VISUALIZING
HUMAN GEOGRAPHY

FIRST EDITION

THE WORLD...

The National Geographic Society (NGS) has been inspiring people to care about the planet since 1888. NGS photographers and cartographers study the world and record it *visually,* making their images and maps ideal resources to help immerse students in the world of Human Geography.

IN YOUR HANDS,

Developed in partnership with the National Geographic Society, *Visualizing Human Geography* integrates photos, maps, illustrations, and video with clear and concise text, to deliver an engaging learning experience. NGS verifies every fact in the book with two outside sources, ensuring accuracy, currency, and effective learning.

TODAY!

A portion of the proceeds of *Visualizing Human Geography* help further the mission of National Geographic: to increase global understanding through education, exploration, research, and conservation.

Experience The Learning

ཨོཾ་མ་ཎི་པདྨེ་ཧཱུྃ

VISUALIZING HUMAN GEOGRAPHY

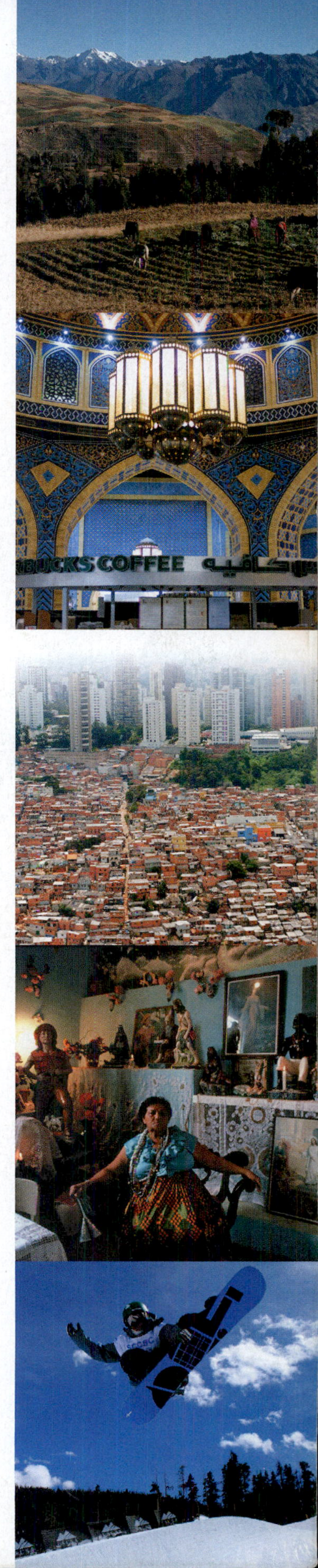

ALYSON L. GREINER
OKLAHOMA STATE UNIVERSITY

WILEY

In collaboration with
The National Geographic Society

VICE PRESIDENT AND EXECUTIVE PUBLISHER Jay O'Callaghan
EXECUTIVE EDITOR Ryan Flahive
DIRECTOR OF DEVELOPMENT Barbara Heaney
MANAGER, PRODUCT DEVELOPMENT Nancy Perry
WILEY VISUALIZING PROJECT EDITOR Beth Tripmacher
ASSOCIATE EDITOR Veronica Armour
WILEY VISUALIZING SENIOR EDITORIAL ASSISTANT Tiara Kelly
EDITORIAL ASSISTANTS Darnell Sessoms, Brittany Cheetham, Sean Boda
ASSOCIATE DIRECTOR, MARKETING Jeffrey Rucker
SENIOR MARKETING MANAGER Margaret Barrett
CONTENT MANAGER Micheline Frederick
SENIOR PRODUCTION EDITOR Kerry Weinstein
SENIOR MEDIA EDITOR Lynn Pearlman
INTERACTIVE PROJECT MANAGER Anita Castro
CREATIVE DIRECTOR Harry Nolan
COVER DESIGN Harry Nolan
INTERIOR DESIGN Jim O'Shea
PHOTO MANAGER Hilary Newman
PHOTO RESEARCHERS Stacy Gold/National Geographic Society, Teri Stratford
SENIOR ILLUSTRATION EDITOR Sandra Rigby
PRODUCTION SERVICES Camelot Editorial Services, LLC

COVER CREDITS Main cover photo: ©Robert Harding Images/Masterfile
Bottom inset photos (left to right): ©Michael S. Yamashita/NG Image Collection; ©Chris Johns/NG Image Collection; ©Alankar Chandra/NG Image Collection; ©Sarah Leen/NG Image Collection; ©Justin Guariglia/NG Image Collection
Back cover inset photo: ©Alankar Chandra/NG Image Collection

This book was set in New Baskerville by Preparé, Inc., printed and bound by Quad/Graphics. The cover was printed by Quad/Graphics.

Library of Congress Cataloging-in-Publication Data

Greiner, Alyson L., 1966-
Visualizing human geography: at home in a diverse world/Alyson Greiner.
p. cm. – (Visualizing series ; 5)
Includes bibliographical references and index.
ISBN 978-0-471-72491-9 (pbk.)
1. Human geography. 2. Geography–Philosophy. 3. Geography–Social aspects. I. Title.
GF41.G735 2010
304.2–dc22

2010037964

ISBN: 978-0471-72491-9
BRV ISBN: 978-0470-91743-5

Printed in the United States of America
10 9 8 7 6 5 4 3

How Is Wiley Visualizing Different?

Wiley Visualizing differs from competing textbooks by uniquely combining three powerful elements: a visual pedagogy, integrated with comprehensive text, the use of authentic situations and issues from the National Geographic Society collections, and the inclusion of interactive multimedia in the *WileyPLUS* learning environment. Together these elements deliver a level of rigor in ways that maximize student learning and involvement. Each key concept and its supporting details have been analyzed and carefully crafted to maximize student learning and engagement.

1. Visual Pedagogy. Wiley Visualizing is based on decades of research on the use of visuals in learning (Mayer, 2005).[1] Using the Cognitive Theory of Multimedia Learning, which is backed up by hundreds of empirical research studies, Wiley's authors select visualizations for their texts that specifically support students' thinking and learning—for example, the selection of relevant materials, the organization of the new information, or the integration of the new knowledge with prior knowledge. Visuals and text are conceived and planned together in ways that clarify and reinforce major concepts while allowing students to understand the details. This commitment to distinctive and consistent visual pedagogy sets Wiley Visualizing apart from other textbooks.

2. Authentic Situations and Problems. Through Wiley's exclusive publishing partnership with National Geographic, *Visualizing Human Geography* has benefited from National Geographic's more than century-long recording of the world and offers an array of remarkable photographs, maps, media, and film from the National Geographic Society collections. These authentic materials immerse the student in real-life issues in human geography, thereby enhancing motivation, learning, and retention (Donovan & Bransford, 2005).[2] These authentic situations, using high-quality materials from the National Geographic Society collections, are unique to Wiley Visualizing.

3. Interactive Multimedia. Wiley Visualizing is based on the understanding that learning is an active process of knowledge construction. *Visualizing Human Geography* is therefore tightly integrated with *WileyPLUS*, our online learning environment that provides interactive multimedia activities in which learners can actively engage with the materials. The combination of textbook and *WileyPLUS* provides learners with multiple entry points to the content, giving them greater opportunity to explore concepts, interact with the material, and assess their understanding as they progress through the course. Wiley Visualizing makes this online *WileyPLUS* component a key element of the learning and problem-solving experience, which sets it apart from other textbooks whose online component is a mere drill-and-practice feature.

Wiley Visualizing and the *WileyPLUS* Learning Environment are designed as a natural extension of how we learn

Visuals, comprehensive text, and learning aids are integrated to display facts, concepts, processes, and principles more effectively than words alone can. To understand why the visualizing approach is effective, it is first helpful to understand how we learn.

1. Our brain processes information using two channels: visual and verbal. Our *working memory* holds information that our minds process as we learn. In working memory we begin to make sense of words and pictures and build verbal and visual models of the information.
2. When the verbal and visual models of corresponding information are connected in working memory, we form more comprehensive, or integrated, mental models.
3. After we link these integrated mental models to our prior knowledge, which is stored in our *long-term memory*, we build even stronger mental models. When an integrated mental model is formed and stored in long-term memory, real learning begins.

The effort our brains put forth to make sense of instructional information is called *cognitive load*. There are two kinds of cognitive load: productive cognitive load, such as when we're engaged in learning or exert positive effort to create mental models; and unproductive cognitive load, which occurs when the brain is trying to make sense of needlessly complex content or when information is not presented well. The learning process can be impaired when the amount of information to be processed exceeds the capacity of working memory. Well-designed visuals and text with effective pedagogical guidance can reduce the unproductive cognitive load in our working memory.

---

[1] Mayer, R. E. (Ed.) (2005). *The Cambridge Handbook of Multimedia Learning.* New York: Cambridge University Press.

[2] Donovan, M.S., & Bransford, J. (Eds.) (2005). *How Students Learn: Science in the Classroom.* The National Academy Press. Available online at http://www.nap.edu/openbook.php?record_id=11102&page=1.

Wiley Visualizing is designed for engaging and effective learning

The visuals and text in *Visualizing Human Geography* are specially integrated to present complex processes in clear steps and with clear representations, organize related pieces of information, and integrate related information with one another. This approach, along with the use of interactive multimedia, provides the level of rigor needed for the course and helps students engage with the content. When students are engaged, they're reading and learning, which can lead to greater knowledge and academic success.

Research shows that well-designed visuals, integrated with comprehensive text, can improve the efficiency with which a learner processes information. In this regard, SEG Research, an independent research firm, conducted a national, multisite study evaluating the effectiveness of Wiley Visualizing. Its findings indicate that students using Wiley Visualizing products (both print and multimedia) were more engaged in the course, exhibited greater retention throughout the course, and made significantly greater gains in content area knowledge and skills, as compared to students in similar classes that did not use Wiley Visualizing.[3]

The use of *WileyPLUS* can also increase learning. According to a white paper titled "Leveraging Blended Learning for More Effective Course Management and Enhanced Student Outcomes" by Peggy Wyllie of Evince Market Research & Communications[4], studies show that effective use of online resources can increase learning outcomes. Pairing supportive online resources with face-to-face instruction can help students to learn and reflect on material, and deploying multimodal learning methods can help students to engage with the material and retain their acquired knowledge. *WileyPLUS* provides students with an environment that stimulates active learning and enables them to optimize the time they spend on their coursework. Continual assessment/remediation is also key to helping students stay on track. The *WileyPLUS* system facilitates instructors' course planning, organization, and delivery and provides a range of flexible tools for easy design and deployment of activities and tracking of student progress for each learning objective.

Figure 1: What a Geographer Sees: Cartographic Scale (Ch. 1)
Through a logical progression of visuals and graphic features such as the arrow and circles, this illustration directs learners' attention to the underlying concept.

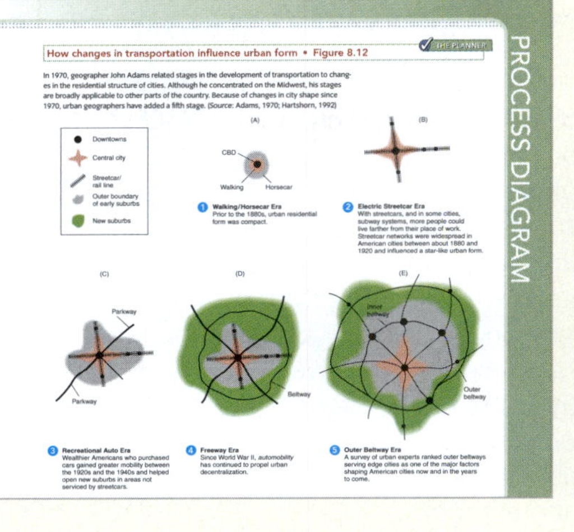

Figure 2: How changes in transportation influence urban form (Figure 8.12)
Textual and visual elements are physically integrated. This eliminates split attention (when we must divide our attention between several sources of different information).

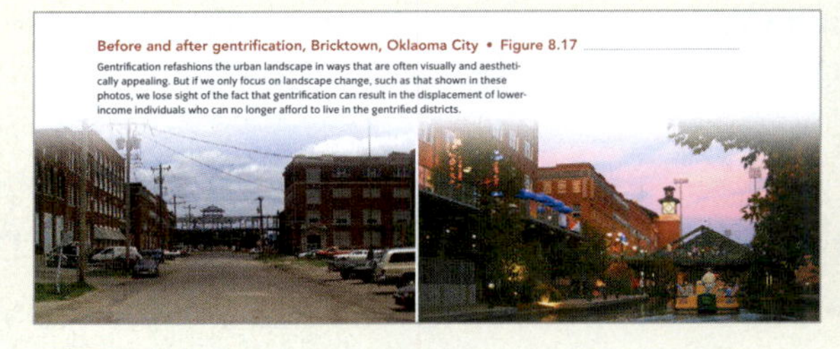

Figure 3: Before and after gentrification (Figure 8.17)
Photos are paired so that students can compare and contrast them, thereby grasping the underlying concept.

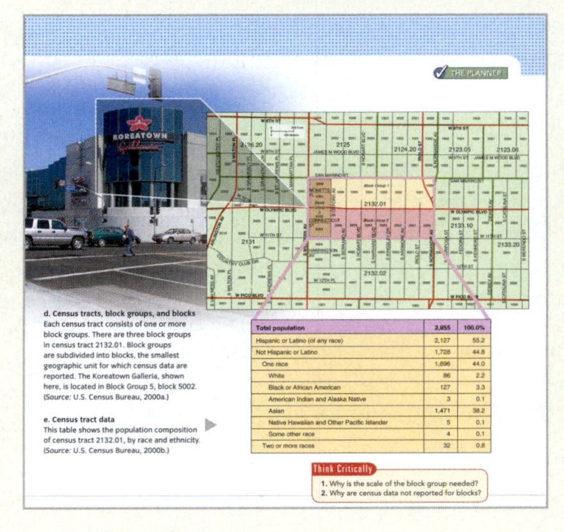

Figure 4: What a Geographer Sees: U.S. Census Geography (Ch. 6)
From reality to abstraction: Linking a photo of a place to its position on census tracts and then showing the data derived from that tract helps students understand how geographic data are produced.

[3] SEG Research (2009). Improving Student-Learning with Graphically Enhanced Textbooks: A Study of the Effectiveness of the Wiley Visualizing Series. Available online at www.segmeasurement.com/

[4] Peggy Wyllie (2009). Leveraging Blended Learning for More Effective Course Management and Enhanced Student Outcomes. Available online at http://catalog.wileyplus.com/about/instructors/whitepaper.html

What Is the Wiley Visualizing Chapter Organization?

Student engagement requires more than just providing visuals, text, and interactivity—it entails motivating students to learn. Student engagement can be behavioral, cognitive, social, and/or emotional. It is easy to get bored or lose focus when presented with large amounts of information, and it is easy to lose motivation when the relevance of the information is unclear. Wiley Visualizing and *WileyPLUS* work together to reorganize course content into manageable learning objectives and relate it to everyday life. The design of *WileyPLUS* is based on cognitive science, instructional design, and extensive research into user experience. It transforms learning into an interactive, engaging, and outcomes-oriented experience for students.

The content in Wiley Visualizing and *WileyPLUS* is organized in learning modules. Each module has a clear instructional objective, one or more examples, and an opportunity for assessment. These modules are the building blocks of Wiley Visualizing.

Each Wiley Visualizing chapter engages students from the start

Chapter opening text and visuals introduce the subject and connect the student with the material that follows.

Global Locator Maps, prepared specifically for this book by the National Geographic Society cartographers, help students visualize where the area depicted in the photo is situated on Earth.

Geographies of Development

BHUTAN'S QUEST FOR GROSS NATIONAL HAPPINESS

Imagine your own Shangri-la—that is, an idyllic place. What place on Earth, if any, comes closest to matching that? Did the country of Bhutan come to mind? Most likely it did not, although in recent years, this small mountainous state nestled between India and China has occasionally been described as a Shangri-la. This designation has less to do with Bhutan's striving to be a perfect place and more to do with its physical setting and its ideology of development.

Until the early 1970s, Bhutan was among the world's most impoverished countries. Then, King Jigme Singye Wangchuck conceived a development strategy that would balance economic growth with environmental protection, Bhutanese cultural traditions, and democratic governance. He envisioned an alternative path to development that, in his words, would bring "gross national happiness."

Bhutan has since invested heavily in education and health care. In the early 1980s, Bhutan had an adult literacy rate of 23% and an infant mortality rate of 163. Today, close to 60% of the adult population is literate, and the infant mortality rate has dropped to 40. In addition, Bhutan has set aside over 30% of its land area—more than any other country—as wildlife sanctuaries, national parks, and nature reserves.

Although Bhutan may not be a Shangri-la and still has room to improve the social well-being of its people, its example is instructive because it highlights a concerted effort to achieve development in a way that is environmentally sustainable and socially conscious.

Global Locator

ASIA

BHUTAN

NATIONAL GEOGRAPHIC

CHAPTER OUTLINE

Chapter Outlines anticipate the content.

CHAPTER PLANNER

- ❑ Study the picture and read the opening story.
- ❑ Scan the Learning Objectives in each section: p. 262 ❑ p. 275 ❑ p. 280 ❑
- ❑ Read the text and study all figures and visuals. Answer any questions.

Analyze key features

- ❑ Geography InSight, p. 274
- ❑ Process Diagram, p. 281
- ❑ What a Geographer Sees, p. 286
- ❑ Video Explorations, p. 288
- ❑ Stop: Answer the Concept Checks before you go on: p. 275 ❑ p. 280 ❑ p. 288 ❑

End of chapter

- ❑ Review the Summary and Key Terms.
- ❑ Answer the Critical and Creative Thinking Questions.
- ❑ Answer What is happening in this picture?
- ❑ Complete the Self-Test and check your answers

Chapter Introductions illustrate key concepts in the chapter with intriguing stories and striking photographs.

Bhutan's national highway zigzags across mountains.

261

The **Chapter Planner** gives students a path through the learning aids in the chapter. Throughout the chapter, the Planner icon prompts students to use the learning aids and to set priorities as they study.

WILEY PLUS

Experience the chapter through a *WileyPLUS* course. The content through *WileyPLUS* transports the student into a rich world of online experience that can be personalized, customized, and extended. Students can create a personal study plan to help prioritize which concepts to learn first and to focus on weak points.

Wiley Visualizing media guides students through the chapter

The content of Wiley Visualizing in *WileyPLUS* gives students a variety of approaches—visuals, words, illustrations, interactions, and assessments—that work together to provide students with a guided path through the content. But this path isn't static: It can be personalized, customized, and extended to suit individual needs, and so it offers students flexibility as to how they want to study and learn the content.

Learning Objectives at the start of each section indicate in behavioral terms the concepts that students are expected to master while reading the section.

Language Diffusion and

LEARNING OBJECTIVES

1. **Explain** how political, economic, and religious forces can affect the diffusion of language.
2. **Identify** factors contributing to linguistic dominance.

What social and geographic factors contribute to the spread, or diffusion, of languages? In our discussion of language families, we learned that the spread of agriculture may have facilitated the historic spread of languages. If we take a broader perspective, we can see that technology and human mobility can contribute significantly to language diffusion. Historically, ships, railroads, and other forms of ...

Every content resource is related to a specific learning objective so that students will easily discover relevant content organized in a more meaningful way.

PROCESS DIAGRAM

Sanctification • Figure 5.13 ✓ THE PLANNER

Sanctification refers to the making of a sacred site after a significant event has occurred there. Here we consider the example of the former site of the Murrah Federal Building in Oklahoma City, which was bombed in an act of domestic terrorism in 1995. This site is today recognized as a national memorial.

Five days after the bombing, a makeshift memorial emerges.

1 The event
The north side of the building in Oklahoma City is blown away by a massive truck bomb on April 19, 1995, killing 168 and injuring over 800.

Ongoing placement of memorials at the site continues to set it apart and affirm its sacredness.

2 Informal consecration: emerging consensus and spontaneous memorials
Consecration involves public activities that set a place apart and contribute to its sacredness. Informal consecration begins as people react to the tragic event and a consensus emerges that the site is worthy of remembrance. Temporary memorials make the site hallowed.

3 Formal consecration: official consensus for a permanent monument
An organization or other entity establishes the memorial boundaries, acquires legal right to use the land, makes plans for a permanent monument, and manages funds for the perpetual upkeep of the site. This is the Oklahoma City National Memorial, with 168 chairs (one for each victim) and a survivors' tree.

4 Final consecration: solemn dedication of the site
Color guard stands at the Field of Chairs during the dedication ...

5 Ritual commemoration
Collective memory of the event at the site is promoted through regular practices, ... annual ... or ...

Process Diagrams provide in-depth coverage of processes correlated with clear, step-by-step narrative, enabling students to grasp important topics with less effort.

Interactive Process Diagrams provide additional visual examples and descriptive narration of a difficult concept, process, or theory, allowing the students to interact and engage with the content. Many of these diagrams are built around a specific feature such as a Process Diagram. Look for them in *WileyPLUS* when you see this icon.

Geography InSights are multipart visual features that focus on a key concept or topic in the chapter, exploring it in detail or in broader context using a combination of photos, diagrams, maps, and data.

Geography InSight
Geographies of language diffusion • Figure 4.8
THE PLANNER
10 JANVIER 2010
VOTEZ
74
FRANCE
PORTUGAL
SPAIN
CHIMOIO
AEROPORTO
SUSSUNDENGA
a. French-language signs in French Guiana remind people to vote
French Guiana became a French colony in the 17th century and has been an overseas department of France since 1946.
b. A waiter in a restaurant in Buenos Aires, Argentina
Buenos Aires means good air or fair winds in Spanish.
c. Portuguese and Bantu language sign in Mozambique
Following its independence from Portugal in 1975, Mozambique changed the name of the city Vila Pery to Chimoio. Education is mainly in Portuguese but Bantu languages, a branch of the Niger-Congo language family, are prevalent across the country.
Generalized Routes of Diffusion of Selected European Languages, circa 1450–1973
French
Portuguese
Spanish
As suggested by this map, European colonization played a major role in the diffusion of certain European languages, including French, Spanish, and Portuguese. Today these languages still influence the linguistic geography of many countries but in varying degrees. In the Philippines, for example, Spanish was an official language until 1973, but English and Filipino (Tagalog) are now more widely spoken.
Language Diffusion and Globalization 107

Maps from the **National Geographic** collection and maps created for this text by NGS cartographers immerse the student in a variety of real-life issues in human geography.

What a Geographer Sees highlights a concept or phenomenon that would stand out to geographers. Photos and figures are used to improve students' understanding of the usefulness of a geographic perspective and to enable students to apply their observational skills to answer questions.

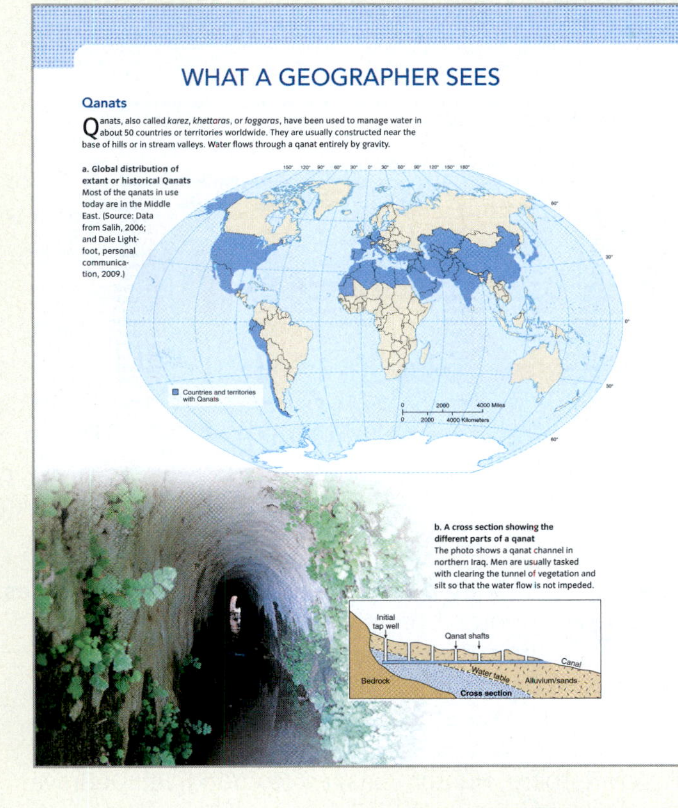

WHAT A GEOGRAPHER SEES
Qanats
Qanats, also called karez, khettaras, or foggaras, have been used to manage water in about 50 countries or territories worldwide. They are usually constructed near the base of hills or in stream valleys. Water flows through a qanat entirely by gravity.
a. Global distribution of extant or historical Qanats
b. A cross section showing the different parts of a qanat
Qanat shafts
Canal
Bedrock
Cross section

Think Critically questions encourage students to analyze the material and develop insights into essential concepts.

In each chapter, the **Video Explorations** feature, researched by Joy Adams of Humboldt State University, showcases one of more than 30 **National Geographic videos** from the award-winning NGS collection. The videos are linked to the text and provide visual context for key concepts, ideas, and terms addressed in the chapter.

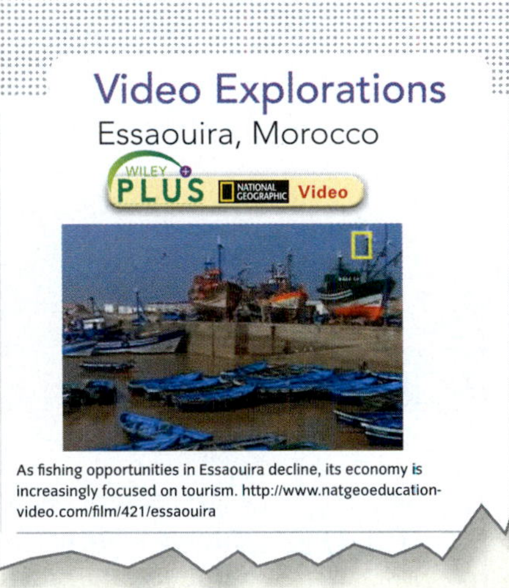

Video Explorations
Essaouira, Morocco
WILEY PLUS NATIONAL GEOGRAPHIC Video
As fishing opportunities in Essaouira decline, its economy is increasingly focused on tourism. http://www.natgeoeducation-video.com/film/421/essaouira

Where Geographers Click showcases a website that professionals use and encourages students to try out its tools.

Where Geographers CLICK

Human Development Reports

http://hdr.undp.org/en/statistics/data/hd_map/

This site provides a very good place to learn more about measures of human development. From here it is possible to view additional maps and access data for specific countries.

Streaming videos are available to students in the context of *WileyPLUS*, and accompanying assignments can be graded online and added to the instructor gradebook.

In concert with the visual approach of the book, **www.ConceptCaching.com** is an online collection of photographs that explores places, regions, people, and their activities. Photographs, GPS coordinates, and explanations of core geographic concepts are "cached" for viewing by professors and students alike. Professors can access the images or submit their own by visiting the website. Caches on the website are integrated in the *WileyPLUS* course as examples to help students understand the concepts.

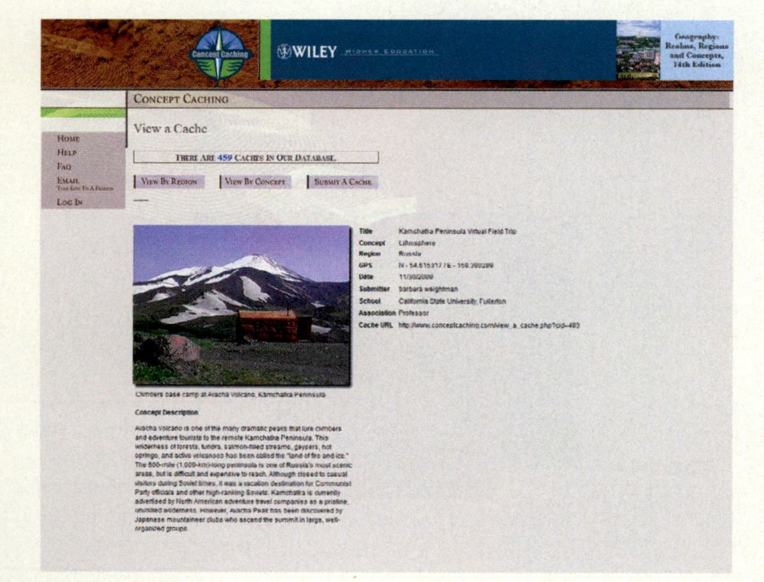

GeoDiscoveries Media Library is an interactive media source of animations, simulations, and interactivities allowing instructors to visually demonstrate key concepts in greater depth.

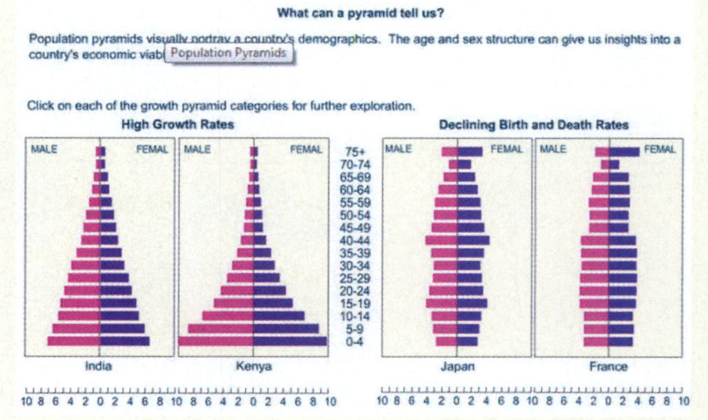

In **Google Earth™ Links, Tours, and Activities**, photos from the *WileyPLUS* eBook are linked from the text to their actual location on the Earth using Google Earth. Tours and activities created by professors engage students with geographic concepts addressed in the text. Contributing professors include Randy Rutberg, Hunter College (New York); Jeff DeGrave, University of Wisconsin-Eau Claire; and James Hayes-Bohanan, Bridgewater State University.

Coordinated with the section-opening **Learning Objectives**, at the end of each section **Concept Check** questions allow students to test their comprehension of the learning objectives.

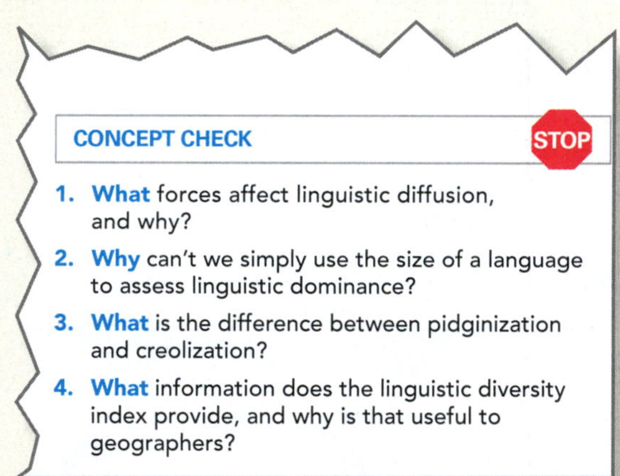
CONCEPT CHECK

STOP

1. **What** forces affect linguistic diffusion, and why?
2. **Why** can't we simply use the size of a language to assess linguistic dominance?
3. **What** is the difference between pidginization and creolization?
4. **What** information does the linguistic diversity index provide, and why is that useful to geographers?

At the end of each learning objective module, students can assess their progress with independent practice opportunities and quizzes. This feature gives them the ability to gauge their comprehension and grasp of the material. Practice tests and quizzes help students self-monitor and prepare for graded course assessments.

Student understanding is assessed at different levels

Wiley Visualizing with *WileyPLUS* offers students lots of practice material for assessing their understanding of each study objective. Students know exactly what they are getting out of each study session through immediate feedback and coaching.

The **Summary** revisits each major section, with informative images taken from the chapter. These visuals reinforce important concepts.

What is happening in this picture? presents an uncaptioned photograph that is relevant to a chapter topic and illustrates a situation students are not likely to have encountered previously.

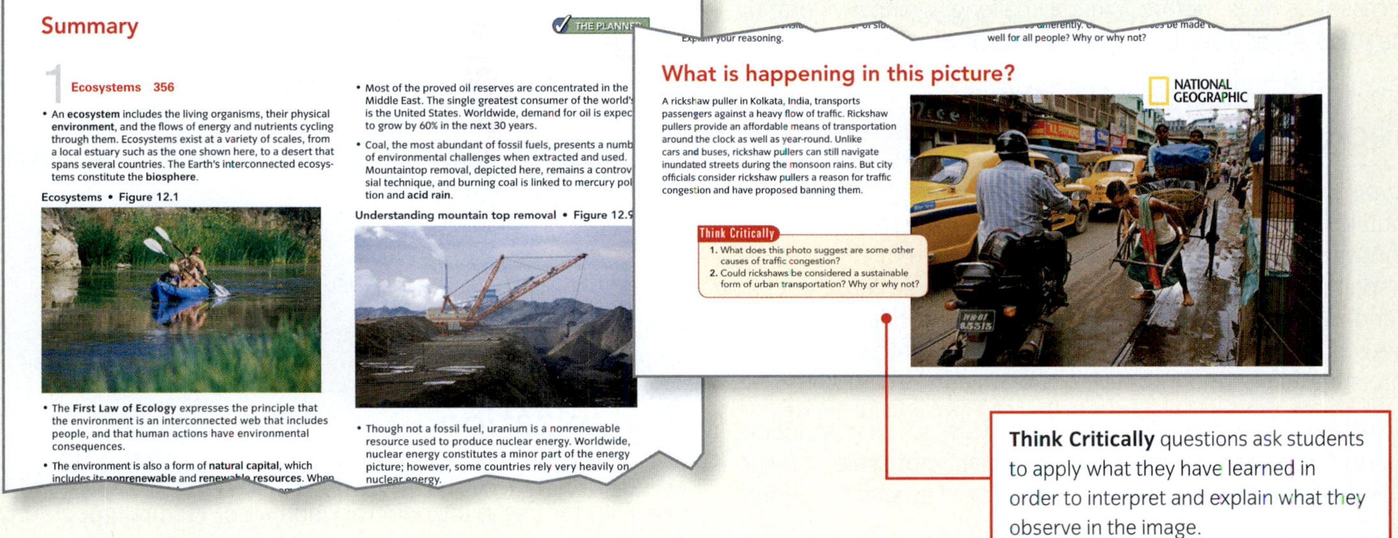

Summary

THE PLANNER

1 Ecosystems 356

- An **ecosystem** includes the living organisms, their physical **environment**, and the flows of energy and nutrients cycling through them. Ecosystems exist at a variety of scales, from a local estuary such as the one shown here, to a desert that spans several countries. The Earth's interconnected ecosystems constitute the **biosphere**.

Ecosystems • Figure 12.1

- The **First Law of Ecology** expresses the principle that the environment is an interconnected web that includes people, and that human actions have environmental consequences.
- The environment is also a form of **natural capital**, which includes its **nonrenewable** and **renewable resources**. When

- Most of the proved oil reserves are concentrated in the Middle East. The single greatest consumer of the world's is the United States. Worldwide, demand for oil is expec to grow by 60% in the next 30 years.
- Coal, the most abundant of fossil fuels, presents a numb of environmental challenges when extracted and used. Mountaintop removal, depicted here, remains a controv sial technique, and burning coal is linked to mercury pol tion and **acid rain**.

Understanding mountain top removal • Figure 12.9

- Though not a fossil fuel, uranium is a nonrenewable resource used to produce nuclear energy. Worldwide, nuclear energy constitutes a minor part of the energy picture; however, some countries rely very heavily on nuclear energy.

Explain your reasoning.

well for all people? Why or why not?

What is happening in this picture?

NATIONAL GEOGRAPHIC

A rickshaw puller in Kolkata, India, transports passengers against a heavy flow of traffic. Rickshaw pullers provide an affordable means of transportation around the clock as well as year-round. Unlike cars and buses, rickshaw pullers can still navigate inundated streets during the monsoon rains. But city officials consider rickshaw pullers a reason for traffic congestion and have proposed banning them.

Think Critically

1. What does this photo suggest are some other causes of traffic congestion?
2. Could rickshaws be considered a sustainable form of urban transportation? Why or why not?

Think Critically questions ask students to apply what they have learned in order to interpret and explain what they observe in the image.

Critical and Creative Thinking Questions challenge students to think more broadly about chapter concepts. The level of these questions ranges from simple to advanced; they encourage students to think critically and develop an analytical understanding of the ideas discussed in the chapter.

Critical and Creative Thinking Questions

1. Will Internet voting ever replace the use of traditional polling places? What geographic and political conditions would be most conducive to such change?
2. List some advantages and disadvantages of majority-plurality and proportional representation systems.
3. Do some fieldwork in the area where you live and identify a relic boundary. What processes led to the creation of that boundary?
4. Not all ethnic groups are nations. Why?
5. Under what circumstances might devolution become a centrifugal force?
6. A political geographer might argue that the Berlin Conference was an exercise in gerrymandering. Explain what is meant by this statement and take a position on it.
7. What similarities and differences are there between the division of Germany after World War II and the division of Cyprus?
8. Review the composition of the UN Security Council. Does it need reform? Why or why not?

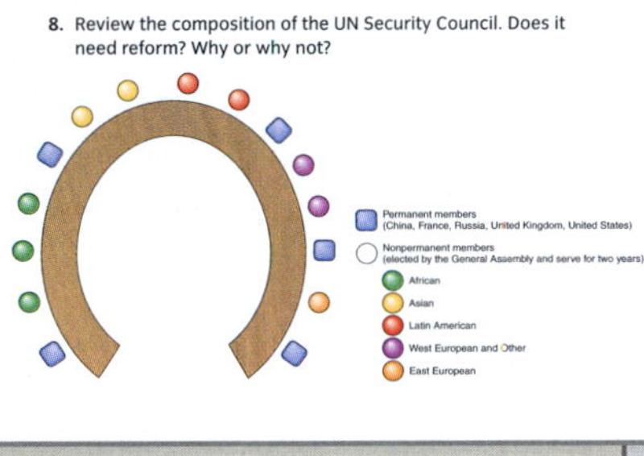

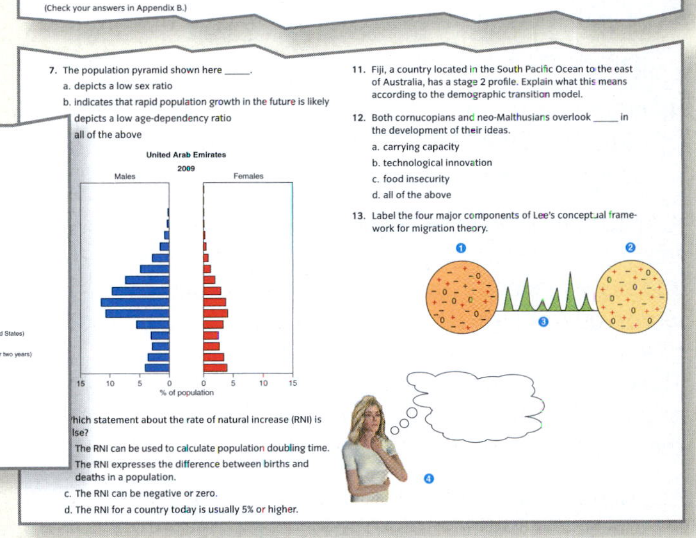

Self-Test

(Check your answers in Appendix B.)

7. The population pyramid shown here ____.
   a. depicts a low sex ratio
   b. indicates that rapid population growth in the future is likely
   depicts a low age-dependency ratio
   all of the above

United Arab Emirates 2009

Males Females

% of population

hich statement about the rate of natural increase (RNI) is lse?
   The RNI can be used to calculate population doubling time.
   The RNI expresses the difference between births and deaths in a population.
   c. The RNI can be negative or zero.
   d. The RNI for a country today is usually 5% or higher.

11. Fiji, a country located in the South Pacific Ocean to the east of Australia, has a stage 2 profile. Explain what this means according to the demographic transition model.
12. Both cornucopians and neo-Malthusians overlook ____ in the development of their ideas.
   a. carrying capacity
   b. technological innovation
   c. food insecurity
   d. all of the above
13. Label the four major components of Lee's conceptual framework for migration theory.

Visual end-of-chapter **Self-Tests** pose review questions that ask students to demonstrate their understanding of key concepts.

Why Visualizing Human Geography?

We live in an ever-changing world in which geographical knowledge is central to the well-being of our communities and society. Perhaps nowhere is the urgency of geographical knowledge made clearer to us than through issues involving the local, national, and global impacts of climate change; the oil spill disaster in the Gulf of Mexico; or the War in Afghanistan. Simultaneously, technological innovations continue to open new horizons in mapping and techniques for visualizing geographic information that enable us to see, explore, and understand local and global processes as never before. What a challenging and invigorating time to be either a student or an instructor of geography.

Geographic literacy

Visualizing Human Geography provides a fresh, new pathway for building geographic literacy and introducing students to the richness of geography, including its many different approaches, perspectives, techniques, and tools. Geographic literacy seeks to endow students with geographic and analytical skills to be creative and capable decision makers and problem solvers. More specifically, geographic literacy includes

1. fostering the skills of spatial analysis so that students gain an understanding of the importance of scale and can evaluate and interpret the significance of spatial variation;
2. enhancing students' comprehension of the interconnectedness of social and environmental dynamics, and the implications of this for people's livelihoods, their use of the Earth, and environmental change;
3. cultivating global awareness in students and exposing them to divergent views so they are prepared and equipped to participate in an increasingly interconnected world; and
4. educating students about the advantages and limitations of tools such as GIS and GPS in the acquisition and use of geographic information.

A fundamental premise guiding the presentation of material in this book is that such key geographical concepts as place, space, and scale cannot be divorced from a study of process. In other words, questions of why and how are vital to our understanding of where activities, events, or other phenomena are located. Thus, every chapter contains at least one Process Diagram in order to show the diverse factors and complex relations among them that drive social and environmental change.

Human geography is well suited to a visually oriented approach for three reasons. First, maps and images are fundamental tools of geographers that help to reveal patterns or trends that might not otherwise be apparent. Second, within the practice of human geography there is a longstanding tradition of studying cultural landscapes for evidence about such processes as diffusion, urbanization, or globalization in order to more fully understand social difference and to assess human use of the Earth. Third, many human geographers are interested in representation, including the kinds of images that are used by different agencies and entities to characterize places, regions, people, and their activities. Therefore, a visual approach enables a more complete instructional use of photographs, maps, and other visually oriented media to explore and evaluate the significance of different representations.

Other features of this book include

- content that reflects the latest developments in geographic thought;
- coverage of geographical models and theory as well as their real-world applications;
- top-notch cartography;
- accurate and up-to-date statistics;
- an appendix devoted to understanding map projections.

Organization

Visualizing Human Geography is a college-level textbook intended for use in introductory human or cultural geography courses. Students need not have had any previous coursework in geography to use this book. The structure of the book is based on a 12-chapter framework suitable for institutions using either the semester or quarter system. The chapters are arranged according to conventional practice. Globalization and gender issues are covered throughout the book. The outline below provides a brief overview of the content of each chapter.

- **Chapter 1, What Is Human Geography?** This foundational chapter introduces students to the discipline of geography and the subfield of human geography. It covers the key concepts of nature, culture, place, space, spatial diffusion, spatial interaction and globalization, and scale. One section of the chapter explains and gives examples of the applications of geographic tools including remote sensing, GPS, and GIS. Students are also introduced to possible careers in geography.
- **Chapter 2, Globalization and Cultural Geography.** This chapter expands on the process of globalization introduced in Chapter 1, then moves to the cultural impacts of globalization such as the diffusion of popular culture and local responses to it. The chapter also explores the commodification of culture through case studies of the diamond industry, representations of indigenous culture, and world heritage. The chapter uses the term *local culture* instead of *folk culture* and examines geographies of local knowledge, including traditional medicine.

- **Chapter 3, Population and Migration.** Such fundamental concepts as population density, fertility, mortality, life expectancy, and their regional differences are discussed and explained in this chapter. Population pyramids, the rate of natural increase, and the demographic transition model are used to examine population change. The chapter also introduces theories about population growth, resource use, food insecurity, and migration, and discusses the patterns of global migration.
- **Chapter 4, Geographies of Language.** Linguistic diversity is an important theme throughout this chapter. Present-day and historical factors help anchor the discussion of the distribution of languages and language families. The relationships among linguistic dominance, status, geographic space, and language endangerment are also covered. The chapter closes with a discussion of dialect geography and toponyms.
- **Chapter 5, Geographies of Religion.** The contrasting geographies of six major religious traditions are discussed in this chapter: Judaism, Christianity, Islam, Hinduism, Buddhism, and Sikhism. The concept of civil religion is introduced and is used to explore the emergence of sacred places and spaces. The chapter addresses the tension between modernism and traditionalism in religion, geographical aspects of religious law, and the origins, diffusion, and globalization of Renewalism. The concept of geopiety provides one way of considering the connections among religion, nature, and landscape.
- **Chapter 6, Geographies of Identity.** Chapters 4, 5, 6, and 7 cover different facets of identity, and this chapter expressly examines race, ethnicity, sexuality, and gender. The chapter treats race as a social construction and examines geographies of racism produced in South Africa during apartheid. The chapter also addresses the complexity of ethnicity, the representation of ethnicity and identity on censuses, theories of ethnic interaction, ethnic conflict, and environmental justice. The section on sexuality and gender challenges students to think about the geographic implications of a heterosexual norm, and the persistence of gender roles and gender gaps.
- **Chapter 7, Political Geographies.** Crucial to this chapter are the development of the state, the geographical characteristics of states, and the geographical implications of centripetal and centrifugal forces as well as separatism and devolution. Discussions of the United Nations and European Union provide contrasting studies of supranational organizations. The topic of global geopolitics is explored through a mix of traditional and contemporary theories as well as globalization and terrorism. Students are introduced to the fundamentals of electoral geography and ways in which cultural landscapes can be used to convey political power and ideologies.
- **Chapter 8, Urban Geographies.** This chapter opens with a discussion of the different types of urban settlements, global patterns of urbanization, the development of megacities and primate cities, and urban hierarchies. The next section of the chapter focuses on models of urban structure. This is followed by a study of the impact of public policy on residential change and urban redevelopment. The extent of urban poverty and the causes of slum formation are detailed, and the chapter closes with a discussion of trends in urban planning.
- **Chapter 9, Geographies of Development.** Students learn what development is, what makes it a normative project, and how it can be measured using development indicators or indexes. The chapter discusses the geography of income inequality, one expression of uneven levels of development, then turns to an examination of the evolution of development theory. Students are introduced to dependency theory, world-system theory, neoliberalism, and poverty-reduction theory, among others, and their geographical ramifications. Students also learn about the technique of poverty mapping.
- **Chapter 10, Changing Geographies of Industry and Services.** This chapter explains distinctions among primary, secondary, tertiary, quaternary, and quinary types of industry. It introduces commodity dependency and staple theory. Students learn about the origins and diffusion of the Industrial Revolution as well as the impact of Fordism and flexible production on manufacturing in the core. The chapter distinguishes between outsourcing and offshoring, and addresses the emergence of newly industrialized economies, export-processing zones, and the globalization of commodity chains. The chapter also examines the process of deindustrialization, characteristics of postindustrial societies, changing patterns of employment in manufacturing, and gender mainstreaming.
- **Chapter 11, Agricultural Geographies.** This chapter follows the chapter on industry because agriculture has been and is still strongly influenced by technological change and systems of industrial production. The chapter identifies three major agricultural revolutions and distinguishes between the Green Revolution and the Gene Revolution. Students are encouraged to think about types of agriculture as agricultural systems, and the global distribution of several examples of subsistence and commercial agriculture is discussed. The chapter also covers the impacts of agriculture on the environment, sustainable agricultural practices, the impact of globalization on agriculture and dietary practices, and the causes of the recent global food crisis.
- **Chapter 12, Environmental Challenges.** The nature and functioning of ecosystems provides a framework for this chapter. A discussion of the concept and process of environmental degradation leads to an examination of Garrett Hardin's work on the tragedy of the commons and an examination of common property resources more broadly. The chapter covers the geographical aspects of the distribution, use, and consumption of all major nonrenewable and renewable energy resources. Students learn about the greenhouse effect, global warming, carbon footprints, and land-use land-cover change. The chapter closes with a discussion of international policies on greenhouse gas reductions.

Also available

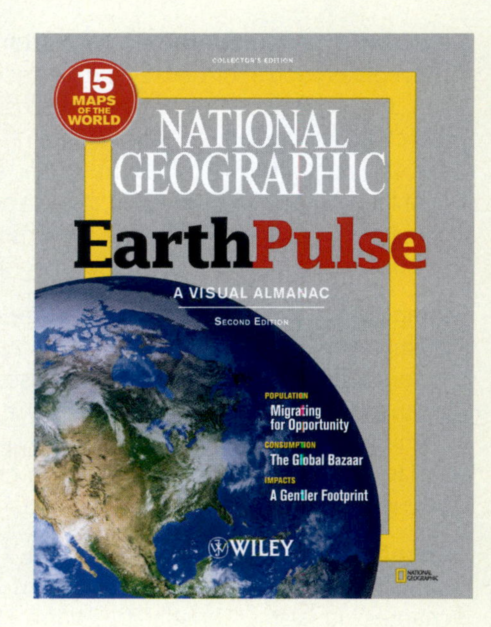

Earth Pulse 2e. Utilizing full-color imagery and National Geographic photographs, *EarthPulse* takes you on a journey of discovery covering topics such as *The Human Condition, Our Relationship with Nature, and Our Connected World*. Illustrated by specific examples, each section focuses on trends affecting our world today. Included are extensive full-color world and regional maps for reference. *EarthPulse* is available only in a package with *Visualizing Human Geography* Contact your Wiley representative for more information or visit www.wiley.com/college/earthpulse.

How Does Wiley Visualizing Support Instructors?

Wiley Visualizing Site

WILEY VISUALIZING™

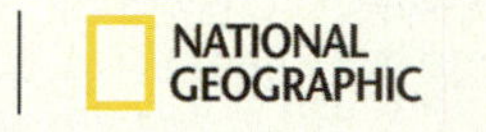

The Wiley Visualizing site hosts a wealth of information for instructors using Wiley Visualizing, including ways to maximize the visual approach in the classroom and a white paper titled "How Visuals Can Help Students Learn," by Matt Leavitt, instructional design consultant. You can also find information about our relationship with the National Geographic Society and other texts published in our program. Visit Wiley Visualizing at www.wiley.com/college/visualizing.

Wiley Custom Select

WILEY Custom LEARNING SOLUTIONS

Wiley Custom Select gives you the freedom to build your course materials exactly the way you want them. Offer your students a cost-efficient alternative to traditional texts. In a simple three-step process create a solution containing the content you want, in the sequence you want, delivered how you want. Visit Wiley Custom Select at http://customselect.wiley.com.

National Geographic Videos

Researched by Joy Adams, Humboldt State University, the **Video Explorations** presented in each chapter of the textbook, are just some of the 30+ NGS videos available to provide visual context for key concepts, ideas, and terms addressed in the textbook. Streaming videos are available to students in the context of *WileyPLUS*, and accompanying assignments can be graded online and added to the instructor gradebook.

Book Companion Site www.wiley.com/college/greiner

All instructor resources (the Test Bank, Instructor's Manual, PowerPoint presentations, and all textbook illustrations and photos in jpeg format) are housed on the book companion site (www.wiley.com/college/greiner). Student resources include self quizzes and flashcards.

PowerPoint Presentations

(available in *WileyPLUS* and on the book companion site)

A complete set of highly visual PowerPoint presentations—one per chapter—is available online and in *WileyPLUS* to enhance classroom presentations. Tailored to the text's topical coverage and learning objectives, these presentations are designed to convey key text concepts, illustrated by embedded text art. Lecture Launcher PowerPoints also offer embedded links to videos to help introduce classroom discussions with short, engaging video clips.

Test Bank (available in *WileyPLUS* and on the book companion site)

The visuals from the textbook are also included in the Test Bank by Carolyn Coulter, Atlantic Cape Community College. The Test Bank has a diverse selection of test items including multiple-choice and essay questions, with at least 20 percent of them incorporating visuals from the book. The Test Bank is available online in MS Word files, as a Computerized Test Bank, and within *WileyPLUS*. The easy-to-use test-generation program fully supports graphics, print tests, student answer sheets, and answer keys. The software's advanced features allow you to produce an exam to your exact specifications.

Instructor's Manual (available in *WileyPLUS* and on the book companion site)

The Instructor's Manual includes creative ideas for in-class activities, discussion questions, and lecture transitions. It also includes answers to Critical and Creative Thinking questions and Concept Check questions.

Guidance is also provided on how to maximize the effectiveness of visuals in the classroom.

1. **Use visuals during class discussions or presentations.** Point out important information as the students look at the visuals, to help them integrate separate visual and verbal mental models.
2. **Use visuals for assignments and to assess learning.** For example, learners could be asked to identify samples of concepts portrayed in visuals.
3. **Use visuals to encourage group activities.** Students can study together, make sense of, discuss, hypothesize, or make decisions about the content. Students can work together to interpret and describe the diagram, or use the diagram to solve problems, conduct related research, or work through a case study activity.

Image Gallery

All photographs, figures, maps, and other visuals from the text are online and in *WileyPLUS* and can be used as you wish in the classroom. These online electronic files allow you to easily incorporate images into your PowerPoint presentations as you choose, or to create your own handouts.

Wiley Faculty Network

The Wiley Faculty Network (WFN) is a global community of faculty, connected by a passion for teaching and a drive to learn, share, and collaborate. Their mission is to promote the effective use of technology and enrich the teaching experience. Connect with the Wiley Faculty Network to collaborate with your colleagues, find a mentor, attend virtual and live events, and view a wealth of resources all designed to help you grow as an educator. Visit the Wiley Faculty Network at www.wherefacultyconnect.com.

How Has Wiley Visualizing Been Shaped by Contributors?

Wiley Visualizing and the *WileyPLUS* learning environment would not have come about without lots of people, each of whom played a part in sharing their research and contributing to this new approach. First and foremost, we begin with NGS.

National Geographic Society

Visualizing Human Geography offers an array of remarkable photographs, maps, illustrations, multimedia, and film from the National Geographic Society collections. Students using the book benefit from the rich, fascinating resources of National Geographic.

National Geographic School Publishing performed an invaluable service in fact-checking *Visualizing Human Geography.* They have verified every fact in the book with two outside sources, to ensure that the text is accurate and up-to-date. This kind of fact-checking is rare in textbooks and unheard-of in most online media.

National Geographic Image Collection provided access to National Geographic's award-winning image and illustrations collection to identify the most appropriate and effective images and illustrations to accompany the content. Each image and illustration has been chosen to be instructive, supporting the processes of selecting, organizing, and integrating information, rather than being merely decorative.

National Geographic Digital Media TV enabled the use of National Geographic videos to accompany *Visualizing Human Geography* and enrich the text. Available for each chapter are video clips that illustrate and expand on a concept or topic to aid student understanding.

National Geographic Maps Group provided access to National Geographic's extensive map collection, or their team of cartographers designed new maps for the text.

Academic Research Consultants

Richard Mayer, Professor of Psychology, UC Santa Barbara. Mayer's *Cognitive Theory of Multimedia Learning* provided the basis on which we designed our program. He continues to provide guidance to our author and editorial teams on how to develop and implement strong, pedagogically effective visuals and use them in the classroom.

Jan L. Plass, Professor of Educational Communication and Technology in the Steinhardt School of Culture, Education, and Human Development at New York University. Plass co-directs the NYU Games for Learning Institute and is the founding director of the CREATE Consortium for Research and Evaluation of Advanced Technology in Education.

Matthew Leavitt, Instructional Design Consultant, advises the Visualizing team on the effective design and use of visuals in instruction and has made virtual and live presentations to university faculty around the country regarding effective design and use of instructional visuals.

Independent Research Studies

SEG Research, an independent research and assessment firm, conducted a national, multisite effectiveness study of students enrolled in entry-level college Psychology and Geology courses. The study was designed to evaluate the effectiveness of Wiley Visualizing. You can view the full research paper at www.wiley.com/college/visualizing/huffman/efficacy.html.

Instructor and Student Contributions

Throughout the process of developing the concept of guided visual pedagogy for Wiley Visualizing, we benefited from the comments and constructive criticism provided by the instructors and colleagues listed below. We offer our sincere appreciation to these individuals for their helpful reviews and general feedback:

Visualizing Reviewers, Focus Group Participants, and Survey Respondents

James Abbott, Temple University
Melissa Acevedo, Westchester Community College
Shiva Achet, Roosevelt University
Denise Addorisio, Westchester Community College
Dave Alan, University of Phoenix
Sue Allen-Long, Indiana University – Purdue
Robert Amey, Bridgewater State College
Nancy Bain, Ohio University
Corinne Balducci, Westchester Community College
Steve Barnhart, Middlesex County Community College
Stefan Becker, University of Washington – Oshkosh
Callan Bentley, NVCC Annandale
Valerie Bergeron, Delaware Technical & Community College
Andrew Berns, Milwaukee Area Technical College
Gregory Bishop, Orange Coast College
Rebecca Boger, Brooklyn College
Scott Brame, Clemson University
Joan Brandt, Central Piedmont Community College
Richard Brinn, Florida International University
Jim Bruno, University of Phoenix
William Chamberlin, Fullerton College
Oiyin Pauline Chow, Harrisburg Area Community College
Laurie Corey, Westchester Community College
Ozeas Costas, Ohio State University at Mansfield
Christopher Di Leonardo, Foothill College
Dani Ducharme, Waubonsee Community College
Mark Eastman, Diablo Valley College
Ben Elman, Baruch College
Staussa Ervin, Tarrant County College
Michael Farabee, Estrella Mountain Community College
Laurie Flaherty, Eastern Washington University
Susan Fuhr, Maryville College
Peter Galvin, Indiana University at Southeast
Andrew Getzfeld, New Jersey City University
Janet Gingold, Prince George's Community College
Donald Glassman, Des Moines Area Community College
Richard Goode, Porterville College
Peggy Green, Broward Community College
Stelian Grigoras, Northwood University
Paul Grogger, University of Colorado
Michael Hackett, Westchester Community College
Duane Hampton, Western Michigan University
Thomas Hancock, Eastern Washington University
Gregory Harris, Polk State College
John Haworth, Chattanooga State Technical Community College
James Hayes-Bohanan, Bridgewater State College
Peter Ingmire, San Francisco State University
Mark Jackson, Central Connecticut State University
Heather Jennings, Mercer County Community College
Eric Jerde, Morehead State University
Jennifer Johnson, Ferris State University
Richard Kandus, Mt. San Jacinto College District
Christopher Kent, Spokane Community College
Gerald Ketterling, North Dakota State University
Lynnel Kiely, Harold Washington College
Eryn Klosko, Westchester Community College
Cary T. Komoto, University of Wisconsin – Barron County
John Kupfer, University of South Carolina
Nicole Lafleur, University of Phoenix
Arthur Lee, Roane State Community College
Mary Lynam, Margrove College
Heidi Marcum, Baylor University
Beth Marshall, Washington State University
Dr. Theresa Martin, Eastern Washington University
Charles Mason, Morehead State University
Susan Massey, Art Institute of Philadelphia
Linda McCollum, Eastern Washington University
Mary L. Meiners, San Diego Miramar College
Shawn Mikulay, Elgin Community College
Cassandra Moe, Century Community College
Lynn Hanson Mooney, Art Institute of Charlotte
Kristy Moreno, University of Phoenix
Jacob Napieralski, University of Michigan - Dearborn
Gisele Nasar, Brevard Community College, Cocoa Campus
Daria Nikitina, West Chester University
Robin O'Quinn, Eastern Washington University
Richard Orndorff, Eastern Washington University
Sharen Orndorff, Eastern Washington University
Clair Ossian, Tarrant County College
Debra Parish, North Harris Montgomery Community College District
Linda Peters, Holyoke Community College
Robin Popp, Chattanooga State Technical Community College
Michael Priano, Westchester Community College
Alan "Paul" Price, University of Wisconsin – Washington County
Max Reams, Olivet Nazarene University
Mary Celeste Reese, Mississippi State University
Bruce Rengers, Metropolitan State College of Denver
Guillermo Rocha, Brooklyn College
Penny Sadler, College of William and Mary
Shamili Sandiford, College of DuPage
Thomas Sasek, University of Louisiana at Monroe
Donna Seagle, Chattanooga State Technical Community College
Diane Shakes, College of William and Mary
Jennie Silva, Louisiana State University

Michael Siola, Chicago State University
Morgan Slusher, Community College of Baltimore County
Julia Smith, Eastern Washington University
Darlene Smucny, University of Maryland University College
Jeff Snyder, Bowling Green State University
Alice Stefaniak, St. Xavier University
Alicia Steinhardt, Hartnell Community College
Kurt Stellwagen, Eastern Washington University
Charlotte Stromfors, University of Phoenix
Shane Strup, University of Phoenix
Donald Thieme, Georgia Perimeter College
Pamela Thinesen, Century Community College
Chad Thompson, SUNY Westchester Community College
Lensyl Urbano, University of Memphis
Gopal Venugopal, Roosevelt University
Daniel Vogt, University of Washington – College of Forest Resources
Dr. Laura J. Vosejpka, Northwood University
Brenda L. Walker, Kirkwood Community College
Stephen Wareham, Cal State Fullerton
Fred William Whitford, Montana State University
Katie Wiedman, University of St. Francis
Harry Williams, University of North Texas
Emily Williamson, Mississippi State University
Bridget Wyatt, San Francisco State University
Van Youngman, Art Institute of Philadelphia
Alexander Zemcov, Westchester Community College

Student Participants

Karl Beall, Eastern Washington University
Jessica Bryant, Eastern Washington University
Pia Chawla, Westchester Community College
Channel DeWitt, Eastern Washington University
Lucy DiAroscia, Westchester Community College
Heather Gregg, Eastern Washington University
Lindsey Harris, Eastern Washington University
Brenden Hayden, Eastern Washington University
Patty Hosner, Eastern Washington University
Tonya Karunartue, Eastern Washington University
Sydney Lindgren, Eastern Washington University
Michael Maczuga, Westchester Community College
Melissa Michael, Eastern Washington University
Estelle Rizzin, Westchester Community College
Andrew Rowley, Eastern Washington University
Eric Torres, Westchester Community College
Joshua Watson, Eastern Washington University

Reviewers of Visualizing Human Geography

Joy Adams, Humboldt State University
Frank Ainsley, University Of North Carolina – Wilmington
Jennifer Altenhofel, California State University – Bakersfield
Jessica Amato, Napa Valley College
Christiana Asante, Grambling State University
Greg Atkinson, Tarleton State University
Timothy Bawden, University of Wisconsin – Eau Claire
Brad Bays, Oklahoma State University
Mark Bonta, Delta State University
Patricia Boudinot, George Mason University
Michaele Ann Buell, Northwest Arkansas Community College
Henry Bullamore, Frostburg State University
Kristen Conway-Gomez, California State Polytechnic University – Pomona
Carolyn Coulter, Atlantic Cape Community College
Christina Dando, Western Kentucky University
Jeff DeGrave, University of Wisconsin – Eau Claire
Ramesh Dhussa, Drake University
Dixie Dickinson, Tidewater Community College – Virginia Beach
Christine Drake, Old Dominion University
James Ebrecht, Georgia Perimeter College
Istvan Egresi, University of Oklahoma
William Flynn, Oklahoma State University
Piper Gaubatz, University of Massachusetts
Jerry Gerlach, Winona State University
Stephen Gibson, Allegany College of Maryland
Charles Gildersleeve, University of Nebraska- Omaha
Jeff Gordon, University of Missouri – Columbia
Margaret Gripshover, University of Tennessee- Knoxville
Joshua Hagen, Marshall University
Helen Hazen, Macalester College
Marc Healy, Elgin Community College
Bryan Higgins, SUNY – Plattsburgh
Juana Ibanez, University of New Orleans
Edwin Joseph, Grand Valley State University
William Laatsch, University of Wisconsin- Green Bay
Heidi Lannon, University of Wisconsin – Oshkosh
Robin Lyons, San Joaquin Delta College
Kenji Oshiro, Wright State University
Siyoung Park, Western Illinois University
Bimal Paul, Kansas State University
Cynthia Pope, Central Connecticut State University
Albert Rydant, Keene State College
James Saku, Frostburg State University
Anne Saxe, Saddleback College
Roger Selya, University of Cincinnati
Dean Sinclair, Northwestern State University
Anne Soper, Indiana University – Bloomington
Christophe Storie, Winthrop University
Tim Strauss, University of Northern Iowa
Ray Sumner, Long Beach City College
Joseph Swain, Arkansas Tech University
Richard Wagner, Louisiana Tech University
William Wheeler, Southwestern Oklahoma State University
Pat Wurth, Roane State Community College
Donald Zeigler, Old Dominion University at Virginia Beach

Survey Respondents

Gillian Acheson, Southern Illinois University
Joy Adams, Humboldt State University
Frank Ainsley, University of North Carolina – Wilmington
Victoria Alapo, Metropolitan Community College
Jennifer Altenhofel, California State University – Bakersfield
Robert Amey, Bridgewater State College
Brian Andrews, Southern Methodist University
Donna Arkowski, Pikes Peak Community College
Christiana Asante, Grambling State University
Michele Barnaby, Pittsburg State University
Steve Bass, Mesa Community College
Sari Bennett, University of Maryland, Baltimore County
Kathryn Besio, University of Hawaii – Hilo
Keith Bettinger, University of Hawaii
Phil Birge-Liberman, Bridgewater State College
Mark Bonta, Delta State University
Fernando Bosco, San Diego State University
Henry Bullamore, Frostburg State University
Rebecca Buller, University of Nebraska – Lincoln
John Burrows, Talladega College
Perry Carter, Texas Tech University – Lubbock
Lisa Chaddock, San Diego City College
Wing-Ho Cheung, Palomar College – San Marcos
Jerry Coleman, University of Southern Mississippi Gulf Coast
Kristen Conway-Gomez, California State Polytechnic University – Pomona
Carolyn Coulter, Atlantic Cape Community College
William Courter, Santa Ana College
G. Nevin Crouse, Chesapeake College
George Daugavietis, Solano Community College
Bruce Davis, Eastern Kentucky University
Jeff DeGrave, University of Wisconsin – Eau Claire
Lorraine Dowler, Pennsylvania State University
Anthony Dutton, Valley City State College
Markus Eberl, Vanderbilt University
Gary Elbow, Texas Tech University
Chuck Fahrer, Georgia College & State University
Johnny Finn, Arizona State University
Roxane Fridirici, California State University – Sacramento
Robert Fuller, North Georgia College & State University
Benjamin Funston-Timms, California Polytechnic State University – San Luis Obispo
Jerry Gerlach, Winona State University
Michael Giammarella, CUNY – Manhattan Community College
Omar Godoy, LACCD – East Los Angeles College
Banu Gokariksel, University of North Carolina
Marvin Gordon, University of Illinois at Chicago
Qian Guo, San Francisco State University
Steve Graves, California State University – Northridge
Angela Gray, University of Wisconsin – Oshkosh
Joshua Hagen, Marshall University
Katherine Hankins, Georgia State University
Timothy Hawthorne, Ohio State University
John Hickey, Inver Hills Community College
Miriam Helen Hill, Jacksonville State University
Larissa Hinz, Eastern Illinois University
Doc Horsley, Southern Illinois University
Ronald Isaac, Ohio University
Ryan James, University of North Carolina – Charlotte
Duncan Jamieson, Ashland University
Wendy Jepson, Texas A & M University – College Station
Chad Kinsella, Kentucky Community & Technical College
Marti Klein, Cypress College
Richard Kujawa, Saint Michael's College
Margareta Lelea, Bucknell University
David Lemberg, Western Michigan University
Anne Lewis, Allegany College of Maryland
Joseph Lewis, Ohio State University
David Liscio, Endicott College
Lee Liu, University of Central Missouri
James Lowry, University of New Orleans
Ronald Luna, University of Maryland
Kerry Lyste, Everett Community College
Taylor Mack, Louisiana Technical University
Michael Madsen, Brigham Young University
Christine Mathenge, Austin Peay State University
Richard McCluskey, Aquinas College
Frank McComb, Georgia Perimeter College – Clarkston
Mark Meo, University of Oklahoma
Diane Meredithn, California State University – East Bay
Silva Meybatyan, University of the District of Columbia
Pam Miller, College of Eastern Utah
Linda Murphy, Blinn Community College
Natalia Murphy, Southern Arkansas University
Hemalatha Navaratne, Borough of Manhattan Community College
Tom Newton, Kirkwood Community College – Iowa City
Kenji Oshiro, Wright State University
Seth Parry, Emmanuel College
Dan Pavese, Wor-Wic Community College
Michael Pesses, Antelope Valley College
Ingrid Pfoertsch, Towson University
Nathan Phillippi, University of North Carolina – Pembroke
Colin Polsky, Clark University
William Price, North Country Community College
Larshale Pugh, Youngstown State University
Melanie Rapino, University of Memphis
Eike Reichardt, Lehigh Carbon Community College
Robert Ritchie, Liberty University
Julio Rivera, Carthage College
Alicia Roe, Inter American University of Puerto Rico – Metropolitan
Karl Ryavec, University of Wisconsin – Stevens Point
James Saku, Frostburg State University
Samuel Sawaya, Sinclair Community College
Andrew Scholl, Wittenberg University
Anita Shoup, CUNY – Hunter College
Steven Silvern, Salem State College
Michael Siola, Chicago State University
Sarah Smiley, Morgan State University
Lisa Stanich, Lakeland Community College
Herschel Stern, Miracosta College
Mary Tacy, James Madison University
Jane Thorngren, San Diego State University

Dan Turbeville, Eastern Washington University
Richard Tyre, Florida State University
David Unterman, University of North Carolina – Greensboro/ Sierra College
Wendy Welch, University of Virginia's College at Wise
Ben Wolfe, Metropolitan Community College – Blue River
Louis A. Woods, University of North Florida
Dawn Wrobel, Moraine Valley Community College
Patricia Wurth, Roane State Community College
Leon Yacher, Southern Connecticut State University
Keith Yearman, College of DuPage
Lei Xu, California State University – Fullerton
Laura Zeeman, Red Rocks Community College
Robert Ziegenfus, Kutztown University of Pennsylvania
William Zogby, Mohawk Valley Community College
Kathleen Zynda, Erie Community College – North Campus

Students and Class Testers

To make certain that *Visualizing Human Geography* met the needs of current students, we asked several instructors to class-test a chapter. The feedback that we received from students and instructors confirmed our belief that the visualizing approach taken in this book is highly effective in helping students to learn. We wish to thank the following instructors and their students who provided us with helpful feedback and suggestions:

Christiana Asante, Grambling State University
Mark Bonta, Delta State University
Patricia Boudinot, George Mason University
Michaele Ann Buell, Northwest Arkansas Community College
Hank Bullamore, Frostburg State University
Chuck Fahrer, Georgia College and State University
Marti Klein, Cypress College
John Kostelnick, Illinois State University
Kerry Lyste, Everett Community College
John Menary, Long Beach City College
Siyoung Park, Western Illinois University
Cindy Pope, Central Connecticut State University
Larshale Pugh, Youngstown State University
Stacey Roush, Montgomery Community College
James Saku, Frostburg State University
Roger Selya, University of Cincinnati
Tim Strauss, University of Northern Iowa
Amy Sumpter, Georgia College and State University
Nicholas Vaughn, Indiana University Bloomington
Pat Wurth, Roane State Community College
Donald Zeigler, Old Dominion University at Virginia Beach

Dedication

To Luis Montes, my husband, whose unwavering support and enthusiasm made this possible and who so unselfishly contributed his time and advice to this project. Thank you for believing in me.

Special Thanks

It has been a tremendous experience working with so many professionals who have been deeply dedicated to this project. I owe a special thanks to Jerry Correa and Anne Smith, for their initial encouragement and infectious enthusiasm about the Wiley Visual Imprint Series. I sincerely appreciate David Kear's incisive comments on an early draft of Chapter 4. In addition, Lindsay Lovier, Kristen Gardner, Courtney Nelson, and Denise Powell expeditiously handled many of the administrative details in the first stages of this project.

I am grateful to Zeeya Merali, who helped generate the art manuscripts for Chapters 2, 3, and 11 and then kindly improved the organization of material in those chapters as well. The art developer, Fred Schroyer, performed amazing feats by transforming my verbal descriptions into coherent and meaningful process diagrams, or other visualizing graphics. I am similarly indebted to Dierdre Bevington-Attardi for her invaluable expertise on map design and map projections.

The photo selection process greatly benefited from the careful management provided by Hilary Newman, Director of Photo Research. Stacy Gold, Research Editor and Account Executive at the National Geographic Society, diligently mined the Society's image collection, and Teri Stratford very patiently endured my exacting requests for specific photos and then willingly embarked on wild goose chases to find them.

Special thanks go to the creative and talented *WileyPLUS* team of Lynn Pearlman, Anita Castro, Alexey Pankov, Ekaterina Kitayeva, Olga Rudnikova, and Mark Lipowicz for their work in creating an online learning and teaching environment for the human geography course.

My thanks also go to Veronica Armour, Assistant Editor; Margaret Barrett, Marketing Manager; and Susan Matulewicz, Senior Marketing Assistant, for the great energy, creativity, and assistance they have provided along the way. Joy Adams, Assistant Professor at Humboldt State University, generously helped identify and recommend videos for the different chapters. Brian Baker, Project Editor, always provided prompt technical assistance and guidance with the videos as well.

Every chapter reflects the wisdom of Developmental Editor Barbara Conover. Her insights as well as her skillful editing enhanced the structure and cohesiveness of the chapters. This project has benefited greatly from the capable direction of Micheline Frederick, Content Manager, and Kerry Weinstein, Senior Production Editor, as well as the production assistance provided by Christine Cervoni of Camelot Editorial Services, LLC, and Assunta Petrone of Preparé, Inc.

I owe a very special thanks also to Nancy Perry, Manager, Product Development, whose contributions to this book are simply innumerable. On a day-to-day basis she suffered through the challenge of working with a new textbook author and did so with remarkable patience and dedication. Development Editorial Assistants, Sean Boda and Brittany Cheetham, also contributed generously of their time to this project, as did Jay Schroyer as he skillfully mapped out the pages of the chapters. In addition, Sandra Rigby, Senior Illustration Editor, provided superb oversight of the art production process, which involved the creation of a host of new pieces and top-notch cartography. I greatly appreciate the swift and efficient permissions assistance provided by Alissa Etrheim.

My students and advisees at Oklahoma State University have been a great source of inspiration for me. I am fortunate to have wonderful colleagues in the Department of Geography who generously provided valuable suggestions, advice, or resources on a moment's notice. I am extremely grateful to my Department Head, Dale Lightfoot, for the support he has shown me throughout the duration of this project. I greatly appreciate the assistance provided by Rebecca Sheehan, who very graciously agreed to teach one of my courses in order to help lighten my teaching load one semester and provided detailed comments on Chapter 6. Sincere thanks also go to Jacqueline Vadjunec, who not only read and commented on several chapters but also suggested ways of improving my examples and definitions and helped me to think through some larger conceptual issues.

Finally, this project would not have been possible without the behind-the-scenes support and assistance provided by numerous other members of the Wiley team. In particular, I thank Beth Tripmacher, Project Editor of the Wiley Visual Imprint, for setting and upholding such high standards. I would also like to thank my Executive Editor, Ryan Flahive; Vice-President and Publisher, Jay O'Callaghan; and Director of Development, Barbara Heaney. I am grateful for their steadfast support and enthusiasm, as well as their efforts on behalf of geography in higher education.

About the Author

Alyson L. Greiner is Associate Professor of Geography at Oklahoma State University. She earned her PhD in Geography from the University of Texas at Austin. She has taught courses on cultural geography, world regional geography, the history of geographic thought, and the regional geography of Europe, Africa, and the Pacific Realm. She regularly teaches undergraduate, graduate, and honors students. In 2000, she received a Distinguished Teaching Achievement Award from the National Council for Geographic Education. Her scholarly publications include *Anglo-Celtic Australia: Colonial Immigration and Cultural Regionalism* (with Terry G. Jordan-Bychkov) and several peer-reviewed journal articles. She is presently the editor of the *Journal of Cultural Geography* and a Regional Councilor for the AAG.

Contents in Brief

Table of Contents

1 What Is Human Geography?

2 Globalization and Cultural Geography

3 Population and Migration

4 Geographies of Language

5 Geographies of Religion

6 Geographies of Identity: Race, Ethnicity, Sexuality, and Gender

7 Political Geographies

8 Urban Geographies

9 Geographies of Development

10 Changing Geographies of Industry and Services

11 Agricultural Geographies

12 Environmental Challenges

INSIGHT FEATURES

Multi-part visual presentations that focus on a key concept or topic in the chapter

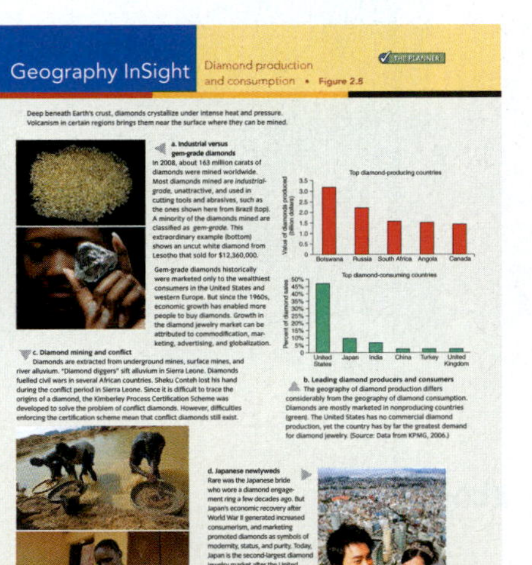

Chapter 1
Remote sensing of post-earthquake damage
A GIS for studying disease incidence

Chapter 2
Diamond production and consumption

Chapter 3
Population control in China
Population pyramids

Chapter 4
Nonspoken languages
Geographies of language diffusion
Language hotspots

Chapter 5
Islam's Five Pillars of practice

Chapter 6
The rise and fall of apartheid

Chapter 7
Café para todos? A model of integration in multinational Spain

Chapter 8
Food deserts
Hybrid city

Chapter 9
The environmental vulnerability index

Chapter 10
Gender mainstreaming in the service sector

Chapter 11
The Green Revolution
Change in the Corn Belt

Chapter 12
Oil reserves

PROCESS DIAGRAMS

A series or combination of figures and photos that describe and depict a complex process

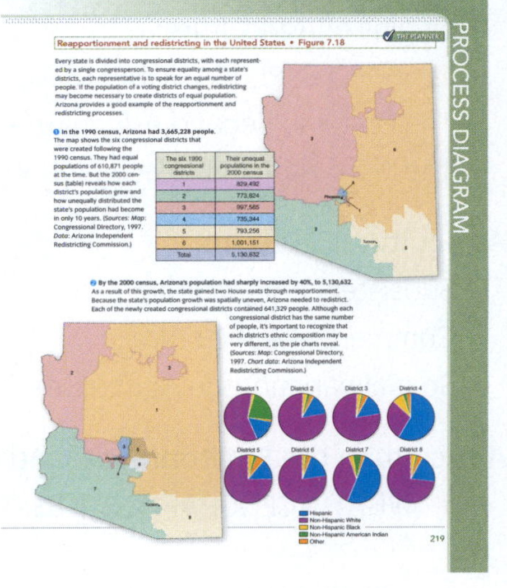

Chapter 1
Understanding hierarchical diffusion

Chapter 2
The diffusion of acupuncture

Chapter 3
Demographic transition model

Chapter 4
Understanding language vitality and endangerment: The example of Yuchi

Chapter 5
Sanctification

Chapter 6
The interaction between race and place

Chapter 7
Reapportionment and redistricting in the United States

Chapter 8
How changes in transportation influence urban form
Slum formation

Chapter 9
Classical model of development

Chapter 10
Manufacturing value added and profit captured in a video iPod

Chapter 11
Four-course crop rotation
Shifting cultivation

Chapter 12
Understanding mountaintop removal

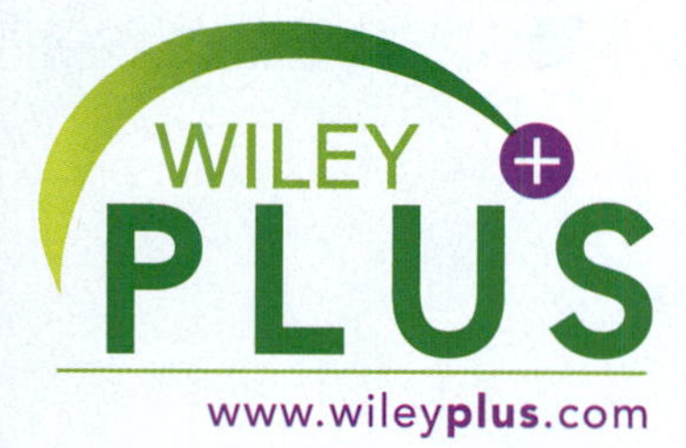
WILEY
PLUS
www.wileyplus.com

1 What Is Human Geography?

GEOGRAPHICAL INQUIRY

Can you find your hometown or city on this image of the Earth at night? Look again at the image. How do the spaces of illumination vary from one continent to another, and how do they vary regionally within continents? What is the relationship among population, urbanization, and night lights? How would you explain the distribution of nighttime illumination? Taking these ideas a step further, what inferences can you make about global interconnectedness, the accessibility of poorly versus well-illuminated places, patterns of human settlement, or the relationship between "light pollution" and environmental modification?

Geographers ask these and similar kinds of questions. Embedded within such questions are concepts relating to location, place, space, region, scale, distribution, and interconnectedness. Thus, geographical inquiry has its roots in a fundamental curiosity about the world. However, there is more to geographical inquiry than simply asking questions. The ability to step back when studying a topic or phenomenon and examine relationships between data in order to generate new insights about how the world works is also important. By doing this, geographical inquiry and analysis contribute to the development of geographical theory—knowledge that advances our understanding of the social, spatial, regional, and ecological facets of our world.

Simply stated, this book is designed to introduce you to geographical inquiry and theory through a perspective that emphasizes people and the spatial variation in their activities around the world. This chapter introduces human geography and illustrates how geographers approach their work, including some of the tools they use.

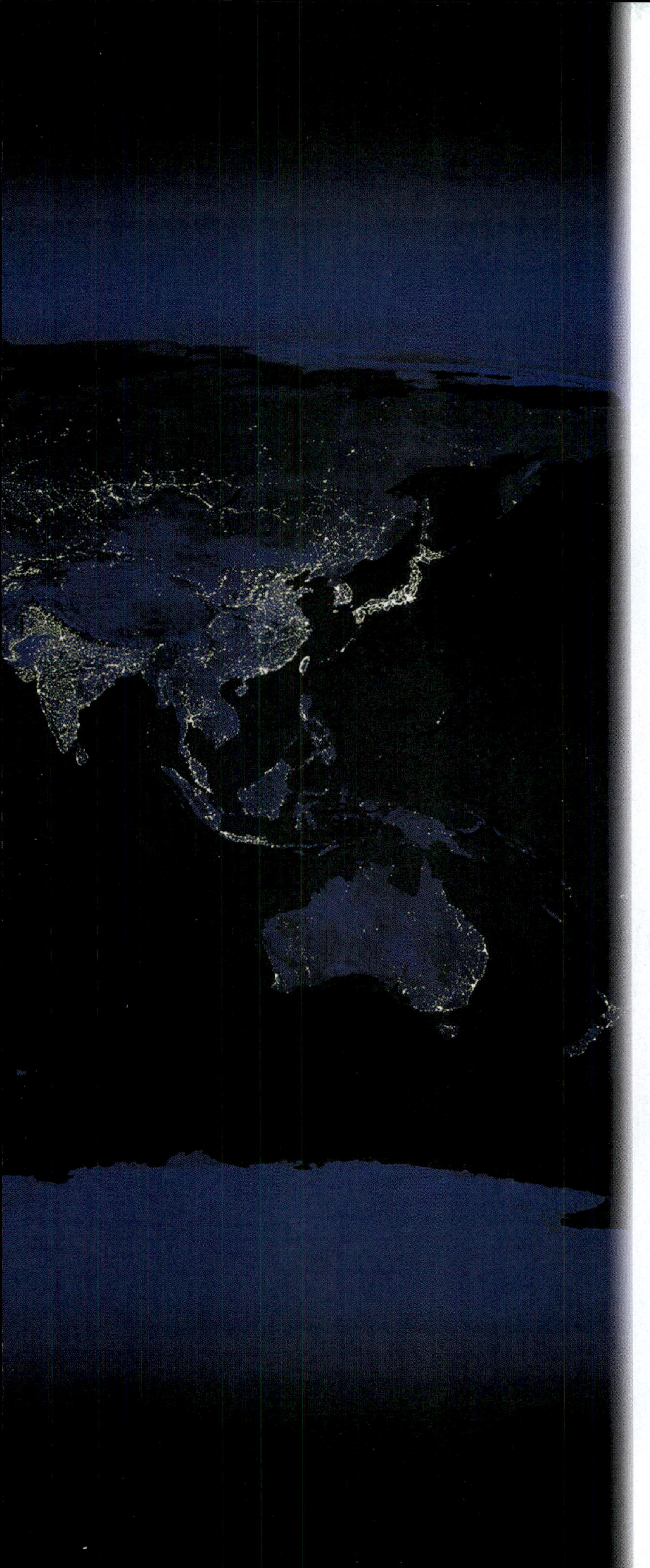

CHAPTER OUTLINE

CHAPTER PLANNER

- ❑ Study the picture and read the opening story.
- ❑ Scan the Learning Objectives in each section: p. 4 ❑ p. 12 ❑ p. 23 ❑
- ❑ Read the text and study all visuals. Answer any questions.

Analyze key features

- ❑ Process Diagram, p. 16
- ❑ What a Geographer Sees, p. 21
- ❑ Video Explorations, p. 23
- ❑ Geography InSight, p. 24 ❑ p. 29 ❑
- ❑ Stop: Answer the Concept Checks before you go on: p. 12 ❑ p. 23 ❑ p. 29 ❑

End of chapter

- ❑ Review the Summary and Key Terms.
- ❑ Answer the Critical and Creative Thinking Questions.
- ❑ Answer What is happening in this picture?
- ❑ Complete the Self-Test and check your answers.

Introducing Human Geography

LEARNING OBJECTIVES

1. **Explain** the basis for the nature-culture dualism.
2. **Review** actor-network theory.
3. **Compare** and contrast formal, functional, and perceptual regions.
4. **Explain** the recent reconceptualization of culture.

We are going to let you in on a little secret: Geography majors go places—in their careers, that is. They also have a lot of fun in the process. This is quite likely because geography is a discipline that encourages people to find a topic or region they are passionate about and explore its many different dimensions. Are you interested in music? Music geographers are needed to understand the globalization of hip-hop as well as its local variations. If you are a sports fan, sports geographers help identify optimal locations for stadiums, golf courses, and other athletic facilities. If your passion is nutrition or health, medical geographers help track and limit the spread of epidemics and study ways to improve people's access to medical care (see *Where Geographers Click*).

Some nongeographers rather naively thought that globalization would make geography irrelevant. Globalization, they claimed, made the world smaller, more accessible, and therefore, easier to know and understand. Meanwhile, geographers politely noted that globalization was not a new phenomenon and that geography had, to the contrary, taken on even greater relevance. For example, understanding the consequences of global climate change on different countries, agricultural production, and coastal populations demands geographic awareness. Similarly, we cannot solve the problem of poverty until we know better its geographic dimensions—where it occurs, how spatially extensive it is, who it affects, and how it is related to access to resources, such as land, water, and housing. Globalization has moved geography to center stage. Simultaneously, improvements and innovations in technology have expanded the geographer's toolbox. These new tools include ways of acquiring data about the Earth with improved GPS receivers, higher resolution satellite imagery, and new ways of visualizing this information with virtual globes such as Google Earth.

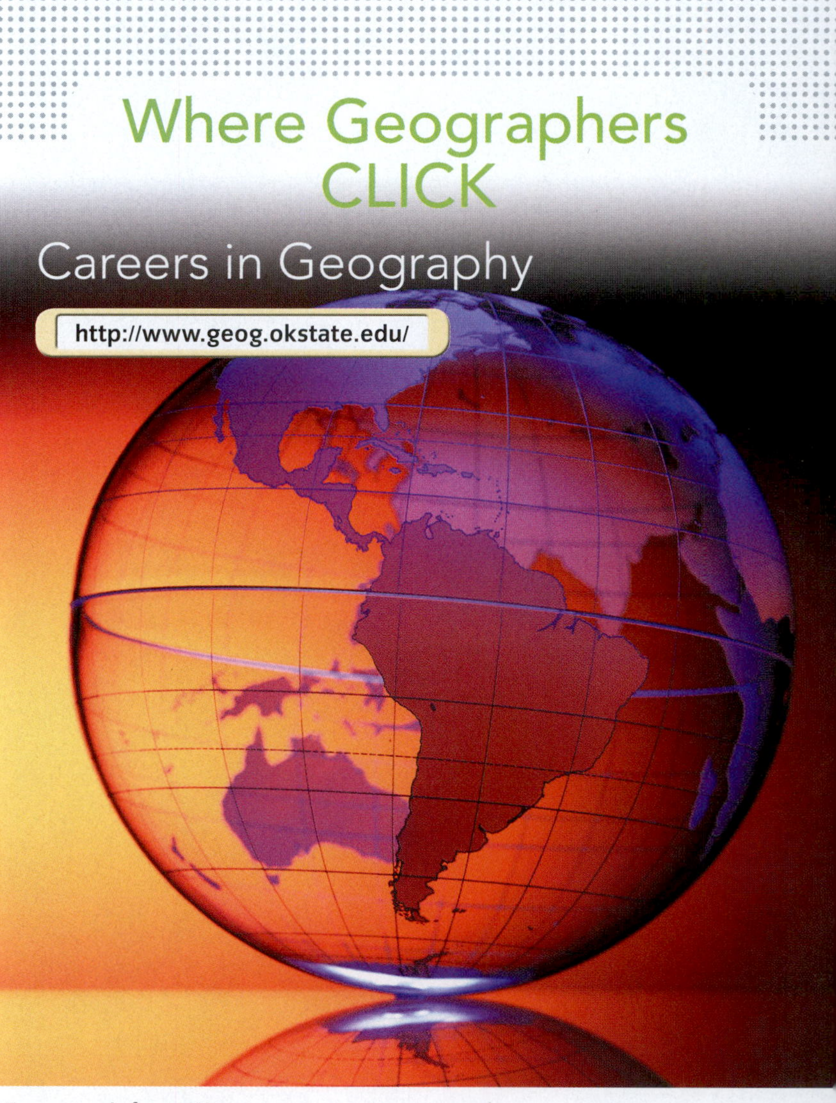

For more information on careers in geography, follow the link and click on the globe.

The word *geography* derives from Greek words (*geo* + *graphia*) meaning *to write about or describe the Earth*. As noted above and in the introduction to this chapter, however, geography is much more than a description of the Earth or a factual listing of countries, their capitals, and resources.

Geography consists of two main branches: physical geography and human geography. Physical geography focuses on *environmental dynamics* (e.g., water quality, soil erosion, forest management) whereas **human geography** focuses on *social dynamics* (e.g., economic development, language diffusion, ethnic identity). Some physical and human geographers focus on *environment-society dynamics* and work on topics that span both branches of the discipline (e.g., vulnerability to environmental hazards, impacts of fossil fuel consumption, social consequences of global climate change). The unity of geography as a discipline stems from a shared philosophy that recognizes the urgency of better understanding the spatial aspects of human and environmental processes and using geographic knowledge to generate solutions to the social and environmental challenges in our world. More specifically, the branch of human geography consists of several different subfields (see **Figure 1.1** on the next page).

human geography
A branch of geography centered on the study of people, places, spatial variation in human activities, and the relationship between people and the environment.

Nature and Culture

What do the words *nature* and *culture* mean to you? At first they seem straightforward, but the longer you think about them the more you realize that they both have a variety of different meanings. For example, nature can refer to the intrinsic qualities of a person, or to the outdoors, and culture can refer to taste in the fine arts or to customary beliefs and practices. Because of this definitional looseness, geographer Noel Castree (2001, p. 5) calls *nature* "a promiscuous concept." The same can be said about *culture*. Nevertheless, these concepts are so fundamental to the practice of geography that we should examine them briefly here.

Very broadly speaking, **nature** is the physical environment; it is external to people and does not include them. People, because of their capacity for intellectual and moral development, are the bearers of **culture**, and it is culture that distinguishes people from nature. When understood in this way, these concepts yield a dualistic framework that sets nature and culture in opposition to one another.

This **nature-culture dualism** has had a significant impact on ways of thinking about social difference. During the 18th century, some European scholars used this distinction between nature and culture to argue that it was the human capacity for culture that made people *superior* to nature. This line of reasoning was subsequently extended and used to rank societies. So, for example, non-Westerners were seen as being closer to nature than so-called civilized and cultured Westerners. Although the origins of these ideas are difficult to unravel, they matter because the way we see human societies in relation to nature and to one another affects not just how we use the environment but also how we interact with others.

Today, many geographers and other social scientists reject the nature-culture dualism because of the way it separates nature from culture. These scholars stress instead that people—in spite of their capacity for culture—are very much a part of nature. This perspective is central to **cultural ecology**, an important subfield within human geography that studies the relationship between people and the natural environment.

When conceptualizing the relationship between people and nature, cultural ecologists and other geographers recognize several different approaches. We discuss four of these next: environmental determinism, possibilism, humans as modifiers of the Earth, and the Earth as a dynamic, integrated system.

Environmental determinism People who take the position that natural factors control the development of human physiological and mental qualities espouse **environmental determinism**. We can trace the intellectual roots of environmental determinism in Western thought to the ancient Greeks, who speculated that human diversity resulted from both climatic and locational factors. For example, plateau environments seemed to produce people who were docile. Similarly, they thought that climatic extremes adversely affected mental capacities. The people with the sharpest minds came from temperate

The scope of human geography • Figure 1.1

Complex linkages between human and environmental systems characterize our world. Human geography is a diverse field that encompasses a number of different subfields (see inset). Though highly simplified, this diagram illustrates some of the components in the web of connections that human geographers study.

Major Subfields of Human Geography

Cultural Ecology
Cultural Geography
Economic Geography
Historical Geography
Political Geography
Population Geography
Urban Geography

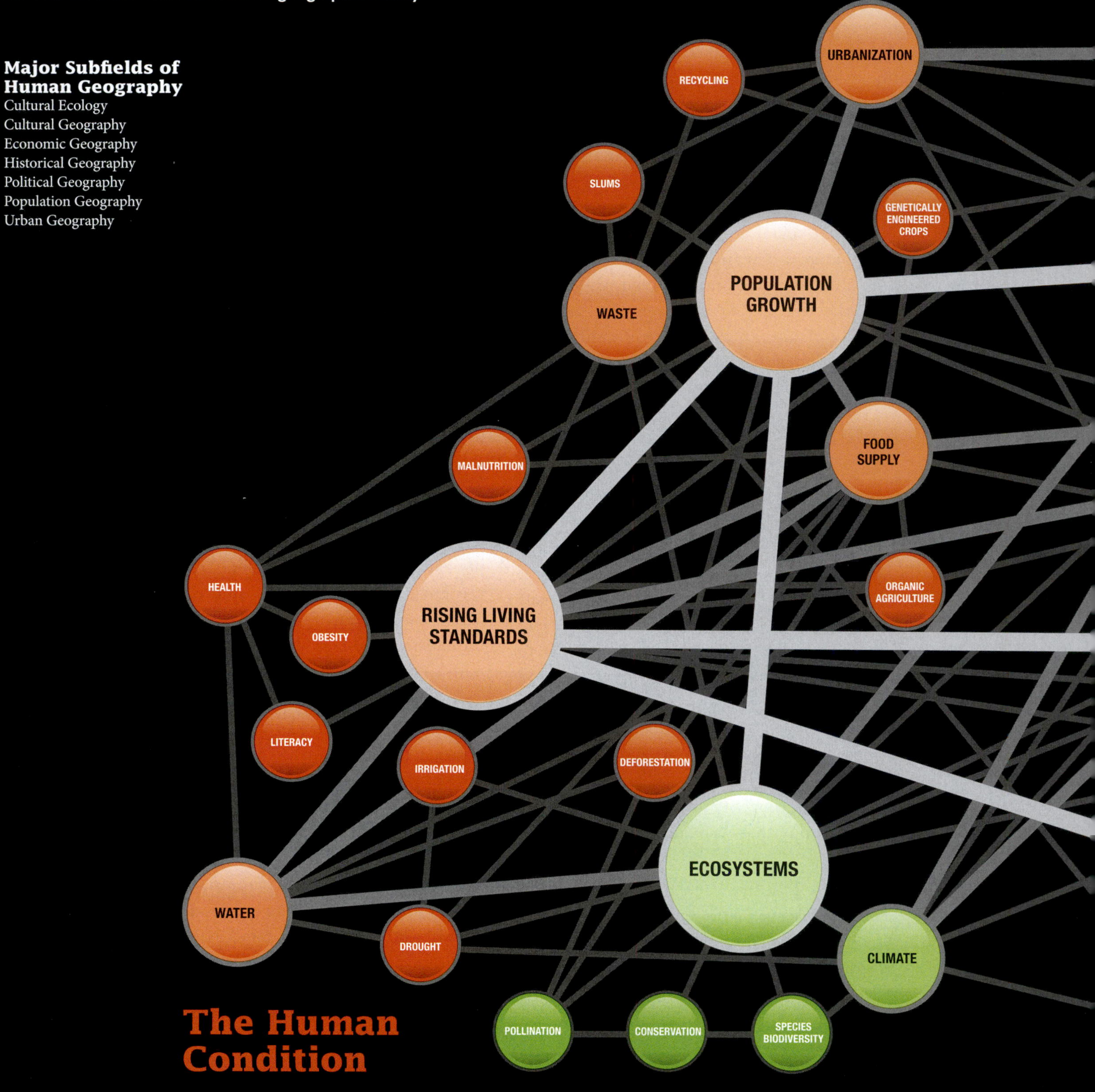

Our Relationship with Nature

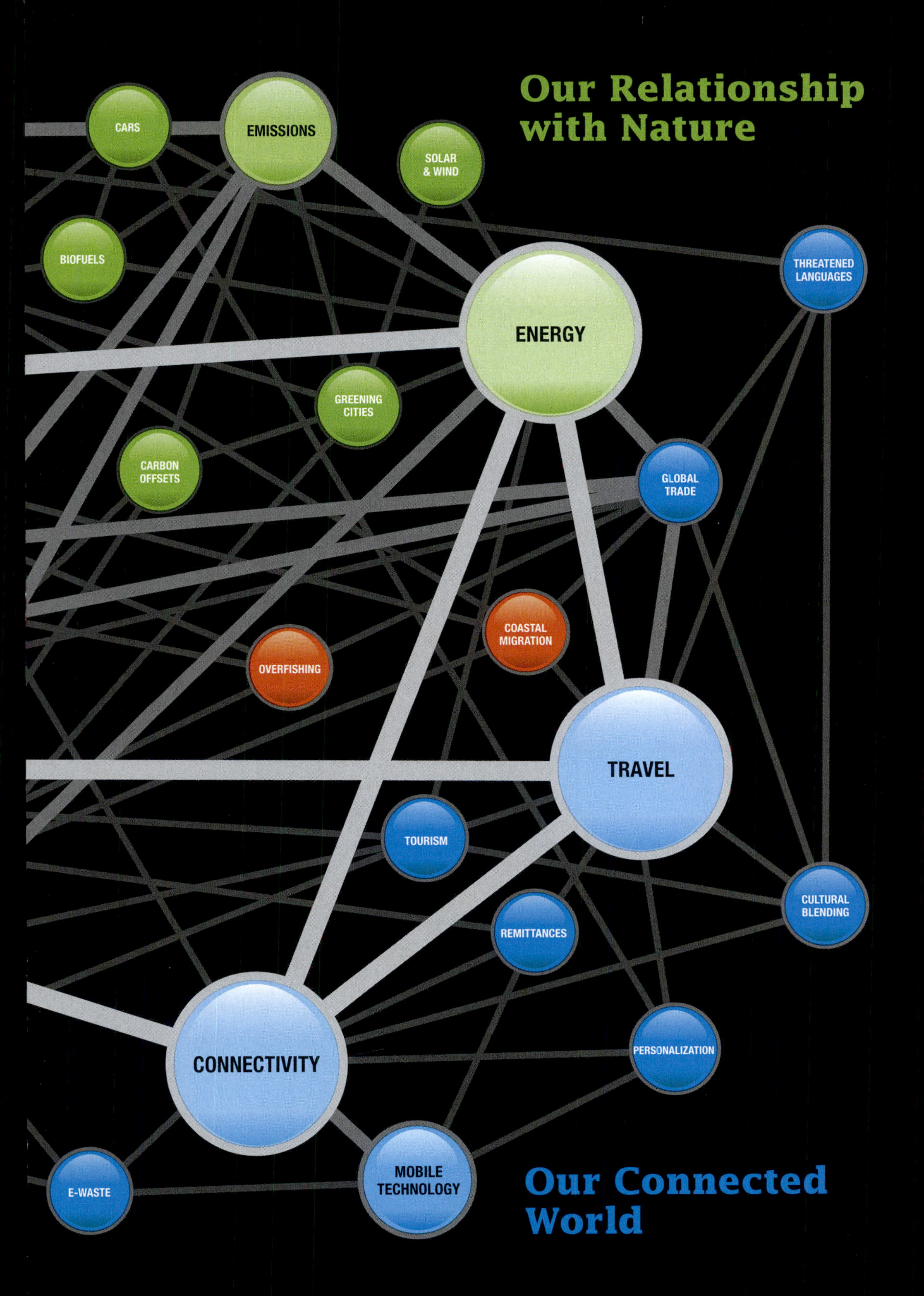

Our Connected World

The four elements and environmental determinism • Figure 1.2

Some ancient scholars thought that all natural phenomena, including people, were made up of the four elements in varying degrees. Environmental determinism attributed cultural difference to human traits that reflected these four elements and were strongly shaped by physical factors, including climate.

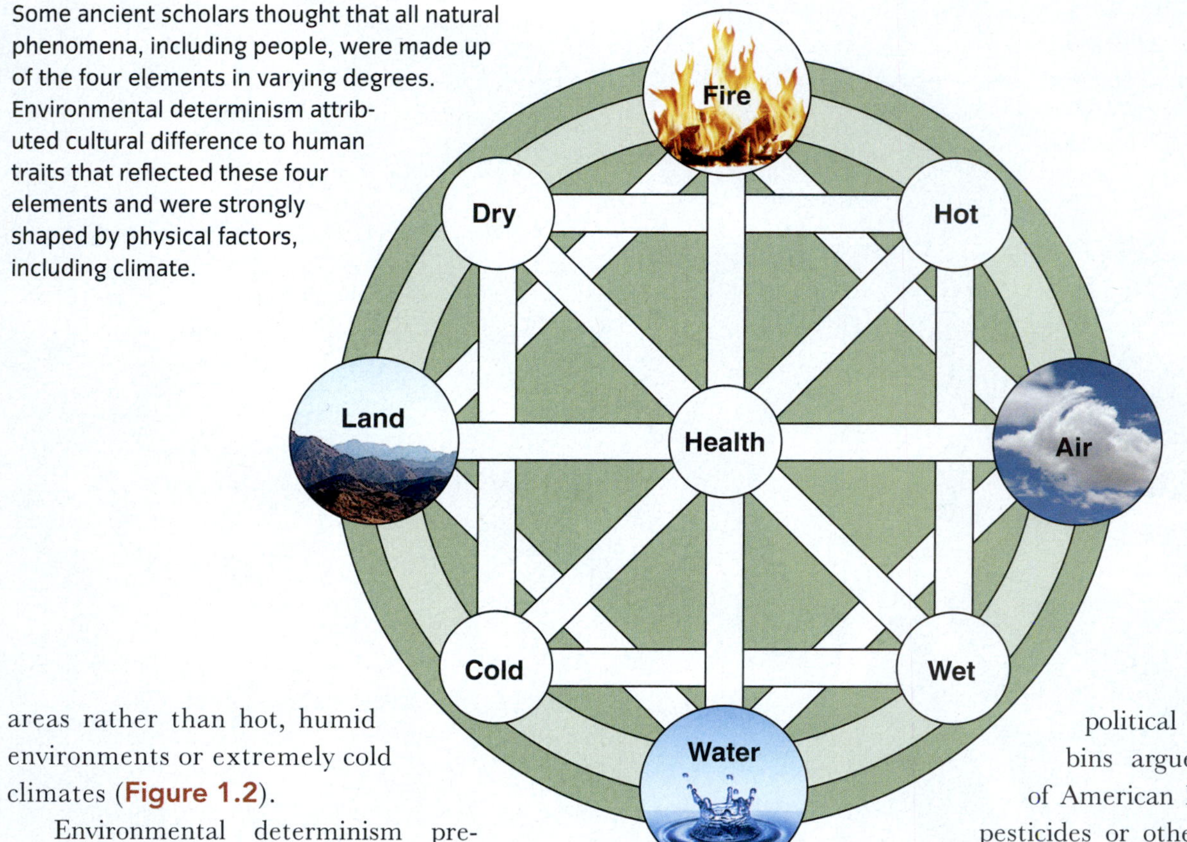

areas rather than hot, humid environments or extremely cold climates (**Figure 1.2**).

Environmental determinism prevailed among American geographers during the early 20th century and then fell quickly into disfavor. Three major criticisms of environmental determinism prompted this change in perspective. First, geographers found overly simplistic the linear, cause-effect relationship that forms the basis of environmental determinism. People, they argued, are more than automatons that simply respond to stimuli, such as the prevailing winds or temperatures in a specific place. Nonenvironmental factors, such as systems of government and law, also help explain human diversity. A second criticism of environmental determinism is that similar natural settings do not produce the same cultural practices or human behavior. Third, environmental determinism tends to contribute to ethnocentric interpretations of sociocultural differences. It is therefore not much of a surprise that some ancient Greek scholars attributed the flourishing of the Greek civilization to the temperate climate of the Mediterranean.

In recent years a radical reinterpretation of environmental determinism has emerged within **political ecology** that involves **actor-network theory**. For example, in his book *Lawn People,* the political ecologist Paul Robbins argues that the decision of American homeowners to apply pesticides or other chemicals to their lawns is the product of multiple interacting factors. These factors include the supply of and demand for lawn chemicals, the importance of property values, community pressure to maintain a well-kept lawn, lawn aesthetics (e.g., ideas about how a lawn should look), and the lawn itself (**Figure 1.3**).

political ecology An offshoot of cultural ecology that studies how economic forces and competition for power influence human behavior, especially decisions and attitudes involving the environment.

actor-network theory A body of thought that emphasizes that humans and nonhumans are linked together in a dynamic set of relations that, in turn, influence human behavior.

Actor-network theory challenges the idea that people have free will. Rather, nonhuman entities gain agency (the ability to exert influence) by virtue of the networks of relations in which they are embedded. As Robbins observes, "the nonhuman world does have an active, ongoing, and crucial role in directing the conditions of the economy and the character of human culture" (2007, p. 137). Unlike environmental determinism, actor-network theory gives agency to natural factors as well as anything human-made (e.g., lawns, machines, or laws) but not in a simplistic cause-effect relationship.

Possibilism Reactions against environmental determinism in the early 20th century gave rise to **possibilism**—the view that people use their creativity to decide how to respond to the

Actor-network theory • Figure 1.3

Actor-network theory acknowledges that our surroundings influence us. Can you see from this diagram how the lawn, the availability of fertilizers, and aesthetics influence human behavior by prompting a homeowner to mow, fertilize, and maintain it? (*Source*: Adapted from P. Robbins, 2007.)

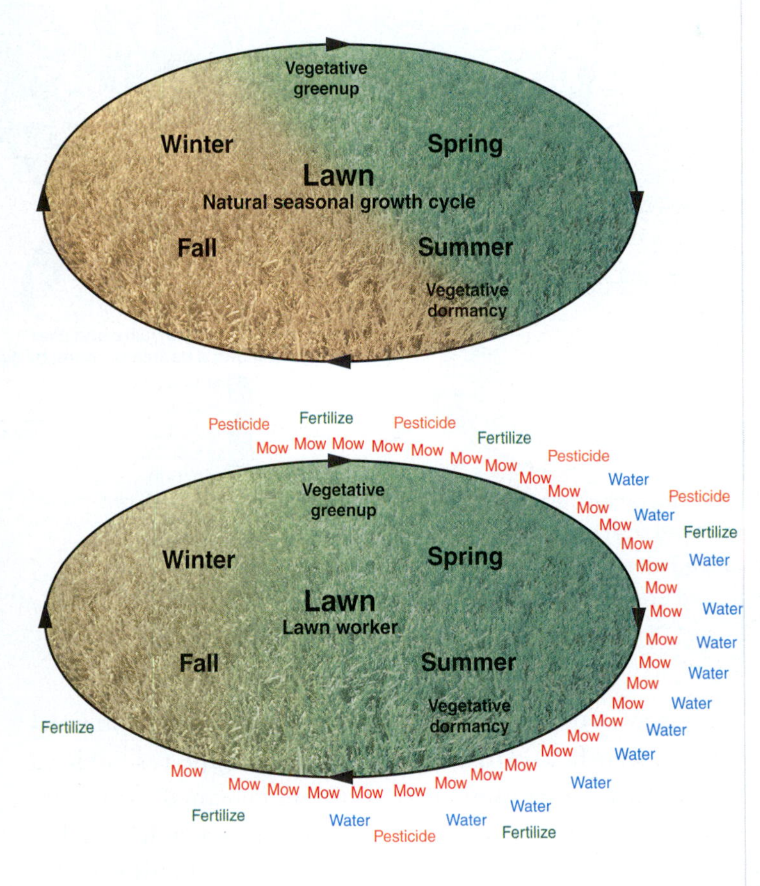

conditions or constraints of a particular natural environment. The word *constraints* is important here because it indicates that the environment is seen as limiting the choices or opportunities that people have. Possibilists, then, do not completely reject the idea of environmental influence; however, they are reluctant to view the environment as the sole or even the strongest force shaping a society. Thus, a possibilist sees technological diversification as one mechanism for expanding the range of choices a society has.

Humans as modifiers of the Earth A different approach to the relationship between people and the environment was advanced by geographer Carl Sauer (1889–1975), beginning in the 1920s. Sauer rejected environmental determinism and emphasized instead human agency, the ability of people to modify their surroundings. He observed that, over time, human activities transform natural landscapes into **cultural landscapes**. Significantly, Sauer's work helped raise awareness of the human role in landscape change. Visually, evidence of humans as modifiers of the Earth is all around us, from our cities to our cultivated agricultural fields (**Figure 1.4**).

An important extension of the humans as modifiers of the Earth approach involves the view that nature is a *social construction*—an invented concept that derives from shared perceptions and understandings. This perspective acknowledges that people shape the natural environment through their practices *and* their ideas about what nature is or should be. A good example of this involves the idea of wilderness in the United States. The environmental historian, William Cronon, has shown that in the 18th century wilderness was equated with wasteland, but by the 19th century wilderness was strongly associated with natural beauty.

Earth as a dynamic, integrated system In this approach, geographers see people as intricately connected with the natural world. Two key principles sum up this approach: (1) the Earth functions as a system made up of diverse components that interact in complex ways; and

An extreme cultural landscape? • Figure 1.4

If your country lacks snow-covered mountains, why not manufacture them? This mountain-themed resort facility is in the United Arab Emirates and features year-round skiing even though outside temperatures rarely dip below 70° Fahrenheit.

Formal, functional, and perceptual regions • Figure 1.5

All kinds of private businesses and governmental and planning agencies make decisions based on spatial information related to these three types of regions.

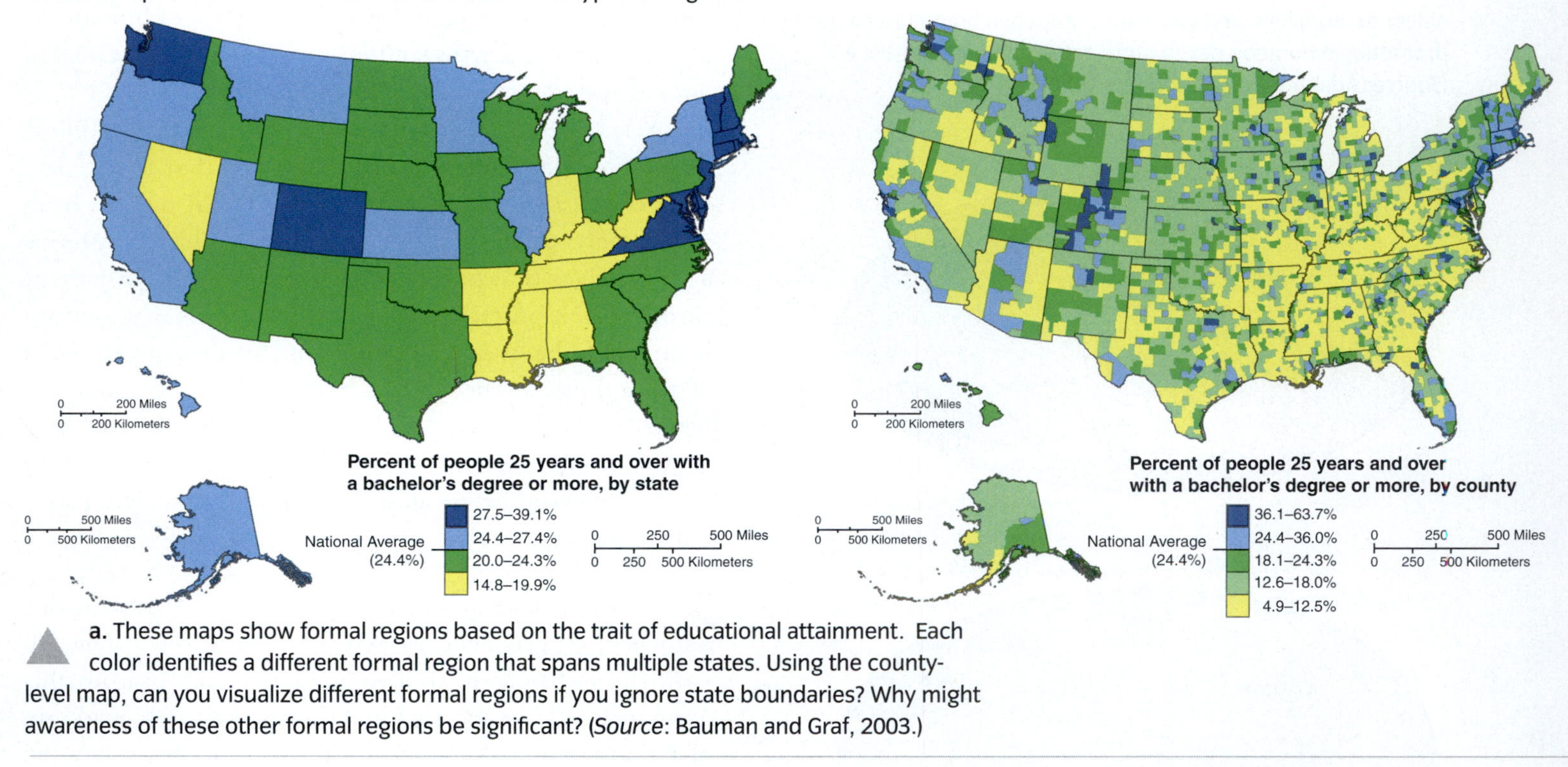

a. These maps show formal regions based on the trait of educational attainment. Each color identifies a different formal region that spans multiple states. Using the county-level map, can you visualize different formal regions if you ignore state boundaries? Why might awareness of these other formal regions be significant? (*Source*: Bauman and Graf, 2003.)

(2) the Earth is constantly changing as a result of natural and human-induced events. We explore these ideas further in Chapter 12.

Cultural Landscapes and Regions

As we have discussed, culture is sometimes used to refer to a person's intellectual improvement through education, particularly the development of an aesthetic appreciation for the arts. In other instances, culture refers to beliefs and practices—such as dietary customs, religious beliefs, and so on—held in common by a group of people. Thus, a cultural group shares certain traits or elements of culture. This understanding of culture guided much of the practice of human geography until the late 20th century. More specifically, two long-standing approaches to the study of culture emphasize reading the cultural landscape and performing regional analysis. The emphasis on cultural landscapes reflects Carl Sauer's influence on geography, especially his view that culture is the driving force for landscape change.

Reading the cultural landscape works from the premise that the cultural landscape constitutes a rich repository of information about cultural beliefs and practices. In other words, the cultural landscape resembles a *palimpsest*—a parchment that, though cleaned, still bears the traces of what was previously inscribed on it. To a human geographer, the visible expressions of culture—for example, the settlement patterns, the structures people build, the architectural styles they choose, and the ways people use land—all provide clues about people's values, identity, and more broadly, their cultures.

Regional analysis involves studying the distinctiveness of regions. This might include understanding how and why the South differs from New England culturally, economically, and politically. Or regional analysis might examine the ways in which the present-day war in Iraq has altered the demographic and religious makeup of the country's provinces and the ramifications of these changes.

Types of regions

Geographers identify three types of regions: formal, functional, and perceptual. A **formal region** is an area that possesses one or more unifying physical or cultural traits. Unlike formal regions, a **functional region** is an area unified by a specific

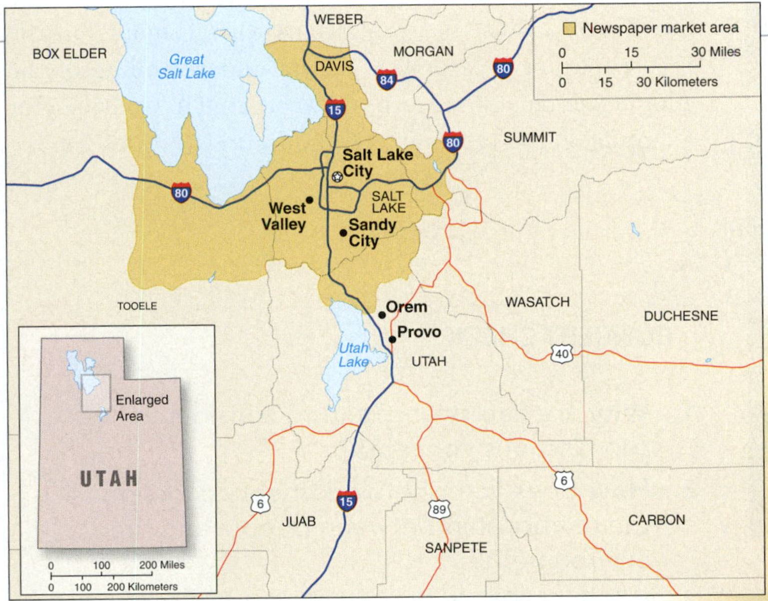

b. Primary market area
A functional region might consist of the main area served by a business. This map shows the functional region for two newspapers: the *Salt Lake Tribune* and the *Deseret Morning News*. How can two newspapers have the same functional region?

c. The perceptual region known as Orange Country surrounds Oklahoma State University. In addition to universities and sports teams, what other factors contribute to the development of perceptual regions?

economic, political, or social activity. Every functional region has at least one node, usually the business, office, or entity that coordinates the activity. For example, each state in the United States constitutes a functional region with its state capital serving as the node. In contrast to both formal and functional regions, **perceptual regions** derive from people's sense of identity and attachment to different areas. The borders of perceptual regions tend to be highly variable since people often have very personal reasons for perceiving an area a certain way (**Figure 1.5**).

Culture reconceptualized Recently, certain geographers have stressed the point that we should think of culture as an abstract concept, not as a material item or collection of cultural traits. According to Don Mitchell, for example, "There's no such *thing* as culture" (emphasis added) (1995, p. 102). By this he means to caution people against trying to limit culture to specific and fixed habits of life. In his view, the visible and tangible expressions of culture are important, but they need to be understood in their dynamic context—in relation to prevailing economic, social, political, and other factors. Similarly, other geographers stress that an understanding of culture that defines the term as a way of life fails to recognize other crucial aspects of culture. Consequently, over the past several decades there has been a significant reconceptualization of **culture** that draws on the following three attributes:

culture A social creation consisting of shared beliefs and practices that are dynamic rather than fixed, and a complex system that is shaped by people and, in turn, influences them.

1. Culture is a social creation that reflects diverse economic, historical, political, social, and environmental factors.
2. Culture is dynamic, not fixed, and can be contested. This is illustrated by the phrase "culture wars."
3. Culture is a complex system. Through interactions with one another, people create and express culture, and in turn, culture shapes and influences people.

The significance of this reconceptualization of culture is that it seeks to make the practice of human geography even more vigorous. For those who work within the reading the landscape approach, this reconceptualization of culture means that sometimes what remains on the landscape provides only a partial understanding of the complex and dynamic forces that created it. Consider, for example, gated residential communities (**Figure 1.6**).

Culture, power, and landscape • Figure 1.6

We can read the cultural landscape to discern that this gated residential community is exclusive. If our approach is informed by a fuller understanding of culture, however, we are better equipped to examine the invisible dimensions of power, identity or class, for example, that also factored in this community's establishment.

CONCEPT CHECK STOP

1. **Why** are nature and culture considered promiscuous concepts?
2. **How** does actor-network theory conceptualize the relationship between people and the environment?
3. **When** might the boundaries of formal and functional regions coincide?
4. **How** has the recent reconceptualization of culture expanded the human geographer's toolbox?

Thinking Like a Human Geographer

LEARNING OBJECTIVES

1. **Contrast** the concepts of place and space.
2. **Distinguish** between spatial variation and spatial association.
3. **Identify** four different types of diffusion.
4. **Explain** the relationship between globalization, spatial interaction, and time-space convergence.
5. **Review** the different scales used in geographical research.

All you need to begin to think like a human geographer is a curiosity about places in the world, whether they are nearby or far away. This curiosity might spur questions similar to those we raised about nighttime illumination in the chapter opener, or it might prompt questions about the connections between different places. Thus, to think like a human geographer is to cultivate a perspective that includes a consideration of one or more of the following: (1) place, (2) space, (3) spatial diffusion, (4) spatial interaction, or (5) scale.

Place

When geographers use the term **place** they are referring to a locality distinguished by specific physical and social characteristics. Every place can be identified by its *absolute location*, or position, reckoned by latitude and longitude on the globe, as well as its **site** and **situation** (**Figure 1.7**).

site The physical characteristics of a place, such as its topography, vegetation, and water resources.

situation The geographic context of a place, including its political, economic, social, or other characteristics.

Places matter because they contribute to the social, political, and economic functioning of our

Site and situation • Figure 1.7

NATIONAL GEOGRAPHIC

a. Physically, Istanbul occupies a hilly site adjacent to a deep harbor and has grown on both sides of the Bosporus Strait, a narrow and strategic waterway that connects the Mediterranean and Black seas. The photo shows the Hagia Sophia, now a museum.

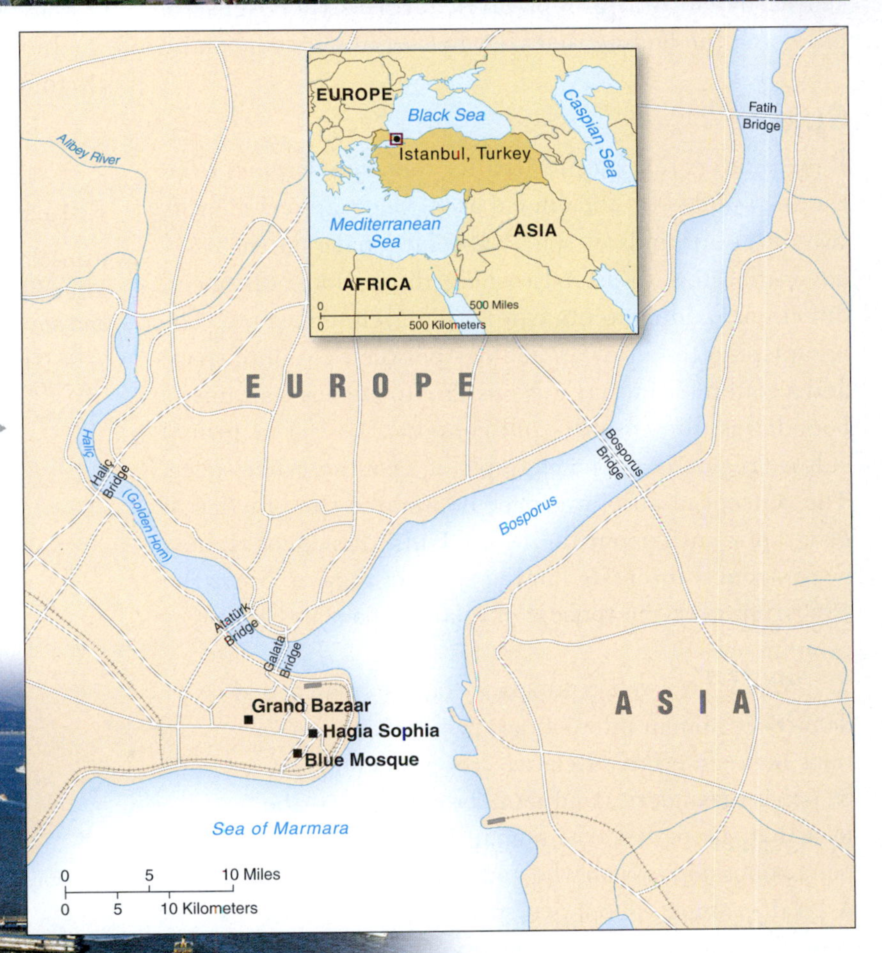

b. These maps depict the situation of Istanbul, Turkey's largest city, in relation to the surrounding bodies of water, the rest of the country, and the neighboring regions of Europe, the Middle East, and North Africa. By virtue of its situation, Istanbul straddles the regions of Europe and Asia.

c. Istanbul's growth as a major port stems from attributes of its site and situation along an important strait. What this photo does not capture, however, is the dynamic nature of a place's situation. Numerous ferries and cargo ships ply the surrounding waters, but a workers' strike or inclement weather can quickly alter Istanbul's situation.

world. Indeed, the tourism industry capitalizes on the fact that no two places are identical and that people enjoy experiencing these differences. Places are also important because they provide anchors for human identity. When you meet someone for the first time and are learning about that person's identity, you typically ask, "What is your name?" and then "Where are you from?" The reverse is also true: Your sense of identity derives in part from your own place-based experiences. Geographers use the term *sense of place* to refer to the complex, emotional attachments that people develop with specific localities. The feeling of belonging is strongly linked to a person's sense of place. Similarly, a part of the collective identity shared by cultural groups often involves their sense of place and the feeling that they belong in a specific place.

Space

If place refers to a specific locality, then **space** refers to either a bounded or unbounded area. Geographers identify two different kinds of space: absolute and relative. *Absolute space* refers to an area whose dimensions, distances, directions, and contents can be precisely measured. Geographers often draw an analogy between absolute space and a container in that it is possible to know a container's boundaries, dimensions, and contents. In fact, a formal culture region is a good example of a containerlike space. The concept of absolute space dominated the practice of geography until about the 1960s. Until then, geographers were strongly interested in the study of regions. Since the 1960s, however, the concept of relative space has gained prominence.

Relative space refers to space that is created and defined by human interactions, perceptions, or relations between events. Relative space is defined less by precise boundaries and more by *contingency*—the idea that the outcome of human interactions and perceptions depends on who and what are involved. A good example of relative space and its contingent character is the space of trade. For trade to occur between two countries, each must be able to supply the products the trading partner needs and enter into a mutual agreement to do so. The contingency of trade, then, depends in part on the countries' ability to continue to supply the desired products and to maintain favorable diplomatic relations. When two countries or businesses engage in trade, they create a relative space of trade that exists between them as long as these contingent conditions are satisfied.

As the trade example shows, political and economic interactions can shape the creation and production of relative space. So, too, can social interactions. In this way, relative space is socially produced. Social networking websites such as MySpace and Facebook provide great examples of this. When you log on and chat with your friends, you are creating and participating in a relative space. It is indeed fascinating, and even a little overwhelming, to think about the millions of relative spaces created not just on the Internet but globally, as people, businesses, and organizations interact. Can you list the different relative spaces you are a part of on a daily basis? More importantly, do you see how the concept of relative space involves horizontal linkages, as well as networks or webs of connections that defy containment (**Figure 1.8**)?

In the course of this discussion we have set up a dichotomy between absolute and relative space. However,

Relative space • Figure 1.8

If absolute space resembles a container, then relative space resembles a network of linked nodes, also referred to as a hub-and-spoke network. In the context of social networking, nodes or hubs represent individuals. You can use the Nexus application for Facebook to generate a map of your network of friends.

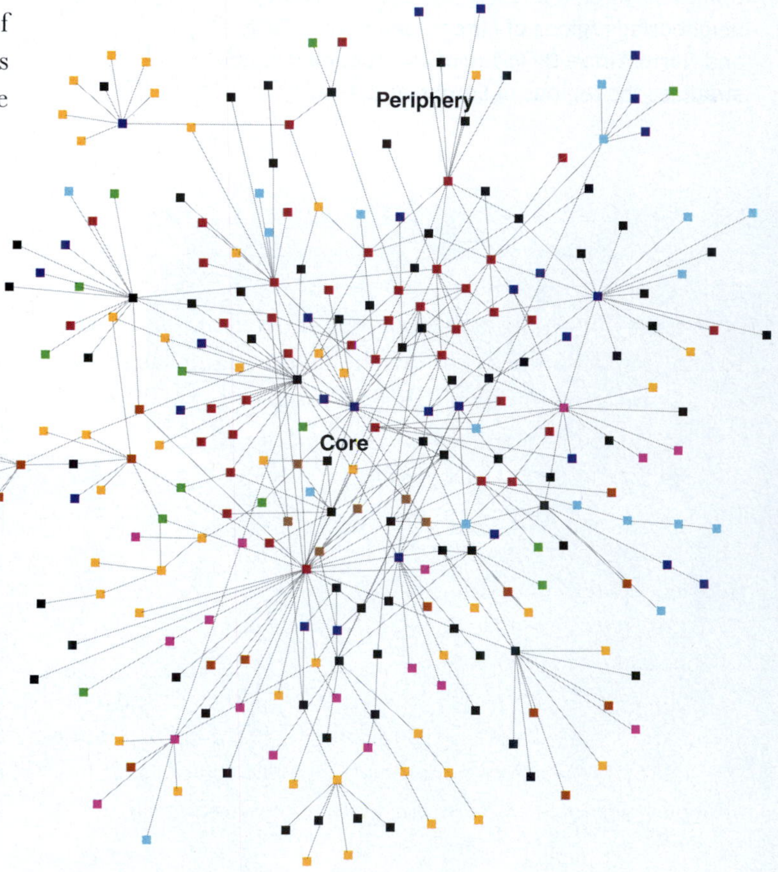

Spatial variation and spatial association • Figure 1.9

a. The spatial variation of closed forests in Vietnam changes markedly from north to south across the country. What are some likely reasons for this?

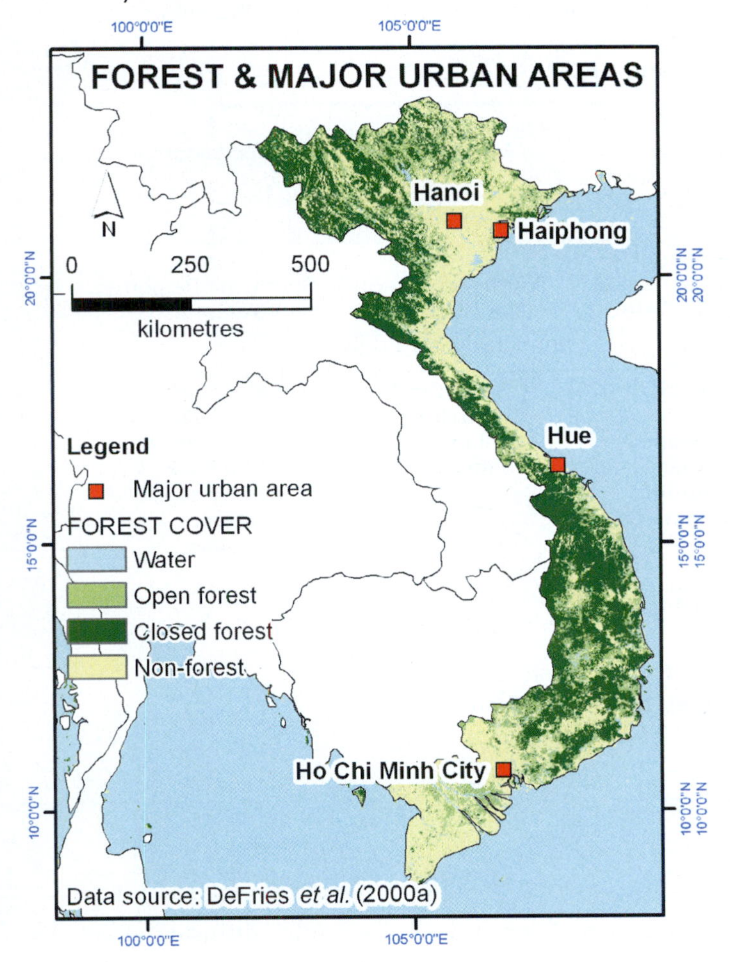

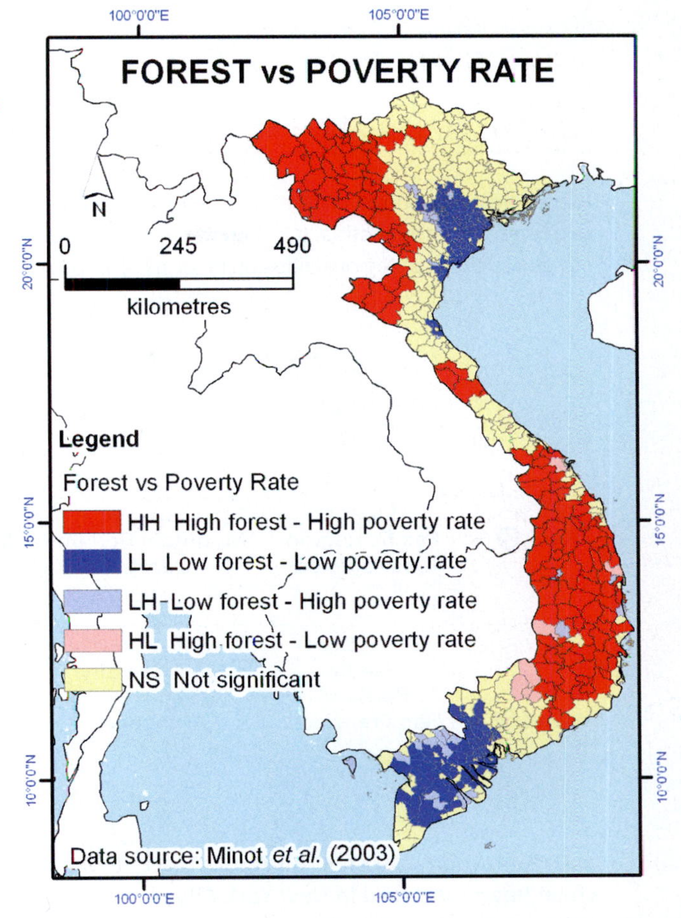

b. There is a strong spatial association between high forest cover and high poverty rates in Vietnam. Deciphering these kinds of relationships is what interests geographers. Just because two or more phenomena have a strong spatial association does not necessarily mean that one phenomenon has caused the other. To understand why high forest cover is spatially associated with high poverty demands an understanding of other factors, including local and global economic forces, political policies, historical developments, and social practices.

the two concepts can and often do overlap. We realize this connection when we think about the relationship between perceptions and space. For example, how does human behavior change when people move from one space to another? How does your own behavior change as you go from home or your dorm to a classroom or to the library? These buildings and rooms have characteristics of both absolute and relative space in that they are bounded, physical spaces but also zones or fields of perception and interaction. The fact that the range of acceptable behaviors changes from one space to another suggests that our perceptions of space can be significantly shaped by many factors, including power relations.

These relations between space and power, or authority, have been informed by the work of French philosopher and historian Michel Foucault (1926–1984). Foucault has shown, for example, that the power relations associated with space have a way of regulating and controlling—or as he calls it "disciplining"—human behavior. Look again at Figure 1.8. Where is the power in this network?

If we hope to gain a fuller understanding of how the world works, then grasping these aspects of space is a key part of that. Consequently, when conducting their work, human geographers adopt and emphasize a *spatial perspective*. That is, they pay particular attention to the variations from one place or space to another in society and environment-society dynamics. **Spatial variation** and **spatial association** are other key concepts geographers use; both concepts build on an understanding of **distribution** (**Figure 1.9**).

spatial variation Changes in the distribution of a phenomenon from one place or area to another.

spatial association The degree to which two or more phenomena share similar distributions.

distribution The arrangement of phenomena on or near the Earth's surface.

PROCESS DIAGRAM

Understanding hierarchical diffusion • Figure 1.10

THE PLANNER

Hierarchical diffusion involves cascading or stair-stepping from one level or rank to another. It is therefore more systematic or structured than contagious diffusion in terms of who is affected and how they hear about a fad or trend. Here is how a fashion innovation—such as a new product line—might diffuse through a company's hierarchy, top-down through the ranks.

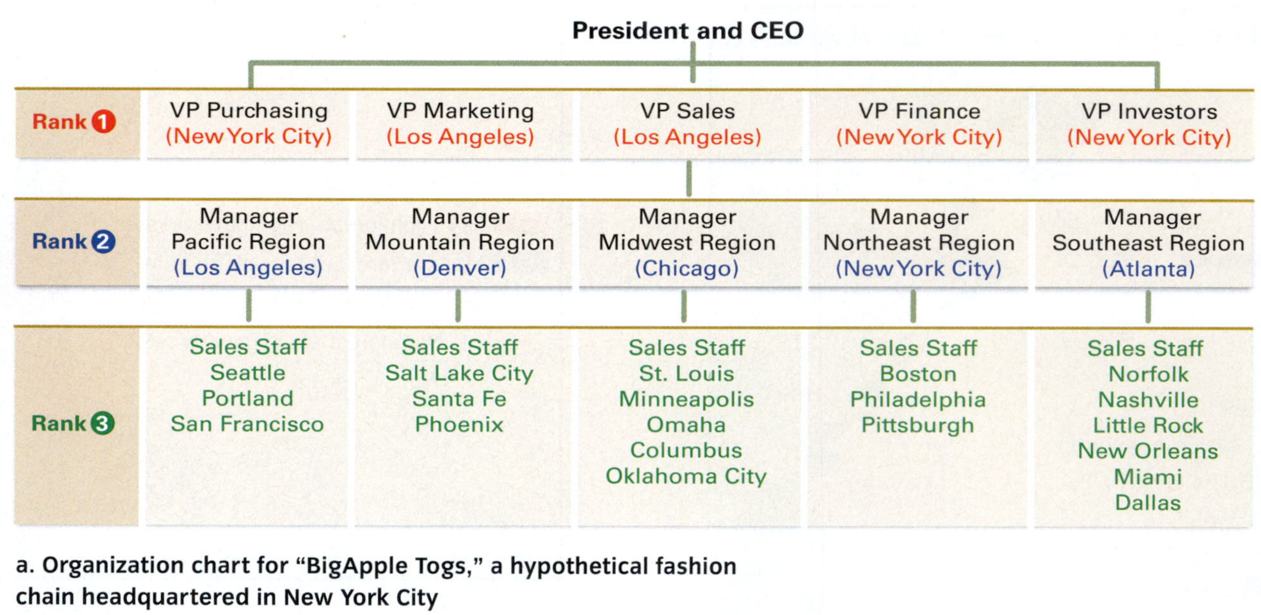

a. Organization chart for "BigApple Togs," a hypothetical fashion chain headquartered in New York City
This structure forms the framework in which hierarchical diffusion of ideas can occur. In addition to the downward diffusion through the ranks of employees, there is also hierarchical diffusion from larger to smaller cities.

Spatial Diffusion

How does fashion, news, gossip, a flu virus, or the latest hi-tech gadget spread through a population and from one place to another? Questions such as this get at the core of **spatial diffusion**. Since spatial diffusion may occur rapidly or slowly, depending on the circumstances, time is always an essential dimension of diffusion.

spatial diffusion The movement of a phenomenon, such as an innovation, information, or an epidemic, across space and over time.

Geographers recognize four different types of diffusion: relocation, contagious, hierarchical, and stimulus. Migration is the most common type of *relocation diffusion*. *Contagious diffusion* occurs when a phenomenon, such as the common cold, spreads randomly from one person to another. In contrast, *hierarchical diffusion* occurs in a top-down or rank-order manner. See **Figure 1.10** for an explanation of hierarchical diffusion.

Stimulus diffusion occurs when the spread of an idea, a practice, or other phenomenon prompts a new idea or innovation. A great deal of stimulus diffusion affects the production and marketing of goods. We can see this readily in the automobile and fast-food industries, for example. The idea behind a successful product often triggers applications of that principle in other settings—whether it is a certain body style on a vehicle or the development of a new kind of fast-food restaurant.

Studies suggest that spatial diffusion often involves a mixture of types. The diffusion of H1N1 flu since April 2009 provides a good example. This flu virus was first detected in Mexico. It spread contagiously within Mexico and to persons in neighboring U.S. states. It then spread to New York City via the relocation diffusion of several students who had vacationed in Mexico. Contagious and relocation diffusion subsequently played a role in the worldwide spread of the disease, which was eventually classified by the World Health Organization as a *pandemic*—an epidemic on a global scale. Not only do the different types of diffusion often work simultaneously, but the presence of *absorbing barriers*—

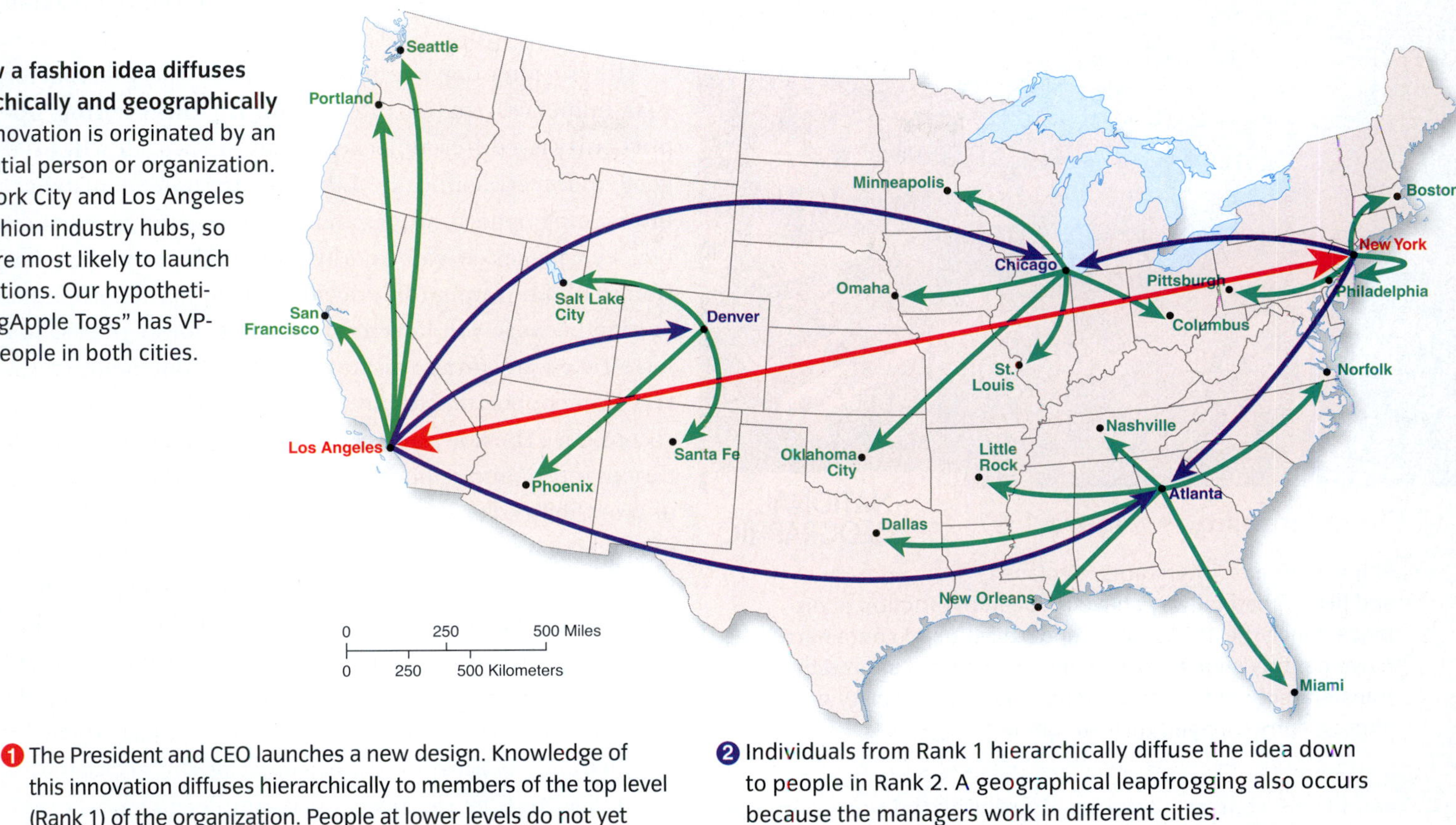

b. How a fashion idea diffuses hierarchically and geographically The innovation is originated by an influential person or organization. New York City and Los Angeles are fashion industry hubs, so they are most likely to launch innovations. Our hypothetical "BigApple Togs" has VP-level people in both cities.

1. The President and CEO launches a new design. Knowledge of this innovation diffuses hierarchically to members of the top level (Rank 1) of the organization. People at lower levels do not yet hear about it because knowledge of the innovation by-passes or leapfrogs them.
2. Individuals from Rank 1 hierarchically diffuse the idea down to people in Rank 2. A geographical leapfrogging also occurs because the managers work in different cities.
3. Individuals from Rank 2 spread the innovation to people in Rank 3.

physical, legal, or other obstacles that stop diffusion—and *permeable barriers* can also affect both the rate and the direction of spatial diffusion.

Spatial Interaction and Globalization

We live in an increasingly globalized world. **Globalization** refers to the greater interconnectedness and interdependence of people and places around the world. Globalization propels and is propelled by **spatial interaction**—the connections and relations that develop among places and regions as a result of the movement or flow of people, goods, or information.

Spatial interaction conveys the interconnectedness of different parts of the world, yet the term was first coined by geographer Edward Ullman in 1954, several decades before the word *globalization* was invented and popularized. For Ullman, a transportation geographer, the study of geography was synonymous with the study of spatial interaction. He identified three factors that influence spatial interaction: complementarity, transferability, and intervening opportunities.

Complementarity exists when one place or region can supply the demand for resources or goods in another place or region. In other words, complementarity provides a basis for trade. Leading coffee producers, such as Brazil, Colombia, and Indonesia, help satisfy the demand for coffee in major consuming regions, such as western Europe and North America, and create a condition of complementarity. Spatial interaction as a result of complementarity can involve short or long distances. Complementarity also exists when people travel from their homes to a movie theater or a gas station.

Complementarity stems from spatial variation. Such spatial variation may relate to the availability of natural resources or to particular economic conditions. For example, countries with scarce coal resources look to coal-rich countries to satisfy their demand for this resource. Economic

Transferability • Figure 1.11 NATIONAL GEOGRAPHIC

Ships, railroads, interstates, airplanes, and high-speed railways have reduced the friction of distance. Since the 1950s, the standardization of containers for movement by ship, tractor-trailer, or train has altered the transferability of freight. How has the transferability of mail or photographs changed through time?

conditions that are associated with spatial variation and lead to complementarity include low costs of production and economies of scale. Low-labor or transportation costs, for example, may make the production of a good less expensive in one place than another, giving that place an economic advantage. Similarly, the ability to create an economy of scale can stimulate complementarity. An *economy of scale* refers to the reduction in the average production cost of an item as a result of increasing the number of items produced. Because certain costs are fixed—for example, the cost of machinery or equipment for an automobile assembly line—lower average costs per vehicle are obtained by producing more of them.

A second factor that influences spatial interaction is **transferability**—the cost of moving a good and the ability of the good to withstand that cost. High-value goods that are not bulky and can be easily transported, such as jewelry, have high transferability. Low-value, bulky goods, such as rocks or hay, have low transferability. In general, goods with low transferability are more likely to be used near their source. Transferability is affected by the *friction of distance*, or the way that distance can impede movement or interaction between places. Historically, distance has deterred spatial interaction, but changes in modes and speeds of transport have reduced the friction of distance (**Figure 1.11**).

Intervening opportunities constitute the third factor that influences spatial interaction. An **intervening opportunity** is a different location that can provide a desired good more economically. Like the friction of distance, intervening opportunities can alter the spatial interaction between places. If you usually stop at the same gas station to fill up your car but decide to frequent another gas station because you have noticed it has lower prices, you have taken advantage of an intervening opportunity. Intervening opportunities are important because they help reconfigure the flows and relations between places. In addition, intervening opportunities point to the importance of accessibility. For geographers, *accessibility* means the ease of reaching a particular place. Different measures of accessibility exist. Accessibility is most commonly expressed in terms of travel time or cost. The greater the accessibility of a place, the lower the travel time to or from it. Public facilities, such as parks and libraries, are considered highly accessible because there are usually no fees to use them.

Distance is an important aspect of the accessibility of a place, but, as the previous example suggests, other aspects can be just as important as or even more important than distance. A business may locate a branch office in a place that is more distant from the market in order to take advantage of lower rents. Alternatively, accessibility can be expressed in terms of a place's *connectivity*—that is, the number and kind of linkages it possesses. Such linkages might include airports, the presence of interstate highways, or the availability of high-speed computer networks. Fiber-optic cables, technology that permits much faster transfer of data compared to copper wires, has helped connect the globe (**Figure 1.12**).

Distance decay Spatial interaction can be affected by **distance decay**. Within cities, population density usually diminishes with increasing distance from the downtown area. Similarly, people are willing to travel a few miles to a grocery store and will do so hundreds of times over the course of a year, but most people are not willing to travel long distances to reach a grocery store. Consequently, distance decay can be an important factor when deciding where to locate certain businesses or public services. It

> **distance decay**
> The tapering off of a process, pattern, or event over a distance.

Spatial interaction and connectivity • Figure 1.12

Spatial interaction occurs in myriad ways as, for example, when you text message a friend, journey from home to work, or transfer funds electronically. Technologies, such as cellular networks, submarine cables, and telephone land lines, facilitate long-distance and international spatial interaction, although the map makes clear the global unevenness of these linkages.

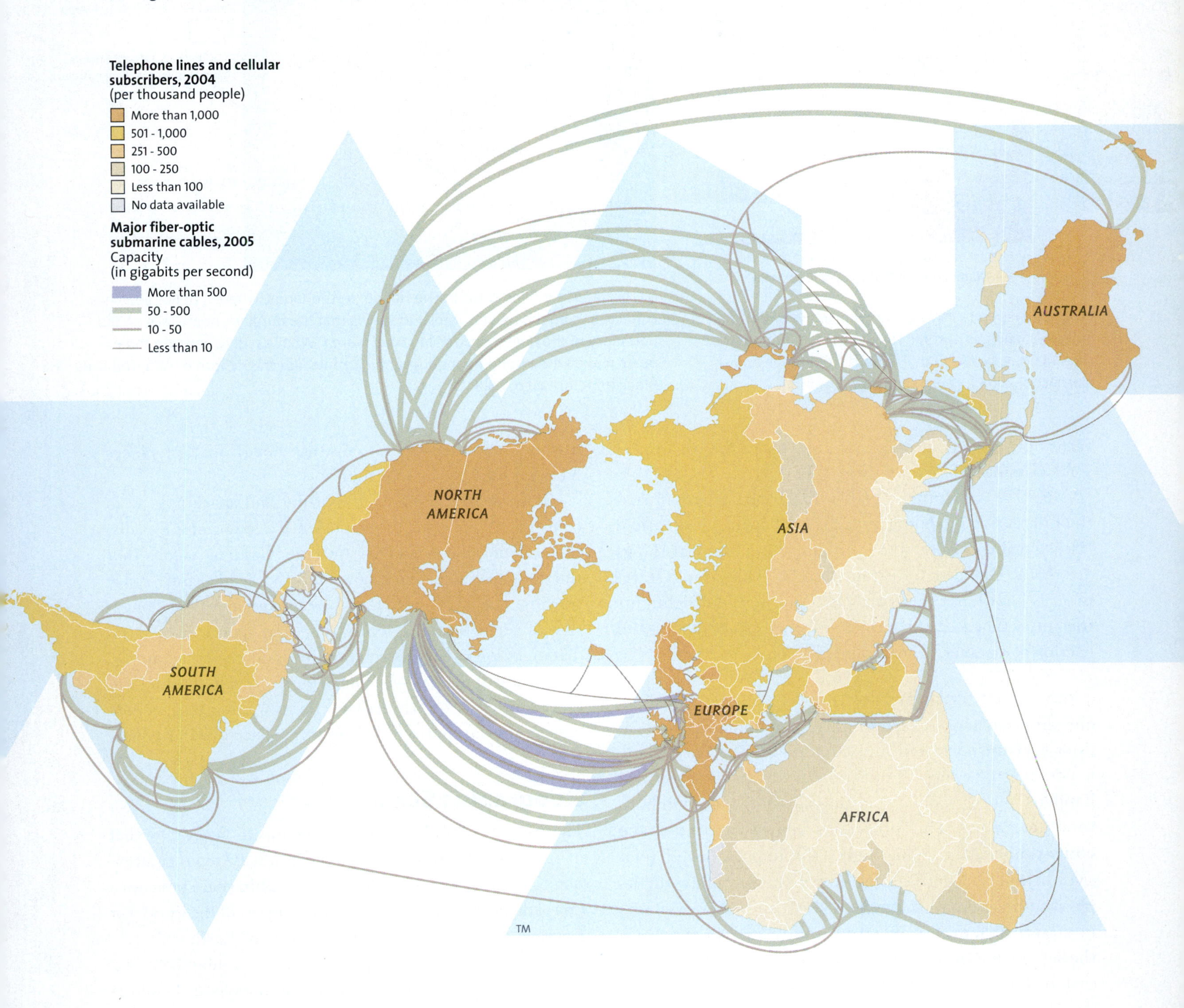

Distance decay and geographic profiling • Figure 1.13

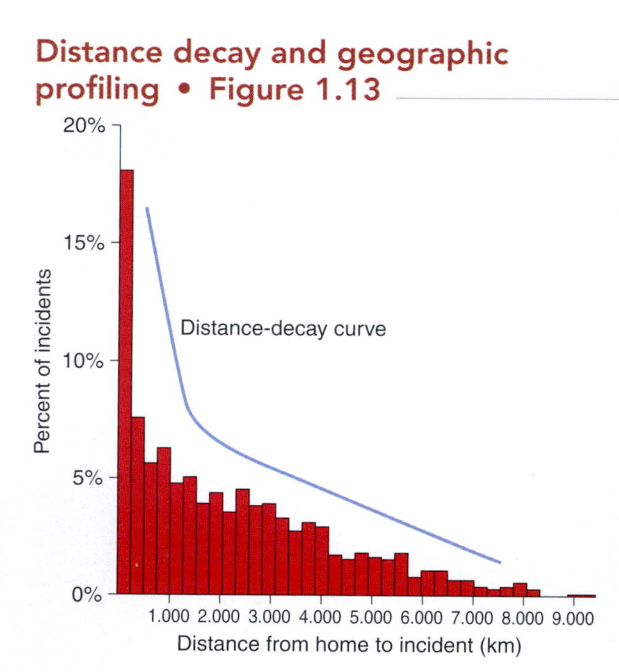

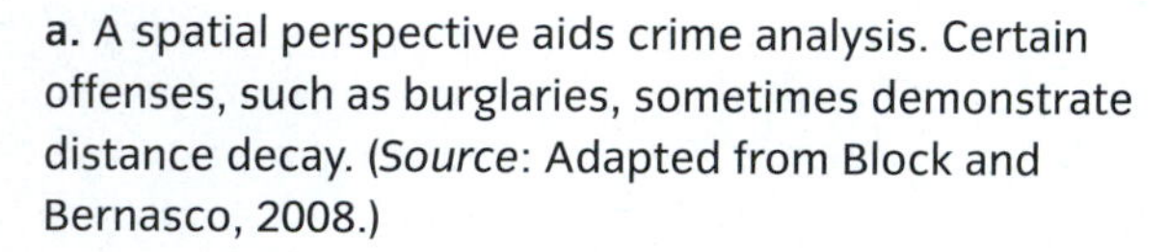

a. A spatial perspective aids crime analysis. Certain offenses, such as burglaries, sometimes demonstrate distance decay. (*Source*: Adapted from Block and Bernasco, 2008.)

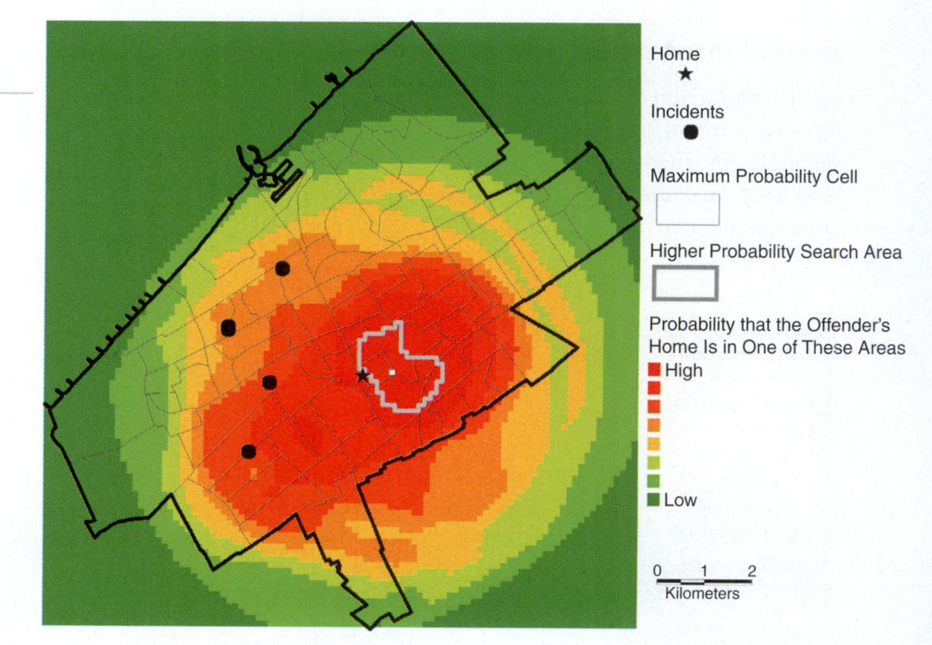

b. This map identifies areas in The Hague, Netherlands, where a police search is expected to be the most fruitful and shows that the burglar's home falls on the boundary of the higher probability search area. What assumptions do the experts make when building a geographic profile like this? (*Source*: Adapted from Block and Bernasco, 2008.)

turns out that distance decay can factor in the patterns of some criminal offenses (**Figure 1.13**).

In 1970, geographer Waldo Tobler, an expert in spatial interaction modeling, made the following observation: "[E]verything is related to everything else, but near things are more related than distant things." This simple statement is known as *Tobler's first law of geography*. It highlights the importance of the principles of distance decay and friction of distance to spatial interaction.

Time-space convergence As we saw previously in our discussion of transferability, technological innovations in transportation and communication have made it possible to reduce the friction of distance. When this happens, places seem to become closer together in both time and space. This process is known as **time-space convergence,** and it highlights the importance of relative distance. Whereas absolute distance refers to the physical measure of separation between points or places in meters or feet, for example, *relative distance* expresses the separation between points or places in terms of time, cost, or some other measure. Globalization does not alter the absolute distance between places, but it can change their accessibility as more places become interconnected. Moreover, globalization can reduce the friction of distance, bringing about a change in our sense of relative distance and making it seem as though distant places have become closer together.

Is it possible that even as time and space appear to converge, social relations experience a lengthening or distanciation? The sociologist Anthony Giddens argues that the same technological innovations that lead to time-space convergence also create *time-space distanciation*, the elongation of social systems across time and space. Such social distanciation occurs as remote interaction—for example, e-mail or cell phones—becomes more prevalent than face-to-face interaction. In his view, even writing is a technological innovation that leads to time-space distanciation.

Geographic Scale

The concept of scale is so fundamental to geography that many geographic works give direct or at least indirect attention to it. In its broadest sense, **geographic scale** provides a way of depicting, in reduced form, all or part of the world. For example, every globe is a scale model of the Earth.

Two classes of geographic scales exist: map or cartographic scale and observational or methodological scale. A *map or cartographic scale* expresses the ratio of distances on the map to distances on the Earth. Geographers also distinguish between large-scale maps and small-scale maps (see *What a Geographer Sees*).

WHAT A GEOGRAPHER SEES

NATIONAL GEOGRAPHIC

THE PLANNER

Cartographic Scale

In making a map, the most basic decisions involve the area to cover and the scale. Mapping a vast area, such as North America, requires a small scale that cannot show much detail. At the opposite extreme, if you show a single town, you can use a very large scale that shows great detail, such as streets and buildings. **Figure a** shows this relationship; **Figure b** explains how we express map scales.

a. Choice of a map's scale controls how much detail can be shown
Note how the detail of Cape Cod changes at these three different scales.

Small scale ⟷ **Large scale**

1 Small-scale map: Eastern North America, with Cape Cod circled

Small-scale maps, such as this one, show larger areas, such as continents or several states, but in less detail.

2 Large-scale map: Northeastern U.S. and neighboring Canada, with Cape Cod circled

This map is three times larger scale than map 1, showing greater detail but a smaller area.

3 Even larger-scale map: Eastern Massachusetts, with Cape Cod circled

This map is about twenty-one times larger scale than map 1, showing much greater detail but a much smaller area.

Verbal scale: 1 inch represents 36,000,000 inches or about 568 miles on the ground	**Verbal scale:** 1 inch represents 12,000,000 inches or about 189 miles on the ground	**Verbal scale:** 1 inch represents 1,750,000 inches or about 28 miles on the ground
Ratio scale: 1:36,000,000	**Ratio scale:** 1:12,000,000	**Ratio scale:** 1:1,750,000
Fractional scale: 1/36,000,000	**Fractional scale:** 1/12,000,000	**Fractional scale:** 1/1,750,000
Graphic scale: 0 KILOMETERS 600 800 1000; 0 MILES 200 400 600 800 1000	**Graphic scale:** 0 KILOMETERS 200 300; 0 MILES 100 200 300	**Graphic scale:** 0 KILOMETERS 30 40 50; 0 MILES 10 20 30 40 50

b. Map scales can be expressed verbally, as a ratio or fraction, or graphically
The advantage of a fraction or ratio scale is that it works for any unit of distance. In the fraction and ratio examples for the Cape Cod map, 1 unit on the map represents 1,750,000 of the same unit on Earth. Geographers frequently use centimeters or inches as units, but they could use any unit, even something whimsical, such as the width of an iPod.

Think Critically

1. What type of map scale is most desirable if you plan to use a copier to reduce or enlarge a map?
2. How does the perspective captured by a map differ from the perspective in a satellite image, and why is this significant?

Observational or *methodological scale* refers to the level(s) of analysis used in a specific project or study. This might include the body, home, neighborhood, city, region, country, or global scale. When geographers talk about the range of observational scale, they say that it extends from small scale (the level of the body) to large scale (the global level). This is the opposite of how they use the terms *small-scale* and *large-scale maps*. With observational scale, the most detailed level of analysis is the body, whereas large-scale maps have the most detail. As with cartographic scale, the choice of observational scale always involves a sacrifice between the area covered and the level of detail of the data.

The body or self constitutes an important scale because it provides a basis for personal and individual identity (**Figure 1.14**). It also helps us see how scale relates to spatial differentiation and social control. In Western society, for example, the home has historically been characterized as a kind of private space and female domain. This contrasts with the realms of politics and work, which have often been characterized as a kind of public space and male domain. In this way, ideas about the female or male body can contribute to the development of segregated spaces. For another example that illustrates the importance of the scale of the body, see *Video Explorations*.

In addition to the different levels of observational scale, it is important to remember that scales are often interdependent. We see this in globalization as things that were once popular on a local, regional, or national scale expand to the global scale. Similarly, we are reminded of the interdependence of scales in the way that the downturn of the U.S. economy in 2008 had global repercussions.

Contested bodies • Figure 1.14

a. One component of the "urban fit" for men includes wearing baggy or sagging pants so that the boxers show.

Crackdown on indecency
It's a crime to wear saggy pants in Flint. Here's the price you could pay:
Punishment for either is 93 days to a year in jail and/or up to $500 in fines.
WARNING
Underwear exposed
DISORDERLY CONDUCT
Underwear exposed, pants below buttocks
INDECENT EXPOSURE
Buttocks exposed
Source: Flint Police Department
MOSES HARRIS/Detroit Free Press

b. The body is personal space and a scale that we control. Or is it? School districts have banned such baggy pants and the city of Flint, Michigan, considers them a punishable offense.

Video Explorations

Teeth Chiseling

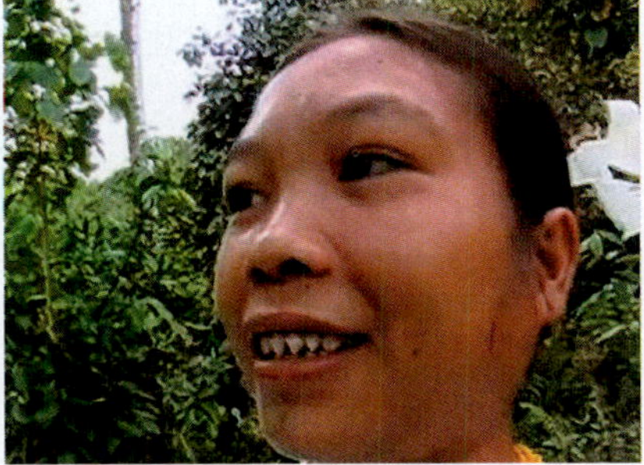

In Indonesia, a chieftain's wife undergoes teeth chiseling to enhance her beauty. To this Sumatran group, beauty is more than skin deep; it is a matter of balance between the soul and body. At what point might cultural standards of beauty become a form of oppression?

CONCEPT CHECK

STOP

1. **Why** does a region resemble absolute space?
2. **What** is meant by a spatial perspective, and how does it relate to the practice of geography?
3. **What** makes hierarchical diffusion more systematic than contagious diffusion?
4. **How** does globalization affect relative distance?
5. **Why** is the body considered a significant scale?

Geographical Tools

LEARNING OBJECTIVES

1. **Explore** how remote sensing works.
2. **Explain** the data structure of a GIS.
3. **Review** some of the applications of remote sensing, GPS, and GIS.

An appealing facet of geography is the ability to use a wide variety of research tools. This includes a mix of exciting and relatively new technologies—such as GPS devices, satellite images, geographic information systems, and interactive maps—as well as more traditional and long-standing research tools, including maps, photographic documentation, archival resources, and interviews. The tool that has been most closely associated with geography is the map, a simple but powerful tool enabling us to visualize parts of the world. Nevertheless, this section focuses primarily on the more recent technological tools, in part because they have significantly expanded the geographer's toolbox. First, however, we need to distinguish between *skills* and *tools*. Skills are a product of our aptitude and learned abilities, whereas tools are the instruments we use to improve procedures or techniques, such as data gathering or visualization.

Like other scholars, geographers seek to cultivate their observational, analytical, and writing skills. Often, it helps to know another language, as well as statistics. Carrying out fieldwork, including extended or repeated visits to a research site or sites, is another skill many geographers hone. Those geographers whose fieldwork involves a great deal of outdoor exploration in remote places where daily luxuries, such as safe drinking water and air conditioning are not readily available, often describe themselves as "muddy-boots geographers."

Remote Sensing

When you scan the road in front of you as you drive, you are, in effect, engaging in a type of **remote sensing**—acquiring information about something that is located at a distance from you. In this case, the human eye acts as a *sensor* that responds to the stimulus of light and transmits certain signals to the brain. For geographers, remote sensing uses instruments or sensors to detect Earth-related phenomena and to provide information about them. As the term *remote sensing* suggests, the sensors are always located at a distance from the subject being studied. With greater reliance on satellite-mounted sensors, the distances between the sensor and

Geography InSight

Remote sensing of post-earthquake damage • Figure 1.15

Innovations in sensor and computer technology have led to improvements in resolution (the detail we can detect), as well as a virtual explosion in the different uses for remotely sensed data. Today, for example, it is possible to obtain remotely sensed images with submeter (less than 3 feet) resolution.

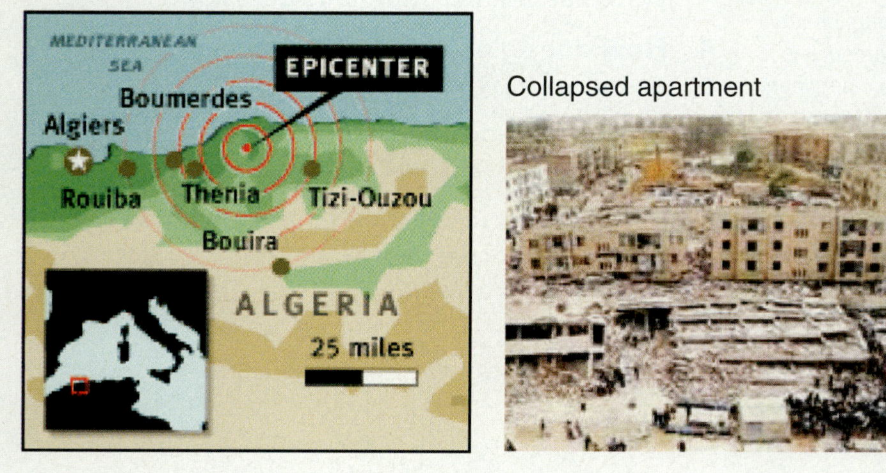

a. The earthquake that struck Boumerdes, Algeria, on May 21, 2003, registered a magnitude of 6.8 on the Richter scale and caused the deaths of more than 2,200 people. Structural damage in the city was severe and spatially variable. (*Source*: Adams et al., 2004.)

b. The availability of high-resolution remotely sensed data with submeter resolution makes possible the rapid identification of structural damage across the affected area, because building collapse produces an identifiable textural signature not present in areas where buildings have not collapsed. Using these textural signatures, geographers can then create maps identifying regions characterized by high percentages of collapsed buildings. (*Source*: Adams et al., 2004.)

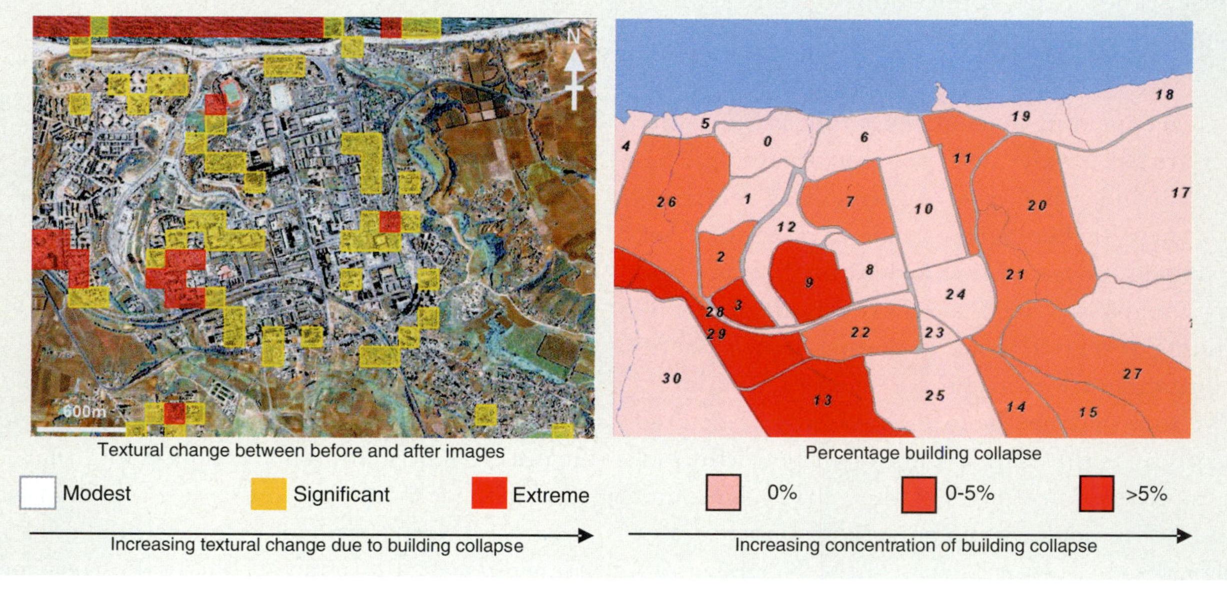

the target of study can be considerable, often spanning thousands of kilometers.

Many of the early applications of remote sensing involved detecting conditions in the natural environment, especially in the area of weather monitoring and forecasting. Human geographers are increasingly making use of remote sensing to study such things as the spatial extent of urban areas or to track oil spills and other

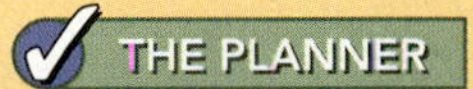

c. The collapse of buildings changes the cultural landscape, in effect creating more edges. Visually, we can see this pattern in the more chaotic and brighter appearance of the images, but the use of software to detect these changes speeds up the analytical process. Identifying high-damage areas is essential to relief coordination following a natural disaster. (*Source*: Adams et al., 2004.)

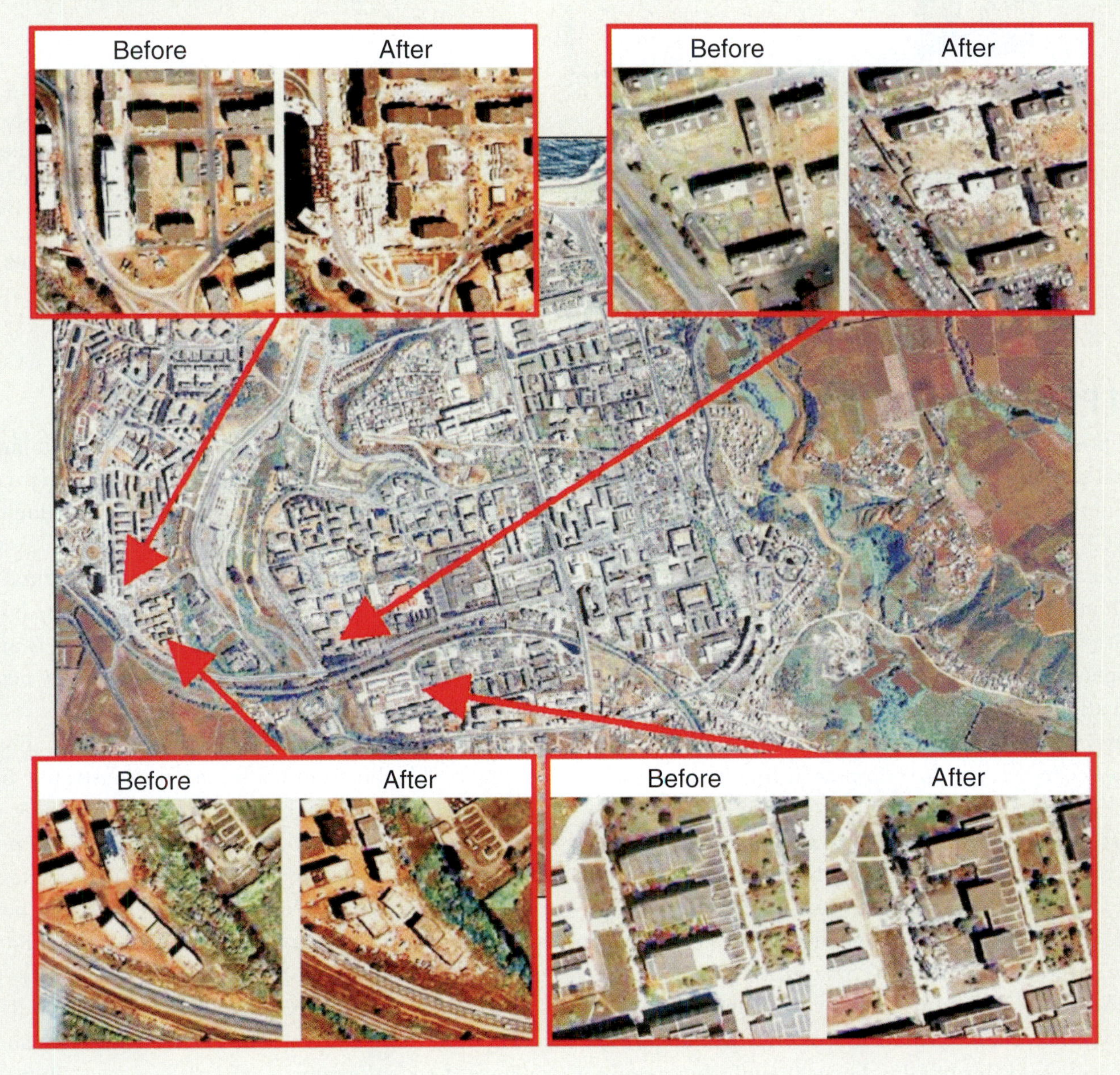

forms of water pollution. Some recent studies suggest that remotely sensed data on nighttime lights, as shown in the chapter-opening photo, might provide a basis for estimating populations in countries that do not have reliable censuses, or even estimating the wealth of a region. Those who study natural disasters are also able to use remote sensing to document and record the extent and damage caused by fires, hurricanes, or other natural hazards (**Figure 1.15**).

GPS applications • Figure 1.16

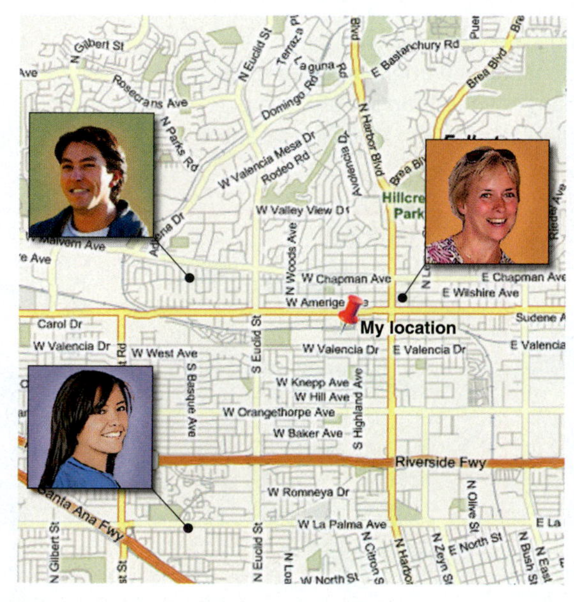

a. Cell phones equipped with GPS receivers make it possible to use location-based services to find friends in your area. Is this kind of geographic awareness a benefit or hindrance to personal security?

b. This diagram shows the time-space paths of a female teenager in one week in Marion County, Indiana. Each disk represents a different day, and the dots represent waypoints (intermediate destinations) collected via a GPS-enabled cell phone as the young woman traveled from home and back. How might time-space paths vary by age, gender, or ethnicity? (*Source*: Wiehe et al., 2008.)

Global Positioning System

A **global positioning system (GPS)** uses a constellation of artificial satellites, radio signals, and receivers to determine the absolute location of people, places, or features on Earth. A GPS receiver uses the time it takes to receive a signal from a satellite to calculate how far away the satellite is. When radio signals are simultaneously transmitted from multiple satellites, it is possible to apply the geometric principles of triangles to determine the latitude, longitude, and altitude of locations on Earth.

When you use a GPS device (which basically functions as a receiver), you are tapping into a system that has been developed and funded by the U.S. Department of Defense. Thus, the term *global positioning system* refers specifically to the system developed by the United States and more generally to the use of multiple satellites as a way of locating things on or navigating between places on Earth. Although the first GPS satellite was put into orbit in the 1970s, GPS did not provide global coverage until 1995. Since then, civilian use of GPS has boomed, and annual global sales of GPS devices regularly amount to several billion dollars.

Like remote sensing, GPS has greatly expedited our ability to acquire data about the Earth. For example, locational information for map features can quickly be acquired and transferred to computers to make or update maps. GPS is regularly used to confirm the legal boundaries of property, to track and inventory different species of plants and animals, and to monitor conditions in agricultural fields. GPS has contributed to the growth in precision farming. For example, GPS can be used to record information on soil types, moisture, or pest infestations at different locations in a field. When this kind of information is combined with GPS-ready agricultural machinery, it is possible to manage pesticide application so that it is applied only where needed and in the smallest amounts possible, preventing waste.

Within the past decade, locational information has become a valuable commodity, as demonstrated by the rapid growth in location-based services. A *location-based service* (LBS) uses the location of a GPS receiver to provide information about nearby businesses and sometimes even people. For example, on some cell phones it is now possible to search for nearby restaurants or ATMs and to find GPS-equipped friends who are in the area (**Figure 1.16**).

GPS technology raises a host of thorny ethical questions. Often there is a fine line between a service and surveillance. Law enforcement officials can use GPS to track the locations of parolees, and parents can use it to know the whereabouts of their kids. Geographers Jerome Dobson and Peter Fisher coined the term *geoslavery* to refer to "a practice in which one entity, the master, coercively or surreptitiously monitors and exerts control over the physical location of another individual, the slave" (2003, pp. 47–48). In what other ways might GPS compromise personal privacy?

Geographic Information Systems

Many people—geographers and nongeographers alike—enjoy poring over paper maps and studying them for the patterns and trends they show. Of course, paper maps do have their limitations. If you need to know the area covered by a lake, for example, doing the calculation manually can be labor intensive and time-consuming. Just imagine trying to manually compute the area covered by lakes in a single country. The emergence of **geographic information systems (GIS)** has its roots in this very issue: how to improve the functionality of maps and the spatial analysis of *georeferenced data*—that is, data tied to locations on Earth.

There are two ways to georeference data: direct and indirect. The most common system used for direct georeferencing is latitude and longitude. For indirect georeferencing, locations may be given by street address, zip code, school district, census tract, or other spatially defined entity (for which latitude and longitude could then be obtained). We can obtain georeferenced data from various sources, including paper maps, satellite imagery, aerial photography, and GPS devices, to name a few. A GIS, then, refers to a combination of hardware and software that enables the input, management, analysis, and visualization of georeferenced data. The usefulness of GIS as a tool stems from its ability to relate different kinds of georeferenced data (**Figure 1.17**).

A GIS can link data, reveal new relationships, and visualize them with maps. In a GIS, maps are interactive, enabling a user to click on a map feature and obtain information about it, to turn different data "on" or "off" for viewing, and to query the data. Another facet of GIS is that

GIS data structure • Figure 1.17

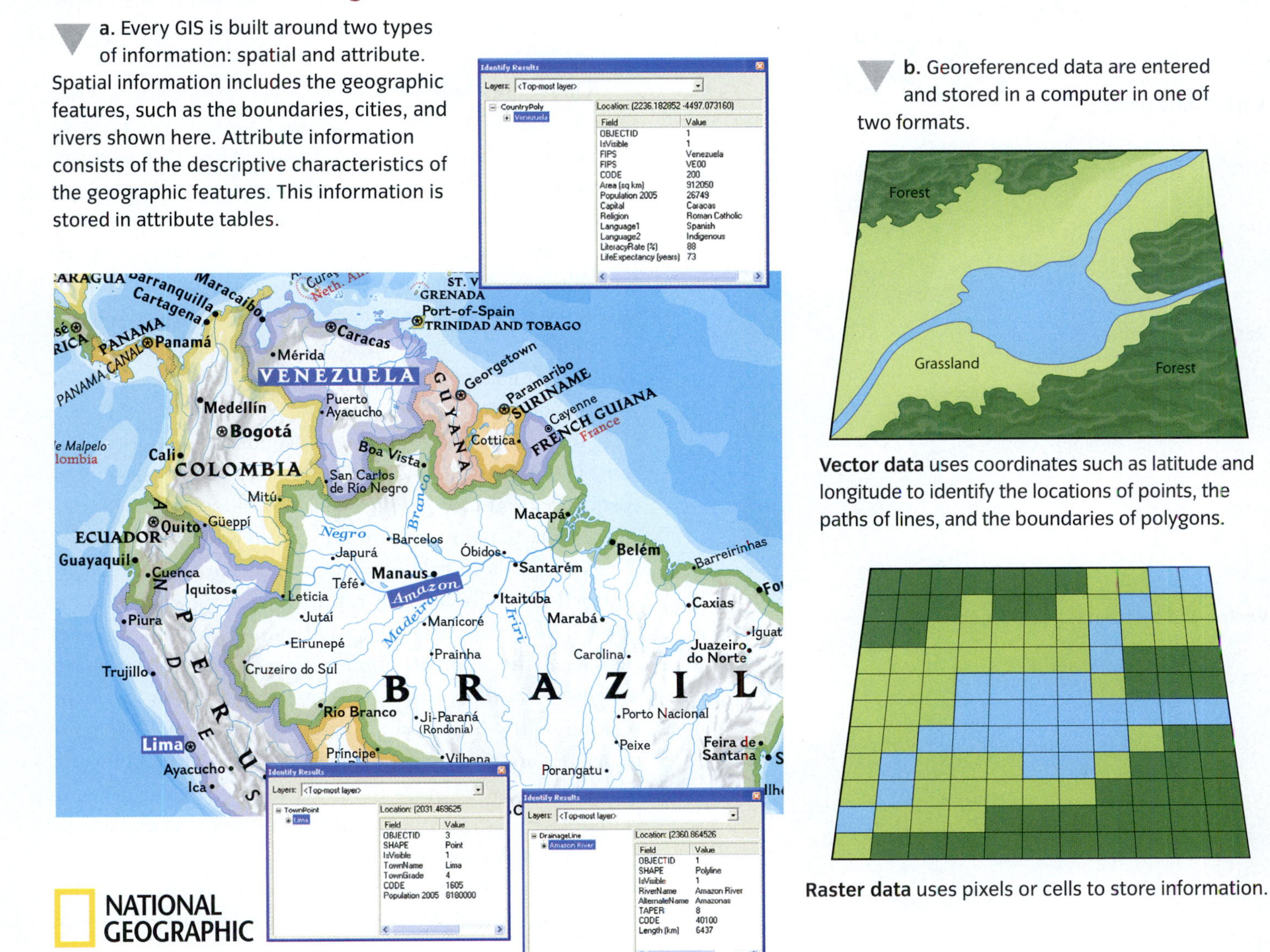

a. Every GIS is built around two types of information: spatial and attribute. Spatial information includes the geographic features, such as the boundaries, cities, and rivers shown here. Attribute information consists of the descriptive characteristics of the geographic features. This information is stored in attribute tables.

b. Georeferenced data are entered and stored in a computer in one of two formats.

Vector data uses coordinates such as latitude and longitude to identify the locations of points, the paths of lines, and the boundaries of polygons.

Raster data uses pixels or cells to store information.

it can accommodate statistical analysis and perform calculations, such as identifying the most optimal route between locations.

One way to conceptualize this is to think of a GIS as a kind of database that stores information in different layers (**Figure 1.18**).

GIS data layers • Figure 1.18

GIS incorporates the ability to combine, through overlays, a wide variety of georeferenced data.

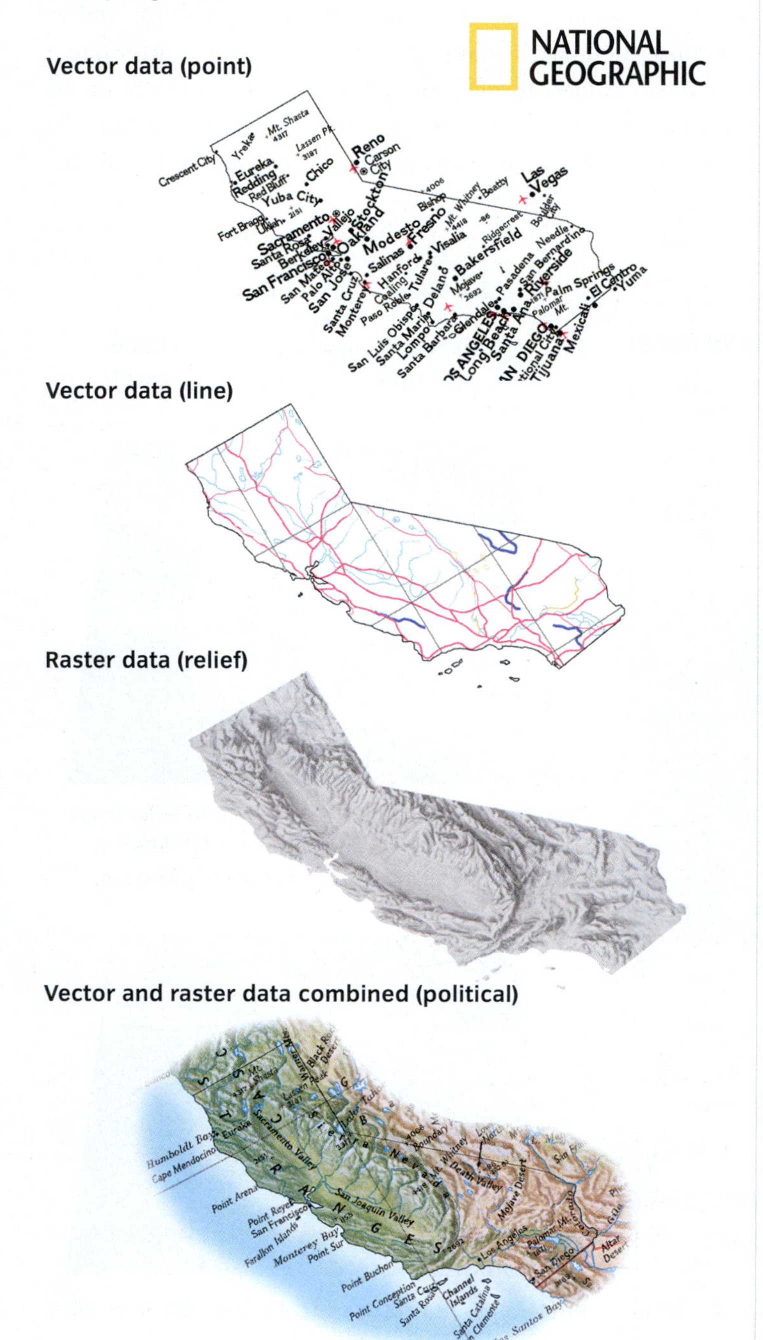

The possible applications for GIS are mind-boggling. For example, GIS has been used in Jamaica to evaluate proposed sites for new schools by examining the terrain and road network in combination with demographic data on the numbers and ages of school children. GIS has also been used to track deforestation in Bolivia over time and, through modeling, to rank and predict areas vulnerable to future deforestation. With support from the Food and Agriculture Organization of the United Nations, water resources, including inland fisheries in Sub-Saharan Africa, are being studied with a GIS to predict fish yield and to help local communities plan for adequate food supplies. In Iowa, GIS has been used to study the spatial associations between demography, ecology, and the incidence of West Nile virus (**Figure 1.19**).

As a tool, GIS has tremendous potential to help solve problems, model social and environmental conditions, and make planning decisions. For students and others who are thinking about career plans, GIS has dramatically transformed employment prospects for geographers by opening up a wide range of job and career options across the public and private sectors.

There are, however, three major criticisms of GIS. One is that to do GIS requires that users have access to the necessary hardware and software. Most GIS software is proprietary and has the greatest functionality on today's newest and fastest computers. Even though the prices for both computers and GIS software have become more affordable, purchasing just the minimum components can still cost a few thousand dollars. In addition, before you can use your GIS, you need data. Some GIS data are publicly available at no charge, but this is not always the case. Thus, you may have to purchase customized data from a GIS services firm or employ personnel to conduct fieldwork to obtain the data. These hardware, software, and data-related costs have prompted people inside and outside the GIS-user community to point out that GIS is still not very accessible.

A second and related criticism is that, given its constrained accessibility, GIS reinforces one power divide in society such that only those individuals and institutions that have the requisite financial resources can purchase and use GIS. This limitation has ramifications for mapmaking and for decisions made based on GIS-derived maps and analysis. On the one hand, access to GIS means we can make more maps than ever before. On the other, it is important to ask, Who is making those maps and whose economic, political, or other interests do they serve?

A third criticism of GIS is that it promotes a detached and strongly Western view of the world. It is entirely

Geography InSight

A GIS for studying disease incidence • **Figure 1.19**

West Nile virus (WNV) is a disease that can be transmitted to humans and animals by certain mosquito species. The first outbreaks of WNV occurred in the United States in 1999. Nationally, Iowa and several other Midwestern states have recorded a high incidence of the disease. These maps illustrate the use of a GIS to study disease incidence. (*Source*: DeGroote et al., 2008.)

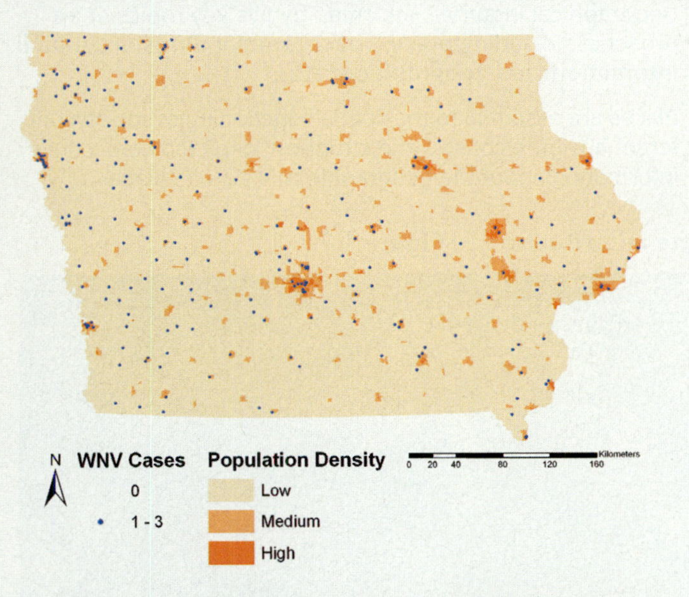

a. This map was created by overlaying a dot map of disease incidence on the map of population density. The scale of the study was conducted at the level of census block groups (a subdivision of a census tract). If WNV occurred in a block group, its center point is shown. (*Source*: DeGroote et al., 2008.)

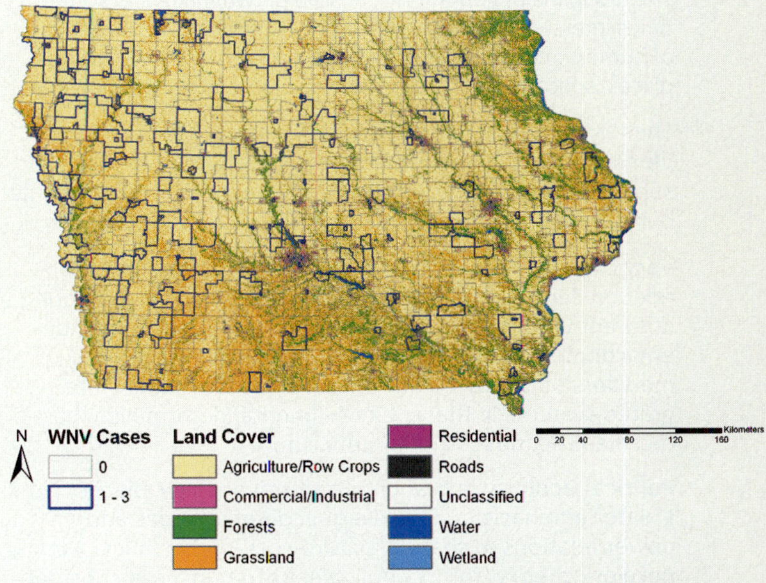

b. Census block groups have been overlaid on a land cover map of Iowa. Darker boundaries indicate block groups recording occurrences of WNV between 2002 and 2006. Using this map and the previous one, what spatial associations can you identify? (*Source*: DeGroote et al., 2008.)

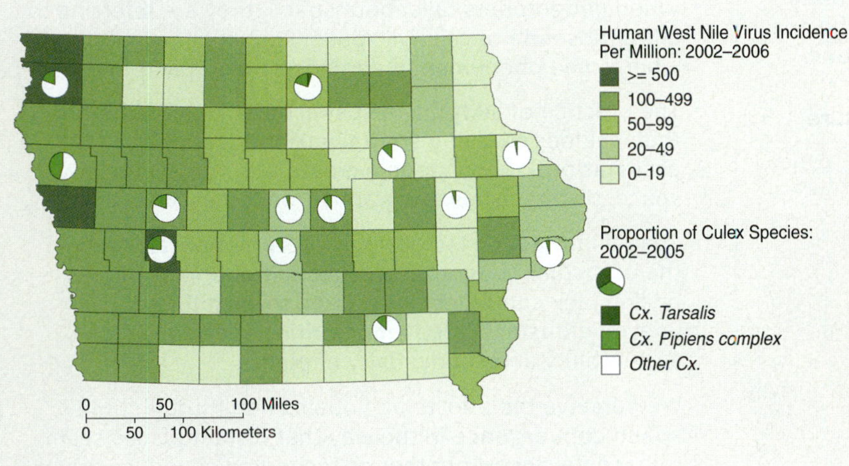

c. Several *Culex* species of mosquito are the most significant transmitters or vectors of the disease. This map shows disease incidence and mosquito species by county. What hypothesis can you suggest to account for these patterns? How might the environment and agricultural practices be involved? (*Source*: DeGroote et al., 2008.)

possible to do GIS by simply sitting in front of a computer and never visiting the site being studied. What impact might this style of work have on the way decisions about a place are made? Although there have been some strides toward incorporating local knowledge into GIS, how places in the world are represented in a GIS remains anchored to concepts of absolute location, defined boundaries, and contemporary political states.

CONCEPT CHECK

STOP

1. **How** does remote sensing incorporate the concept of scale?
2. **How** does a GIS incorporate and make use of georeferenced data?
3. **What** are some applications and limitations of remote sensing, GPS, and GIS?

Summary

1 Introducing Human Geography 4

- The discipline of geography consists of two main, and sometimes intersecting, branches: physical and human. The scope of **human geography** is broad and encompasses the study of places, spatial variation, and human-environment relationships.
- The terms ***nature*** and ***culture*** rank among the most complex words in the English language. Dualistic thinking that separates nature from culture has shaped Western thought, but many geographers reject the **nature-culture dualism**.
- Very broadly defined, **cultural ecology** focuses on the relationships between people and the environment. Four different ways of understanding that relationship include **environmental determinism**, **possibilism**, humans as modifiers of the Earth, and the Earth as a dynamic, integrated system. Of these, environmental determinism has received the most strident criticism.
- **Political ecology**, a branch of cultural ecology, places greater emphasis on the role of economic forces and power relations in shaping nature-society dynamics. **Actor-network theory**, for example, seeks to reinsert a consideration of environmental influence in such studies.
- **Culture** can refer to shared beliefs and practices of a group. Reading the **cultural landscape** and performing **regional analysis** have long been associated with this conceptualization of culture. Regional analysis frequently involves the study and mapping of **functional regions**, like the one shown here, as well as **formal** and **perceptual regions**.

Formal, functional, and perceptual regions • Figure 1.5

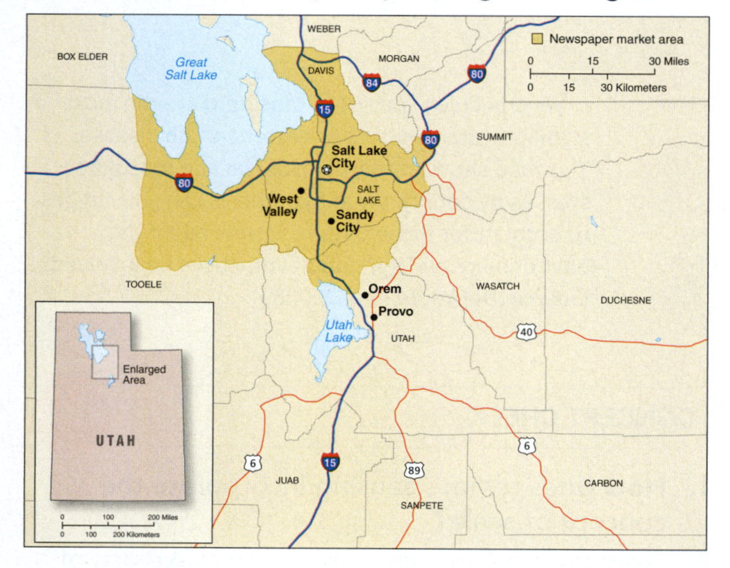

- There has been a significant reconceptualization of culture within human geography. This reconceptualization sees culture not only as a collection of cultural traits but also as a social construction that is dynamic and contested.

2 Thinking Like a Human Geographer 12

- Geographical inquiry is informed by five key topics or approaches, including **place**, **space**, **spatial diffusion**, **spatial interaction**, and **geographic scale**.
- Places are essential to the spatial functioning of society. Different attributes of **site** and **situation,** depicted here, convey information about the geographic context of a place.

Site and situation • Figure 1.7

- When geographers talk about space, they are referring to either absolute or relative space. Spatial diffusion occurs when some phenomenon spreads across space.
- The practice of adopting a spatial perspective, whether that includes studying **spatial variation**, **spatial association**, or how people create and perceive relative space, distinguishes geography from other fields of study.
- Spatial interaction increasingly fuels **globalization**, but the particular geographies of globalization are shaped by complex contingencies related to such things as uneven diffusion, complementarities of trade, and the accessibility and connectivity of places.
- We perceive the effects of globalization through **time-space convergence** in the way that distant places seem to become closer together as technologies reduce travel time and cost. Although the geographic scale of the globe remains unchanged, time-space convergence affects relative distance.

3 Geographical Tools 23

- From the paper map to the interactive map, the geographer's toolbox continues to expand through technological advances related to **remote sensing**, **GPS**, and **GIS**. Remote sensing has extended the visual horizons of geography by enabling us to see and detect things not visible to the naked eye. GIS has improved our ability to examine spatial associations and visualize them, especially by creating maps such as the ones shown here, and overlaying them.

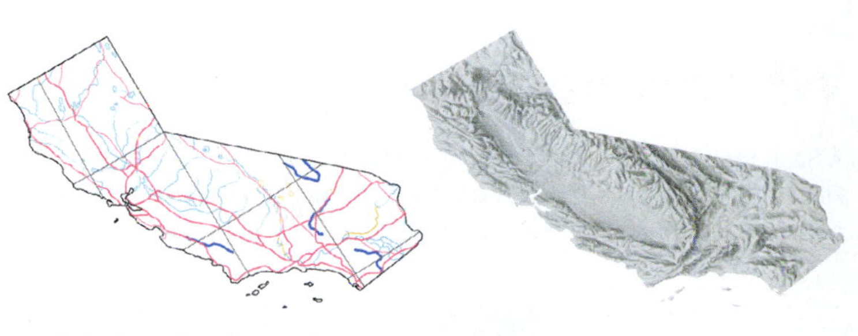

GIS data layers • Figure 1.18

- These exciting technologies do raise serious ethical questions about privacy and surveillance. GPS has been criticized for its potential to contribute to geoslavery.

Key Terms

- actor-network theory 8
- complementarity 17
- cultural ecology 5
- cultural landscape 9
- culture 11
- distance decay 18
- distribution 15
- environmental determinism 5
- formal region 10
- functional region 10
- geographic information system (GIS) 27
- geographic scale 20
- global positioning system (GPS) 26
- globalization 17
- human geography 5
- intervening opportunity 18
- nature 5
- nature-culture dualism 5
- perceptual region 11
- place 12
- political ecology 8
- possibilism 8
- regional analysis 10
- remote sensing 23
- site 12
- situation 12
- space 14
- spatial association 15
- spatial diffusion 16
- spatial interaction 17
- spatial variation 15
- time-space convergence 20
- transferability 18

Critical and Creative Thinking Questions

1. Applying what you have learned about diffusion, is it feasible to close borders between countries when an epidemic appears to be intensifying and becoming global in scale?
2. Do national parks and protected areas reflect a nature-culture dualism? Explain your reasoning.
3. Do you agree with the actor-network theory? Discuss your answer.
4. Propose a GIS project and identify the spatial and attribute data you would need to conduct it.
5. Plan a research project that would enable you to cartographically depict the boundaries of a perceptual region.
6. Keep a geographical diary in which you record the times and the places you go during a week. (You can also do this by collecting waypoints if you have a GPS-enabled cell phone or other GPS receiver.) Use the Internet to find a suitable base map and plot out your time-space paths, using this figure as an example. If you know how, you could even make a mash-up and include photos of your favorite places.

What is happening in this picture?

A flash mob protest assembles in Terminal 5 at Heathrow Airport in London. Flash mobs are large groups, often coordinated via social networking sites, that gather in a public place for a specific purpose.

Think Critically

1. How does a flash mob challenge taken-for-granted notions of space?
2. What type of diffusion drives flash mob formation?

Self-Test

(Check your answers in Appendix B.)

1. Which of the following statements about place is FALSE?
   a. Studies of place may begin with a consideration of site characteristics.
   b. Every place has a unique absolute location.
   c. The situation of places can change.
   d. Sense of place is related to the ability to navigate.

2. Which of the following is most closely associated with relative space?
   a. a GPS receiver
   b. trade between two cities
   c. site
   d. formal regions

3. _____ diffusion, shown here, involves _____, where certain individuals or places are skipped because of their rank or status.
   a. Stimulus; bypassing
   b. Contagious; overlapping
   c. Stimulus; randomization
   d. Hierarchical; leapfrogging

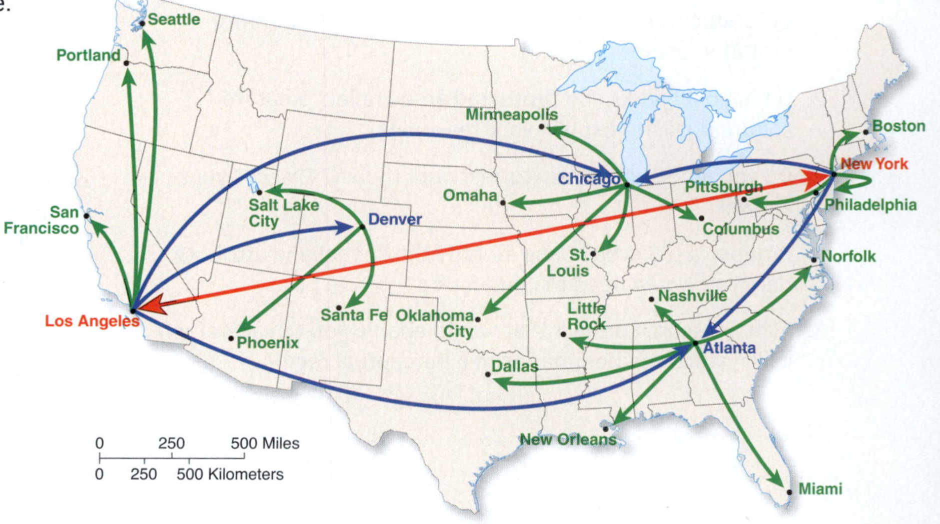

4. As discussed in the chapter, key factors that influence spatial interaction include all of the following *except* _____.
   a. transferability
   b. intervening opportunities
   c. complementarity
   d. relative distance

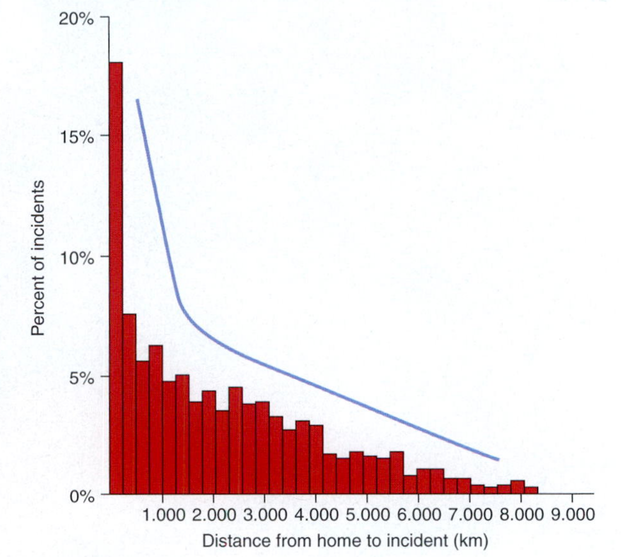

5. What spatial process is illustrated here?
   a. distance decay
   b. connectivity
   c. accessibility
   d. hierarchical diffusion

6. Globalization and time-space convergence affect our perception of _____.
   a. culture
   b. relative distance
   c. regions
   d. absolute distance

7. When we think about how different scales affect the ways we use space, we are building on an awareness of _____.
   a. mathematical scale
   b. cartographic scale
   c. observational scale
   d. representative scale

8. Which of the following situations is *not* likely to involve remote sensing?
   a. measuring the extent of an oil spill
   b. overlaying different mapped datasets
   c. identifying a pest infestation in an agricultural field
   d. locating new settlements in rural areas

9. GPS is associated with all but one of the following. Which item does *not* belong?
   a. absolute location
   b. navigation
   c. location-based services
   d. indirect georeferencing

10. Which of the following statements does *not* describe a characteristic of a GIS?
    a. A GIS can relate settlement density to elevation.
    b. A GIS can use raster or vector data.
    c. A GIS can use directly but not indirectly georeferenced data.
    d. A GIS links attribute data to spatial data.

11. As discussed in the chapter, which of the following is *not* a major criticism of GIS?
    a. If all the data are on the computer, GIS users may feel no need to know a place firsthand.
    b. Because of its reliance on state or regional data, GIS can reinforce conventional views of society.
    c. GIS data may not be publicly available for a specific place or project.
    d. GIS is a recognized low-cost solution to decision making and planning.

12. Consider this statement: "Houses constructed with steep roofs or heavy thatch roofs are just two examples of responses to wet environmental conditions." Which viewpoint does it best express?
    a. possibilism
    b. Earth as a dynamic system
    c. environmental determinism
    d. cultural ecology

13. A good example of a functional region is _____.
    a. Red Sox Nation
    b. a wealthy residential community
    c. the area served by a TV station
    d. an area with a high percentage of college graduates

14. An approach that uses the cultural landscape as a clue to people's values and priorities is _____.
    a. reading the landscape
    b. political ecology
    c. regional analysis
    d. actor-network theory

15. This billboard illustrates _____.
    a. the presence of a permeable barrier
    b. the concept of contagious diffusion
    c. the existence of a perceptual region
    d. the presence of an absorbing barrier

THE PLANNER

Review your Chapter Planner on the chapter opener and check off your completed work.

2 Globalization and Cultural Geography

TATTOOING AND GLOBALIZATION

If you have a tattoo, what does it mean to you? If you don't have one, do you want one? Some people consider tattoos a form of body decoration and artwork, whereas others find them unattractive. These differences of opinion can extend beyond personal preference and affect larger groups. Until 2006, for example, tattoo parlors were banned from Oklahoma even though they were legal in other states.

Tattooing has a long history and is geographically widespread. It has been practiced by Samoans for more than 2,000 years. To Samoans, the tattoo is a mark of distinction. Boys are often tattooed at puberty with designs extending from their waist down to their knees. Thigh tattoos are common among women. Among New Zealand Maori males, full facial tattoos employing spiral designs are not unusual, and some Maori women have tattoos on their chins, although these practices were more widespread in the past.

The late-18th-century Pacific voyages of Captain Cook helped increase the popularity of tattooing among Westerners. Even as tattooing became common among European and American sailors, who favored the armband style of tattoo, Christian missionaries working in the Pacific region sought to curb the practice because they considered it unholy.

Tattooing is a good example of a globalized cultural practice that has become extremely popular in Western countries. Despite the globalization of tattooing, diverse local practices endure though they have changed over time. Tattooing therefore reveals the fascinating interplay between events at local and global scales.

CHAPTER OUTLINE

CHAPTER PLANNER

- ❑ Study the picture and read the opening story.
- ❑ Scan the Learning Objectives in each section: p. 36 ❑ p. 40 ❑ p. 44 ❑ p. 50 ❑
- ❑ Read the text and study all figures and visuals. Answer any questions.

Analyze key features

- ❑ Geography InSight, p. 46
- ❑ Video Explorations, p. 52
- ❑ Process Diagram, p. 53
- ❑ What a Geographer Sees, p. 56
- ❑ Stop: Answer the Concept Checks before you go on: p. 40 ❑ p. 44 ❑ p. 50 ❑ p. 59 ❑

End of chapter

- ❑ Review the Summary and Key Terms.
- ❑ Answer the Critical and Creative Thinking Questions.
- ❑ Answer What is happening in this picture?
- ❑ Complete the Self-Test and check your answers.

Contemporary Globalization

LEARNING OBJECTIVES

1. **Account** for the development of contemporary globalization.
2. **Distinguish** between multinational corporations and foreign direct investment.
3. **Identify** some of the geographical aspects of foreign direct investment.

In Chapter 1, we defined the concept of **globalization** as those processes contributing to greater interconnectedness and interdependence among the world's people, places, and institutions. Some of the clearest expressions of globalization can be found in the geography of the foods we eat and the clothes we wear. In the ingredients listed on some brands of trail mix, for example, we learn that the raisins are a product of the United States, Chile, Argentina, South Africa, and Mexico; and the almonds and cashews are products of Africa, India, and Brazil. Garment labels on certain tee-shirts and jeans proclaim "Made of 100% U.S.A. materials. Assembled in Costa Rica." You may not be someone who scrutinizes the labels on the things you buy, and that is partly the point. We have grown so accustomed to the global sourcing of the products we use every day that we do not realize how much a part of our lives globalization has become or what makes globalization such a significant development.

Globalization stems from the expansion of capitalism and international trade. If we approach globalization purely from the standpoint of scale, we see that globalization is not an entirely new process. The trade in spices from Asia and Africa to Europe illustrates a connectedness that developed at the global scale as early as the 15th century. However, *contemporary globalization*—the focus of this chapter—is a more recent development that has been underway since the 1960s and has been especially rapid since the 1980s and 1990s. Even though the concept of globalization has long existed, use of the word *globalization* did not become commonplace until the 1980s.

Contemporary globalization differs significantly from historical examples of globalization because of the greater degree of financial, political, and cultural interdependence that now exists. The spice trade extended horizontal or international connections between places. Since the 1960s, however, globalization has involved both an ongoing **horizontal expansion** via rapid flows of goods, people, and ideas between places, and simultaneously a kind of **vertical expansion** as well, especially through the development of policies, such as trade agreements that formalize linkages and strengthen them like a deep root system. Thus, we can think of globalization as a process that both widens and deepens connectedness. As we saw in Chapter 1, globalization is both a cause and an effect of spatial interaction (**Figure 2.1**).

Five major factors have encouraged globalization:

1. The quest for global markets associated with *capitalism*. This includes searching for locations where goods can be produced and distributed efficiently.
2. Technological advances, especially in the areas of transportation, telecommunication, and digital computers.
3. Reduced business costs, such as lower costs for long-distance transportation.
4. An increase in the flows of financial **capital**, as a result of trade and international investments.
5. Policy, including laws and institutional arrangements, that supports the four previously mentioned factors.

capital Financial, social, intellectual, or other assets that are derived from human creativity and are used to create goods and services.

One policy development in the expansion of globalization was the creation of the World Trade Organization (WTO) in 1995 (from its precursor, the General Agreement on Tariffs and Trade [GATT]). The WTO's primary purpose is to establish and enforce the rules of trade. Today more than 150 countries belong to the WTO.

International tourism and globalization • Figure 2.1

Tourists walk across a bridge in a Costa Rican rainforest. Growth in international tourism highlights the importance of spatial interaction, flows, and mobility to globalization. Worldwide, the number of international tourist arrivals exceeded 900 million in 2007, a 70% increase since 1995.

Globalization is marked by large flows of capital around the world. Multinational corporations have emerged as major players in the global economy, and foreign direct investment has increased. A **multinational corporation (MNC)** (also called a transnational corporation, or TNC) owns offices or production facilities in one or more other countries. Some business experts more narrowly define an MNC as a corporation that derives at least a quarter of its revenue from its foreign operations. For example, General Electric (GE) is headquartered in the United States and has various business operations in more than 100 countries. In recent years more than half of the company's revenue has come from its overseas business endeavors.

A few statistics help underscore the importance of MNCs in the global economy. Today there are 82,000 MNCs with 810,000 foreign affiliates, up from 37,000 MNCs and 170,000 foreign affiliates in the early 1990s. Some of the largest economic entities in the world are MNCs. In 2008, each of the three largest MNCs—Royal Dutch Shell, Exxon Mobil, and Walmart Stores—earned revenues in excess of $400 billion. The revenues for each of these MNCs exceeds the value of all goods and services produced in many countries, including Denmark, Greece, Ireland, Argentina, Malaysia, Egypt, and Nigeria among others (**Figure 2.2** on the next page).

Multinational corporations transfer money from their home countries to foreign or host countries to finance their overseas business activities. This process is known as **foreign direct investment (FDI)**. For instance, purchasing or constructing manufacturing plants or equipment in the host country is a form of FDI. There is considerable disagreement over the impacts of FDI. On the one hand, FDI increases the flow of cash in a country and can help promote economic activity, raise employment, and lead to the transfer of knowledge, technology, and infrastructure. On the other hand, FDI can make it difficult for local companies

Multinational corporations • Figure 2.2

The distribution of MNCs and their operations in other countries, or foreign affiliates, demonstrate the globalization of business.

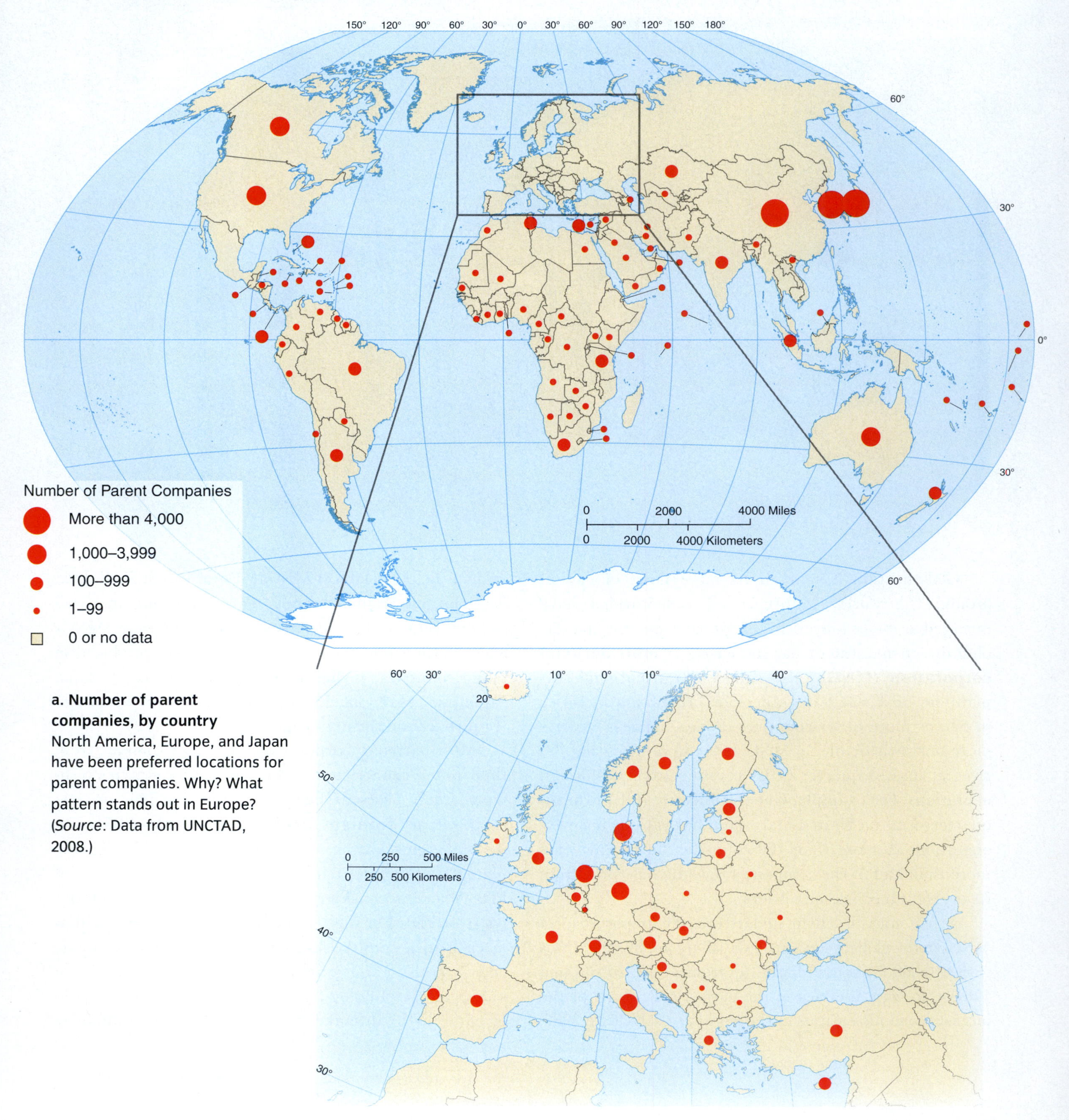

a. Number of parent companies, by country
North America, Europe, and Japan have been preferred locations for parent companies. Why? What pattern stands out in Europe? (*Source*: Data from UNCTAD, 2008.)

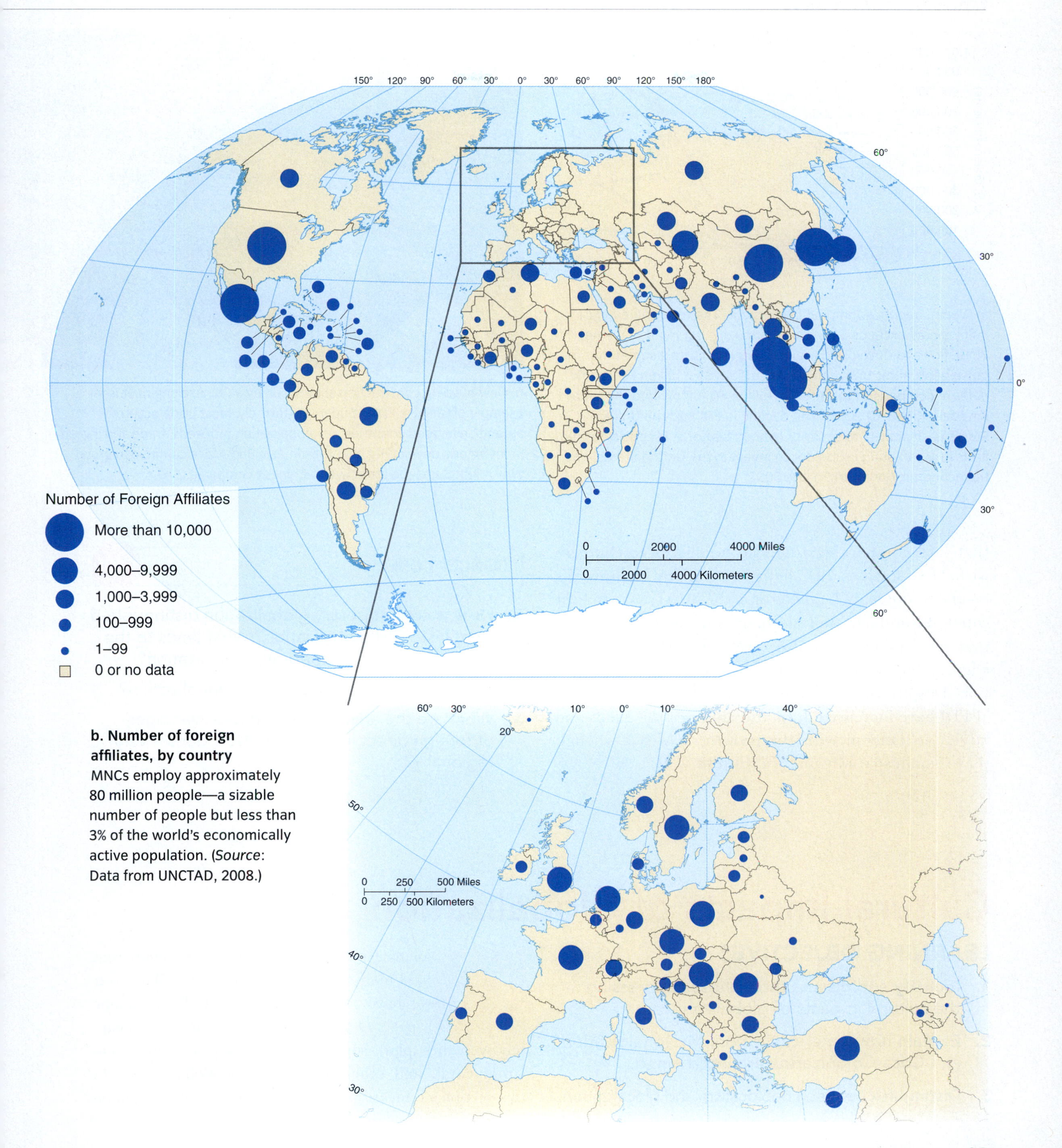

b. Number of foreign affiliates, by country
MNCs employ approximately 80 million people—a sizable number of people but less than 3% of the world's economically active population. (*Source*: Data from UNCTAD, 2008.)

Flows of foreign direct investment (FDI) into different regions • Figure 2.3

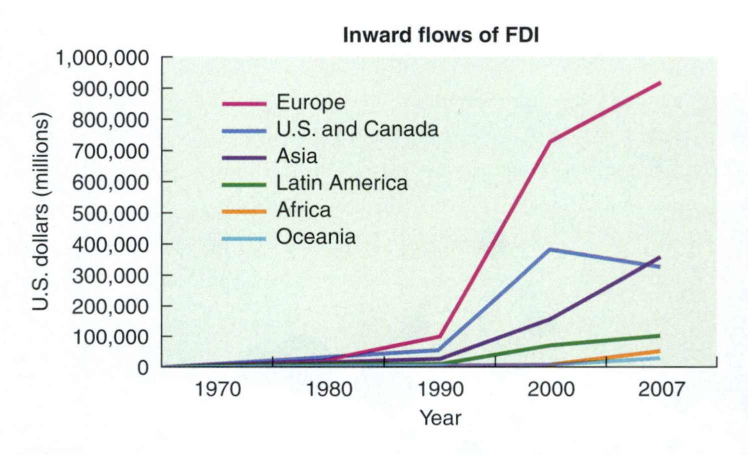

a. Wealth begets additional investment. The total value of FDI inflows now exceeds $1.8 trillion annually, but most of it—nearly 70%—flows to developed regions. Regional totals sometimes disguise geographic unevenness: Portugal's $5.6 billion FDI inflow pales in comparison to the $224 billion FDI inflow into the United Kingdom (UNCTAD 2008). The decline in FDI into the United States is attributed to security concerns after 9/11, decisions by firms in Japan and Europe not to invest in the United States, and the economic downturn in late 2007.

b. New building construction supervised by a Chinese engineer is an example of Chinese FDI in Sudan. Historically, most trade occurred among developed countries. As globalization proceeds, trade and investment among developing countries in Asia and Africa continue to grow. Why might Sudan be attractive to Chinese companies?

that lack comparable financial resources to compete with MNCs. Since FDI helps a company improve its business, the main benefactor is the MNC, not the host country. Similarly, there is no guarantee that a country that receives FDI will also receive transfers of knowledge or technology, in part because of patent protection. Despite the debate over the impact of FDI, flows of FDI have increased dramatically since the 1990s and largely in tandem with the growth of MNCs, though they remain spatially uneven (**Figure 2.3**).

CONCEPT CHECK

STOP

1. **How** is contemporary globalization distinguished from historical globalization? What leads to the emergence of contemporary globalization?
2. **How** is capital associated with globalization?
3. **What** are the advantages and disadvantages of foreign direct investment? How does it vary regionally?

Cultural Impacts of Globalization

LEARNING OBJECTIVES

1. **Identify** and discuss three theses addressing the cultural impacts of globalization.
2. **Explain** how Americanization, McDonaldization, and Coca-Colonization are intertwined.
3. **Distinguish** between neolocalism and glocalization.

Globalization is a complex phenomenon that has many different dimensions. Thus far we have focused on some of the economic aspects of globalization. In subsequent chapters we will explore some of the political, industrial, environmental, and other dimensions of globalization. In this section we investigate the impacts of globalization on cultural difference through an examination of some facets of **popular culture**. Popular culture encompasses products that are mass-produced—for example music, video games, TV shows, cars, clothing—as well as widely held attitudes about preferred forms of leisure, recreation, and entertainment. Popular culture is heavily influenced by mass media, including TV broadcasting, the motion picture industry, the Internet, and traditional forms of book and newspaper publishing. Rapid change, as seen in the way fads come and go, is a characteristic of popular culture.

popular culture The practices, attitudes, and preferences held in common by large numbers of people and considered to be mainstream.

Views about the cultural consequences of globalization are related to our understanding of spatial diffusion. In Chapter 1 we discussed spatial diffusion and noted that it often occurs via a combination of methods, for example via *hierarchical diffusion* and *contagious diffusion*. In some instances, *reverse hierarchical diffusion*—in which diffusion occurs in a "bottom-up" manner—also plays an important role (**Figure 2.4**).

Reverse hierarchical diffusion • Figure 2.4

Few globally successful retail establishments have small-town origins, yet Walmart does. The company was launched in the small town (population 6,000) of Rogers, Arkansas, in 1962.

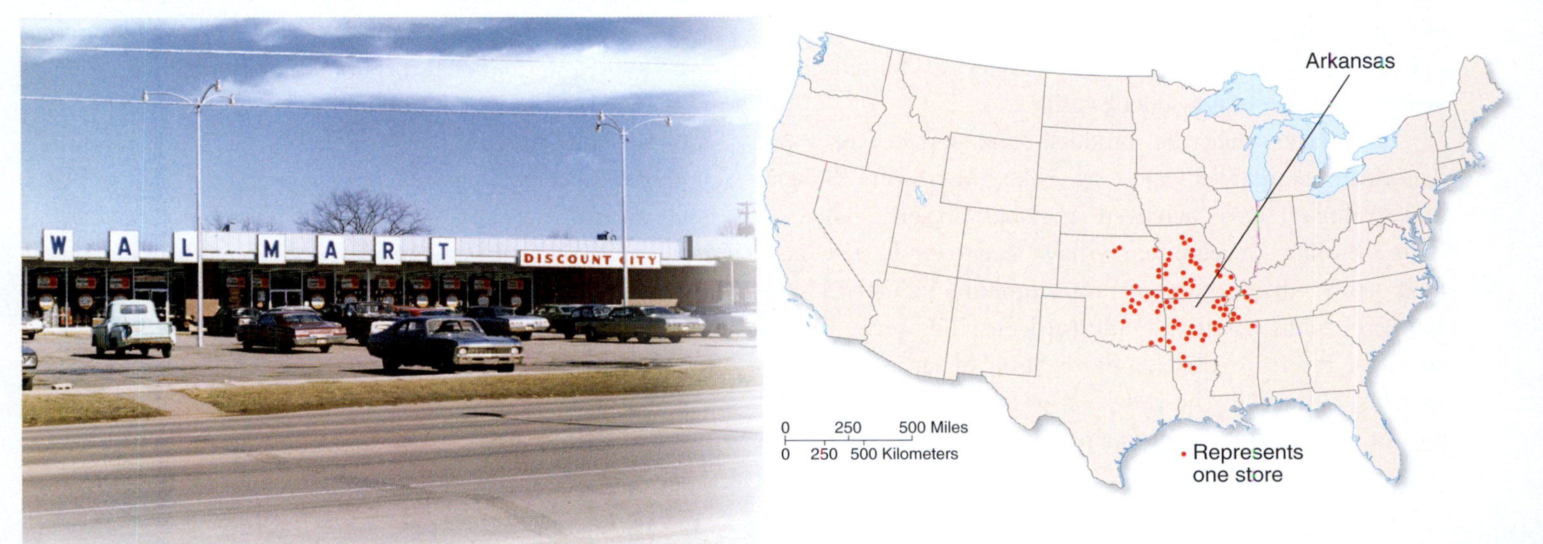

a. The first Walmart store and Walmart store openings: 1970–1974
The map reveals the small-town locations of store openings soon after the company's establishment. (*Source*: Graff and Ashton, 1993.)

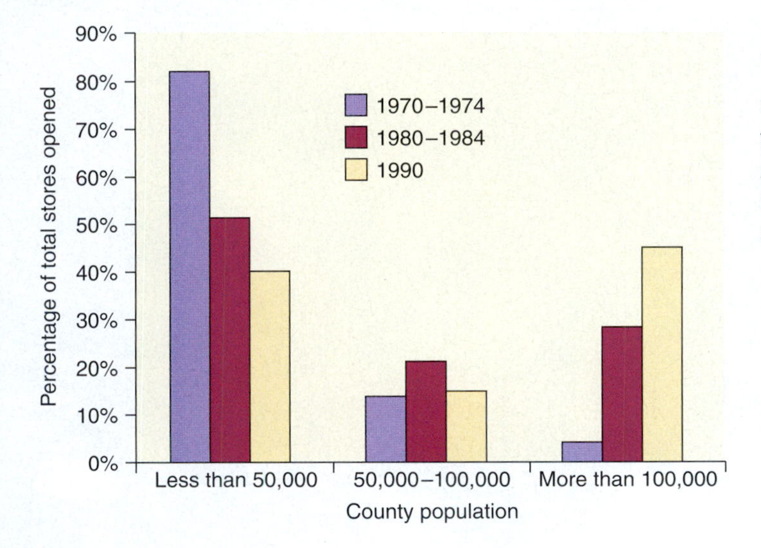

b. Walmart store openings by county size
Initially, Walmart spread to other nearby small towns. It was not until the late 1980s and 1990s that new stores concentrated on the larger markets associated with major metropolitan areas. Approximately what percentage of stores did Walmart open in counties with more than 100,000 people from 1970 to 1974 and in 1990? (*Source*: Data from Graff and Ashton, 1993.)

Social scientists have proposed a simple but useful framework to help us make sense of the cultural consequences of globalization. This framework consists of three key ideas or theses: the homogenization thesis, the polarization thesis, and the glocalization thesis. According to the homogenization thesis, globalization makes cultural tastes, beliefs, and practices converge and become more alike. The global diffusion of fast-food franchises, restaurants, hotel chains, and big-box retailers, such as Walmart, is often cited as evidence of the homogenization thesis. In the 1970s, before the word *globalization* was coined, geographer Edward Relph developed the term **placelessness** to draw attention to the loss of the unique character of different places and the increasing standardization of places and cultural landscapes. In the 1990s, writer and cultural critic James Howard Kunstler lamented what he saw as the emergence of a "geography of nowhere," primarily because of so much standardization in the American landscape. The epitome of "nowhere" is the cookie-cutter residential subdivision where every house looks alike.

It is clear that the products and services of some American companies, such as Nike, Microsoft, Google, and Walmart, have achieved a global presence and visibility. Similar developments have occurred in the fast-food, soft drink, and entertainment industries—the source of many American cultural icons. The term *McDonaldization* refers to the standardization of eating habits—specifically through the provision of fast food eaten on the go, out of styrene packages or paper wraps. In addition, the global reach of the Cartoon Network—broadcast in over 160 countries—now surpasses both McDonald's and Coca-Cola.

The homogenization thesis is tied to the notion that economic processes shape cultural practices. These economic processes are largely attributed to the expansion of capitalism, and although the United States is not the sole driver of globalization, the country exerts a great deal of economic and cultural influence. Consequently, homogenization is widely understood to bring **Americanization**. The pervasive presence of American products involves the expansion of multinational corporations. Some proponents of Americanization point out that it is much more than just a diffusion of goods; it also involves the spread of values and attitudes, such as consumerism, freedom, and individualism. Critics of Americanization note that even though MNCs do employ local workers and managers, the bulk of the profits accrue to the multinational corporation, not the foreign country or the local economy where their franchises or branch plants are located. Thus, they see the diffusion of American corporations and brands as an expression of American economic, political, and cultural hegemony or dominance. The term *Coca-Colonization* implies that the hegemony of MNCs creates a set of power relations similar to those that exist between a country and its colonies, producing a kind of imperialism (**Figure 2.5**).

Americanization The diffusion of American brands, values, and attitudes throughout the world.

According to the polarization thesis, globalization contributes to a heightened sense of sociocultural identity that serves to fragment people and trigger social

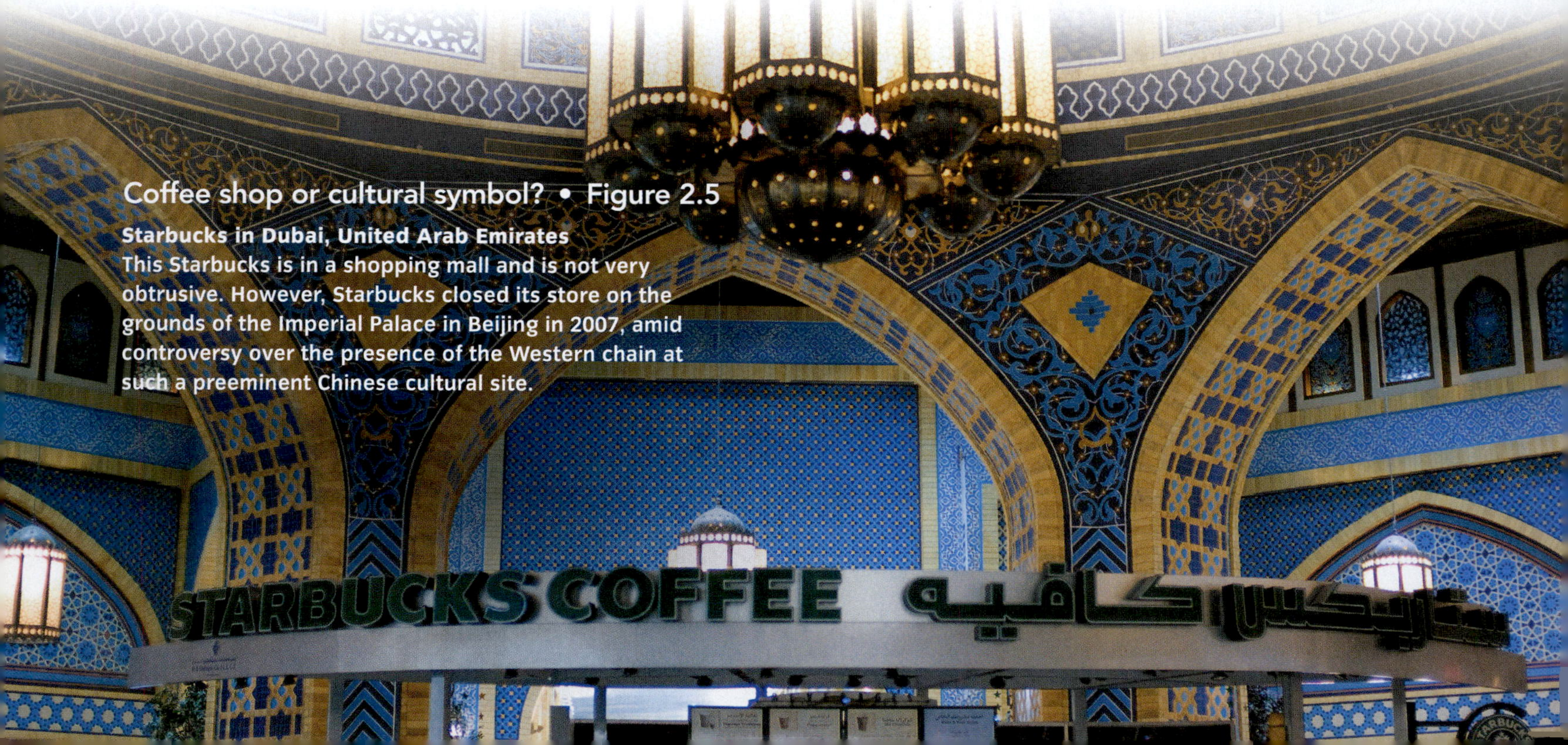

Coffee shop or cultural symbol? • Figure 2.5

Starbucks in Dubai, United Arab Emirates
This Starbucks is in a shopping mall and is not very obtrusive. However, Starbucks closed its store on the grounds of the Imperial Palace in Beijing in 2007, amid controversy over the presence of the Western chain at such a preeminent Chinese cultural site.

STARBUCKS COFFEE ستاربكس كافيه

Cultural flows and neolocalism • Figure 2.6

a. Beliefs, practices, and products do not simply flow from the Western world to other parts of the globe—they also flow out from non-Western places. Japanese anime (from *animation*) and cosplaying (costume role-playing of anime characters) have become very popular in North America, Europe, and Australia.

b. Local ordinances established by area communities or collective preferences can resist some of the homogenizing forces of globalization. Residents on Chicago's South Side have resisted attempts by Walmart to open a supercenter.

disorder and instability instead of creating a standardized global culture. Those who subscribe to the polarization thesis believe that rather than homogenizing the world, globalization has unleashed powerful separatist forces that have heightened concerns about security not just for individuals but also for countries. They point to the fact that numerous wars and struggles over identity in the Balkans, the Caucasus, and Africa have coincided with the spread of globalization, as has the prevalence of global terrorism. Similarly, globalization has made possible and, with the Internet, has even facilitated the extension of cross-border criminal networks. Although the homogenization and polarization theses are quite popular in the media, many scholars find that they tend to oversimplify how globalization works because they depict globalization as having a single and fixed outcome. However, many scholars doubt that we will witness a complete homogenization of culture or landscape. They argue that human creativity tends to resist uniformity and that there will always be some people who refuse to conform. Furthermore, scholars stress that globalization is not unidirectional from the West to the rest of the world. The diffusion of tattooing discussed at the start of the chapter is an example of cultural practices spreading to the West. Reggae spread from Jamaica around the world. The existence of Mexican restaurants in Vietnamese cities, or Indian, Thai, and Ethiopian restaurants in London, Toronto, or Sydney, also illustrates the multidirectional nature of globalization.

Similarly, globalization has as much to do with connectivity as it does with polarization. As the previous discussion of the polarization thesis shows, globalization not only creates homogenizing forces, it can also stimulate local awareness. **Neolocalism** is a term coined by geographer Wes Flack to describe a renewed interest in sustaining and promoting the uniqueness of a place (**Figure 2.6**).

The glocalization thesis provides a third framework for understanding the cultural consequences of globalization. **Glocalization** occurs when a multinational corporation alters its business practices to reflect local preferences. From a geographic standpoint, glocalization is the result of the bridging of the local

glocalization The idea that global and local forces interact and that both are changed in the process.

Glocalization • Figure 2.7

In India, where cows are sacred to Hindus and Muslims avoid consuming pork, the menu often features the McVeggie, a beefless burger, the Chicken Maharaja Mac, and a variety of other vegetarian and nonvegetarian options but no beef or pork products.

and global scales to create what is sometimes called the *local-global nexus*. Stated differently, there is always a dynamic relationship between local and global forces such that local forces become globalized and global forces become localized. Glocalization is a business strategy as well as a conceptual framework about the impacts of globalization. Globalization expert Jan Nederveen Pieterse (2009, p. 52) draws attention to the local-global nexus with the maxim that "all business is local" (Figure 2.7).

CONCEPT CHECK STOP

1. **What** are the strengths and weaknesses of the homogenization and polarization theses?
2. **How** are Americanization, McDonaldization, and Coca-Colonization similar to and different from one another?
3. **What** evidence supports the glocalization thesis?

The Commodification of Culture

LEARNING OBJECTIVES

1. **Discuss** how commodification can shape cultural practices and meanings.
2. **Explain** how representation and commodification relate to the notion that culture is contested.
3. **Explain** what is meant by heritage dissonance.

s we discussed in Chapter 1, the term *culture* is a social creation consisting of shared beliefs and practices that are dynamic rather than fixed. Culture is a complex system that is shaped by people and, in turn, influences them. Manifestations or expressions of culture take both material and nonmaterial forms. **Material culture** includes the tangible and visible artifacts, implements, and structures created by people. Furniture, dwellings, musical instruments, and tools are all examples of material traits. **Nonmaterial culture** is not tangible and is associated with oral traditions and behavioral practices. Examples of nonmaterial traits include recipes, songs, knowledge, or philosophies shared by word of mouth, and the way we behave in certain circumstances as when we greet people. **Cultural geography** is a branch of human geography that emphasizes human beliefs and activities

and how they vary spatially, utilize the environment, and change the landscape.

Cultural geographers are extremely interested in the **commodification** of culture. Slavery commodifies human beings, while online dating services commodify the procedures for meeting and getting to know prospective partners. Plasma collection centers and sperm banks reflect the commodification of bodily fluids. Seeing what's available on eBay quickly reveals the extent of commodification.

commodification The conversion of an object, a concept, or a procedure once not available for purchase into a good or service that can be bought or sold.

consumption Broadly defined, the use of goods to satisfy human needs and desires.

Commodification is closely associated with **consumption**. Cultural geographers note that consumption both influences and is influenced by culture. Your cultural background influences the things that you use, but these items also shape the interactions you have with other people. Commodities affect social relationships. Prenuptial agreements signed by couples before the formal marriage ceremony provide a good illustration of how commodities shape social relationships.

Commodities also affect social relationships because they communicate meaning. Think, for example, about the different social messages that your clothing, jewelry, car, house, or apartment send about you. Commodities become symbols of ideas and values. People who share the same cultural background share the same codes for interpreting these symbols and their meanings.

Advertising, Commodification, and Cultural Practice

Advertising is one of the main forces affecting patterns of consumption locally as well as globally. Simply stated, advertising is designed to influence consumer behavior, and it does so by the clever manipulation of images, text, symbols, and slogans. In this section we explore some of the connections between geography, advertising, and consumption through developments involving diamonds. This example highlights some of the ways in which commodification can shape cultural practices.

Do you agree that gems have intrinsic value? Globally there are enough people who do to sustain a lucrative market for them. Among Westerners, diamonds have long been considered the most precious of all gems. The average amount spent by grooms on a diamond engagement ring now exceeds $3,000, double what they would have paid if they had made their purchase in 1995. If they shop at Tiffany's, the renowned luxury diamond retailer, they can expect to pay, on average, $10,000 for a diamond engagement ring. But diamonds are not scarce, so how can they command such high prices? The answer is a complicated mix of economics, geography, history, and marketing.

Think about the diamond commercials and advertisements you have seen. If you believe that "a diamond is forever," you have accepted one of the diamond industry's most familiar and durable advertising slogans, first launched in the United States by De Beers in 1947. The diamond industry cultivates the idea that diamonds are associated with everlasting love, will retain their value over time, and are symbols of beauty, preciousness, happiness, and status. Also, diamond marketers have strongly conventionalized—even ritualized—the act of giving a diamond engagement ring.

The history of De Beers is instructive in understanding the commodification of diamonds. De Beers began as a mining company in the late 19th century in what is now South Africa. The Englishman Cecil Rhodes established De Beers and sought to control the worldwide diamond industry by acquiring the diamond mines in the region. Under the direction of the Oppenheimer family in the 20th century, De Beers established a diamond **cartel** that controlled the supply of diamonds and therefore the demand for them.

cartel Entity consisting of individuals or businesses that control the production or sale of a commodity or group of commodities—often worldwide.

De Beers solicited contracts with other diamond producers to bring them into the cartel; as a result, for the better part of the 20th century the company controlled between 66 and 80% of the global supply of rough, unprocessed diamonds. This control gave the company the power to decide when to sell diamonds and also to determine the selling price. Even though diamonds are not scarce and can actually be manufactured, the De Beers cartel succeeded in creating the illusion that diamonds are in short supply by carefully controlling the quantities of diamonds they made available for sale.

In the late 1980s and 1990s, a number of events helped to break the De Beers monopoly. New diamond mines opened in Russia and Canada. Russia only participated in the cartel during some years, while Canada did not participate at all. Diamond mining companies in Australia and Angola also remained outside the cartel.

Geography InSight

Diamond production and consumption • Figure 2.8

Deep beneath Earth's crust, diamonds crystallize under intense heat and pressure. Volcanism in certain regions brings them near the surface where they can be mined.

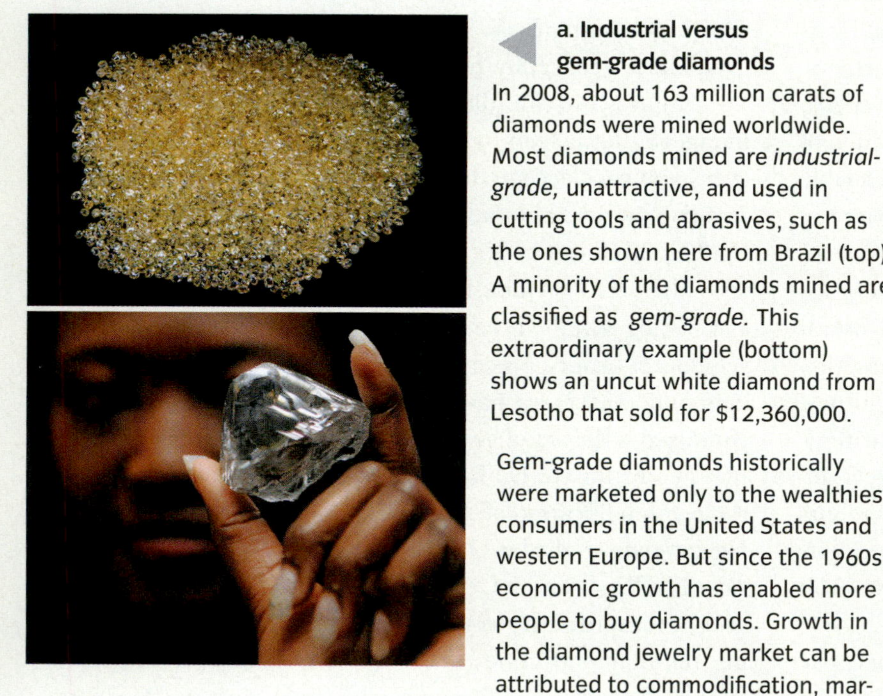

a. Industrial versus gem-grade diamonds
In 2008, about 163 million carats of diamonds were mined worldwide. Most diamonds mined are *industrial-grade,* unattractive, and used in cutting tools and abrasives, such as the ones shown here from Brazil (top). A minority of the diamonds mined are classified as *gem-grade.* This extraordinary example (bottom) shows an uncut white diamond from Lesotho that sold for $12,360,000.

Gem-grade diamonds historically were marketed only to the wealthiest consumers in the United States and western Europe. But since the 1960s, economic growth has enabled more people to buy diamonds. Growth in the diamond jewelry market can be attributed to commodification, marketing, advertising, and globalization.

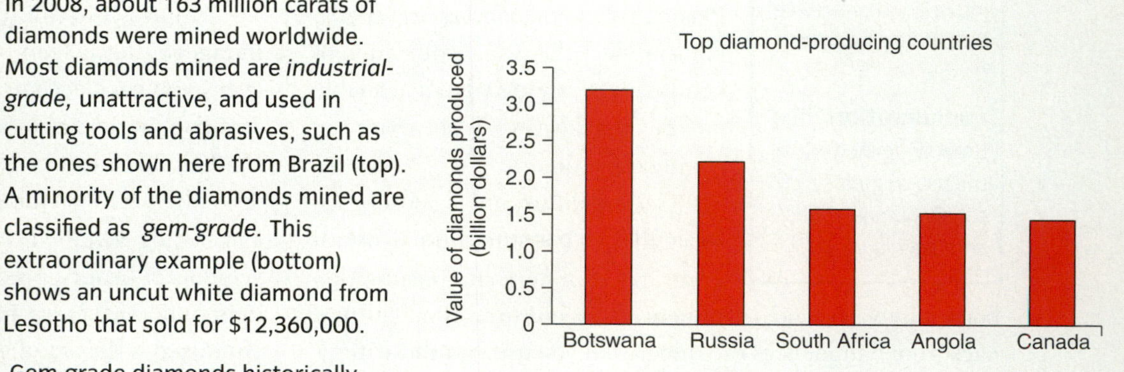

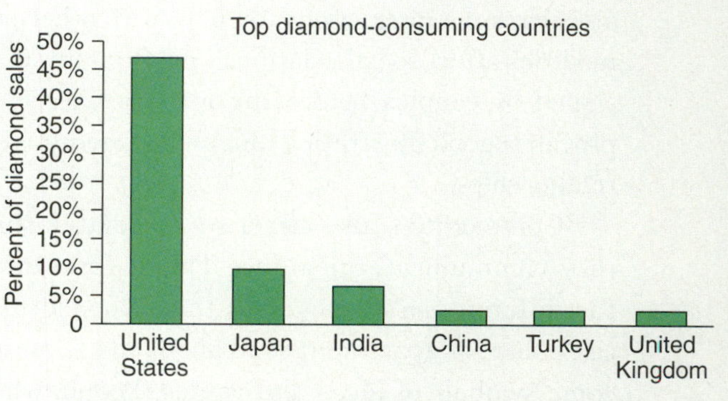

b. Leading diamond producers and consumers
The geography of diamond production differs considerably from the geography of diamond consumption. Diamonds are mostly marketed in nonproducing countries (green). The United States has no commercial diamond production, yet the country has by far the greatest demand for diamond jewelry. (*Source:* Data from KPMG, 2006.)

c. Diamond mining and conflict
Diamonds are extracted from underground mines, surface mines, and river alluvium. "Diamond diggers" sift alluvium in Sierra Leone. Diamonds fuelled civil wars in several African countries. Sheku Conteh lost his hand during the conflict period in Sierra Leone. Since it is difficult to trace the origins of a diamond, the Kimberley Process Certification Scheme was developed to solve the problem of conflict diamonds. However, difficulties enforcing the certification scheme mean that conflict diamonds still exist.

d. Japanese newlyweds
Rare was the Japanese bride who wore a diamond engagement ring a few decades ago. But Japan's economic recovery after World War II generated increased consumerism, and marketing promoted diamonds as symbols of modernity, status, and purity. Today, Japan is the second-largest diamond jewelry market after the United States.

Developments in Angola were among the first to draw attention to the problem of **conflict diamonds** or *blood diamonds*, but diamonds also factored in bitter conflicts in Sierra Leone, Liberia, and the Democratic Republic of the Congo. In response to public concerns, many diamond retailers now certify that the diamonds they sell are not conflict diamonds (**Figure 2.8**).

conflict diamonds Diamonds sold to finance wars or terrorist activities.

Sports, Representation, and Commodification

The discussion of diamonds shows how successful some enterprises are at shaping cultural practices, both locally and globally. Commodification is not limited to material culture, however. Commodification also influences nonmaterial culture and increasingly involves indigenous and local communities, in part because it presents opportunities for commercial transactions and economic gain. The Maori are the indigenous peoples of New Zealand. One facet of Maori nonmaterial culture that has been heavily commodified within the past decade is the **haka**, a collective, ritual dance (**Figure 2.9**).

The commodification of the haka has been hotly contested for two key reasons. The first reason is that, as Maori scholars have pointed out, the haka is not actually a Maori war dance and the rugby players do not use authentic haka moves. The second reason involves ownership and control of Maori culture. In 2000, some Maoris began to seek a greater share of the profit that both Adidas and the New Zealand Rugby Football Union (NZRFU) were making on contracts for the televising of All Blacks games. Representatives of the NZRFU commented that their use of the haka was not for commercial purposes. When Maoris raised concerns over violations of copyright in the commercial and also on billboards, lawyers for NZRFU argued that it was not possible to prove who created the haka or had ownership of it.

If you are not Maori, the controversy over the haka may seem insignificant. Nevertheless, there are at least two crucial points to take from this discussion. The first is that culture is contested. Culture is so contested because people derive a good part of their sense of identity from it. The second and related point is that the controversy over the haka involves questions of ownership as well as cultural boundaries. For example, who should have a say in the use of cultural symbols, such as the haka or the representation of peoples and their cultural practices? Also, are there

The haka • Figure 2.9

Before a match, members of the New Zealand rugby team, the All Blacks, gather in the middle of the field and perform the haka, an ancient ritual dance consisting of a series of loud chants, body slapping, forceful movements of the arms and legs, and jumping. The goal is to psych up the members of the All Blacks team and intimidate their opponent.

The dissonance of heritage • Figure 2.10

a. Hundreds of Native Americans in Oklahoma protested the centennial celebration of the state in 2007. To them the celebrations recalled a heritage rooted in treaties with their nations that the U.S. government failed to uphold and in the taking of Native American land for the purposes of white settlement.

b. Robben Island, the notorious prison for South Africa's anti-apartheid protestors, including Nelson Mandela, former president of South Africa and Nobel Peace Prize winner. To the indigenous San people, some of whom were also imprisoned there, the place symbolizes oppression by European colonialists. What might the heritage of Robben Island mean to former prison staff?

limits to the aspects of culture that can or should be commodified? If so, who sets those limits? These issues are all the more complicated by the friction between Maoris and New Zealanders of European descent, and the long-term prejudice the Maori have encountered since colonial days.

The Heritage Industry

Questions about commodification and authenticity also surround the **heritage industry**, which has experienced considerable growth since the 1980s. This growth is closely associated with a shift in the meaning of heritage. Geographer David Lowenthal, for example, has tracked the evolution of this term. Until very recently, heritage referred to property transmitted to an heir. It also included cultural practices relating to inheritance and the management of one's estate through wills and trusts. Today, however, **heritage** more commonly refers to any contemporary use of the past. For example, even radio stations that play "classic rock" are using a kind of heritage.

> **heritage industry** Enterprises, such as museums, monuments, and historical and archaeological sites that manage or market the past.

The creation of heritage attractions frequently involves the commodification of the past. The heritage industry, in effect, packages the past for sale. In order for it to have broad appeal, the idea of heritage has to be simplified, sanitized, and made entertaining. These views underscore the fact that heritage is characterized by **dissonance**—the quality of being inconsistent. Heritage dissonance expresses the idea that the meaning and value of heritage vary from group to group.

In their book, *Dissonant Heritage: The Management of the Past as a Resource in Conflict*, J. E. Tunbridge and G. J. Ashworth identify two main reasons that heritage is always dissonant. The first reason involves the opposing uses of heritage. For example, many heritage sites are also sacred sites. Allowing tourists at these sacred sites has the potential to lead to conflict. The second reason involves the *particularism* of heritage. That is, the meanings and uses of heritage are group-specific. Simultaneously inclusive and exclusive, heritage creates an inconsistency or dissonance in its meaning. The dissonance of heritage is clearly revealed in multicultural societies where heritage is frequently contested (**Figure 2.10**).

World Heritage

World or **global heritage** refers to sites perceived to have outstanding universal value for all of humanity. Geographer Douglas Pocock has noted that identifying extraordinary sites dates to at least the Classical Greek identification of the Seven Wonders of the World. The movement to protect and preserve world heritage, however, gathered its momentum in the 1960s and 1970s and has been led by UNESCO, the educational, scientific, and cultural organization of the United Nations.

In 1972, UNESCO adopted the Convention Concerning the Protection of the World's Cultural and Natural Heritage. This paved the way for the creation of a World Heritage Committee (WHC). One of the charges of the WHC was to generate a World Heritage List of cultural and natural sites deemed to possess outstanding universal value. In 1992, the list was modified to incorporate cultural landscapes that reflect different facets of the relationship between people and the environment.

Pocock argues that adding cultural landscapes to the World Heritage List marked a significant shift to a broader vision of heritage that is not limited to material evidence. For example, a particular landscape may be associated with significant religious, social, or other meanings regardless of the presence of tangible artifacts. Since this change in 1992, a World Heritage property might be a cultural site, a natural site, a cultural landscape, or a mixture of these (**Figure 2.11**).

The World Heritage List has been celebrated for its role in raising awareness about global cultural resources and

World Heritage sites • Figure 2.11

a. The event that linked UNESCO with world heritage was the impending destruction of the Abu Simbel temples by the formation of Lake Nasser behind the Aswan High Dam in Egypt. UNESCO considered Abu Simbel, built in the 13th century BCE, an irreplaceable cultural resource and in the 1960s arranged to disassemble the two temples and reassemble them on land above the waters of Lake Nasser.

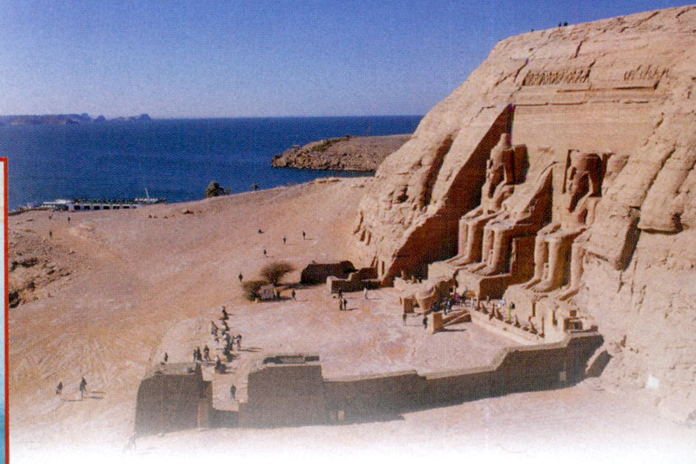

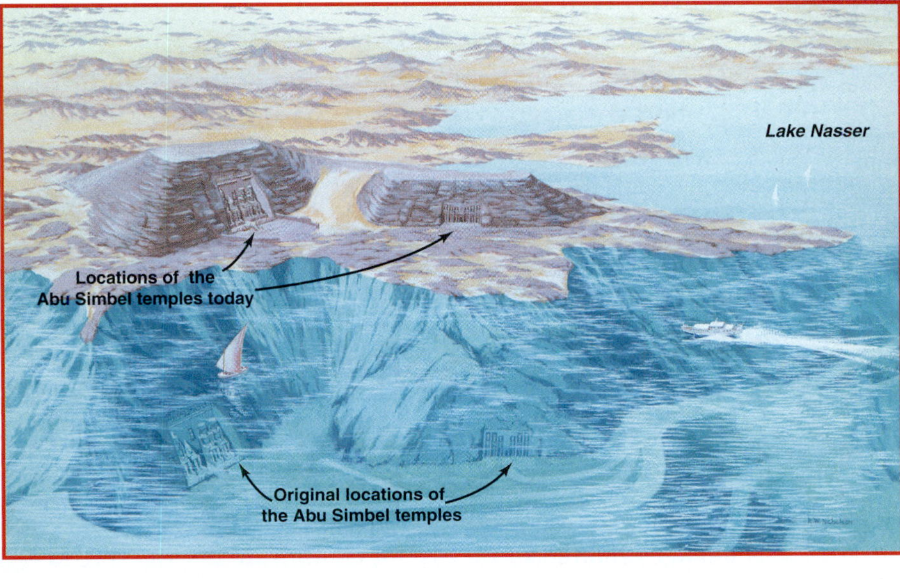

b. The first cultural landscape recognized in the World Heritage List was New Zealand's Tongariro National Park. Originally included as a natural site on the World Heritage List, in 1993 its listing was revised to recognize it as a cultural landscape with major religious significance to the Maoris.

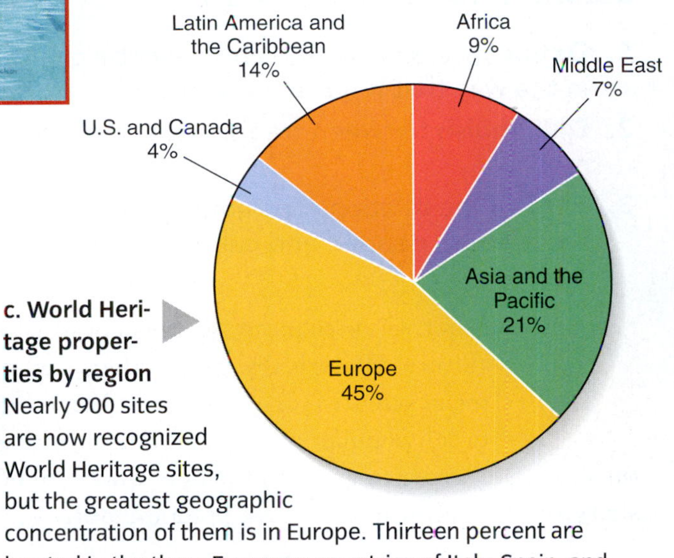

c. World Heritage properties by region Nearly 900 sites are now recognized World Heritage sites, but the greatest geographic concentration of them is in Europe. Thirteen percent are located in the three European countries of Italy, Spain, and France. (*Source*: Data from UNESCO, 2009.)

Where Geographers CLICK

UNESCO World Heritage List

http://whc.unesco.org/en/list

At this Web site, you can view a map of World Heritage sites and learn more about them.

sometimes for triggering the development of new tourist sites (see *Where Geographers Click*). But the World Heritage List has also been criticized for four main reasons. First, it reflects a Eurocentric bias with an overrepresentation of sites in Europe, including buildings and properties associated with Christianity. Second, when World Heritage sites are identified, tourists often expect that the site will be accessible to them, regardless of any competing uses of the site or their state of repair. Third, these sites can be very expensive to maintain. Fourth, critics question whether there really can be such a thing as global heritage. They base this criticism on the point, made earlier, that heritage is group-specific.

CONCEPT CHECK STOP

1. **How** does the history of the De Beers cartel illustrate the connections between commodification and cultural practice?
2. **Why** is the commodification of the haka a contentious issue?
3. **Why** is the World Heritage List both celebrated and criticized? How has it changed the way heritage is defined?

Cultural Geographies of Local Knowledge

LEARNING OBJECTIVES

1. **Define** local knowledge and describe changes in the way that it has been viewed.
2. **Distinguish** between traditional and allopathic medicine.
3. **Explain** the relationships among local knowledge, gender, and cultural ecology.

Earlier we introduced the concept of popular culture. Historically, geographers and other social scientists have drawn a distinction between popular culture and folk culture, with *folk culture* referring to groups of people whose members share similar cultural traits, live predominantly in rural areas, and whose livelihood is minimally connected to the global market economy. The basis for this distinction between folk and popular culture stems from social changes related to the rise of capitalism and the spread of industrialization.

When this two-part classification was initially conceived in the 19th century, the labels "folk society" and "urban society" were used. At the time, folk society was synonymous with preindustrial society and urban society was associated with development. Thus, the term *folk* is problematic because of the way it suggests a less advanced group of people and makes those groups classified as folk cultures seem like quaint curiosities representative of a bygone era. A folk culture was understood to include those people who represented the common folk (as opposed to the elites) and whose cultural artifacts were handmade, not mass-produced. But this practice of excluding the customs of local elites introduces a bias and also calls into question the usefulness of the term *folk culture*. Because of these criticisms and in light of ongoing globalization, especially the greater interconnectedness of the global economy, we prefer instead to speak of **local culture**.

local culture The practices, attitudes, and preferences held in common by the members of a community in a particular place.

How do local cultures make decisions about natural resources? What factors do they take into consideration when planting or harvesting, preparing food, or caring for the sick? The complicated answers to these kinds of questions depend on the people we are talking about and on the knowledge they possess, as well as on the political, economic, environmental, and social circumstances in which they live. For a long time scholars have recognized that people acquire, share, use, and transmit knowledge in many different ways. But it is only recently that researchers have begun to understand and appreciate the importance of local ways of knowing. In this section we explore cultural geography through a discussion of local knowledge.

Local Knowledge

Geographers and other scholars use the term **local knowledge** to refer to the collective knowledge of a community that derives from the everyday activities of its members. Three characteristics help clarify the meaning of local knowledge:

1. Local knowledge is usually transmitted orally and is rarely written down. In many cases, oral transmission is supplemented by activities or stories that help to demonstrate a procedure or reinforce a particular practice.
2. Local knowledge is dynamic and continuously evolving—changing to reflect the acquisition of new observations and information.
3. Local knowledge does not exist as a single, monolithic entity. Rather, numerous reservoirs of local knowledge are retained by different individuals and groups within a community. It may be more suitable to talk about *local knowledges* instead of local knowledge.

In the past, Western thinkers assumed that local knowledge and land-use practices were outmoded, even inferior. Geographer Jim Blaut argues that these kinds of prejudices were part of the European **rationality doctrine**, prevalent during the lengthy period of European colonialism, when non-Europeans were perceived to be irrational and childlike. By the 1950s and 1960s, Blaut notes, Europeans attributed a scientifically deficient mentality to their colonial subjects. This vision contributed to the emergence of **diffusionism**—the view that the diffusion of Western science, technology, and practices to other peoples would enable them to advance socially and economically. Ironically, diffusionists—many of whom were geographers—overlooked the importance of place as well as the relevance of local knowledge.

rationality doctrine The attitude and belief that Europeans were rational and non-Europeans, especially colonized peoples, were irrational.

Most researchers today share the view that awareness of the local context—including local knowledge—is essential. Because local knowledge provides a framework for individual and community problem solving, not only on a day-to-day basis but also over the long term, it contributes to and informs **sustainable development**—an approach to resource use and management that meets economic and social needs without compromising the resources for future generations. Moreover, an awareness of local knowledge provides insights on the **social capital** of a particular community, which affects how decisions on matters such as resource use are made.

social capital The social ties, networks, institutions, and trust that members of a group use to achieve mutual benefits.

Geographies of Traditional Medicine

Traditional medicine is an important reservoir of local knowledge that varies considerably from one place to another. Those who practice traditional medicine carry different titles, including healer, herbalist, bone setter, and spiritualist. A healer may acquire her

traditional medicine Medical practices, derived from a society's long-established health-related knowledge and beliefs, that are used to maintain or restore well-being.

THE PLANNER

Video Explorations

Leeches for Curing Illness

WILEY PLUS NATIONAL GEOGRAPHIC Video

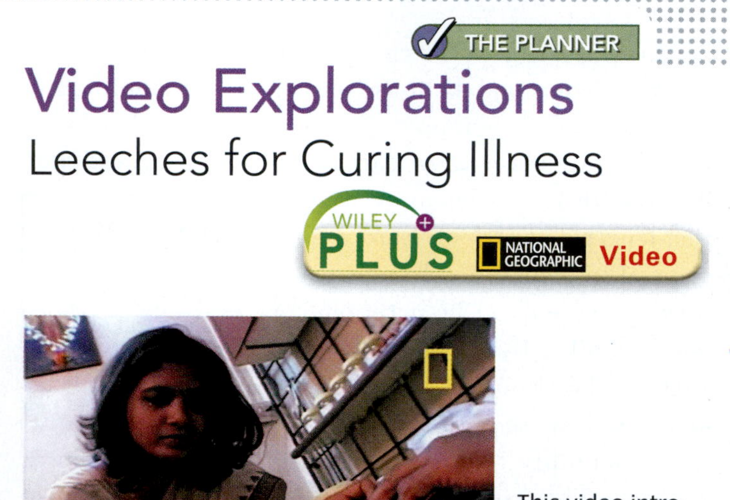

This video introduces Ayurvedic medicine, a holistic form of medicine common in India.

or his knowledge from an experienced elder, or may be trained at a university. Despite its name, traditional medicine can actually be conventional, particularly if it is widely practiced by the general populace or forms the basis of a country's health care system. Alternatively, traditional medicine might only be practiced in special circumstances.

Most approaches to traditional medicine share two attributes: They tend to be holistic and personal. A **holistic approach** to medicine sees health as encompassing all aspects—physical, mental, social, and spiritual—of a person's life. A **personal approach** to medicine means that it is quite possible for two people to have the same symptoms but receive different treatments. Traditional medicine is usually contrasted with **allopathic medicine**, sometimes called modern or Western medicine. The oldest form of medicine practiced by humankind is traditional medicine. Allopathic medicine, by contrast, developed in conjunction with advances in biology and chemistry, the rise of the experimental method, and the invention of new technologies, including the microscope.

allopathic medicine Medical practice that seeks to cure or prevent ailments with procedures and medicines that have typically been evaluated in clinical trials.

Traditional medicine in India grew out of *Ayurvedic* beliefs and practices (in Sanskrit *ayurveda* means "the science of life") based on the sacred Hindu text, the Vedas. It emphasizes the maintenance of harmony between a person and her or his environment, through such techniques as the use of herbal and oil treatments (see *Video Explorations*).

A similar ancient theory exists in Chinese traditional medicine and involves the *yin* and *yang*—opposing forces that, if out of balance, can result in illness. An important facet of Chinese traditional medicine is **acupuncture**, which has been practiced for at least 2,500 years (**Figure 2.12**).

acupuncture An ancient form of traditional Chinese medicine that promotes healing through the insertion of needles into the body at specific points.

Traditional medicine is geographically widespread; however, since many countries do not keep statistics on the practice or use of traditional medicine, it is difficult to draw comparisons between traditional medicine and allopathic medicine. Also, because there are very few systems of licensing practitioners of traditional medicine, there is no easy method to keep track of them. Studies conducted in Zambia, for example, indicate that there are 44 traditional healers for every allopathic doctor in the country. In addition, the position of traditional medicine within a country's health care system varies considerably.

Traditional medicine is fully and officially part of the national health care system in just four countries of the world: China, North Korea, South Korea, and Vietnam. In China, for example, the country's constitution grants traditional medicine legal authority to be practiced and used. It is official policy in China that allopathic and traditional medicine should coexist. Thus, a patient can visit either kind of doctor, and health insurance will provide full coverage for the visit, diagnosis, and treatment. Similar practices exist in North Korea, South Korea, and Vietnam, where the national health care system recognizes and supports both traditional and allopathic systems of medicine. In Japan, by contrast, most medical schools train physicians to practice allopathic medicine, though it is often possible for students to take classes in *kampo*, Japanese traditional medicine.

In Western countries the term *complementary and alternative medicine* (CAM) is now preferred to *traditional medicine*. CAM is sometimes described as the adoption of traditional medicine within an allopathic system. The United States is much like Japan in its long-standing emphasis on the training of allopathic practitioners. However, in 1998 the United States created the National Center for Complementary and Alternative Medicine within the National Institutes of Health in order to support and advance research on CAM. Recent studies

PROCESS DIAGRAM

THE PLANNER

The diffusion of acupuncture • Figure 2.12

a. The precise origins of acupuncture within China are not known
One of the first references to the use of needles for therapeutic purposes appears in a text called *The Yellow Emperor's Classic of Internal Medicine*, which was compiled between about 480 and 220 BCE. These models, labeled in Chinese, show the body's acupuncture points. Acupuncture is now one of the most recognized forms of traditional medicine in the West. How did knowledge of acupuncture diffuse?

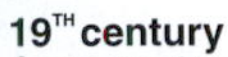

1 Several centuries BCE
Acupuncture originates in ancient China.

2 6TH century CE
Knowledge of acupuncture diffuses to Korea and Japan with the Buddhist religion. A Chinese medical text was given to Japan in 552 CE and ten years later a Chinese acupuncturist visited the country. Subsequent visits to China by Japanese monks studying medicine also helped diffuse knowledge of acupuncture.

3 8TH–10TH centuries
Acupuncture diffuses to the area of present-day Vietnam via trade.

4 10TH–16TH centuries
The diffusion of acupuncture to the West likely followed the Silk Road, an ancient trade route. Tenth-century Persian medical texts mention acupunture. French Jesuits working as missionaries in East Asia during the 16th century brought back information on acupuncture and helped its diffusion to Europe.

5 19TH century
Acupuncture probably arrived in North America as early as the 1800s via at least two pathways: Chinese immigrants who continued to use acupuncture, and information transfer with the medical community.

To France

The Silk Road

From China

b. Following its arrival in the West, acupuncture was marginalized for a long time by practitioners of mainstream, allopathic medicine
Only since the 1970s has acupuncture become more widely accepted and available within the United States. In the 1980s, acupuncture became eligible for coverage under some insurance plans.

A woman receives acupuncture.

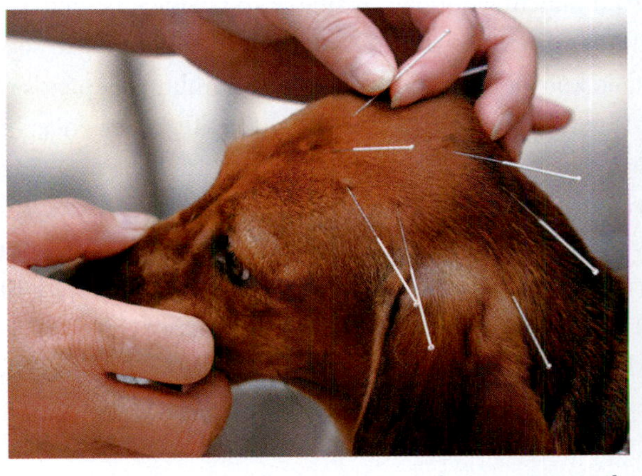

Even pets may undergo acupuncture for treatment of certain ailments.

The global resurgence of traditional medicine • Figure 2.13

a. Products for complementary and alternative treatments on drug store shelves
CAM may include the use of herbal medicines, now estimated to have a global market value of $60 billion. CAM is increasingly a part of medical practice in countries where allopathic medicine has dominated. Yearly spending on CAM exceeds $2 billion each in the United States, Canada, and the United Kingdom.

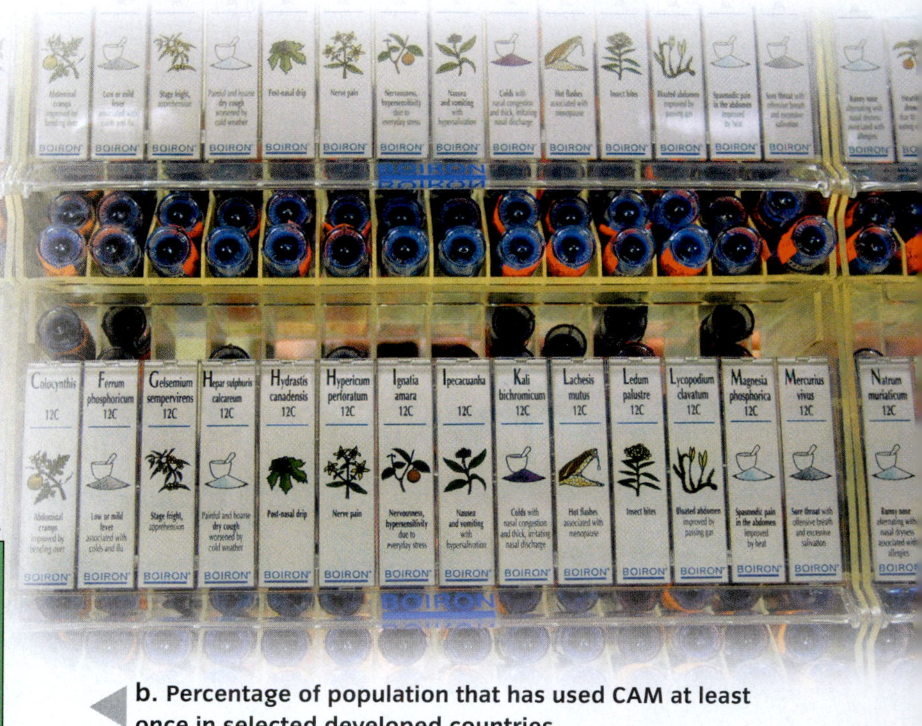

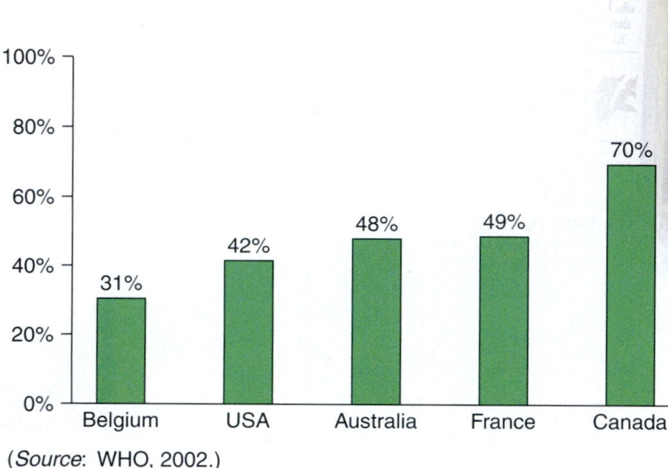

(*Source*: WHO, 2002.)

b. Percentage of population that has used CAM at least once in selected developed countries
Even in countries where allopathic practices prevail, people indicate a willingness to try CAM. What might account for the difference between the United States and Canada?

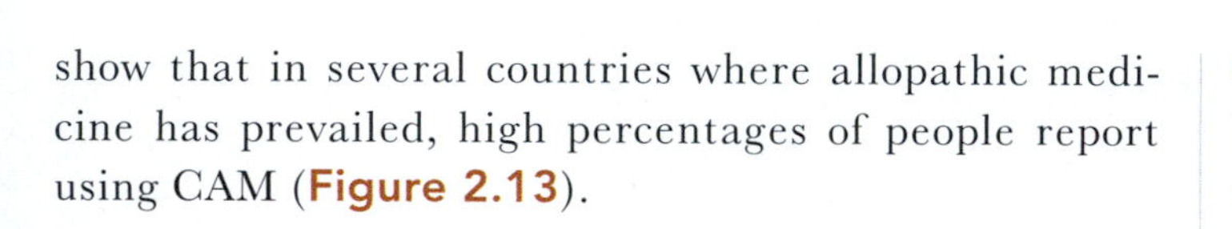

show that in several countries where allopathic medicine has prevailed, high percentages of people report using CAM (**Figure 2.13**).

Cultural Ecology and Local Knowledge

Local knowledge embodies a great deal of information about the environment. Indeed, approximately 75% of the plant-derived pharmaceuticals available today have been developed from plants used in traditional medicine. As we have seen, the specific knowledge an individual accumulates depends on her or his gender, age, status, wealth, and life experience. Significantly, the local knowledge possessed by one group may not be readily shared with members of another group. Often, an important reason for this is that the knowledge accumulated by certain members of the community is considered specialized knowledge. Therefore, cultural practices shape the ways in which local knowledge is acquired and shared.

Gender and traditional medicine In the case of traditional medicine, women, men, or both may be the repositories of the specialized knowledge depending on where we are and the community involved. A study of South African villages found that men, especially those 50 years of age or older, tended to have more extensive knowledge of plant resources for medicines than did the women. Within Talaandig communities in the Philippines, however, women accumulate more knowledge about lowland herbs, whereas men accumulate more knowledge about forest herbs. The reason for this gender-based distinction involves daily work and roles: Men spend more time in forest environments engaged in agricultural pursuits. In towns in Amazonia, however, women typically are the primary repositories of knowledge of medicinal plants. They frequently cultivate such plants in their yards and gardens, and they share knowledge about the plants with other women.

Water resources, local knowledge, and gender It is not unusual for local knowledge about a given topic, such as medicinal plants or water resources, to reveal an intimacy of awareness about place and ecology. San communities in the Kalahari Desert of southern Africa, for example, once utilized *sip-wells* as part of their strategy for obtaining water. They learned that water, though not visible on the desert surface, collects in and below the sands in certain locations. By using straws from plant parts, they could sip water from the sands and store the water in ostrich eggs for use at a later time.

Both women and men play important roles in managing water resources, although their influence tends to vary by scale and location. Broadly speaking, men exert more influence at the regional, national, and international scales, whereas women have more influence at the household scale and, depending on the place, sometimes also at the village and community scales. Across Sub-Saharan Africa, about one-fourth of women in rural households require 30 minutes or more a day to collect water. However, in some instances women spend the better part of a day collecting water for household use. Where manual or mechanized pumps have been installed at well sites, a number of cultural (and ecological) changes have occurred. Use of the pumps reduces the amount of time spent collecting water. Moreover, maintaining the pumps—a job initially allocated to men—is increasingly performed by women who are more invested in pump maintenance and water sanitation because of their role in managing daily and household water needs. In the eastern regions of Nepal, however, some well pumps have been installed in places that lack adequate privacy for women to bathe—a reflection of the fact that women were not included when information about water usage was obtained.

In parts of the Middle East, North Africa, Central Asia, and Mediterranean Europe, one facet of the local knowledge about water resources involves **qanats**. These ancient systems have long provided villages with water for agriculture and household consumption. Construction and maintenance of the qanats as well as decisions about how much water to allocate to different families and for different purposes have historically been carried out by men, while women have been responsible for water collection. To understand how qanats work, see *What a Geographer Sees* on the next page.

qanat A system of water supply that uses shaft and tunnel technology to tap underground water resources.

Vernacular architecture The study of **vernacular architecture** has been closely associated with human geography for well over a century. Cultural geographers maintain that vernacular architecture provides valuable insights on human use of space—whether in terms of the layout of a house or building, or of the design of a village or settlement. According to Paul Oliver (1997, p. xxiii), vernacular architecture is an architecture "of the people." More specifically, vernacular structures

vernacular architecture The common structures—dwellings, buildings, barns, churches, and so on—associated with a particular place, time, and community.

> are customarily owner—or community—built, utilizing traditional technologies. All forms of vernacular architecture are built to meet specific needs, accommodating the values, economies and ways of living of the cultures that produce them.

Vernacular dwellings and structures often incorporate locally available materials and resources and are adapted to environmental conditions as well as cultural practices and needs. In China, building traditions usually reflect awareness of cosmic forces, which the geographer Ronald Knapp refers to as **mystical ecology**. One example of mystical ecology includes **feng shui**, the art and science of situating settlements or designing cultural landscapes in order to harmonize the cosmic forces of nature with the built environment. The concepts of *yin* and *yang* that we discussed in relation to acupuncture are also relevant to *feng shui*. The cosmic force or energy known as *chi* or *qi* (pronounced *chee*) not only controls the interaction of *yin* and *yang* but is also believed to be visibly expressed in the physical landscape. Thus, an auspicious site has positive *chi*. Assessing the characteristics of a particular site according to

mystical ecology The interrelationship between an awareness of cosmic forces and human use of the environment.

WHAT A GEOGRAPHER SEES

Qanats

Qanats, also called *karez*, *khettaras*, or *foggaras*, have been used to manage water in about 50 countries or territories worldwide. They are usually constructed near the base of hills or in stream valleys. Water flows through a qanat entirely by gravity.

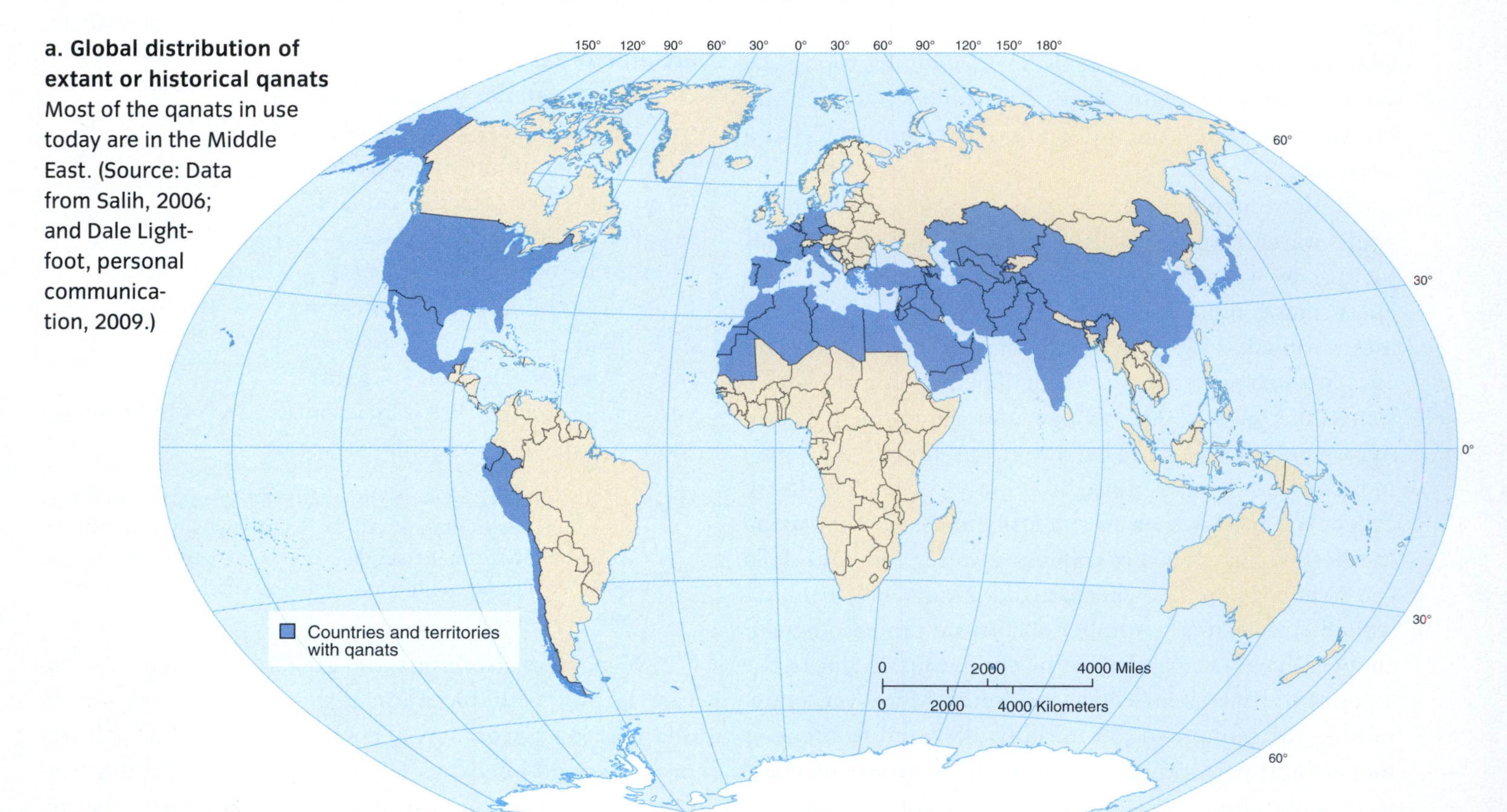

a. Global distribution of extant or historical qanats Most of the qanats in use today are in the Middle East. (Source: Data from Salih, 2006; and Dale Lightfoot, personal communication, 2009.)

b. A view inside a qanat and as a cross section The photo shows a qanat channel in northern Iraq. Men are usually tasked with clearing the tunnel of vegetation and silt so that the water flow is not impeded.

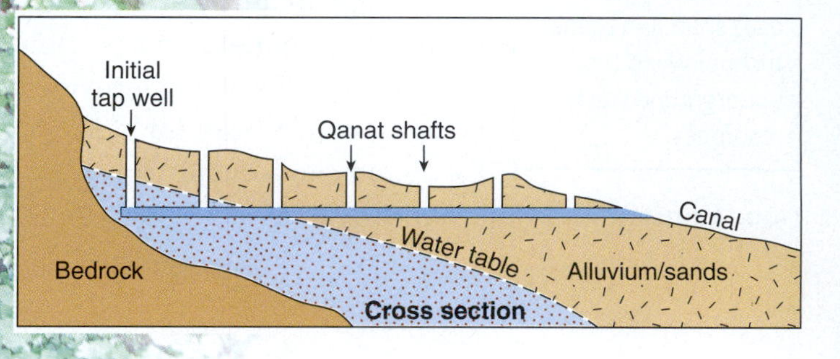

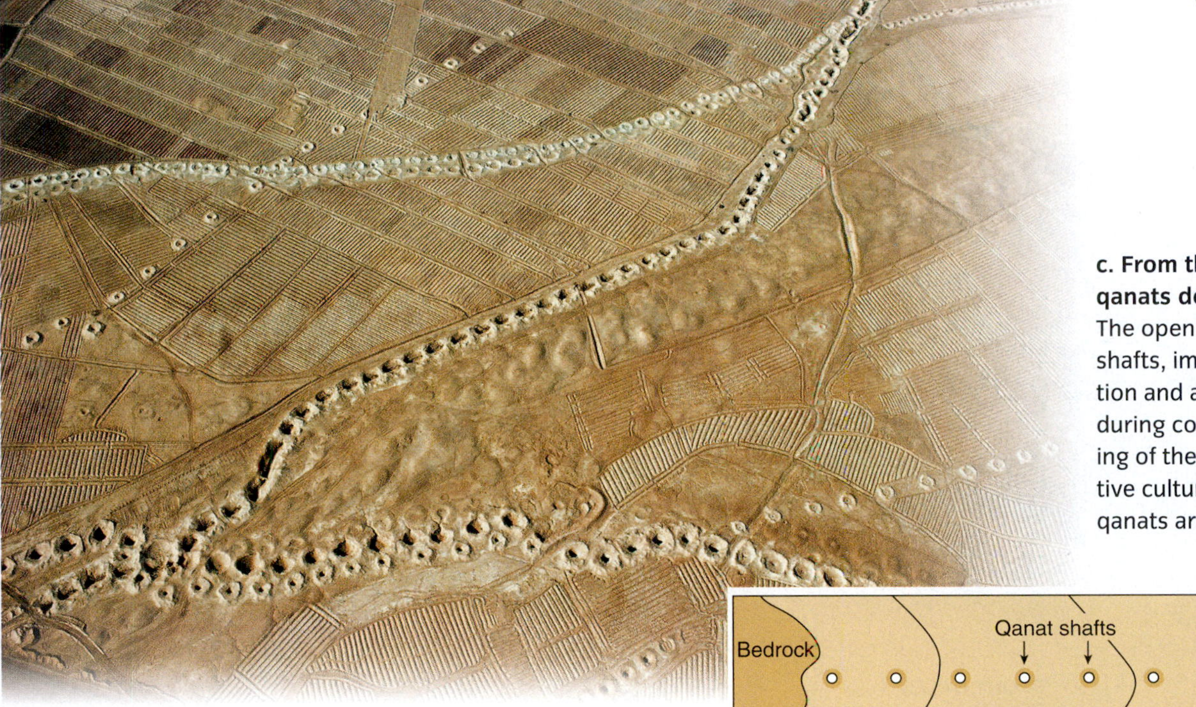

c. From the air, qanats dot the landscape The openings of the qanat shafts, important for ventilation and access to the tunnel during construction and cleaning of the qanat, create distinctive cultural landscapes. These qanats are in western China.

d. Diminishing lifeline The Kunaflusa qanat in northern Iraq still provides water to the villagers, though the flow of water has diminished considerably. This change is mainly the result of other water withdrawals for agriculture from the same underground water source that feeds this qanat and of drought.

Think Critically

1. What social, political, or other factors might contribute to qanat abandonment?
2. What is a problem with using a modern political map (as in **a**) to indicate the location of qanats?

Architecture and the environment • Figure 2.14

a. Mongolian nomads dwell in buildings known as *ger*, or *yurts*, with walls consisting of a circular lattice frame and a roof supported by slender poles. The entire structure is covered with felt and can be collapsed for transport to a new location.

FRONT GATES
DESIGN To be in a straght line parallel to the octagonal square in front of them.
REDESIGN To form an angle of 12 degrees to ensure prosperity.

Theme Park

MAIN WALKWAY
DESIGN To be straight.
REDESIGN Angled to prevent positive chi from flowing into the sea.

DISNEY'S HOLLYWOOD HOTEL
DESIGN Outside wall was supposed to be parallel to a berm alongside it.
REDESIGN Hotel location was shifted to form an angle so that it would receive positive chi.

HONG KONG DISNEYLAND HOTEL
DESIGN Incorporates motifs of the five elements as defined by feng shui: wood, earth, metal, fire, and water.

b. Even popular architecture may be designed in accordance with feng shui principles. Disney altered some aspects of the design of Hong Kong Disneyland after consulting with feng shui experts.

c. This church in Pereira, Colombia, is built from "Giant Bamboo" or guadua (*Guadua angustifolia*). Guadua, which grows in Colombia and Ecuador, has been used for housing by the indigenous people for centuries. Guadua is increasingly favored as an earthquake-resistant building material.

feng shui principles is complex and includes the lay of the land and the location of water sources. Indeed, *feng shui* literally means *wind and water*. Worldwide, vernacular architecture displays tremendous variety in building techniques, styles, and materials and often influences current building practices (**Figure 2.14**).

CONCEPT CHECK

STOP

1. **How** have Western views of local knowledge changed over time?
2. **How** has the status of traditional medicine changed over time? How does its status differ around the world?
3. **Why** are gender differences relevant to resource management?

Summary

THE PLANNER

1 Contemporary Globalization 36

- **Globalization** extends the interconnectedness and interdependence of people and places in the world. Although globalization existed before the term was coined, geographers recognize that contemporary globalization differs from its historical precedents. This is because the world is experiencing a widening and deepening of economic, cultural, and political interconnectedness not seen before.
- **Capital** refers to those assets that stem from human creativity and are used to create goods and services. One of the hallmarks of globalization is the increased circulation of financial capital.
- Today, **multinational corporations** (MNCs) are a significant component of the global economy. They not only dominate global trade, but they also influence the flow of financial capital especially through **foreign direct investment**, as shown here.

Flows of foreign direct investment (FDI) into different regions • Figure 2.3

2 Cultural Impacts of Globalization 40

- Globalization's impacts on society are complex and diverse. When examining the cultural impacts of globalization, geographers and other social scientists have identified three key theses: the homogenization thesis, the polarization thesis, and the glocalization thesis.
- While diffusion can occur in two or more directions, the fact that many American companies and products have achieved a global presence raises concerns about **Americanization**.
- **Glocalization** and **neolocalism** challenge the view that the cultural impacts of globalization flow only or primarily from the Western world to the rest of the world.

Glocalization • Figure 2.7

3 The Commodification of Culture 44

- Culture can be expressed in tangible or intangible ways, giving rise to the distinction between **material culture** and **nonmaterial culture**. **Cultural geography** is the branch of human geography that studies the impact of human beliefs and activities on other groups and the environment.
- **Commodification** converts an object, a concept, product, or procedure formerly not available for purchase into a good or service that can be bought or sold. Commodification is related to **consumption,** which is encouraged by advertising. As noted in the discussion of diamonds, advertising can even influence cultural practices. In the case of the **haka**, shown here, advertising helped commodify an aspect of Maori **nonmaterial culture** and demonstrates the contested nature of culture.

The haka • Figure 2.9

- It can also be said that the **heritage industry** involves the commodification of the past. The particularism of **heritage** means that it is always characterized by **dissonance**. The identification of **world (or global) heritage** sites has been both celebrated and criticized.

4 Cultural Geographies of Local Knowledge 50

- **Local knowledge** encompasses the collective knowledge of a community that derives from the everyday activities of its members. In every community there are multiple repositories of local knowledge. These are the individuals and groups who retain specialized knowledge for different activities.
- Historically, European views of non-Europeans and especially colonized peoples were influenced by the **rationality doctrine** and **diffusionism**.
- **Traditional medicine** is a kind of local knowledge that is often contrasted with Western, or **allopathic,** ways of practicing medicine. Traditional medicine is widely practiced around the world, and some facets of it, such as **acupuncture** (see photo), have become more mainstream and globalized.

The diffusion of acupunture • Figure 2.12

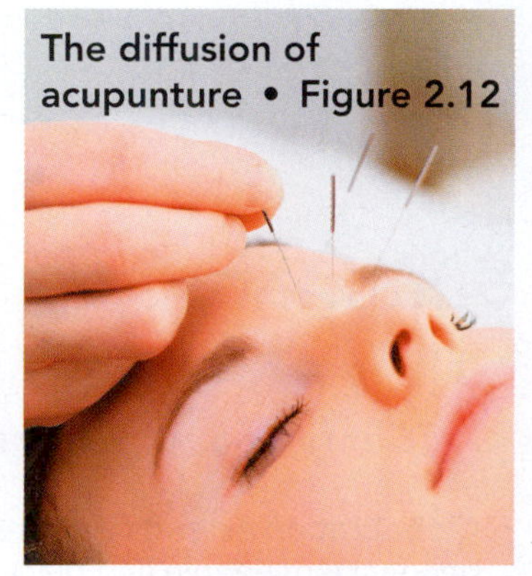

- Local knowledge often reveals an intimate awareness and understanding of the environment. A more complete understanding of **qanats**, **vernacular architecture**, and **mystical ecologies** such as **feng shui** demands an awareness of local knowledge and the fact that gender differences can play a major role in shaping the specific knowledge an individual possesses.

Key Terms

- acupuncture 52
- allopathic medicine 52
- Americanization 42
- capital 36
- cartel 45
- commodification 45
- conflict diamonds 47
- consumption 45
- cultural geography 44
- diffusionism 51
- dissonance 48
- feng shui 55
- foreign direct investment (FDI) 37
- globalization 36
- glocalization 43
- haka 47
- heritage 48
- heritage industry 48
- holistic approach 52
- horizontal expansion 36
- local culture 51
- local knowledge 51
- material culture 44
- multinational corporation (MNC) 37
- mystical ecology 55
- neolocalism 43
- nonmaterial culture 44
- personal approach 52
- placelessness 42
- popular culture 41
- qanat 55
- rationality doctrine 51
- social capital 51
- sustainable development 51
- traditional medicine 51
- vernacular architecture 55
- vertical expansion 36
- world (or global) heritage 48

Critical and Creative Thinking Questions

1. Using what you have learned in this chapter, would you say that consumerism is an expression of democracy? Explain your reasoning.
2. In what ways do different genres of music express resistance to or acceptance of forces of globalization?
3. The United States has contributed very little to the globalization of sports. Why?
4. Is distance education a form of commodification? Explain your reasoning.
5. What is meant by the commodification of nature? Give examples to support your interpretation.
6. What are some consequences of commodifying public goods such as water?
7. Study the residential architecture in your hometown. Can you identify any vernacular styles? If so, what patterns do you observe in the distribution of those styles?
8. To what extent is local knowledge, such as vernacular architecture and as seen in this photo, a kind of heritage?

What is happening in this picture?

People from the country of Myanmar, formerly known as Burma, protest the South Korean multinational Daewoo International Corporation and its recent foreign direct investment in a natural gas project off the country's coast.

Think Critically

1. Why are many people around the world suspicious of foreign direct investment?
2. More broadly, why is there dissent against globalization?

Self-Test

(Check your answers in Appendix B.)

1. Which factor below is NOT considered central to the development of globalization?
   a. the development of laws that encourage trade
   b. technological advances
   c. growth in advertising and consumption
   d. capitalism's quest for global markets

2. To a country receiving foreign direct investment, benefits may include all but one of the following. Which item does not belong?
   a. FDI supports small, local businesses.
   b. FDI improves a country's cash flow.
   c. FDI can increase employment.
   d. FDI can transfer technology.

3. Placelessness is most closely associated with which thesis about the impacts of globalization?
   a. the homogenization thesis
   b. the polarization thesis
   c. the glocalization thesis
   d. the Americanization thesis

4. The spread of Walmart stores shown in this diagram illustrates _____.
   a. glocalization
   b. neolocalism
   c. hierarchical diffusion
   d. reverse hierarchical diffusion

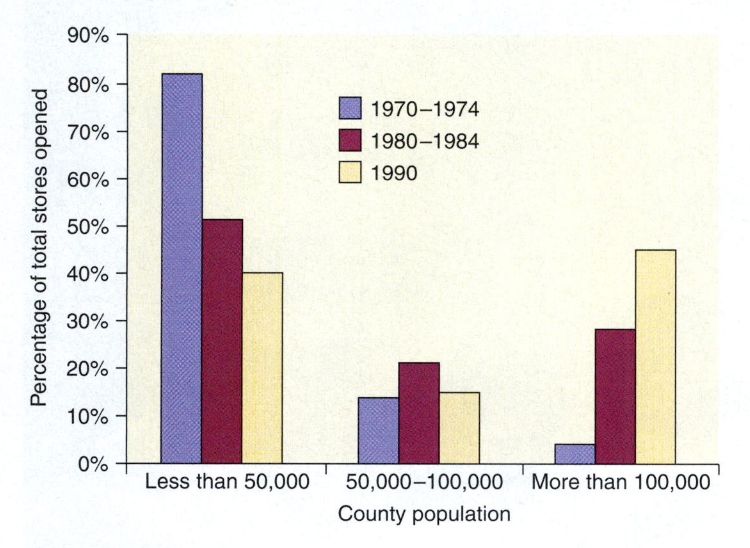

5. _____ is a pejorative term that implies that culture flow is unidirectional; _____ counters the view that homogenization is inevitable.
   a. Neolocalism; globalization
   b. Polarization; dissonance
   c. Coca-Colonization; glocalization
   d. Americanization; reverse hierarchical diffusion

6. This McDonald's menu from India is an example of _____.
   a. globalization and glocalization
   b. glocalization and diffusionism
   c. glocalization and neolocalism
   d. cultural dissonance and Americanization

7. List two limitations of the term *folk culture*:

   ____

   ____

8. Based on the discussion of the commodification of diamonds, which of the following statements is FALSE?
   a. The demand for diamonds is generally greatest in countries that do not mine them.
   b. The De Beers cartel helped ritualize and spread the practice of giving diamonds as gifts.
   c. The De Beers cartel used advertising to create an illusion of diamond scarcity.
   d. Blood diamonds have little to do with commodification.

9. Which of the following is *not* a characteristic of vernacular architecture?
   a. The design of vernacular architecture seems at odds with its surroundings.
   b. Vernacular architecture can tell us about human use of space.
   c. Vernacular architecture is an architecture of the people.
   d. Vernacular architecture can reveal shared beliefs such as mystical ecologies.

10. Which of the following is *not* a criticism leveled against the World Heritage List?
   a. There is no such thing as global heritage.
   b. Listed sites are not always made available to the public.
   c. Listed sites display a Eurocentric bias.
   d. Profit from tourism to listed sites is not returned to the local communities.

11. Which of the following terms cannot be used to describe local knowledge?
   a. oral
   b. evolving
   c. monolithic
   d. nonmaterial

12. Using the accompanying map as a guide, briefly explain the diffusion of acupuncture.

13. _____ long affected development policies and contributed to _____. However, _____ is needed to provide insights on the _____ of communities.
   a. Dissonance; sustainable development; allopathic knowledge; health care
   b. The rationality doctrine; diffusionism; local knowledge; social capital
   c. Traditional knowledge; sustainable development; social capital; dissonance
   d. Local knowledge; health care; the rationality doctrine; traditional knowledge

14. Allopathic medicine is _____ and _____; traditional medicine is _____ and _____.
   a. expensive; requires formal training; inexpensive; never requires formal training
   b. holistic; impersonal; informal; spiritual
   c. widespread; personal; indigenous; targets specific ailments
   d. nonholistic; associated with the West; holistic; individual

15. In terms of qanat systems, what is a primary function of the circular features shown in this picture?
   a. to capture water for underground storage
   b. to provide access for cleaning
   c. to provide a place to install a mechanized pump
   d. They don't have a particular function.

THE PLANNER

Review your Chapter Planner on the chapter opener and check off your completed work.

3 Population and Migration

WHEN IS A PLACE OVERPOPULATED?

Can you imagine living in a cemetery? In a section of Cairo the homes of more than 500,000 people are located in a cemetery called al-Qarafa. This historic cemetery contains hundreds of tombs, mausoleums, and shrines built from the late 14th century onward to commemorate Muslim religious leaders and dynastic families. They were designed as much for the living as for the dead, with many tombs containing one or two extra rooms for family members to spend time in, remembering the deceased.

Since World War II, Cairo's population has grown explosively, climbing from about 2 million people in 1947 to more than 12 million today. But housing construction has not kept pace, and many people have taken up residence in al-Qarafa's vacant rooms or have made makeshift dwellings between the tombs.

Media reports usually characterize al-Qarafa as an overcrowded slum—an illustration of Cairo's overpopulation. The residents, however, have mixed feelings. While thankful to have a place to live, they are also frustrated by the lack of basic services such

as electricity, running water, and sewage systems. Many resent the social stigma of living there. Still others consider the use of the cemetery for housing a desecration.

Is Cairo overpopulated? Are overcrowding and overpopulation the same thing? To what extent does crowding relate to quality of life? What circumstances prompt people to migrate within a country or to a new country? In this chapter we explore the topics of population and migration through spatial, ecological, and regional perspectives.

CHAPTER OUTLINE

CHAPTER PLANNER

- ❑ Study the picture and read the opening story.
- ❑ Scan the Learning Objectives in each section: p. 66 ❑ p. 73 ❑ p. 78 ❑ p. 81 ❑
- ❑ Read the text and study all visuals. Answer any questions.

Analyze key features

- ❑ Geography InSight, p. 71 ❑ p. 74 ❑
- ❑ Video Explorations, p. 73
- ❑ Process Diagram, p. 77
- ❑ What a Geographer Sees, p. 86
- ❑ Stop: Answer the Concept Checks before you go on: p. 73 ❑ p. 78 ❑ p. 80 ❑ p. 91 ❑

End of chapter

- ❑ Review the Summary and Key Terms.
- ❑ Answer the Critical and Creative Thinking Questions.
- ❑ Answer What is happening in this picture?
- ❑ Complete the Self-Test and check your answers.

Population Fundamentals

LEARNING OBJECTIVES

1. **Define** arithmetic density and physiological density.
2. **Distinguish** between the total fertility rate and replacement level fertility.
3. **Account** for recent changes in global fertility.
4. **Identify** the geographic dimensions of China's one-child policy.

Al-Qarafa and the city of Cairo provide a useful starting point for thinking about the relations between people, space, place, and more broadly—**geodemography**. As its name suggests, *geodemography* brings together geography and *demography*, the statistical study of characteristics of human populations. Four basic dimensions of geodemography are population distribution, population density, fertility, and mortality. We examine these in turn.

geodemography The study of the spatial variations among human populations.

Population Distribution and Density

As alluded to in the chapter opener, people are unevenly spread across the city of Cairo. The world's 6.8 billion people are also unevenly distributed (**Figure 3.1**).

Global population distribution • Figure 3.1

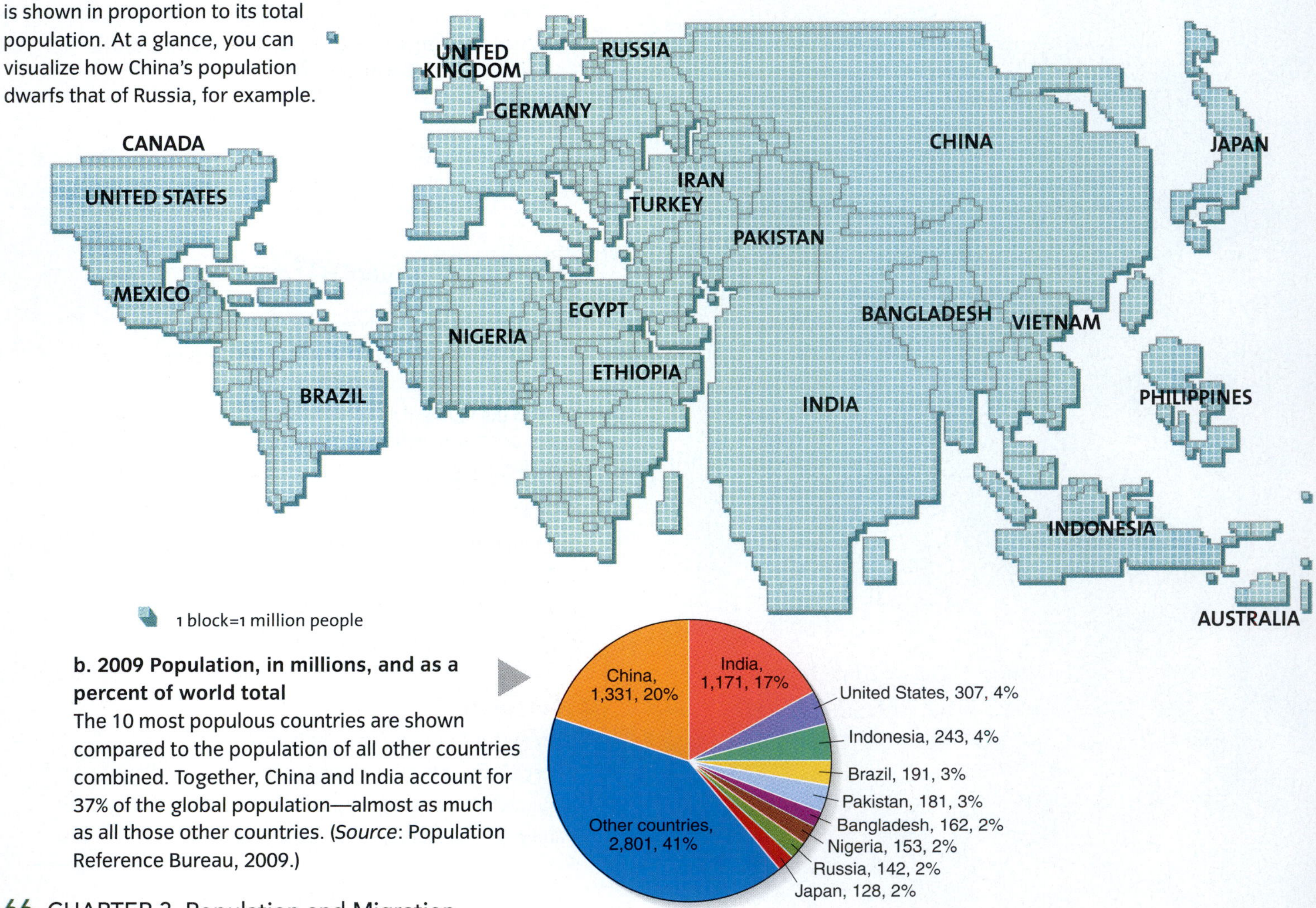

a. Population cartogram
The uneven distribution of population among countries is highlighted here. Unlike conventional maps that depict the land area of countries, on this cartogram the size of a country is shown in proportion to its total population. At a glance, you can visualize how China's population dwarfs that of Russia, for example.

b. 2009 Population, in millions, and as a percent of world total
The 10 most populous countries are shown compared to the population of all other countries combined. Together, China and India account for 37% of the global population—almost as much as all those other countries. (*Source*: Population Reference Bureau, 2009.)

Density expresses the number of people, structures, or other phenomena per unit area of land. When geodemographers want to know about the pressure a population exerts on the land, they calculate the population density. More specifically, geodemographers use two different measures of population density: **arithmetic density** and **physiological density**. To calculate the arithmetic density, you divide the total population by the total land area. One limitation of the arithmetic density is that it does not distinguish between types of land in an area, some of which may be inhospitable. The physiological density takes into account how much of the land is **arable land**, that is, how much of the land can be used for agriculture. To calculate the physiological density, you divide the total population by the total area of arable land. See **Figure 3.2** on the next page to understand how and why arithmetic and physiological densities vary from country to country.

arithmetic density The number of people per unit area of land.

physiological density The number of people per unit area of arable land.

Look again at **Figure 3.2a**. Globally, the pattern of densely and sparsely settled regions suggests that people tend to prefer to live in certain natural settings, such as on level land and away from deserts. Nearly 70% of the world's people live within 400 kilometers (250 miles) of a coast and on just 10% of the Earth's land. It is also important to keep in mind that the global distribution of population is increasingly urban. Indeed, cities are home to half of all people on the planet.

Fertility

Populations are dynamic, and two of the most important causes of population change include births and deaths. In its broadest sense, *fertility* refers to the ability to produce offspring. More narrowly, however, fertility refers to the births within a given population. Fertility and *mortality*—the incidence of death within a given population—are affected by biological, social, economic, political, and cultural factors.

Geodemographers use two important measures of fertility: the **crude birth rate (CBR)** and the **total fertility rate (TFR)**. Of these measures, the CBR is the most familiar, but it is also a more general measure. Thus, it is "crude" in the sense that it reflects childbearing trends within society as a whole rather than by specific age group. Globally, the CBR ranges from 7 in the small European principality of Monaco to 53 in the West African country of Niger. The CBR for the world is 20, while that for the United States is 14.

crude birth rate (CBR) The annual number of births per 1,000 people.

total fertility rate (TFR) The average number of children a woman is expected to have during her childbearing years (between the ages of 15 and 49), given current birth rates.

The TFR helps population experts to gauge family size and predict future population trends. When the TFR is 2.1, a population is said to be at the **replacement level**—the fertility rate necessary for a population to replace itself. When the replacement level is achieved, the population will eventually stabilize. (It may seem that the replacement level should be 2.0 exactly. It is slightly higher than this because some females will die before they have children.)

Factors influencing fertility Biological factors and patterns of reproductive behavior directly affect fertility. For example, women who breastfeed after giving birth become temporarily infertile because of hormonal changes. Certain diseases or a poor diet also affect a woman's ability to conceive or carry a pregnancy to term. The use of contraception directly reduces fertility. Higher fertility rates are associated with women who become sexually active earlier and who marry at younger ages.

Numerous cultural, social, economic, and political factors affect patterns of reproductive behavior, and in turn, influence fertility. It is difficult to pinpoint a single factor that controls fertility; rather, complex associations among factors are often involved. For example, gender roles often contribute to the low status of women, and in many societies women do not have the prerogative to make decisions about family planning. Across the developing world, children are considered an economic investment because they will be able to work and contribute to the family's income. Although Americans tend to expect their parents to have retirement savings or pensions, in many developing countries children may be the sole source of financial support for aging parents.

Poverty also tends to be associated with higher fertility, but the relationship between poverty and fertility is complicated because the poor are also more likely to have lower levels of education. Studies now indicate that literacy

Population densities • Figure 3.2

a. Dot map of population density

This map depicts arithmetic densities and helps distinguish densely from sparsely settled regions. Identify the South Asian, East Asian, and European population clusters on the Eurasian landmass. Respectively, these clusters are home to about 25%, 23%, and 11% of the global population. Most people in the South Asian cluster live in the countryside, whereas most people in the European cluster are city dwellers. About half of East Asia's population is rural.

150° 120° 90° 60° 30° 0° 30° 60° 90°

c. Crowded subway

In Tokyo, subway workers push people onto trains, ensuring high arithmetic densities in each subway car during the rush-hour commute.

NATIONAL GEOGRAPHIC

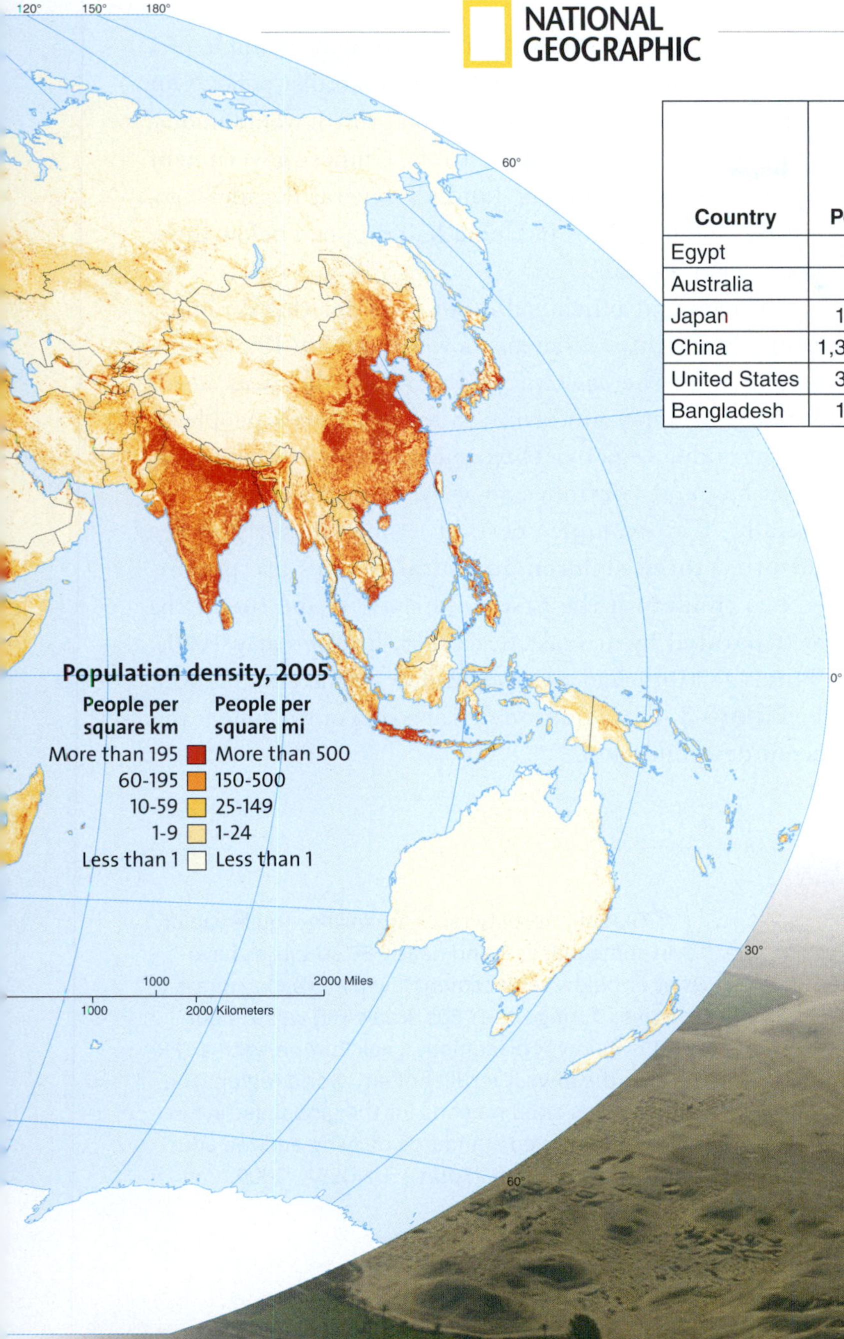

Country	Total Population	Land Area (square km)	Arithmetic Density (people per square km)	Arable Land (percent)	Arable Land (square km)	Physiological Density (people per arable square km)
Egypt	78,600,000	995,000	79	3	29,850	2,633
Australia	21,900,000	3,000,000	7	7	210,000	104
Japan	127,600,000	375,000	340	12	45,000	2,836
China	1,331,400,000	9,326,000	143	15	1,398,900	952
United States	306,800,000	9,162,000	33	18	1,649,160	186
Bangladesh	162,200,000	134,000	1,210	55	73,700	2,201

b. Arithmetic and physiological densities for selected countries

The higher the physiological density is, the greater the pressure that a population exerts on land that is used for agriculture. Bangladesh records high arithmetic and physiological densities even though the country has a high percentage of arable land. (*Source*: Population Reference Bureau, 2009; *CIA World Factbook*, 2009.)

Population density, 2005

People per square km	People per square mi
More than 195	More than 500
60-195	150-500
10-59	25-149
1-9	1-24
Less than 1	Less than 1

1000 2000 Miles
1000 2000 Kilometers

d. Edge of arable land in Egypt

In addition to irrigating, what other strategies exist for increasing arable land in different environments?

is linked to delays in marriage and childbirth, as well as lower rates of teen pregnancy. Because of these complex factors, fertility varies depending on the region, country, or social group being studied. **Figure 3.3** illustrates some of this variation.

Population control policies Governments can influence fertility by introducing policies to promote or limit population growth. Such population control policies are, respectively, *pro-natalist* or *anti-natalist*, and usually involve incentives to change reproductive behavior and, consequently, fertility. Although France's TFR of 2.0 is not high by global standards, the country has one of the highest fertility rates in Europe. Many experts attribute this to pro-natalist policies that are family-friendly, support two-career parents, and are based on gender equality. For example, France gives tax concessions to families with three or more children, subsidizes day care, permits parental leave, and prevents women from being fired while on maternity leave.

China likely provides the most familiar example of a country that has implemented an anti-natalist policy. Out of concern that explosive population growth would undermine the country's development, the Chinese government began to promote smaller families several decades ago. By 1979, fertility rates in China had dropped below three children per woman.

Government officials did not consider this dramatic decline in fertility sufficient, however, and in 1979 China implemented the *one-child policy*. This policy was never intended to apply uniformly to all of China's people or its geographic regions. Thus, the implementation of the policy has and continues to vary demographically and spatially. For example, certain ethnic minorities are permitted three children and rural couples are permitted two children if the first child is a girl and the births are separated by at least five years. By the early 1990s, China's fertility had dropped below replacement level. See **Figure 3.4** for additional discussion of the policy and recent developments.

Global dimensions of fertility • Figure 3.3

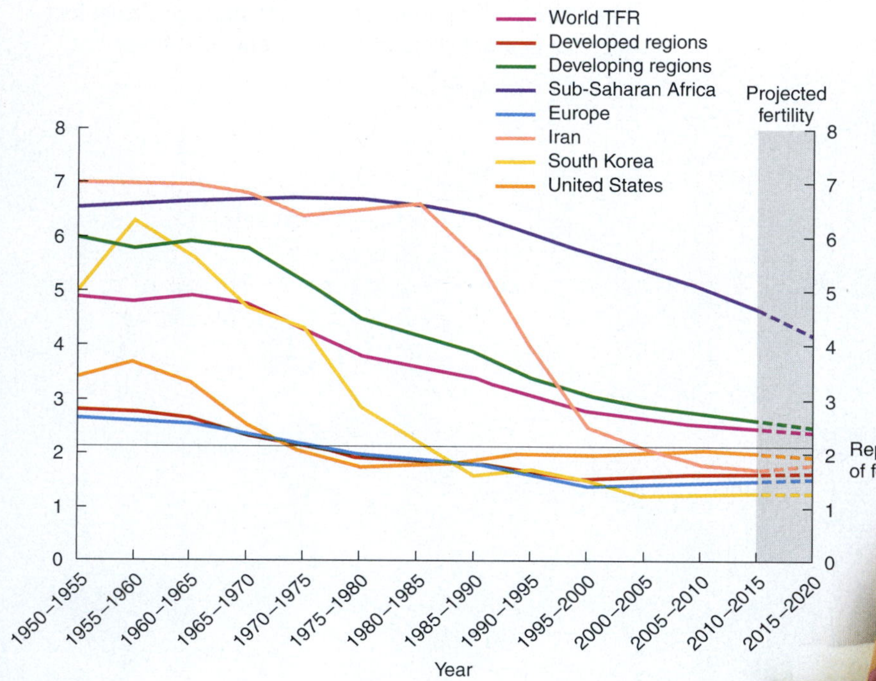

a. Globally, fertility rates are falling—quite rapidly in some places—and nearly 80 countries have TFRs at or below replacement level. Iran's example is instructive: During the 1950s, Iran's TFR was 7.0 but today it is below replacement level. Europe, with a TFR of 1.5, has the lowest fertility of any world region. The two most often cited reasons for the global decline in fertility are improved standards of living and the education of more women. (*Source*: UNDESA, 2009.)

b. Sub-Saharan Africa records some of the highest fertility rates in the world, often in excess of 6.0. In Ghana, shown here, the TFR dropped from 6.4 to 4.0 between 1998 and 2008. Among Ghanaian women with high school or postsecondary education, the TFR falls to 2.1.

Geography InSight

Population control in China • Figure 3.4

China's one-child policy has important geographical dimensions. The policy was conceived at the national scale but has always been implemented at the local scale. For example, local areas were assigned birth quotas, and couples had to request permission to have their first child.

a. Translated from Chinese, this billboard from the 1970s reflects the government's "later, longer, fewer" message, a voluntary anti-natalist strategy prior to the one-child policy. It promoted marriage at a later age, longer periods of time between pregnancies, and fewer pregnancies.

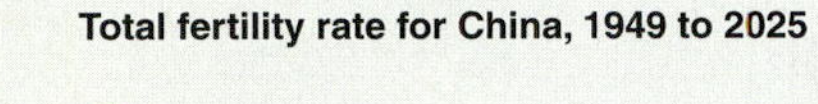

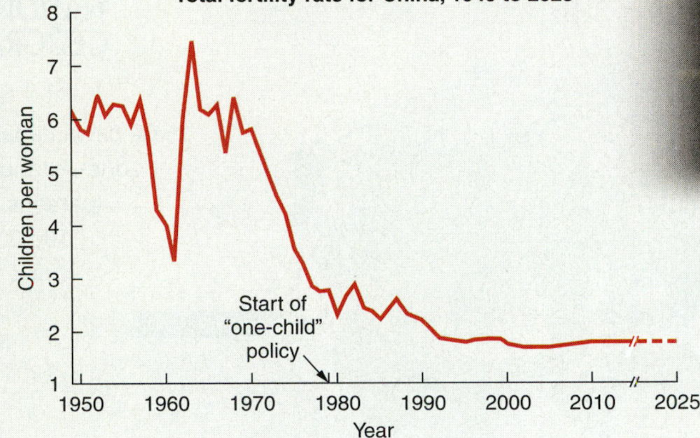

b. The steep fertility decline underway prior to 1979 leads some experts to question whether the one-child policy was necessary. (*Source*: Post–2000 data and projections from U.S. Census Bureau International Data Base, 2009.)

c. Enforcement of the one-child policy has, until recently, been strictest in urban areas. Faced with an aging population, Shanghai in 2009 began encouraging couples, who themselves were from one-child families, to have two children.

Mortality

Like fertility, patterns of mortality also affect population change. A standard measure of mortality is the **crude death rate (CDR)**. Today, no country has a death rate in excess of 23. Globally, some of the lowest death rates—1 to 2 deaths annually per 1,000 people—have been recorded in the Middle Eastern countries of Qatar and Kuwait. In contrast, some of the highest death rates occur in the African countries of Lesotho (23) and Sierra Leone (20), a result of the ongoing AIDS epidemic and poor access to medical care.

crude death rate (CDR) The annual number of deaths per 1,000 people.

Natural and social factors can increase mortality. For example, natural disasters can lead to temporary increases in the death rate. Prior to Hurricane Katrina, the death rate for Orleans Parish, where New Orleans is located, was 11.3. More than a year after the storm, the death rate was 14.3—an increase of more than 26%. Many medical officials believe that the higher death rate was the result of ongoing stress associated with the personal, financial, and psychological hardships Katrina wrought.

The state of the health care system is another factor that affects death rates. Poor countries often cannot afford to provide vaccinations or purchase medicines needed to manage preventable diseases. Countries affected by wars and violent social upheaval are likely to experience increased death rates.

Although poorer countries often have high death rates, having a high death rate does not necessarily mean that a country is less developed. A country that has a high proportion of elderly people will, typically, have a high death rate regardless of the country's wealth. For example, Denmark has a much higher proportion of elderly people than Honduras and its death rate is 10, whereas the death rate in Honduras is 5. Thus, we should remember that the death rate cannot tell us about the quality of life or health in a country. For that, demographers examine life expectancy and the rate of infant mortality.

Life Expectancy and Infant Mortality

Life expectancy is the average length of time from birth that a person is expected to live given current death rates. Women have slightly longer life expectancies than men. Globally, the life expectancies for both men and women have risen substantially in the past century, from about 29 years in 1900 to 67 years today. Geodemographers have recently identified places called *blue zones* that have exceptionally long-lived populations (**Figure 3.5**).

Life expectancy can decline if the level of poverty increases or if a country experiences a social upheaval. Following the breakup of the Soviet Union, Russia experienced a decline in its life expectancy, especially among men. In 1990 life expectancy for Russian men was 64 years, by 2000 it had fallen to 59 years; today it is 61 years. By contrast, the life expectancy for Russian women is 74 years. The social and economic trauma of Russia's reforms, as well as high rates of alcoholism among men, help account for these differences.

Globally, the impact of HIV/AIDS on life expectancies varies. In wealthier countries, better access to medical care and pharmaceuticals has helped bring life expectancies of persons living with HIV/AIDS close to normal. Elsewhere, however, HIV/AIDS has reduced life expectancies by as much as 20 years. Lesotho and Botswana, southern African countries, have life expectancies of 40 and 49 years, respectively. (See *Video Explorations*.)

The second key measure that indicates the quality of life in a particular population is the **infant mortality rate**—the number of deaths of infants under one year of

Life expectancy and blue zones • Figure 3.5

NATIONAL GEOGRAPHIC

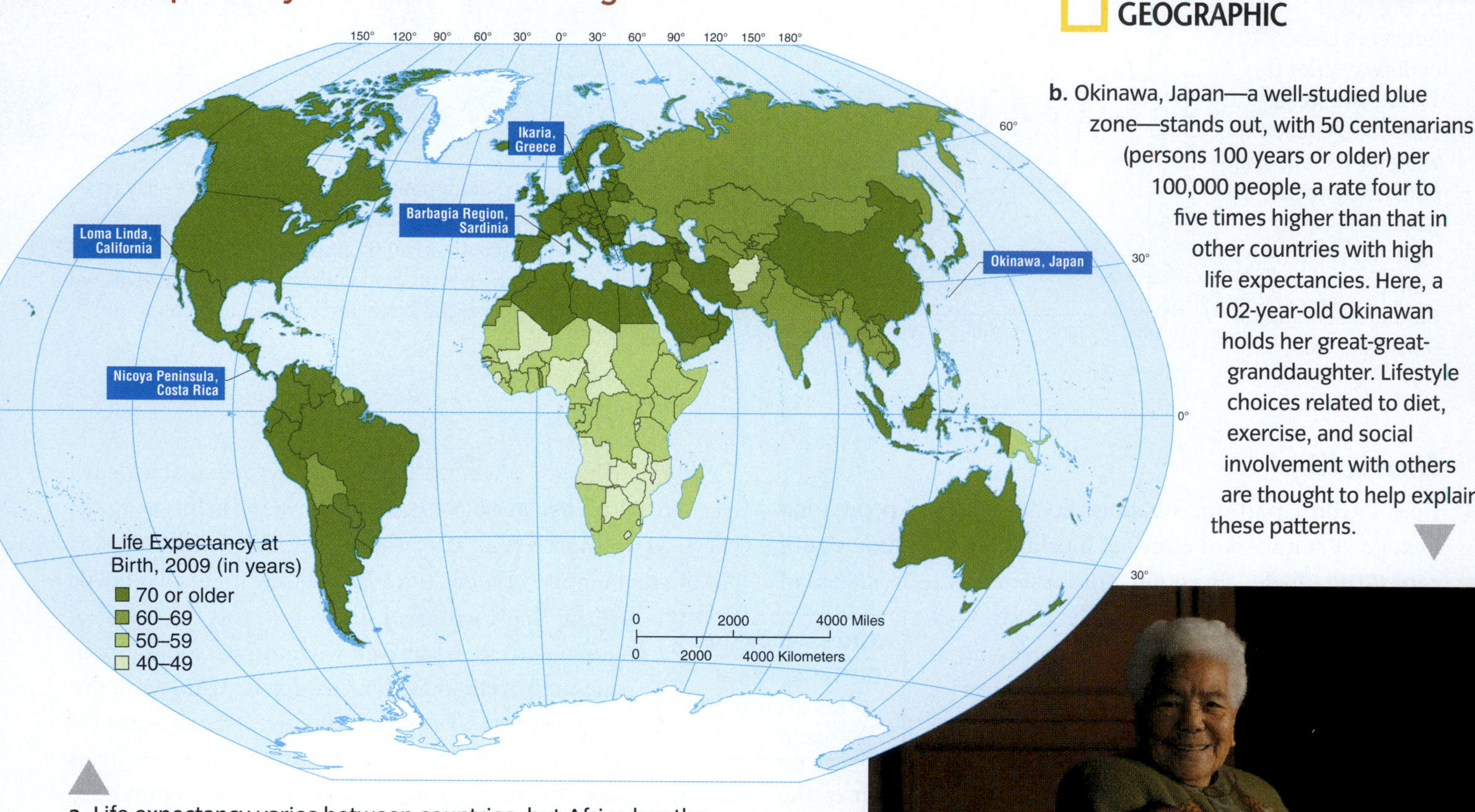

a. Life expectancy varies between countries, but Africa has the lowest life expectancies of any continent. The five labels identify *blue zones*; the one in Loma Linda corresponds to a Seventh-Day Adventist community.

b. Okinawa, Japan—a well-studied blue zone—stands out, with 50 centenarians (persons 100 years or older) per 100,000 people, a rate four to five times higher than that in other countries with high life expectancies. Here, a 102-year-old Okinawan holds her great-great-granddaughter. Lifestyle choices related to diet, exercise, and social involvement with others are thought to help explain these patterns.

THE PLANNER

Video Explorations

AIDS

WILEY PLUS NATIONAL GEOGRAPHIC Video

AIDS is an acronym for Acquired Immune Deficiency Syndrome, a disease caused by the HIV virus. In Sub-Saharan Africa, HIV infection is spread primarily via heterosexual intercourse. Across the region, mortality from AIDS remains high, with significant long-term social and demographic consequences.

age per 1,000 live births. High infant mortality rates signal that inadequacies exist in the health care given to pregnant women and to newborn babies. At the start of the 20th century the infant mortality rate for the United States was over 100. Today the infant mortality rate is about 6, on a par with the rate in the European Union. In many Sub-Saharan African countries struggling with malnutrition, unreliable access to safe water, and/or civil strife, infant mortality rates approach and sometimes exceed 100.

CONCEPT CHECK

STOP

1. **What** are two different measures of density, and why do geodemographers need them?
2. **What** is the significance of replacement level fertility?
3. **What** has happened to global fertility and why?
4. **How** has the implementation of China's one-child policy varied geographically?

Population Composition and Change

LEARNING OBJECTIVES

1. **Identify** the three basic shapes of population pyramids.
2. **Explain** how to calculate the age-dependency ratio.
3. **Summarize** the factors that may contribute to an imbalanced sex ratio.
4. **Identify** the components used to measure population change.
5. **Describe** the differences among the four stages in the demographic transition.

We can describe every population in terms of its composition—that is, in terms of the characteristics of the specific groups within it. Take, for example, the student population of your college or university. We can ask: How many students are between the ages of 18 and 22? How many students work part-time? What percent of the student body commutes? What is the ratio of men to women? For geodemographers, this kind of information creates a demographic picture of the variations within a given population.

Significantly, population composition also provides valuable clues about how a population is likely to change in the future. For example, college recruiters target areas with large populations of high school students. Similarly, the types of television shows and advertisements change from afternoon to evening as the composition of the viewing population changes. One tool that gives us a glimpse of the composition of a population is the population pyramid.

Population Pyramids

The vertical axis of a **population pyramid** depicts age cohorts—groups of people born in the same time span. Most population pyramids use five-year cohorts and begin with the youngest age group, which includes infants and children up to 4 years of age. The age cohorts on the left of the vertical axis usually represent males, while the age cohorts on the right side of the vertical axis represent females. The horizontal axis indicates the percentage of the total population. Population pyramids can be grouped into three categories with

population pyramid Bar graph that shows the age and gender composition of a population.

Geography InSight

THE PLANNER

Population pyramids • Figure 3.6

Population pyramids help to visualize the age and gender composition of a population. Their shape derives mainly from the birth rate, and they can be grouped into three categories: rapid growth, slow growth, and population decline. Population pyramids can help predict future changes in the structure of a country or region's population, but they are just as valuable for helping us to visualize some of the demographic impacts of significant past events.

a. Rapid population growth
The population pyramid for the Philippines closely resembles a pyramid. The broad base indicates that birth rates have been and continue to remain high, and that there is a high percentage of young people. These characteristics signal that the country's population is increasing rapidly. Photo shows Filipinos at a job fair. (*Source*: U.S. Census Bureau International Data Base, 2009.)

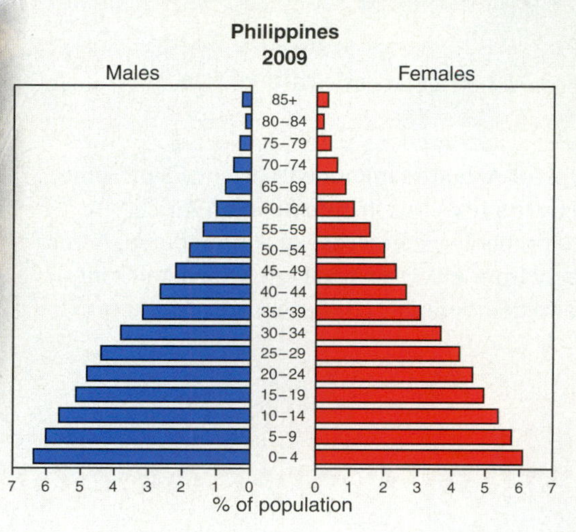

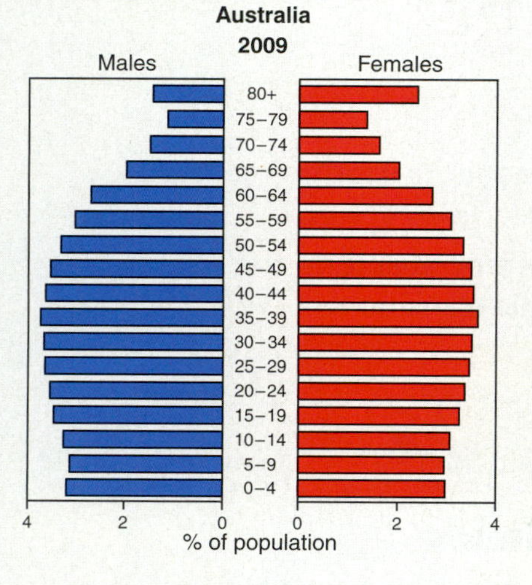

b. Slow population growth
Australia's population pyramid has a narrowing base, indicating that birth rates have fallen and that the momentum for future population growth is not nearly as high as in the Philippines.

c. Population decline
Japan's pyramid is the most top-heavy, with 22% of its population over the age of 65. By 2050 the number of Japanese elderly is projected to outweigh the number of children by 4 to 1. What are some potential economic and social consequences of such a ratio?

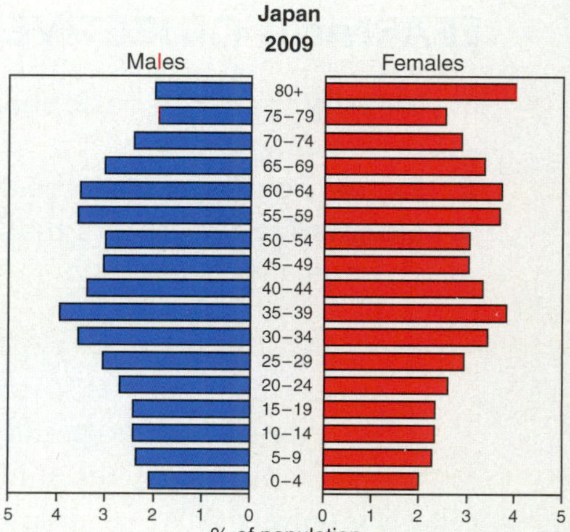

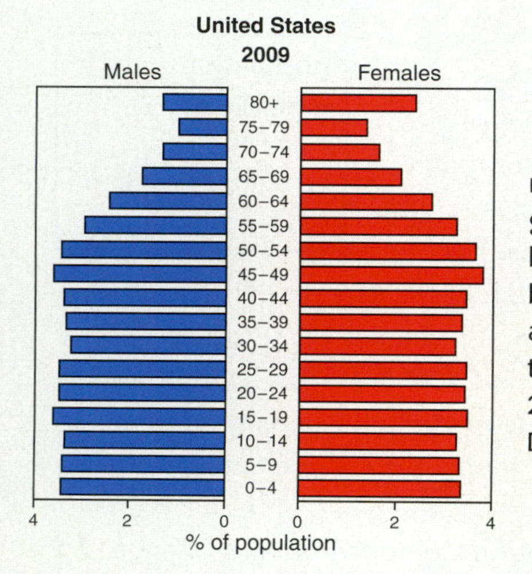

d. Baby booms
The shape of a country's population pyramid can be transformed within a lifetime or a generation. Find the bulge that includes persons born between 1946 and 1964; it represents the baby boom generation born in the United States after World War II. The other bulge corresponds to the New Boomers, born between 1983 and 2001. (*Source*: U.S. Census Bureau International Data Base, 2009.)

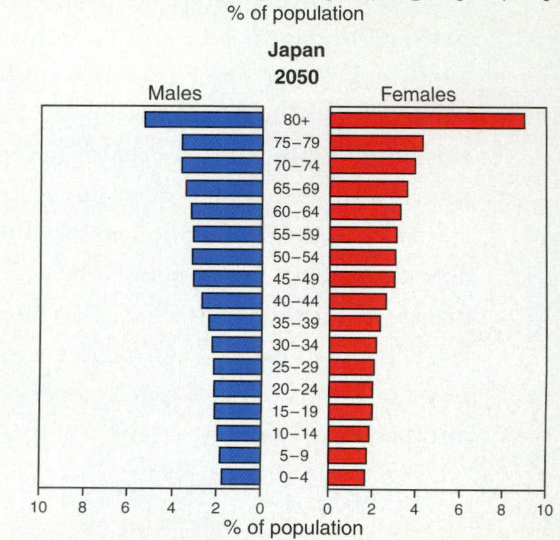

distinct shapes: rapid growth, slow growth, and population decline. See **Figure 3.6** for a discussion of these categories and follow the link in *Where Geographers Click* to view some animations.

Age-Dependency Ratio

You can use the population pyramid in **Figure 3.6a** to estimate that 35% of the people in the Philippines are below the age of 15. What percentage of the country's population is over the age of 65?

Demographers pay close attention to the number of people younger than 15 or over 65. People in these groups are termed *age dependents* because most of them do not work on a full-time basis. The **age-dependency ratio** helps countries to predict and plan for how their society will change. Countries with a youthful population need to ensure that there is sufficient classroom space and that there will be jobs available as these youths age.

> **age-dependency ratio** The number of people under the age of 15 and over the age of 65 as a proportion of the working-age population.

To calculate the age-dependency ratio, you divide the number of age dependents by the working-age population (the number between ages 15 and 64) and multiply the result by 100. Look again at Figure 3.6a. In the Philippines 39.3% of the population (38,500,000) are age dependents, and 60.7% (59,400,000) are of working age. The Philippines has an age-dependency ratio of 65:100—that is, for every 100 people between the ages of 15 and 64 there are 65 age dependents, a high age-dependency ratio. The high age-dependency ratio in the Philippines stems from the youthfulness of the population—there are nearly nine times as many young dependents as there are elderly dependents. An age-dependency ratio of 100 means that the working-age and age-dependent populations are equal.

Where Geographers CLICK

U.S. Census Bureau International Data Base

http://www.census.gov/ipc/www/idb/informationGateway.php

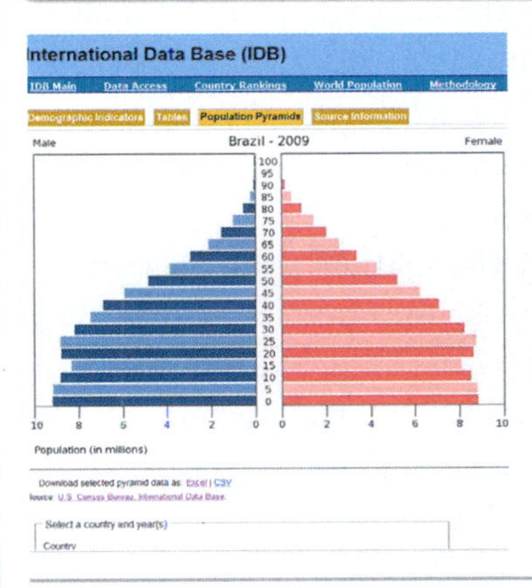

You can look at population pyramids for most of the world's countries at this website. You have to first select a country, click submit, and then select the population pyramids tab. With a little practice, you can compare population pyramids for different years and set the population pyramids in motion.

Sex Ratio

Another way to examine the composition of a population is with the **sex ratio**. In normal conditions, more males are born than females, giving a slight imbalance. The natural sex ratio is 105:100, meaning 105 boys are born for every 100 girls.

> **sex ratio** The proportion of males to females in a population.

Diverse forces can create disparities between the number of men and women in a population. Men have higher mortality rates than women in virtually every age cohort. Women live longer, and you can see this in the population pyramids in Figure 3.6. War usually has a disproportionate impact on the male population; immigration can also affect the sex ratio (**Figure 3.7**).

Sex ratio in the United Arab Emirates • Figure 3.7

a. The United Arab Emirates recruits laborers from other countries for their industries. Foreigners constitute more than 80% of the country's labor force. South Asians, shown here, are heavily concentrated in construction jobs.

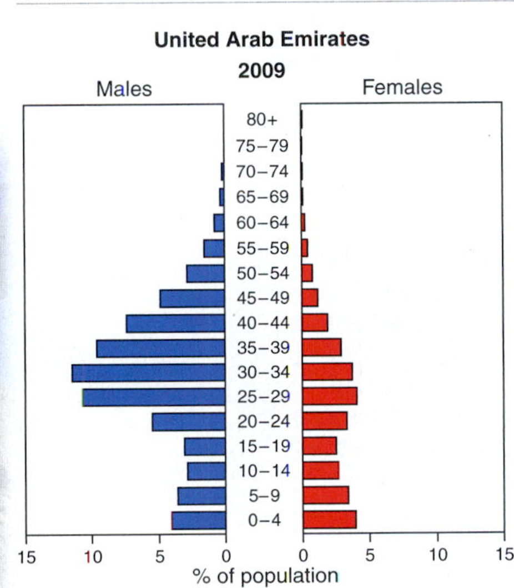

b. The population pyramid reveals how reliance on migrant labor has massively altered the country's sex ratio. For the entire population, the sex ratio is 219:100; for 15- to 64-year-olds, the sex ratio rises to 274:100.

Some Asian countries, including India and China, have high sex ratios. India's sex ratio is 113 but rises to 123 in some northwestern states. In 2005, China's sex ratio was 120, and there were 32 million more men than women under the age of 20. In China's rural provinces, where couples are permitted a second child if their first is a girl, the sex ratio for second births exceeded 160. In both countries, a strong cultural preference for boys is partly responsible for these disproportions, but in China the one-child policy also likely plays a role. The extremely imbalanced sex ratio for second births points to the practice of sex selective abortions to ensure at least one boy. What are some possible long-term effects of such uneven sex ratios?

Geodemographers track the spatial variations that result from population growth or decline. To do this, they need to know about changes over a given time within a population. In the next section, we will concentrate on the changes to the population through births and deaths. Later in the chapter, we will look at changes to the population through immigration (in-migration) and emigration (out-migration).

Rate of Natural Increase

A population experiences natural increase when the number of births exceeds the number of deaths. To calculate the **rate of natural increase (RNI)** geodemographers subtract the death rate from the birth rate and convert that figure to a percentage. For example, in 2009 the birth rate for the world was 20 per 1,000 and the death rate was 8 per 1,000. The difference is 12 per 1,000, which, when converted to a percentage, yields a rate of natural increase of 1.2%. This figure may seem low, but we should keep in mind that even a small percentage increase translates into a large number of persons. In 2009, the world population grew through the addition of about 80 million people.

rate of natural increase (RNI) The percentage of annual growth in a population excluding migration.

The rate of natural increase can be zero, as in Austria, where both the birth rate and the death rate are 9 per 1,000. A negative rate of natural increase indicates that the death rate is higher than the birth rate and there is a natural *decrease* in the population. With a birth rate of 12 and a death rate of 15, Russia presently has a negative rate of natural increase. **Figure 3.8** shows how the rate of natural increase varies around the world.

Geodemographers use the rate of natural increase to determine the number of years it takes a population to double. The **population doubling time** helps relate current population trends to future population size. To calculate the population doubling time, you divide the number 72 by the rate of natural increase. This calculation gives a population doubling time for the world of 60 years (72 ÷ 1.2). If the rate of natural increase remains

Rates of natural increase around the world • Figure 3.8

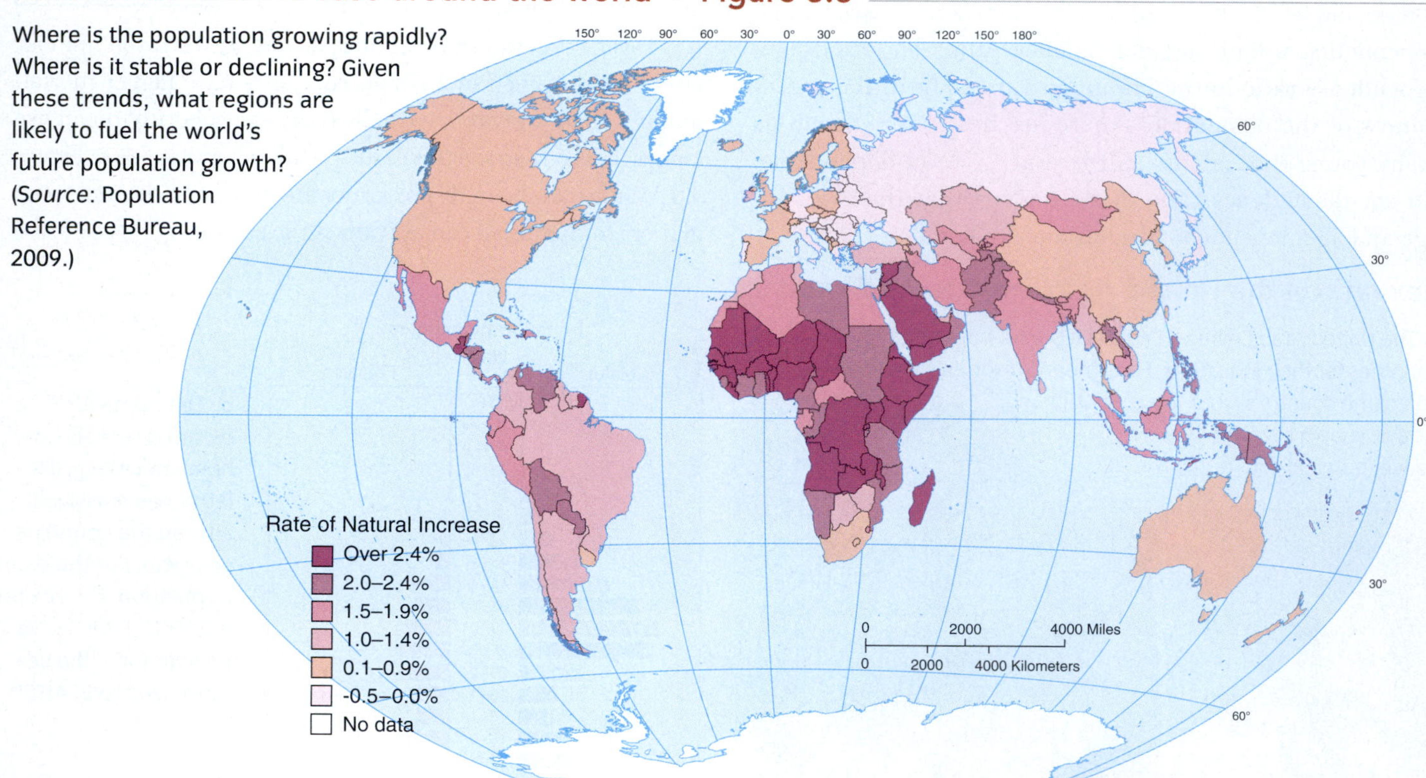

Where is the population growing rapidly? Where is it stable or declining? Given these trends, what regions are likely to fuel the world's future population growth? (*Source*: Population Reference Bureau, 2009.)

PROCESS DIAGRAM

THE PLANNER

Demographic transition model • Figure 3.9

The demographic transition model is derived from population trends in western Europe before, during, and after the Industrial Revolution. It consists of four stages.

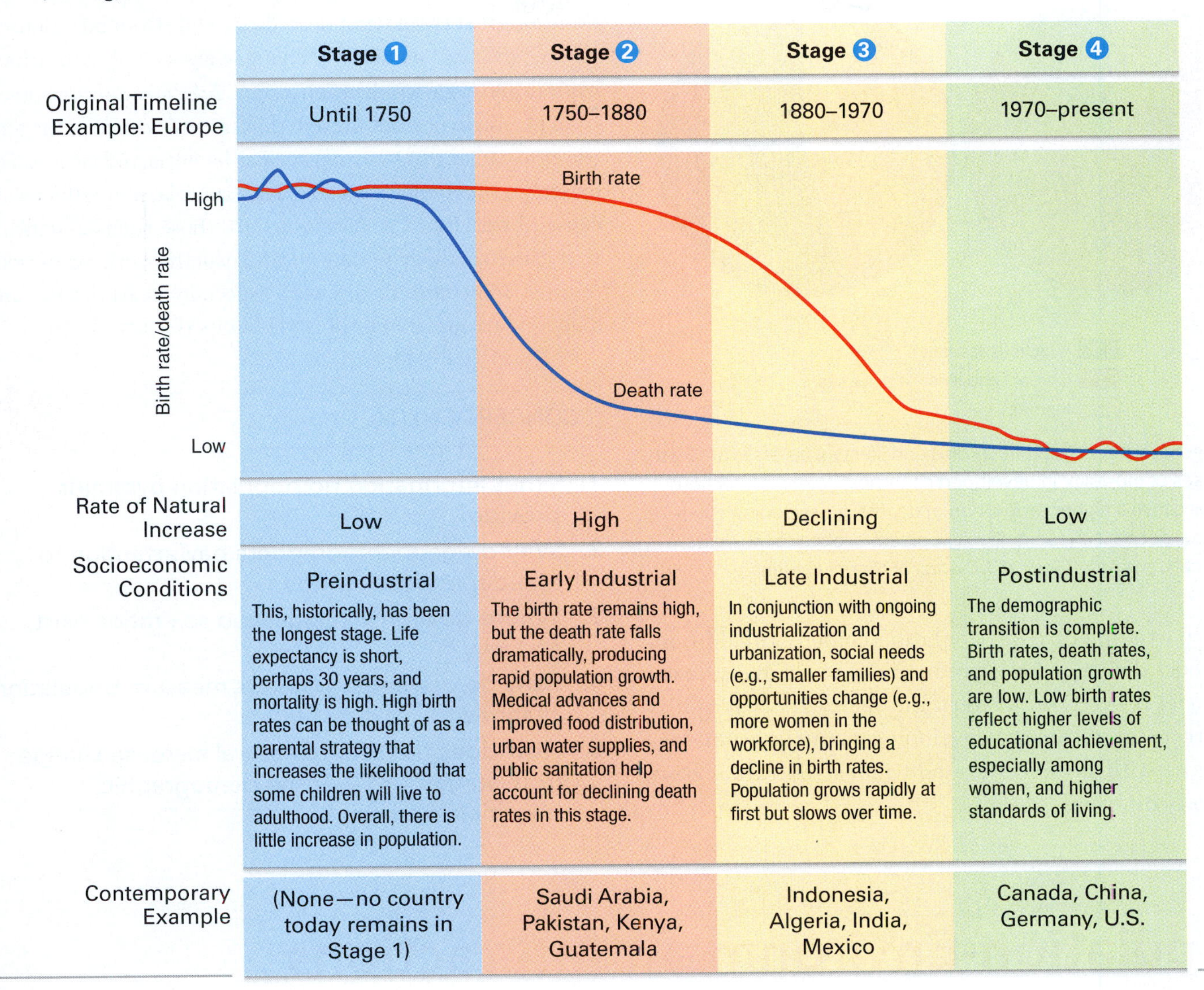

the same, the global population will reach 13 billion people in 2067. The current rate of natural increase for Ethiopia is 2.4%, giving it a population doubling time of just 30 years. In contrast, Spain has a rate of natural increase of 0.2% and a population doubling time of 360 years. Why does the population doubling time matter?

Stage 1
Stage 2
Stage 3
Stage 4
Original Timeline Example: Europe
Until 1750
1750–1880
1880–1970
1970–present
High
Low
Birth rate/death rate
Birth rate
Death rate
Rate of Natural Increase
Low
High
Declining
Low
Socioeconomic Conditions
Preindustrial
This, historically, has been the longest stage. Life expectancy is short, perhaps 30 years, and mortality is high. High birth rates can be thought of as a parental strategy that increases the likelihood that some children will live to adulthood. Overall, there is little increase in population.
Early Industrial
The birth rate remains high, but the death rate falls dramatically, producing rapid population growth. Medical advances and improved food distribution, urban water supplies, and public sanitation help account for declining death rates in this stage.
Late Industrial
In conjunction with ongoing industrialization and urbanization, social needs (e.g., smaller families) and opportunities change (e.g., more women in the workforce), bringing a decline in birth rates. Population grows rapidly at first but slows over time.
Postindustrial
The demographic transition is complete. Birth rates, death rates, and population growth are low. Low birth rates reflect higher levels of educational achievement, especially among women, and higher standards of living.
Contemporary Example
(None—no country today remains in Stage 1)
Saudi Arabia, Pakistan, Kenya, Guatemala
Indonesia, Algeria, India, Mexico
Canada, China, Germany, U.S.

Demographic Transition Model

The **demographic transition model** grew out of several studies of population trends in Europe. Broadly, this model relates changes in the rate of natural increase to social change as a result of urbanization and industrialization. More specifically, it describes a common demographic shift from high birth and death rates to low birth and death rates over time (Figure 3.9).

Using what you know about the relationship between the rate of natural increase and the demographic transition, you can refer again to Figure 3.8 to identify other countries that have stage 2, 3, or 4 profiles.

More important than assigning a country to a specific stage is that we understand why a country has a certain demographic profile. As we have seen, economic, social, and political factors affect demographics. In turn, these factors are bound up with development, a topic we discuss in greater detail in Chapter 9. Moreover, the demographic transition model has serious limitations. Because it does not take migration into account, it presents

Epidemiological transition • Figure 3.10

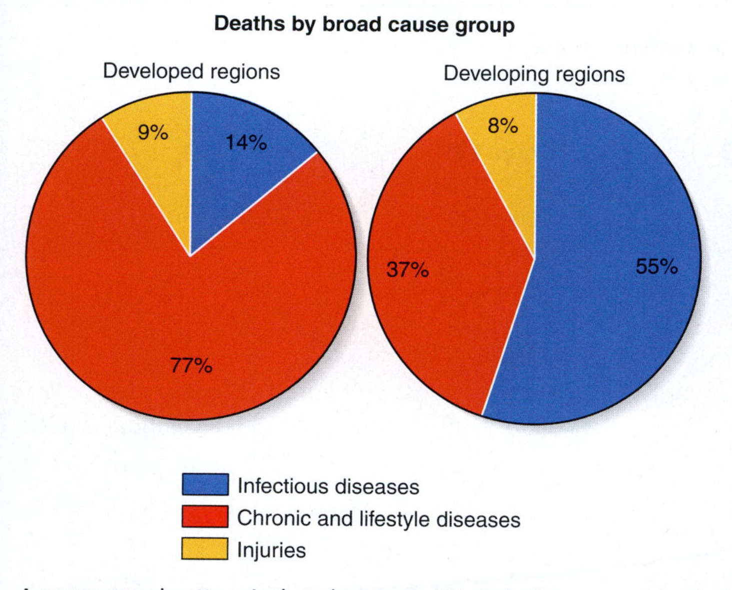

A very general pattern is that chronic and lifestyle diseases are leading causes of mortality in developed countries, whereas infectious diseases cause a higher proportion of deaths in developing countries. How does the present obesity epidemic relate to the epidemiological transition? (*Source*: *UC Atlas of Global Inequality*, 2009.)

only a partial picture of population change. In addition, the model does not have predictive value. Finally, since the model is based on western Europe's experience, it is not directly applicable to developing countries where experiences with urbanization and industrialization have been very different.

For decades now, demographers and medical geographers have observed that when countries experience the demographic transition, a parallel shift occurs in the kinds of diseases that affect mortality in the population. The **epidemiological transition** describes a shift from infectious to chronic diseases as lifestyle changes associated with urbanization and industrialization occur. *Infectious diseases* spread from person to person through the transmission of pathogens (disease-causing organisms) such as bacteria and viruses. For example, tuberculosis is caused by bacteria, and influenza is caused by a virus. *Chronic diseases* are those such as arthritis that cause the body to deteriorate, usually over long periods of time. Some chronic diseases, especially heart disease and diabetes, are affected by lifestyle habits (**Figure 3.10**).

CONCEPT CHECK

1. **What** information do population pyramids provide?
2. **Why** do geodemographers pay attention to age-dependency ratios?
3. **Where** do highly imbalanced sex ratios exist, and why?
4. **How** do geodemographers measure population change?
5. **How** does the rate of natural increase change from stages 1 to 4 of the demographic transition, and why?

Population-Environment Interactions

LEARNING OBJECTIVES

1. **Summarize** Malthusian population theory.
2. **Contrast** the neo-Malthusian and cornucopian theories.
3. **Define** food insecurity.

The chapter opener described the crowded conditions in al-Qarafa, one of Cairo's cemeteries. From a more global perspective, is there a limit to the number of people the Earth can support? How are population size, quality of life, and the state of the environment related? These complex questions have long challenged many thinkers. The study of **population ecology** examines the impacts populations have on their environments as well as the ways in which environmental conditions affect people and their livelihoods. In this section, we discuss some past and present theories that address different population-environment interactions.

Malthusian Population Theory

One person whose ideas have received a great deal of attention for more than two centuries is the English economist Thomas Malthus who, in 1798, published *An Essay on the Principle of Population*. Malthus wondered whether society could be perfected so that people everywhere could enjoy prosperity and well-being, but he came to the conclusion that rapid population growth was a major cause of human poverty and misery.

Malthus was writing at a time when England was in stage 2 of the demographic transition and its population was rapidly increasing. Perhaps these factors influenced his argument that the food supply increased arithmetically—from 1 to 2 to 3 to 4, for example—but that population increased geometrically—from 1 to 2 to 4 to 8 to 16, and so on. The result, he claimed, was that the number of people in a country would quickly exceed their food supply. According to Malthus, *positive checks* such as famine and disease would then spread, raise mortality, and reduce the population. In essence, population size was held in check by a country's food-producing capacity. To avoid such dire events, Malthus, who was also an Anglican clergyman, argued that people would need to implement voluntary *preventive checks* such as postponing marriage and practicing sexual restraint.

Neo-Malthusians and Cornucopians

Malthus's views are provocative and, to many people, extremely pessimistic. Nevertheless, more than two centuries after the publication of his *Essay* Malthus's work still prompts debate about population size and rates of population growth. Significantly, after World War II, neo-Malthusian thought shaped the population control strategies adopted in many parts of the developing world. **Neo-Malthusians** are people who share the same general views of Malthus. They argue that, since the world's resources are limited, there is also a natural limit to the number of people the Earth can support at a comfortable standard of living. This limit is the world's **carrying capacity**.

One of the major criticisms of Malthusian population theory is its assumption that the environment is the main determinant of population size because environmental conditions constrain food production. Malthus simply could not imagine that food production could be significantly expanded. Moreover, he neglected to consider the influence of political factors on the distribution of food.

One of Malthus's critics was the development theorist Ester Boserup, who, beginning in the 1960s, argued that with more people to provide labor, food production could actually be increased. Boserup's argument provides a good example of **cornucopian theory**—a theory positing that human ingenuity will result in innovations that make it possible to expand the food supply. (The word *cornucopia* means "horn of plenty.") Boserup was by no means the first cornucopian, but her work presented an important challenge to neo-Malthusianism (**Figure 3.11**).

Food Insecurity

One of the key issues that both the cornucopians and neo-Malthusians overlooked is that of **food insecurity**—when people do not have physical or financial access to basic foodstuffs. Conditions leading to food insecurity include poverty, population growth, war and civil strife, natural resource constraints and environmental degradation, and natural disasters. Chronic problems with food insecurity in East and West Africa have prompted the development of the Famine Early Warning System (FEWS), an initiative led by

Population problem or promise? • Figure 3.11

Neo-Malthusians contend that population growth needs to be strictly controlled in order to avoid a future characterized by great loss of human life, environmental catastrophes, and poverty, whereas cornucopians consider people to be the most valuable resource. Some cornucopians consider population control misguided.

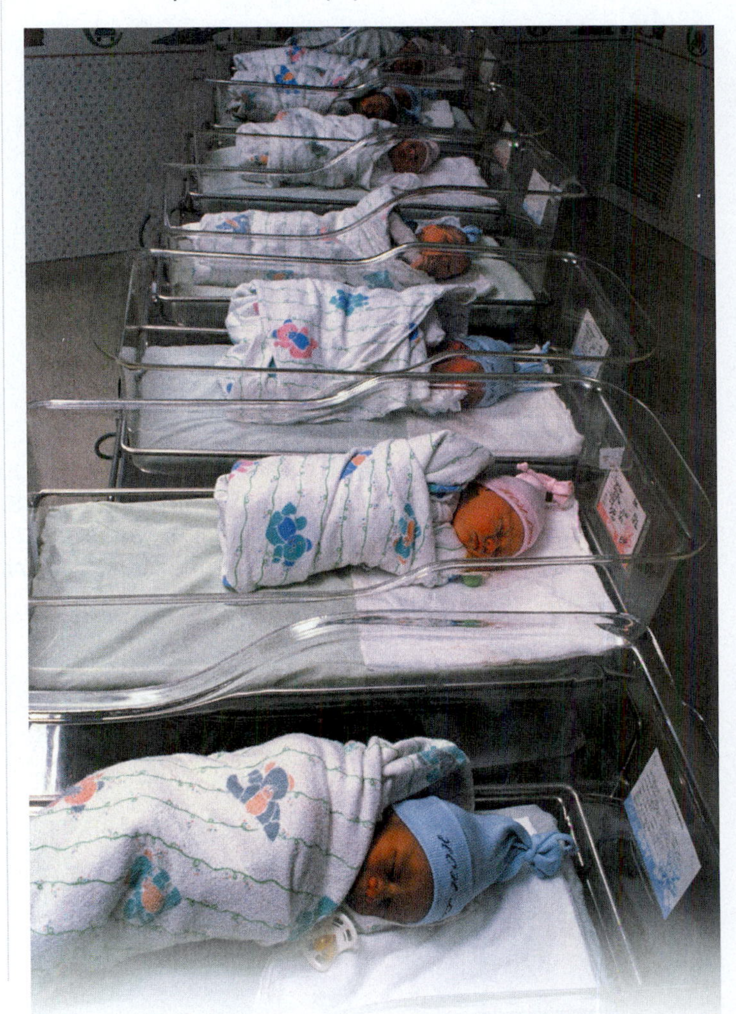

Food insecurity • Figure 3.12

a. Combining satellite data about environmental conditions, food price information, and urban and rural settlements, the FEWS maps areas of food security and insecurity in order to identify populations susceptible to famine. Food insecurity is not just a concern of other countries; it exists in different magnitudes in all U.S. states.

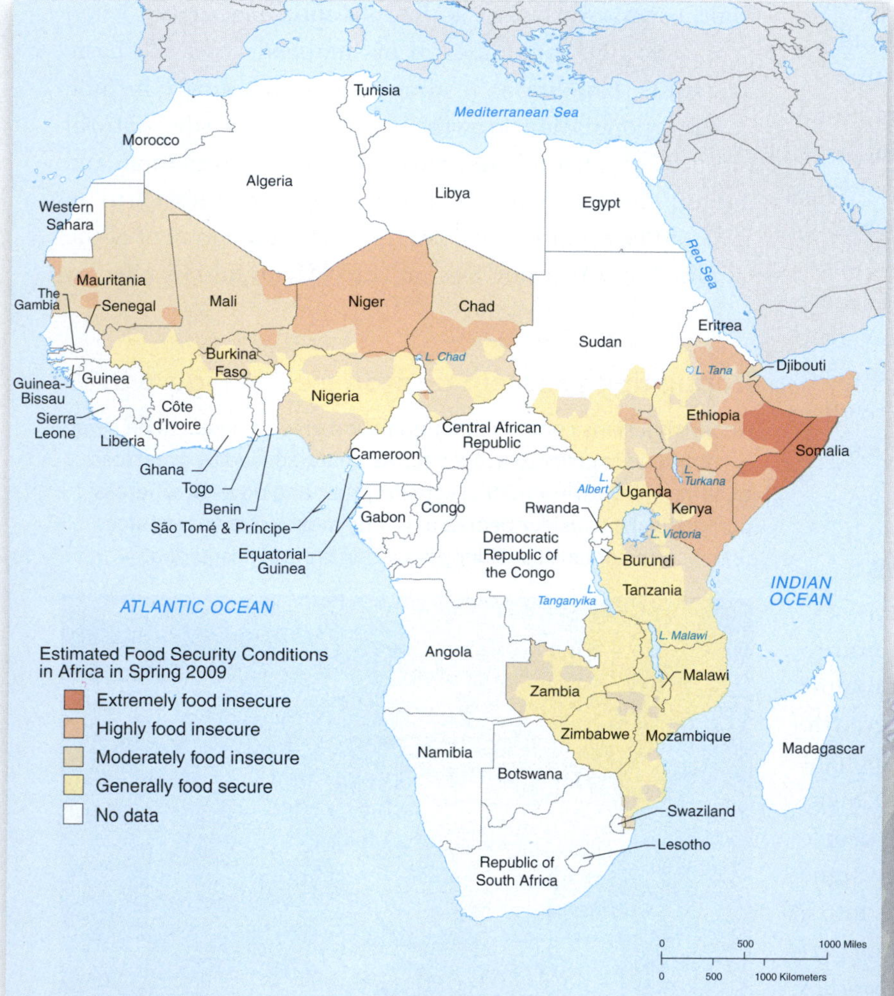

b. Stunted growth is a sign of chronic food insecurity and malnutrition. Iodine deficiencies also impair health. Food insecurity in North Korea is most severe in the northeastern provinces, where level and arable land are in short supply. North Korea's food insecurity stems from poverty but is complicated by environmental conditions and the policies of its repressive government. (*Source*: DPRK, 2005.)

c. North Korean workers bag donated wheat. The cost of fertilizer has been beyond the means of this impoverished country. Officially, U.S. policy is not to use food aid for strategic purposes. However, the United States often does stipulate that the distribution of such aid be monitored. Does this make food a political tool?

the U.S. Agency for International Development. Since hunger is an expression of food insecurity, another approach to population-environment studies involves the political ecology of hunger. Geographer Michael Watts has long argued that famine and hunger have multiple causes, requiring an understanding of environmental conditions and political policies (**Figure 3.12**).

CONCEPT CHECK

1. **What** principle of population did Malthus advance, and how has it been criticized?
2. **What** are the implications of neo-Malthusian and cornucopian perspectives for resource use?
3. **Why** is food insecurity a geographic issue?

Migration

LEARNING OBJECTIVES

1. **Identify** Ravenstein's principles of migration.
2. **Explain** Lee's migration theory.
3. **Explain** how transnationalism relates to migration.
4. **Distinguish** between an unauthorized immigrant, an asylum seeker, a refugee, and an internally displaced person.

Thus far, we have focused our discussion of population change on the relationship between births and deaths in a given population. There is, however, another important factor that contributes to population change: **migration**. Geographers distinguish between migration and **circulation**. For example, if you spend the academic year at college but return home during the summers, your movement is classified as a kind of circulation rather than a migration. Geographers consider migration and circulation two kinds of *spatial mobility*.

migration The long-term or permanent relocation of an individual or group from one place to another.

circulation The temporary, often cyclical, relocation of an individual or group from one place to another.

Migration always involves **emigration**, the out-migration or departure of people from a location, and **immigration**, the in-migration or arrival of people to a location. **Net migration** accounts for changes to the population of a location (e.g., state, province, country, or region) as a result of immigration and emigration:

$$\text{Net migration} = \text{Number of immigrants} - \text{Number of emigrants}.$$

Population change in a region can then be calculated using the **demographic equation**, defined as natural increase plus net migration over a specified period of time. The following equation shows how to calculate population change from 2008 to 2009:

$$\text{2009 Population} = \text{2008 Population} + \underbrace{(\text{Births} - \text{Deaths})}_{\text{Natural Increase}} + \text{Net Migration}$$

Forced and Voluntary Migration

We can group most migrations into one of two categories: forced or voluntary. **Forced migration** occurs when a person, group, government, or other entity insists that another individual or group must relocate. The people being forced to move have no say about where they are moving, when, or other conditions of the move. The Five Civilized Tribes were forced from their homelands in the southeastern United States to Indian Territory—now Oklahoma—in the 19th century. During the late 1990s, Germany forcibly returned several thousand refugees from the war in the former Yugoslavia to Bosnia. The most notorious example of forced migration is the trans-Atlantic African slave trade, which resulted in the forcible relocation of nearly 12 million Africans to destinations in the Caribbean and North, Central, and South America between 1450 and 1900.

In contrast, a **voluntary migration** is a long-term or permanent move that stems from choice. Decisions about voluntary migration may be made during times of considerable hardship and migrants may have limited options, but there is still an element of choice involved in the decision that distinguishes a voluntary from a forced migration. Most migration is voluntary migration and shares several characteristics, which were called the *laws of migration* when first proposed in 1885 by the British demographer E. G. Ravenstein. Ravenstein enumerated several laws, and some included multiple ideas that can be distilled into the following seven principles:

1. Most migrations cover short distances and do not cross international boundaries.
2. Migrants moving to towns and cities create gaps or open spaces that are filled by migrants moving from more distant places.
3. Migration involves two opposite processes: dispersion (the departure of migrants from a place of origin) and absorption (the arrival of migrants in a place of destination).
4. Migration flows produce counterflows.
5. Urban areas are common destinations of long-distance migrants.

6. Urban residents tend to be less likely to migrate than rural residents.
7. Women migrate more than men within their country of birth, whereas men more frequently migrate beyond their country of birth.

Ravenstein was one of the first researchers to draw attention to gender differences in migration. Recent trends, however, suggest that women increasingly participate in international migration and may do so as frequently as men. Ravenstein also observed that, as the distance traveled during a move increases, the number of migrants decreases. Thus, he highlighted the role of distance decay in migration flows and noted that all migrations, through the processes of absorption and dispersion, involve spatial interaction. Still, an important question remains unanswered: What specific factors influence people's decision to move? These are the subject of the next section.

Push and Pull Factors

The decision to migrate is a complex one. People must consider many things, such as the opportunities a new place presents, the challenges of living and working away from one's home community, the cost of the move, as well as how and when the move will occur. Often, social networks—systems of personal ties and communication—play an important role in migration.

push factors
Unfavorable conditions or attributes of a place that encourage migration.

pull factors
Favorable conditions or attributes of a place that attract migrants.

All voluntary migrants confront a combination of **push factors** and **pull factors.** In 1966, the social scientist Everett Lee revisited Ravenstein's work in order to develop a theory of migration. An important part of Lee's theory centered on migration as a decision-making process that is compelled by the personal perception of many different variables (**Figure 3.13**).

Patterns of Global Migration

International migration occurs when people cross international boundaries and take up long-term or permanent residence in another country. Despite the fact that international migration attracts a great deal of media attention, most people in the world will never move away from the country in which they were born. Today, there are roughly 168 million international migrants, who represent 3% of the global population. Why is this percentage so small? One reason is that international migration typically introduces additional complexities, including citizenship status, the need for a passport and/or visa, and other costs. International migration is also more stringently regulated today than ever before.

Migration occurs on a global scale when migrants cross an ocean. This type of international migration is a relatively recent phenomenon that has grown in conjunction with European colonization and globalization. As we have seen, the trans-Atlantic slave trade raised global migration to a new level in terms of the sheer numbers of people involved. Then, in the 19th and early 20th centuries, Europe experienced the greatest outpouring of individuals ever, when—as the result of political strife, famine, and other factors—approximately 20 million Europeans crossed the Atlantic Ocean between 1880 and 1914.

Globally, about 60% of all international migrants move from developing countries to developed countries and about 40% of international migrants move from one developing country to another developing country. Three world regions—Asia, Africa, and Latin America—are major sources of international migrants. Because more migrants leave these regions than enter them, they are said to have *net emigration*. Conversely, North America, Europe, and Oceania are regions of *net immigration*. That is, more people are admitted to these places than depart from them in a year.

Northern America Both the United States and Canada are major destination countries, regularly admitting more than half of all immigrants in the world. Today most immigrants to Northern America come from Asia and Latin America. This pattern differs from the historical one—from about 1750 to 1950—in which Europeans dominated the flow of immigrants to Northern America.

The United States and Canada set quotas—maximum limits—on the number of immigrants admitted, as do numerous other countries. Prior to 1965, the United States set quotas for individual countries. These quotas demonstrated a preference for European immigrants, but country-specific quotas were abolished by the Immigration and

Migration theory • Figure 3.13

Everett Lee's migration theory still informs studies of migration today. He developed a simple conceptual framework to help comprehend the diverse factors that affect the decision to migrate. He grouped these factors into four categories, illustrated in the diagram.

1 Area of origin
Positive, negative, and neutral factors shape the nature of people's attachment to a place.

2 Area of destination
Positive, negative, and neutral factors also influence the attractive pull that destinations have on people. Prospective migrants typically have only partial information about the area of destination since they do not know it firsthand.

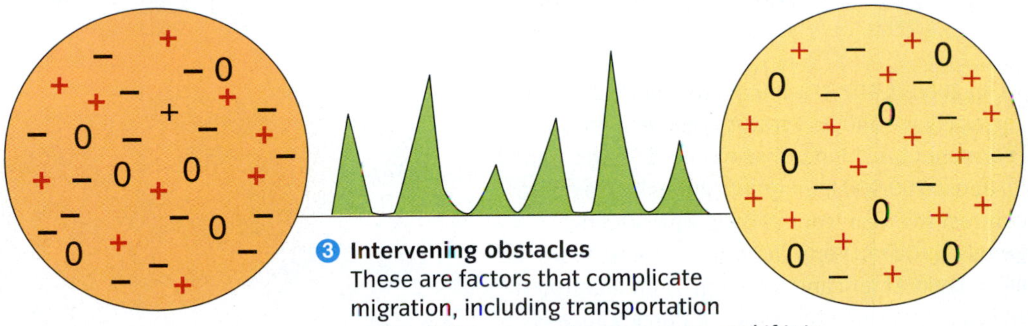

3 Intervening obstacles
These are factors that complicate migration, including transportation costs, distance, moving expenses, and if it is an international migration, the ability to get a passport or visa. Intervening obstacles vary from person to person. They also vary in terms of how difficult they are to overcome.

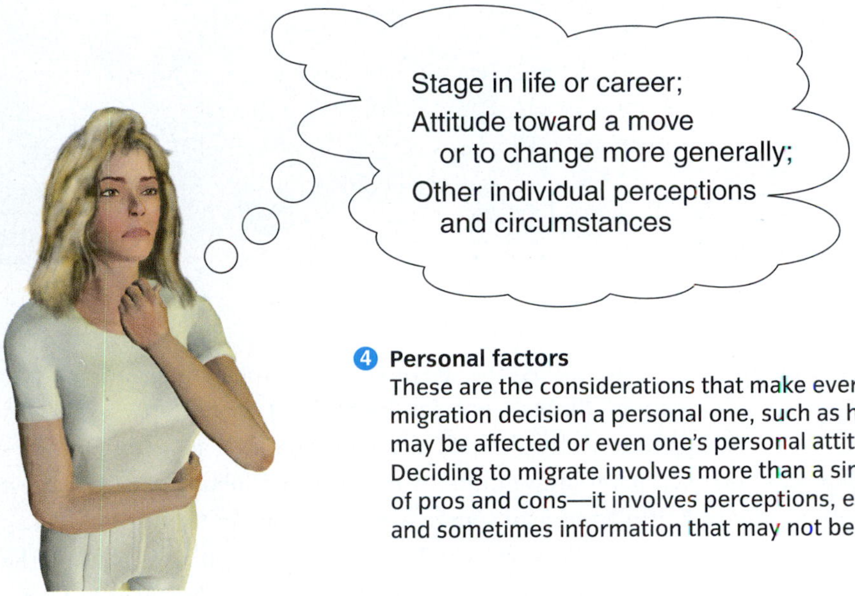

Stage in life or career;
Attitude toward a move or to change more generally;
Other individual perceptions and circumstances

4 Personal factors
These are the considerations that make every migration decision a personal one, such as how children may be affected or even one's personal attitude toward change. Deciding to migrate involves more than a simple weighing of pros and cons—it involves perceptions, emotions, and sometimes information that may not be entirely accurate.

Nationality Act Amendments of 1965. Since that time the United States has used category-based quotas, including family-sponsored and employment-based immigrants.

In the United States **authorized immigrants** are legal permanent residents, also called green-card holders. U.S. statistics count new arrivals and status adjusters (those already in the country) as immigrants. An employer can sponsor for legal permanent residence someone who arrives on a temporary work visa. If approved, this person is counted as a status adjuster. In contrast, **unauthorized immigrants**, also called undocumented or illegal immigrants, are people who come to the United States on a temporary visa but remain in the country after their visa expires, or they cross the border without being detected.

Immigration to the United States and Canada • Figure 3.14

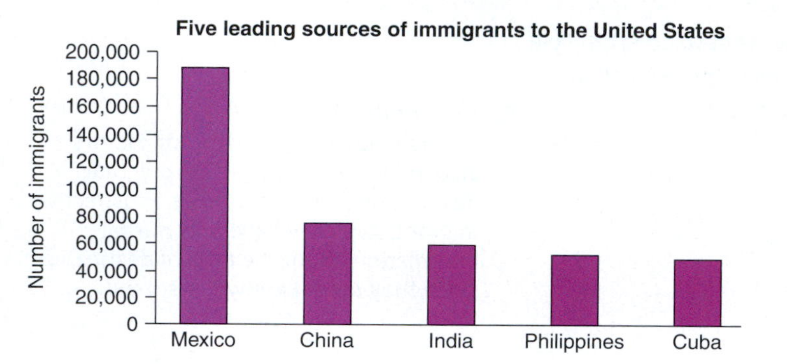

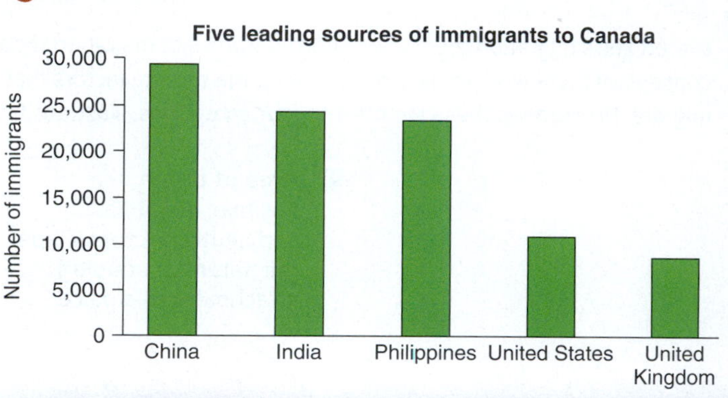

a. In 2008, 1,107,126 immigrants (466,558 new arrivals plus 640,568 status adjusters) gained legal permanent residence in the United States. Canada admits about 250,000 immigrants annually. Compare and contrast the main source countries sending immigrants to Canada and the United States. More than 10 million Latin Americans have legally immigrated to the United States since the 1960s. (*Source*: Citizenship and Immigration Canada, 2009 and U.S. Department of Homeland Security, 2009.)

b. The controversial fence along the U.S.-Mexico border attempts to block undocumented immigrants, thought to number 11.9 million, or 4% of the population in 2008. Many migration specialists maintain that the flow of these illegal immigrants will continue until the root causes—the economic and political push and pull factors—are addressed.

Figure 3.14 illustrates some of the trends in migration to the United States and Canada.

Latin America Prior to 1950, Latin America was a leading destination for immigrants, with the largest migration streams flowing from Spain, Portugal, and Italy. Immigrants also came from Japan, Germany, Russia, and a host of other countries. Since 1950, however, Latin America has become a region of emigration, largely because of political and economic instability. For example, civil unrest in some Central American countries in the late 1980s contributed to heightened levels of emigration from the region. The number of emigrants from El Salvador, for instance, peaked in 1990 when some 80,000 left for the United States. Since that time the numbers of El Salvadorans migrating to the United States has dropped to about 20,000 per year.

Mexico is the leading source of immigrants to the United States, and in 2008 the United States officially admitted 190,000 Mexican immigrants, nearly four times as many immigrants as came from Cuba, the next largest Latin American source area (refer again to **Figure 3.14a**). Some of the world's highest net emigration rates occur in the Caribbean, mostly to the United States, but also to Canada. Limited economic opportunities, especially in the small island states of the Caribbean, are important push factors. The reasons for emigration are quite different for the people of Colombia, however, where violence and armed confrontations associated with a decades-long uprising continue to wrack the country. This internal strife has uprooted more than 3 million people, forcing them to move from the countryside into towns and cities for their safety. It also continues to sustain emigration from the country. **Figure 3.15** illustrates some of these aspects of Latin American migration.

transnationalism In migration studies, the process by which immigrants develop and cultivate ties to more than one country.

Beginning in the 1990s, a number of scholars began questioning the impact of globalization on international migration, specifically within the context of Latin American migration. What these scholars found is that **transnationalism** is often a key aspect of an immigrant's identity. Globalization and the greater connectivity among places facilitate the development of transnationalism.

From the standpoint of geography and geodemography more specifically, transnationalism is significant because it demonstrates that migration involves a system of circulation. Ravenstein hinted at this when he noted that migrations are not just one-way movements of people but also trigger counterflows. Perhaps the strongest evidence of these counterflows can be seen in **remittances**,

Latin American migration patterns • Figure 3.15

a. Blue circles on the map indicate that a country loses more people to emigration than it admits. Some of these migrations have been strongly channelized. Most Dominicans, for example, have moved to New York City and most Cubans to southern Florida. (*Source*: United Nations Population Division, 2009.)

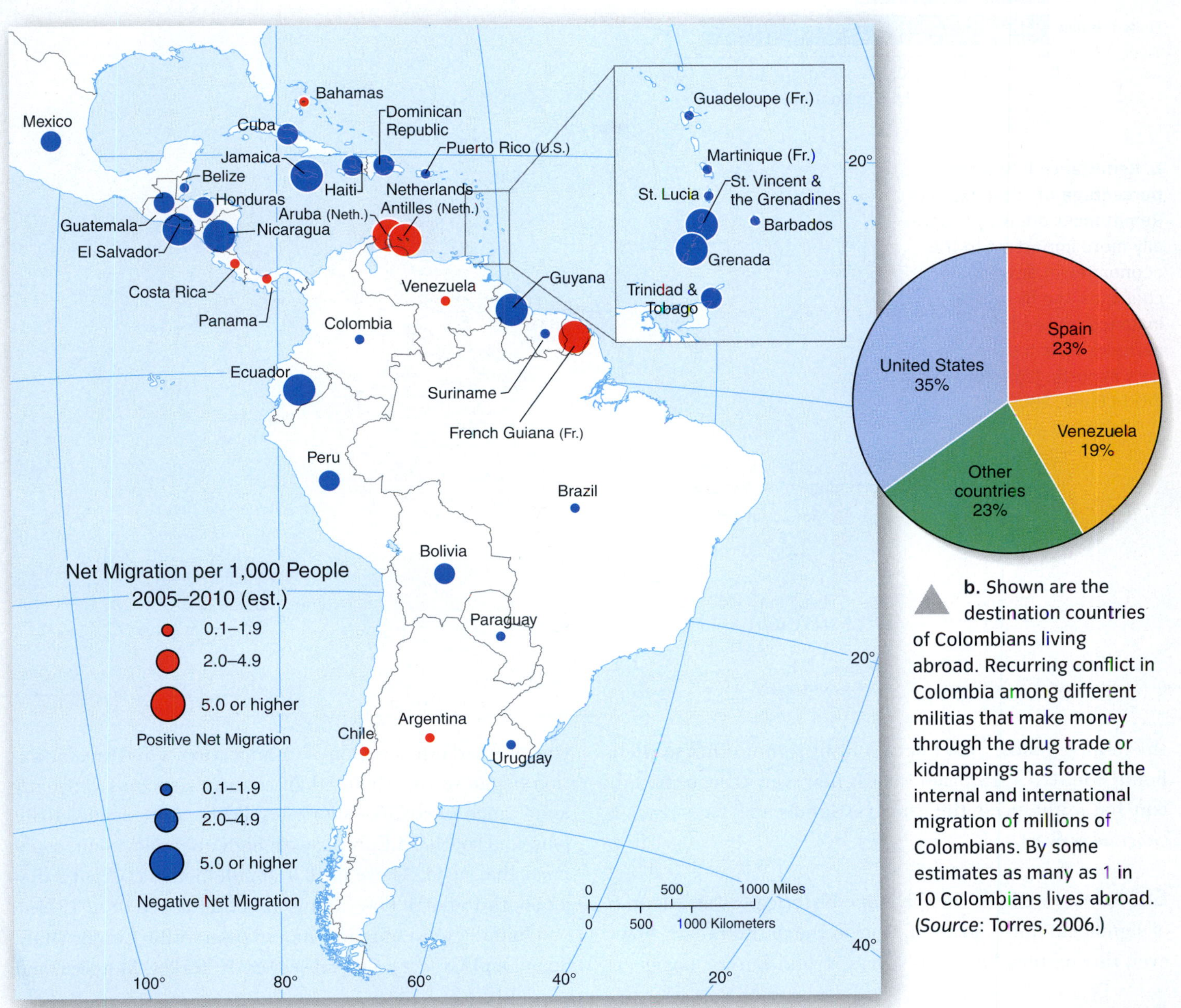

b. Shown are the destination countries of Colombians living abroad. Recurring conflict in Colombia among different militias that make money through the drug trade or kidnappings has forced the internal and international migration of millions of Colombians. By some estimates as many as 1 in 10 Colombians lives abroad. (*Source*: Torres, 2006.)

WHAT A GEOGRAPHER SEES

Economic and Sociocultural Transnationalism

In Chapter 2 we learned that economic transnationalism or multinationalism involves the establishment of branch offices of a corporation in other countries. With respect to migration, however, economic transnationalism focuses on the financial and monetary connections between an immigrant and her or his home country. When immigrants create political, social, or family-based ties that are rooted in the values and practices of their home country and community, they forge a kind of sociocultural transnationalism. Geographers can detect different kinds of transnationalism by tracking remittances and analyzing changes in the cultural landscape.

Top five remittance-sending countries

Country	US$ (billion)
Germany	15.0
Saudi Arabia	16.1
Switzerland	19.0
Russia	26.1
United States	47.2

a. Leading remittance-sending countries
In 2008, migrants sent approximately $433 billion in remittances, compared with $132 billion sent in 2000. The amounts that are remitted change frequently and decline in periods of economic recession. The United States and Russia are, respectively, the leading remittance-sending countries. (*Source*: World Bank, 2009.)

b. Remittance inflows as a percentage of total exports
Remittances are proportionally more important to the economies of developing countries and in several instances exceed the earnings from a country's exports. (*Source*: World Bank, 2009.)

Remittance Inflows as a Percentage of All Exports
- More than 100%
- 20–75%
- 6–19%
- 1–5%
- Less than 1%
- No data

the money, goods, or services sent by immigrants to their home countries. In other cases, however, transnationalism has influenced the cultural landscape (see *What a Geographer Sees*).

Europe As we have seen, Europe historically was a source of significant out-migration. Within the past 50 years, however, this pattern has been reversed, and Europe has been transformed into a region of in-migration. This transformation began in the 1960s when countries such as Germany and France faced labor shortages. People were needed to fill jobs, and southern Europe, with high unemployment, could meet that need. These push and pull factors created a major south-to-north flow of migrants that lasted until 1974.

Initially, most migrants moved from within Europe (Italy, Spain, and Greece), but subsequently Turkey, Morocco, and

THE PLANNER

c. Brazilian transnational landscapes in the United States Higher wages constitute a powerful pull factor. A Brazilian bakery in Marietta, Georgia.

d. Transnational landscapes in Brazil In the city of Piracanjuba, one of the important source communities for Brazilians in Marietta, the landscape suggests that ideas about architecture may have been influenced by Brazilians who have lived in Georgia or other parts of the U.S. South.

Think Critically

1. What are some other ways immigrants might express transnationalism?
2. Can you think of any circumstances in which a migration does not produce a counterflow?

Algeria became important source countries. These migrants came largely as *guest workers*, receiving temporary permits to live and work in the host country. Once their employment ended, however, many guest workers did not return home. Subsequent *chain migrations* brought family members and others from their community or town to join them.

refugee One who flees to another country out of concern for personal safety or to avoid persecution.

This pattern of immigration took on a different form in the 1980s for two reasons. First, the fall of communism meant that controls on the movement of East Europeans were loosened. Second, millions of foreigners, including many Bosnian **refugees** from the war that developed as Yugoslavia broke apart, sought **asylum** in various European countries.

asylum Protection from persecution granted by one country to a refugee from another country.

Asylum applications in Germany • Figure 3.16

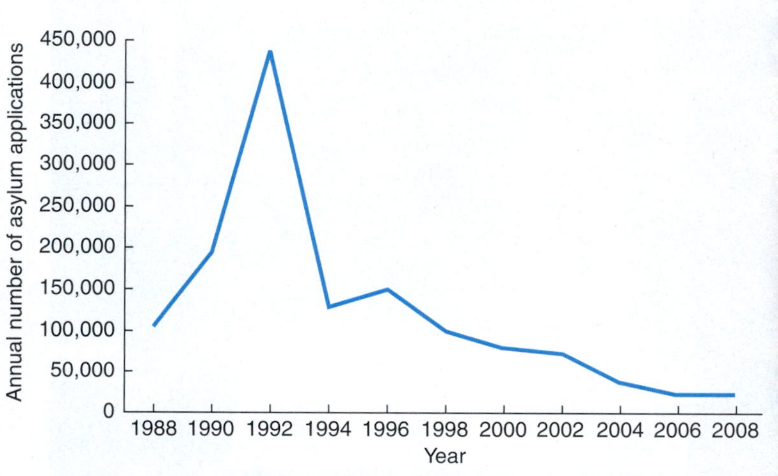

a. After peaking at 438,191 in 1992, the number of asylum applications in Germany has plummeted and in 2008 was just over 21,000. By comparison, the United States received about 49,000 asylum applications in 2008. (*Source*: UNHCR, 2009; Juchno, 2007.)

b. The Turkish market in Berlin, Germany
Two of the leading sources of Germany's asylum applications at the end of the 20th century were Turkey and the former Yugoslavia.

People fearing persecution in their home country can apply for asylum when they enter another country—at a border, in an airport, or in an embassy of the country in which they wish to seek asylum—and a judge will hear their case. Most applications for asylum are rejected. Nonetheless, while a case is being reviewed or appealed, the state is responsible for housing the asylee and meeting their basic daily needs.

Germany provides a good example of a country impacted by the flow of asylum seekers in the 1990s. In most of the decade, Germany received more asylum applications than any other European country. Why was Germany so popular with asylum seekers? An important reason was that Germany provided a fairly generous living allowance for asylum applicants. However, the German government reformed its asylum laws when public sentiment turned against asylum seekers and immigrants in general. Since 1993, Germany has applied the *safe country policy* in which it can deny asylum to people who have passed through another safe country on their way to Germany. In practice, the first safe country asylees pass through is the country where they should seek asylum. As a result of this policy, Germany has the right to return the asylum seeker to that country. **Figure 3.16** depicts the changing trends in applications for asylum in Germany.

Immigration into Europe adds between 300,000 and 500,000 newcomers each year, and most European countries have had positive net migration recently. As discussed earlier in the chapter, Europe is characterized by natural population decline. In 2009, for example, deaths in Europe exceeded births, giving the region a natural decrease of −0.3 per 1,000 population. For that same year, however, in-migrants exceeded out-migrants and the net migration rate for Europe was 1.2 per 1,000 population. Thus, population growth in Europe is being fueled by immigration, not by natural increase.

This pattern of immigration raises three key issues. First, the flow of immigrants into Europe is spatially uneven, with Southern Europe the preferred destination. In particular, Cyprus, Spain, Italy, and Portugal have recorded some of the largest population increases from immigration. Second, Europe has not historically been a region of immigration, and many countries have witnessed the resurgence of anti-immigrant political parties. Third, although immigration into Europe is expected to continue, it is not likely to solve the problem of Europe's declining population. Current estimates indicate that France, Germany, Italy, and the United Kingdom would need to admit nearly 700,000 immigrants just to keep their population at the level it was in 1995. There is, however, little public support in Europe for such high levels of immigration.

Africa Africans make up about 9% of international migrants. Their movement from one African country to another and their emigration from the continent have been shaped by colonialism. Mines and plantations established by European colonialists drew heavily on African labor and established enduring patterns of migration. In southern Africa, for example, employment in the coal, copper, diamond, and gold mines has long shaped the flow of people, and in West Africa a well-established migration stream carries agricultural laborers from interior to coastal regions or, increasingly, into cities. Western Europe and North America have been the main destination regions of emigrating Africans. The movement of Nigerians to the United Kingdom or of Algerians to France reflects the colonial connections established between these places.

Virtually all developing regions confront challenges associated with **brain drain**—the migration of skilled professionals (for example, teachers, engineers, doctors) to another country, where they can obtain a higher paying job and better quality of life. Experts agree, however, that Africa has one of the most serious problems with brain drain. Annually, more than 10% of the health professionals in some African countries emigrate. Such departures compromise the ability of many African countries to deal with the HIV/AIDS epidemic and other public health concerns.

Conflict in Africa continues to have a major impact on migration, prompting forced migrations and displacing large numbers of people. In 1994 the conflict in Rwanda prompted approximately 2 million people to flee into neighboring portions of the Democratic Republic of the Congo (formerly Zaire) and Tanzania. Many of these refugees have since returned to Rwanda, but their mass migration is the largest in recent history. Sudan is estimated to have nearly 5 million **internally displaced persons**, the result of ongoing civil unrest.

> **internally displaced persons** People forcibly driven from their homes into a different part of their country.

Other important migration trends in Africa include a growing south-to-north flow of migrants and an increase in the proportion of women who are migrants. Since the 1980s there has been a more significant flow of Sub-Saharan Africans into North Africa. Three countries—Morocco, Tunisia, and Libya—have become a destination as well as a temporary stop for migrants on their way to Europe (**Figure 3.17**).

African migration • Figure 3.17

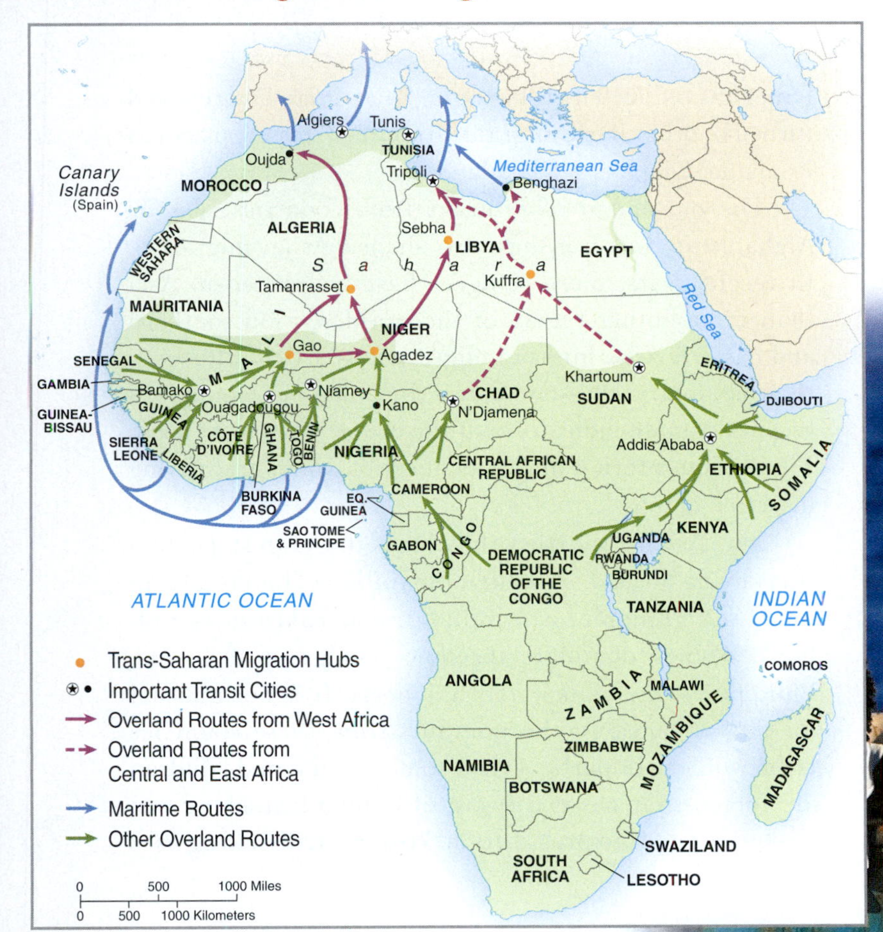

Common routes used by Sub-Saharan Africans. Each journey presents a different set of perils, whether crossing the Sahara Desert or plying the waters of the Atlantic Ocean or Mediterranean Sea often in small fishing boats, like this one arriving in Malta.

Foreign-born • Figure 3.18

Country	Percent Foreign-Born
Qatar	81
United Arab Emirates	70
Kuwait	69
Bahrain	38
Saudi Arabia	27
Oman	26
Iran	3
Yemen	2
Iraq	1

(*Source*: UN Population Division, 2009.)

a. Immigration has substantially altered the demographic composition of several of the Middle Eastern countries around the Persian Gulf, with the result that some of them have very high percentages of foreign-born individuals. In comparison, the foreign-born population in the United States is about 12%. (*Source*: United Nations Population Division, 2009.)

b. Israel is another Middle Eastern country with a high percentage of individuals who are foreign-born (28%), but for different reasons. Israel—much like the United States, Canada, Australia, and New Zealand—has been a country of immigration since its establishment. By law, any Jew can immigrate to Israel and become a citizen. On different occasions since the 1980s, the Israeli government has flown Ethiopian Jews to Israel, shown here, through a kind of assisted immigration.

Asia Asia supplies about 25% of the world's migrants, the largest percentage from any world region. Most of these migrants move from one Asian country to another. Within the past 10 to 15 years, an eastward flow of Asians to countries such as Malaysia, Singapore, and Indonesia has emerged. Asia is also a major source region for immigrants to the United States, Canada, and Europe.

The reasons for Asian migration are complex and varied, but many Asians move in search of jobs, creating sizable streams of labor migration. Beginning in the 1970s, oil-producing Middle Eastern countries with expanding economies attracted millions of migrants, mostly other Asians, who sought work in the oil fields and in construction (**Figure 3.18**).

Like other Middle Eastern countries, Israel has relied on foreign laborers—usually Palestinians—in agriculture and construction. Following the start of the *intifada*, or Palestinian uprising, in 1987, however, restrictions were placed on the movement of Palestinians, many of whom commuted to work in Israel. Because of security concerns, Israel has reduced its reliance on Palestinian laborers and turned to other labor migrants from such diverse places as Romania, Ghana, and Thailand.

The movement of refugees from countries such as Afghanistan and Iraq has also shaped Asian migration. In recent years, more refugees have originated in Asia than in any other region of the world. Within southern and eastern Asia, internal migration—particularly rural-to-urban migration—dominates. In China, for example, a recent estimate indicates that the country possesses some 130 million internal migrants, most of whom are moving in search of jobs.

The Asian countries that have the largest populations living abroad are China (40 million), India (20 million), the Philippines (8 million), and Pakistan (4 million). Women dominate the flow of migrants from the Philippines, Sri Lanka, and Indonesia. It is not unusual for 60% or more of the migrants from these countries to be women. In parts of Asia some women and children are forced to move as a result of human trafficking and are often sold into prostitution. **Human trafficking** uses force, violence, or coercion to recruit people for work in exploitative conditions. We lack comprehensive statistics on the global volume of human trafficking, but estimates indicate that between 2 and 4 million people are trafficked annually. Southeast Asia is the leading region from which victims are trafficked, and they are moved to destinations including Japan, Thailand, Malaysia, and Cambodia, among other countries. Human traffickers use many different routes; one of these routes takes women and children from Bangladesh and Nepal into India and Pakistan, with some being transported to the Middle East. Young Vietnamese women are also frequently trafficked across the border into China.

CONCEPT CHECK

1. **How** did Ravenstein relate absorption and dispersion to migration?
2. **What** are intervening obstacles, and how do they factor in Lee's migration theory?
3. **Why** is transnationalism associated with migration, and in what ways might transnationalism be expressed?
4. **What** similarities and differences exist between Germany's experience with asylum seekers and concerns in the United States about unauthorized immigrants?

Summary

THE PLANNER

1 Population Fundamentals 66

- Geodemography is the study of the spatial variations of human populations. People are unevenly distributed on the world's landmasses and tend to be concentrated in coastal lowlands.
- Two important geodemographic measures of population density are **arithmetic density**, depicted here, and **physiological density**, which takes the amount of **arable land** into consideration.

Population densities • Figure 3.2

- A variety of cultural, economic, and political factors influence the patterns of birth and death around the world. The **crude birth rate (CBR)** provides one measure used to track birth-related trends in populations. Another measure with more predictive power is the **total fertility rate (TFR)**.

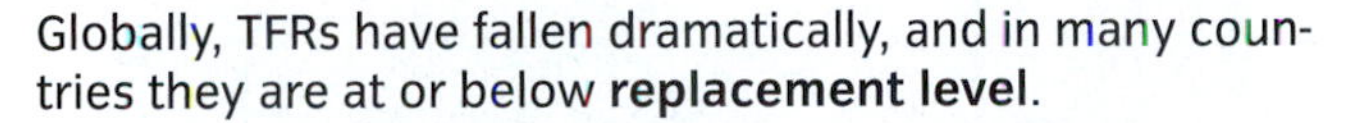

Globally, TFRs have fallen dramatically, and in many countries they are at or below **replacement level**.

- The **crude death rate (CDR)** measures the mortality within a population. Information about the quality of life of a population can be gleaned from its **life expectancy** and **infant mortality rate**.

2 Population Composition and Change 73

- **Population pyramids,** like this one, help show the age and gender makeup of a particular population. Population pyramids can be grouped into one of three categories: rapid population growth, slow population growth, and population decline.

Population pyramids • Figure 3.6

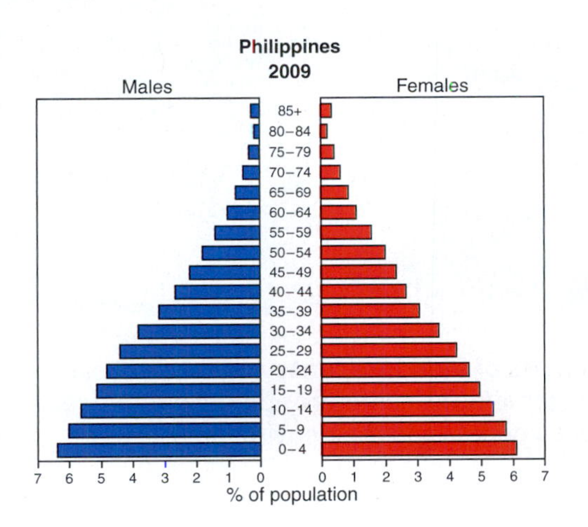

- Two important dimensions of the composition of a population include the **age-dependency ratio** and the **sex ratio**.
- Geodemographers use the **rate of natural increase (RNI)** to indicate the rate at which a population grows or declines in a year. RNIs can also be used to calculate the **population doubling time** for a particular population.
- The relationship between the birth rate, death rate, and RNI forms an essential part of the **demographic transition model.** Changing patterns of infectious and chronic diseases often parallel the demographic transition. These trends are expressed in the **epidemiological transition**.

3 Population–Environment Interactions 78

- Geodemography includes the study of population–environment interactions, or **population ecology**. **Neo-Malthusians** emphasize the concept of a **carrying capacity** for Earth. **Cornucopian theory** considers population growth, as represented by this photo, a stimulus for technological innovations.

Population problem or promise? • Figure 3.11

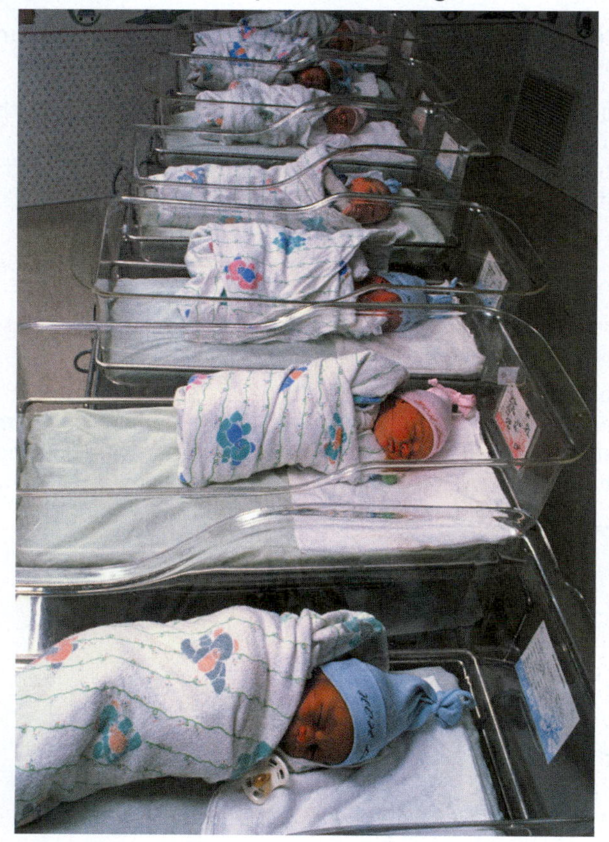

- Awareness of **food insecurity** and the political ecology of hunger affords a different approach to the relationship between people and their environment.

4 Migration 81

- **Migration** and **circulation** are different forms of spatial mobility. The demographic equation takes into consideration natural population change as well as **emigration** and **immigration**. There are two broad types of migration: **forced migration** and **voluntary migration**. Although numerous examples of forced migration exist, most migrants move on a voluntary basis. Ravenstein's principles of migration describe the salient characteristics of voluntary migrations.
- Deciding to move is a complicated process that usually involves weighing **push** and **pull factors**. Everett Lee's theory of migration provides a conceptual framework for understanding the factors affecting the decision to migrate.
- Globalization has increased the potential for **international migration**, but the percentage of people who move internationally remains very small. One notable international flow draws Sub-Saharan Africans north and then across the Mediterranean Sea to Europe, often via small boats (see photo).

African migration • Figure 3.17

- Latin America has become a region of net emigration, yet Asia supplies the largest share of world's emigrants. Like most countries, the United States distinguishes between **authorized** and **unauthorized immigrants**.
- One result of migration is **transnationalism**. **Remittances** constitute a form of economic transnationalism and are proportionally more important to the economies of developing countries.
- Migrations compelled by civil strife can lead to large numbers of **asylum** seekers, **refugees,** or **internally displaced persons**. Another aspect of migration that is less visible but no less significant, is **human trafficking**.

Key Terms

- age-dependency ratio 75
- arable land 67
- arithmetic density 67
- asylum 87
- authorized immigrant 83
- brain drain 89
- carrying capacity 79
- circulation 81
- cornucopian theory 79
- crude birth rate (CBR) 67
- crude death rate (CDR) 71
- demographic equation 81
- demographic transition model 77
- emigration 81
- epidemiological transition 78
- food insecurity 79
- forced migration 81
- geodemography 66
- human trafficking 90
- immigration 81
- infant mortality rate 72
- internally displaced persons 89
- international migration 82
- life expectancy 72
- migration 81
- neo-Malthusian 79
- net migration 81
- physiological density 67
- population doubling time 76
- population ecology 78
- population pyramid 73
- pull factors 82
- push factors 82
- rate of natural increase 76
- refugee 87
- remittance 85
- replacement level 67
- sex ratio 75
- total fertility rate (TFR) 67
- transnationalism 85
- unauthorized immigrant 83
- voluntary migration 81

Critical and Creative Thinking Questions

1. Do you think it would be feasible to establish limits on the population size or density of the world's largest cities? Why or why not?
2. Is there a relationship between the age structure of a population and crime rates? Explain your answer.
3. What would a population pyramid for a cemetery look like? Speculate on the shape of the population pyramid for Arlington National Cemetery in Virginia, where military personnel and their families are buried, and explain your reasoning.
4. What evidence does this photo provide to challenge a conventional view of age dependents?

5. Discuss how Malthusian views might affect public policy, including welfare programs.
6. If you or your family has moved, reflect on the experience and identify the specific push, pull, and personal factors as well as any intervening obstacles that came into play. If you haven't moved, find a friend who has and explore the factors that influenced his or her move.
7. Reflect on the following questions: Is the right of reproduction a basic human right? If so, do anti-natalist policies violate it?
8. What relationships might exist between education, fertility, and food security? Explain your reasoning.

What is happening in this picture?

NATIONAL GEOGRAPHIC

A giant male figure has been carved into a chalk ridge near Cerne Abbas, England. The giant is more than 300 years old, and, though its origins are obscure, it is a recognized historic site. Conventional lore considers it an emblem of fertility.

Think Critically

1. Many societies have fertility rituals. How are they similar to and different from present-day pro-natalist population control policies?
2. Can you identify some ways in which depictions of fertility vary by cultural group?
3. Do media depictions of fertility give equivalent representation to men and women?

Self-Test

(Check your answers in Appendix B.)

1. What is a population cartogram?
   a. a map that shows population density
   b. a map that shows the size of a country in proportion to its population
   c. a pie chart showing the world's population by country
   d. a graph showing population growth or decline over time

2. To calculate the physiological density of your state, you would _____.
   a. divide the area of arable land in the state by the area of arable land in the country
   b. divide the state's farming population by the area of land they own
   c. divide the area of arable land in the state by the total land area of the state
   d. divide the state's population by the area of arable land in the state

3. Which of the following is not associated with low fertility?
   a. use of contraception
   b. education
   c. younger age at marriage
   d. literacy

4. Identify two geographic dimensions of the implementation of China's one-child policy.

5. _____ and _____ are useful indicators of the quality of life in a population.
   a. Infant mortality rate; life expectancy
   b. Total fertility rate; life expectancy
   c. Infant mortality rate; crude death rate
   d. Total fertility rate; crude death rate

6. A high sex ratio in a population means that _____.
   a. men significantly outnumber women
   b. women significantly outnumber men
   c. there will be a lot of age-dependents
   d. the rate of natural increase will also be high

7. The population pyramid shown here _____.
   a. depicts a low sex ratio
   b. indicates that rapid population growth in the future is likely
   c. depicts a low age-dependency ratio
   d. all of the above

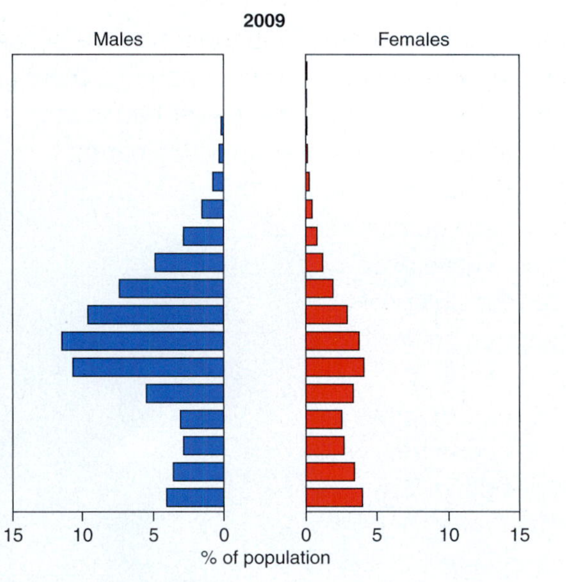

8. Which statement about the rate of natural increase (RNI) is false?
   a. The RNI can be used to calculate population doubling time.
   b. The RNI expresses the difference between births and deaths in a population.
   c. The RNI can be negative or zero.
   d. The RNI for a country today is usually 5% or higher.

9. Brazil has a crude birth rate of 18 per 1,000 people and a crude death rate of 6 per 1,000 people. What is the population doubling time for Brazil?
   a. 12 years
   b. 48 years
   c. 60 years
   d. 72 years

10. Using what you have learned about the demographic transition model, draw the lines for the birth rate and death rate on this diagram.

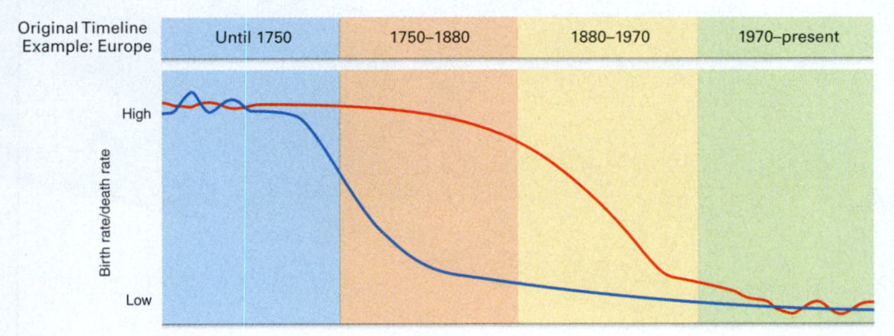

11. Fiji, a country located in the South Pacific Ocean to the east of Australia, has a stage 2 profile. Explain what this means according to the demographic transition model.

12. Both cornucopians and neo-Malthusians overlook _____ in the development of their ideas.
   a. carrying capacity
   b. technological innovation
   c. food insecurity
   d. all of the above

13. Label the four major components of Lee's conceptual framework for migration theory.

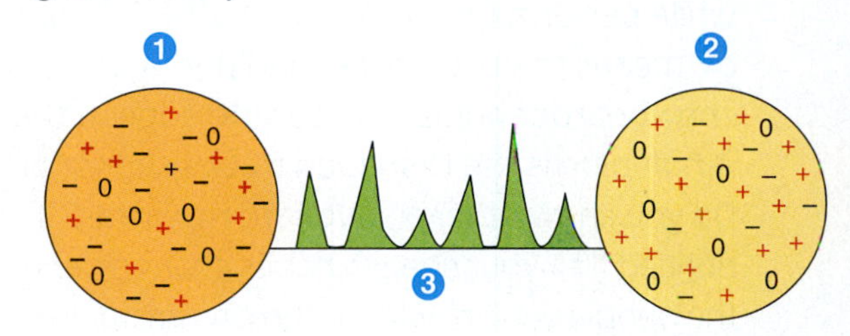

14. In terms of migration patterns, what do Asia, Africa, and Latin America have in common?
   a. They are regions of net emigration.
   b. They are favored destinations of asylum seekers.
   c. They have been only minimally affected by remittances.
   d. They are regions of net immigration.

15. Brain drain and significant south-to-north flows of migrants are shaping _____.
   a. North America
   b. Europe
   c. Africa
   d. the Middle East

THE PLANNER

Review your Chapter Planner on the chapter opener and check off your completed work.

4 Geographies of Language

THE LANGUAGE OF SNOWBOARDING

How well do you know the language of snowboarding? Here, a snowboarder catches some big air while performing an Indy grab off a half-pipe. *Catching air* means to go airborne, and during the *Indy grab* the snowboarder holds the toe-side edge of the snowboard. A *half-pipe* is the U-shaped track used for aerial tricks. Do you know other snowboarding words? If you have noticed that your speech habits change depending on the people you are with or the circumstances you are in, then you already have some familiarity with one important aspect of language—its *situational* quality. Saying that language is situational points out another characteristic of language—its *flexibility*.

Languages change, and the changes can be tracked over time and space. Some language changes happen very quickly, from one generation to the next or even over shorter time spans. Even though in one sense you speak the same language as your parents, there are definite differences. If you have ever had to stop in the middle of a conversation with a parent or grandparent and explain what you mean when you use a certain word, you have experienced the ability of a language to be flexible and to change. The meanings of words can vary over time, and how words are used can differ from one place to another.

Language plays a major role in society because it provides a basis for communication, shapes peoples' identities, and reflects their relationship with place.

CHAPTER OUTLINE

CHAPTER PLANNER

- ❑ Study the picture and read the opening story.
- ❑ Scan the Learning Objectives in each section:
  p. 98 ❑ p. 105 ❑ p. 117 ❑
- ❑ Read the text and study all figures and visuals. Answer any questions.

Analyze key features

- ❑ Geography InSight, p. 100 ❑ p. 106 ❑ p. 112 ❑
- ❑ Process Diagram, p. 116
- ❑ Video Explorations, p. 117
- ❑ What a Geographer Sees, p. 122
- ❑ Stop: Answer the Concept Checks before you go on:
  p. 105 ❑ p. 117 ❑ p. 124 ❑

End of chapter

- ❑ Review the Summary and Key Terms.
- ❑ Answer the Critical and Creative Thinking Questions.
- ❑ Answer What is happening in this picture?
- ❑ Complete the Self-Test and check your answers.

Languages in the World

LEARNING OBJECTIVES

1. **Distinguish** among the terms *language, dialect,* and *mutual intelligibility*.
2. **Identify** the major language families and contrast the distribution of them.
3. **Explain** one theory about the origins of language families.

Globalization calls to mind the interconnectedness of the world, but that interconnectedness depends, in large measure, on our ability to communicate. Although we use language almost continuously throughout the day—in our speech, our thoughts, and even in our dreams—we tend to take it for granted. That is, we sometimes forget how important it is to the overall functioning of society and as a marker of who we are.

Language is a key component of culture. When two or more people speak the same language, communication occurs because the speakers know the symbols and their meanings, and how to put the symbols together to make more complicated meanings. Of course, language does not always facilitate communication; instead, it is sometimes the source of misunderstandings.

language A system of communication based on symbols that have agreed-upon meanings.

dialect A particular variety of a language characterized by distinctive vocabulary, grammar, and/or pronunciation.

Deciding what a language is presents some complications as well. **Dialects** are not usually treated as separate languages. In addition, some languages are so similar that they have a high degree of **mutual intelligibility**. For example, people who speak Serbian can also understand Croatian, and vice versa, even though Serbian and Croatian are considered separate languages.

All human languages express thought and are capable of expressing all kinds of thought, from simple thoughts to the most complex. But different languages express thought in different ways. Each language has its own particular strategies for communicating, strategies that, in many cases, developed over thousands of years. These strategies include rules of grammar, specialized vocabularies, and distinctive systems of pronunciation. For example, the pronunciation of words in some of the

Linguistic differences • Figure 4.1

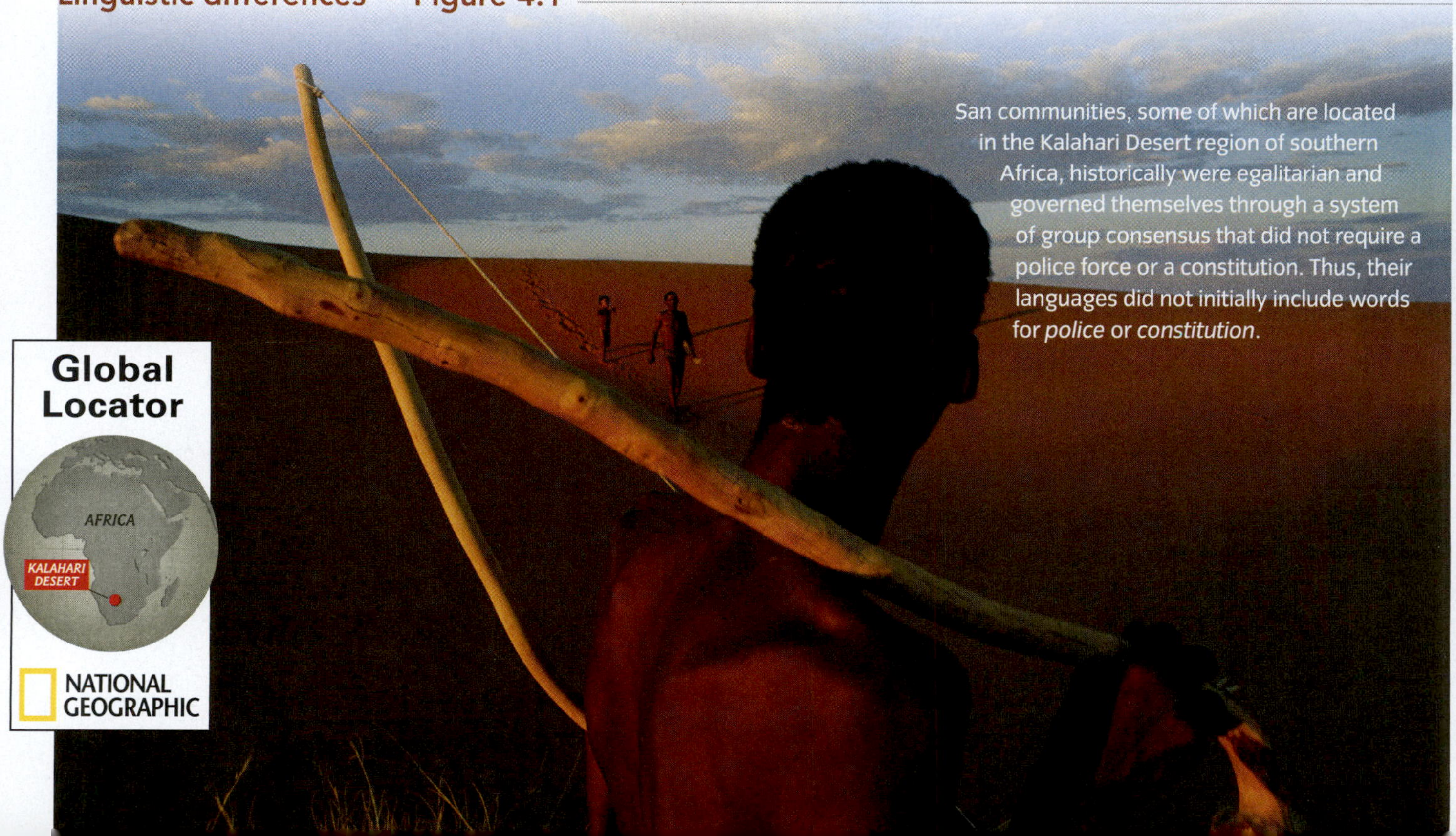

San communities, some of which are located in the Kalahari Desert region of southern Africa, historically were egalitarian and governed themselves through a system of group consensus that did not require a police force or a constitution. Thus, their languages did not initially include words for *police* or *constitution*.

Global Locator

AFRICA

KALAHARI DESERT

NATIONAL GEOGRAPHIC

The language of greetings • Figure 4.2

Body language is highly situational and can vary from one cultural group to another. Greetings are one kind of body language. Businesspeople bow in greeting at a crosswalk in Tokyo's financial district. In Saudi Arabia a traditional greeting between male friends and relatives involves a handshake and a kiss on each cheek.

languages of southern Africa incorporates the use of click sounds. In the Hausa language, spoken in Nigeria, and in Chinese, the meaning of a word can change depending on the rise or fall in the pitch of one's voice.

Social factors influence language use and development. At one time the Japanese language did not have words for *fork* or *coffee*, but as these items entered their culture, words to express them developed. The Yiddish word *farpotchket* means "to make something worse when you were trying to fix it." There is no word for this idea in English. The absence of words in a language does not imply any inability of that language to express thought (**Figure 4.1**).

Types of Language

Although we tend to think of language as spoken and written systems of communication, certain types of language cannot be spoken, and other languages developed orally and were never written. *Sign languages* are nonspoken languages used to communicate with people whose hearing or speech is impaired. Sign languages are based on body movements, especially hand motions (signs that stand for words and concepts) instead of sounds. Sign languages differ from country to country—there is no universal sign language. For example, a person who uses American Sign Language will not be able to communicate with someone who uses British Sign Language because they use different signs for the same words.

Sign language is one kind of *body language*—a communication system based on gestures, facial expressions, and other body movements. But sign language is also a true language, a formalized system capable of expressing all types of thought. In contrast, the body language most of us use on a daily basis consists of the gestures used in greetings, in signaling one's emotions, and in specific situations such as asking for the check in a busy restaurant. Unlike the gestures of sign language, these situational gestures do not make up a fully developed system of communication—they cannot communicate all kinds of thought (**Figure 4.2**).

Touch or *tactile language* constitutes another kind of nonspoken language. The most familiar touch language is Braille, which uses raised dots to represent letters, numbers, and other symbols. Some nonspoken languages, such

Geography InSight

Nonspoken languages • Figure 4.3

Specialized languages might use simplified pictures and symbols to express an idea, whereas tactile languages communicate via sense of touch.

a. One specialized language uses a semaphore alphabet consisting of colored flags held in various positions to signal different letters. The peace sign was created from the semaphore letters "n" and "d"—the first letters of the words *nuclear disarmament*.

b. In Malaysia a sign reminds visitors, who might not read English, that public displays of affection are not permitted. Another specialized language has emerged to meet the needs of travelers and tourists.

c. A blind person reads a tactile-relief political map that is also in Braille. Geographic education for the visually impaired often uses textural differences to communicate spatial information.

as computer programming languages, exist only in a written form. These and other *specialized languages* are designed for specific purposes, not necessarily for everyday communication among people (**Figure 4.3**).

Natural languages are those that have emerged and evolved within living or historic human communities. Languages that are intentionally constructed by people for international communication or fictional purposes are called *artificial languages*. J. R. R. Tolkien invented the Elvish languages featured in *The Lord of the Rings* trilogy. Some artificial languages have been designed for the purpose of creating a *universal language*, a language that could be understood and used by everyone in the world. The most well-known example of an artificial language is Esperanto. Invented by a Polish doctor in the late 19th century, Esperanto has a very simple and regular grammar that is intended to be much easier to learn than the grammar of natural languages. Estimates suggest that between 1 million and 15 million people may know Esperanto with varying degrees of fluency. Today it is possible to find newspapers, journals, and Bibles published in Esperanto (**Table 4.1**).

Esperanto: A language without borders? Table 4.1

A portion of President Obama's 2009 inaugural address in English and Esperanto
Esperanto uses a twenty-eight letter Latin alphabet. Twenty-two of the letters are the same as those used in English. Does the use of the Latin alphabet conflict with the claims that the language is universal?

English	Esperanto
My fellow citizens: I stand here today humbled by the task before us, grateful for the trust you've bestowed, mindful of the sacrifices borne by our ancestors. . . .	Miaj samlandaj civitanoj: Mi staras antaŭ vi hodiaŭ, humiliĝinta de la tasko antaŭ ni, dankema pro la fido donacita de vi, konscia de la oferojn de niaj prapatroj. . . .
. . . Forty-four Americans have now taken the presidential oath. The words have been spoken during rising tides of prosperity and the still waters of peace. Yet, every so often, the oath is taken amidst gathering clouds and raging storms. At these moments, America has carried on not simply because of the skill or vision of those in high office, but because we, the people, have remained faithful to the ideals of our forebears and true to our founding documents.	. . . Kvardek kvar Usonanoj estis ĵuranta kiel prezidanto. La vortoj estis laŭtparolitaj dum leviĝintaj tajdoj de prospero, kaj dum la trankvilaj akvoj de paco. Tamen de tempo al tempo oni faras la ĵuron inter kunigantaj nuboj kaj tempestaj ŝtormegoj. Dum ĉi tiaj momentoj, Usono persistis—ne nur pro vizieco aŭ lerteco de altranguloj, sed ĉar Ni la Popolo restis fidela al la idealoj de niaj prapatroj, kaj lojale apogis niajn fundamentajn dokumentojn.
So it has been; so it must be with this generation of Americans.	Tiel estis. Tiel devos esti per ĉi tiu generacio de Usonanoj.
(*Source*: The White House Blog http://www.whitehouse.gov/blog/inaugural-address/.)	(*Source*: Translation provided by Esperanto-USA: http://esperanto-usa.org/en/node/1308.)

Languages by Size

It is generally agreed that there are about 6,900 different languages in the world today. When we group the languages of the world into categories based on their estimated number of speakers, striking patterns emerge: there are a lot of small languages but just a handful of very large languages (**Figure 4.4**).

From a historical perspective, the emergence of these very large languages is a recent phenomenon that highlights an important change in the linguistic geography of

Languages by size • Figure 4.4

a. Languages from very large to very small
More than half of the world's languages have fewer than 10,000 speakers. There are, however, a few hundred languages (counted as "unknown") for which we lack reliable counts or estimates of the number of people who speak them. (*Source*: Lewis, 2009.)

Size	Number of speakers	Number of languages	Percent of the world's languages	Percent of global population
Very large	100,000,000 to 999,999,999	9	0.13026	40.79696
Large	1,000,000 to 99,999,999	380	5.50007	53.29407
Medium	10,000 to 999,999	2,719	39.35447	5.77057
Small	100 to 9,999	3,052	44.17427	0.13818
Very small	Fewer than 100	472	6.83167	0.00022
Not classified	Unknown	277	4.00926	—
TOTAL		6,909	100	100

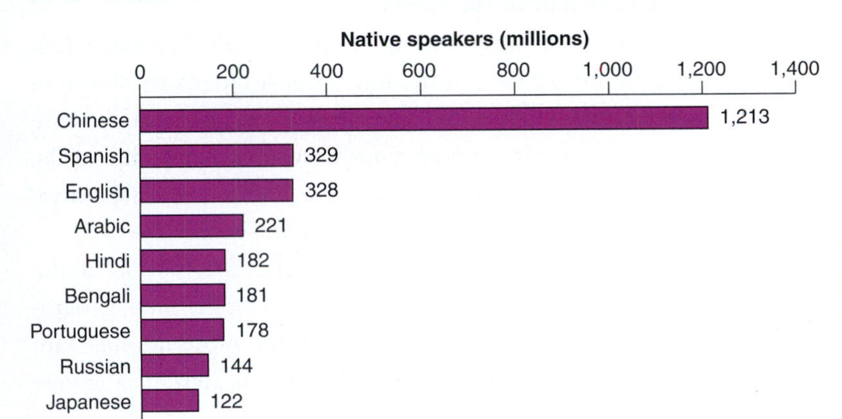

b. Estimated number of native speakers, in millions
These are the nine very large languages. Hindi and Bengali are languages primarily associated with India. How do these very large languages compare and contrast in terms of their global distribution?

Comparing language families • Figure 4.5

Language families are diverse: Some contain more than 1,000 languages, whereas others may contain just a few. Thus, one way to distinguish "major" language families is to identify those that proportionally have the most languages.

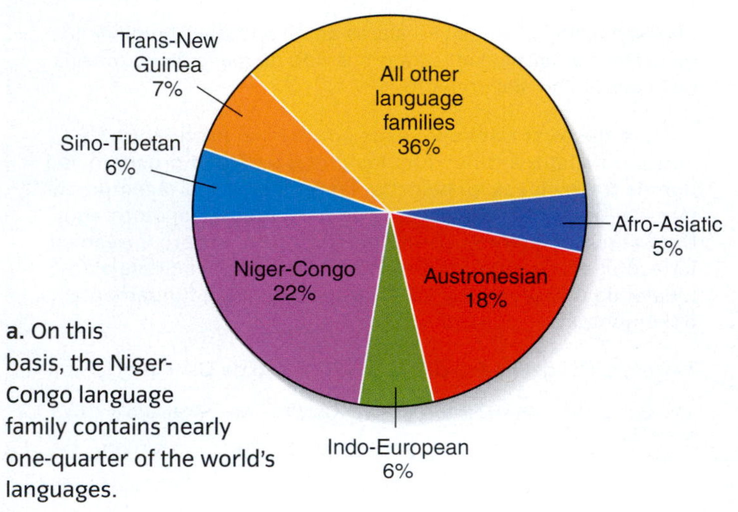

a. On this basis, the Niger-Congo language family contains nearly one-quarter of the world's languages.

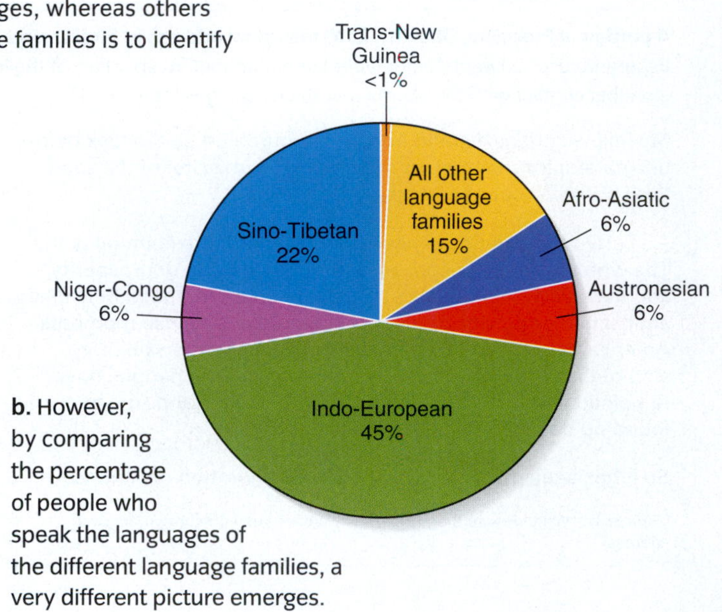

b. However, by comparing the percentage of people who speak the languages of the different language families, a very different picture emerges.

Language family	Example language(s)
Indo-European	English, Hindi
Sino-Tibetan	Mandarin Chinese, Burmese
Afro-Asiatic	Arabic, Hebrew
Niger-Congo	Yoruba, Zulu
Austronesian	Tagalog, Bahasa Indonesia
Trans-New Guinea	Tetum

c. This table provides selected examples of languages from each of the major language families. Yoruba is spoken in Nigeria, Zulu in South Africa, Tagalog in the Philippines, and Tetum in East Timor.

the world. Their expansion has prompted some linguistic geographers to question whether the very small languages can survive, a topic we will return to later.

Language Families

To this point we have seen how languages can be classified by type or size. For geographers and linguists, however, another important system of classification involves the historical relationships among languages. For instance, do all languages derive from one source, or did multiple languages develop independently? This question remains unsettled and highly controversial. Language probably existed at least 30,000 years ago or even much earlier, but we do not know its date of birth. Much of what we do know about language development comes from surviving artifacts and written texts, but there are many gaps in this historical record because most early languages were never written down. Despite these complexities, scholars continue to pursue the study of the evolution of languages and the relationships among them. Why? One major reason is that knowledge of language development contributes to our understanding of past societies, contacts among them, and patterns of human migration.

It is clear that most languages share a distant historical and genetic relationship with one or more other languages. Terms such as **language family** and *branch* express these relationships. The world's languages have been classified into more than 90 different language families, but 6 are considered major language families (**Figure 4.5**).

> **language family** A collection of languages that share a common but distant ancestor.

Almost half of the world's people speak languages belonging to the Indo-European language family. In fact, of the nine largest languages shown in Figure 4.4b, all but three (Chinese, Arabic, and Japanese) are Indo-European. In later sections we will explore some of the consequences of the growth of these large languages, but we want to turn now to some facets of language distribution.

Since it is impractical to show the distribution of the world's several thousand languages on a single map, geographers instead make and use maps of the more familiar language families. Maps showing the distribution of language families instantly prompt questions about where language families originated—their **hearths**—and how they spread. Among geographers and linguists a prevalent, though controversial, theory posits that the rise of agriculture and the subsequent spread of farming populations transformed the distribution of the world's languages and language families. This transformation is thought to have occurred through absorption by intermarriage and language replacement as the language of the farming population spread. Compared to hunters and gatherers, agricultural societies gained numerical dominance through their ability to support more people and sociopolitical dominance through their ability to organize large armies. This dominance facilitated the diffusion of the languages spoken by the farming populations. In most cases, these processes of language spread occurred over thousands of years. For example, the spread of the Niger-Congo language family is associated with the expansion of Bantu farmers who migrated east toward Lake Victoria about 4,000 years ago and subsequently spread south (**Figure 4.6**).

hearth A place or region where an innovation, idea, belief, or cultural practice begins.

Figure 4.6 shows that the Austronesian language family covers a vast maritime area. The spread of farming populations, especially rice cultivators and seafaring merchants, played a role in the spread of Austronesian languages. The origins of the language family date to about 3000 BCE in Taiwan, but did not reach the island of Madagascar until about the first millennium CE with the arrival of Indonesian merchants.

Of the language families, the Indo-European language family has the largest number of speakers and the widest geographical distribution (refer again to Figure 4.6). As we have seen, about half of the world's people speak Indo-European languages, and the language family is divided into multiple branches. The Italic languages, more popularly referred to as Romance languages, form one branch. All of the Romance languages developed from Latin, an Italic language that became the language of the citizens of Rome in about the 6th century BCE. The subsequent rise and expansion of the Roman Empire played a major role in spreading the Latin language across much of southern and western Europe. Within the Roman Empire two classes of Latin existed: a standardized written form that came to be known as *Classical Latin*

Language families and some hearths • Figure 4.6

The origins of the different language families are often not precisely known either geographically or historically. This is especially the case in the Americas where the extermination of native populations and the loss of their languages as a consequence of European conquest complicate the historical record. Keep in mind also that some of the patterns on this map—for example, the spread of the Niger-Congo language family—are traced to events that occurred thousands of years ago, whereas some are just a few centuries old. Identify a pattern on the map that is the result of more recent developments.

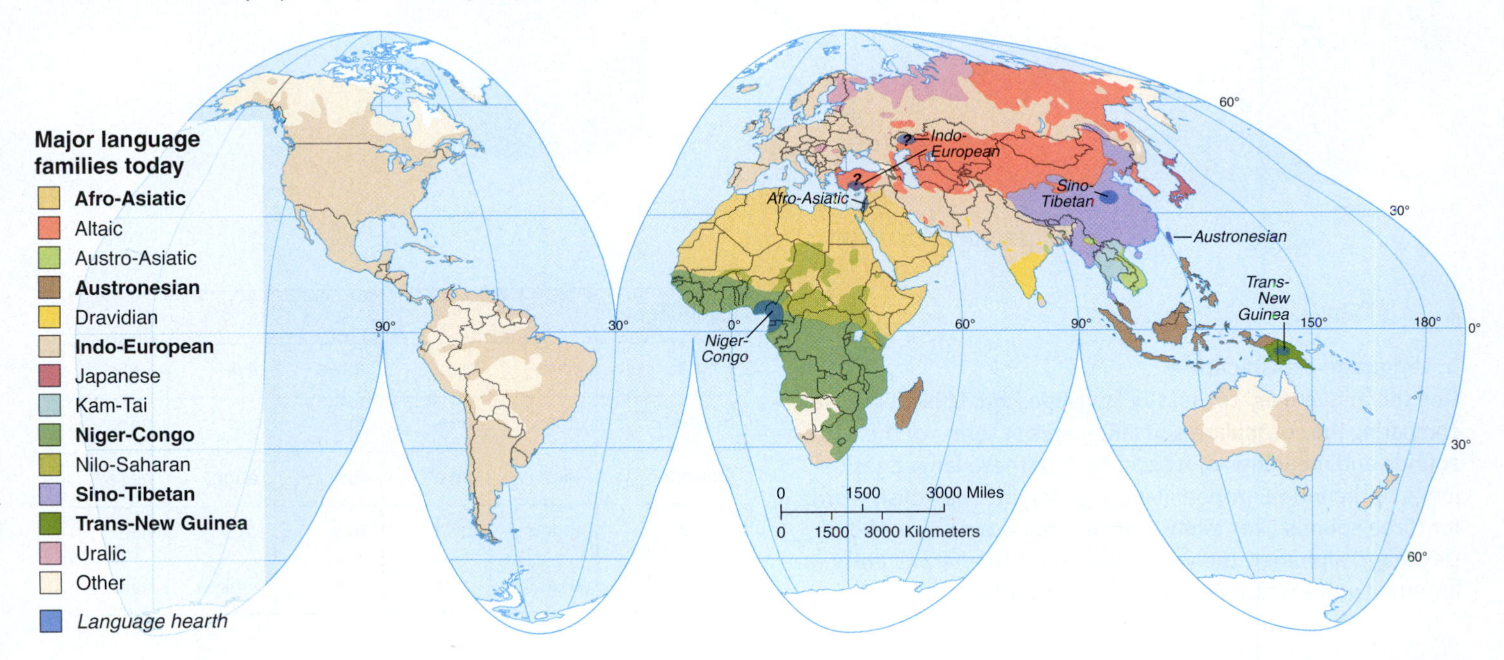

The Indo-European language family tree • Figure 4.7

The prefix *proto-* identifies an ancestral language. Note that the presence of a language on the tree does not necessarily mean that it exists today. All the Anatolian languages, for example, are extinct.

a. Indo-European language family tree
Within large language families such as this one, some languages are more closely related than others. These are grouped into branches that may be further divided into subgroups or individual languages.

b. Language classification
One technique used to classify languages into families involves comparing the vocabularies of languages in terms of the sounds and meanings of words. Each of these languages belongs to the Indo-European language family. The Polish word for *three* sounds like "tshay." Since Hindi is written in the Devanagri alphabet, the words here have been *transliterated* into the Latin alphabet to facilitate comparison.

Language	Term				
English	one	two	three	eye	nose
Spanish	uno	dos	tres	ojo	nariz
German	eins	zwei	drei	auge	nase
Polish	jeden	dwa	trzy	oko	nos
Hindi	eka	doe	tina	amkha	naka

and a nonstandardized, spoken form called *Vulgar Latin*, the language of the common person. Lacking standardization, Vulgar Latin varied from place to place within the Roman Empire—that is, numerous dialects of Vulgar Latin existed. This spatial variation in the spoken forms of Vulgar Latin contributed to the gradual evolution and emergence of the different Romance languages (**Figure 4.7**).

Scholars think that the Indo-European language family may have originated in the region to the north of the Black and Caspian Seas or in Anatolia, a region in Turkey, perhaps about 5000 BCE (Figure 4.6). Notice, however, that the Turkish language is not an Indo-European language. In the 11th century the Seljuk Turks, whose empire stretched from Central Asia, conquered Anatolia and introduced the language that would become the precursor of modern Turkish.

CONCEPT CHECK STOP

1. **What** is the difference between a language and a dialect?
2. **What** are the six major language families, and where are they distributed?
3. **Where** might the Indo-European language family have originated, and how might it have spread?

Language Diffusion and Globalization

LEARNING OBJECTIVES

1. **Explain** how political, economic, and religious forces can affect the diffusion of language.
2. **Identify** factors contributing to linguistic dominance.
3. **Distinguish** among pidgin languages, creole languages, and lingua francas.
4. **Relate** the concept of language endangerment to linguistic diversity.

What social and geographic factors contribute to the spread, or diffusion, of languages? In our discussion of language families, we learned that the spread of agriculture may have facilitated the historic spread of languages. If we take a broader perspective, we can see that technology and human mobility can contribute significantly to language diffusion. Historically, ships, railroads, and other forms of transportation opened physical spaces for language diffusion, and today the Internet continues to open new virtual spaces in which language can diffuse.

Political, economic, and religious forces can also shape language diffusion. The rise of the British Empire contributed to the expansion of English. As a result of European colonization and immigration to overseas destinations, there are now more English speakers outside than inside the United Kingdom, more French speakers outside than inside France, more Spanish speakers outside than inside Spain, and more Portuguese speakers outside than inside Portugal (**Figure 4.8** on the next page).

Economic forces influence the diffusion of languages in various ways. For example, in many countries, tourism and foreign business are important sources of revenue. Being able to accommodate international tourists and conduct business in English or another European language not only creates opportunities for expanding a country's economy but also shapes language diffusion. For example, Mongolia's new English-language program seeks to make its citizens bilingual in Mongolian and English in order to attract outsourced jobs (see Chapter 10). Even on an individual level, the perception that fluency in another language will improve one's ability to land a job or earn a higher salary can influence language spread. This remains a powerful force behind the decision by many immigrants to learn the language of their new land.

Religion also influences the spread of language. Muslims whose first language is not Arabic study the Arabic language in order to be able to read the Qur'an in its original language. Historically, the diffusion of the Arabic language has been closely associated with the diffusion of Islam.

When studying language spread, linguistic geographers also consider the contexts in which a language is used. For example, in a given place one language may be used at home, another in school, and still another for business. Being aware of the different uses of languages and the spaces or settings in which they are spoken helps us understand how languages become dominant.

Geography InSight

Geographies of language diffusion • **Figure 4.8**

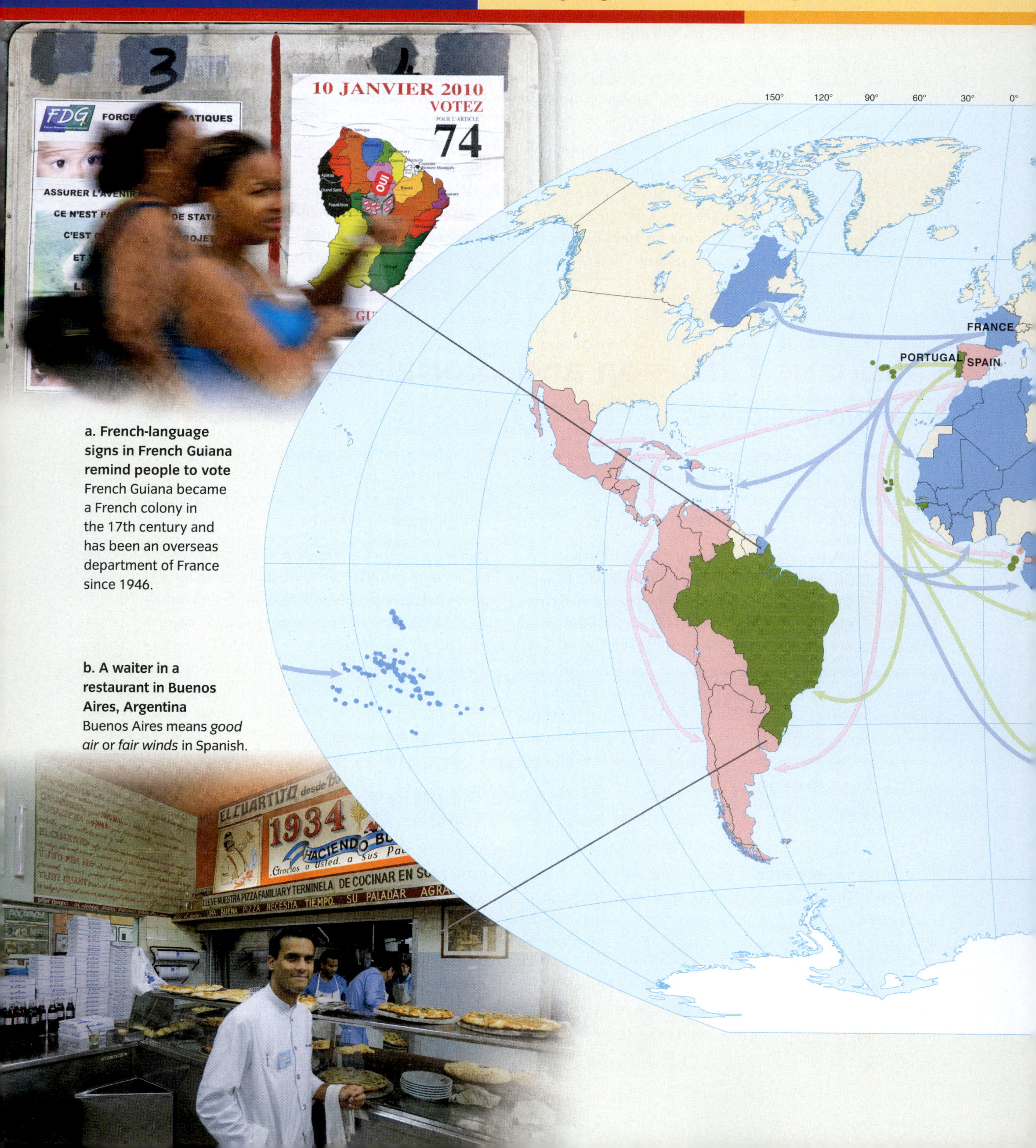

a. French-language signs in French Guiana remind people to vote French Guiana became a French colony in the 17th century and has been an overseas department of France since 1946.

b. A waiter in a restaurant in Buenos Aires, Argentina Buenos Aires means *good air* or *fair winds* in Spanish.

c. Portuguese and Bantu language sign in Mozambique Following its independence from Portugal in 1975, Mozambique changed the name of the city Vila Pery to Chimoio. Education is mainly in Portuguese but Bantu languages, a branch of the Niger-Congo language family, are prevalent across the country.

Generalized Routes of Diffusion of Selected European Languages, circa 1450–1973

- French
- Portuguese
- Spanish

As suggested by this map, European colonization played a major role in the diffusion of certain European languages, including French, Spanish, and Portuguese. Today these languages still influence the linguistic geography of many countries but in varying degrees. In the Philippines, for example, Spanish was an official language until 1973, but English and Filipino (Tagalog) are now more widely spoken.

Linguistic Dominance

> **linguistic dominance**
> A situation in which one language becomes comparatively more powerful than another language.

Sheer numbers affect **linguistic dominance**, but size is not everything. Chinese, for example, with more than a billion speakers, commands the largest speech community in the world, but the geographic range of Chinese is far more restricted than that of English. On the world stage, therefore, English is considered a more dominant language than Chinese. This status shows that linguistic dominance is sometimes more a result of economic and political power than of size.

The association of a language with an independent country is also important. There are about 200 independent political states in the world but about 6,900 languages. In other words, there is what we might call a *language gap*. That is, a majority of the world's languages are not directly associated with the functions of a state. Such *stateless languages* are not used for government functions and are rarely taught in schools. Although they are used in the daily lives of their speakers and are very much a part of people's identity, these uses may not confer the same kind of status on them.

Identification of an official language is often among the first acts of a newly independent country. An **official language** is one that a country formally designates for use in its political, legal, and administrative affairs. This designation is usually made in the country's constitution. A country can designate more than one official language; thus, countries can be officially unilingual, bilingual, trilingual, and so on. However, not all countries have an official language (see **Figure 4.9**).

Official languages • Figure 4.9

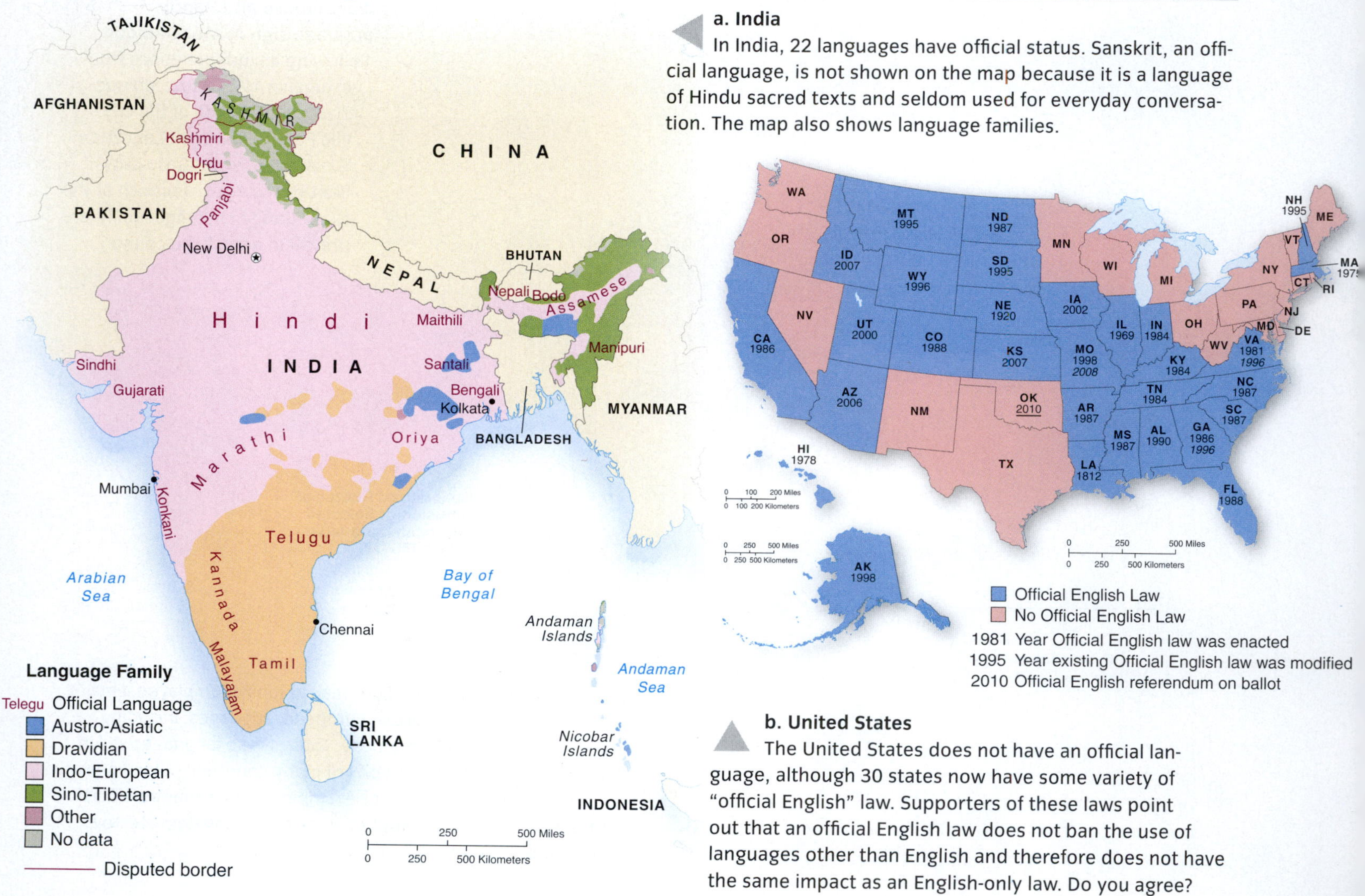

a. India
In India, 22 languages have official status. Sanskrit, an official language, is not shown on the map because it is a language of Hindu sacred texts and seldom used for everyday conversation. The map also shows language families.

b. United States
The United States does not have an official language, although 30 states now have some variety of "official English" law. Supporters of these laws point out that an official English law does not ban the use of languages other than English and therefore does not have the same impact as an English-only law. Do you agree?

Linguistic borrowing • Figure 4.10

Contact among speakers of different languages commonly results in linguistic borrowing.

a. Selected loanwords and their origins
The list includes direct loans (e.g., *luau*) and indirect loans (e.g., *hammock*, which entered Spanish and French and then English).

Loanwords	Language
fjord, ski	Norwegian
beef, naive	French
bandit, duet	Italian
peninsula, ultimate	Latin
landscape, cruise	Dutch
glen, slogan	Scottish Gaelic
hammock, hurricane	Carib
dungarees, jungle	Hindi
caravan, candy	Persian
giraffe, sofa	Arabic
catamaran, curry	Tamil
ukulele, luau	Hawaiian
boomerang, koala	Dharuk (Australian Aboriginal)
wok, hoisin	Cantonese
karaoke, tsunami	Japanese
cola	Temne (West Africa)
okra	Igbo (West Africa)

b. The diffusion of ethnic food items, such as fajitas, often contributes to linguistic borrowing.

c. The Japanese word karaoke has been borrowed into English as well as Mandarin Chinese. When pronounced, the Chinese symbols sound like *karaoke*.

International political and economic institutions such as the United Nations (UN) and the European Union (EU) can also influence linguistic dominance. Languages gain status from being selected as official languages for organizations. The UN, for example, recognizes six official languages—English, French, Spanish, Russian, Arabic, and Chinese. To avoid favoring one language over another, the EU recognizes 23 official languages, and all EU documents must be produced in each of these languages.

Language Dynamics

Languages change over time and from one place to another. As noted at the start of this chapter, snowboarders have developed a specialized vocabulary for their sport. New technologies and innovations frequently stimulate vocabulary change as words are invented to express new concepts or to name new things (e.g., *blog* or *spam*), but they can also enable new ways of using language, such as texting.

Often, vocabulary change occurs when one linguistic community borrows words from another language. These borrowed words are called **loanwords** (**Figure 4.10**).

loanword A word that originates in one language and is incorporated into the vocabulary of another language.

Pidgin and Creole Languages

One of the most significant forces affecting language change is human mobility or migration. Mobility can fragment linguistic communities, paving the way for new language usages to develop. The emergence of American English, Australian English, and South African English as distinct from British English provides a good example of this facet of language change.

Just as geographic separation can create conditions favorable for linguistic change, contact between members of different speech communities can also result in language

Creole language in Papua New Guinea • Figure 4.11

The sign reads: "Work on road, all cars must stop when you see the red sign." The name of Papua New Guinea's creole language is Tok Pisin, literally *Talk Pidgin*, a reflection of the language's pidgin origins. Some words are Polynesian in origin, others derive from the languages indigenous to Papua New Guinea, and still others reflect German and English influences—a result of colonization. Tok Pisin is now used for some governmental functions and in journalism.

> **pidgin language**
> A language that combines vocabulary and/or grammatical practices from two or more languages that have come in contact.

change. When people who speak different languages come into contact and need to communicate, they might create a **pidgin language**. This process of creating a common language by people who do not share one is known as *pidginization*. Pidgin languages typically have specialized and limited functions because they develop in response to particular circumstances.

Pidgin languages demonstrate creative and adaptive linguistic mixing. They tend to be oral languages, though some can be written, and they are rarely the first language a person learns. Pidgin languages endure as long as the contact situations in which they emerged are sustained. For example, the pidgin language Tay Boi was used for communication between the French and the Vietnamese from the 1860s to the 1950s, when Vietnam was a French colony.

What is Spanglish? There is little agreement among scholars about this. Some linguists consider Spanglish to be a pidgin language that has grown out of the contact between Spanish-speakers and English-speakers in the United States, in regions of Mexico adjacent to the United States border, and in Puerto Rico where Spanish and English are recognized locally as official languages. Words such as *chatear* (to chat), *lonchear* (to lunch), *mapo* (map), and *cuora* (quarter) illustrate the hybridization common in Spanglish. Other linguists contend that Spanglish is a kind of *code switching,* or a linguistic technique in which a speaker alternates between languages during a single sentence or conversation; for example: La fiesta por mi abuelita es domingo, so I will arrive on Friday. (*The party for my grandmother is Sunday, so I will arrive on Friday*.)

We take the position that code switching is a fundamental dimension of Spanglish, and as a result, Spanglish is not as specialized and limited as most pidgins. In this respect it might be useful to think of Spanglish as intermediary between a pidgin language and a **creole language**. This is not to say, however, that pidgins always develop into creoles because they don't. Nevertheless, *creolization* describes a process of linguistic change in which the functions and use of a pidgin language expand. For example, Hawaiian Creole English, which formed during the early 20th century, is based on a pidgin language that was used by the ethnically and linguistically diverse population of Hawaii. This population included native Hawaiians, Americans, and immigrant Chinese, Japanese, and Portuguese, many of whom worked on sugar and pineapple plantations. Contact among such linguistically diverse groups gave rise to Hawaiian Pigdin English, which was eventually taught to children as a first language. This practice helped to expand the language and extend its use beyond the immigrant communities, leading to the development of Hawaiian Creole English. In

> **creole language**
> A language that develops from a pidgin language and is taught as a first language.

Papua New Guinea, one of the most linguistically diverse countries in the world, Tok Pisin has emerged as a creole language that is now learned by many children as their first language (**Figure 4.11**).

Lingua Francas

As we have discussed, contact among people who speak different languages can result in the emergence and use of a pidgin language. Another option is to identify a **lingua franca**. The Hausa language is a lingua franca used in northern Nigeria, Niger, and neighboring regions of West Africa. Swahili is a lingua franca spoken in the East African countries of Kenya, Tanzania, Uganda, and Burundi. Russian is still used as a lingua franca in Uzbekistan, Turkmenistan, and a few other former Soviet republics. However, it is not clear how long this will last since most of these countries have identified other languages as their official languages, and some have even discouraged the use of Russian through de-Russification policies.

lingua franca A language that is used to facilitate trade or business between people who speak different languages.

Today there is a growing consensus that English has become a global lingua franca. Consider the prevalent use of English in certain international contexts. For example, English is used for communication at sea or in the air and dominates in the areas of science, medicine, technology, and international business. Every day around the world hundreds of millions of people whose first language is not English use English to communicate.

Could English eventually become a universal language through ongoing globalization? According to most experts, probably not, for two important reasons: First, as we learned in Chapter 2, globalizing forces can generate different local outcomes. The global diffusion of English provides a good example of this effect (**Figure 4.12**).

A second reason we will likely not witness the development of English as a universal language involves the spaces or domains of language use and the nature of human identity. Although English prevails in certain spaces, for example, as a language of commerce and medicine, the number of people who learn English as a second or third language exceeds the number of people who learn it as their first language. This suggests that there are spaces—households and local communities, for example—that are more resistant to the diffusion of English. Moreover, some people neither want to nor desire to speak English, preferring instead to use their first language. Nevertheless, globalization and the growth of English and other languages as very large languages are altering the geography of linguistic diversity, a topic we discuss in the next section.

English in global and local context • Figure 4.12

No other language has the international standing or the global reach of English. Although this diagram shows the influence of American and British English on the global spread of the language, there is not a single version of English. Rather, the varieties of English that exist are highly localized. So, for example, Nigerian English differs from Pakistani English.

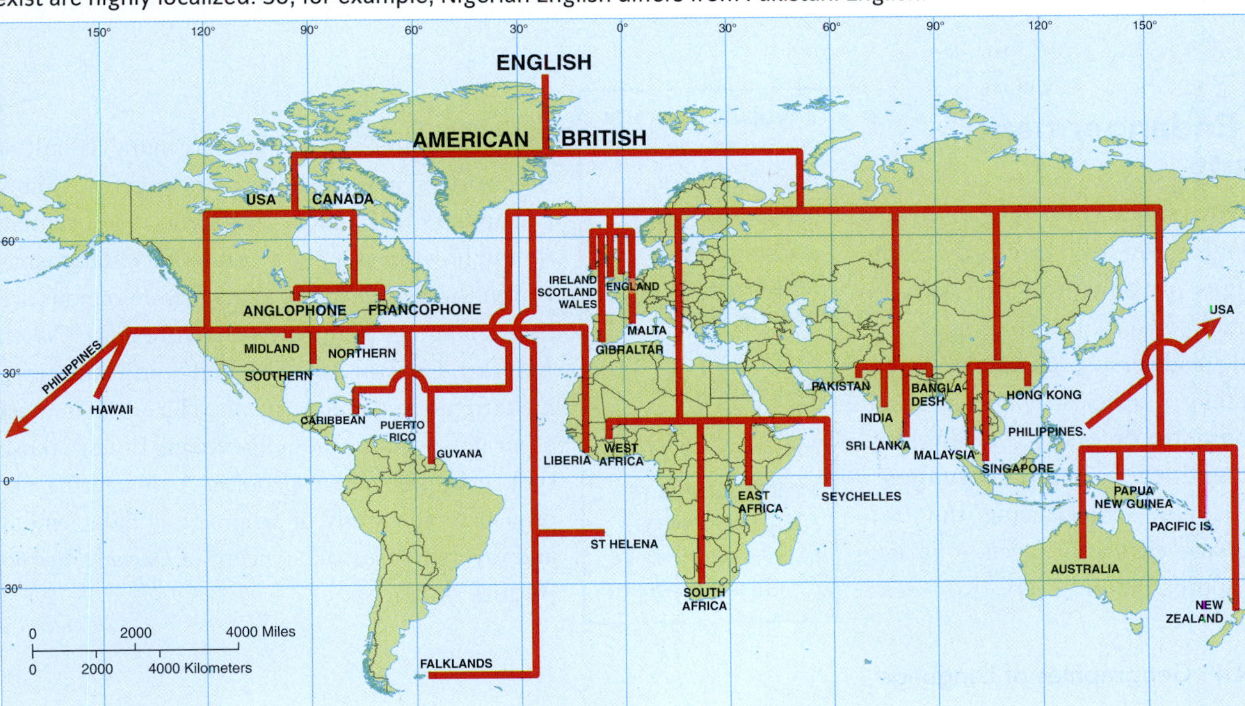

Geography InSight

Language hotspots • **Figure 4.13**

a. The concept of language hotspots was developed by Greg Anderson and David Harrison of the Living Tongues Institute for Endangered Languages. A language hotspot exists when three factors converge: high language endangerment, high linguistic diversity, and languages that are poorly documented. To calculate linguistic diversity, they divide the number of language families in an area by the total number of languages.

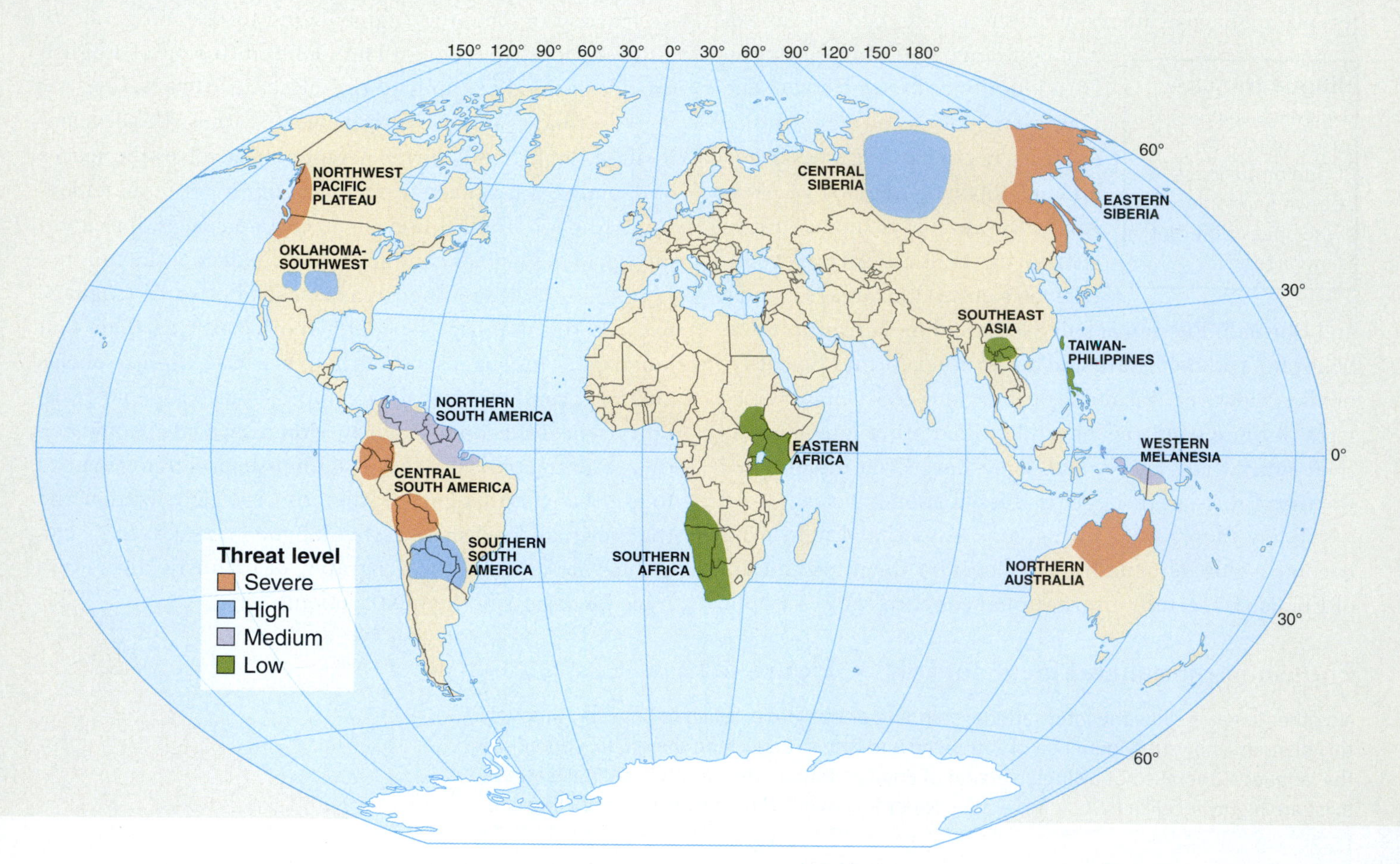

Language Endangerment and Diversity

During the 1990s, researchers began to use biological analogies when characterizing the state of languages in the world. Just as wildlife ecologists and conservation biologists use the concepts of biodiversity, endangered species, and extinction of species, language scholars speak of **linguistic diversity**, **endangered languages**, and **extinct** (or dead) **languages**.

Our world is now experiencing the fastest rate of language extinction ever—one language dies out approximately every two weeks.

linguistic diversity The assortment of languages in a particular area.

endangered language A language that is no longer taught to children by their parents and is not used for everyday conversation.

extinct language A language that has no living speakers; also called a dead language.

Some estimates suggest that as many as half of the world's languages are endangered. Many researchers fear that if this trend is not halted, we might witness a mass extinction of languages within the next 50 or 60 years. Three regions are losing languages quickly—the Americas (North America, Central America, and South America), Eastern Siberia, and Australia. Here, and in most other hotspots, the languages being lost are those spoken by the native peoples. A pioneering approach to the study of language endangerment and diversity uses the concept of *language hotspots* (**Figure 4.13**).

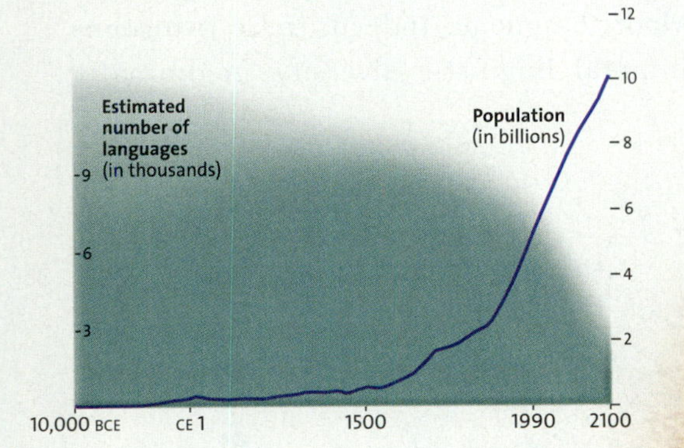

b. Globally, the historical trend is for linguistic diversity to decline as global population rises and for large languages to spread, usually at the expense of smaller languages.

c. Researchers in northern Australia consult with Charlie Mangulda, the last known speaker of Amurdag, an Aboriginal Australian language. In general, the smaller the speech community, the more urgent is the task of ensuring the survival of its language.

The extinction of languages is one factor that influences the distribution and mixture of languages in the world. Mapping language hotspots thus is an important strategy for identifying those areas that possess languages that we know little about and where those languages are at risk of becoming extinct—that is, areas where research is needed.

If geographers and other experts want to know how linguistically diverse a country or region is based on the size of its population and the number of different languages spoken within its borders, they compute a **linguistic diversity index (LDI)**. Values for the LDI can range from 0.00 to 0.99. Countries with LDIs at or close to zero have no or very little linguistic diversity; thus, two people selected at random will probably speak the same first language. Countries with LDIs close to 1.0 possess considerable linguistic diversity, so that two randomly selected people will probably speak different first languages.

linguistic diversity index (LDI) A measure that expresses the likelihood that two randomly selected individuals in a country speak different first languages.

Keep in mind that the LDI does not simply reflect the number of languages spoken in a country. If it did, all countries with a large number of languages would also have high LDIs. Consider the examples of Brazil and Mexico. Brazil has nearly 200 languages, and Mexico has nearly 300 languages, but both have very low LDIs because most of their citizens speak the predominant language—Portuguese in Brazil and Spanish in Mexico (**Figure 4.14**).

You might be wondering what happened to the hotspot in northern Australia (refer again to Figure 4.13). It does not appear on the map in Figure 4.14b because of differences in the way linguistic diversity is calculated. Remember that the LDI conveys the likelihood that two randomly selected people will speak different first languages. Aboriginal Australians constitute just 3% of the country's population, so, for the country as a whole, the LDI is quite low. A similar situation exists in New Zealand with the Maori language. Indeed, these two countries possess minimal linguistic diversity, in dramatic

Geography of the linguistic diversity index • Figure 4.14

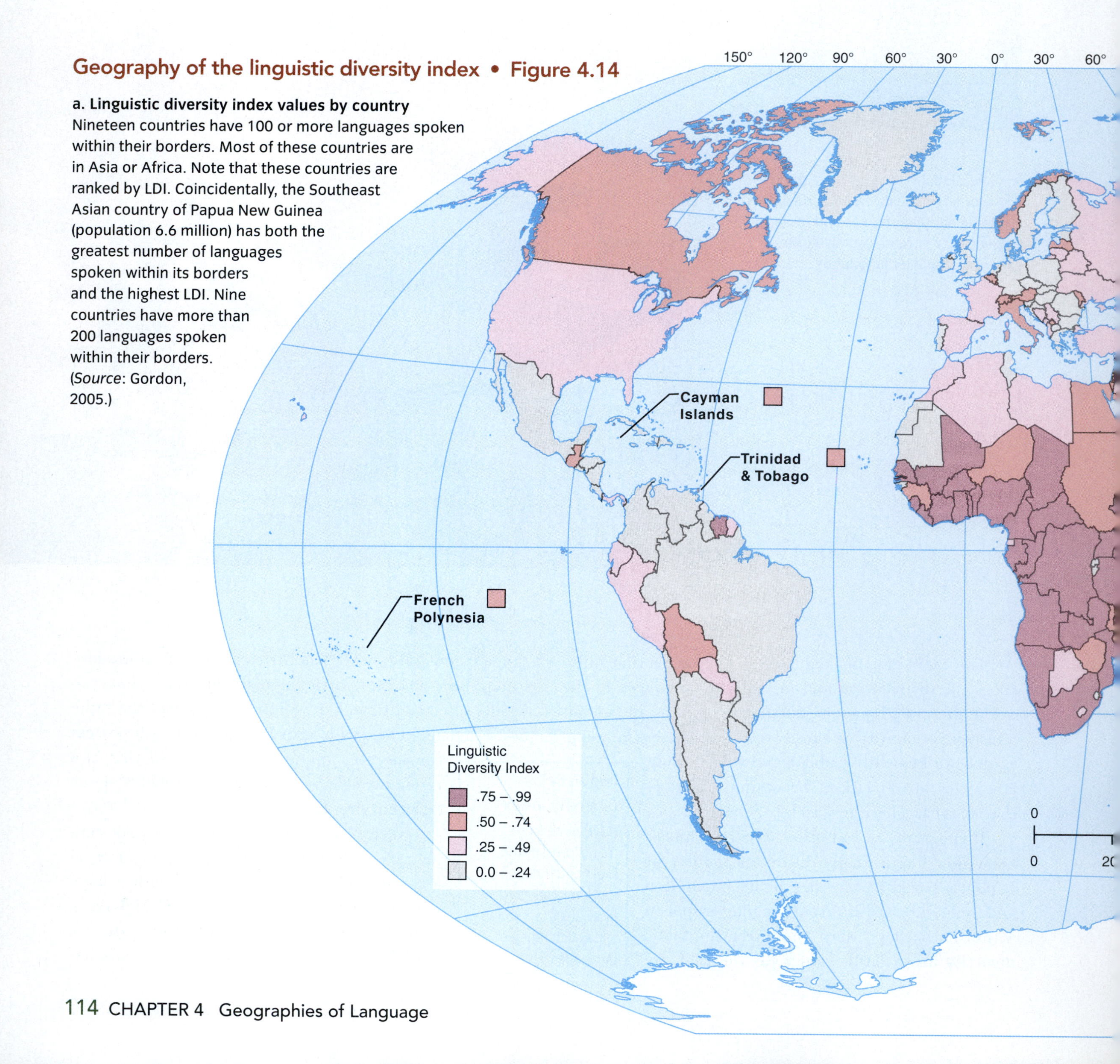

a. Linguistic diversity index values by country
Nineteen countries have 100 or more languages spoken within their borders. Most of these countries are in Asia or Africa. Note that these countries are ranked by LDI. Coincidentally, the Southeast Asian country of Papua New Guinea (population 6.6 million) has both the greatest number of languages spoken within its borders and the highest LDI. Nine countries have more than 200 languages spoken within their borders. (*Source*: Gordon, 2005.)

contrast to the general pattern across Southeast Asia and the Pacific islands.

Other regions of low linguistic diversity include much of Mesoamerica (Mexico and Central America), South America, and Europe. The presence of nation-states—countries where the boundaries of a national group match the boundaries of the country—in Europe helps to explain the generally low LDIs in that region. European colonization is an important factor in the low LDIs in Mesoamerica and South America, as well as in Australia and New Zealand. In precolonial times, these regions possessed a diverse mixture of indigenous languages, but many of these languages have been in decline since the colonial period, and several of them

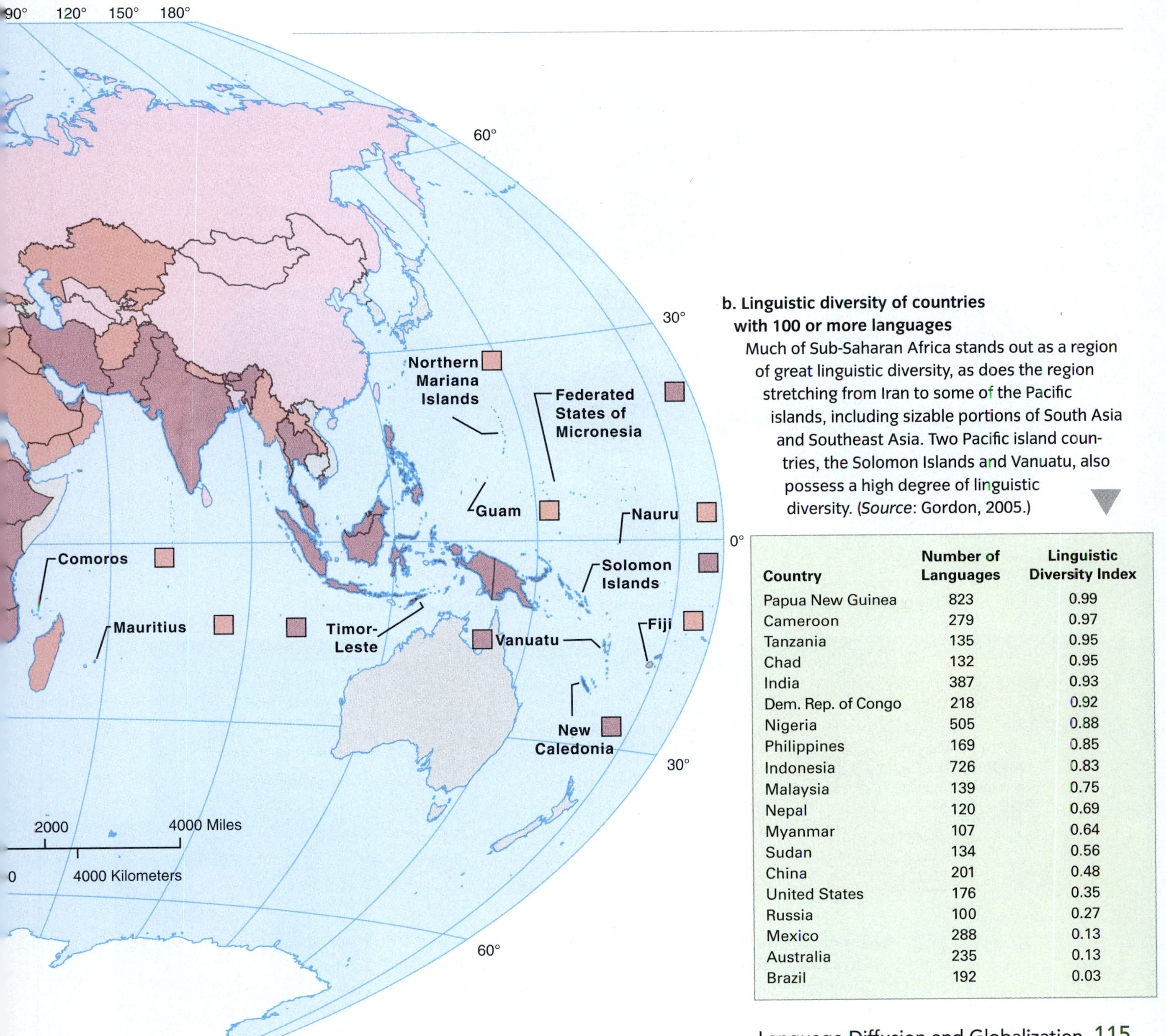

b. Linguistic diversity of countries with 100 or more languages
Much of Sub-Saharan Africa stands out as a region of great linguistic diversity, as does the region stretching from Iran to some of the Pacific islands, including sizable portions of South Asia and Southeast Asia. Two Pacific island countries, the Solomon Islands and Vanuatu, also possess a high degree of linguistic diversity. (*Source*: Gordon, 2005.)

Country	Number of Languages	Linguistic Diversity Index
Papua New Guinea	823	0.99
Cameroon	279	0.97
Tanzania	135	0.95
Chad	132	0.95
India	387	0.93
Dem. Rep. of Congo	218	0.92
Nigeria	505	0.88
Philippines	169	0.85
Indonesia	726	0.83
Malaysia	139	0.75
Nepal	120	0.69
Myanmar	107	0.64
Sudan	134	0.56
China	201	0.48
United States	176	0.35
Russia	100	0.27
Mexico	288	0.13
Australia	235	0.13
Brazil	192	0.03

PROCESS DIAGRAM

THE PLANNER

Understanding language vitality and endangerment: The example of Yuchi • Figure 4.15

Yuchi is a Native American language spoken in Oklahoma. Thousands once spoke it in all domains or spaces of social interaction. Since the 1800s, however, Yuchi has declined as language shift occurred, and today it has been almost entirely replaced by English. Only a handful of Yuchi speakers remain.

Language shift occurs when speakers of a language change their speech behavior—for example, by acquiring another language and altering the geography, or spaces, where their original language was used. Yuchi speakers not only learned English, but the spaces where English is used expanded as the spaces of Yuchi use contracted. The social pressures that cause language shift are complex.

1 Pressures at different scales affect language vitality and endangerment

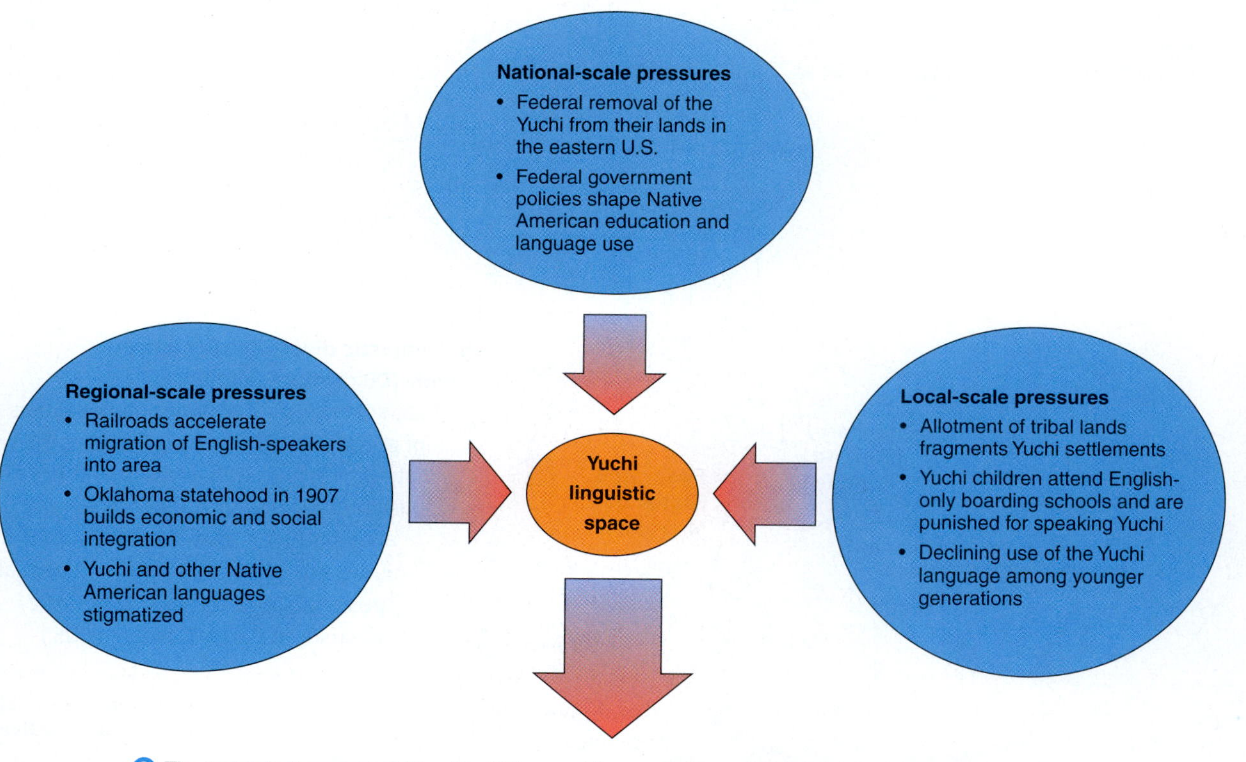

National-scale pressures
• Federal removal of the Yuchi from their lands in the eastern U.S.
• Federal government policies shape Native American education and language use
Regional-scale pressures
• Railroads accelerate migration of English-speakers into area
• Oklahoma statehood in 1907 builds economic and social integration
• Yuchi and other Native American languages stigmatized
Yuchi linguistic space
Local-scale pressures
• Allotment of tribal lands fragments Yuchi settlements
• Yuchi children attend English-only boarding schools and are punished for speaking Yuchi
• Declining use of the Yuchi language among younger generations

2 The result is language shift as Yuchi (orange) loses vitality and English (blue) grows dominant.

Pre-1830s—Yuchi language vitality
Yuchi is the first language acquired and is used for all purposes—home, commerce, government, education, religion.

1830s–1940s—Yuchi language gradually becomes subordinate to English
National and regional forces confine Yuchi's linguistic space while the spaces of English use expand.

Since the 1950s—Yuchi language endangerment
English dominates most linguistic spaces and Yuchi is no longer taught to children.

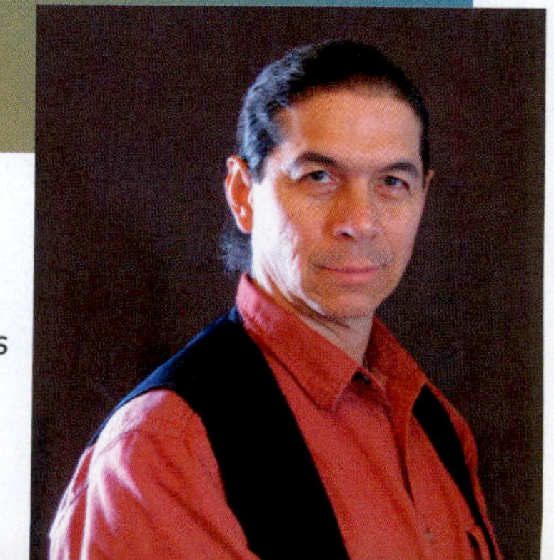

3 The Yuchi attempt to revitalize their language.
Richard A. Grounds, a Yuchi/Seminole and Director of the Yuchi Language Project, works with today's few elderly Yuchi speakers to preserve their language, customs, and traditions.

have become extinct. **Figure 4.15** illustrates some of the social forces that can contribute to language endangerment, using the example of the Yuchi language.

For a glimpse of some of the work involved in preserving a language, see *Video Explorations*.

CONCEPT CHECK

STOP

1. **What** forces affect linguistic diffusion, and why?
2. **Why** can't we simply use the size of a language to assess linguistic dominance?
3. **What** is the difference between pidginization and creolization?
4. **What** information does the linguistic diversity index provide, and why is that useful to geographers?

Video Explorations

Enduring Voices Expeditions

Members of the Enduring Voices Project travel to locations in northern Australia and northeastern India to record endangered languages. In addition to those shown in the video, what are some other strategies for language preservation?

Dialects and Toponyms

LEARNING OBJECTIVES

1. **Define** the term *isogloss*.
2. **Identify** three major dialect regions that exist in the United States.
3. **Distinguish** between prestige and standard dialects.
4. **Explain** what toponyms are and what information they can provide.

Beyond its function as a communication system, language also is a marker of cultural and personal identity. We identify different cultural groups around the world on the basis of their language. And on an individual level, our language—the way we speak, the words we use, our dialect—defines who we are and who we are not. Dialects, the various forms of a single language, provide important clues about the construction of linguistic boundaries and the relationships between language and identity. In the following section we explore some facets of the dialect geography of the United States.

Dialect Regions

We know that dialects can differ in their grammar, their vocabulary, and their pronunciation. When someone speaks a dialect of our language that has a different pronunciation from our own dialect, we perceive that person as having an **accent**. Most people think that they do not speak with an accent, but the truth is that we all have an accent—we simply don't hear our own accent as an accent. In addition, in the course of our lives, each of us develops peculiarities in pronunciation—our personal accent.

Although most linguistic geographers find individual differences in pronunciation fascinating, they usually concentrate their study on the spatial patterns of dialect usage. This field of study is called **dialect geography**. Hans Kurath helped pioneer the study of dialect geography in the United States beginning in the 1930s. He was specifically interested in the patterns of word usage along the East Coast. By mapping word usage, he was able to identify **isoglosses** (**Figure 4.16**).

isogloss A line that marks a boundary of word usage.

Differences in pronunciation help distinguish among the dialect regions in the United States. One striking difference involves the pronunciation of "r" sounds. With the Northern accent, if the "r" follows a vowel, the "r" sound is lost. For example, the word *car* sounds like "cah," while the word *skirt* sounds like "skuht." The Southern accent similarly drops the "r" after vowels and between them as well (*more* sounds like "mo"). Pronunciation of "i" as if it sounded like "ah" also characterizes the Southern accent. The word *fire* sounds like "fahr" and *I'm fine* sounds like "Ahm fahn." The Southern practice of pronouncing one-syllable words as if they had two syllables, as in "thiyus" (*this*), and "wayell" (*well*), creates the Southern drawl.

Unlike either the Northern or the Southern accent, the Midland accent keeps "r" sounds after vowels. The Midland accent forms the basis of *network standard*—a way of speaking commonly utilized by TV and radio announcers with national audiences that emphasizes accent reduction through the avoidance of regionally distinctive pronunciation practices. Jay Leno has purposely acquired the network standard dialect over the course of his career. However, if you listen closely to Leno's pronunciation, you might be able tell that he is originally from the Northern dialect region by the way he drops some of his "r" sounds. (Leno was born in Boston.) To explore these and other dialects, see *Where Geographers Click*.

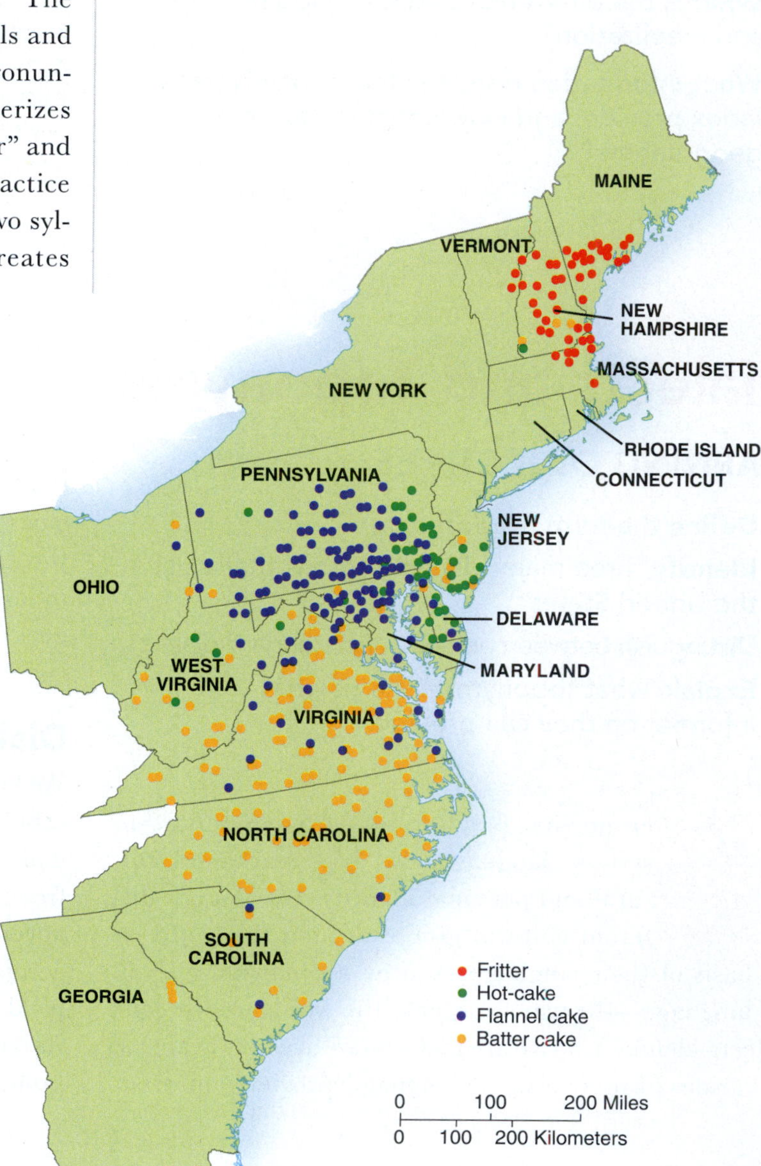

Kurath's word geography • Figure 4.16

a. This map, based on research conducted during the 1930s and '40s, shows the regions where other terms for pancake prevailed. Where would you draw the isoglosses for *hot-cake*, *batter cake*, and *flannel cake*? Because no two words have the exact same distribution, the isoglosses of two words never exactly coincide.

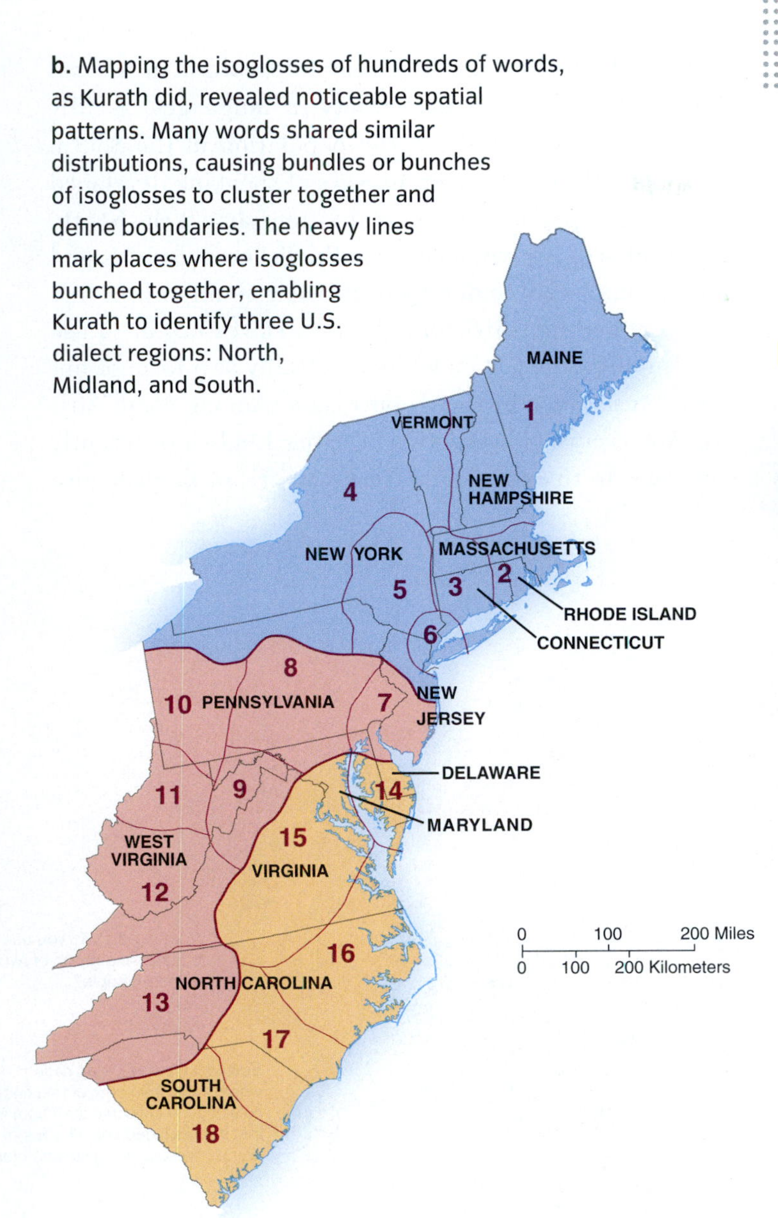

b. Mapping the isoglosses of hundreds of words, as Kurath did, revealed noticeable spatial patterns. Many words shared similar distributions, causing bundles or bunches of isoglosses to cluster together and define boundaries. The heavy lines mark places where isoglosses bunched together, enabling Kurath to identify three U.S. dialect regions: North, Midland, and South.

Where Geographers CLICK

Dictionary of American Regional English

http://www.dare.wisc.edu/?q=node/1

DICTIONARY OF
AMERICAN REGIONAL ENGLISH

- Home
- About DARE
- 100 Entries
- DARE News & Newsletters
- Fred Cassidy Tribute
- Index
- Participate
- Purchase
- Educational Resources

Navigation

- Search

DARE

Like other dictionaries, the *Dictionary of American Regional English (DARE)* is arranged alphabetically by headword, from A to Z. What is different about *DARE* is that it shows *where* people use the words that are included. We all know, for example, that Americans have many names for the kind of sandwich that includes meats, cheeses, lettuce, tomatoes, etc., served in a long bun. What *DARE* can tell you (and can often illustrate through the use of maps based on fieldwork) is where the words *hero, hoagie, grinder, sub, torpedo, Cuban*, etc. are the local terms for this sandwich.

Donate to the dictionary

Word of the Month

hopping John

It can also tell you where people use the words *darning needle, ear cutter, eye stitcher, mosquito* (or *skeeter*) *hawk, sewing needle, snake doctor*, or *snake feeder* (among other terms) for a dragonfly.

And what about the words people use for the strip of grass between the sidewalk and the street? *Boulevard, devil strip, grass plot, neutral ground, parking, parking strip, parkway, terrace, tree bank, tree*

WHAT IS *DARE?*

The *Dictionary of American Regional English (DARE)* is a multi-volume reference work that documents words,

DARE, short for the *Dictionary of American Regional English*, is a leading reference for people who want to know about American dialects. There is not yet an online version of this work, but the DARE website has word quizzes and podcasts so you can test your knowledge of word usage and listen to different accents. Go to the website and click on **Educational Resources**. In addition, go to http://dare.wisc.edu/?q=node/17 for an explanation of DARE's unique population-adjusted maps showing the distributions of various words and phrases offered by informants during the 1965–1970 fieldwork for the *Dictionary*.

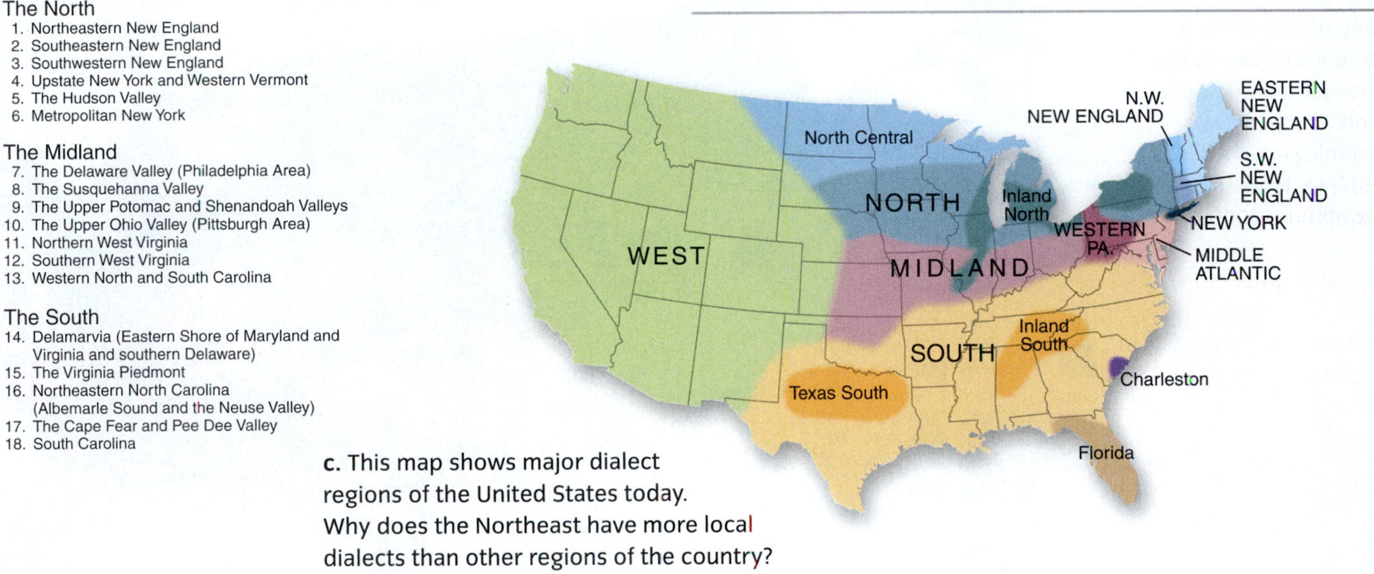

c. This map shows major dialect regions of the United States today. Why does the Northeast have more local dialects than other regions of the country?

Along with pronunciation, regional patterns of word usage also help to distinguish between these dialect regions. Perhaps one of the most distinctive Southern usages is saying "y'all," when referring to a group of people (**Figure 4.17**).

African American English

The Northern, Midland, and Southern dialect regions reflect the European origins of the settlers along the East Coast. Kurath's work, however, did not give much attention to the speech patterns of African Americans. Yet, from the 1600s into the 1800s the importation of African slaves, who spoke a wide variety of languages, greatly altered the composition of the population in the South. Consequently, many different ways of speaking developed among slave populations on plantations across the South. The origins of African American English are complex and only incompletely understood. In their communities and with one another, African American slaves began to use English differently from whites, partially as a form of linguistic resistance to their oppression. Among some African Americans, the practice of using English differently continues to this day, and African American English, also

Regional patterns of word usage • Figure 4.17

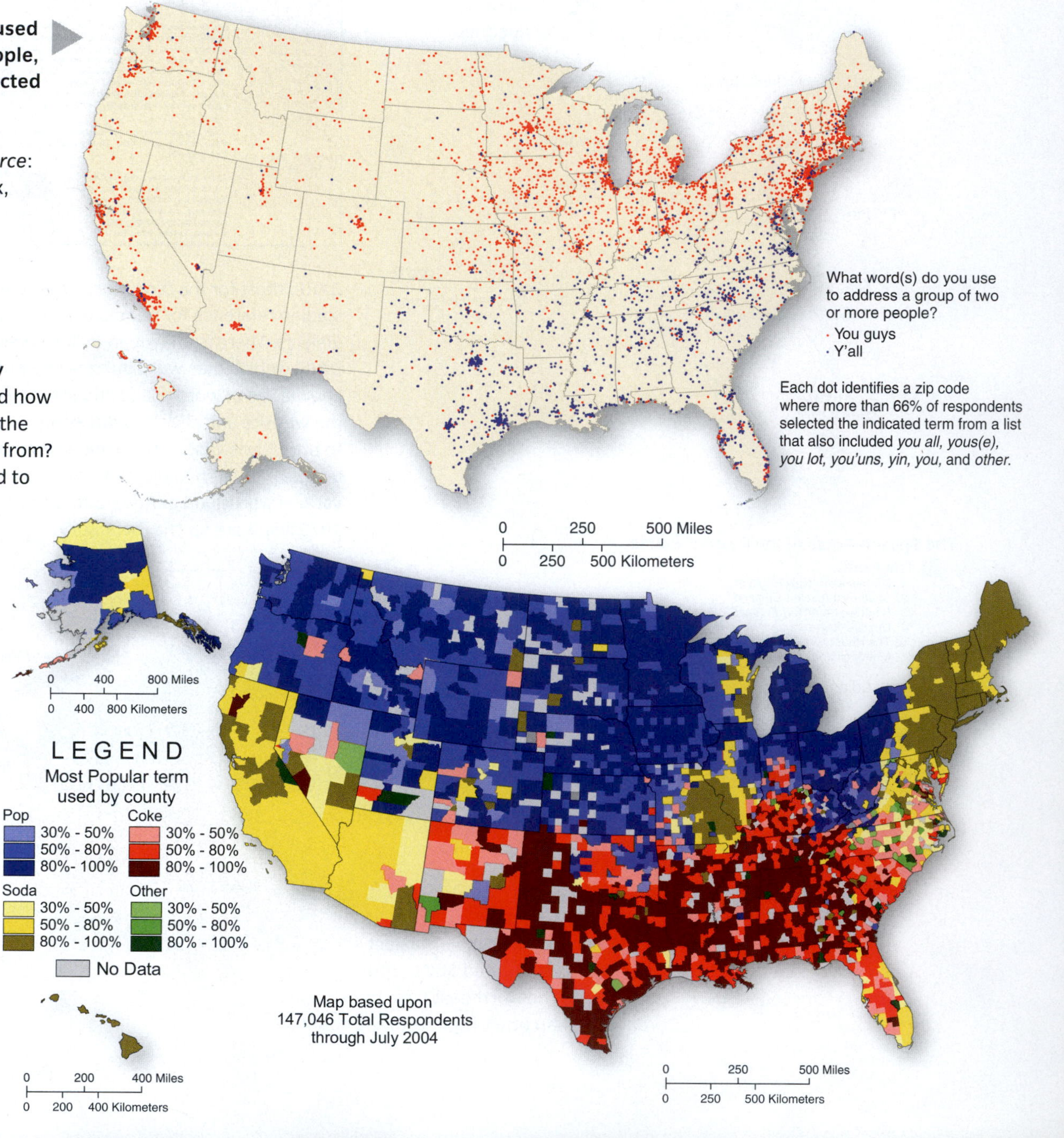

a. Distribution of words used to refer to a group of people, based on a survey conducted by Bert Vaux
Where would you draw an isogloss on this map? (*Source*: Data courtesy of Bert Vaux, University of Cambridge.)

b. Generic names of soft drinks, by county
What word do you use, and how well does it correlate with the dialect region that you are from? Once, Southerners referred to soft drinks as *cold drinks*. Today, many Southerners use *Coke* to refer to any kind of soft drink, a practice influenced by the fact that the Coca-Cola Company was established in Atlanta. (*Source*: *Web Atlas of Oklahoma*, 2005.)

called African American Vernacular English or Black English Vernacular, constitutes another dialect of American English. Because of its emergence in the South, it shares some features with the Southern dialect, such as not pronouncing the "g" in words ending with *-ing* and dropping the "r" when it occurs between vowels.

Other pronunciation practices associated with African American English include loss of the "l" sound after vowels (*help* sounds like "hep"). When words end in two or more consonant sounds, the final consonant is often simplified (*meant* sounds like "men"). African American English also possesses some distinctive grammatical practices. One of these involves the absence of an "es" or "s" on third-person present tense verbs. Thus, instead of using "he goes"/"she goes"/"it goes," speakers of African American English say "he go"/"she go"/ "it go." Note that this grammatical practice makes English more regular. Another grammatical practice associated with African American English involves the use of the verb *to be*. *Be* is used to convey that something happens regularly: "She be workin' every day." To indicate that something is taking place at this moment, *be* is not used: "She workin' right now."

African American English possesses a distinctive vocabulary. Some scholars interpret this as a linguistic strategy that slaves used to help them communicate with one another but not be understood by their owners. In addition, many African American expressions have become very widely adopted and used. Some of these words include *jazz*, *funky*, *chill out*, *high-five*, *phat*, and *bling-bling*. Because of the international popularity of hip-hop, the vocabulary of African American English actively shapes patterns of language use not only in the United States but also around the world (**Figure 4.18**).

Although African American English emerged in the South, it has diffused widely across the country. In a large-scale movement of population called the *Great Migration*, beginning about 1916 and continuing until 1970, more than 6 million African Americans moved from the South to such Northern and Midwestern cities as New York, Philadelphia, Chicago, and Detroit. Social pressures, including residential segregation, forced many of these migrants into the inner cities, where they formed African American neighborhoods. One result of this movement is that the geography of African American English today is largely Southern, but beyond the South it is also highly urban. Of course, not all African Americans use African American English, and today some whites, Hispanics, and others have adopted aspects of African American English.

Hip-hop artist Jay-Z • Figure 4.18

Shawn Carter is more popularly known by his stage name, Jay-Z, said to be derived from the word *jazzy*. The lyrics of many African American hip-hop artists illustrate highly creative linguistic practices.

Chicano English

Kurath's work made important contributions to the study of dialect patterns in the United States in the 1930s. Since that time, however, immigration has altered the linguistic geography of the country. *Latino* (or *Latina*) usually refers to someone who is from Latin America and has a Spanish- or Portuguese-speaking background. Although the term *Latino English* is used to identify the variety of English spoken by these people, the term is slightly misleading because there is no set of linguistic practices that members of this diverse group share.

Mexicans constitute the largest Latino population in the United States. Chicano English (the word *Chicano* derives from *Mexicano*) refers to the dialect that has emerged within that population. This dialect is frequently associated both with Mexican Americans and with Mexicans who learn English as a second language. A distinctive pronunciation in Chicano English is using an "s" sound instead of a "z" sound, so *crazy* and *his* sound like "craysee" and "hiss." Grammatically, Chicano English tends to omit the word *have* in certain instances. Instead of "I have been on campus all day" you might hear "I been on campus all day."

WHAT A GEOGRAPHER SEES

Toponyms, or What Is in a Name?

Are you aware of the political and symbolic meanings in toponyms? Geographers consider the naming of a place a fundamental way of taking possession of it and promoting unity among citizens. But renaming a place can also serve as a strategy for dispossessing people of land they have historically claimed, or for weakening a people's attachment to a place. To a geographer, toponyms are often more than mere names.

Not all toponyms reflect the past in this way, and some are interesting just for their oddity. Perhaps you have heard of Hot Coffee, Mississippi, or Truth or Consequences, New Mexico.

a. This monument commemorates the defense of Leningrad ("Lenin's City") against the Nazis in World War II. When the Soviet Union dissolved in 1991, the citizens voted to restore the pre-Soviet name, St. Petersburg. This deeply symbolic act removed the name of Lenin, founder of the Soviet Union, and re-honored Peter the Great, who helped modernize Russia.

Because of the long period of contact in the borderlands between the United States and Mexico, many Spanish vocabulary words have diffused into English. You know a number of these—*enchiladas, frijoles, tortillas*—largely because of the popularity of Mexican-style food. But words such as *rodeo* and *mesa* also reflect other Spanish contributions to the English vocabulary. As with African American English, not all Mexican Americans use Chicano English.

Standard Dialects

When different dialects of a language are spoken in an area, one may be designated or become accepted as the **standard dialect**, the norm or authoritative model of language usage. You will, for example, sometimes hear references to Standard British English or Standard Russian. The selection of a standard dialect may reflect the dominance of a given way of speaking. It is more likely, however, to reflect a way of speaking that is associated with high socioeconomic or educational status or political power. Thus, the standard dialect may be perceived by some as a *prestige dialect*. A phrase, such as "the Queen's English," not only refers to the dialect that became the standard in the United Kingdom but also associates that dialect with status, power, and authority. In the United States, network standard (see page 118) comes the closest to being a standard dialect. Use of a standard dialect brings a consistency to the way a language is written and spoken. Standard dialects are preferred in business, government, education, and mass media, such as television and radio.

Beginning in the 1950s, government officials in China selected Northern Mandarin as the country's official language and standard dialect and have been promoting its use ever since, despite the fact that millions of Chinese use other dialects, such as Cantonese and Shanghainese. The phrase "Mandarin policy" refers to the government's active efforts to spread Northern Mandarin for the sake of national unity.

b. European colonization often erased local toponyms, supplanting them with the names of European kings, queens, heroes, or other familiarities. In colonial India, numerous toponyms were changed mainly by the British. Since the 1990s, however, some of India's colonial toponyms have themselves been expunged. Colonial place-names such as Bangalore, Bombay, Calcutta, among others, became Bengaluru, Mumbai, and Kolkata, respectively, as shown on this current map.

c. Like place-names, street names also have practical, symbolic, and commemorative value. They not only help us locate places, they also convey political values. Consider, for example, Independence Avenue in Washington, D.C., or the way state names have been incorporated as street names there. Street names also serve as memorials to the past. The first street to commemorate Martin Luther King, Jr., was established in Chicago in 1968. There are now more than 700 MLK streets in cities and towns across the country, most of which are in the South.

Think Critically

1. How might residents in a place express resistance to toponymic change imposed by a government or other group?
2. How might toponyms reflect spatial thinking?

Sometimes people mistakenly claim that using a standard dialect is the only correct or proper way to use a language. This mistake arises from not knowing how a standard dialect gets selected, and it contributes to a negative stereotyping of nonstandard dialects. For example, the Southern dialect of American English has long been negatively stereotyped. But imagine if the Southern dialect had been selected as the standard dialect—our ideas of "proper" English would be entirely different. Chicano English and African American English have been even more negatively stereotyped than the Southern dialect. Such prejudice is one of the ways in which language can be used to create and reinforce social barriers between people. When people label certain ways of speaking as "right" or "wrong," "proper" or "improper," they are making highly subjective judgments based on ignorance about how language works. Dialects represent different ways of using a language, not ways that are right or wrong.

Toponyms

As the previous discussion suggests, language use is closely associated with identity as well as social and political power. We can see this in the selection of **toponyms**, because the names given to places can make powerful statements about a group's sense of belonging in, attachment to, or control of a place (see *What a Geographer Sees*).

toponym A place-name.

Geographers study toponyms not only for the insights they provide about territorial possession and political power, but also because they can provide clues about settlement history. A glance at a map of Quebec, for example, reveals Native American, French, and English-derived place-names—a reflection of its peopling. Many English-language toponyms consist of identifiable generic and specific elements. The generic element identifies the feature (mountain, city, and so on), and the specific element provides more descriptive information. In the toponym Black Mesa, *Black* is the generic element and *Mesa* is the specific element.

What do cultural groups choose to name? What can toponyms tell us about past land-use practices in a place or environmental change? Geographer J. L. Delahunty's study of yew tree toponyms in Ireland identified 70 unique place-names that contain a reference to yew trees. Examples include Youghal (yew wood) from the Irish *Eochaill* and Ture (yew) from the Irish *Iubhar*. Delahunty's research suggests a past cultural preference for yew trees, a more abundant distribution of yew trees in the past, or both.

Through the study of spatial patterns of language use, geographers make languages visible. Maps serve as a valuable tool in this respect. Maps that show isoglosses, languages used, or the distribution of language families draw out the visible dimensions of linguistic geography. Signage, too, is important. From our gestures and dialects to our toponyms, we etch our language and our identities into our surroundings, demonstrating the interconnectedness of people, place, and environment.

CONCEPT CHECK

STOP

1. **What** is an isogloss, and how are isogloss bundles used?
2. **How** do accents and vocabulary differ in the three main dialect regions of the Eastern United States?
3. **Why** might a country promote a standard dialect?
4. **Why** are toponyms important to linguistic geography?

Summary

1 Languages in the World 98

- **Language** gives people a voice and shapes their identity. Language is a defining element of culture. Geographers study the spatial patterns of languages and **dialects**, beginning with an understanding of **mutual intelligibility**, the kinds of languages that exist, their sizes, and relationships.
- **Language families**, such as the Indo-European language family depicted here, reflect genetic relationships among languages. There are 6,900 languages spoken in the world today. When these languages are classified into families, six stand out because of the large percentage of languages and speakers that they contain.

The Indo-European language family tree • Figure 4.7

- More than half of the languages that exist today are classified as "small" or "very small" languages, but they are spoken by only a few hundred thousand people. Conversely, just nine of the world's languages are classified as "very large languages"—yet they are spoken by almost half of the world's people.

2 Language Diffusion and Globalization 105

- Technology, especially changes in transportation, and human migration influence language diffusion. In addition, the spread of language may depend on political, economic, and even religious forces.
- Contact among speakers of different languages can lead to the invention or borrowing of words, shown here, or the emergence of **pidgin languages**, **creole languages**, or **lingua francas**.

Linguistic borrowing • Figure 4.10

- The size of a language and the situations in which it is spoken have consequences not only for **linguistic dominance**, but also for language survival, **endangerment**, and **language extinction**. The death or extinction of languages reduces linguistic diversity. The **linguistic diversity index** is used to express the likelihood that two randomly selected individuals in a country will speak different first languages.

3 Dialects and Toponyms 117

- **Dialect geography** has long been a part of the spatial study of languages and frequently relies on **isoglosses** to help us understand how vocabulary usage varies from place to place.
- African American English and Chicano English are two highly dynamic and contested dialects of American English. The identification of a **standard dialect** gives an element of authority and legitimacy to one way of speaking, even though no dialect is inherently better or worse than another.
- Whether recorded on maps or signs, **toponyms** (see photo) provide telling clues about our priorities, preferences, and cultural practices.

What a Geographer Sees

- Equally fascinating are the events or circumstances that prompt people to change or erase toponyms.

Key Terms

- accent 117
- creole language 110
- dialect 98
- dialect geography 118
- endangered language 112
- extinct language 112
- hearth 103
- isogloss 118
- language 98
- language family 102
- lingua franca 111
- linguistic diversity 112
- linguistic diversity index (LDI) 113
- linguistic dominance 108
- loanword 109
- mutual intelligibility 98
- official language 108
- pidgin language 110
- standard dialect 122
- toponym 123

Critical and Creative Thinking Questions

1. Use the photo below to develop a proposal to create a tactile map of your campus or neighborhood for those who are visually impaired.

2. Can you think of American English words not mentioned in this chapter that have local or regional usages? Where are they used?
3. Identify some challenges associated with counting the number of speakers of a language.
4. It has been said that network standard is a geographically neutral dialect. Do you agree? Can a dialect be socially neutral? Explain your reasoning.
5. How might governmental and educational policies prevent linguistic dominance?
6. Some scholars argue that a language must possess a literary tradition to be counted as a fully separate language. Others claim that a language should have status as an official language. What are the advantages and disadvantages of these approaches to language identification?
7. What do the toponyms where you live reveal about settlement patterns, politics, or commemoration more generally?
8. Linguistically, is the world becoming more alike or different? Explain your answer.
9. This chapter has not discussed slang. What is slang, and why is it a controversial subject? What would you say is the difference between slang and linguistic creativity? In what linguistic spaces is slang used?

What is happening in this picture?

These students in Tejgadh, India, are making dictionaries of the languages that they first learned to speak. Until recently, these languages existed only as oral languages and were not written. Thousands of such languages exist around the world today.

Think Critically

1. What are some advantages of having a written language?
2. Are linguistic dominance and linguistic discrimination related?
3. Can linguistic dominance occur without discrimination?

Self-Test

(Check your answers in Appendix B.)

1. Find the false statement about language families.
   a. Within a single language family, mutual intelligibility will exist among the speakers of the various languages.
   b. Language families may have originated within farming populations.
   c. Language families can include extinct languages.
   d. Sino-Tibetan is considered a major language family.

2. What is surprising about the indigenous languages of Madagascar?
   a. They are "click" languages.
   b. They are unrelated to any other known languages.
   c. They probably diffused to Madagascar from Indonesia.
   d. None of the above is correct.

3. Accents _____.
   a. are situational, and dialects are not
   b. include distinctive vocabulary and pronunciation
   c. are mainly associated with the Southern dialect of American English
   d. can serve as markers of personal or social identity

4. An example of an artificial language intended to be universal is _____.
   a. Esperanto
   b. Elvish
   c. Latin
   d. body language

5. The geography of African American English _____.
   a. was once southern but is now mainly northern in its distribution
   b. was initially southern but has become increasingly urban
   c. is associated mainly with the East and West coasts
   d. has remained confined largely to the South

6. Fill in the table with the correct language family.

Language family	Example language(s)
	English, Hindi
	Mandarin Chinese, Burmese
	Arabic, Hebrew
	Yoruba, Zulu
	Tagalog, Bahasa Indonesia
	Tetum

7. List three characteristics of African American English.

____

____

____

8. Find the false statement about pidgin languages.
   a. They demonstrate adaptive linguistic mixing.
   b. They are the native language of people who live in multilingual communities.
   c. They are most likely to be oral languages rather than written languages.
   d. They are considered contact languages that have limited or specialized functions.

9. Creolization _____.
   a. relates to the establishment of a lingua franca
   b. describes the practice of creating and changing toponyms
   c. refers to the process by which a pidgin language develops into a first language
   d. explains the spread of the Indo-European language family around the world

10. What do the words *fajita*, *mesa*, and *rodeo* have in common?
   a. They are toponyms.
   b. They are isoglosses.
   c. They help identify speakers of Chicano English.
   d. They are loanwords.

11. This graph shows that _____.
   a. the number of languages declines as population increases
   b. the number of languages increases as population increases
   c. the number of languages and the population size are unrelated
   d. in AD 1 the number of languages and the population were both low

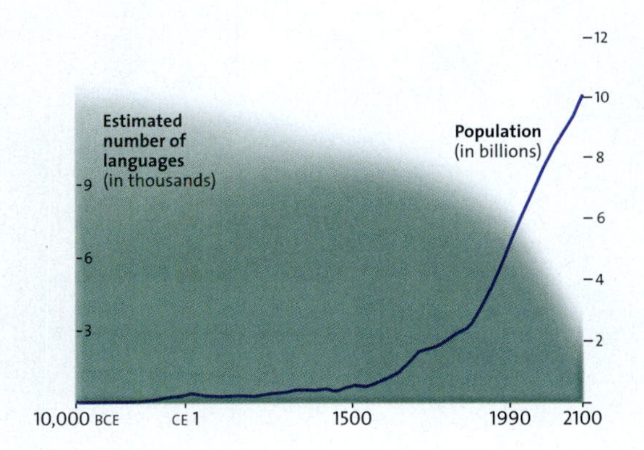

12. The linguistic diversity index expresses _____.
   a. the likelihood that two randomly selected individuals in a country will speak the same first language
   b. the ratio between the number of languages spoken in a country and the area of that country
   c. the ratio of the number of multilingual individuals in a country and the country's total population
   d. the likelihood that any randomly selected Web page will be in a language other than English

13. Why do Mexico and Brazil have such low linguistic diversity indexes (0.13 and 0.03, respectively)?

____

____

14. Find the false statement about language extinction.
   a. The world is experiencing the highest rate of language extinction ever.
   b. Of the world's regions, Europe is experiencing very high rates of language extinction.
   c. When speech communities are small, the potential for language extinction is usually very high.
   d. One language becomes extinct approximately every two weeks.

15. List at least one that has contributed to the contraction of Yuchi linguistic space at each of the three scales.

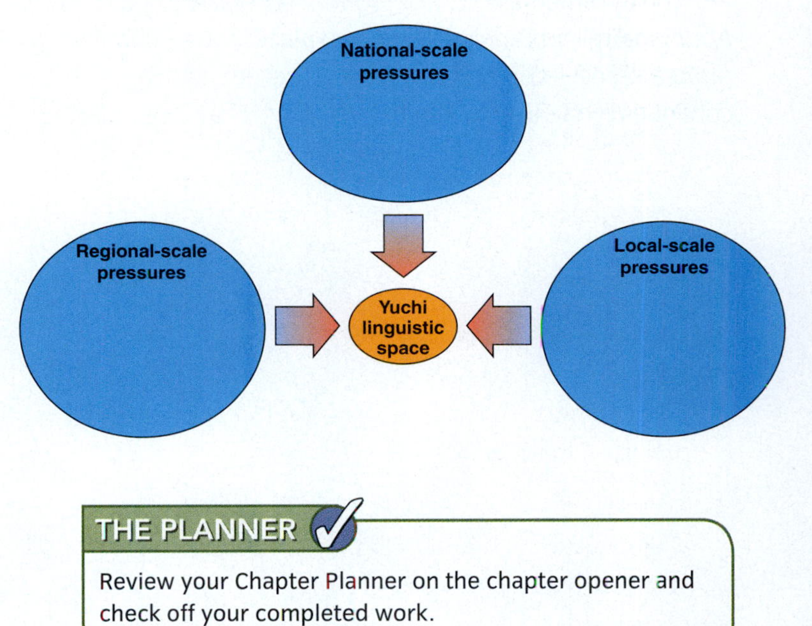

THE PLANNER

Review your Chapter Planner on the chapter opener and check off your completed work.

5 Geographies of Religion

THE ABORIGINAL DREAMTIME

The Aboriginal inhabitants of Australia share a belief in a mythic creation that they call the *Dreamtime* or *Dreaming*—the time when the Earth was transformed from a featureless and inactive realm into the world as we know it. Possessing powerful energies, Ancestral Spirits crossed the Australian continent. Along the way they created physical features such as mountains, plateaus, and river valleys. They also established boundaries between different Aboriginal groups.

By singing, the Ancestral Spirits gave names to the land, to its plants, animals, lakes, and deserts. The Australian continent, therefore, is inscribed with a kind of musical score consisting of the tracks or pathways taken by the Ancestral Spirits during the Dreamtime. Aboriginals call these routes *dreaming tracks*.

The Dreamtime is associated with two sacred realms in Aboriginal religious beliefs: time and place. The sacredness of Dreamtime—encompassing past, present, and future—continues today. In addition, the actions of the Ancestral Spirits during the Dreamtime made the world a sacred place. When their creative work was finished, they retired into the Earth, becoming part of the land, sea, and sky. But the Ancestral Spirits have not died. They are present everywhere—in estuaries, streams, rocks, and trees. The spiritual energies of the ancestral beings can be harnessed through the performance of certain religious rituals. Aboriginals celebrate and commemorate events in the Dreamtime through song, dance, body painting, and other forms of art. Dance factors prominently in the rituals of Australian Aboriginals. Here, in the accompanying photo, an Aboriginal dances in a ceremony called a *corroboree*.

CHAPTER OUTLINE

CHAPTER PLANNER

- ❑ Study the picture and read the opening story.
- ❑ Scan the Learning Objectives in each section: p. 130 ❑ p. 138 ❑ p. 141 ❑ p. 152 ❑
- ❑ Read the text and study all visuals. Answer any questions.

Analyze key features

- ❑ Video Explorations, p. 130 ❑ p. 150 ❑
- ❑ Geography InSight, p. 134
- ❑ Process Diagram, p. 147
- ❑ What a Geographer Sees, p. 154
- ❑ Stop: Answer the Concept Checks before you go on: p. 137 ❑ p. 141 ❑ p. 151 ❑ p. 155 ❑

End of chapter

- ❑ Review the Summary and Key Terms.
- ❑ Answer the Critical and Creative Thinking Questions.
- ❑ Answer What is happening in this picture?
- ❑ Complete the Self-Test and check your answers.

Religion in Global Context

LEARNING OBJECTIVES

1. **Distinguish** between animistic and syncretic religions.
2. **Identify** characteristics of universalizing and ethnic religions.
3. **Contrast** the distributions of the Abrahamic faiths.
4. **Identify** similarities and differences among Buddhism, Hinduism, and Sikhism.

Religion is a system of beliefs and practices that help people make sense of the universe and their place in it. A religion can be very personal, or it can be highly institutionalized. Religion may involve the worship of the divine or supernatural. A religion may be **monotheistic**, **polytheistic**, or **atheistic**. However, it is important that we keep in mind that such labels yield only a simplistic way of comprehending what a religion means to its adherents. **Animistic religions**, like those of the Australian Aboriginals and many other indigenous belief systems, incorporate veneration of spirits or deities associated with natural features—rocks, mountains, trees, or rivers, for example. The environment is, in other words, an inspirited realm. Many religions are also **syncretic**—that is, they demonstrate a notable blending of beliefs and practices, usually as a result of contact between people who practice different religions. Certain African and Roman Catholic traditions became fused as a result of the African slave trade, giving rise to the syncretic religions of Santeria in Cuba and Candomblé in Brazil (**Figure 5.1**). For an introduction to Cuban Santeria, see *Video Explorations*.

monotheistic The belief in or devotion to a single deity.

polytheistic The belief in or devotion to multiple deities.

atheistic The belief that there is no deity.

Video Explorations

Santeria

WILEY PLUS NATIONAL GEOGRAPHIC Video

Santeria draws on the beliefs of West Africans, especially Yorubans, who were brought to the New World as slaves. Devotion to orishas is a fundamental aspect of Santeria, as is animal sacrifice.

A religion might also provide an explanation of the beginning of the world, or **cosmogony**. Dreamtime, discussed in the chapter opener, is a cosmogony shared by Australian Aboriginals. Cosmogonies are important because they can influence people's sense of belonging and attachment to place. Similarly, a religion may be associated with a code of behavior, morals, or ethics. For their adherents, then, religions represent or express certain truths.

Candomblé • Figure 5.1

One expression of syncretism in Candomblé links worship of orishas, powerful animistic deities that manifest different qualities and energy of the divine, with Catholic saints. Candomblé recognizes a priesthood, in which the head priest is usually a woman—technically a priestess—called the *Mãe-de-Santo*, or mother of saints. Here, a Mãe-de-Santo sits in her consulting room in Bahia, Brazil.

Distribution of major religions • Figure 5.2

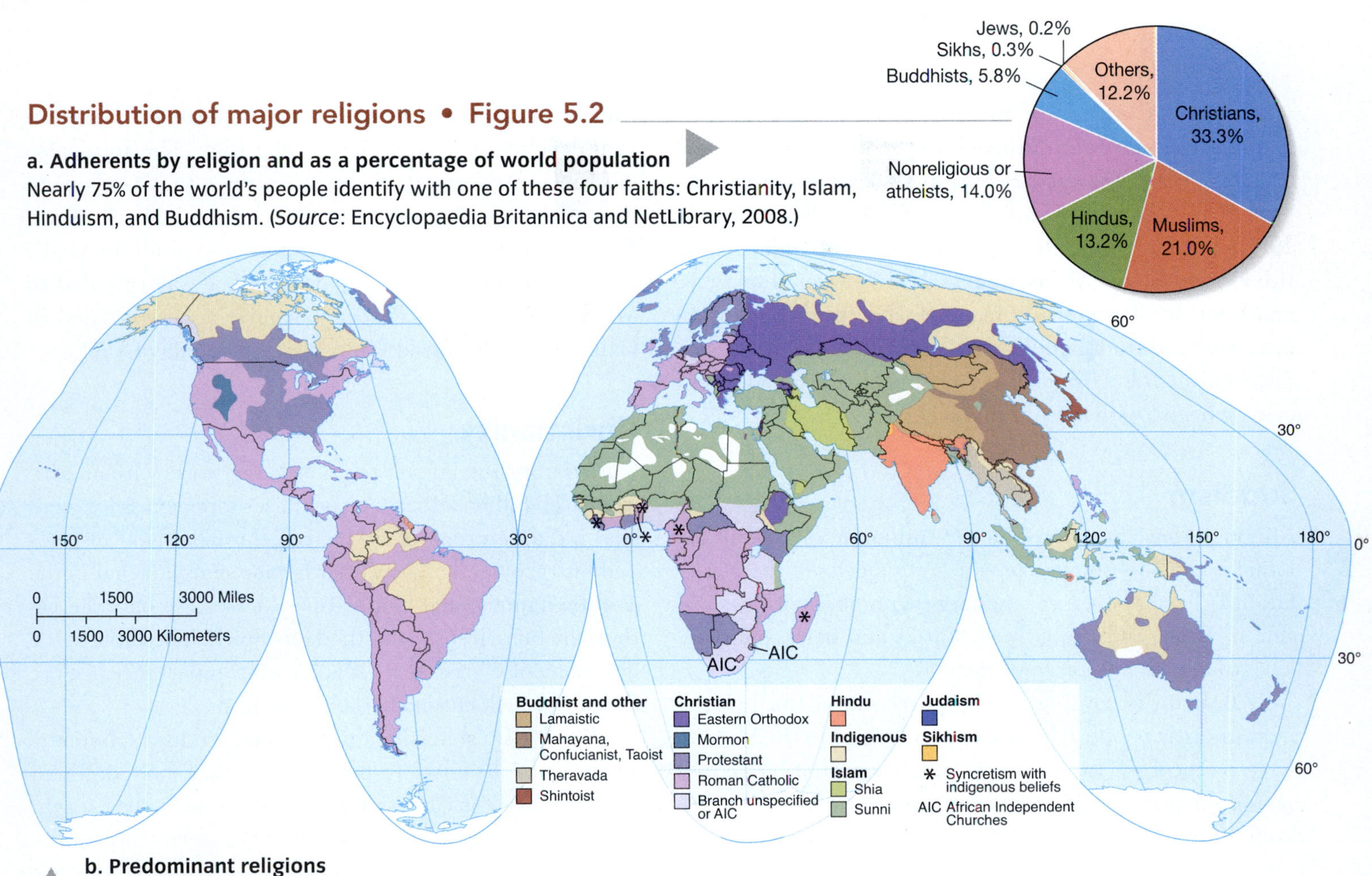

a. Adherents by religion and as a percentage of world population
Nearly 75% of the world's people identify with one of these four faiths: Christianity, Islam, Hinduism, and Buddhism. (*Source*: Encyclopaedia Britannica and NetLibrary, 2008.)

b. Predominant religions
Christianity's global spread stems from European conquest, colonization, and the ongoing work of missionaries. Although we call the Middle East a Muslim realm, 40% of all Muslims live in the Asian countries of Indonesia, Pakistan, India, and Bangladesh. Hinduism is the largest faith in India, but the country has more than twice as many Muslims as Egypt. Buddhism is widespread across East Asia and often mixes with other religious traditions, including Confucianism, which some scholars classify as a system of ethics or a civil religion.

Like language, religion is another facet of culture. Religion shapes the identity of a person, or an entire community. It helps people define who they are, how they behave, and how they interpret the world. Religious behavior may include the practice of **rituals** such as prayer, the maintenance of dress codes, or the celebration of religious festivals. **Piety** (or piousness) means to be deeply devoted to a religion.

Religions can be loosely grouped into two broad categories: **universalizing** and **ethnic**. Distinctions between these categories mainly involve how a religion acquires adherents or followers, and at what scale. Christianity, Islam, Buddhism, and Sikhism (a religion of northern India) are universalizing religions. Universalizing religions are closely associated with a key individual who established the religion.

ritual Behavior, often regularly practiced, that has personal and symbolic meaning.

universalizing religion A belief system that is worldwide in scope, welcomes all people as potential adherents, and may also work actively to acquire converts.

ethnic religion A belief system largely confined to the members of a single ethnic or cultural group.

Membership in ethnic religions is usually conferred by birth, and ethnic religions rarely use missionaries to increase the number of adherents. Ethnic religions include Judaism, Hinduism, Shintoism (a religion of Japan), and many of the belief systems of the world's indigenous peoples. Although the oldest religions in the world are ethnic religions, many of them are challenged by the growth and expansion of universalizing religions.

Globally, religion remains a profoundly influential factor in people's lives. When religious notions, symbols, and rituals infuse the political culture of an area, we refer to this as **civil religion**. For example, on the Great Seal of the United States, the Latin phrase *Annuit Cœptis* states: "He [God] has favored our undertakings." (This is visible on the back side of a dollar bill.) We will discuss civil religion in later sections, but first we need an introduction to some of the major faiths mapped in **Figure 5.2**.

Judaism, Christianity, and Islam are sometimes classified as *Abrahamic faiths*. Although the specific details differ, each of these faiths has a historical association with Abraham, who is thought to have lived in the Middle East in the 19th century BCE. In contrast, Hinduism and Buddhism are two *Vedic faiths*. The Vedas are India's oldest sacred writings and influenced the development of Hinduism. Buddhism later diverged from Hinduism. Neither Abrahamic nor Vedic, Sikhism draws from both Islam and Hinduism. Our discussion begins with the Abrahamic faiths.

Judaism

Although there are more than 13 million Jews worldwide, Israel is the only country in which a majority of the population is Jewish. Even so, the largest numbers of Jews reside in Israel and the United States, and both countries have roughly 5 million Jews each.

Abraham is considered the patriarch of the Jews, a monotheistic people who trace their origins to the Middle East. As recorded in the Torah, which is part of Judaism's sacred scripture and is sometimes referred to as the Hebrew Bible, Moses led the Jews out of slavery in Egypt. This event is called the Exodus and constitutes a significant development in the Jewish tradition. The Torah also describes a covenant or an agreement that God made with Abraham. According to this covenant, the Jews are "chosen people" selected to uphold and abide by God's law. At Mount Sinai, in Egypt, this law was revealed to Moses. After a long desert journey the Jews settled in Canaan, the Promised Land, in what is today Israel.

Christianity

Christianity, the largest religion in the world with an estimated 2.3 billion adherents, promises forgiveness for one's sins and an eternal life in heaven through belief in Jesus and his resurrection. Most Christians share a belief in the Trinity; namely, that God is three persons in one: the Father, the Son (Jesus), and the Holy Spirit. The Bible is the holy book of Christians. Of particular importance are the Gospels, which chronicle the life of Jesus.

Since its establishment Christianity has splintered. There is not one form of Christianity; rather, many different varieties exist. As Christianity spread across Europe, it split into Western and Eastern branches. Western Christianity,

Christianity and its shift south • Figure 5.3

a. Composition of Christianity

Slightly more than half of Christians who affiliate with a church are Catholics. Followers of independent churches are also proportionally important. Many of these churches began in the late 20th century from grassroots movements in Africa and Latin America and see themselves as distinct from the European-derived Christian churches classified as Catholic, Protestant, or Orthodox. The category "Other Christians" includes Latter-day Saints and Unitarians, among others. The accompanying photo shows a member of an African independent church being baptized at a beach in Durban, South Africa.

Stated Christian affiliation*, by percent**	
Roman Catholicism	54%
Protestantism	22%
Independent churches	21%
Eastern Orthodoxy	11%
Other Christians	2%

*A significant number of Christians do not specify an affiliation.

**Percentages sum to 110% because some Christians are affiliated with two churches and have been counted twice.
(*Source*: Encyclopaedia Britannica and Net Library, 2008.)

centered on Rome and referred to as Roman Catholicism, recognized the authority of the pope. Eastern Christianity, or Eastern Orthodoxy as it came to be called, was based in the city of Constantinople (present-day Istanbul) and did not recognize the papacy. Instead, Eastern Orthodoxy gave rise to 15 independent churches including, for example, the Greek Orthodox Church and Russian Orthodox Church. In the 16th century, Western Christianity was split by the Protestant Reformation, which took issue with Catholic practices.

The Roman Catholic, Protestant, and Eastern Orthodox branches form the three conventional branches within Christianity. About 80% of all Christians belong to one of these three branches. However, the emergence of numerous independent churches constitutes a noteworthy trend. Globally, the distribution of Christianity has changed significantly over time (**Figure 5.3**).

Islam

After Christianity, Islam is the second largest religion. More than one-fifth of the world's people call themselves Muslims. Islam is also the fastest-growing religion in the world. Geographically, Islam is the dominant religion in a belt that stretches from North Africa across the Middle East, Central Asia, and into South Asia.

Muhammad, the founder of Islam, was born in about 570 CE in Mecca, in what is now Saudi Arabia. On several occasions while meditating, Muhammad received revelations from God, whom Muslims call Allah; these revelations were communicated to him via the angel Jibril (Gabriel). Muhammad began to share with others what had been revealed to him, especially belief in the unity of Allah. He met resistance in part because of enduring polytheistic beliefs in the region.

Muslims believe that the Qur'an, Islam's holy book, records the word of God as it was revealed to Muhammad. Muslims recognize many of the people in the Old and New Testaments, including Abraham, Moses, and

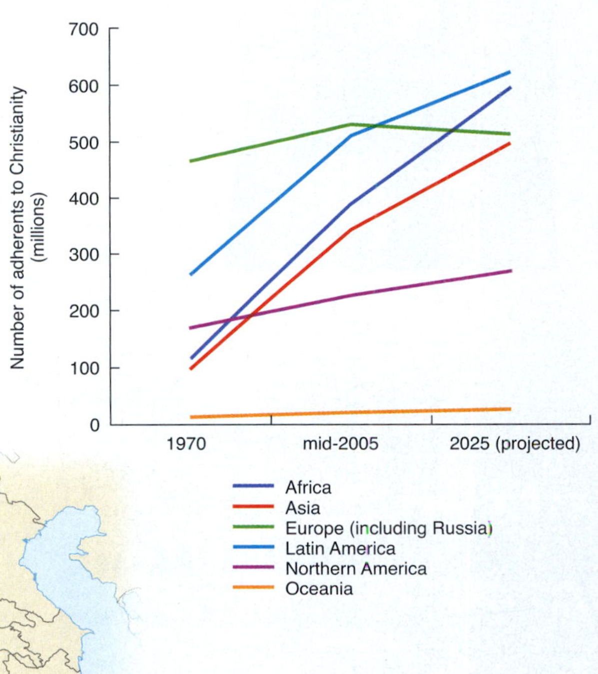

b. Adherents to Christianity, 1970–2025
For many centuries, Christianity was a religion associated mainly with Europe and then with the Americas. In the past several decades, however, Christianity has gained large numbers of adherents in Latin America, Africa, and Asia as a result of missionary activity, the popularity of independent churches, and population growth. (*Source*: Barrett, Johnson, and Crossing, 2005.)

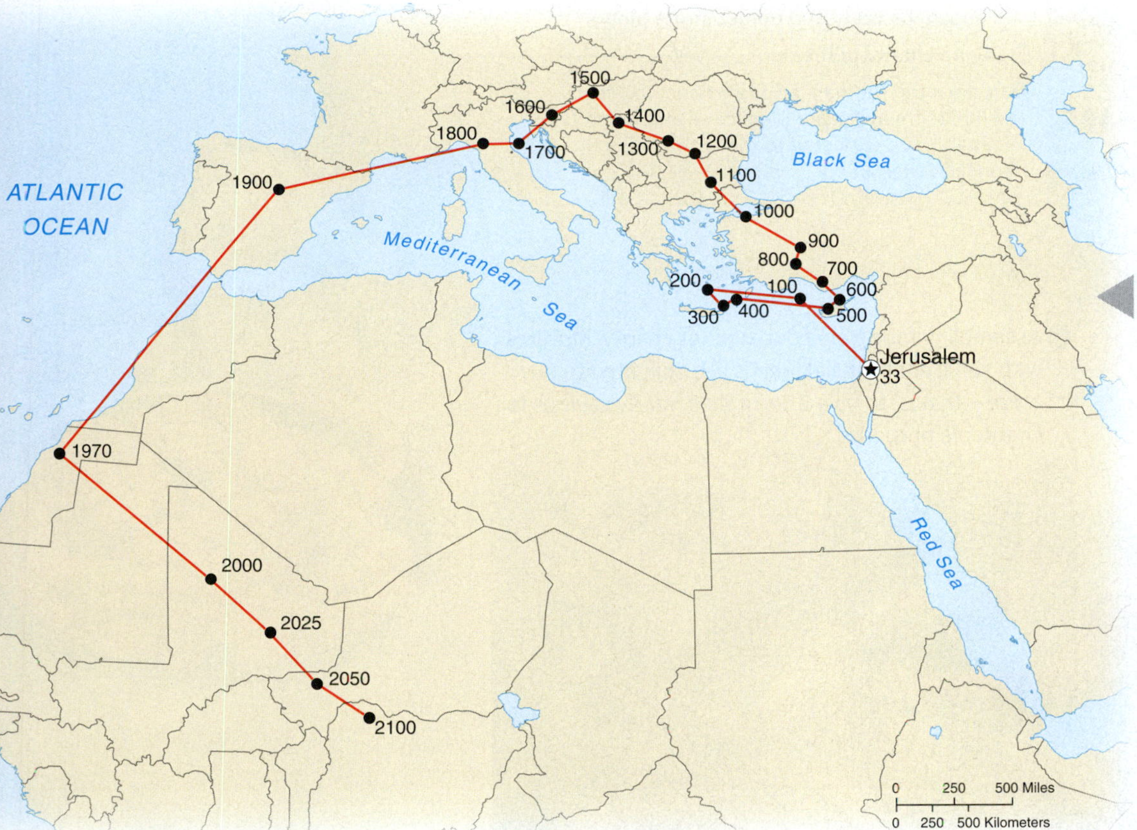

c. Christianity's global center of gravity, 33–2100 CE
An equal number of Christians lives east, west, north, and south of the center of gravity. A recent study projects that the center of gravity will be located in northern Nigeria by 2100, with more than three times as many Christians living in Latin America, Africa, and Asia than in Europe and the Americas. Map shows present-day boundaries. (*Source*: Johnson and Chung, 2004.)

Geography InSight

Islam's Five Pillars of Practice • Figure 5.4

Islam, like many other religions, is often described as a way of life. The Five Pillars of Practice express and embody this lived dimension of the religion. Of course, being a Muslim involves more than achieving these five pillars, but they serve as major tenets of the faith.

1 The *shahadah*, or profession of faith, in Arabic over the gateway to Topkapi Palace in Istanbul, Turkey.

2 Muslims pray near the ticketing area of the Minneapolis-St. Paul International Airport. Muslims customarily face Mecca when they pray and kneel on a prayer rug. *Qibla*, in Arabic, means the direction of Mecca. When traveling, Muslims can use a qibla compass to determine which way to face. Alternatively, some mobile devices now come with built-in qibla compasses.

1 Making the *shahadah*, a profession of faith expected of all Muslims. The shahadah states: There is no god but Allah, and Muhammad is his prophet.

2 Praying five times a day. All Muslims are expected to pray before dawn, after noon, in the late afternoon, at sunset, and at night.

3 Fasting during Ramadan, one of the months in the Islamic lunar calendar. Muslims fast from sunrise to sunset during this month in order to cleanse the body and demonstrate piety.

4 Giving charity to the poor.

5 Performing the *hajj*, or pilgrimage, to Mecca. All Muslims who are physically and financially capable are expected to make the hajj at least once in their life.

3 A Muslim vendor sells food outside a mosque in Shanghai, China, during Eid al-Fitr, the festival that marks the end of Ramadan.

5 The hajj is an annual pilgrimage to Mecca that Muslims participate in simultaneously. Since Islam uses a lunar calendar, the dates for the hajj, as for Ramadan, vary. Recently, the hajj has drawn more than 2 million Muslims.

4 Assembling food items donated for charity. Muslims with sufficient financial means are obligated to pay *zakat*—that is, to give 2.5% of their yearly savings for charitable purposes.

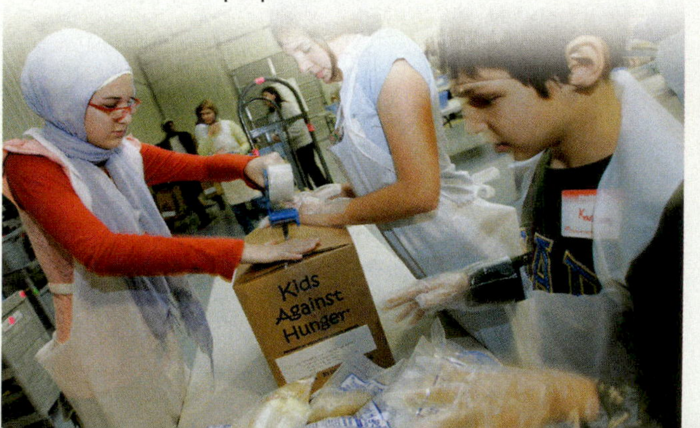

Jesus, for example, but they consider Muhammad to be the final prophet of God. Although faith is very important to Muslims, expressing their faith through actions remains an essential part of Islam. The Five Pillars of Practice is a set of rituals central to the religion (**Figure 5.4**).

Islam consists of two main branches, Sunnis and Shiites. The Sunni branch is the largest and most geographically widespread branch of Islam. About 80% of all Muslims are Sunnis. No more than 15% of Muslims are Shiites, although they make up a majority of the population in four countries: Iran, Iraq, Azerbaijan, and Bahrain. Other smaller branches account for the remainder of the Muslim population.

The Sunni and Shia branches emerged following Muhammad's death in 632 CE when disagreements arose about who should succeed him. Those Muslims who felt that Muhammad's immediate successor should be someone from his family became known as the Shia, or Shiites, to distinguish themselves from the others, called the Sunnis, who accepted someone outside Muhammad's family as the first successor. Other differences separate Sunnis from Shias as well. For example, Sunnis use the term *imam* to refer to a religious leader, especially one who leads group prayer. Shia Muslims, however, consider imams to be divinely inspired.

Hinduism

Sanatana dharma, meaning "eternal truth," is the name some Hindus use for their religion. The terms *Hindu* and, later, *Hinduism* came to be used by outsiders to refer to the people and their religion in the region that would become India. Approximately 900 million people in the world identify Hinduism as their religion, making it the largest ethnic religion in the world. Most Hindus live in South Asia, specifically India.

Hinduism includes a great diversity of religious beliefs and practices, but some common elements exist. Hindus view existence as cyclical such that souls are immortal and subject to reincarnation. The process of reincarnation brings spiritual suffering and is controlled by *karma*, the influence of past thoughts and actions. Hindus strive to attain *moksha*, or release from the cycle of death and rebirth. Moksha has been described as a state of freedom or bliss.

In the Hindu cosmogony, Brahman is the supreme spiritual source and sustainer of the universe, variously understood as an absolute and eternal force as well as a supreme being. Hinduism includes a vast number of gods and goddesses, and Hindus believe these deities express different qualities of Brahman (**Figure 5.5**).

Shiva statue • Figure 5.5

One prominent Hindu deity is Shiva, the destroyer of evil. Located in Bengaluru, India, this statue depicts a meditating Shiva at Mount Kailash, his heavenly home. His four arms are symbolic of supreme power.

Buddhism • Figure 5.6

a. A devotee stands in front of a statue of Buddha in Sri Lanka. In contrast to Islam, which forbids depictions of Muhammad and God, representations of the Buddha are common where Buddhism is practiced.

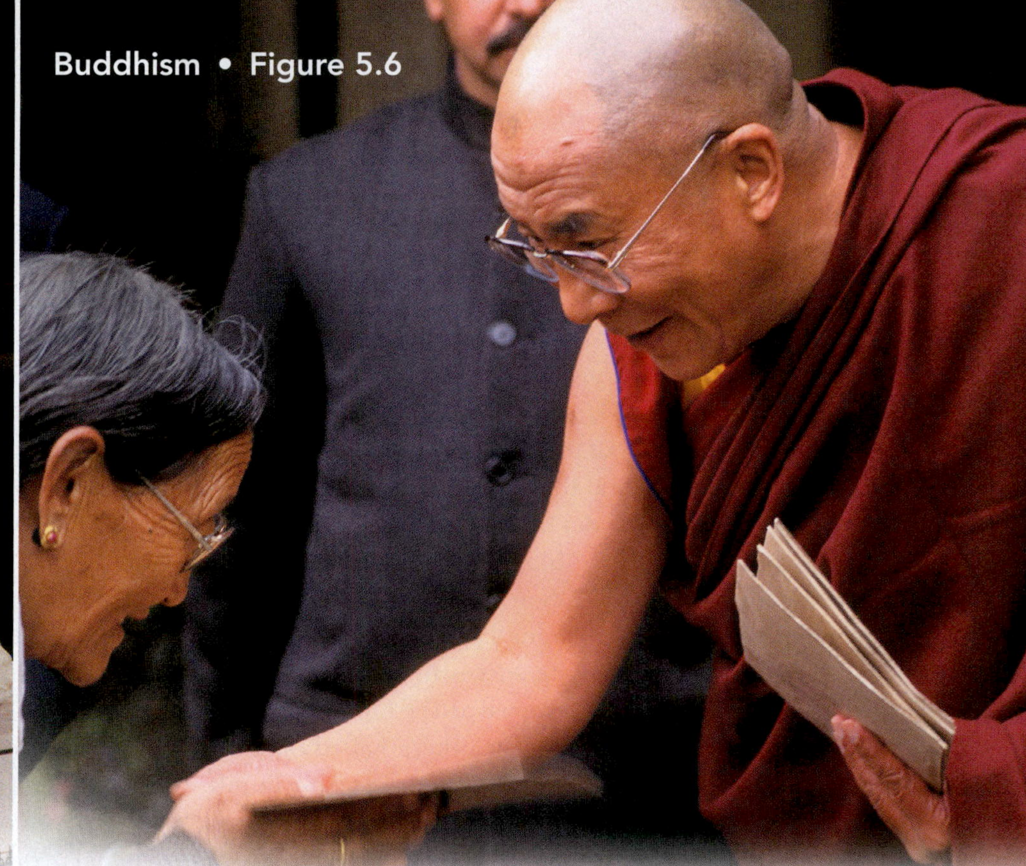

b. The Dalai Lama, shown here greeting Tibetans, serves as the spiritual authority of Tibet, but he has lived in exile since China invaded and occupied the region in 1959. Tibetan Buddhists believe that the Dalai Lama is the reincarnation of the bodhisattva of mercy and that when he dies his spirit will be reincarnated and enter a child.

Buddhism

Buddhism remains closely associated with East and Southeast Asia. In China, Japan, Hong Kong, Taiwan, and Singapore, Buddhism is prevalent but mixes with other local traditions, including Confucianism.

The founder of Buddhism, Siddhartha Gautama (6th century BCE), was raised a Hindu prince and rather sheltered from the sufferings of the sick and the problem of poverty. Tradition maintains that he was troubled by the suffering he saw and, out of compassion, gave up his life of comfort and sought a way to end human suffering. Siddhartha Gautama attained enlightenment while meditating. This marks his transformation into the *Buddha*, literally the *enlightened one*. After this, he began to teach and acquire disciples. His teachings are recorded in different documents, the oldest of which is the *Tripitaka*.

Buddhists believe that suffering is linked to reincarnation, and they seek to attain *nirvana*, escape from the cycle of death and rebirth. The Buddha's teachings centered on the Four Noble Truths:

1. Life brings suffering.
2. Desire causes this suffering.
3. This suffering can be overcome and nirvana can be attained.
4. Disciplining the mind and body by practicing proper thinking and behavior ends this suffering and leads to nirvana.

There are three major branches within Buddhism: Theravada, Mahayana, and Tantrayana Buddhism. Theravada means "Way of the Elders," and this form of Buddhism cultivates a monastic approach that emphasizes the study of Buddhist scripture and the practice of disciplined behavior. Meditation forms an important part of this effort. Theravada Buddhism maintains a greater presence in the island country of Sri Lanka as well as the mainland Southeast Asian countries of Thailand, Myanmar, Cambodia, and Laos. Theravada Buddhists account for fewer than 40% of all Buddhists.

Mahayana, meaning "Great Vehicle," broadened the appeal of Buddhism. Mahayana Buddhists believe that the Buddha is a compassionate deity and that Mahayana Buddhism provides a way for believers to be saved from the cycle of rebirth. Devotion to *bodhisattvas* ("Buddhas-to-be") also distinguishes this branch of the religion. Bodhisattvas are celestial

c. A Zen Buddhist priest meditates at the "dry landscape" garden of Zuiho-in at Daitoku-ji Temple. Zen Buddhism developed as a movement within Mahayana Buddhism and encourages followers to cultivate the Buddha within oneself. Gardens like this aid meditation.

beings that represent the qualities of the Buddha and help others achieve enlightenment. Mahayana Buddhism is most closely associated with China, Korea, Japan, and Vietnam. Slightly more than half of all Buddhists are Mahayanists.

The smallest branch of Buddhism, Tantrayana Buddhism, emerged in Tibet and spread to Mongolia. Syncretic in character, Tibetan Buddhism fused aspects of Mahayana Buddhism and native Tibetan beliefs. Indigenous influences in Mongolia have shaped the practice of Buddhism there. For many decades, however, Communist rule suppressed religious practices in both places. **Figure 5.6** illustrates some other dimensions of Buddhism.

Sikhism

Some 23 million adherents practice Sikhism, making it the smallest of the major universalizing religions. A *guru* is an inspired religious teacher, and the word *Sikh* means *disciple*. Guru Nanak (1469–1538) founded Sikhism. Sikhs believe that Guru Nanak experienced a divine revelation, after which he began teaching and establishing Sikh communities.

Sikhism emerged in northern India and bears the influences of Hinduism and Islam. For example, Sikhism teaches belief in and worship of one creator god and yet emphasizes the importance of karma. The holy book for Sikhism is the *Guru Granth Sahib*, also called the *Guru Granth*. It consists of a compilation of hymns revealed to Guru Nanak and several other gurus. Sikhs call this holy book a *guru* because they consider it to be the source of spiritual authority today.

CONCEPT CHECK

STOP

1. **What** is a syncretic religion? Explain using a specific example.
2. **How** does the concept of scale relate to universalizing and ethnic religions?
3. **How** has the global geography of Christianity recently changed? Why?
4. **Where** are Buddhism and Hinduism prevalent, and how are the two faiths similar and different?

Religious Hearths and Diffusion

LEARNING OBJECTIVES

1. **Distinguish** between primary and secondary hearths.
2. **Define** diaspora.
3. **Relate** the spread of religion to different types of diffusion.

Every religion emerges in a hearth. For many adherents to a specific religion, places or sites within the hearth often acquire sacred qualities because of their association with significant events in the growth and development of the religion. Geographers recognize two categories of religious hearths: primary and secondary. *Primary hearths* are those places or regions where a wholly new religion develops. In contrast, *secondary hearths* are the places or regions where a religion fragments internally to form a new branch. Remarkably, two primary hearths, the Semitic and the Indic, have generated more religions—and more enduring religions—than any other regions (**Figure 5.7**).

Religions of the Semitic Hearth

As the birthplace of a religion, the hearth usually represents its dominant symbolic center even though the hearth may or may not be home to many of that religion's adherents. The region known historically as Palestine, and which today roughly coincides with Israel, constitutes the hearth of Judaism. However, the opportunity for Jews to live in the hearth has been much contested. The term **diaspora** developed as a specific reference to the dispersion of the Jewish population that occurred in the sixth century BCE when the Babylonians sacked Jerusalem and exiled the Jews to Babylonia, now part of Iraq. Then, following the Roman destruction of Jerusalem in 70 CE, the Jews were again expelled from Palestine. By some estimates, as many as 2 to 5 million Jews lived outside of Palestine in the first century CE, many taking residence in Europe and North Africa. Nazi atrocities committed prior to and during World War II killed 6 million Jews in Europe. These horrific events prompted many Jews to leave

> **diaspora** The scattering of a people through forced migration.

Primary hearths • Figure 5.7

Judaism, Christianity, and Islam emerged in the Semitic hearth, whereas Hinduism, Buddhism, Sikhism, and another South Asian faith, Jainism, emerged in the Indic hearth. In both the Semitic and Indic hearths, these developments played out over many centuries. Why does the identification of religious hearths sometimes become a contested practice?

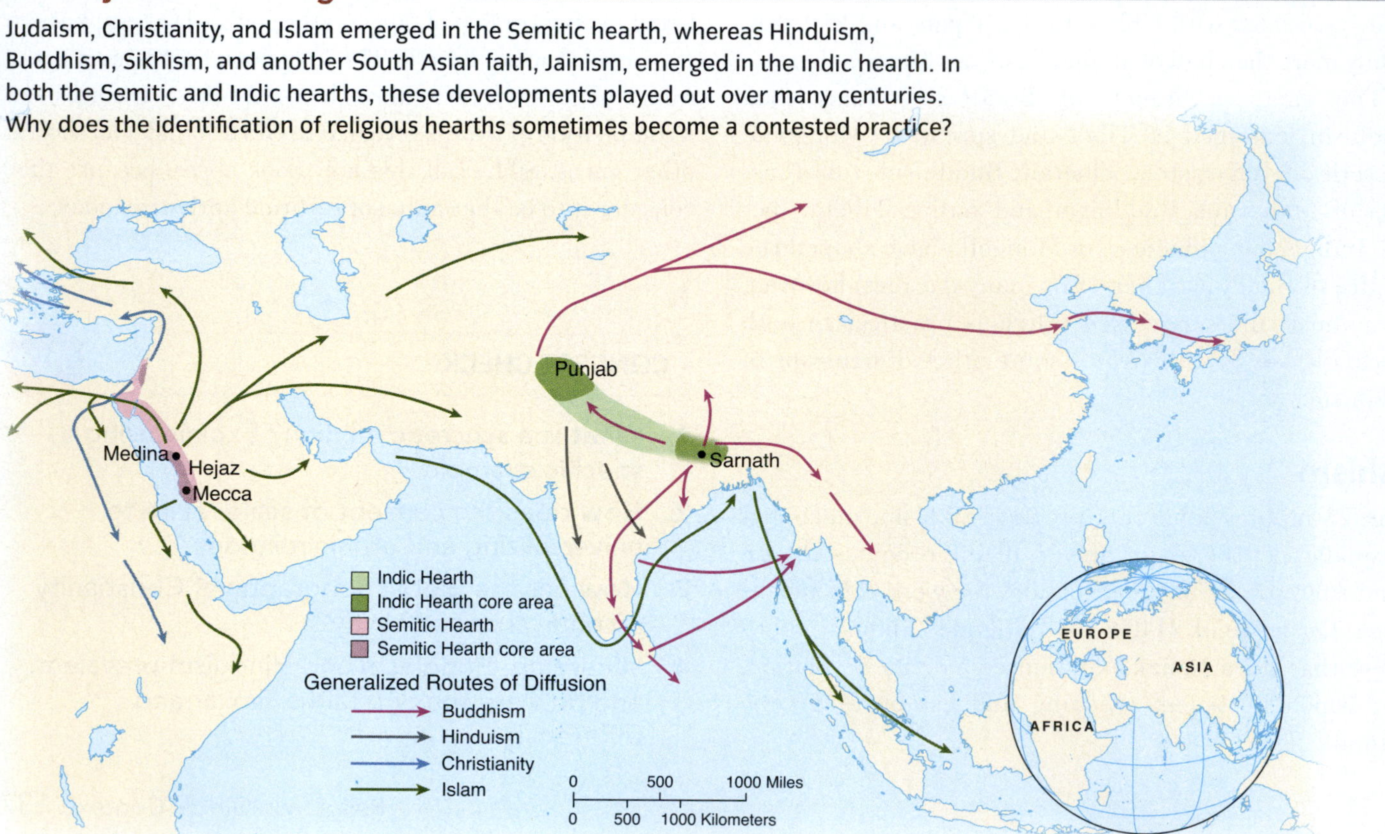

Christianity in the United States: diffusion and fragmentation • Figure 5.8

Colonists from Europe, followed by later waves of immigrants, transferred and implanted their religious beliefs in the Americas, reshaping the religious landscape in the process.

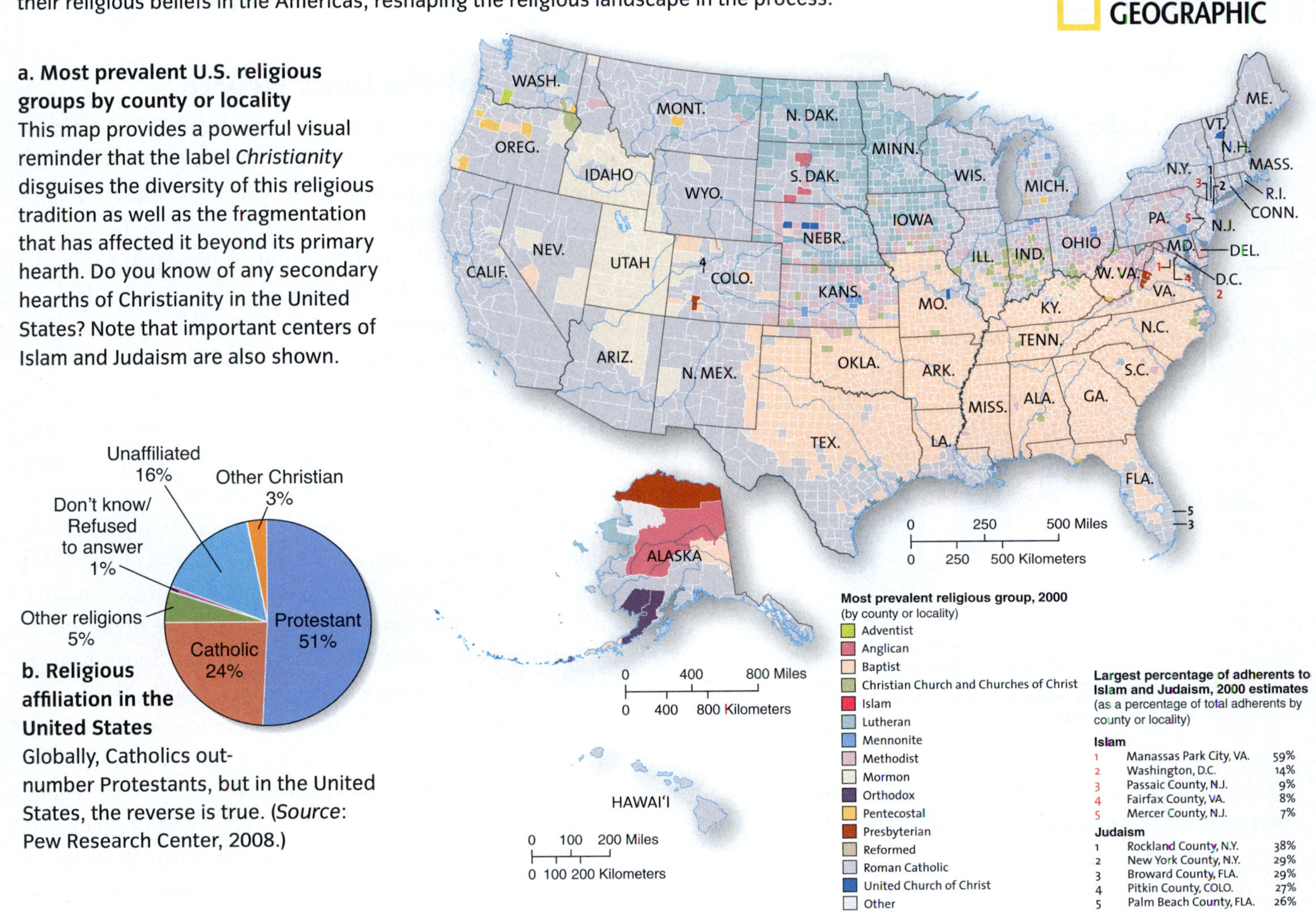

a. Most prevalent U.S. religious groups by county or locality
This map provides a powerful visual reminder that the label *Christianity* disguises the diversity of this religious tradition as well as the fragmentation that has affected it beyond its primary hearth. Do you know of any secondary hearths of Christianity in the United States? Note that important centers of Islam and Judaism are also shown.

b. Religious affiliation in the United States
Globally, Catholics outnumber Protestants, but in the United States, the reverse is true. (*Source*: Pew Research Center, 2008.)

Europe for Palestine and the Americas and also renewed the Jewish desire for a country of their own. Amid much controversy, in 1948 the Jewish state of Israel was carved out of land in the Middle East.

Palestine, the region of the Middle East where Jesus was born, began to teach, and gained disciples, constitutes the hearth of Christianity. Thus, Christianity and Judaism share similar but not identical hearths. At the local scale, for example, the geography of the hearth of Christianity becomes defined by those places closely associated with key events in the religion's formative period.

Contagious, relocation, and hierarchical diffusion all factored in the spread of Christianity. Jesus and the 12 disciples initially spread Christianity in contagious fashion through their personal interactions with non-Christians. Following the crucifixion of Jesus, the apostle Paul traveled as a missionary to regions bordering the Mediterranean Sea, visiting parts of Syria, Turkey, Greece, and Italy and spreading the Christian religion via relocation and contagious diffusion. In 313 CE, the Roman Emperor Constantine converted to Christianity. Subsequently, Christianity became the official religion of the Roman Empire. This development set in motion a form of hierarchical diffusion in which the religion of the emperor became the religion of the people. The rise of Christianity in cities such as Alexandria, Damascus, Constantinople, and Rome before it became established in the countryside surrounding these places shows another type of hierarchical diffusion, from large cities to smaller towns and villages. European colonialism as well as past and present missionary activity have carried Christianity far beyond its hearth, and in the process Christianity has become highly fragmented. See **Figure 5.8** for one depiction of this fragmentation.

Where Geographers CLICK

Pew Forum on Religion & Public Life: U.S. Religious Landscape Survey

http://religions.pewforum.org/maps.

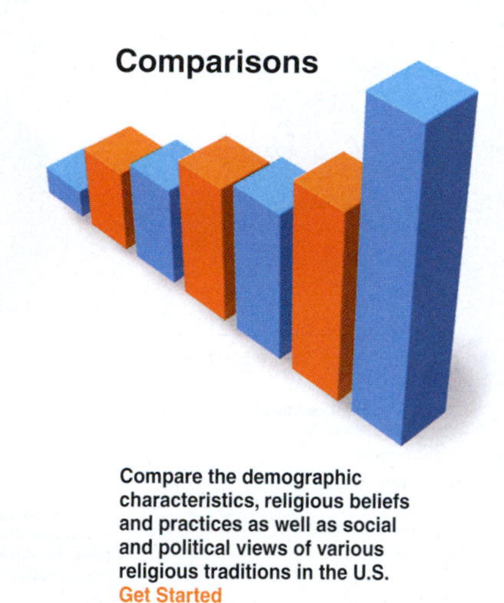

Comparisons
Compare the demographic characteristics, religious beliefs and practices as well as social and political views of various religious traditions in the U.S.
Get Started

At this site, you can view and compare maps of the United States that show affiliation with different religious traditions.

See *Where Geographers Click* to explore additional maps of religious beliefs in the United States.

Unlike Judaism and Christianity, Islam has its origins in territory that is now part of Saudi Arabia. More specifically, the Hejaz region where Mecca and Medina are situated constitutes the primary hearth of Islam. The origin and initial diffusion of Islam are closely linked to urban places. After its establishment in the 6th century, Islam spread rapidly and hierarchically, moving along established trading routes across North Africa and the Middle East. Commerce, military conquest, and scholarship contributed to the spread of Islam. Muslim scholars established libraries and academies in North Africa. By the early 8th century, Islam stretched from Spain across Central Asia and into India. Relocation diffusion was also involved. Muslim traders reached Southeast Asia in the 13th century, laying the foundation for the growth of Islam in present-day Indonesia and Malaysia.

Religions of the Indic Hearth

One of the world's oldest religions, Hinduism began to take shape some 4,000 years ago. Although the hearth of Hinduism is difficult to pinpoint precisely, most scholars situate it in the Punjab, a region spanning the border between northern India and Pakistan (refer again to Figure 5.7). Unlike universalizing religions, the origin of Hinduism is generally associated with the upper classes rather than the influence of one key individual.

Hinduism diffused hierarchically and in conjunction with Sanskrit, the language of the *Vedas*. From the Punjab, Hinduism spread through the Ganges River Valley and south across the subcontinent. Relocation diffusion has carried Hinduism beyond South Asia. About 4 million Hindus live in Indonesia, mainly on the island of Bali, and about 1 million Hindus live in the United States. Smaller Hindu communities have developed a notable presence on the Caribbean island of Trinidad and in the South American countries of Suriname and Guyana. South Asian Indians were brought to these places as indentured laborers after the African slave trade ended.

Compared to Hinduism, Buddhism is a more recent faith that dates to the 6th century BCE. Siddhartha Gautama delivered his first sermon at Sarnath, near Varanasi, helping to establish the hearth of Buddhism in the eastern lowlands of the Ganges River in India. Following his enlightenment, the Buddha continued to travel and teach for more than 40 years, playing a role in the diffusion of the religion. His disciples also helped spread the religion within India. Buddhism spread through Central Asia along trade routes, entering China during the 1st century CE., In India and China, Buddhist monks frequently settled together, creating monastic communities called *sangha*. These communities often became centers of learning and continued the diffusion of Buddhism. In the 4th century CE, Buddhism diffused from China to Korea. From Korea, Buddhism spread to Japan, becoming established there in the 6th century. Buddhism was introduced to Tibet from northern India in the 7th century.

Sikhism, a third religion of the Indic hearth, dates to the 16th century CE and has its origins in the Punjab, the same region where Hinduism emerged. Compared

Sikhism and Khalistan • Figure 5.9

The creation of India and Pakistan in 1947 led to the division of the Punjab, Sikhism's hearth. Many Sikhs desire their own independent country, Khalistan, and celebrate Khalistan Day (*Vaisakhi*) as a way of expressing their religious and political commitment to this goal. This parade is in Washington, D.C.

to other universalizing religions, however, Sikhism has not spread as extensively. Nevertheless, the British colonization of South Asia contributed to the relocation diffusion of Sikhism and the emergence of diasporic Sikh communities. For example, Sikhs were recruited to work as police officers in Hong Kong, as soldiers in the British army in what is now Malaysia, and as railroad workers in East Africa, among other places (**Figure 5.9**).

CONCEPT CHECK

1. **Why** is it important to identify primary religious hearths?
2. **What** is a diaspora, and how is the concept related to the geography of religion?
3. **How** has hierarchical diffusion factored in the spread of Christianity and Hinduim?

Religion, Society, and Globalization

LEARNING OBJECTIVES

1. **Explain** why certain sacred spaces in Jerusalem are contentious.
2. **Summarize** the process of sanctification.
3. **Describe** the geography of Renewalism.
4. **Distinguish** between religious fundamentalism and Islamic traditionalism.

At the start of this chapter we discussed the close connection between religion and culture. Indeed, many adherents describe their religion as a "way of life." Of particular interest to geographers are the ways in which religion colors how people understand and interpret the world as well as their place in it. As Roger Stump (2008, p. 23) has observed, "religious groups do not simply exist in space; they also imagine and construct space in terms related to their faith." In the following sections, we explore some of the ways that religion shapes individual and collective views about territory, identity, and society. We begin with a consideration of sacred space.

Sacred Space

Sacred space is space that has special religious significance and meaning that makes it worthy of reverence or devotion. Sacred space includes specific places and sites that are recognized for their sanctity; however, a sacred space need not be territorially defined. The performance of religious rituals, prayer for example, often creates a highly personal sacred space.

Ideas about sacred space can become a source of disagreement and even conflict among people. For the Mescalero Apache in New Mexico and west Texas, some sacred spaces include regions where they collect natural

Jerusalem's many sacred spaces • Figure 5.10

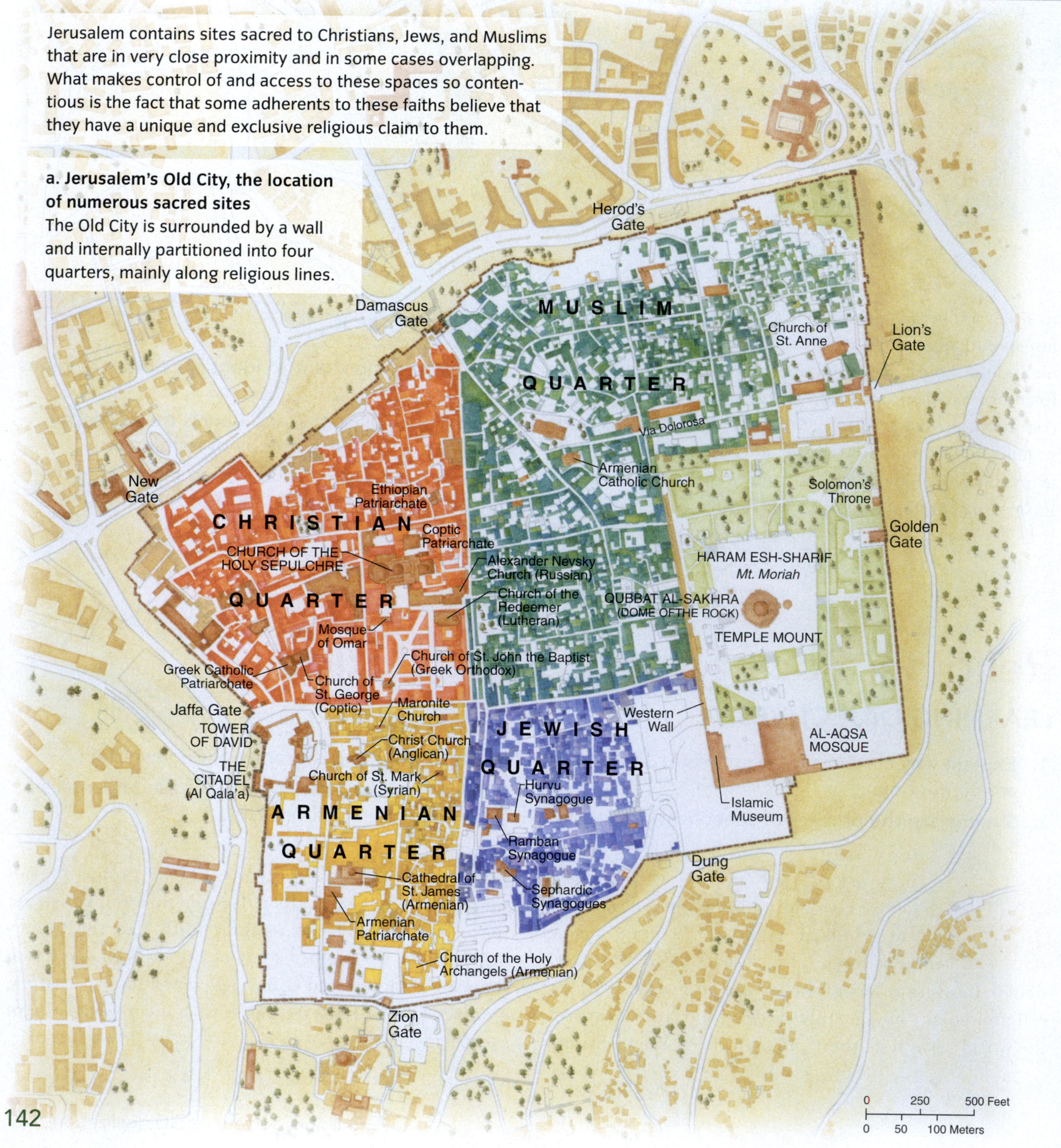

Jerusalem contains sites sacred to Christians, Jews, and Muslims that are in very close proximity and in some cases overlapping. What makes control of and access to these spaces so contentious is the fact that some adherents to these faiths believe that they have a unique and exclusive religious claim to them.

a. Jerusalem's Old City, the location of numerous sacred sites
The Old City is surrounded by a wall and internally partitioned into four quarters, mainly along religious lines.

resources such as sage or mesquite fruits. When these sacred spaces exist on land that has since become federal land or highway easements, the Mescalero Apache have been charged fees, fined, or harassed for collecting these resources.

To protect their sacredness, two of the holiest cities in Islam, Mecca and Medina, are closed to non-Muslims. In an effort to resist globalization and specifically Americanization, al-Qaeda has argued that all of Saudi Arabia constitutes an Islamic sacred space. To al-Qaeda members, the presence of non-Muslim Americans in the country constitutes a perceived desecration of this sacred space. One of the most complex and contentious examples of sacred space involves the city of Jerusalem (**Figure 5.10**).

NATIONAL GEOGRAPHIC

b. The Temple Mount, as it is known to Jews, or Haram al-Sharif—the Noble Sanctuary—to Muslims
This contested site is sacred to both groups. Muslims believe the Dome of the Rock marks the site from which Muhammad ascended to heaven. After Mecca and Medina, Jerusalem ranks as the third holiest site in Islam. Jews pray at the Western, or Wailing Wall, below the Dome of the Rock. For Jews, Jerusalem's Temple Mount is sacred because it housed the First and Second Temples of the Jews, which were destroyed by the Babylonians and Romans, respectively. All that remains of the Second Temple is the Western Wall. Another source of contention between Muslims and Jews involves the al-Aqsa Mosque. It is the second oldest Islamic mosque but was built atop the ruins of the Second Temple (refer again to **a**).

c. The Church of the Holy Sepulchre
For Christians, the Church of the Holy Sepulchre marks the site of the crucifixion, burial, and resurrection of Jesus. Space within the church is heavily contested by different Christian communities, including the Greek Orthodox, Roman Catholic, Armenian, and others, and disputes among them are not uncommon.

Pilgrimage A **pilgrimage** is a journey to a sacred place or site for religious reasons. Some pilgrimages, such as the hajj for Muslims, are obligatory, whereas others are voluntary. Pilgrims conduct these journeys for a variety of reasons: to purify the soul, to demonstrate devotion, to fulfill a vow, to contemplate or gain proximity to the divine, to seek divine or supernatural assistance, to give thanks, or to perform an act of penance. Although specific reasons for and patterns of pilgrimage vary from one religion to another and even from one person to another, the geographical study of pilgrimage includes the identification of major and minor pilgrimage sites and the patterns of circulation. Major destinations can be distinguished from minor destinations on the basis of their perceived holiness, the number of pilgrims they attract, and the *catchment area*—the size and kinds of areas sending pilgrims to a particular site. Going on a pilgrimage constitutes a kind of ritual and factors prominently in many faiths (**Figure 5.11**).

Community, identity, and scale Religion can provide a strong basis for community and individual identity. In Islam the idea of a Muslim community is often expressed at different scales, from the global to the individual. The Arabic term *umma*, meaning "community of believers," refers to the worldwide population of Muslims and imagines this group of people as a community whose identity transcends doctrinal differences and geographic separation. *Dar al-Islam,* or House of Islam, usually refers to countries that are majority Muslim, but depending on the context it might also include Muslims who live in non-Muslim countries. A third scale involves the mosque. Weekly communal worship at a local mosque contributes to the sense of a shared identity among Muslims, as does planning and accumulating the resources to construct a mosque. Finally, when Muslims fast during Ramadan or when Muslim women veil themselves, the body becomes the site of an important scale of identity.

Pilgrimage in selected religions • Figure 5.11

NATIONAL GEOGRAPHIC

a. Pilgrimage varies in the different Christian traditions. In general, pilgrimage is virtually absent in Protestantism and most pronounced in Catholicism, which possesses an elaborate network of sacred pilgrimage sites. Major destinations include the Holy Land—encompassing places such as Bethlehem and Jerusalem—as well as Rome, Italy. For many Catholics, sites associated with appearances of the Virgin Mary receive special veneration and are also major pilgrimage destinations. Most of these sites are outside Christianity's hearth. Annually, more than 6 million pilgrims flock to Lourdes, France, shown here. The basilica was built above a grotto reported to be the site of apparitions of the Virgin Mary and miracle cures.

How do other religions express a sense of community or identity at different scales?

Religion and settlement We have already seen how beliefs about sacred space can cultivate strong attachments between people and territory and can lead to competing ideas about how a particular area or site is used or who claims it. However, other distinctive associations between religious communities and territory can also develop. Diasporic religious communities—those that have become displaced from their religious hearth—provide a good example. For instance, some diasporic communities develop powerful ideas about specific territories, such that these spaces come to be perceived not only as a holy land but also as a homeland for the community of believers.

To escape persecution, the Mormons migrated beyond the frontier in the United States in the 19th century and into territory that would become Utah. There they sought to create an ideal community, a settlement that they called *Zion* and that is known today as Salt Lake City. To the Mormon settlers, Zion represented a new Jerusalem and the start of their endeavor to build the Kingdom of God on Earth. Thus, Salt Lake City developed as a holy city for Mormons in accordance with principles established by Joseph Smith, the religion's founder. The city plan, including its square blocks, wide streets, and a designated Temple Square, linked their religious ideals with a distinctive urban form.

For the Jews, another diasporic religious community, religion and settlement have long been closely interconnected. As we have discussed, Jewish people have a special relationship with the land of Israel, which they believe was promised to Abraham and his descendants by God. Thus, they refer to this area as the Promised Land, and the desire of eventually returning to it became

b. Buddhism does not require that its adherents go on pilgrimage, but the religion has still developed a tradition of pilgrimage. Major pilgrim destinations are concentrated in the hearth area and are associated with significant events in the Buddha's life: his birthplace, his enlightenment, his first sermon, and his death. These pilgrims pray and meditate at the Bodhi Tree, or bo, where Siddhartha Gautama attained enlightenment.

c. Among Hindus, the Ganges River is revered for its energy and for its ability to sustain life and remove negative karma. A number of important pilgrimage sites have developed in close proximity to or along the banks of this river at places like this in Hardiwar, where people can immerse themselves.

and continues to be an aspiration of many Jews. During the 19th century, a political movement known as *Zionism* developed. One of the goals of this movement was to create a religious and political homeland for Jews. This goal was partially realized in 1948 with the creation of the state of Israel; however, this development brought other issues to the forefront (**Figure 5.12**).

Sacred places and civil religion We have seen that sacred sites can become powerful magnets for pilgrims and that ideas about sacred space more broadly can influence settlement patterns. But how does a place become a sacred place? The answer involves the process of **sanctification**. Since the concept of sacredness also applies in the realm of civil religion, as, for example, in the way societies venerate certain political leaders, veterans, and symbols of states such as flags, we can use civil religion to understand how sanctification occurs. Through the study of landscapes of tragedy, geographer Kenneth Foote has demonstrated how sacred sites develop and gain significance within the context of civil religion. **Figure 5.13** provides an adaptation of his explanation of sanctification.

Israeli-Palestinian conflict • Figure 5.12

a. Establishment of the state of Israel in 1948 brought the Zionist dream for a Jewish homeland, reminiscent of the Kingdom of Israel that existed in biblical times, closer to fruition. Simultaneously, however, Palestinians, a largely Muslim population living in the area, were made stateless.

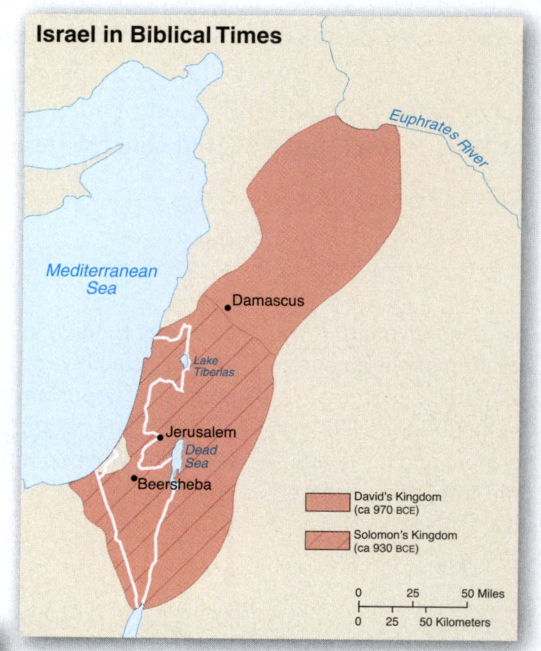

b. The Zionist dream was more fully realized when Israel gained control of the West Bank—part of the biblical lands known as Judea and Samaria and to which Jews have a deep attachment—and access to the Western Wall in Jerusalem after the Six-Day War.

c. The political status of Jerusalem and the West Bank remains highly contested because both Palestinians and Jews feel they have legitimate claims to them. Some Jews have built settlements in the West Bank, though they are considered illegal by the United Nations. Religion is a significant factor in the Israeli-Palestinian conflict, but questions of citizenship and water rights are also involved.

Sanctification • Figure 5.13

Sanctification refers to the making of a sacred site after a significant event has occurred there. Here we consider the example of the former site of the Murrah Federal Building in Oklahoma City, which was bombed in an act of domestic terrorism in 1995. This site is today recognized as a national memorial.

1 The event
The north side of the building in Oklahoma City is blown away by a massive truck bomb on April 19, 1995, killing 168 and injuring over 800.

Five days after the bombing, a makeshift memorial emerges.

2 Informal consecration: emerging consensus and spontaneous memorials
Consecration involves public activities that set a place apart and contribute to its sacredness. Informal consecration begins as people react to the tragic event and a consensus emerges that the site is worthy of remembrance. Temporary memorials make the site hallowed.

Ongoing placement of memorials at the site continues to set it apart and affirm its sacredness.

3 Formal consecration: official consensus for a permanent monument
An organization or other entity establishes the memorial boundaries, acquires legal right to use the land, makes plans for a permanent monument, and manages funds for the perpetual upkeep of the site. This is the Oklahoma City National Memorial, with 168 chairs (one for each victim) and a survivors' tree.

4 Final consecration: solemn dedication of the site
Color guard stands at the Field of Chairs during the dedication ceremony that dignifies this sacred site.

5 Ritual commemoration
Collective memory of the event at the site is promoted through regular practices, such as an annual ceremony, or events like this Memorial Marathon.

Tradition and Change

All religions confront the pressure to change. Indeed, as societies change economically and politically, so do people's attitudes, beliefs, and values. Of the many developments propelling such change, three deserve attention: modernism, secularization, and globalization.

Modernism refers to an intellectual movement that encourages scientific thought, the expansion of knowledge, and belief in the inevitability of progress. The roots of modernism date to the European Enlightenment of the 1700s. To some religious groups, however, modernism is perceived as dangerous because of its potential to contribute to **secularization** by reducing the scope or influence of religion. Similar concerns have been raised about globalization and its capacity to secularize society. In the early 20th century, for example, Pope Pius X formally condemned modernism because of the way it challenged basic Catholic beliefs such as the authority of the Bible. To this day, the tension between tradition and change can still be seen in the way the Vatican resists expanding women's reproductive rights and allowing women to serve as priests.

One area where Hinduism has encountered tension between tradition and change involves castes. The **caste system** refers to a hierarchical form of social stratification historically associated with Hinduism. The ancient Hindu literature known as the *Vedas* (1500–1200 BCE) contains the oldest written description of this system. The *Vedas* identify four social classes called *varnas*, ranked on the basis of purity. The caste system is hereditary, and children are born into the varna of their parents. In the past there was a strong connection between a person's varna and the occupations a person could expect to enter. Later developments introduced a fifth category of people who were considered so impure that they fell outside the varna system altogether—literally outcasts. These came to be known as the "untouchables," and today they are called "Dalits." Although the caste system has been abolished by law and is less meaningful in urban settings, it still affects social relations and interaction in rural areas (**Figure 5.14**).

Resistance to change is sometimes expressed through various kinds of fundamentalism. **Religious fundamentalism** involves an interpretation of the principles of one's faith in such a way that they come to shape all aspects of private and public life. Although the media often report on violent acts of fundamentalist groups, most fundamentalist groups do not espouse violence or terrorism. Moreover, fundamentalist movements have affected many religions, including Christianity, Hinduism, Islam, and Judaism. Christian fundamentalists consider the Bible to be infallible and to provide clear direction on political, moral, and social issues. In the United States, Christian fundamentalists oppose such practices as the teaching of evolution and the legalization of abortion.

Islamic traditionalism, or Islamism, is a movement that favors a return to or preservation of traditional, premodern Islam and resists Westernization and globalization. The al-Qaeda attacks on 9/11 specifically targeted the World Trade Center and the Pentagon

Caste system • Figure 5.14

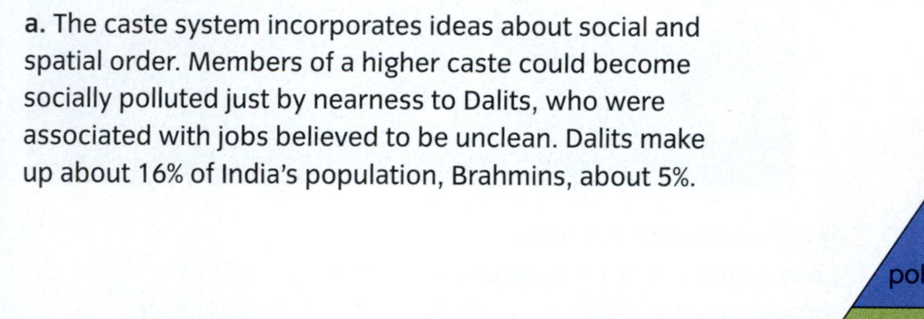

a. The caste system incorporates ideas about social and spatial order. Members of a higher caste could become socially polluted just by nearness to Dalits, who were associated with jobs believed to be unclean. Dalits make up about 16% of India's population, Brahmins, about 5%.

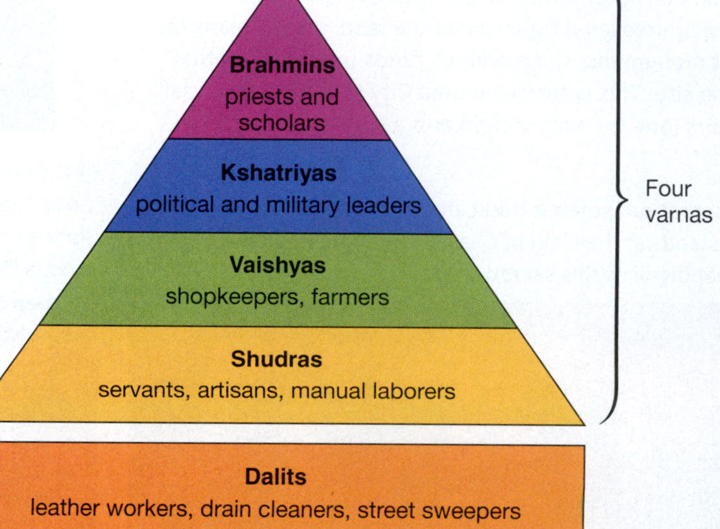

in part because of their association with Westernization, globalization, and the perceived secularization of the world. Islamic traditionalism has existed for many years, but it is only recently that militants within this religious movement have adopted terrorism to help them accomplish their goals.

What is jihad? **Jihad** is popularly understood to mean "holy war," but a preferred translation yields the phrase "utmost struggle." This phrase has two meanings. It can be understood as a personal struggle to uphold the tenets of the faith or as a defense of Islam from threats posed by nonbelievers. In contrast to conventional Islam, some Islamic traditionalists take a very literal interpretation of this second sense, claiming that it justifies the use of terrorism as a way for Muslims to defend their faith. However, most Muslims reject this view.

Religious Law and Social Space

Religion clearly forms a significant part of the everyday functioning of society. Religion has a major presence in numerous dimensions of private and public space. Institutions such as the legal system, education, and government can reflect and be influenced by faith. These close relationships between religion and society may seem odd or unusual to most Americans, who are accustomed to the separation of church and state. Yet, in a number of states, Sunday-closing laws, or *blue laws*—laws that prohibit the sale of certain nonessential merchandise such as alcohol, tobacco, or even motor vehicles—still exist. Blue laws derive from Sabbath-day laws that were once common in England and were designed to encourage proper observance of the day of rest.

Observance of the Sabbath is stipulated in the Ten Commandments and forms a part of the Judeo-Christian tradition but certainly is not unique to it. For Jews, the Sabbath extends from Friday to Saturday evening. As with Christians, there are variations among Jews in how the Sabbath is observed. Orthodox Jews, for example, eschew use of electricity and of motor vehicles on the Sabbath. Jewish law, called *halacah*, goes beyond the Ten Commandments and includes principles given in the Torah as well as other Jewish teachings. Dietary practices associated with keeping kosher, such as not mixing meat and milk products, stem from the divine law of the Torah.

Muslims use the term **sharia** to refer to Islamic law derived from the Qur'an, the teachings of Muhammad, and other sources. Sharia addresses different dimensions of people's lives, such as marriage, divorce, inheritance, and the status of women. Across the Muslim world many different interpretations of sharia exist. Broadly speaking, modernists see the need for sharia to be flexible and open to different interpretations in order to apply in today's society, whereas traditionalists favor narrower, more literal interpretations. Some of the strictest interpretations of sharia are associated with the Taliban. In Afghanistan between 1996 and 2001 and in the Swat Valley of Pakistan in 2009, Taliban forces used sharia to justify their tyrannical control over local citizens, including prohibiting girls from

b. Shown here is a mass conversion to Buddhism in Mumbai, India. As it developed, the caste system sanctioned discrimination. Nowhere was this clearer than in the treatment of the Dalits, who were excluded from Hindu temples and other public facilities. Understandably, many Dalits and other lower caste Hindus have converted to Buddhism.

The status of Islamic law (sharia) • Figure 5.15

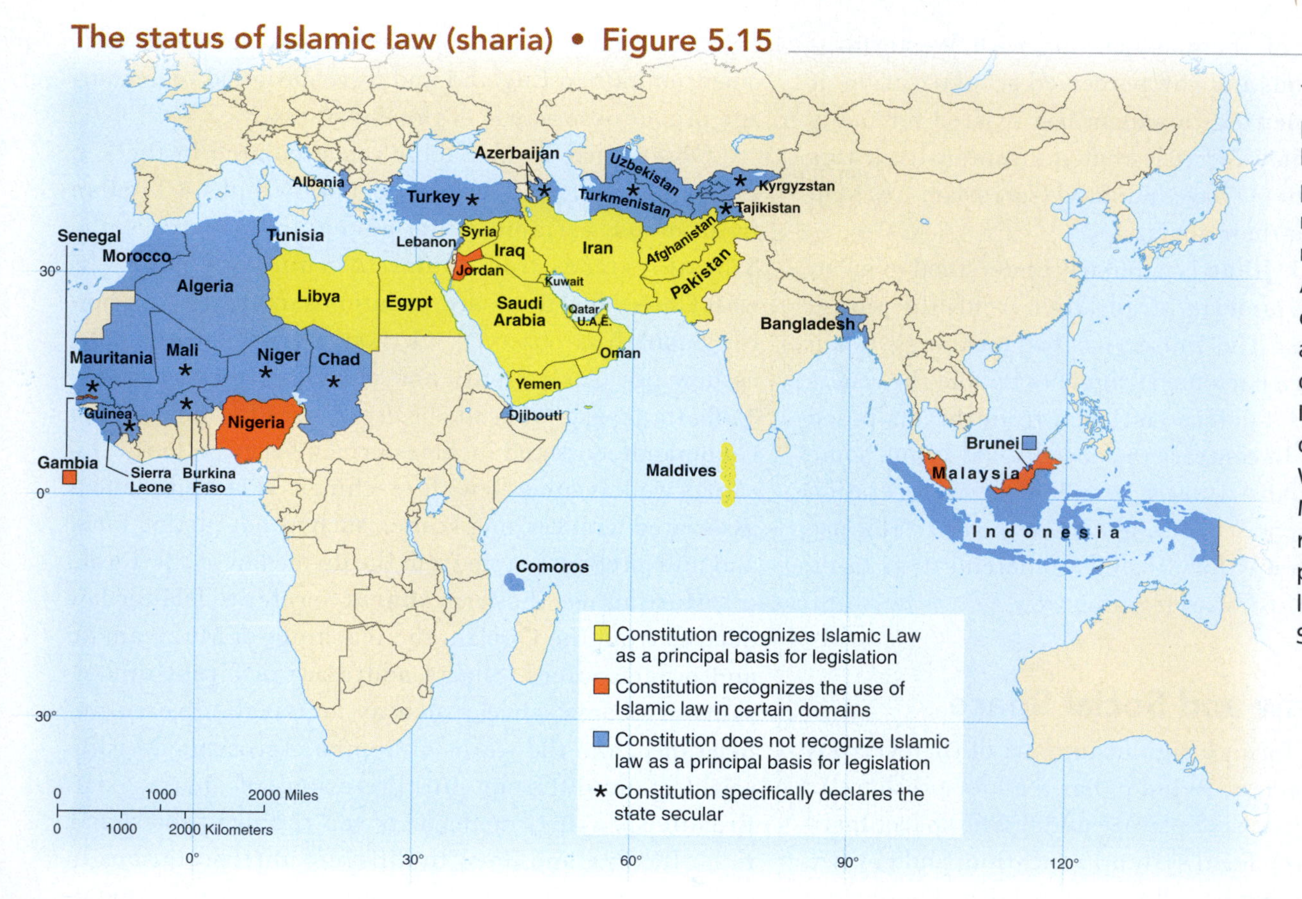

This map shows a core-periphery pattern and highlights a range of perspectives on the relationship between religion and the state. According to Iran's constitution, for example, all laws must be based on Islam. In Malaysia, Islamic law applies only to Muslims. The West African country of Mauritania is an Islamic republic but does not privilege sharia in its legislation. (*Source*: Stahnke and Blitt, 2005.)

attending school. See **Figure 5.15** for a depiction of the spatial variation in the status of sharia across the Muslim world.

Globalization of Renewalism

The fastest growing branch within Christianity is **Renewalism**, a broad term that includes the Pentecostal and Charismatic movements. Examples of Pentecostal denominations include the Assemblies of God and the Church of God in Christ. Pentecostalism emphasizes interacting with and being "filled with the Holy Spirit" (Acts 2:4 NAS). For example, speaking in tongues, faith healing, and the performance of miracles are considered manifestations of the Holy Spirit. Charismatics are Catholics, Protestants, or Orthodox Christians who share a belief in the *charisms*, or gifts, of the Holy Spirit but do not belong to Pentecostal denominations. Note that Christian fundamentalists are not Renewalists and do not share the same beliefs about the gifts of the Holy Spirit.

Uncontrolled bodily movements such as speaking in tongues and altered states of consciousness, have a long historical relationship with different belief systems. For an Indonesian example, see *Video Explorations*.

Video Explorations

Self-Stabbing

In Bali, where Hinduism has mixed with animism, religious devotees undergo a self-stabbing ritual while in a trance. Trance states are common in the rituals of certain faiths, including Renewalism, as for example when adherents speak in tongues. In rituals of this sort, how important is place?

Megachurches • Figure 5.16

a. With 800,000 members, Yoido Full Gospel Church, a Pentecostal church in Seoul, Korea makes Lakewood Church in Houston, Texas, the largest megachurch in the U.S. with an average weekly attendance that exceeds 43,000, seem small. However, some 7 million television viewers also watch Lakewood Church services. What is the attraction of a megachurch?

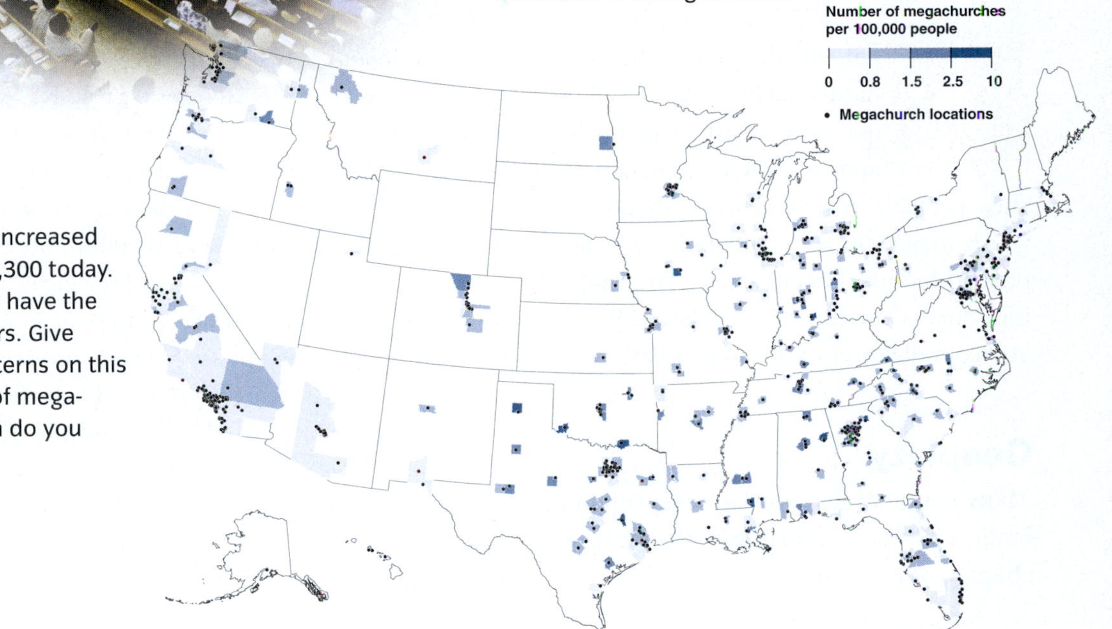

b. The number of megachurches has increased from about 10 in 1970 to more than 1,300 today. California, Texas, Georgia, and Florida have the megachurches with the most members. Give some reasons to help explain the patterns on this map. If you mapped the distribution of megachurches in a single city, what pattern do you think you might find?

Renewalists now constitute the second largest group of Christians, after Catholics. Estimates indicate that about 26% of all Christians are Renewalists and that this proportion will increase to 30% by 2025. In 1970, by contrast, only about 6% of all Christians were Renewalists. The diffusion of Renewalism has been rapid in Latin America, Sub-Saharan Africa, and parts of Asia.

Four factors help account for the success of Renewalism. First, Renewalist missionaries have competently transcended linguistic and cultural barriers. Translating the Bible into local languages, establishing local contacts to serve as liaisons with missionaries, and encouraging local people to give testimonies in their own language have been a key part of this movement. A second and related factor is that Renewalism eschews hierarchies—anyone can receive the gift of the Holy Spirit—and has been accessible to women. Third, the emphasis on spiritual healing and personal experience with the Holy Spirit parallels beliefs about spirituality in many of the local religious traditions in Africa, Asia, and the Americas. Fourth, Renewalist churches have been quick to adopt technology and incorporate popular music. One practice that has clouded the Renewalist movement, however, is the *prosperity gospel*—the idea that one's physical and economic well-being are direct consequences of the financial contributions a member makes to the church. Although exploitation of this sort has occurred, it should not be seen as representative of all Renewalist churches.

megachurch A church with 2,000 or more members that follows mainline or Renewalist Christian theologies.

In the United States, so-called mainline Catholic and Protestant denominations are losing members to Renewalist churches, and **megachurches** have become very popular, especially those with Renewalist theologies (**Figure 5.16**).

CONCEPT CHECK

STOP

1. **How** are pilgrimage and sacred space related?
2. **Why** is sanctification associated with civil religion?
3. **How** is Islamic traditionalism different from religious fundamentalism?
4. **What** is Renewalism, and why has it been successful?

Religion, Nature, and Landscape

LEARNING OBJECTIVES

1. **Distinguish** among geopiety, environmental stewardship, and religious ecology.
2. **Trace** connections between different religions and their expression on the landscape.

What can the geography of religion reveal about nature-society relationships? One answer to this question might focus on how people interpret and view their surroundings. Religion can influence people's environmental perception. These influences, which may stem from religious doctrine, are frequently expressed in views about nature. Alternatively, we can explore the topic of nature-society relationships through a consideration of religious imprints on the land or religious landscapes.

Geopiety

Many connections exist among religion, nature, and the landscape. The Aboriginal Dreamtime discussed in the chapter opener shows that, to the Aboriginals, the Earth is imbued with holiness. A number of religious holidays (originally *holy days*) and rituals in different religions are associated with seasonal changes, including, for example harvest festivals and solstice celebrations. Similarly, the designation of sacred forests, mountains, and rivers attests to other ways in which religion and nature are linked.

Geographer J. K. Wright coined the term **geopiety** to reflect the religious-like reverence that people may develop for the Earth. Geographer Yi-fu Tuan subsequently extended the definition of geopiety in order to encompass the strong attachments that people associate with both sacred and secular places. Two important points follow from this. First, a person need not be religious to experience geopiety. Second, nature can be an important dimension of civil religion. Each member of the Boy Scouts of America (BSA), for example, pledges "to do my duty to God and my country." Being reverent is also a Scout characteristic, and in their activities and service, the BSA has long cultivated a certain appreciation for and understanding of the outdoors.

Environmental awareness • Figure 5.17

a. Indonesians plant mangrove trees to commemorate Earth Day
In the 1980s, the United Nations Environment Program planned an Environmental Sabbath—a single day in June for people to contemplate the relationship between their beliefs and the Earth. The idea did not catch on, although Earth Day celebrations have. What religious beliefs or practices directly affect the environment?

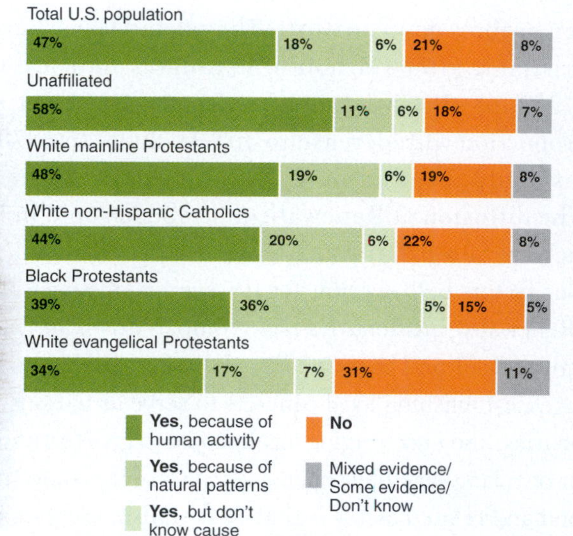

b. Opposing views on climate change
How do different faiths communicate environmental or ecological issues, such as climate change, to their adherents? (*Data from*: Pew Research Center survey conducted April 23–27, 2008, among 1,502 American adults. Results for other religious groups are not reported because of the small sample sizes.)[1]

[1] Full question wording available in the permissions credits section at the end of the book. **Note:** Statistical analysis indicates that political ideology and partisanship, not religious affiliation, are the main determinants of U.S. public opinion on climate change. A full analysis of public opinion on global warming is available in an October 2009 poll report by the Pew Research Center for the People & the Press.

At what point do the connections between religion and nature provide a basis for **religious ecology**—that is, an awareness of the interdependency between people and nature? In a provocative and controversial article in 1967, the historian Lynn White Jr. argued that Judeo-Christian views played a part in contributing to the world's ecological problems because certain biblical scriptures emphasized people's dominion over the Earth. Given the extent of environmental change the world over, it is clear that it is not possible to lay the blame for ecological problems on a specific religion.

Nevertheless, the greater legacy of White's article may be that it drew attention to the relationship between religion and the environment, and specifically the fact that religion can affect how people perceive and use the environment. Indeed, many scholars firmly agree with White (1967, p. 1204) on one point—namely, that "what people do about their ecology depends on what they think about themselves in relation to things around them." Today a number of Christians emphasize **environmental stewardship**—the idea that they should be responsible managers of the Earth and its resources; many other faiths incorporate environmental ethics as well (**Figure 5.17**).

Religion and Landscape

Buildings constructed for religious uses make some of the most distinctive cultural landscapes. These buildings serve important practical and symbolic purposes. The concept of the church varies in Christianity. For Protestants, the church building is not sacred, but it serves as a place for adherents to gather and worship together. For Roman Catholics and Eastern Orthodox, however, the church is a sacred place—the house of God. Mosques are the gathering places for congregational worship in Islam, and Muslims do not consider mosques sacred. Rather, they are buildings where the Muslim community assembles to hear a sermon and to pray.

Buddhists and Hindus generally do not gather for congregational worship on a specific day of the week as do Christians, Jews, and Muslims. Rather, worship tends to be more individual, though some Buddhist temples do have large halls used for instruction or other special events. Hindu temples are considered sacred space and, more specifically, an architectural expression of God. Devout Hindus usually visit a temple once a week; however, most Hindus keep a shrine in their homes for daily worship or *puja* (**Figure 5.18**).

Since customs associated with the disposal of the dead vary from one religion to another, an interesting and sometimes more subtle landscape expression of religion involves *deathscapes*. For example, many Christians share the practice of aligning the graves on an east-to-west axis. Some Buddhists also bury the dead in cemeteries and select specific grave sites based on their *feng shui*, how well they harmonize with the cosmic forces. Although some Christians and Muslims construct mausoleums—large tombs built to accommodate above-ground burials—Judaism requires in-ground burial of the dead.

Theravada Buddhists, Hindus, and Sikhs normally cremate the deceased. As traditionally practiced in India, cremation involves placing the body on a pyre, a large wooden structure that is set on fire. Two of the ghats—the steps down to the Ganges River in Varanasi—are used for cremation. Hindus believe that to be cremated so close to the Ganges brings *moksha*, release from the cycle of death and rebirth. For illustrations of some deathscapes, see *What a Geographer Sees* on the next page.

Hindu temple • Figure 5.18

This is an exceptional example of a Hindu temple in India's Tamil Nadu state. The innermost sanctuary contains a shrine to Lord Vishnu, the god who preserves the universe.

WHAT A GEOGRAPHER SEES

Deathscapes

Necrogeographers study the cultural and spatial variations in the disposal of the dead. Often their work involves mapping data as well as reading the landscape for clues about cultural practices and their change through space and over time. Geographers see cemeteries as fascinating hybrid spaces.

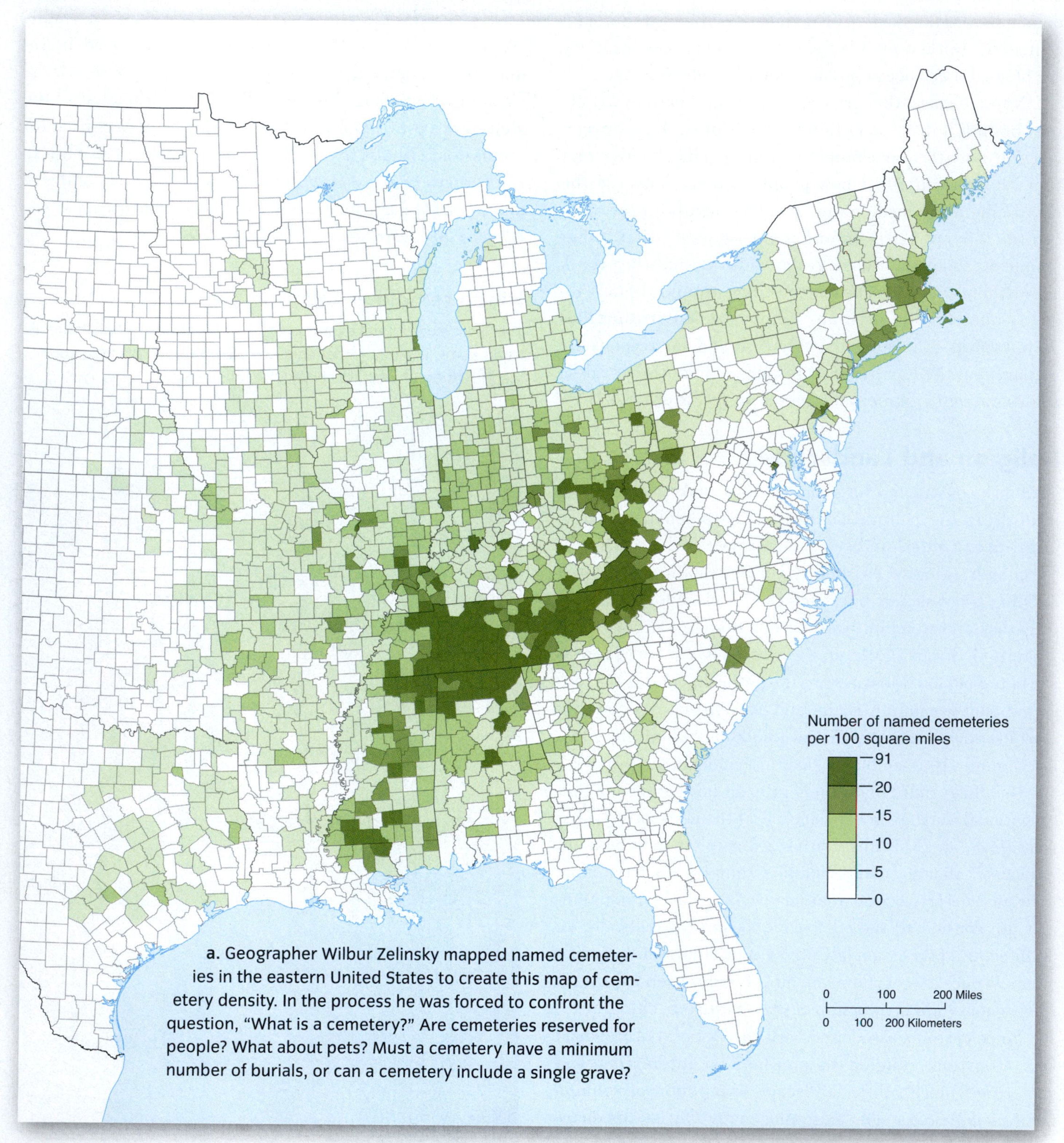

a. Geographer Wilbur Zelinsky mapped named cemeteries in the eastern United States to create this map of cemetery density. In the process he was forced to confront the question, "What is a cemetery?" Are cemeteries reserved for people? What about pets? Must a cemetery have a minimum number of burials, or can a cemetery include a single grave?

b. Eklutna Cemetery in Alaska
Russian Orthodox and Native American practices fused here, as demonstrated by the spirit houses, each fronted with an Orthodox cross. In Orthodox Christianity, burial grounds are sanctified space and the large Orthodox cross shown here testifies to that belief.

c. Holywell Cemetery and wildlife site in Oxford, England
In urban areas, cemeteries can provide valuable natural spaces for people as well as for flora and fauna. To promote biodiversity, this cemetery is not mowed or treated to keep grass and weeds at bay.

d. A columbarium storing a person's cremated remains
This is a high-rise columbarium in Hong Kong. Local customs have traditionally favored in-ground burial in sites with auspicious feng shui. However, land scarcity makes such use of space economically impractical.

e. A mega-mausoleum
This is gravesite of William Andrews Clark Jr., founder of the Los Angeles Philharmonic Orchestra.

Think Critically

1. What factors help explain the patterns on map **a**?
2. How is a kind of hybridity expressed in photo **e**?

CONCEPT CHECK STOP

1. **What** is geopiety, and how is it expressed in religious and nonreligious contexts?
2. **How** do religions vary in their use of space and landscape?

Summary

1 Religion in Global Context 130

- **Religion** refers to the beliefs and practices that people use to understand the universe and their place in it. **Monotheistic, polytheistic,** and **atheistic** religions exist, but these terms provide at best a very rough approximation of a specific faith. Adherents of **animistic religions** believe supernatural forces and deities are present in the natural surroundings. The Candomblé religion in Brazil (see photo) is an example of a religion that is **syncretic.**

Candomblé • Figure 5.1

- Religious beliefs often provide the basis for different **cosmogonies**, which may be celebrated or reenacted through **rituals**. Shared rituals help reinforce a sense of community, belonging, and identity among adherents. Such rituals can also influence individual and group behavior and make visible a person's **piety**.
- Religions can be classified as either universalizing or ethnic. **Universalizing religions** include Buddhism, Christianity, and Islam. Hinduism, Judaism, and the religions of many indigenous peoples are **ethnic religions**. **Civil religion** takes shape when certain aspects of religion become woven into the political culture of an area.
- Christianity grew out of Judaism, and Buddhism grew from Hinduism. Islam is the youngest of the Abrahamic faiths, and Sikhism developed syncretically from the blending of aspects of Hinduism and Islam. Most religions have experienced fragmentation into different branches, creating distributions that are complex and fascinating.

2 Religious Hearths and Diffusion 138

- All religions have a hearth, but the precise details of their development are not always known. Geographers distinguish between primary and secondary hearths. The Semitic and Indic primary hearths have witnessed the emergence of the world's most prevalent and influential religions.
- Geographers and other scholars identify, map, and study religious hearths because they help us to know how religions develop and change over time and across space. The geography and history of some religions, such as Judaism and Sikhism (see photo), have been shaped by **diasporas**.

Sikhism and Khalistan • Figure 5.9

- Religions are dynamic belief systems. No religion is uniform or immune to fragmentation, though some religions have fragmented more than others. In general, the spread of a religion over space is likely to increase the potential for fragmentation.

3 Religion, Society, and Globalization 141

- The identification of **sacred space** demonstrates that religion influences how people perceive and understand the world. **Pilgrimages** involve a journey to and encounter with sacred space; thus they affect patterns of human circulation and mobility. People create sacred places and sites through the process of **sanctification**. Conflict can occur when the sacred spaces of different religions overlap.
- The tension between tradition and change shapes religious beliefs and practices. **Modernism** can be perceived as a challenge to the authority of a religion or even as a form of **secularization**. Globalization and secularization are related, but globalization does not always result in secularization. Indeed, the continued importance of sacred space helps illustrate this.
- In Hinduism, the **caste system**, now abolished, highlights the tension between tradition and change. In Christianity and Islam such tensions are often expressed through **religious fundamentalism**. Since 9/11, **Islamic traditionalism** has received a great deal of attention, but fundamentalism can affect any religion. From Sabbath-day observance to **sharia**, the Abrahamic faiths in particular illustrate some ways in which religious law provides parameters for the management of social space.

- Within Christianity, **Renewalism** has grown very rapidly in the past few decades and has become a global phenomenon. **Megachurches**, such as this one, now rival mainline Protestant and Catholic churches for members.

Megachurches • Figure 5.16

4 Religion, Nature, and Landscape 152

- Religion provides another lens through which to explore nature-society relationships including **religious ecology** and geopiety. **Geopiety** includes the reverence people may have for the Earth as well as the place attachments they develop. **Environmental stewardship** positions people as responsible managers of the Earth and its resources.
- The visible imprint of religion on the landscape varies from place to place and from one religion to another. Religious structures provide visible clues about beliefs and practices in an area. Customs for disposing of the dead also vary by religion and give rise to **deathscapes,** shown here.

What a Geographer Sees

Key Terms

- animistic religion 130
- atheistic 130
- caste system 148
- civil religion 131
- cosmogony 130
- diaspora 138
- environmental stewardship 153
- ethnic religion 131
- geopiety 152
- Islamic traditionalism 148
- jihad 149
- megachurch 151
- modernism 148
- monotheistic 130
- piety 131
- pilgrimage 144
- polytheistic 130
- religion 130
- religious ecology 153
- religious fundamentalism 148
- Renewalism 150
- ritual 131
- sacred space 142
- sanctification 146
- secularization 148
- sharia 149
- syncretic religion 130
- universalizing religion 131

Critical and Creative Thinking Questions

1. Describe a religious landscape in your hometown. Discuss the forces that created it.
2. Why is Protestantism much more fragmented than Roman Catholicism? Why has Islam experienced comparatively less fragmentation than Christianity?
3. Who are the Amish and Mennonites? Do some research to learn of their origins, diffusion, and way of life.
4. View the movie *Bend It Like Beckham* and discuss its representation of Sikhism.
5. Research and then discuss the geography of cremation globally and in the United States.
6. How has the religious symbolism on graves changed over time? Test your hypothesis by visiting a cemetery to gather some data.
7. Why has the celebration of Earth Day been more successful than the celebration of an Environmental Sabbath?
8. What parallels can be drawn between religious practices and proper treatment of the American flag? Is this evidence of civil religion? Explain your reasoning.
9. Think about pilgrimage from the standpoint of spatial interaction. To what extent does the concept of distance decay apply? Discuss, giving specific examples.
10. Note that evidence of at least two different religious traditions is reflected in this photograph. What are they? Is this a religious landscape? Why or why not?

What is happening in this picture?

A Tibetan woman carries a prayer stone to a sacred site. Called *mani* stones, these are hand-carved and inscribed with sacred symbols or Buddhist scripture. The stone wall beside her has also been decorated with religious art.

Think Critically

1. Can virtual pilgrimages, via the Internet, fulfill the same purpose as a personal visit to a sacred site?
2. What are some ways that religion influences gender roles?

Self-Test

(Check your answers in Appendix B.)

1. A religion that has incorporated influences of one or more other faiths is _____.
   a. animistic
   b. polytheistic
   c. syncretic
   d. universalizing
2. Civil religion _____.
   a. draws on religious law
   b. relates religion to politics
   c. promotes atheism
   d. emphasizes fundamentalism
3. Cosmogonies _____.
   a. are initiation rites
   b. stem from the natural landscape
   c. may influence environmental perception
   d. None of the above statements is correct.
4. Which is not one of Islam's Five Pillars of Practice?
   a. completing a pilgrimage to Medina at least once
   b. making a profession of faith
   c. fasting during Ramadan
   d. praying five times a day
5. Explain the significance of this mass conversion to Buddhism that occurred in Mumbai, India.

6. What happens in a secondary religious hearth?
   a. A new religion develops.
   b. A minor pilgrimage destination develops.
   c. Adherents experience a renewal or an awakening of their faith.
   d. A religion splits internally.

7. Find the *false* statement about Buddhism.
   a. Tantrayana Buddhism is closely associated with Japan.
   b. Buddhism is a universalizing religion.
   c. Mahayana Buddhists see the Buddha as a compassionate deity.
   d. Theravada Buddhism tends to be more monastic.

8. Identify the step in the process of sanctification illustrated by this photograph.

____

9. Give an example of hierarchical diffusion in Christianity.

10. Zionism is important to the geography of religion because _____.
    a. it explains why and how diasporic communities form
    b. it demonstrates the spread of secularization
    c. it shows the emergence of a secondary religious hearth
    d. it illustrates the bonds between religious groups and territory

11. The emergence of Sunni and Shia branches in Islam _____.
    a. was about differences over sharia
    b. involved a dispute over Muhammad's successor
    c. was shaped by linguistic differences among Muslims
    d. was about the infallibility of the Qu'ran

12. Religious fundamentalism _____.
    a. endorses terrorism
    b. means that religious principles guide all aspects of an adherent's life
    c. is only associated with Islam
    d. All of the above statements are correct.

13. Which point did Lynn White make?
    a. Any space can be a sacred space.
    b. People around the world should recognize an environmental sabbath.
    c. Judeo-Christian views contributed to environmental abuse.
    d. Jihad means "utmost struggle."

14. This picture illustrates _____.
    a. geopiety
    b. hybridity
    c. resistance to change
    d. None of the above statements is correct.

15. What is Renewalism?
    a. It is a form of secularization.
    b. It is a branch of Judaism.
    c. It refers to the spread of Islamic traditionalism.
    d. It is the fastest growing branch within Christianity.

THE PLANNER

Review your Chapter Planner on the chapter opener and check off your completed work.

6 Geographies of Identity: Race, Ethnicity, Sexuality, and Gender

IDENTITY, GENDER, EQUITY

"Hi, my name is John, and I'm from planet Earth. I've lived in Texas and in Malaysia, but I'm not from those places." So began a series of student introductions in a recent college seminar. When you are in similar situations, how do you construct your identity?

Who you are is a deceptively complicated question that cannot be answered absolutely because every person's identity is contingent. That is, our sense of who we are depends on where we are geographically and temporally, whom we are with, and what we are doing. Identity is both individual and collective. As John's response indicates, our identity is also a matter of scale, and we can express our identity in relation to the individual, local, regional, national, or even planetary scale.

Our identity matters because it affects social relations, and these create or construct different kinds of social spaces that can foster or inhibit opportunities. For example, in the United States, the space of professional sports leagues has predominantly been a masculine space. One exception is the WNBA, the most prominent professional sports league for women, which began in 1996. The WNBA is a legacy of Title IX, legislation enacted in 1972, which requires that all educational programs receiving federal funding provide equal opportunities for men and women. How has the professionalization of women's athletic leagues changed the landscape of sports? What consequences might this have for the way society constructs identities for women and for the geography of gender?

CHAPTER OUTLINE

CHAPTER PLANNER

❑ Study the picture and read the opening story.

❑ Scan the Learning Objectives in each section:
p. 162 ❑ p. 166 ❑ p. 170 ❑ p. 176 ❑ p. 182 ❑

❑ Read the text and study all visuals.
Answer any questions.

Analyze key features

❑ Process Diagram, p. 167

❑ Geography InSight, p. 168

❑ What a Geographer Sees, p. 172

❑ Video Explorations, p. 182

❑ Stop: Answer the Concept Checks before you go on:
p. 165 ❑ p. 169 ❑ p. 175 ❑ p. 181 ❑ p. 186 ❑

End of chapter

❑ Review the Summary and Key Terms.

❑ Answer the Critical and Creative Thinking Questions.

❑ Answer What is happening in this picture?

❑ Complete the Self-Test and check your answers.

Race and Racism

LEARNING OBJECTIVES

1. **Explain** why using race as a classification system is problematic.
2. **Distinguish** between a social construction and an ideology.
3. **Contrast** the geography of the trans-Atlantic slave trade and the geography of human trafficking.

The topics of this chapter—race, ethnicity, sexuality, and gender—are among the most complex and sensitive of any discussed in this book. To ignore these topics, however, would be to overlook a crucial aspect of human geography involving the ways that different groups have modified space and place. Thus, this chapter seeks to balance a celebration of the diversity that makes the world such a rich human tapestry with an examination of some unfortunate developments, too, such as racism, ethnic conflict, and gender inequalities.

You may be wondering why such diverse topics as race, sexuality, and gender inequality are included in a single chapter. These topics have a great deal in common, especially from the standpoint of spatial inclusion and exclusion, or who is deemed to belong in a place and who is excluded from it. We will explore these points in greater detail below, but we begin by considering two basic principles about identity that will guide subsequent discussions: (1) Human identity is dynamic and contingent, and therefore cannot be easily categorized; (2) Efforts to classify people into groups tend to exaggerate the differences among people, especially visible differences.

What Is Race?

Race refers to the highly influential but mistaken idea that one or more genetic traits can be used to identify distinctive and exclusive categories of people. To understand why race is a mistaken idea, we need to see how the concept developed. Initially, four major groups of people inhabiting the Earth were identified: African, American (referring to Native Americans), Asiatic, and European. The Swedish naturalist Carolus Linnaeus developed these categories in the 18th century in conjunction with his system for naming and classifying plants and animals.

Although Linnaeus called these four groups of people "varieties," his method of classification was based on the single trait of skin color, which, to him, appeared to vary by geographic region. Some subsequent scholars insisted there were as few as three races, whereas other scholars identified 30 races or more.

These disagreements over the number of different races highlight a fundamental problem with the concept of race: the boundaries between races are always arbitrary. Geography and biology can help us understand this problem. First, physical traits in people tend to change gradually over space, like transition zones rather than sharp boundaries. Second, no two physical traits (e.g., skin color or hair type) have the same spatial distribution. Thus, both geographical and biological facts demonstrate why race is a mistaken idea. Simply stated, racial categories are subjective and arbitrary, and have no geographical or biological basis. Indeed, when people are divided into so-called races, the amount of genetic variation *within* a single race turns out to be much greater than the variation *among* the races.

The prevailing view among academics is that race is best thought of as a **social construction**—an idea or a phenomenon that does not exist in nature but is created and given meaning by people (Figure 6.1).

How Has Racism Developed?

Racism includes the belief that genetic differences produce a hierarchy of peoples, from the most superior to the most inferior. Racism fuels prejudice, discrimination, and the hatred of others. It results in people being excluded and disadvantaged, as well as emotionally and physically abused. Racism can be thought of as an **ideology** that has been exploited by certain groups, such as the Nazis, at different times.

racism Intolerance of people perceived to be inherently or genetically inferior.

ideology A system of ideas, beliefs, and values that justify the views, practices, or orientation of a group.

Constructions of race • Figure 6.1

Children born to interracial couples often come to know that race is a social construction; from firsthand experiences they learn that people, including their close friends, apply different racial identities to them. For example, U.S. President Barack Obama's mother was a Kansan of English ancestry, and his father was a Kenyan. Some people construct his racial identity as biracial while others construct it as black.

Scholars agree that events associated with the European colonization and settlement of the Americas during the 16th and 17th centuries contributed significantly to the development of racism. Historians have shown that the practice of separating whites and blacks, in particular, became emphasized in colonial Virginia around the time of Bacon's Rebellion in 1676 when poor whites and blacks formed an alliance against white elites. Bacon's Rebellion called attention to the fact that cooperation among whites and blacks could threaten elite control. Thus, racism and slavery provided mechanisms that could be used to maintain social control and power.

It was during the Enlightenment, a European intellectual movement of the late 17th and 18th centuries that probed the relationship between society, nature, and religion, that racial difference became strongly associated with inferiority. One of the more influential ideas to shape collective perceptions in Europe and European colonial outposts was that of the *great chain of being* (**Figure 6.2**).

The great chain of being contributed to an ideology of racism that helped Europeans make sense of the world, but in ways very different from the system of classification developed by Linnaeus. More specifically, the concept of the great chain of being had a far-reaching legacy that can be summarized in the following three points. First, it helped to *naturalize* the idea of human difference. The great chain of being was understood not as a social construction, but rather as a natural construction—a reflection of a God-given hierarchical ordering of the world that included human races. Second, the great chain of being supported views that emphasized the differences between people and provided a way to link those differences to ideas about genetic and intellectual inferiority as well as inequality. Third, whiteness became a standard against which others were measured.

Great chain of being • Figure 6.2

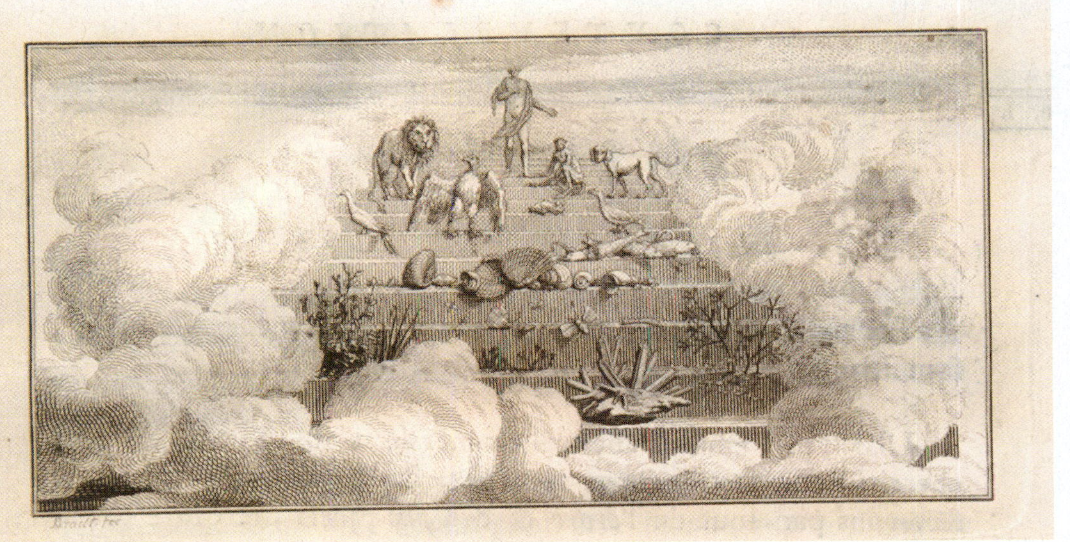

The great chain of being is based on the idea that there is a natural and ranked order of life in the world, stretching from aquatic life at the bottom up to the heavens. If the broad classes of living things were ranked, it followed that the members of those classes could be ranked as well. Visible differences including skin color were used to distinguish among different ranks and races of people (Bonnet, 1779–1783, IV, p. 1).

During the era of European imperialism, this ideology of racism diffused widely and was used not only to justify European dominance and colonization, but also to reinforce European power. The ideology of racism supported the idea that social inequality was a natural phenomenon and that the place for whites was at the top of the social hierarchy. In the colonies, Europeans were typically the administrators, managers, and settlers who controlled the colonized people. Residential areas in colonial towns were segregated on the basis of race (see Chapter 8). This ideology of racism helped to perpetuate slavery. Although slavery is illegal today, it still persists as one kind of human trafficking. **Human trafficking** occurs when people are forcibly and/or fraudulently recruited for work in exploitative conditions—for example, as child soldiers or prostitutes.

Historical and contemporary geographies of slavery • Figure 6.3

The enslavement of people perceived to be of a low status or class is an ancient practice known to have existed in many societies around the world. Only in the Americas, however, did a form of slavery so strongly based on visible physical difference emerge.

a. Generalized origins and destinations of African slaves

The map, with arrows resembling on- and off-ramps, shows the highway that the Atlantic Ocean became as a result of the slave trade between the 16th and 19th centuries. Brazil and the Caribbean were the leading destinations of African slaves. Two factors that sustained the trans-Atlantic slave trade were the strong demand for labor in the Americas and the increasingly common view that racial categories were natural. Note that numbers on the map are estimates. (*Source*: Eltis and Richardson, 2009.)

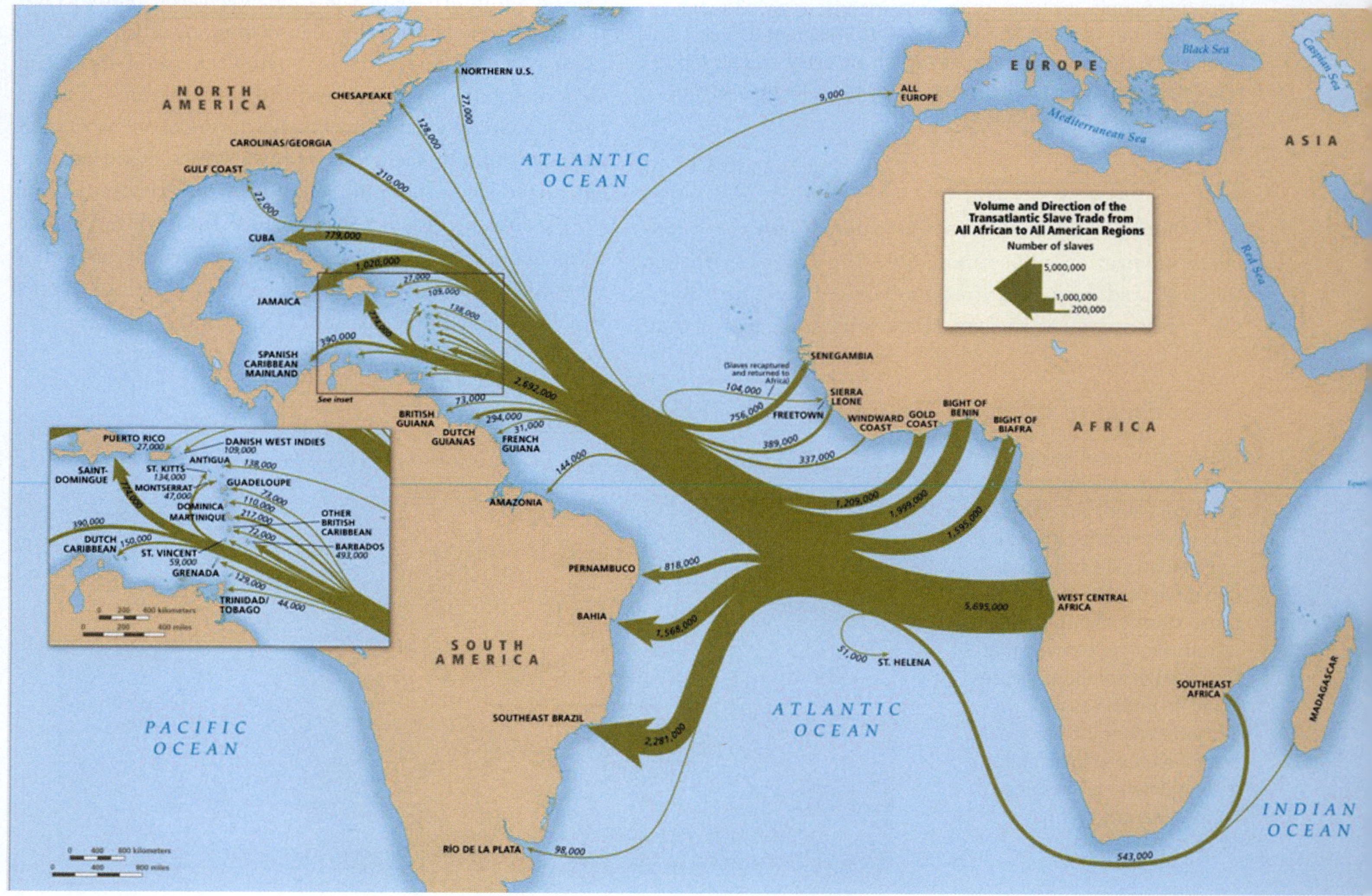

The geography of the trans-Atlantic slave trade was built on well-established routes between Africa and the Americas. Because slavery was legal for centuries, it was visible and remains so in historical documents. In contrast, the geography of human trafficking is global in scale, affecting every region of the world, but it is much less visible. Human trafficking has been called "the underside of globalization" (**Figure 6.3**).

CONCEPT CHECK STOP

1. **What** are the geographic and biological problems associated with using race as a classification system?
2. **How** did racism become an ideology?
3. **How** has the geography of slavery changed over time?

b. Bonded and forced labor
In India, this young girl and her family were trafficked and work as bonded laborers at a brick-making factory. Bonded labor means that the employer has a long-term lease for your services because of a debt that is owed. Often a family's debt is passed on from one generation to the next, creating a vicious cycle of debt slavery.

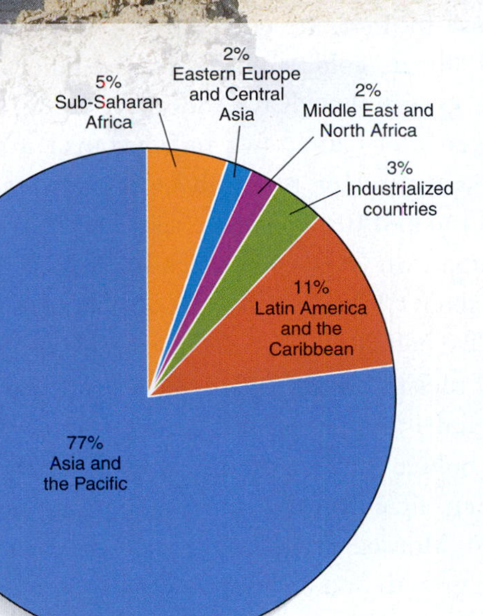

c. Estimated number of forced laborers, by region
Forced labor is regionally most prevalent in Asia, in part because of enduring systems of bonded labor. Perhaps 12 million people work as forced laborers worldwide—as many people as were transported to the Americas as slaves. Workers are recruited with the promise of high wages, but once employed those promises are broken and laborers are forced to work with threats of penalty or harm. (*Source*: ILO, 2005.)

Geographies of Race and Racism

LEARNING OBJECTIVES

1. **Understand** what is meant by institutional discrimination.
2. **Summarize** the relationship between race and place in Vancouver's Chinatown.
3. **Examine** the spatial consequences of the rise and fall of apartheid in South Africa.

Although academics agree that there is no biological basis for race, neither the term nor the concept has fallen out of use. Ideas about race as well as the practice of discrimination more broadly have tended to be persistent and, in some cases, institutionalized. Geographically, race and racism have influenced the spatial organization of people and their activities. In this section we examine how **institutional discrimination**, often grounded in racist views, influenced the development of two specific places: Chinatown in Vancouver, Canada, and the country of South Africa.

institutional discrimination A situation in which the policies, practices, or laws of an organization or government disadvantage people because of their cultural differences.

Race and Place in Vancouver's Chinatown

Chinese immigrants came to the Vancouver area in the mid-1800s as gold prospectors and laborers. In a classic study, geographer Kay Anderson (1987) examined the emergence of Chinatown in Vancouver at the end of the 19th century. Her research demonstrates how attitudes about racial difference gave rise to the strong association of Chinatown with negative characteristics and how these ideas about Chinatown influenced city policies (**Figure 6.4**).

City policies in Vancouver institutionalized an ideology of racism that affected social relations between white Canadians and those of Chinese origin. Over the years, those policies have been removed or altered, and relations between the cultural groups, though not perfect, have improved. Moreover, the many contributions of the Chinese to the growth and development of Vancouver are increasingly acknowledged.

1 Race makes place

Observable differences reinforced the perceived distance between whites mainly of European descent and Chinese. Thus, Chinatown was believed to be an unsanitary place and a center of vice in spite of evidence to the contrary. For example, a number of Chinese operated laundries, the "business of cleanliness," according to Anderson (1987, p. 586), and Chinatown was not a center of epidemics.

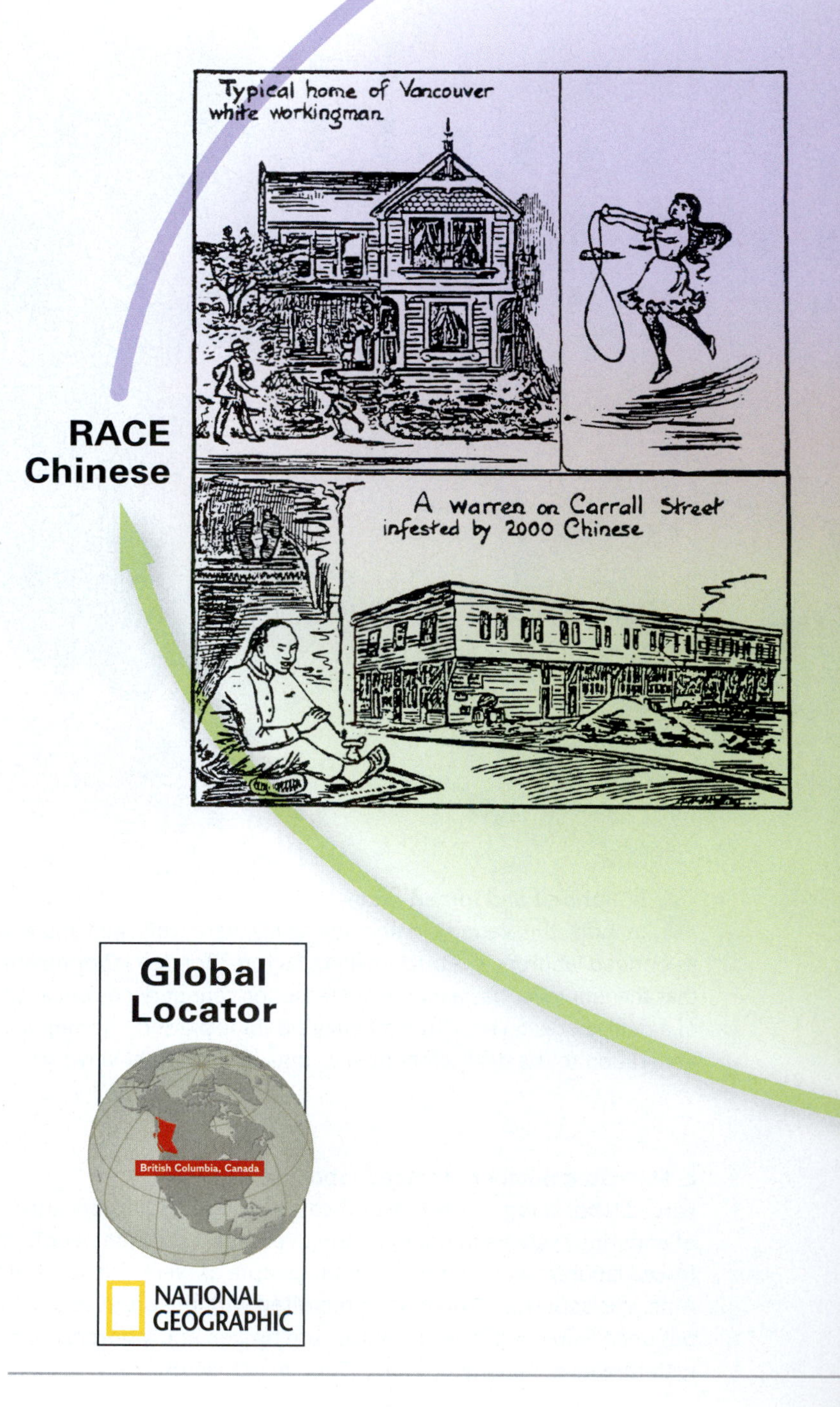

THE PLANNER

The interaction between race and place • Figure 6.4

One of the key contributions of Kay Anderson's research on Vancouver's Chinatown from the 1880s to the 1920s is her demonstration of the mutually reinforcing processes of how race makes place and place makes race. This shows how an arbitrary classification of Chinese as *Orientals* provided a rationale for discrimination against them.

Today, Vancouver's Chinatown is widely perceived to be a popular destination and downtown attraction.

Cultural Difference

Including such characteristics as appearance, language, religion, and other customs

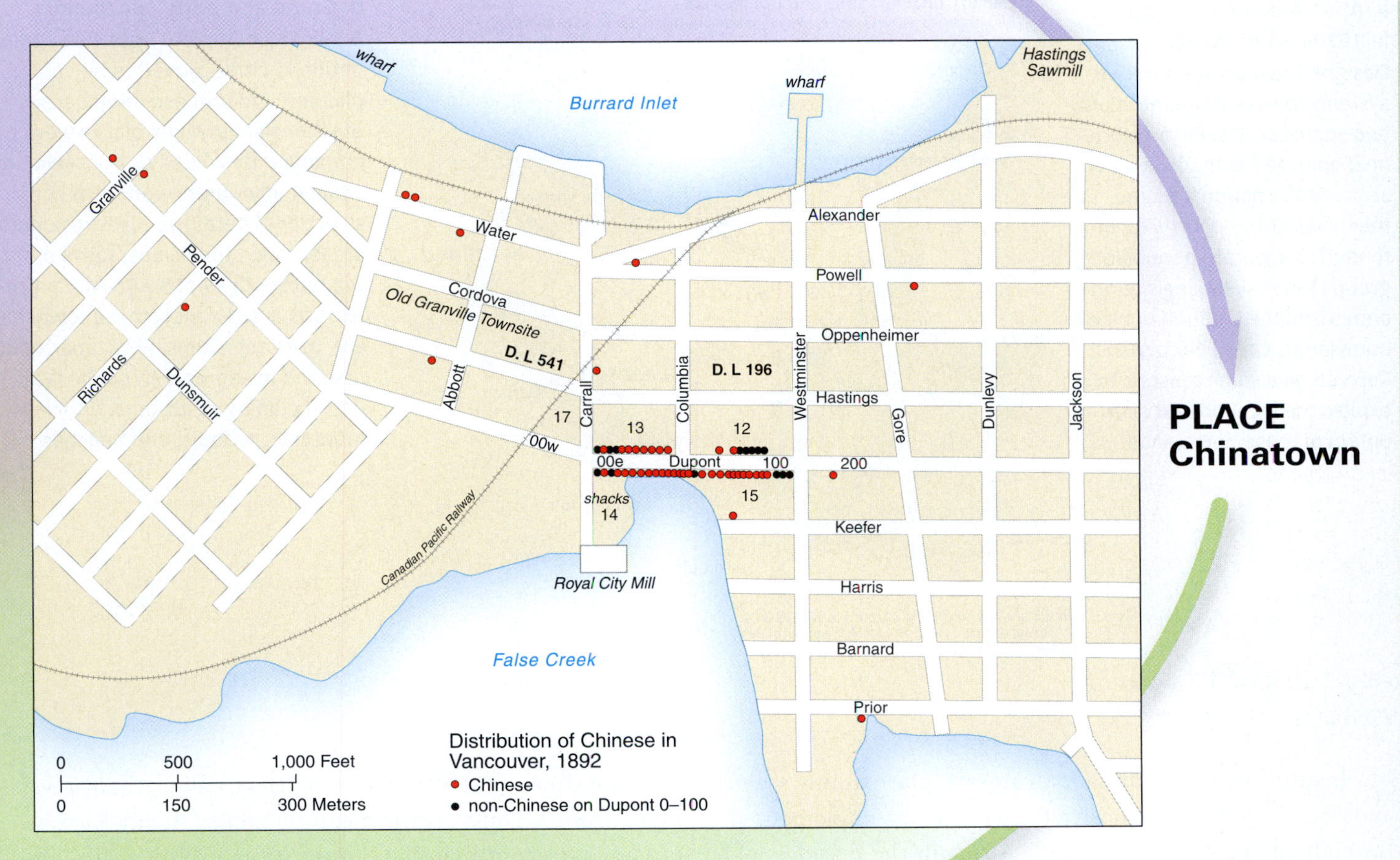

PLACE Chinatown

Institutional Discrimination

Examples of some anti-Chinese city practices:

- Restricting locations of Chinese-operated laundries.
- Treating Chinatown like a factory, a place needing frequent inspection because of its perception as a public nuisance.
- Targeting Chinese residences or buildings deemed unsanitary for destruction.

2 Place makes race

The stigmatization of Chinatown as a socially troubled area influenced the actions of Vancouver's civic leaders, who sought to contain and control Chinatown's "problems" through city rules and regulations. To say that place makes race means that perceptions of Chinatown reinforced racialized views of the Chinese as a different breed of people who needed to be spatially separated and monitored.

Geography InSight The rise and fall of apartheid • Figure 6.5

In theory, apartheid was to produce a society that was segregated on a racial and territorial basis. Every racial group was to have its own geographic space. Underscoring all of this was the myth of racial purity, a myth shared by many Afrikaner Nationalists. Physical separation of the races would therefore protect their racial purity and enable each group to pursue its separate cultural and economic development. In reality, however, no group had the same access to or control of resources as white South Africans did.

a. Apartheid on the map in 1970s South Africa
Designed as a comprehensive system, apartheid operated on two spatial scales. *Petty apartheid* operated at the individual scale. At the national scale there was *grand apartheid*, intended to separate population groups territorially. Legislation authorized the creation of black homelands, called Bantustans. Can you give some reasons to explain the location and distribution of these homelands?

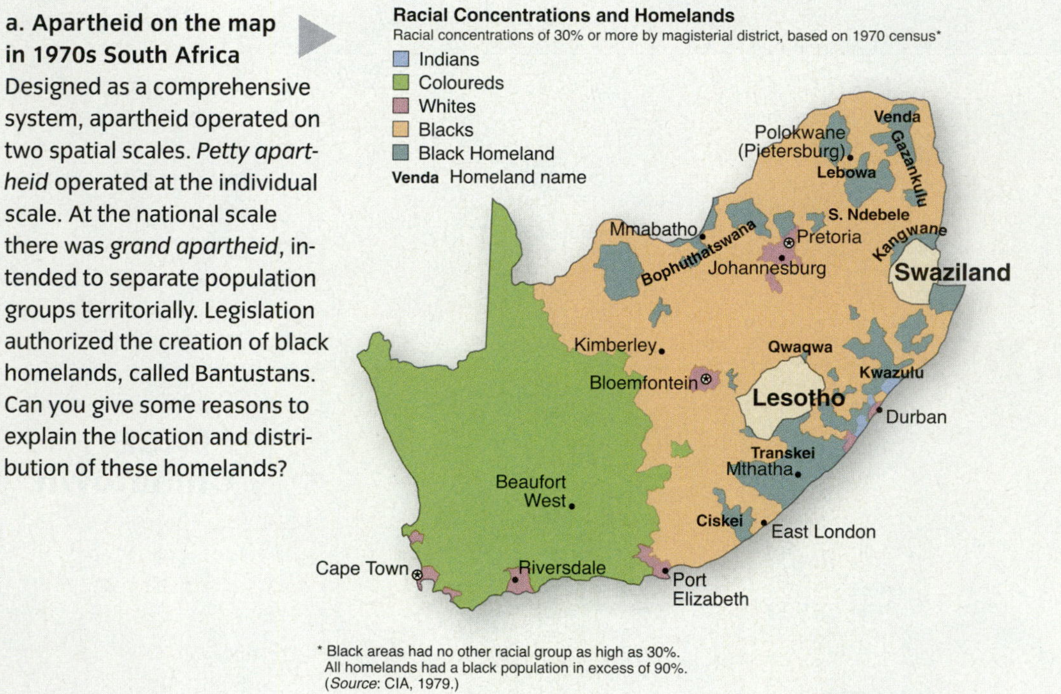

b. Grand and petty apartheid
Achieving complete physical segregation in cities proved to be complicated and unreasonable because of labor needs. Many blacks were employed by whites as housekeepers, while many others worked in construction. Therefore, black residential areas, called *townships*, formed on the outer edges of South African cities. Still in existence, these townships are, in effect, slums. Petty apartheid resulted in segregated public facilities, including restrooms, drinking fountains, schools, and beaches.

Institutional discrimination can take place at any scale, and one of the most extensive forms of it was implemented throughout the country of South Africa in the decades following World War II. As the civil rights movement gathered momentum in the United States and the push to end segregation spread, events in South Africa moved that country in the opposite direction and led to the implementation of apartheid. The word **apartheid**, meaning *apartness*, refers to the government-sponsored policy of racial segregation and discrimination that came to define South Africa and regulated the social relations and opportunities of its people.

Geographies of Apartheid

Southern Africa was already home to numerous different African cultural groups when Dutch settlers arrived in Cape Town in 1652. These settlers called themselves *Boers*, from a Dutch term meaning *farmer*. Today, persons of Dutch descent living in South Africa usually call themselves *Afrikaners* (from *Afrikaans*, the name of the Dutch-derived language they speak).

Toward the end of the 18th century, however, a number of British settlers had also moved into the area. To avoid British rule, most of the Boers eventually migrated away from the Cape Town vicinity and took control of lands in the interior. Beginning in the middle part of the 19th century, the British shipped Indian laborers from their colonies in South Asia to the area around Durban to work on the sugar plantations. In South Africa the term *coloured* (retaining the British spelling) came to identify a person of mixed ancestry who did not fit either

c. Waiting to vote in post-apartheid free elections In 1991 the legal framework of apartheid was repealed. Three years later all South African adults were allowed to vote in national elections for the first time ever, creating long lines at polling stations such as this one. Nelson Mandela, a key figure in the anti-apartheid movement, was elected president.

Population Group	Percent Unemployed
Black African	27.6%
Coloured	21.4%
Indian/Asian	10.9%
White	4.4%

d. Apartheid's deep legacy South Africa continues to implement programs to reduce poverty, but unemployment—especially among blacks—remains very high. (*Source*: *Statistics South Africa*, 2008.)

the African, Asian Indian, or European racial categories. This categorization was different from that in the United States, where the term *colored* was mainly used to refer to blacks.

When South Africa achieved independence in 1910, it formalized a system of white minority rule. At the time, the population composition of the country was about 69% black, 20% white, 8% coloured, and 3% Asian. Then, in 1948, the Afrikaner Nationalist Party gained power. One of the main components of its platform was *baaskap*, a term meaning to keep blacks in their place. According to many Afrikaners, previous governments had not done enough to segregate the races. Apartheid, then, was the mechanism to help achieve baaskap, and it became official state policy in 1948. See **Figure 6.5** to learn more about how apartheid altered South Africa's human geography.

CONCEPT CHECK STOP

1. **How** can racism or discrimination become institutionalized?
2. **How** and why did race and place mutually affect one another in Vancouver's Chinatown?
3. **How** might institutional discrimination alter the spatial arrangement of a city or country?

What Is Ethnicity?

LEARNING OBJECTIVES

1. **Summarize** the relationship between ethnicity and othering.
2. **Identify** different components of ethnicity.
3. **Identify** a regional pattern associated with use of the terms *race* and *ethnicity* on censuses.
4. **Explain** the concept of discourse.

The word *ethnicity* stems from the Greek *ethnos*, meaning people. In a broad sense, ethnicity involves the formation and maintenance of individual and collective identities. Therefore, we take the terms *ethnicity* and *ethnic identity* to mean the same thing, and we will use the terms synonymously. Ethnicity is another complex concept, but it becomes a little easier to comprehend when we emphasize two fundamental points about it. First, ethnicity is about people constructing a sense of social belonging. Second, this process of belonging involves **othering**—the act of differentiating between individuals and groups such that distinctions are made between "me" and "you" and between "us" and "them."

Defining and Characterizing Ethnicity

ethnicity The personal and behavioral basis of an individual's identity that generates a sense of social belonging.

ethnic group People who share a collective identity that may derive from common ancestry, history, language, or religion, and who have a conscious sense of belonging to that group.

Understanding **ethnicity** provides a starting point for identifying and studying **ethnic groups**. Although the terms *nationality* and *ethnic group* are closely related, they are not synonymous. *Nationality* expresses a person's affiliation with a country, usually in terms of citizenship.

When speaking of ethnicity, geographers recognize that it has different facets. For example, ethnicity has internal, personal components as well as external, behavioral components. A personal component of ethnicity includes who we think and feel we are. However, sense of identity may or may not stem from a person's ancestry. A person whose father is Peruvian and whose mother is Japanese, for example, might consider herself to be ethnically Japanese, ethnically Peruvian, a hybrid of both, or none of these. Ethnicity is, therefore, subjective.

One's ethnic identity also depends on how an individual's identity has formed over time. At various times in a person's life, an individual may choose to embrace or reject an ethnic identity. In addition, the process of ascription can strongly influence one's identity. **Ascription** occurs when people assign a certain quality or identity to others, or to themselves (called *self-ascription*). A person's ethnic identity derives from processes of ascription and self-ascription. For example, when Bülent married Leyla, a Kurd, the Turkish community that Bülent belonged to did not approve of the marriage and ascribed to him a Kurdish identity. Bülent now considers himself a Kurd and is accepted as one by the Kurdish community. Ethnicity is, therefore, both subjective and flexible. In the chapter opener we noted that identity is contingent. As a form of identity, a person's ethnicity is contingent on the circumstances and people that shape our lives.

In addition to personal components of ethnicity, there are also behavioral components. These typically include practices that mark who we are, such as language, religious beliefs, and customary traditions associated with dietary preferences, styles of dress, dance, music, or art. But it is often the case that the personal and behavioral components of ethnicity are mutually reinforcing. In other words, one's sense of ethnic identity can depend on or be defined by the practice of certain customs. Another aspect of ethnicity relates to indigenous or native peoples and their identity (Figure 6.6).

There is no agreed-upon definition of indigenous peoples, but three characteristics are commonly cited: (1) ancestral ties to pre-colonial or pre-settler societies; hence indigenous peoples are sometimes called *first peoples*; (2) self-identification of and acceptance by others as a member of an indigenous group; and (3) status in society as a nondominant (e.g., numerically, economically, or politically) group. Thus, indigenous peoples are ethnic groups, but they may prefer to identify themselves by different names, such as aboriginals, tribes, or native peoples. Moreover, their sense of identity is also highly contingent as demonstrated in the following quote:

> [W]hen we are introduced to a man in the village of Mishongnovi on Second Mesa in Arizona . . . we are told his name and that he is a member of the Coyote Clan. When he goes on business to the nearby town of Window Rock, capital of the Navajo Nation, he specifies that he is a Hopi; at a lecture he delivers in Chicago he claims to be Native American and at the Palais Wilson in Geneva,

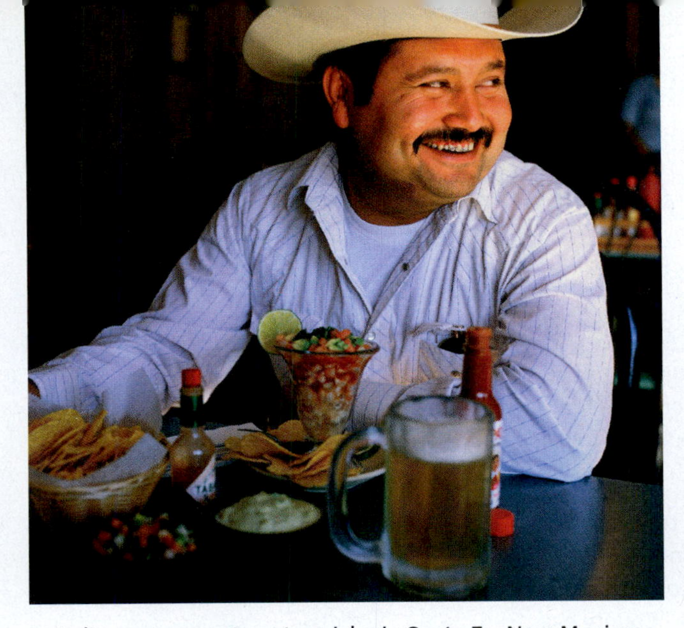

a. A diner prepares to eat ceviche in Santa Fe, New Mexico. Ceviche is popular among some Latinos. The dish is customarily made with raw seafood marinated in a mix of lime juice, chili peppers, and other seasonings. To what extent are your food preferences an expression of your ethnicity?

Aspects of ethnicity • Figure 6.6

b. Bolivian president Evo Morales takes part in a purification ritual prior to his formal inauguration. Morales (center) is an Aymara Indian and Bolivia's first indigenous president. What components of ethnicity are visible here?

as he sits between a Dayak woman from Kalimantan, Indonesia, and an Ogiek man from Kenya while attending an international human rights conference, he identifies himself, and is identified by others, as indigenous. (Levi and Maybury-Lewis, 2010, p. 4)

Ethnicity, Race, and the Census in Global Context

Like race, ethnicity is a subjective social construction that defies the use of fixed categories that neatly divide and classify people. Although collecting data on ethnicity proves to be extremely difficult, many governments insist on identifying different ethnic or racial groups and enumerating them. During the apartheid era in South Africa, government officials required this enumeration so that they could maintain social and political control. In countries such as Brazil, Canada, the United Kingdom, and the United States, such enumerations help to identify minority groups and prevent discrimination. For example, the U.S. Equal Credit Opportunity Act is a civil rights law requiring the collection of information about the ethnic or racial identity of loan applicants as a way to monitor fairness in lending practices.

Even though many countries collect data in order to chart ethnic or racial affiliation, there is considerable spatial variation in the use of terms such as *ethnicity* and *race* around the world. The sociologist Ann Morning has produced a fascinating study of the geographic differences in the terminology and meaning of words, including *ethnicity*, *race*, *indigenous group*, and *nationality* on official censuses. Her study reveals that most countries do enumerate their populations on the basis of some form of national, ancestral, or ethnic identity. The most common term used in such questions is *ethnicity*. Globally, the United States is exceptional in that its census asks three separate questions: one about race, one about ethnicity, and one about ancestry. "More specifically," Morning explains, "census usage of *race* is found almost entirely in the former slaveholding societies of the Western Hemisphere and their territories" (Morning, 2008, p. 248). See **Table 6.1**.

Census terminology Table 6.1

Countries or territories with census questions about race*

World Region	Country or territory
Africa	Zambia
Caribbean	Anguilla (U.K.)
	Bermuda (U.K.)**
	Jamaica
	Saint Lucia
	Puerto Rico (U.S.)
	Virgin Islands (U.S.)
North America	United States
South America	Brazil
Oceania	American Samoa (U.S.)
	Guam (U.S.)
	Northern Mariana Islands (U.S.)
	Solomon Islands

* Based on a sample of 138 official censuses. (*Source*: Morning, 2008.)

** Bermuda, in the mid-Atlantic, is usually grouped with the Caribbean Islands.

WHAT A GEOGRAPHER SEES

U.S. Census Geography

Many geographers use data from the U.S. Census Bureau in their work. Effective use of this information requires knowledge of the geographic areas for which data are reported. These areas include the following scales: national, state, county (or parish or borough), and city block. We will start at the state scale and use California as an example.

a. California counties We'll look at Los Angeles County, home to 9.8 million people, or approximately 27% of the state's population.

b. Census tracts of Los Angeles County Counties are divided into census tracts. Los Angeles County has some 2,000 census tracts. The size of a census tract varies, but each contains, on average, 4,000 people.

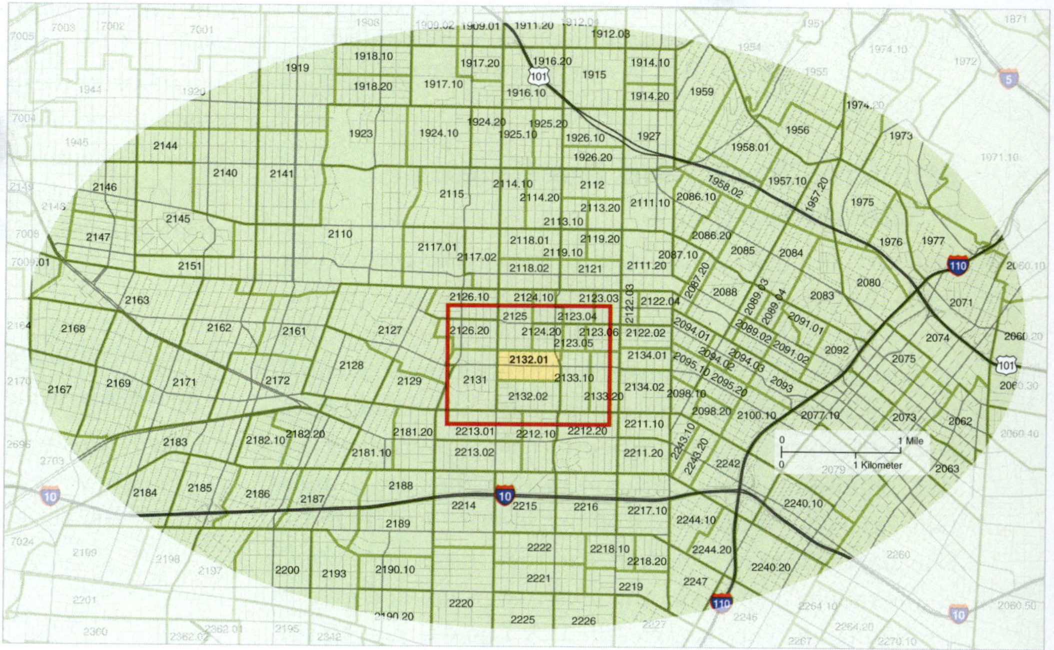

c. Census tracts near downtown LA Every census tract is assigned a unique number. Census tract 2132.01 is located in the area known as Koreatown. (*Source*: U.S. Census Bureau, 2000d.)

Ethnicity, Race, and the Census in the United States

The United States Census Bureau is the government agency responsible for enumerating the country's population. A crucial reason for the census is to ensure proper political representation (see Chapter 7) and to plan for the provision of public services. The Census Bureau collects a wide range of data about the population and the economy, such as age, education, household size, employment, and commuting patterns in order to better understand important trends. Human geographers and other researchers greatly appreciate the volume and diversity of data collected by the Census Bureau. Indeed, many geographical studies would not be possible without U.S. Census data. For an introduction to U.S. Census geography, see *What a Geographer Sees*. For access to U.S. Census data, follow the link in *Where Geographers Click*.

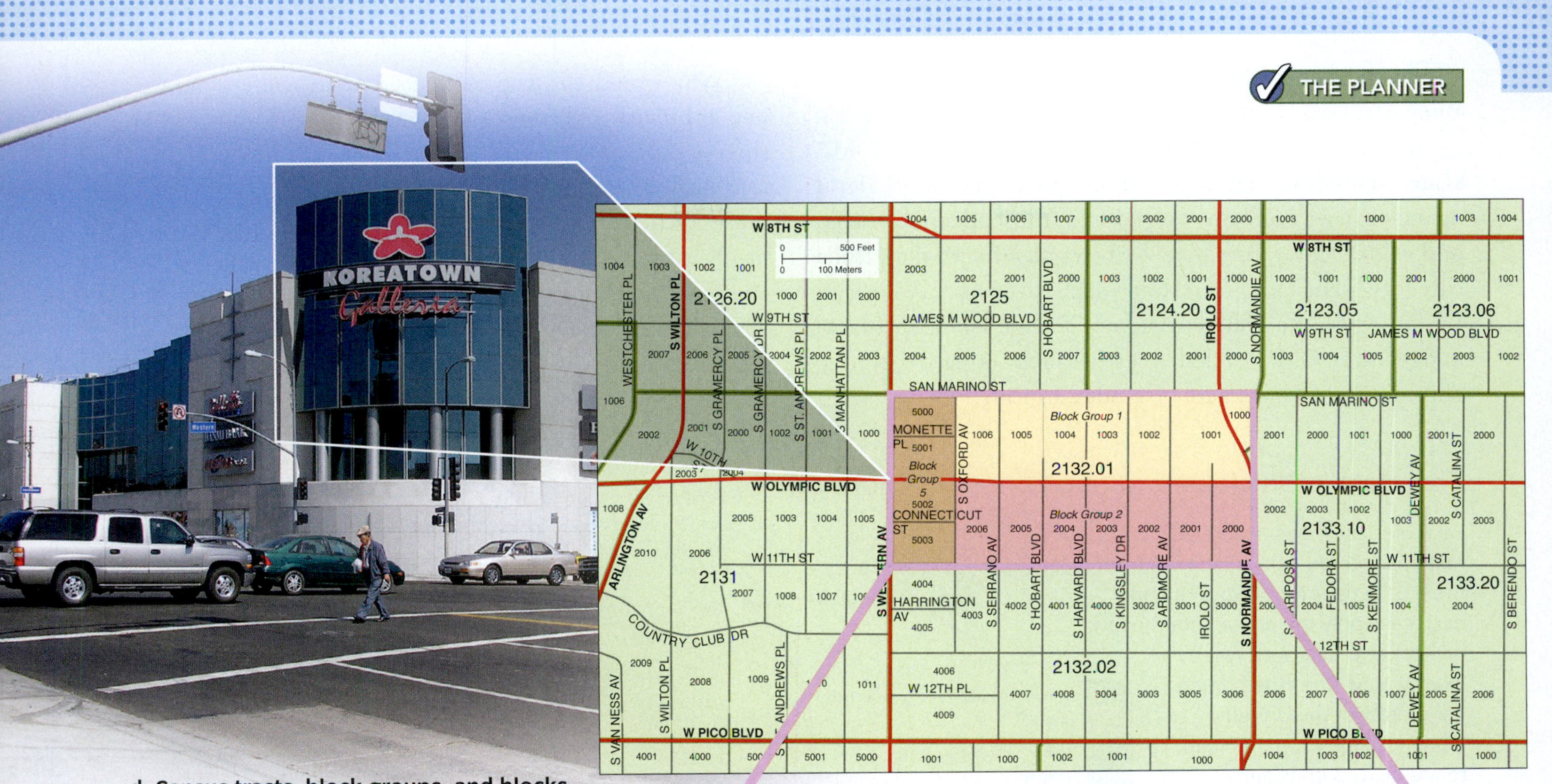

d. Census tracts, block groups, and blocks Each census tract consists of one or more block groups. There are three block groups in census tract 2132.01. Block groups are subdivided into blocks, the smallest geographic unit for which census data are reported. The Koreatown Galleria, shown here, is located in Block Group 5, block 5002. (*Source:* U.S. Census Bureau, 2000a.)

e. Census tract data This table shows the population composition of census tract 2132.01, by race and ethnicity. (*Source:* U.S. Census Bureau, 2000b.)

Total population	3,855	100.0%
Hispanic or Latino (of any race)	2,127	55.2
Not Hispanic or Latino	1,728	44.8
One race	1,696	44.0
White	86	2.2
Black or African American	127	3.3
American Indian and Alaska Native	3	0.1
Asian	1,471	38.2
Native Hawaiian and Other Pacific Islander	5	0.1
Some other race	4	0.1
Two or more races	32	0.8

Think Critically

1. Why is the scale of the block group needed?
2. Why are census data not reported for blocks?

Where Geographers CLICK

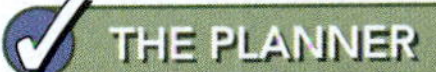

American Factfinder

http://factfinder.census.gov/home/saff/main.html?_lang=en

This is an online mapping resource provided by the U.S. Census Bureau. You can use this website to view and create thematic maps of different kinds of census data. From the link above, click on Maps and then Thematic Maps.

In the United States the Office of Management and Budget (OMB) is charged with the supervision of federal agencies, including the Census Bureau. In 1977 the OMB issued Directive 15, which mandated that all federal agencies use uniform procedures for collecting racial and ethnic data. Directive 15 recognized a single ethnic category—Hispanic—and four racial categories: American Indian or Alaska Native, Asian or Pacific Islander, Black, and White. OMB added a fifth racial category, Native Hawaiian or Other Pacific Islander, in 1997.

Why would the U.S. Census Bureau collect data on one ethnicity but multiple races? One answer is that this practice represents an artifact of the guidelines established by Directive 15. A more critical response, however, involves the role of **discourse**. For example, a conversation is a kind of a discourse, and the language that we use—or fail to use—reveals the meanings we give to

discourse Communication that provides insight on social values, attitudes, priorities, and ways of understanding the world.

Prevalent ethnicities and ancestries • Figure 6.7

Each of these maps is derived from responses to questions (shown) from the U.S. Census form. Note that in each case only the prevalent ethnicity, race, or ancestry is shown by county. How does the human geography of the United States change from one map to the next? Significantly, the U.S. Census is based on self-reported data. How do you respond to the different census questions? (*Source*: Suchan et al., 2007; U.S. Census Bureau, 2000c.)

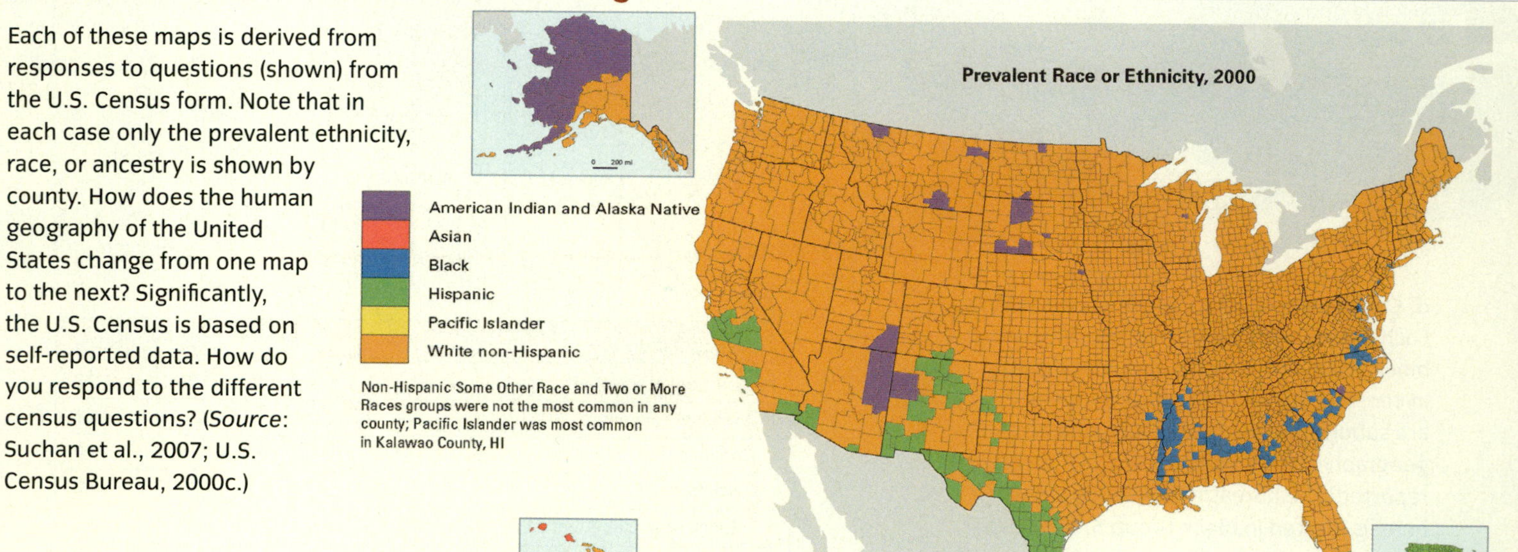

Is Person 1 Spanish/Hispanic/Latino? *Mark* ☒ *the* ***"No"*** *box if* ***not*** *Spanish/Hispanic/Latino.*

☐ **No,** not Spanish/Hispanic/Latino
☐ Yes, Puerto Rican
☐ Yes, Mexican, Mexican Am., Chicano
☐ Yes, Cuban
☐ Yes, other Spanish/Hispanic/Latino — *Print group.*

What is Person 1's race? ***Mark*** ☒ ***one or more races*** *to indicate what this person considers himself/herself to be.*

☐ White
☐ Black, African Am., or Negro
☐ American Indian or Alaska Native — *Print name of enrolled or principal tribe.*

☐ Asian Indian
☐ Japanese
☐ Native Hawaiian
☐ Chinese
☐ Korean
☐ Guamanian or Chamorro
☐ Filipino
☐ Vietnamese
☐ Samoan
☐ Other Asian — *Print race.*
☐ Other Pacific Islander — *Print race.*

☐ Some other race — *Print race.*

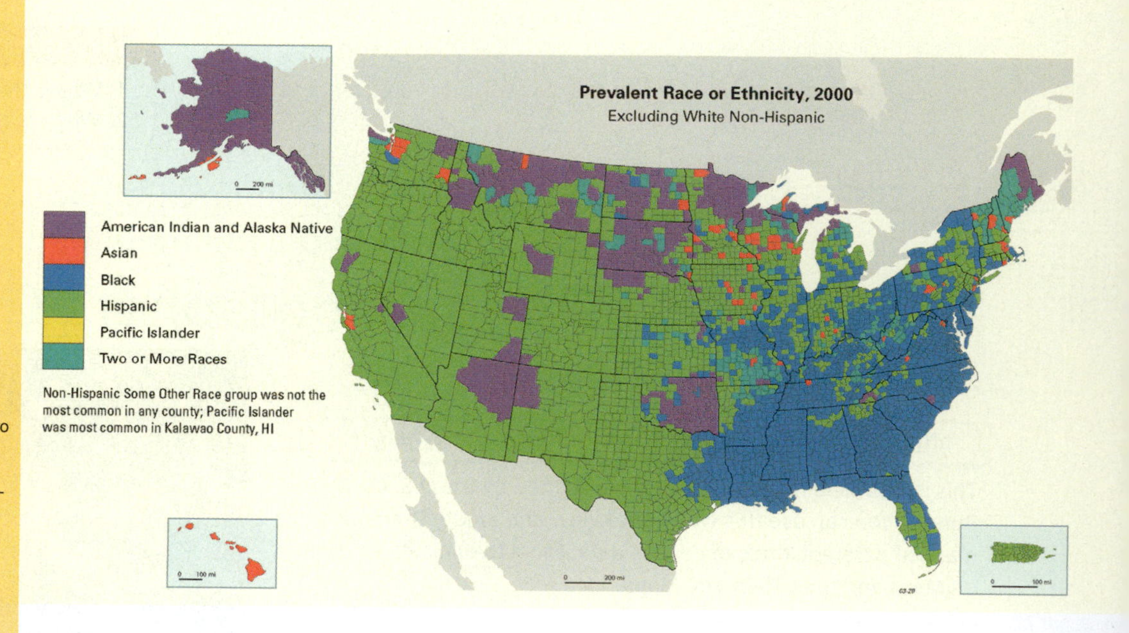

places, people, and events. Institutions, including legal systems and entities such as the U.S. Census Bureau, also create discourses.

The way in which the U.S. Census gathers information about race and ethnicity is part of a broader discourse of identity that has developed around two important issues. The first issue is that a person's race can be objectively defined. The second and related point is that the U.S. Census uses ethnicity in a way that does not capture the sense of social belonging that ethnicity involves. In other words, one's membership in a racial or ethnic group is still based on skin color (for example, white or black) and ancestry (for example, Chinese or Native Hawaiian). Bear in mind that these issues are contrary to some of the main characteristics of race and ethnicity already presented in this chapter.

The discourse of the U.S. Census, then, reinforces and legitimizes certain ways of thinking about race and ethnicity even though they may be misleading or inconsistent. In fact, the U.S. Census has often been inconsistent in its classification of the population by ethnicity and race. Some scholars have argued, for example, that the census has been anchored to a discourse that draws on a binary vision created by the categories of "white" and "nonwhite." The "nonwhites" are further subdivided into blacks, American Indians, Chinese, Asian Indians, and so on, but "white" remains a singular category that is not subdivided.

The Census Bureau clearly plays an influential role in shaping a discourse of identity in the United States. Specifically, it contributes to the erroneous assumption that race is real. The Bureau's actions and policies carry authority because it is the key institution charged with enumerating and classifying people. According to Kenneth Prewitt, former director of the Census Bureau, "[T]he public face of America's official racial classification is its census" (Prewitt, 2005, p. 6). Thus, it is instructive to critically appraise the procedures the Census Bureau uses so that we are better informed users of its data (**Figure 6.7**).

CONCEPT CHECK

STOP

1. **How** does othering relate to ethnicity?
2. **What** are some different components of ethnicity? Give some examples.
3. **How** does census collection of data on ethnicity and race vary globally?
4. **How** and why does the U.S. Census contribute to a discourse of identity?

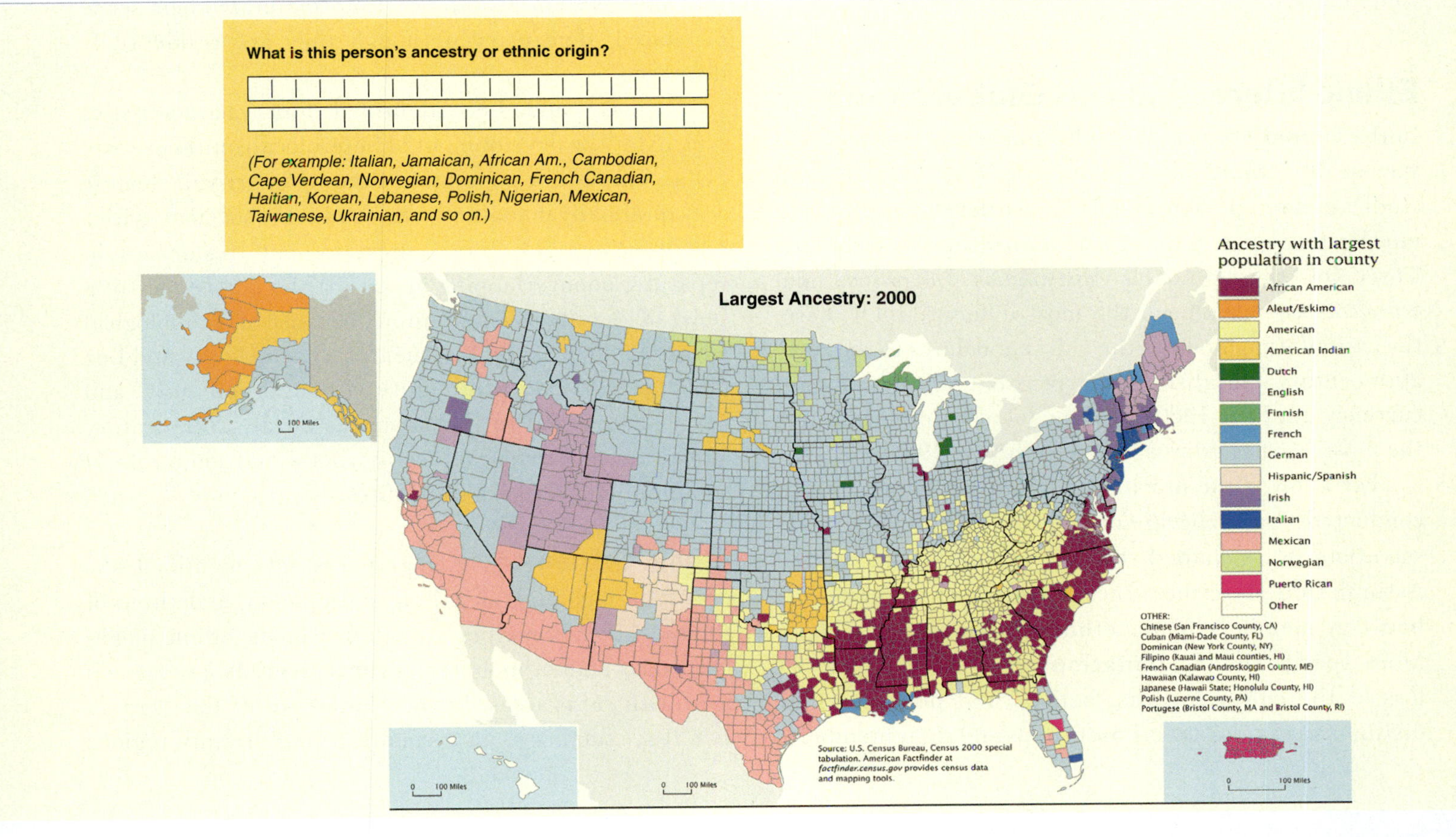

Ethnicity in the Landscape

LEARNING OBJECTIVES

1. **Identify** three models of ethnic interaction.
2. **Distinguish** between ethnic islands, ethnic neighborhoods, and ethnoburbs.
3. **Explain** symbolic ethnicity.
4. **Explain** the reasons for ethnic conflict and violence in Sudan.
5. **Define** environmental justice.

thnic geography is a subfield of human geography that studies the migration and spatial distribution of ethnic groups, ethnic interaction and networks, and the various expressions or imprints of ethnicity in the landscape. The study of such ethnic imprints, or **ethnoscapes**, has traditionally focused on identifying and documenting distinctive examples of material culture including religious buildings, community centers, murals, and ethnic restaurants. Lately, the study of ethnic imprints has broadened to include radio and television stations that cater to particular ethnic groups, as well as Internet sites that support ethnic communities.

ethnoscape A cultural landscape that reveals or expresses aspects of the identity of an ethnic group.

Ethnic Interaction and Globalization

In the United States, the study of ethnic geography, and specifically ethnic interaction, tends to overlap with studies of immigration. Scholars have developed different models to depict and help explain ethnic interaction. Three of these models—*assimilation*, *pluralism*, and *heterolocalism*—have been the most influential. Of these, the assimilation model is the oldest and dates to the early 20th century. Pluralism is nearly as old but did not gain currency until the 1960s. The model of heterolocalism is the most recent and was first proposed in 1998.

For a long time much scholarship on ethnic groups conducted by geographers, historians, and social scientists was shaped by the assimilation model. Assimilation describes the outcome of interactions between members of an ethnic group and outsiders. More specifically, **assimilation** refers to the gradual loss of the cultural traits, beliefs, and practices that distinguish immigrant ethnic groups and their members. The assimilation model promotes the view of society as a melting pot. The **pluralism** model, by contrast, builds on the premise that members of immigrant ethnic groups resist pressures to assimilate and retain those traits, beliefs, and practices that make them distinctive. The pluralism model gives rise to the tossed salad metaphor for society, and geographically to the idea of *ethnic enclaves*—areas with a notable concentration of members of an ethnic group.

In a groundbreaking study, geographer Wilbur Zelinsky and the sociologist Barrett Lee coined the term *heterolocal*, literally meaning *different place*, and argued that globalization had so fundamentally altered the patterns and consequences of ethnic interaction that a new model was needed. The concept of **heterolocalism** means that members of an ethnic group maintain their sense of shared identity even though they are residentially dispersed.

Zelinsky and Lee identified four characteristics crucial for the development of heterolocalism. These are: (1) an immigrant population that, upon arrival, clusters only minimally if at all; (2) the conduct of social activities (shopping, employment, entertainment, residence) in separate, nonoverlapping areas; (3) the persistence of a sense of identity as a community because of technological advances, such as the Internet, or what Zelinsky and Lee (1998, p. 285) call "community without propinquity" and (4) a history tied to the processes of late-20th-century globalization. Cartographically, what would a heterolocal community look like? See **Figure 6.8**.

Ethnic settlements Geographers have identified several different kinds of ethnic settlements, and three of the more common types are ethnic islands, ethnic neighborhoods, and ethnoburbs. **Ethnic islands** are associated with rural areas. In the United States they vary in size from smaller than a county to multicounty regions

Heterolocalism • Figure 6.8

a. Distribution of ethnic Vietnamese by census tract
Some 40,000 Vietnamese live in and around Washington, D.C., but at the level of the census tract nowhere do they make up more than 18% of the population. By contrast, they accounted for less than 1% of the population in most census tracts. (*Source*: Airriess, 2007.)

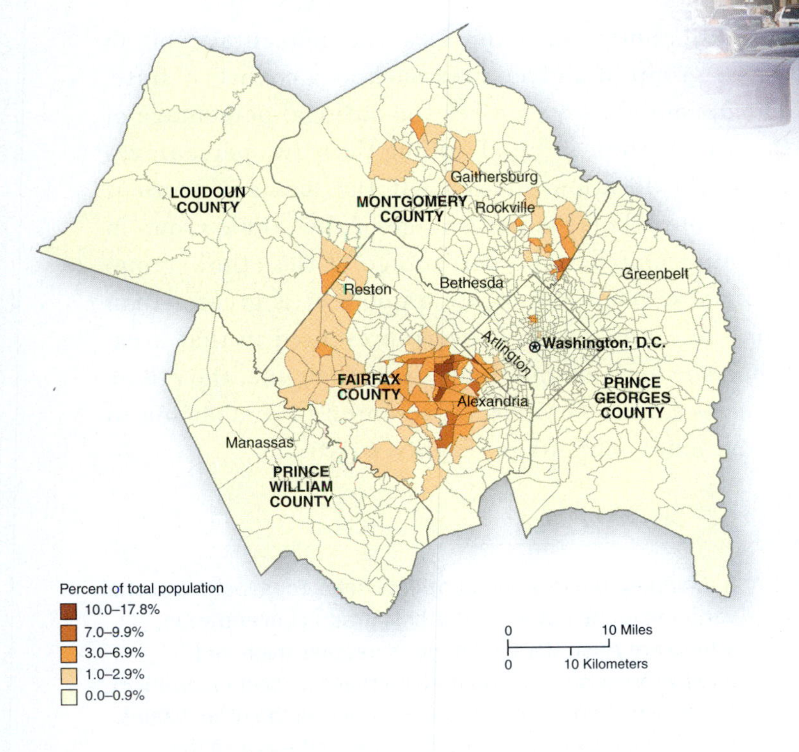

b. Eden Center in Falls Church, Virginia
This shopping center caters to a predominantly Vietnamese clientele and those with a taste for Vietnamese and other Asian cuisine. The clock tower resembles one in Saigon, and the yellow flag is the flag of the former South Vietnam.

that may extend into several states. In terms of population they may have fewer than 100 residents or several thousand (**Figure 6.9**).

In contrast to ethnic islands, **ethnic neighborhoods** develop in urban areas but may vary in scale from a few city blocks to sizable districts within a city. Such places as Chinatown and Little Italy are examples of ethnic neighborhoods. A *ghetto* is a type of ethnic neighborhood. The first ghettos were involuntary settlements created in the Middle Ages when Jews were forced to live in one district of the city. In the United States, the term *ghetto* refers to inner-city neighborhoods with predominantly African American populations. As impoverished and economically marginalized areas, ghettos—like Chinatowns—have historically been racially stigmatized spaces.

Ethnic islands • Figure 6.9

The Hopi and Navajo Reservations (with respective populations of approximately 7,000 and 180,000) have an unusual geography such that a portion of Navajo territory is surrounded by the Hopi Reservation, which in turn is surrounded by the Navajo Reservation, creating an ethnic enclave within an ethnic enclave.

Ethnoburbs are multiethnic suburban settlements associated mainly with large metropolitan areas. Two forces play a role in the formation of ethnoburbs: economic globalization and social stratification. Greater economic globalization and changes in immigration policies have diversified the pool of new immigrants to the United States and contributed to the multiethnic composition of ethnoburbs. Also, many of these immigrants are well educated and work in jobs that are linked to highly globalized sectors of the economy, including international trade and banking. Some of these immigrants even shuttle frequently between the United States and their country of origin, maintaining global as well as local social networks. Significantly, ethnoburbs are mosaics. In contrast to ethnic neighborhoods, ethnoburbs function as both commercial and residential areas. Moreover, the population within ethnoburbs is not only ethnically diverse, but it is also economically stratified, with low-skill, low-wage workers as well as more affluent residents.

ethnoburb
A multiethnic residential, commercial, or mixed suburban cluster in which a single ethnic group is unlikely to form a majority of the population.

Ethnic groups and location quotients When geographers want to know how the proportional presence of an ethnic group in a region compares to the proportional presence of that same ethnic group in the country, they use the **location quotient**. The location quotient (LQ) can be calculated using the following formula:

$$LQ = \frac{\text{(Population of ethnic group in an area/Total population of that area)}}{\text{(National population of ethnic group/National population)}}$$

The numerator expresses the percentage of an ethnic group in a given area such as a county or state. The denominator expresses the national percentage of that same ethnic group. Dividing these two percentages yields the location quotient. If the location quotient equals one, then the percentage of the ethnic group in the state matches the percentage of the ethnic group nationally. When the location quotient is greater than one, the ethnic group is overrepresented in that area. When the location quotient is less than one, the ethnic group is said to be underrepresented in comparison to the national percentage (**Figure 6.10**).

Location quotients • Figure 6.10

a. The table presents population data for Australia and its indigenous populations, the Aboriginals and Torres Strait Islanders. Nationally, they constitute just 2.5% of the population, and only in the Northern Territory do they make up more than one-quarter of the population. (*Source*: Australian Bureau of Statistics, 2009.)

Australian state or territory	Indigenous population*	Total population*	Percentage indigenous	Location quotient for indigenous populations
New South Wales	152,685	6,816,087	2.24%	0.90
Victoria	33,517	5,126,540	0.65%	0.26
Queensland	144,885	4,090,908	3.54%	1.42
South Australia	28,055	1,567,888	1.79%	0.72
Western Australia	70,966	2,059,381	3.45%	1.38
Tasmania	18,415	489,951	3.76%	1.50
Northern Territory	64,005	210,627	30.39%	12.16
Australian Capital Territory	4,282	334,119	1.28%	0.51
Australia	517,043	20,697,880	2.50%	

* Numbers are estimates.

b. Mapping the location quotients helps to reveal the spatial variation in the patterns of ethnic group concentration. Where are Australia's indigenous peoples over- and underrepresented? If you noticed that the location quotient for the Northern Territory is well above all the other values, then you have discovered one of the limitations of this measure. Although the location quotient can be less than one, it can never be lower than zero. This means that the location quotient is not as effective at capturing the degree of underrepresentation of a particular group.

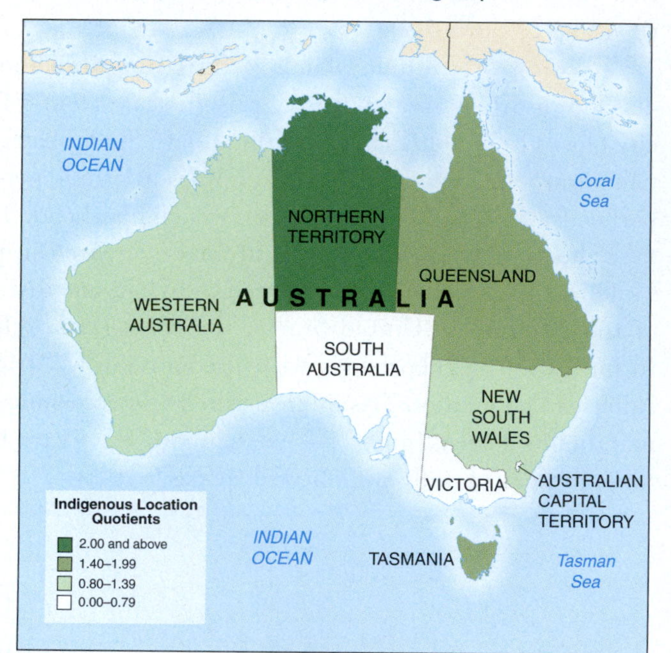

Other Ethnic Imprints

On a spectrum of visibility, ethnic imprints may be subtle, highly visible, or virtually anywhere between these two possibilities. For example, Joseph Sciorra used yard shrines and sidewalk altars as markers of Italian American residences in New York City. Ethnic radio and television are other examples of ethnic imprints. Since 1998 Asian Indians in the United States with access to satellite television have been able to tune into Zee TV, a station based in India which broadcasts in Hindi and other Indian languages.

Ethnic restaurants are one of the most visible ethnic imprints, but even within this category a great deal of variation exists. For example, Chinese and Mexican restaurants are widespread in the United States, but in the United Kingdom Indian and Chinese restaurants are more prevalent. Why do you suppose that Pakistani, Bangladeshi, Filipino, or Nigerian restaurants are less common even in countries with diverse immigrant populations?

Public festivals also provide a highly visible expression of ethnicity and identity. In the Crown Heights neighborhood of Brooklyn, for example, immigrants from several of the Caribbean Islands hold a yearly festival they call the West Indian Carnival. Like Mardi Gras, Carnival developed as a pre-Lenten celebration in which participants design and wear elaborate costumes. Carnival was originally a winter festival held indoors, but in Brooklyn today it takes place outdoors on Labor Day and attracts more than 3 million visitors. Some now refer to the festival as the "Labor Day Carnival" (**Figure 6.11**).

To some observers, festivals are at best contrived and staged examples of ethnicity that generate business and tourist revenue. But this interpretation neglects to consider the complexity of ethnic identity and the role of **symbolic ethnicity**—the way in which a collection of symbols imparts meaning and identity to members of an ethnic group. These symbols can be thought of as expressions of ethnicity and often include such things as flags, music, styles of dress, and cuisine.

Participation in or attendance at festivals such as Carnival, then, is a component of symbolic ethnicity and may in fact be very meaningful in terms of reaffirming a sense of belonging individually and collectively. Ethnic geographers agree that it is important to study the evolution of different traditions such as Carnival in order to understand how they have changed over time and what they mean in the context of symbolic ethnicity, rather than simply dismiss them as contrived or inauthentic.

West Indian Carnival • Figure 6.11

Participants in the annual Carnival parade in Crown Heights, Brooklyn. Once home primarily to persons of European descent, the Crown Heights area of Brooklyn is experiencing ethnic succession. Today Crown Heights has a sizable population of immigrants from the Caribbean and a smaller population of Orthodox Jews.

Ethnic Conflict

No discussion of ethnicity would be complete without a consideration of ethnic conflict. The phrase *ethnic conflict* is commonly used to characterize conflicts, including the crisis in Darfur (a region portrayed as troubled by ethnic and tribal rivalries), the breakup of Yugoslavia (described as a product of historical ethnic animosities among the Croats, Serbs, Bosnians, and others), and the unrest in Sri Lanka (cited as an ethnic conflict between Sinhalese and Tamils).

Many human geographers, however, find the label *ethnic conflict* to be problematic because it tends to suggest a single cause for conflict: ethnic difference or ethnic hatred. Most conflicts, however, have multiple causes such as political exclusion or disputes over land or access to other resources. Events in the Darfur region of Sudan provide a good example of the complex and multifaceted nature of conflict.

Conflict in Sudan • Figure 6.12

a. Sudan is Africa's largest country, and the Darfur region is comparable in size to Texas. Since 2003, more than 400 villages have been destroyed, approximately 2 million people have been displaced, and between 100,000 and 400,000 people have been killed. Beyond Darfur, how might oil also factor in the tensions between northern and southern Sudan?

b. A Sudanese man and member of a rebel group looks at the charred remains of a house in a Darfur village that was burned by janjaweed militia. Some scholars note that globalization has made it easier to obtain weapons—an additional factor that has affected this conflict.

The origins of the conflict in Darfur are rooted in tensions that have grown between sedentary farmers and nomadic pastoralists over access to water and land. In Darfur (meaning land of the Fur people), the Fur, Masalit, and Zaghawa are sedentary farmers of African descent, whereas the pastoralists are Arab-descended. All groups are Muslim, though indigenous and Christian religions are practiced elsewhere in Sudan, mainly in the southern parts of the country. The conflict is also linked to a desire among many southern Sudanese for self-rule. Historically, disputes among these groups in Darfur were settled through indigenous systems of mediation. Only recently has the conflict in Darfur become charged with ethnic and racial overtones.

A Sudanese military coup in 1989 brought Omar Hassan al-Bashir to power. In an effort to consolidate power and control over Sudan's southern regions, his regime has used ethnic difference to stoke hatred. During the civil war, the Bashir regime armed Arab groups and directed them to wage war on non-Arab southern Sudanese. When a separate rebellion broke out in Darfur in 2003, the Bashir regime employed a similar strategy of organizing and equipping Arab militias to quell the rebellion. Known as the *janjaweed* (a word for armed men riding horses or camels), these government-backed militias perpetrated **ethnic cleansing** (**Figure 6.12**).

ethnic cleansing
The forced removal of an ethnic group from an area.

Proximity to environmental burdens in the San Francisco Bay area • Figure 6.13

Certain large industrial facilities (called toxic release facilities) are required to report to the Environmental Protection Agency amounts of toxins they emit into the air. NIMBY—not in my backyard—expresses how most people feel toward the siting of such facilities.

a. Population by race/ethnicity and proximity to a toxic release facility

How does the composition of population groups change with increasing distance from the toxic release facility? (*Source*: Pastor et al., 2007.)

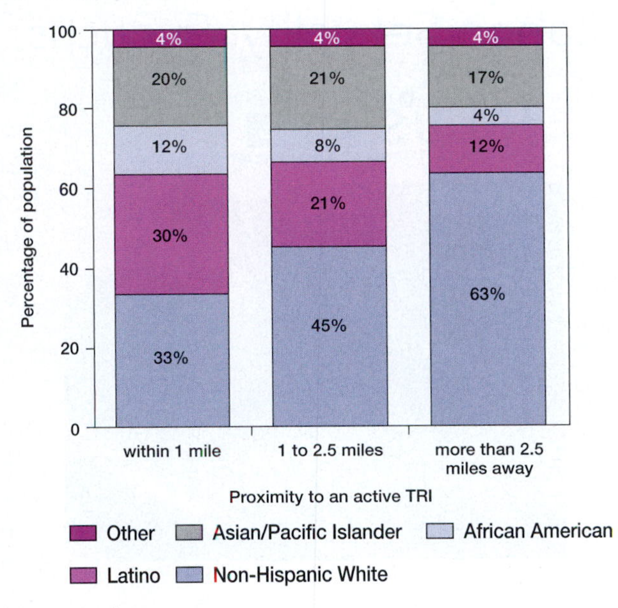

b. Households within one mile of a toxic release facility by income, race/ethnicity

Compare this graph to the one in (a). What do these data suggest in terms of the distribution of environmental burdens in the San Francisco Bay area and the interconnectedness of geography, race or ethnicity, and income? What other factors need to be considered? (*Source*: Pastor et al., 2007.)

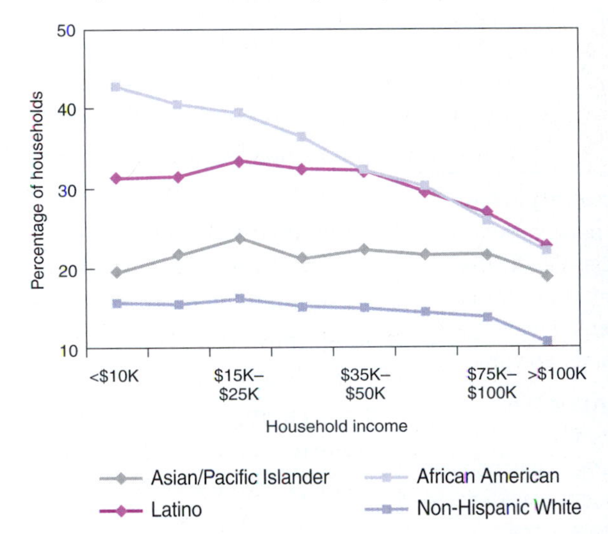

Environmental Justice

Another factor that can contribute to tension among ethnic groups is the lack of equity in the distribution of hazardous sites or facilities. *Equity* means without favoritism or bias. The movement for environmental justice originated in the 1980s amid a growing awareness that certain groups had little involvement in decisions that affected the locations of toxic waste sites, industrial hazards, and other undesirable uses of land. According to the U.S. Environmental Protection Agency, "**Environmental justice** is the fair treatment and meaningful involvement of all people regardless of race, color, national origin, or income with respect to the development, implementation, and enforcement of environmental laws, regulations, and policies." Since 1994 all federal agencies have adopted the principle of environmental justice.

At the global scale, the disposal of toxic waste, including electronic or e-waste generated from discarded computers, TVs, and other electronic devices, is a major area of concern for advocates of environmental justice. Considerable volumes of contaminated waste are exported, sometimes illegally, to developing countries for recycling or disposal. When the recycling or disposal is not effectively regulated, however, workers can be exposed to dangerous acids and environmental contamination can occur.

Environmental justice and its counterpart environmental injustice are inherently geographical because they involve understanding what groups of people and which places are disproportionately burdened by exposure to environmental hazards (see **Figure 6.13**).

CONCEPT CHECK

STOP

1. **What** are three models of ethnic interaction, and how are they different?
2. **How** is globalization related to the formation of ethnoburbs?
3. **What** is symbolic ethnicity? Give an example to illustrate it.
4. **What** circumstances have shaped the conflict in Sudan?
5. **How** can geography inform the study of environmental justice?

Sexuality and Gender

LEARNING OBJECTIVES

1. **Distinguish** between sexuality and gender.
2. **Explain** how ideas about sexuality can affect the use of space.
3. **Explore** the geographical variation in gender roles and gender gaps.

Two other important facets of identity include **sexuality** and **gender**. Both are social constructions; however, the distinction between the sexes has been based on a dichotomy once thought to be exclusive. Biologically, females have two X chromosomes and males have one X and one Y chromosome. But we now know that human biology is not always this straightforward. For example, some individuals inherit only one X chromosome (XO), or an extra X or Y chromosome (XXY or XYY, respectively). None of these chromosome patterns fits either the female (XX) or male (XY) categories. Thus, it is clear that human sexuality cannot be reduced to a simple binary or twofold classification of individuals into females and males. Moreover, the problems of classifying people according to their chromosomal makeup resemble the problems we mentioned earlier of attempting to classify people into separate and distinct racial categories.

sexuality A basis for personal and social identity that stems from sexual orientation, attitudes, desires, and practices.

gender The cultural or social characteristics society associates with being female or male.

The term **gender role** conveys the idea that there are certain social expectations, responsibilities, or rights associated with femininity and masculinity. Notice how a dominant discourse in society privileges a binary view about sexuality and gender. This discourse specifically reinforces the thinking that there should be a correspondence between the two. In other words, society expects that a person who is female will behave in ways that reinforce and reflect a female gender role and a feminine identity. However, a person's sexual identity may or may not be related to her or his biological or chromosomal makeup. **Transgendered** persons, for example, do not identify with the gender assigned them at birth (see *Video Explorations*).

THE PLANNER

Video Explorations

Taboo Sexuality: Eunuchs

WILEY PLUS NATIONAL GEOGRAPHIC Video

A eunuch is a castrated male, but in India the term *eunuch* or, more commonly, *hijra* also encompasses people who are born male but identify as women. Thus, eunuchs are referred to as a third sex or third gender in India. Paradoxically, they are considered to be endowed with spiritual powers, yet they are also discriminated against. The video highlights a group of eunuchs in their struggle to make a life in India, where society privileges the conventional view that the existence of two genders, male and female, is "normal."

Sexuality, Identity, and Space

As individuals, our identity may be shaped as much by our sexuality as by our gender, ethnicity, ancestry, and other unique biographical details. Until the late 1970s, however, geographers ignored the ways in which sexuality contributed to identity, promoting instead the fiction that the world could best be understood through approaches that privileged a **heterosexual norm**—a binary vision of the sexes based on clearly defined masculine and feminine gender roles. Today a number of geographers study the ways in which diverse sexualities influence the configuration and use of space.

The relationship between sexuality and space makes sense if we think about how language is often used to describe sexuality. For example, we talk about gays and lesbians as "coming out." The heterosexual norm promotes the idea that those who are not heterosexual are socially deviant and need to suppress, ignore, or hide their identity and occupy or use different social spaces. This norm has geographic ramifications because it can affect certain kinds of space, including public space. **Public space** can be thought of as a kind of commons—a space open and accessible to anyone. Sometimes the enforcement of heterosexual norms turns public space into space that is actually not public. One of the clearest examples of this involves the conflict over the issue of gays in the military. Instead of allowing gays and lesbians to be open about their identity, the U.S. military adopted the "don't ask/don't tell" policy. This is still the military's official policy, and it means that gays and lesbians are not to speak about their sexual orientation. The policy effectively imposes limitations on free speech in a public space.

public space A kind of commons; a space intended to be open and accessible to anyone.

During the 1950s and 1960s, police sometimes raided private establishments frequented by gay, lesbian, and transgendered individuals. In these instances the police monitored private spaces in the name of public decency. Local statutes against solicitation, public lewdness, or vagrancy, for example, were used to justify arresting the patrons of gay bars. This kind of harassment and institutional discrimination worked to forge a sense of shared identity among gay, lesbian, bisexual, and transgendered individuals and to launch the gay rights movement following the Stonewall Rebellion in New York City in 1969. The Stonewall Rebellion occurred when customers at the Stonewall Inn, a gay bar, resisted and fought against police during a raid of the bar. These events helped geographers and other scholars realize that sexuality can often influence a person's political identity (**Figure 6.14**).

Geography and Gender

Janet Townsend, a geographer at the University of Durham, provided one of the most succinct statements about the relationship between geography and gender when she observed that "gender matters to geography and geography matters to gender at all places and scales" (1991, p. 25). She went on to explain that a thorough understanding of a particular place must involve an examination of the ways in which gender intersects with and affects the lives and activities of people in that place. One way to accomplish this is to consider the spatial variation in gender roles.

Same-sex marriage • Figure 6.14

The status of same-sex marriage and other legal alternatives such as domestic partnerships and civil unions demonstrates the politically charged nature of sexuality. Most U.S. states have both a law and a constitutional amendment that define marriage as the union of a man and a woman and, in effect, prohibit same-sex marriage. Same-sex marriage is legal in five states and the District of Columbia. (*Source*: Data from Vestal, 2009; HRC 2010a; HRC 2010b.)

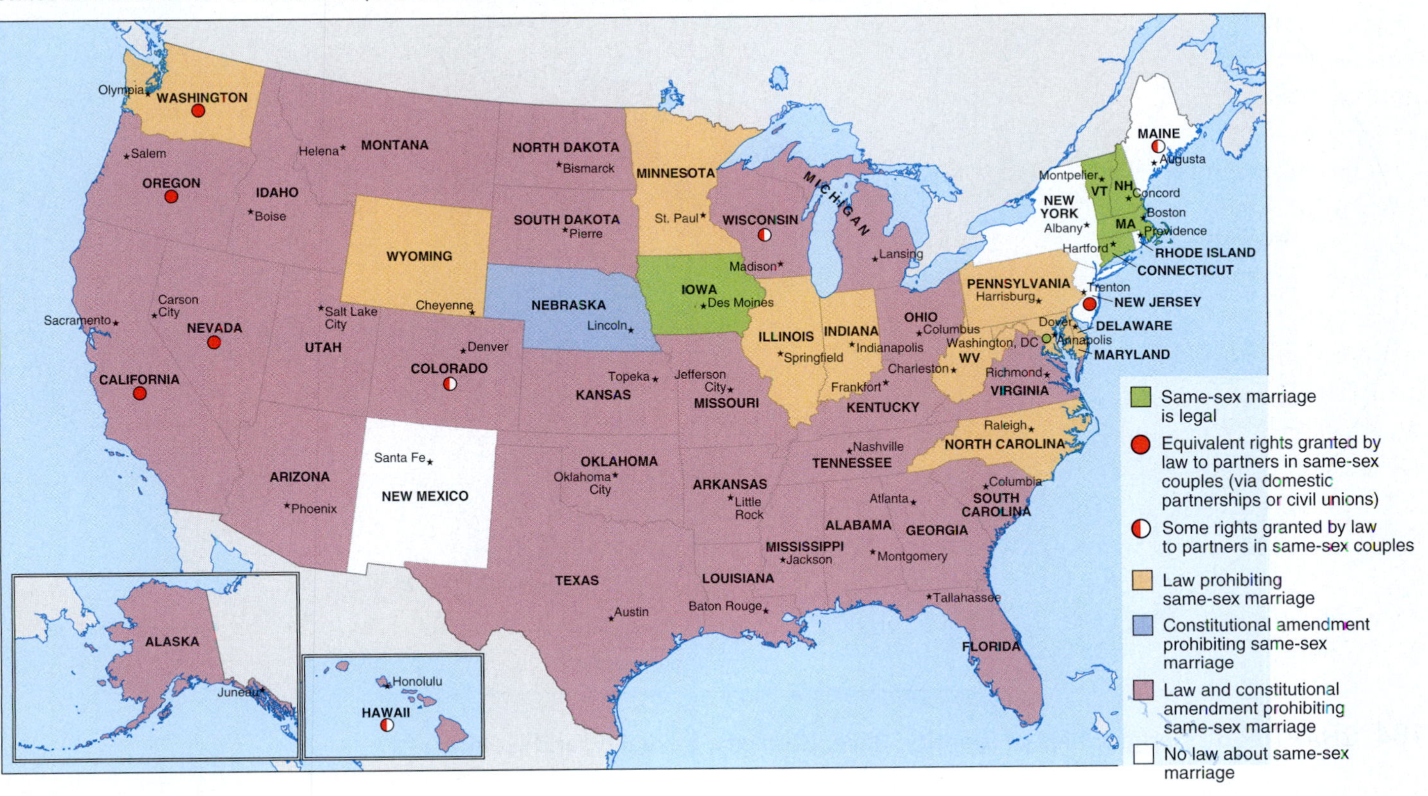

Gender roles vary geographically and can influence the division of labor. In rural Tanzania, for example, men are expected to be the main economic providers and women the caretakers of families. In contrast, different gender roles in Ghana ensure that women manage a wide variety of activities, including wholesaling and retailing in urban markets. Significantly, gender roles can change over time. In the United States, for example, gender roles changed during World War II as more women began working outside of the home in factories and in jobs once considered "men's work."

The factors that influence gender roles are numerous but include ideas (sometimes mistaken) about the skills women and men possess, as well as attitudes about empowerment and who should have decision-making authority. The mass media and other religious, educational, political, and corporate institutions strongly reinforce gender roles. Even in the United States the slowness to provide college-level education for women and the fact that it has required Title IX legislation to help prevent discrimination as mentioned in the chapter opener, reveal enduring and deep-seated views about gender roles.

Elsewhere, long-standing customary practices within families and communities continue to shape gender roles. Some Muslim and Hindu women, for example, observe *purdah*, which refers to the practice of wearing clothes that cover all of the body. Purdah also includes the practice of socially segregating women and men. Purdah developed not only as a means of ensuring that women were not seen by men unrelated to them, but also as a way of defining acceptable behavior and gender roles. For example, in Saudi Arabia, women are not permitted to drive and, if traveling alone, they must have permission from a male relative to board an airplane. Moreover, in Saudi Arabia, segregation of men and women is sanctioned by law. This has given rise to banks that have

Gender gap index and gender disparities • Figure 6.15

a. Sweden has the narrowest gender gap (0.81), and Yemen has the widest (0.45). To calculate the index, male and female enrollments in school are examined, as are labor participation rates. How might you explain the large gender gap index for Japan? (*Source*: World Economic Forum, 2007.)

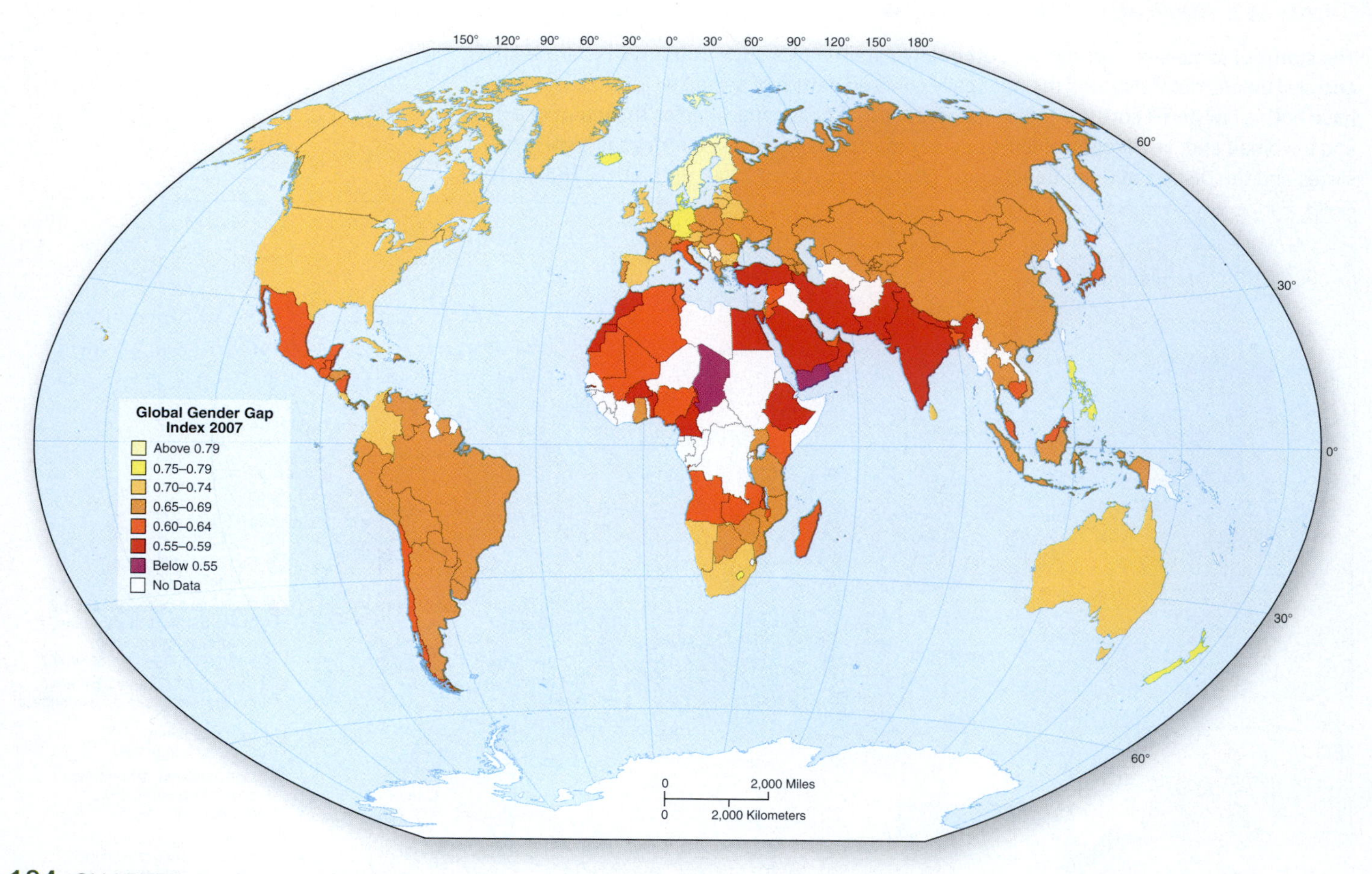

separate branches for women, segregated universities, and segregated workplaces. Because of this segregation, it has only been within the past 20 years that Saudi women have been able to pursue careers in such fields as architecture and journalism.

gender gap A disparity between men and women in their opportunities, rights, benefits, behavior, or attitudes.

Gender roles do not always result in men and women working in separate spheres or spaces, however. In Tamil Nadu, India, for example, both men and women work in agriculture and participate in the same activities. However, men spend comparatively more time plowing the fields, whereas women spend more time selecting seeds, weeding, and preparing crops for storage.

The persistence of gender roles can contribute to the development of stereotypes about what constitutes "men's work" or "women's work." When such stereotypes develop, they can create social barriers that constrain the opportunities available and contribute to a **gender gap**. Many researchers pay close attention to gender gaps because they are good indicators of gender-based inequalities. For example, even though women constitute 40% of the global labor force, they earn just 25% of the income generated in the world. As a whole, women are less likely to work in salaried jobs, and when they do, they do not earn as much as men, creating a gender gap in wages.

One way of examining the gender gap is to use the gender gap index—a measure developed in 2006 to help assess how effective different countries have been at closing the gender gap. The index is based on a mixture of economic, political, educational, and health-related data. Countries that have a value close to one have a narrow gender gap (a value of one would indicate gender equality rather than a gender gap), while countries with values close to zero have a wide gender gap (**Figure 6.15**).

b. Bangladeshi women carry a banner proclaiming "gender equality needs equal rights to property and resources." In parts of Asia, women cannot inherit property, but they can inherit debt.

c. In Saudi Arabia, women can own cars, but they cannot drive them. In a remarkable protest against the ban in 1990, women drove cars in Riyadh. In commemoration of that day, this woman takes advantage of a loophole in the law and drives an ATV .

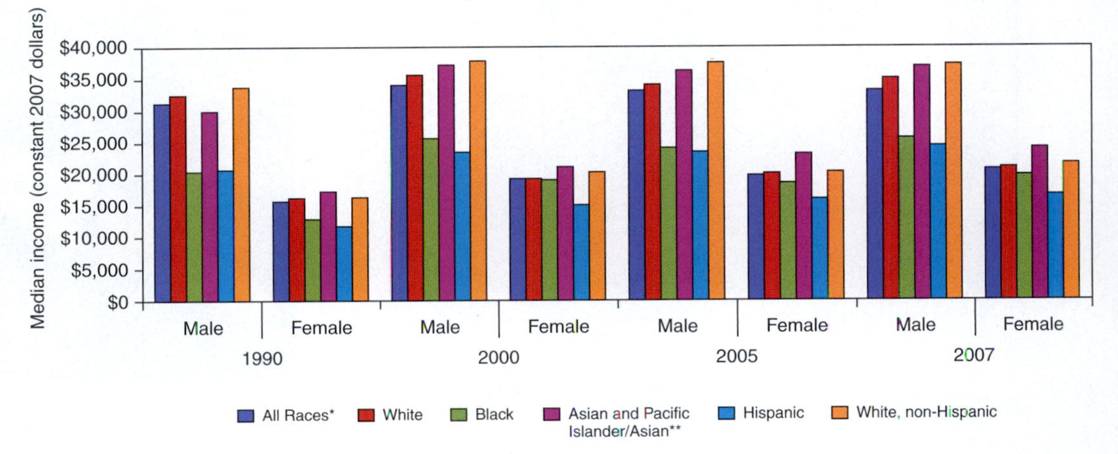

d. In a list of values ordered from low to high, the median is the value in the middle; half of the values are higher than it, and half of the values are lower than it. Based on median income data for the U.S., shown here, a persistent income gap not only separates men from women but also separates blacks and Hispanics, both male and female, from other racial groups. In 2007, the highest median income for women (among Asians) was still below the lowest median income for men (among Hispanics). (*Source*: U.S. Census Bureau, 2010.)

The presence of a gender gap matters because it points to the institutionalization of status differences between women and men in society. Status is a way of socially ranking and valuing certain types of knowledge and skills that people possess. Status takes on a geographic dimension when it affects how the spaces and places of home, school, and work are organized and controlled. We continue our discussion of geography and gender in several other chapters, including Development, Industry, and Agriculture.

CONCEPT CHECK STOP

1. **What** are the limitations of binary views of sexuality and gender?
2. **How** can norms about sexuality affect public space?
3. **Why** is an understanding of gender important to the study of geography?

Summary

1 Race and Racism 162

- The concept of **race** developed as a way to classify people into biologically distinct groups, or races. On a genetic basis, however, humans are so similar that identifying distinct racial categories is an entirely arbitrary exercise. Since nature does not provide a basis for classifying people into racial categories, race is best thought of as a **social construction**.
- **Racism** is rooted in perceptions that people who are different from us are inferior to us. During the Enlightenment in Europe, a complex **ideology** of racism developed in conjunction with ideas about the great chain of being, shown here. **Human trafficking** is also tied to perceptions of human difference and represents an enduring form of slavery.

Great chain of being • Figure 6.2

2 Geographies of Race and Racism 166

- **Institutional discrimination** disadvantages people because of their cultural differences. Geographers recognize that institutional discrimination and racism can have a significant impact on the spatial functioning and configuration of cities and countries.
- In certain instances, ideas about race and place mutually reinforce one another and can lead to institutional discrimination, racism, or both.
- From 1948 to 1991, **apartheid**, a system of racial segregation, was official policy in South Africa. Apartheid affected the use of space not only at the national scale but also, as the photo shows, at the individual scale.

The rise and fall of apartheid • Figure 6.5

3 What Is Ethnicity? 170

- **Ethnicity** refers to the personal and behavioral basis of an individual's identity, which generates a sense of social belonging among members of an **ethnic group**. As illustrated in the photo, the preparation of food and the kinds of foods consumed might be two personal components of ethnicity. Others might include language, religion, or styles of dress or music. An individual's ethnicity is contingent and flexible, and stems from processes of **othering** and **ascription**.

Aspects of ethnicity • Figure 6.6

- Many countries use their censuses to collect data about the ethnicity of their citizens. Geographically, use of the term *race* on census forms is most commonly associated with the Americas, prompting interest in how census questions contribute to **discourses** of identity.

4 Ethnicity in the Landscape 176

- **Ethnic geography** includes the study of **ethnoscapes**. In the study of ethnic interaction, three models have prevailed: the **assimilation** model, the **pluralism** model, and the **heterolocalism** model. Heterolocalism occurs when members of an ethnic group who, upon arriving in a place, disperse or cluster minimally while maintaining a sense of ethnic identity, such as the Vietnamese in metropolitan Washington, D.C. (See map.)

Heterolocalism • Figure 6.8

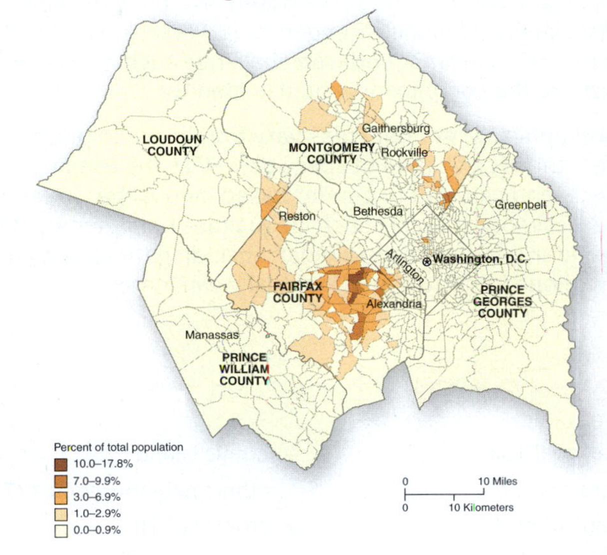

- When studying ethnic settlements, geographers distinguish among **ethnic islands**, **ethnic neighborhoods**, and **ethnoburbs**.
- The **location quotient** is a measure that helps to determine how the presence of an ethnic group in a region compares to that same group's presence nationally.
- Ethnic imprints provide landscape expressions of the presence of ethnic groups. Ethnic festivals are one kind of ethnic imprint and help to shed light on the ways in which **symbolic ethnicity** shapes the identity of an individual or a group.
- Ethnic tension may be a factor that contributes to ethnic conflict, but few conflicts can be explained solely on the basis of ethnic strife, even when **ethnic cleansing** occurs.
- Concern about the inequitable exposure to environmental hazards associated with the locations of toxic waste sites and toxic release facilities prompted the **environmental justice** movement.

5 Sexuality and Gender 182

- Convention teaches us that there are only two sexes: females and males. On a biological basis, however, human **sexuality** is similar to race in that it cannot always be made to fit into separate categories. These points remind us that a person's identity—whether sexual, ethnic, or gendered—is not biologically determined, nor is it fixed.

- The configuration and use of space tend to reflect dominant beliefs about **gender** and sexuality. In addition, ideas about sexuality have important geographical and political implications.
- Globally, a **heterosexual norm** prevails. This norm is rooted in a binary vision of the sexes based on clearly defined masculine and feminine gender roles. Persons who identify themselves as **transgendered** challenge such norms and illustrate the complexity of human identity.
- Geographically, **gender roles** vary from place to place, can affect the organization of space as well as the use of **public space**, and can lead to gender-based inequalities in terms of access to resources. Such inequalities help perpetuate a **gender gap** that many women, including those shown here, hope to narrow and eventually eliminate.

Gender gap index and gender disparities • Figure 6.15

Key Terms

- apartheid 168
- ascription 170
- assimilation 176
- discourse 174
- environmental justice 181
- ethnic cleansing 180
- ethnic geography 176
- ethnic group 170
- ethnic island 176
- ethnic neighborhood 177
- ethnicity 170
- ethnoburb 178
- ethnoscape 176
- gender 182
- gender gap 185
- gender role 182
- heterolocalism 176
- heterosexual norm 182
- human trafficking 164
- ideology 162
- institutional discrimination 166
- location quotient 178
- othering 170
- pluralism 176
- public space 183
- race 162
- racism 162
- sexuality 182
- social construction 162
- symbolic ethnicity 179
- transgender 182

Critical and Creative Thinking Questions

1. Use the U.S. Census Bureau website and the map shown here to identify the census tract, block group, and block where you live. Describe the ethnic composition of your census tract or several census tracts in your county.
2. Use the U.S. Census Bureau website to find the population data for your state and county and then calculate the state and county location quotients for one or two ethnic groups of your choice. What geographical observations can you make about your findings?
3. Do some additional reading on ghettos and Indian reservations. How are they similar and different?

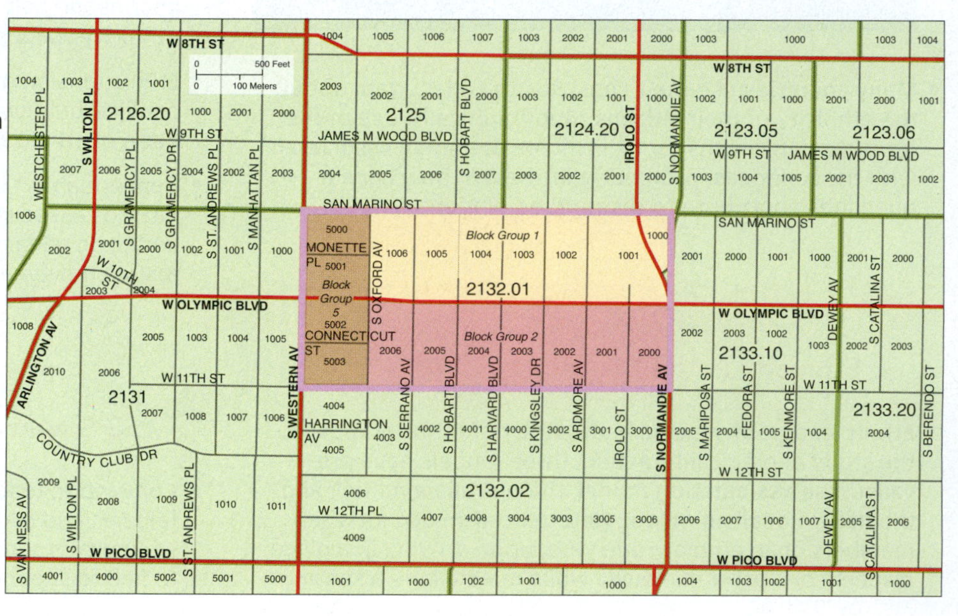

4. Some scholars have argued that in the United States one discourse about wilderness is that it is an ideal promoted by and for white people to the exclusion of people of color. What evidence can you provide for or against this position?
5. Do some fieldwork in your community. What ethnic imprints can you identify? What spatial patterns do you notice, and how might you account for them?
6. Cities with sizable gay communities are increasingly promoting them as a kind of alternative ethnic space and another tourist attraction. What are the implications of this for the gay communities?
7. Some scholars argue that gender segregation and strongly defined gender roles are empowering for women. Do you agree? Why or why not?
8. If poor people move into an area where a toxic waste site is located, has environmental injustice occurred? Explain your reasoning.

What is happening in this picture?

In Papua New Guinea, Huli men participating in certain ceremonies wear elaborate wigs constructed from their hair and adorned with feathers from birds. Boys who want to become warriors in Huli society are segregated from women and learn the art of beautifying their hair.

Think Critically

1. What are the implications of this for our understanding of masculinity, gender roles, and geography?
2. Do you think there is a relationship between the visibility of an ethnic group and the cohesiveness of its members?

Self-Test

(Check your answers in Appendix B.)

1. Which of the following is not associated with the historical development of a European ideology of racism?
   a. the great chain of being
   b. a hierarchical order of nature
   c. skin color
   d. overlapping racial categories

2. _____ exists when policies, practices, or laws disadvantage people because of their differences.
   a. Ideology
   b. Institutional discrimination
   c. Interaction between race and place
   d. Apartheid

3. Which statement about apartheid is false?
   a. It contributed to the impoverishment of many South Africans.
   b. It was implemented at different scales.
   c. It contributed to the development of Bantustans.
   d. It was most effectively implemented in urban areas.

4. The phrase "the underside of globalization" is a reference to _____.
   a. human trafficking
   b. the widening gender gap
   c. institutional discrimination
   d. environmental injustice

5. What types of census geography are shown here?
   a. census tracts and census groups
   b. block groups and census blocks
   c. census areas and census tracts
   d. counties and block groups

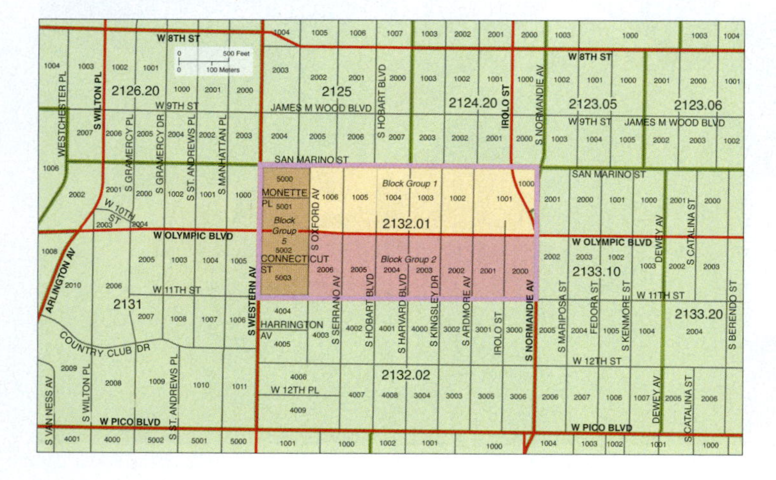

6. Briefly explain the fundamental problem with the concept of race:

   ____

   ____

7. _____ occurs when people assign a specific quality or identity to others.
   a. Othering
   b. Racism
   c. Ascription
   d. Social construction

8. Which characteristic of ethnicity does this photo illustrate?
   a. Ethnicity is subjective.
   b. Ethnicity is inherited.
   c. Ethnicity is expressed behaviorally.
   d. Ethnicity is fixed.

9. As discussed in the chapter, the U.S. Census is unusual because _____.
   a. its questions are discourse-free
   b. it separates race from ethnicity
   c. it is conducted every year
   d. each state conducts a separate census

10. Provided sufficient corroborating evidence, this chart might indicate _____ .
    a. an ethnic enclave
    b. a ghetto
    c. environmental injustice
    d. pluralism

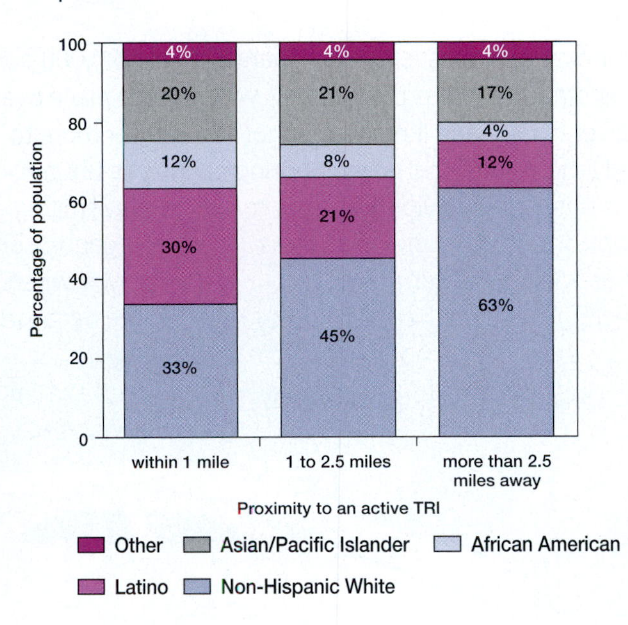

11. Which is not a characteristic of heterolocalism?
    a. community without propinquity
    b. living in one area but shopping in another
    c. globalization
    d. ethnic enclaves of immigrant communities

12. Which is not a characteristic of an ethnoburb?
    a. It is mainly a commercial space.
    b. It is ethnically diverse.
    c. It is socially stratified.
    d. It forms in a suburban location.

13. Which statement about Darfur is false?
    a. It is a region where ethnic cleansing has occurred.
    b. It is rich in oil.
    c. The conflict there has only recently become charged with ethnic hatred.
    d. Disputes in the region were customarily settled via mediation.

14. The social expectation that certain kinds of work are "men's work" is a good example of _____ .
    a. a gender gap
    b. a heterosexual norm
    c. a gender role
    d. All of the above statements are correct.

15. Purdah _____ .
    a. refers to the practice of socially segregating women and men
    b. refers to the practice, among certain women, of wearing clothes that cover the entire body
    c. may affect public as well as private space
    d. All of the above statements are correct.

THE PLANNER

Review your Chapter Planner on the chapter opener and check off your completed work.

7 Political Geographies

VOTING PRACTICES AROUND THE WORLD

Election practices vary considerably around the world. In remote parts of Mongolia, the *ger*, or traditional tent of nomadic herders, is converted into a polling station, and it is not uncommon for these herders to travel to the polls on horseback. Some countries have election weeks instead of an election day. Election officials in India, for example, must staff and operate some 900,000 polling stations across the country. Voting takes place over several days, and it may take another week or more to tally the results. In the neighboring country of Bhutan, families—not individuals—cast votes, thereby limiting popular participation in elections. If you are Ugandan or Guatemalan and happen to be out of the country when elections are held, you forfeit your right to vote because these countries do not permit absentee voting.

Most Americans probably did not think much about the mechanics of ballot design until the 2000

presidential election, when the use of punch card ballots in Florida came under scrutiny because of problems determining how votes had been cast, crucial to the final election result. Until recently, voters in the Philippines wrote the candidates' names on the ballot, but such a practice requires that voters be literate (and have a good memory). Alternatively, many countries design ballots so that even those who cannot read or write may still cast a vote for each of the political parties. Some ballots even show photographs of the different candidates.

Global Locator

MONGOLIA

NATIONAL GEOGRAPHIC

CHAPTER OUTLINE

CHAPTER PLANNER

- ❑ Study the picture and read the opening story.
- ❑ Scan the Learning Objectives in each section: p. 194 ❑ p. 201 ❑ p. 209 ❑ p. 212 ❑ p. 218 ❑ p. 221 ❑
- ❑ Read the text and study all visuals. Answer any questions.

Analyze key features

- ❑ What a Geographer Sees, p. 204
- ❑ Video Explorations, p. 206
- ❑ Geography InSight, p. 208
- ❑ Process Diagram, p. 219
- ❑ Stop: Answer the Concept Checks before you go on: p. 201 ❑ p. 208 ❑ p. 212 ❑ p. 217 ❑ p. 221 ❑ p. 223 ❑

End of chapter

- ❑ Review the Summary and Key Terms.
- ❑ Answer the Critical and Creative Thinking Questions.
- ❑ Answer What is happening in this picture?
- ❑ Complete the Self-Test and check your answers.

Key Concepts in Political Geography

LEARNING OBJECTIVES

1. **Define** sovereignty.
2. **Distinguish** between a state and a nation.
3. **Identify** some of the impacts of colonialism on the political geography of Africa.

t is sometimes said that **political geography** exists because people are territorial. Most political geographers consider human **territoriality** to be more than an instinctive, biological response. Instead, they see it as a complex form of behavior that is shaped by diverse social and cultural factors related to human identity.

Territoriality can be expressed by individuals and by groups. The concept of personal space helps us understand territoriality on an individual level. Personal space is

political geography The study of the spatial aspects of political affairs.

territoriality Strong attachment to or defensive control of a place or an area.

Political map of the world • Figure 7.1

The 194 countries shown on this map provide one powerful expression of human territoriality. The use of such territorially defined political units has become the dominant mode of political organization around the world only within the past four centuries. Throughout most of human history, people organized themselves in different ways, for example, according to class, kinship, or as subjects owing allegiance to a king, an emperor, or other ruler.

the space around our bodies that we consider to be an extension of ourselves and therefore "our space." People differ a great deal in terms of how expansive their personal space is but are likely to feel uncomfortable when others encroach on it. The presence of political states (countries), connected with a specific territory, shows that territoriality exists among groups of people (**Figure 7.1**).

The Development of the State and Its Sovereignty

How did the political map of the world come about? Although the answer to this question goes beyond the scope of this chapter, the question points to one of the key concepts in political geography: the development of the state. The practice of using territory as a basis for political organization stems from ideas about **sovereignty** that gained prominence following the Peace of Westphalia in 1648. This peace settlement included a series of treaties signed at the end of the Thirty Years' War, a long religious and territorial conflict in Europe. Since the mid-17th century, then, sovereignty has meant that states are distinct territorial units, that

sovereignty Supreme authority of a state over its own affairs and freedom from control by outside forces.

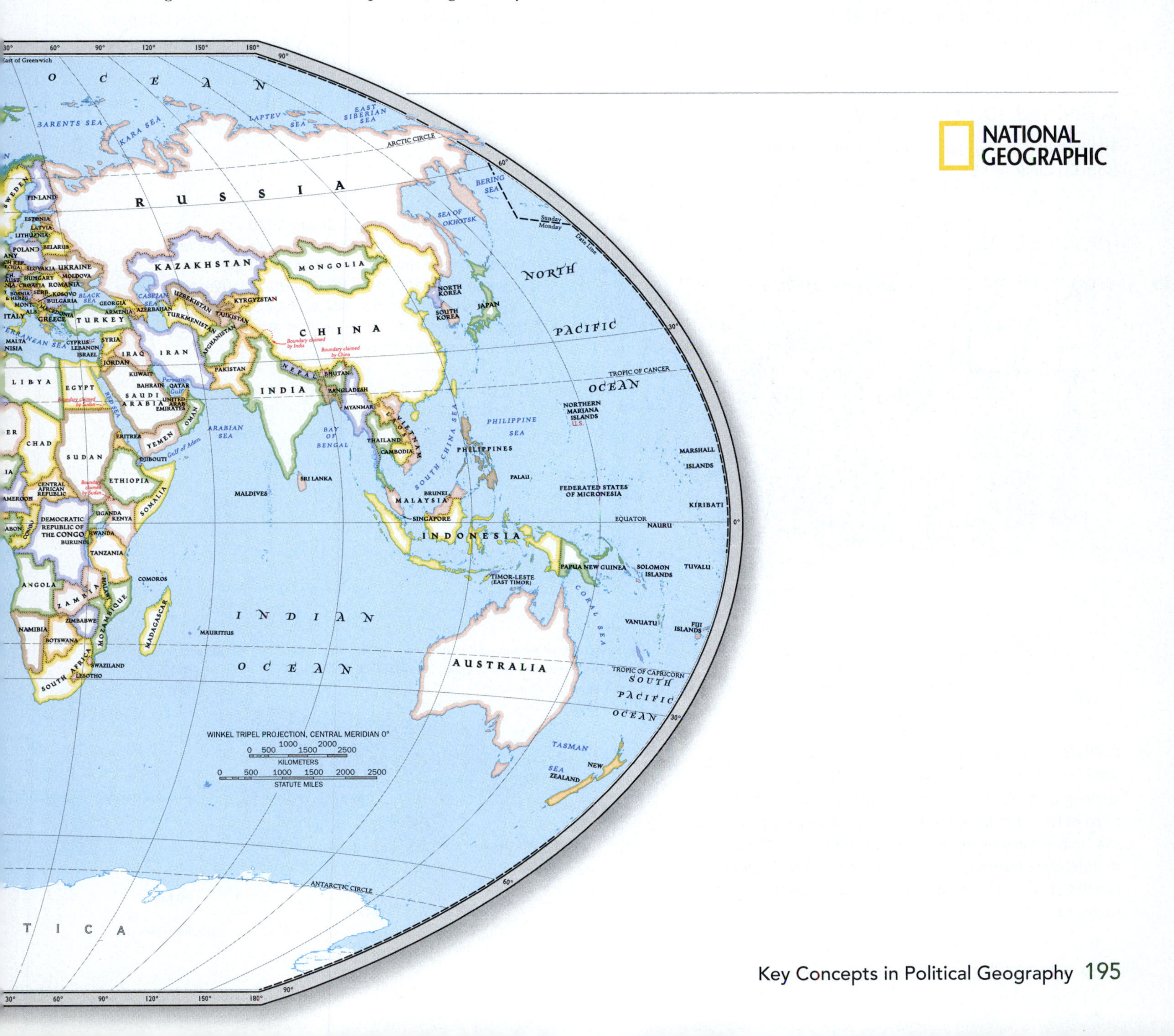

one state has no business interfering in the affairs of another state, and that states are expected to recognize the existence of other states.

For political geographers, the terms *state* and *country* mean the same thing; *state* is considered a more formal term. Therefore, a **state** exists when the following four conditions are met:

1. It consists of a specific territory with defined boundaries.
2. Its territory includes a permanent population.
3. It is recognized as a state by other states.
4. It has a government with supreme authority over its domestic and international affairs.

state An internationally recognized political unit with a permanently populated territory, defined boundaries, and a government with sovereignty over its domestic and international affairs.

Sovereignty is a fundamental aspect of statehood, but even though we have carefully defined sovereignty, the fact of the matter is that people, acting on behalf of states, use it to suit their own purposes. Thus, questions of sovereignty can become a basis for political disputes. For example, different perspectives exist concerning the status of Taiwan. The origins of this dispute date to the 1940s, when civil war broke out between the Communists and Nationalists in China. Following the war, the People's Republic of China was established. In defeat, the Nationalists retreated to Taiwan in 1949. Since that time, China and Taiwan have developed very different economic and political systems. Although the Taiwanese

A divided state • Figure 7.2

The Republic of Cyprus gained its independence from Britain in 1960.

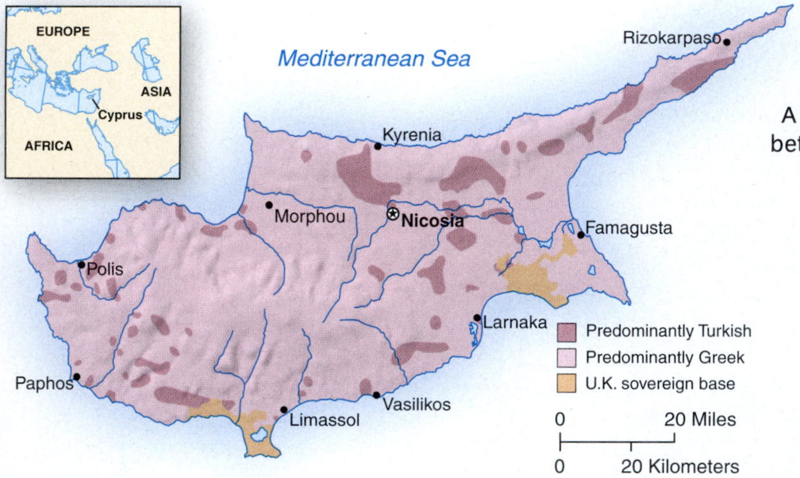

a. Ethnic distribution, 1970
The country's independence, however, did not heal the rifts that had developed between Greek Cypriots (78% of the population) and Turkish Cypriots (18% of the population). A degree of residential separation already existed between these groups by the 1970s.

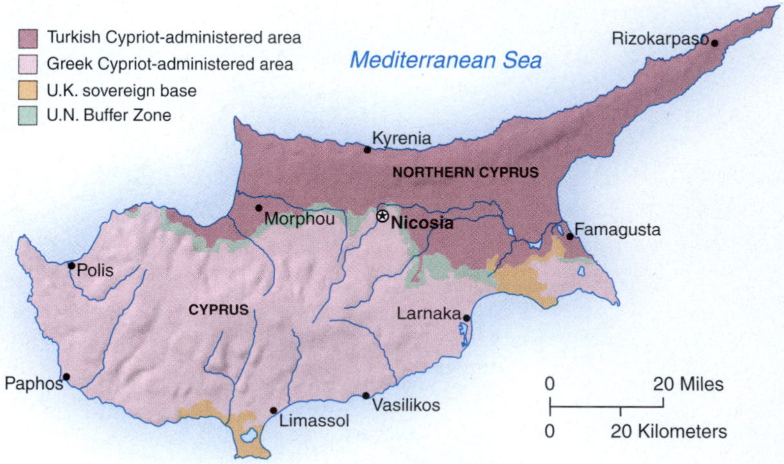

b. Division of Cyprus
After a coup in 1974 led by some Greek Cypriots seeking unification with Greece, Turkey invaded the northern third of the island. Partition followed (compare to **a**), with the United Nations maintaining a buffer zone. Thousands of people were internally displaced. The Turkish Republic of North Cyprus was established in 1983, but only Turkey has recognized it.

have never declared their independence from China, the government of Taiwan represents itself as the Republic of China—a continuation of the government that existed in China before its civil war.

The Chinese government asserts sovereignty over Taiwan and considers the island to be its twenty-third province, but it does not control the island's political affairs. Thus, it can be said that Taiwan has de facto (actual) sovereignty because it manages its own affairs, but Taiwan lacks de jure (legal) sovereignty because the international community does not recognize it as a full-fledged state.

Different conceptions of sovereignty also make a simple question like "how many states are there?" more complicated than it may seem. We might reasonably ask whether Cyprus, a sovereign state divided since the 1970s between a Turkish-dominated North and a Greek-dominated South, should be counted as one state or two (**Figure 7.2**).

Nations and States

Thus far we have talked about states, but we have said very little about the people within them, especially those who see themselves as belonging to a **nation**. What gives a nation its shared sense of identity? Often it is a mixture of various historical, cultural, economic, or political circumstances. Certainly the Palestinian nation has been shaped by its long struggle to gain independence and statehood, among other factors. For the French-

nation A sizable group of people with shared political aspirations whose collective identity is rooted in a common history, heritage, and attachment to a specific territory.

c. Nicosia divided
When Cyprus was divided, so was the capital city, Nicosia. Ledra Street, shown here, was blocked to prevent movement between the Turkish and Greek Cypriot parts of the city.

d. Potential resolution
In April 2008, Ledra Street was reopened in a symbolic move to build support for reunification of the island. Like sovereignty, the issue of which states are counted as states involves some subjectivity.

speaking Québécois in Canada, a shared language and experience as a minority group have shaped their national identity.

In popular usage, the terms *nation* and *state* are frequently used interchangeably, but political geographers and other scholars are careful to note that the terms are not synonymous. Simply stated, a nation refers to a people and a state refers to a political unit. *Nationalism*, then, is the expression of loyalty to and pride in a nation, whereas *patriotism* is the expression of love for and devotion to one's state.

In a **multinational state**, the population consists of two or more nations. Most countries in the world are multinational states, including, for example, Brazil, Canada, China, Indonesia, Mexico, Nigeria, Spain, Sudan, and the United Kingdom. A **nation-state** exists when the boundaries of a nation coincide with the boundaries of the state and the people share a sense of political unity. A narrow understanding of the nation-state concept means that a nation-state possesses a fairly homogeneous population. For example, Icelanders make up 94% of the population of Iceland, and Japanese account for 99% of the population of Japan. But very few of the states in the world meet this strict definition of nation-state. Thus, a broader understanding of the nation-state concept helps us see that even a multinational state can develop an identity as a nation-state by socially, economically, and politically integrating its people. Let's consider the United States. Because of its Native American nations, such as the Chickasaw Nation and the Navajo Nation among others, the United States is multinational in terms of its population composition. Nevertheless, the United States functions as a nation-state through the creation of a political identity that sees the American nation and the state of the United States as identical and indivisible.

For a variety of political, economic, and social reasons, some multinational states are simply not able to successfully integrate the nations inside their borders. This is demonstrated by, among other events, the collapse of three multinational states in the 1990s: the Union of Soviet Socialist

The breakup of a multinational state • Figure 7.3

Some factors contributing to the fall of the Soviet Union were economic crises such as food shortages, poor industrial performance, German reunification, and the related fall of communism in eastern Europe. When Russia declared its sovereignty in 1990, that act significantly undermined the political legitimacy of the Soviet Union. Home to a diverse population, the Soviet Union fractured largely along internal political and national lines as 15 new states came into existence in 1991.

NATIONAL GEOGRAPHIC

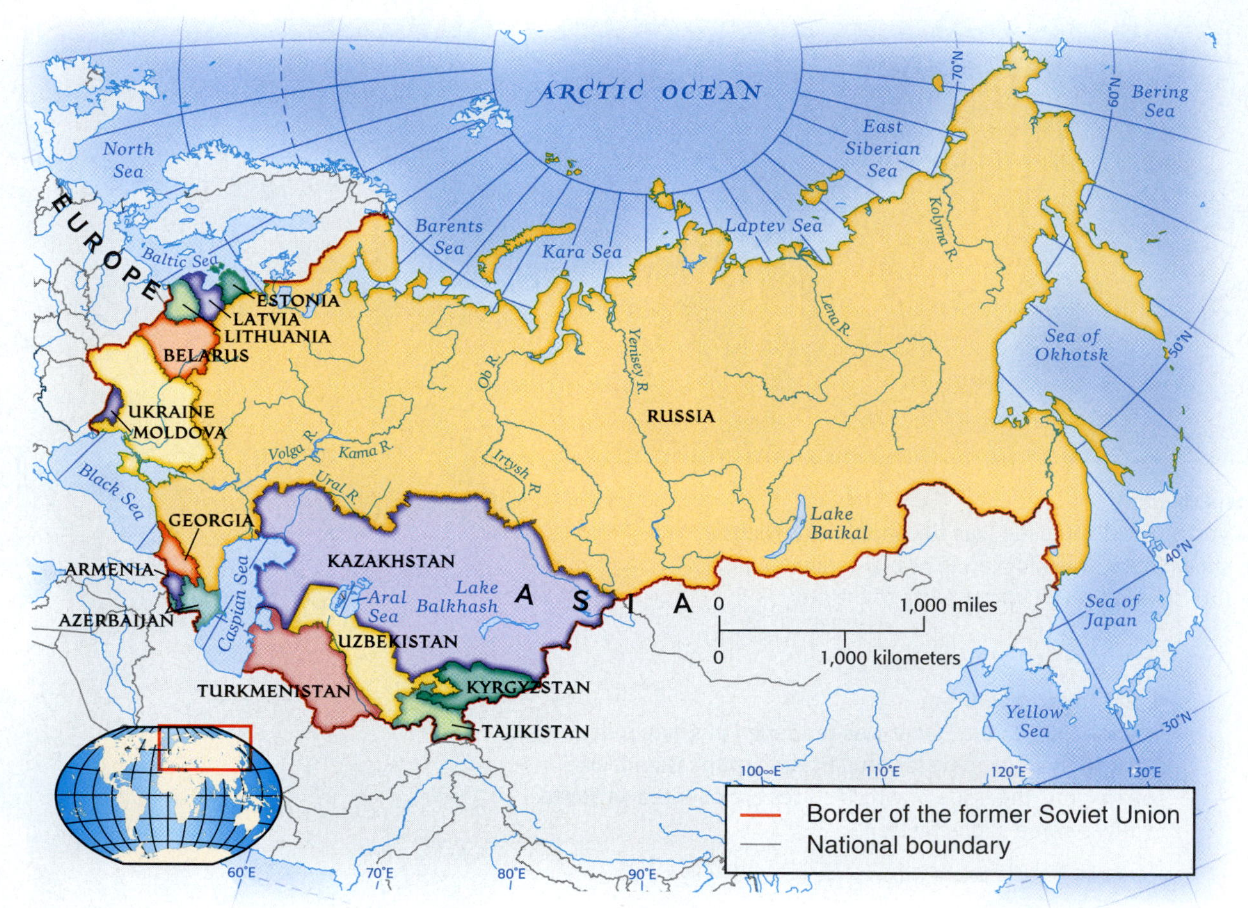

Republics (U.S.S.R.), Yugoslavia, and Czechoslovakia. Since 1991, 24 new states have been created from the breakup of these three states. The splitting up of the U.S.S.R. alone accounted for 15 of these new states, of which Russia is the largest (see **Figure 7.3**).

Imperialism and Colonialism

The dual processes of **imperialism** and **colonialism** have contributed to the creation of many of the world's multinational states largely because gaining access to and control of additional territory usually brings different national groups into contact. Imperialism and colonialism are closely connected, though the terms are not necessarily interchangeable.

imperialism One state's exercise of direct or indirect control over the affairs of another political society.

colonialism A form of imperialism in which a state takes possession of a foreign territory, occupies it, and governs it.

States use imperialism and colonialism as strategies to extend their power over other lands and peoples. In the 15th century, news of Portuguese explorations along the coast of Africa, and Portuguese and Spanish ventures in the Americas, prompted the Netherlands, Britain, and France to seek additional territory beyond their borders. Other European states such as Belgium, Germany, and Italy followed suit, and European colonies were eventually established in Africa, Asia, the Americas, and the Pacific. Although most of the colonies in the Americas had gained independence by 1825, sizable portions of Africa and Asia were still controlled by Europeans in 1914. With colonies on every continent (except Antarctica), the British created the largest colonial empire in history. France built the second most extensive empire, assembling colonies across West Africa, Southeast Asia, and the Pacific. The establishment of European colonies in Africa provides a stark example of one of the largest land grabs in history, often referred to as the "scramble for Africa."

a. As the rupture occurred, the boundaries of the internal Soviet republics became the boundaries of the newly independent states. For example, the Uzbek Soviet Socialist Republic became the new state of Uzbekistan. Tellingly, the names of these new states, like the names of the republics before them, evoke the diverse human mosaic that the Soviet Union was—for embedded in each is also the name of a prominent ethnic group: the Russians, Uzbeks, Tajiks, and so on.

b. In 2008, the Georgian military attacked locations in South Ossetia ostensibly to protect Georgians living in the region, and the Russians responded by sending troops into the region. The photo shows Georgians protesting this intervention outside the Russian Embassy in Paris. Expressions of Georgian nationalism have at times invoked the idea of a homogeneous nation-state and alienated other national groups in the country by claiming that "Georgia is for Georgians."

c. Banners carried by Latvian nationalists reveal their enduring opposition to Russia and Russian dominance associated with the former Soviet Union. Significantly, however, about 30% of the Latvian population is Russian. In fact, all of the states created in the breakup of the Soviet Union are themselves multinational.

Colonial legacies in Africa • Figure 7.4

Colonialism transformed Africa. Among other impacts, it gave rise to a number of multinational and landlocked states, shaped new political identities, and built new road, rail, and trade networks.

a. Africa in 1914

The political map of Africa, as we know it today, began to take shape during the colonial period, and the first major boundary-making exercise by colonial powers took place in conjunction with the 1884–1885 Berlin Conference.

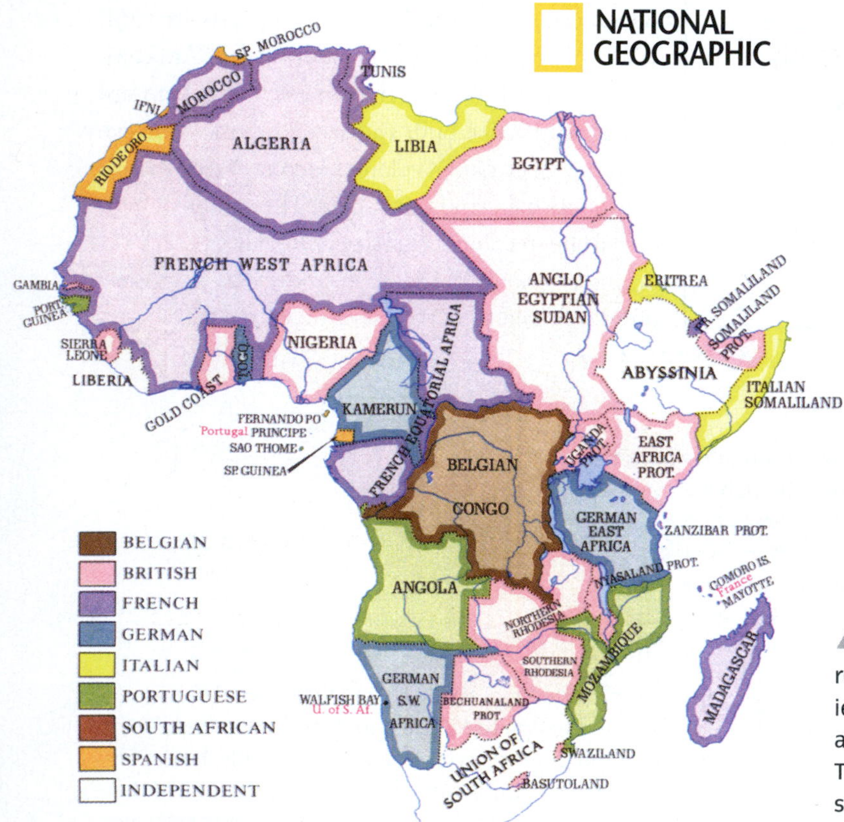

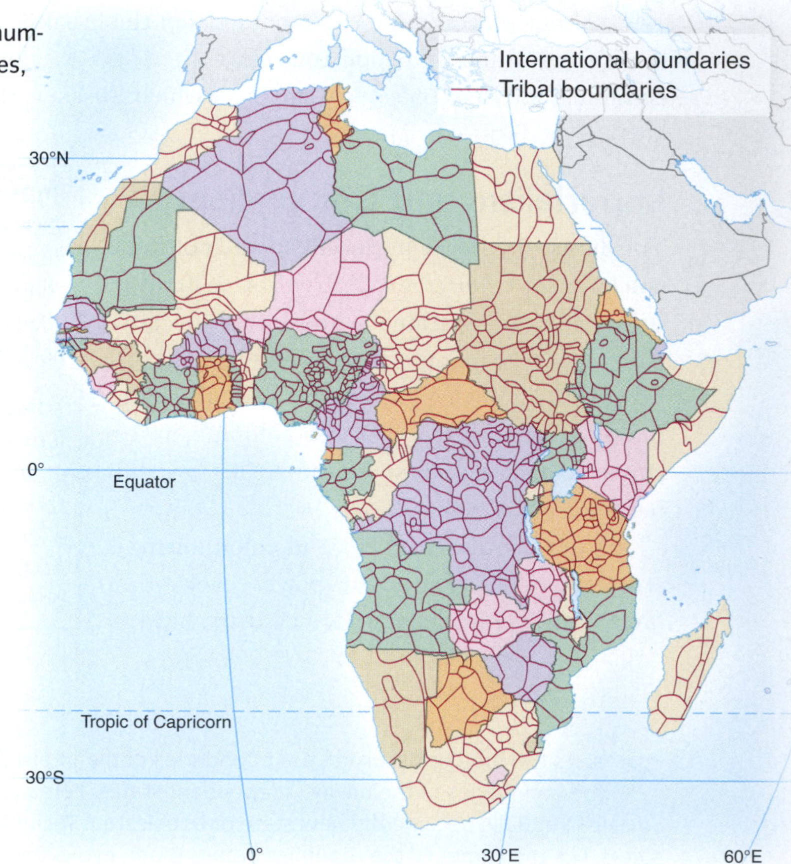

b. Africa's states and one depiction of its cultural groups

The present-day boundaries of African states more closely resemble those of the European colonies than they do the boundaries of the African peoples. But notice that this statement implies that a country should be a nation-state, in the strict sense of the term. The human diversity of the world, glimpsed via this map of Africa, suggests that the idea of creating homogeneous nation-states is impractical, if not impossible.

c. Rwandan refugees and a refugee camp

In some places, like Rwanda, European colonialists stoked animosities between local groups that have continued to the present. Rwanda descended into a horrific civil war in the 1990s. In 1994, approximately 250,000 refugees entered neighboring Tanzania in just 24 hours, prompting the establishment of substantial refugee camps. The adoption of a new constitution in 2003 ensuring a balance of power between Hutus and Tutsis has helped stabilize the country, but there are still thousands of refugees scattered across the region.

The Berlin Conference, held in 1884–1885, began the process of formalizing modern political boundaries in Africa. Representatives from the leading European powers, as well as the United States and Russia, met to discuss the partitioning of the African continent. Not a single African was present at the conference. In a boardroom in Berlin, Europeans made decisions about their control of territory in Africa almost as if they were playing a game of Monopoly in which the object was to control the most (and most valuable) land.

Boundaries separating British, French, Belgian, Portuguese, Spanish, Italian, and German spheres of influence in Africa were superimposed. That is, they were often drawn as straight lines with little awareness of or concern for the different ethnic groups living there. In East Africa, for example, the Somali people lived in lands that were partitioned between the British, French, and Italians. In numerous other instances tens and sometimes hundreds of different ethnic groups were combined in a single colony.

Each of the European powers administered its colonies in different ways, but a guiding philosophy of managing colonies was rooted in the racist belief that Africans were inferior. Sometimes colonial administrators used ethnic differences to create animosities between certain Africans. In Rwanda, for example, the Germans and later the Belgians exploited differences between Tutsis and Hutus. The Belgians showed favoritism to the Tutsis, however, allowing some of the men to attend school and rewarding them with jobs handling the day-to-day affairs of the colony. In 1962, following a Hutu rebellion, the Belgians granted independence to Rwanda, which came into existence as a multinational state with a population that was seriously divided (**Figure 7.4**).

Before and after World War II there was a surge in efforts by Africans and other colonized peoples to attain independence and achieve **self-determination**, the ability to choose their own political status. Globally, colonialism fell out of favor. On the continent of Africa, for example, some 32 former colonies gained their independence between 1960 and 1970. Even though European colonialism was waning, the framework for the political organization of space that exists today was cemented because the newly independent countries often retained their colonial boundaries.

CONCEPT CHECK STOP

1. **How** did the concept of sovereignty change after the Peace of Westphalia, and why is this significant?
2. **What** are the four criteria that define a state, and how is a state different from a nation?
3. **How** did colonialism affect the political map of Africa?

Geographical Characteristics of States

LEARNING OBJECTIVES

1. **Explain** how boundaries affect access to resources.
2. **Compare** and contrast centripetal and centrifugal forces.
3. **Identify** two systems of internal spatial organization.
4. **Define** devolution.

Earlier, in our discussion of Africa, we touched on boundaries and the role they played in transforming the political spaces of the African continent. Boundaries are regulatory devices that not only sanction territorial possession, but also help identify the contents—the people, natural resources, and territory—of states. In this section, we first examine different types of boundaries, visualizing the ways boundaries simultaneously secure, divide, and configure political space. Then, in subsequent sections, we consider other forces that affect the spatial functioning of states such as pressures of political integration or belonging, and political separation.

Exclusive economic zones • Figure 7.5

EEZs extend up to 200 nautical miles from shore, as shown for Australia. Coastal and island states have a reduced form of sovereignty—the "sovereign right"—to manage the resources in the waters and on the ocean floor in that area within their exclusive economic zones. This contrasts with their territorial seas, where they have full sovereignty. Note that Australia is one of several countries that assert sovereignty over part of Antarctica.

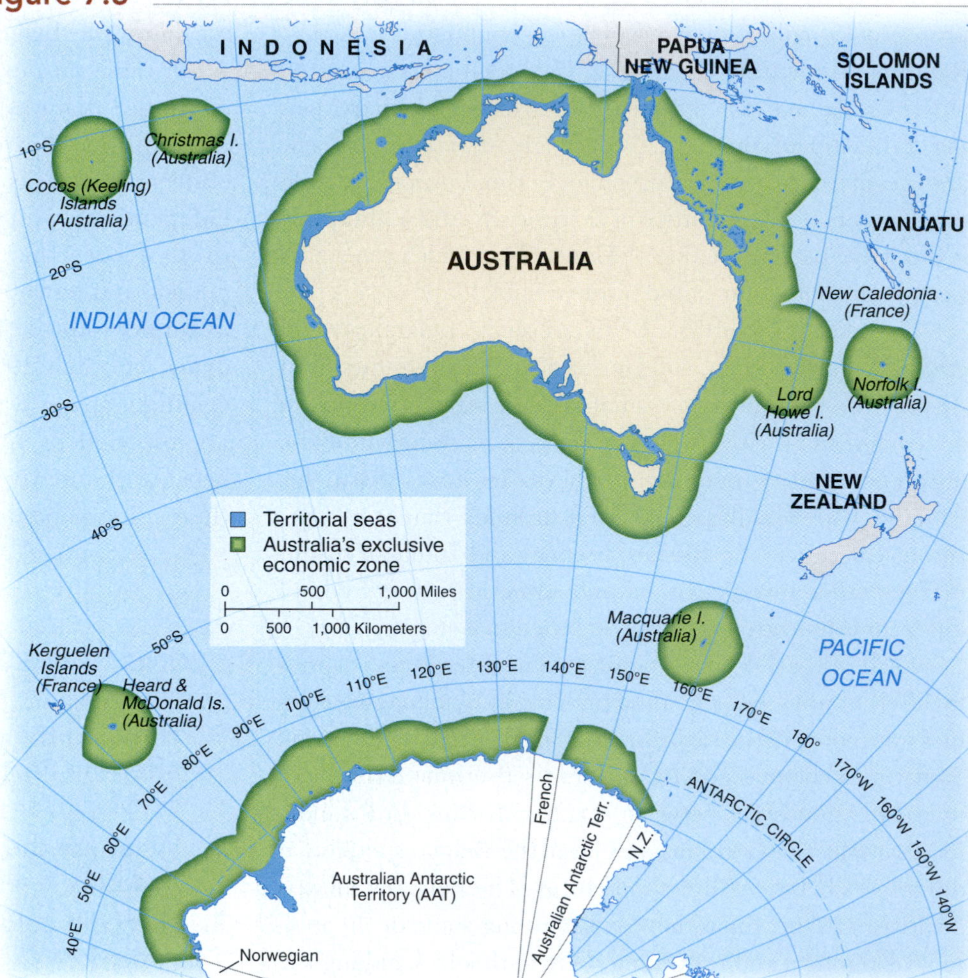

Boundaries

Every state consists of a defined territory marked by at least one **boundary**. We tend to think of boundaries as lines that stretch horizontally through space—the way we see them on maps—but boundaries are better understood as having a vertical extent, dividing the airspace above the ground and the rocks and resources below ground. The boundaries of coastal states extend offshore. The waters enclosed by these boundaries are considered part of the territory of the state and are called **territorial seas**. By international convention, territorial seas rarely exceed 19 km (12 mi). Over the past several decades, desire for rights to marine resources has led to the development of *exclusive economic zones* (EEZs) for coastal and island states (Figure 7.5).

boundary
A vertical plane, usually represented as a line on a map, that fixes the territory of a state.

By convention, boundaries are defined in legal documents, are drawn or delimited on maps, and may also be demarcated on the ground with signs, posts, fences, or other markers. Not all boundaries are marked or demarcated on the ground. A boundary might not be demarcated because it is disputed. This is the case with portions of the boundary between India and China. Sometimes even boundaries that are not disputed are not demarcated as, for example, in lightly populated areas, such as the high-altitude zones along the Andes Mountains between Chile and Argentina.

All political boundaries are human creations, but they often make use of physical features. A *physiographic boundary* follows a natural feature, such as a river or mountain range. For example, the boundary between Bulgaria and Romania follows the Danube River for much of its length. In mountainous areas, physiographic boundaries usually follow the crest, the line connecting the highest points. When a river is used, the boundary may be placed along one bank, in the middle of the river, or along the deepest part of the river channel. One problem with using rivers as boundaries is the potential for a river to substantially shift its course (Figure 7.6).

Fixing a boundary • Figure 7.6

We expect boundaries to be fixed or stable, but rivers are dynamic. This variability was a longstanding problem between El Paso, Texas, and Ciudad Juarez, Mexico, where the Rio Grande (called the Rio Bravo in Mexico) separates the two countries. The issue was finally resolved in the 1960s when a portion of the river was canalized—forced to flow in a concrete-lined channel in order to keep it from changing course and causing people to find themselves on a different side of a political boundary.

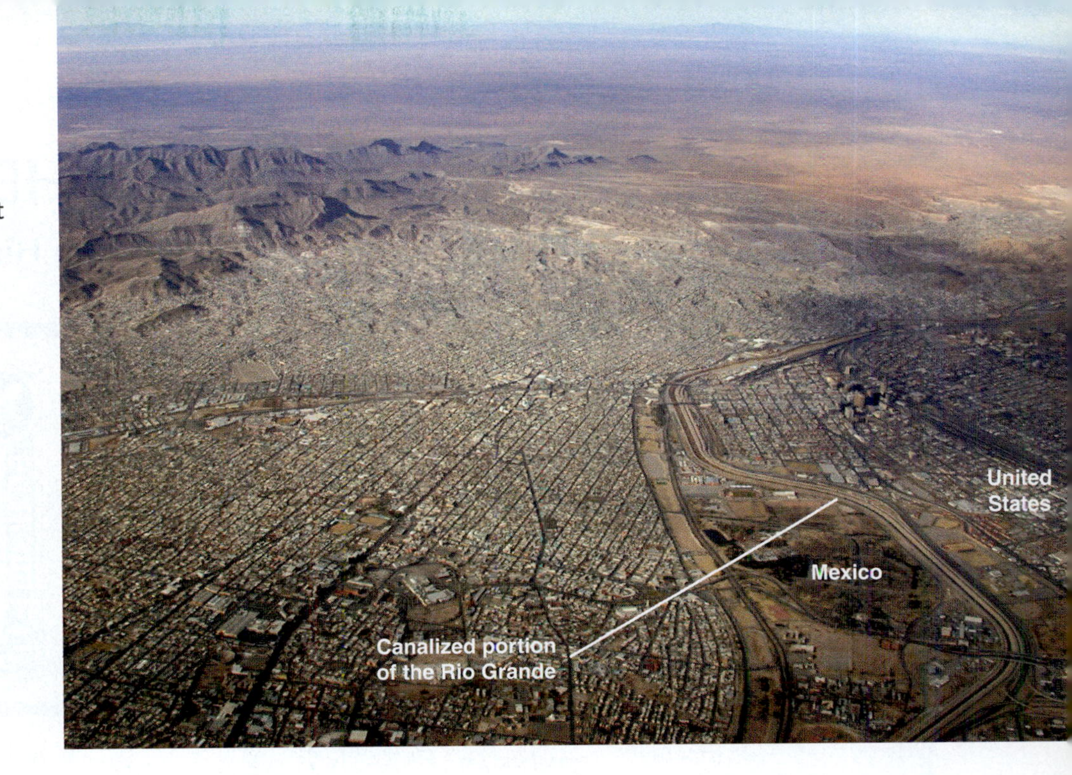

Geometric boundaries are drawn as straight lines and sometimes follow lines of latitude or longitude. As we have discussed, the concentration of geometric boundaries in Africa stems from the Berlin Conference. West of the Great Lakes, the boundary between the United States and Canada follows the 49th parallel, and the straight line between Alaska and the Yukon Territory follows the 141st meridian.

Ethnographic boundaries may be based on one or more cultural traits such as religion, language, or ethnicity. In South Asia, the boundary drawn between India and Pakistan was conceived as an ethnographic boundary separating Hindus from Muslims. Linguistic boundaries are common in Europe. For example, the boundary between Spain and Portugal is linguistic, as is the boundary separating Bulgaria from Greece.

A *relic boundary* is one that used to exist but is no longer recognized as an official boundary and, therefore, is no longer formally defined or delimited. Relic boundaries result from changes in the ways that geographic space is administered over time. Perhaps the most familiar example is the Great Wall of China (**Figure 7.7**)

Relic boundary • Figure 7.7

NATIONAL GEOGRAPHIC

This section of the Great Wall of China, located about 40 miles from Beijing, dates to the 15th century, but other parts of it were constructed at various times between the 7th century BCE and the 16th century CE to protect the Chinese Empire from intrusions by nomadic peoples. The impressive structure served not only as a boundary but also as a fortification.

WHAT A GEOGRAPHER SEES

THE PLANNER

The Making of a Boundary on Hispaniola

Politically, the Caribbean island of Hispaniola is occupied by two states: the Dominican Republic on the east and Haiti on the west. Spain asserted a claim to the island following the voyages of Christopher Columbus, but the island later came under French control. Although Haiti and the Dominican Republic achieved independence, respectively, in 1804 and 1844, disputes developed over the location of the boundary.

a. This map shows the present location of the international boundary, which dates from a 1929 agreement negotiated between Haiti and the Dominican Republic. Prior to this agreement, however, the location of the boundary was a source of contention.

b. At the turn of the 20th century, the Dominican Republic recognized a very different boundary. When the Dominican Republic issued this stamp in 1900, it provoked a political crisis. As you can see, the Dominican Republic had pushed the boundary significantly to the west and claimed possession of about three-fourths of the island, much to the consternation of the Haitian government as well as Haitians living in the disputed area. The Dominican Republic eventually relinquished its claims, paving the way for the 1929 treaty.

c. Can you spot the international boundary in this satellite image?

Global Locator

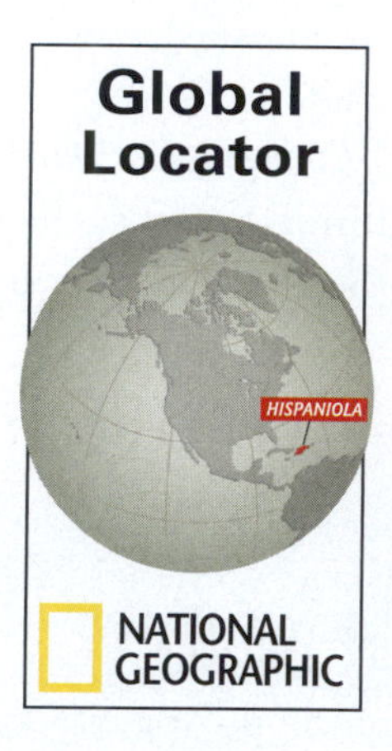

Think Critically

1. What might cause such a striking difference between the two countries, as shown in image **c**?
2. How is the boundary an ethnographic boundary?

Many boundaries have complex origins and in reality often reflect a consideration of multiple factors, including competing political concerns (see *What a Geographer Sees*).

Territorial Extent and Configuration

States come in a variety of shapes and sizes (refer again to Figure 7.1). The smallest state in the world, Vatican City,

State	Land Area (sq km)	Area (sq mi)
Russian Federation	17,098,242	6,601,665
China	9,984,670	3,855,101
United States	9,826,675	3,794,099
Canada	9,596,961	3,705,406
Brazil	8,514,877	3,287,611
Australia	7,741,220	2,988,901
India	3,287,263	1,269,219
Argentina	2,780,400	1,073,518
Kazakhstan	2,724,900	1,052,089
Sudan	2,505,813	967,499

The 10 largest states in the world Table 7.1

(*Source*: Data from CIA, 2009.)

covers just 44 hectares (109 acres) within the city of Rome. It is a *microstate*, a political state that is extremely small in total land area. At the opposite extreme, Russia is the largest state in the world, more than one and a half times the size of the United States (see **Table 7.1**). Antarctica is the only landmass that is not part of any state, though some countries have claims to parts of it (refer again to Figure 7.5).

On the basis of their shape, states can be classified as compact, elongated, prorupt, fragmented, or perforated. **Figure 7.8** illustrates these shapes.

A rupture in the territory of a state may result in the creation of an **enclave** or an **exclave**. Vatican City is not only a microstate, it is also an enclave state because it is completely surrounded by territory belonging to Italy. As shown in Figure 7.8, Lesotho perforates the territory of South Africa, making Lesotho an enclave state. Look again at Figure 7.8. Is Swaziland also an enclave state? Strictly speaking, the answer

enclave Territory completely surrounded by another state but not controlled by it.

exclave Territory that is separated from the state to which it belongs by the intervening territory of another state.

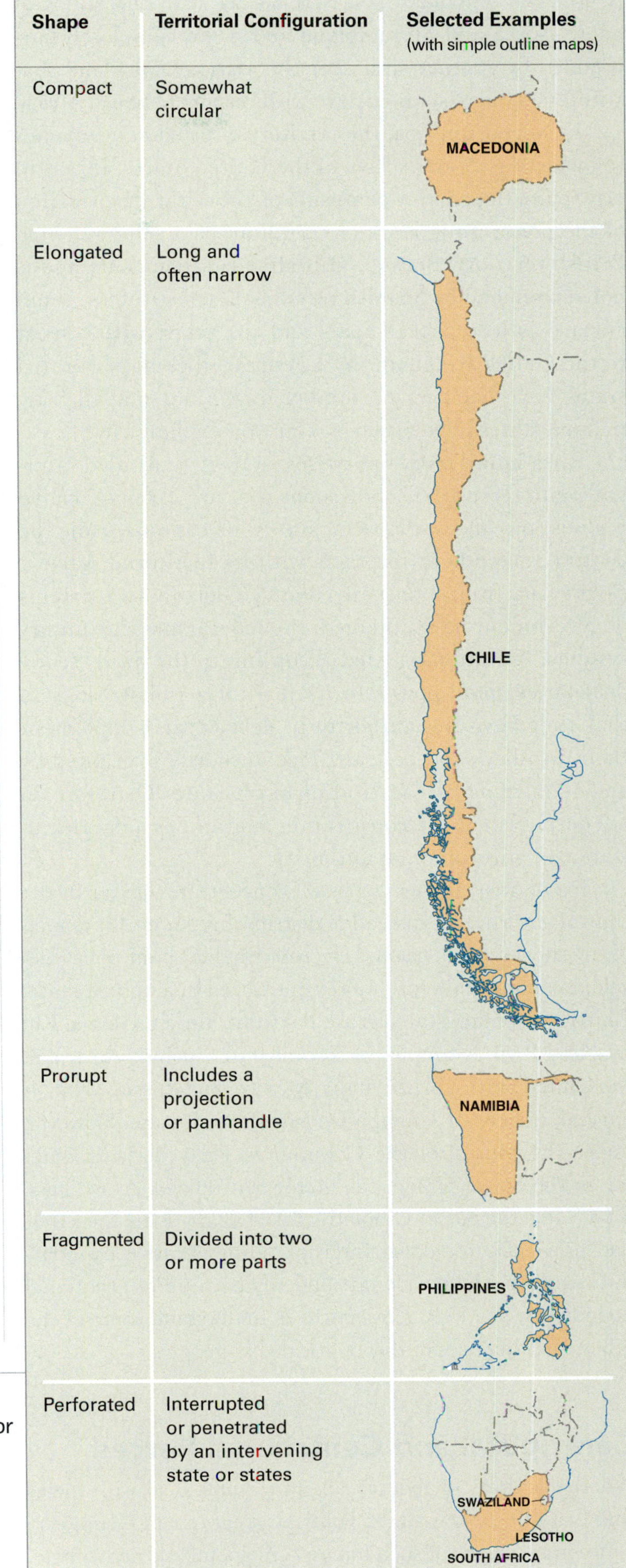

Shape
Territorial Configuration
Selected Examples
(with simple outline maps)
Compact
Somewhat circular
MACEDONIA
Elongated
Long and often narrow
CHILE
Prorupt
Includes a projection or panhandle
NAMIBIA
Fragmented
Divided into two or more parts
PHILIPPINES
Perforated
Interrupted or penetrated by an intervening state or states
SWAZILAND
LESOTHO
SOUTH AFRICA

Shapes of states • Figure 7.8

Geographers recognize five basic shapes of states, but sometimes a state exhibits characteristics of more than one of these shapes. Note, for example, that Chile is both elongated and fragmented, and that South Africa has two perforations and is prorupt. For political geographers, awareness of the configuration of a state's territory, the nature of its topography, the characteristics of its boundaries, and its relations with its neighbors are just some factors that affect the security of a state.

is no since Swaziland shares a border with the states of South Africa and Mozambique and is not completely surrounded by another state. For this reason, Swaziland constitutes a kind of semi-enclave with respect to South Africa.

In North America, the territory of Canada intervenes to make Alaska an exclave of the United States. In southern Spain, the narrow peninsula of Gibraltar (the location of the picturesque Rock of Gibraltar) is a self-governing British territory. Because Gibraltar is a peninsula and is not surrounded by Spanish territory, it constitutes a semi-enclave with respect to Spain and an exclave with respect to the United Kingdom. As a general rule, all perforated states have enclaves or semi-enclaves, but not all fragmented states have exclaves. Can you explain why?

For administrative purposes, states are divided internally into territorial subdivisions that are variously called regions, provinces, districts, states, or cantons. How the central government interacts with its territorial subdivisions varies from country to country. Globally, two systems of government have emerged: the federal and the unitary systems. A state organized according to the *federal system* distributes some power to its territorial subdivisions so that they have the authority to develop and implement their own laws and policies. This authority is granted by and detailed in the constitution of the state. Therefore, in a federal system the constitution guarantees a decentralization of governmental authority.

In contrast, the *unitary system* concentrates power in the central government instead of distributing some of it among its territorial subdivisions. Decision making and policy development are therefore more centralized in a unitary state. China is a unitary state, as are Pakistan, Turkey, Ghana, Kenya, and Peru. Examples of federal states include Australia, the United States, Brazil, India, Mexico, and Russia. While it existed, the Soviet Union was also a federal state. However, under the control of the Communist Party decision making in the Soviet Union was highly centralized. As we have seen, when the Soviet Union fractured apart, some lines that had previously served as internal, administrative boundaries within the Soviet Union suddenly became international boundaries. See *Video Explorations* to understand some of the other consequences of this political change.

Centripetal and Centrifugal Forces

Whether unitary or federal, all states must deal with forces that can affect their unity. Political geographers distinguish between two kinds of such forces: centripetal and centrifugal.

Video Explorations

Estonia—Identity, Religion, and Politics

THE PLANNER

WILEY PLUS NATIONAL GEOGRAPHIC Video

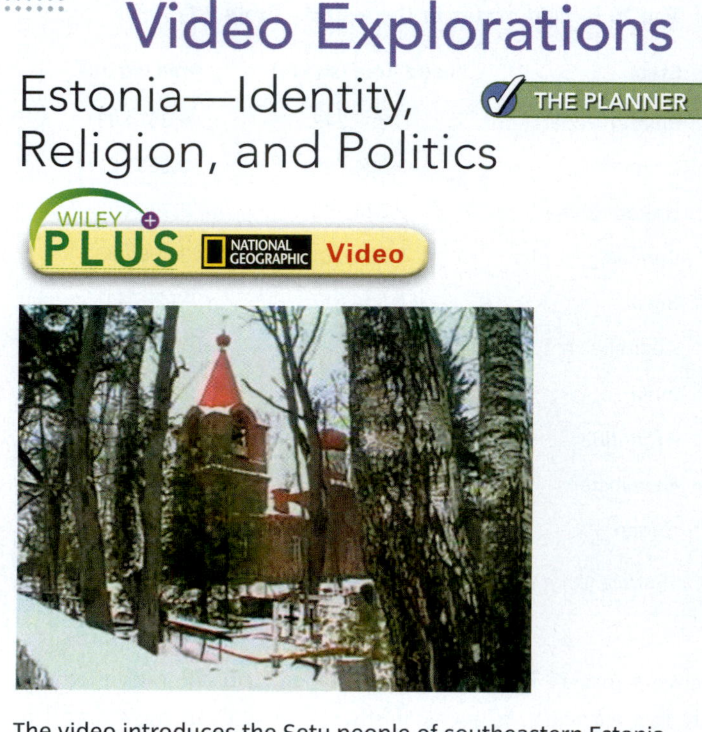

The video introduces the Setu people of southeastern Estonia and explores the impact of political change on them and their local landscapes.

The events on 9/11 are considered a **centripetal force** for the United States because they contributed to a collective sense of being under attack and to a shared sense of grief. The cultural diversity of some states, where different groups struggle for access to resources, can be a **centrifugal force**, especially if one group feels that it is not treated equally. In addition to cultural diversity, economic disparities within a country's population can have a centrifugal effect. Similarly, a serious economic downturn can erode people's support for the government. Government policies that exclude one or more groups of people or diminish their political voice, can contribute to rebellions or insurgencies.

centripetal force An event or a circumstance that helps bind together the social and political fabric of a state.

centrifugal force An event or a circumstance that weakens a state's social and political fabric.

Centripetal forces include certain policies and practices of governmental and nongovernmental institutions. Schools as well as the armed forces inculcate patriotism and build support for a state. Even international athletic competitions can unite a country's people.

Writing in 1940 when World War II was underway, the political geographer Richard Hartshorne drew attention to the importance of centripetal forces and specifically the value of the raison d'être, or reason for being.

A *raison d'être* is the idea, belief, or purpose that justifies the existence of a state. It is also the most significant centripetal force. The raison d'être for Pakistan was to create a Muslim-majority state in South Asia. The raison d'être for Israel was to create a homeland for the Jews. The absence of a raison d'être can be equally detrimental for the cohesiveness of a state. Yugoslavia was created as a multinational state whose citizens were separated by linguistic and religious differences and who did not share a raison d'être. The dictatorial rule of President Tito (from 1953 to 1980) held its diverse peoples together. After Tito's death, the country broke apart largely along ethnic, religious, and linguistic lines.

Whether a specific event is a centripetal or centrifugal force depends on one's perspective and may even change over time. For example, the 1989 reunification of Germany was widely perceived as a powerful centripetal force when it happened. However, the challenges of bridging the social divide between East and West Germans and the financial burden of reunification have, for a number of citizens, contributed to a greater awareness of the centrifugal forces associated with reunification.

It is impossible to list all of the kinds of centripetal and centrifugal forces. Depending on the situation, religion might be a centripetal force in one instance but highly divisive in another instance. The same can be said for issues involving language, immigration, election results, and national identity (**Figure 7.9**).

Separatism and Devolution

States change over time, and so do the centripetal and centrifugal forces that affect them. The stability of a state depends on how effectively it manages these forces. Even very stable states continually face challenges to their unity. **Separatism**, the desire of a nation to break apart from its state, is one of these challenges. A nation's sense of identity and the perception that it is different from other groups in the state can contribute to separatist sentiment. Separatism is closely connected with calls for greater *autonomy*, or self-government. When a state transfers some power a self-identified community within it, the process is called **devolution**. Devolution is one mechanism that states use to help accommodate separatist pressures.

Centrifugal and centripetal forces in Serbia • Figure 7.9

Kosovo is a region in the country of Serbia. The population of Kosovo consists of an Albanian majority and a Serb minority. Kosovo's declaration of independence from Serbia in 2008 has had both centrifugal and centripetal consequences: centrifugal in that it has challenged the unity of the Serbian state, but centripetal in that it has renewed Serb attachment to the region.

a. With a sign proclaiming "Kosovo is Serbia," protestors march in opposition to Kosovo's declaration of independence.

b. A Kosovar Albanian teacher shows her class the new Kosovo flag.

Geography InSight

Café para todos? A model of integration in multinational Spain • Figure 7.10

THE PLANNER

From the late 1930s until 1975, Spain was ruled by the dictator Francisco Franco. His idea of nationalism followed the idea of the nation-state strictly. He banned the use of the Euskera and Catalan languages and forbade the use of national symbols such as the Basque and Catalan flags. Franco's treatment of both groups fortified their national identity. It also contributed to the formation of Euskadi Ta Azkatasuna, or ETA, a Basque group that has often resorted to terrorism. How has Spain integrated these nations since?

a. Spain ratified a new constitution in 1978 that recognized the unity of the Spanish nation but also guaranteed autonomy to the "nationalities and regions integrated in it." This has been called the "café para todos" (*coffee for all*) model. The constitution fashioned the autonomous communities system and paved the way for the creation of the country's 17 autonomous regions (see map). As autonomous regions, the Basque Country and Catalonia control their own police forces, education, health care, and other services.

Atlantic Ocean
Autonomous Region
Greater Basque Country
FRANCE
Cantabria
Basque Country
Asturias
Galicia
Navarra
ANDORRA
La Rioja
Castile and León
Aragon
Catalonia
Madrid
Madrid
PORTUGAL
SPAIN
Extremadura
Castile-La Mancha
Valencia
Balearic Islands
Murcia
Mediterranean Sea
Andalusia
Ceuta
Melilla
Note: The Canary Islands, also an autonomous region, are not shown.
0 100 Miles
0 100 Kilometers

b. Basque separatists at a rally carrying a banner with a map of the Basque region. Note that there is a Basque population in southwestern France and that part of France is included in the territory referred to as "Greater Basque Country." In 2008, the Basque government advanced a proposal for Basque self-determination—a sign that some Basques desire separation from Spain. The proposal was overturned by the Spanish government.

c. Catalans march in support of the Catalan Charter. They carry Catalan flags, and their placards read "som una nació" (*we are a nation*). Approved in 2006, the Charter devolves still more power to Catalonia in areas of immigration, taxation, and transportation. Catalonia is one of the wealthiest regions in Spain and has often used the economic importance of the region as a bargaining chip to gain additional autonomy.

Spain is a multinational state (population 47 million) that has confronted separatist movements led by two of its nations: the Basques (population about 2.2 million), and the Catalans (population about 7.4 million). The national identity of the Basques stems in part from their language, called Euskera, and a shared history as a self-governing people. Catalan nationalism derives from similar factors, including the Catalan language and a tradition of autonomy. See **Figure 7.10** to learn more about these movements and how Spain has dealt with them.

CONCEPT CHECK

1. **What** is meant by the vertical extent of a boundary, and why is it significant?
2. **How** might a state's shape contribute to centrifugal or centripetal forces?
3. **What** are the main differences between unitary and federal systems of government?
4. **How** have devolution and multinationalism shaped Spain's political geography?

Internationalism and Supranational Organizations

LEARNING OBJECTIVES

1. **Explain** how internationalism and supranational organizations are related.
2. **Distinguish** between the General Assembly and the Security Council of the United Nations.
3. **Summarize** the key events leading to the establishment of the European Union.
4. **Explain** the challenges Turkey faces in terms of becoming a member of the EU.

Although separatism is a potentially destabilizing force, it is in some ways countered by the spread of **internationalism**, the development of close political and economic relations among states. The growth of supranational political organizations provides the clearest expression of internationalism. A **supranational organization** consists of multiple states that agree to work together for a common economic, military, cultural, or political purpose, or a combination of several of these. The United Nations (UN) is a supranational organization promoting global peace and security. The Association of Southeast Asian Nations (ASEAN), the Commonwealth of Independent States (CIS), and the European Union (EU) are other supranational organizations whose member states cooperate for political and economic purposes. The North Atlantic Treaty Organization (NATO) is a military alliance including several North American and European countries. Each of the supranational organizations mentioned here—indeed most of the supranational organizations that exist today—were formed after World War II.

The benefits of membership in a supranational organization vary depending on the purpose of the organization but typically include improved political security or enhanced trading opportunities. Membership in a supranational organization, however, has a cost, and that cost is associated with the loss of a portion of a state's sovereignty. The existence of supranational organizations introduces a tension between internationalism and sovereignty. Therefore the benefits of membership need to be perceived as outweighing any sacrifices to a state's sovereignty. Participation in a supranational organization indicates a willingness to be a team player—to support the decisions and policies of the organization rather than to take actions that support only the goals of one's own state.

The United Nations

The **United Nations** was founded in 1945 as a supranational organization charged with promoting peace in the world. The mission of the UN includes building and sustaining cooperative relations among states and, when conflict arises, using diplomacy to negotiate peaceful solutions. The experience of two devastating world wars plus the desire to avoid a third one contributed to international support for an organization like the UN. Almost every country in the world is represented in this organization. At present, Vatican City does not have full membership but maintains permanent observer status because of its expressed desire to remain neutral on certain issues. Although Kosovo declared its independence in 2008, this Balkan territory's status remains disputed, and it is not a member of the UN. (This explains why there are 192 members of the UN but 194 sovereign states in the world.)

Headquartered in New York City, the UN has many different constituent parts. We can distinguish between specialized agencies and principal organs. Some specialized UN agencies include the World Health Organization and the Food and Agriculture Organization. The Security Council, General Assembly, and International Court of Justice (ICJ) are principal organs of the UN. Located in the Hague, Netherlands, the ICJ resolves international legal disputes. The General Assembly, which consists of all members of the UN, controls its budget and oversees the activities of the other branches of the organization.

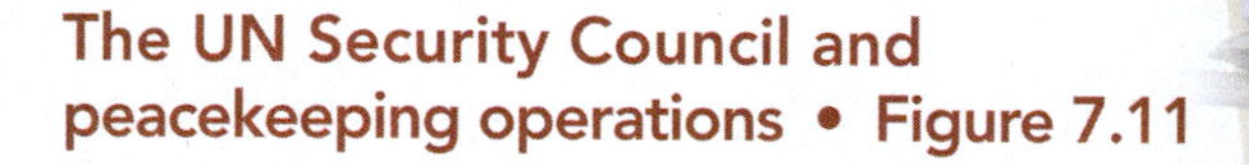

The UN Security Council and peacekeeping operations • Figure 7.11

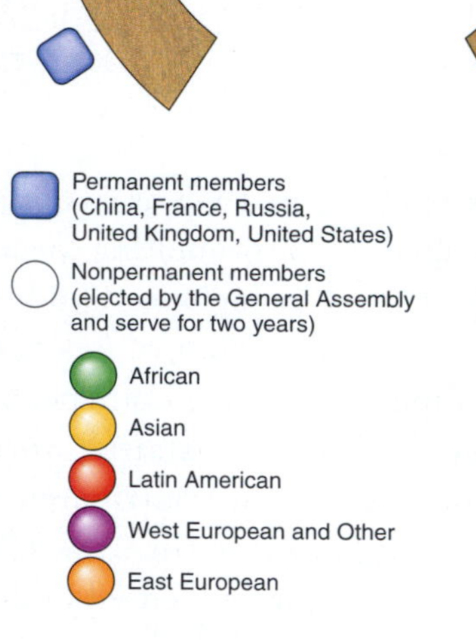

Permanent members (China, France, Russia, United Kingdom, United States)
Nonpermanent members (elected by the General Assembly and serve for two years)
- African
- Asian
- Latin American
- West European and Other
- East European

a. The composition of the Security Council
The UN Charter is vague on how equitable geographic representation of nonpermanent members should be achieved. In practice, seats are allocated according to the regional groups shown. The "other" in the West European group includes Australia and New Zealand.

b. UN peacekeepers at work
Serving in a humanitarian capacity, UN peacekeepers from Bolivia distribute water and meals to the residents of Cité Soleil, Haiti, after the devastating earthquake in January 2010. UN peacekeeping forces are provided by the members of the General Assembly.

The real power for the day-to-day maintenance of international peace and security lies with the Security Council. Depending on the situation, the Security Council might recommend sanctions against a country or might recommend that peacekeeping forces be deployed. In order for recommendations made by the Security Council to be acted upon, there must be nine affirmative votes from nonpermanent members and unanimous support from the permanent members. In 2003, for example, the United States and the United Kingdom sought UN support for the invasion of Iraq to overthrow Saddam Hussein. UN support was not forthcoming because three members of the Security Council —China, France, and Russia—opposed it. **Figure 7.11** provides an overview of the Security Council.

The European Union

Whereas the UN is a supranational organization that is global in scale and focuses primarily on issues of international security and well-being, the **European Union** (EU) is regional in scale and came into existence in order to enhance economic cooperation in western Europe. Five important developments contributed to the establishment of the EU.

1. Creation, in 1944, of Benelux, an association made up of Belgium, the Netherlands, and Luxembourg. These three small western European countries realized that economically they could lower their costs of production if they cooperated with one another to remove tariffs and ease restrictions on the movement of goods among them.
2. Implementation of the Marshall Plan following World War II, which stimulated the rebuilding of Europe and encouraged regional cooperation.
3. Establishment of the separate European Coal and Steel Community (ECSC) in 1952. The ECSC worked to remove barriers on the movement of coal and steel. The Benelux countries joined France, West Germany, and Italy as members of the ECSC.
4. Acceptance of the Treaty of Rome in 1957, which created the European Economic Community (EEC),

sometimes also called the Common Market. The Treaty of Rome committed its signatories to still greater economic union, through the creation of a single common market to enable the unrestricted movement of goods, people, services, and capital among them. The six countries that belonged to the ECSC were also founding members of the EEC.

5. Implementation of the Treaty of Brussels, or "Merger Treaty," in 1967. This treaty amended the Treaty of Rome and provided a framework for political cooperation, including a European parliament. To reflect its broader mission, the EEC was renamed the European Community (EC).

The Treaty of Rome incorporated the idea that the EEC would attract other countries sharing the same economic and political ideals. By 1981, six other countries had become members. Then, in 1992, these 12 member countries signed the Treaty of European Unity (also called the Maastricht Treaty) in the Netherlands. From this point forward, the term *European Union* has been used as the name of the supranational organization. Since 1993 the EU has admitted another 15 countries as members, expanding its membership to a total of 27 countries (**Figure 7.12**).

Turkey has applied for membership in the EU but has not yet been admitted for economic, demographic, cultural, and political reasons. Economic reasons include the country's large amount of debt and problems with high inflation. At current rates of natural increase, Turkey's population is expected to surpass that of Germany in the next few years. If admitted to the EU, Turkey would likely become the most populous member. It would also be the only member whose population is predominantly Muslim.

EU members and the euro zone • Figure 7.12

The EU now has a population of 495 million and a gross domestic product of nearly $15 trillion, making it a leading economic power. The passage of the Treaty of European Unity has enabled a greater degree of economic and political integration. Major progress toward monetary union occurred in 1999 when the EU adopted a single currency, the euro. The 16 countries that have since replaced their currency with the euro collectively form a region known as the euro-zone. Why is membership in the EU more widespread than the euro zone? Why has use of the euro expanded to nonmembers?

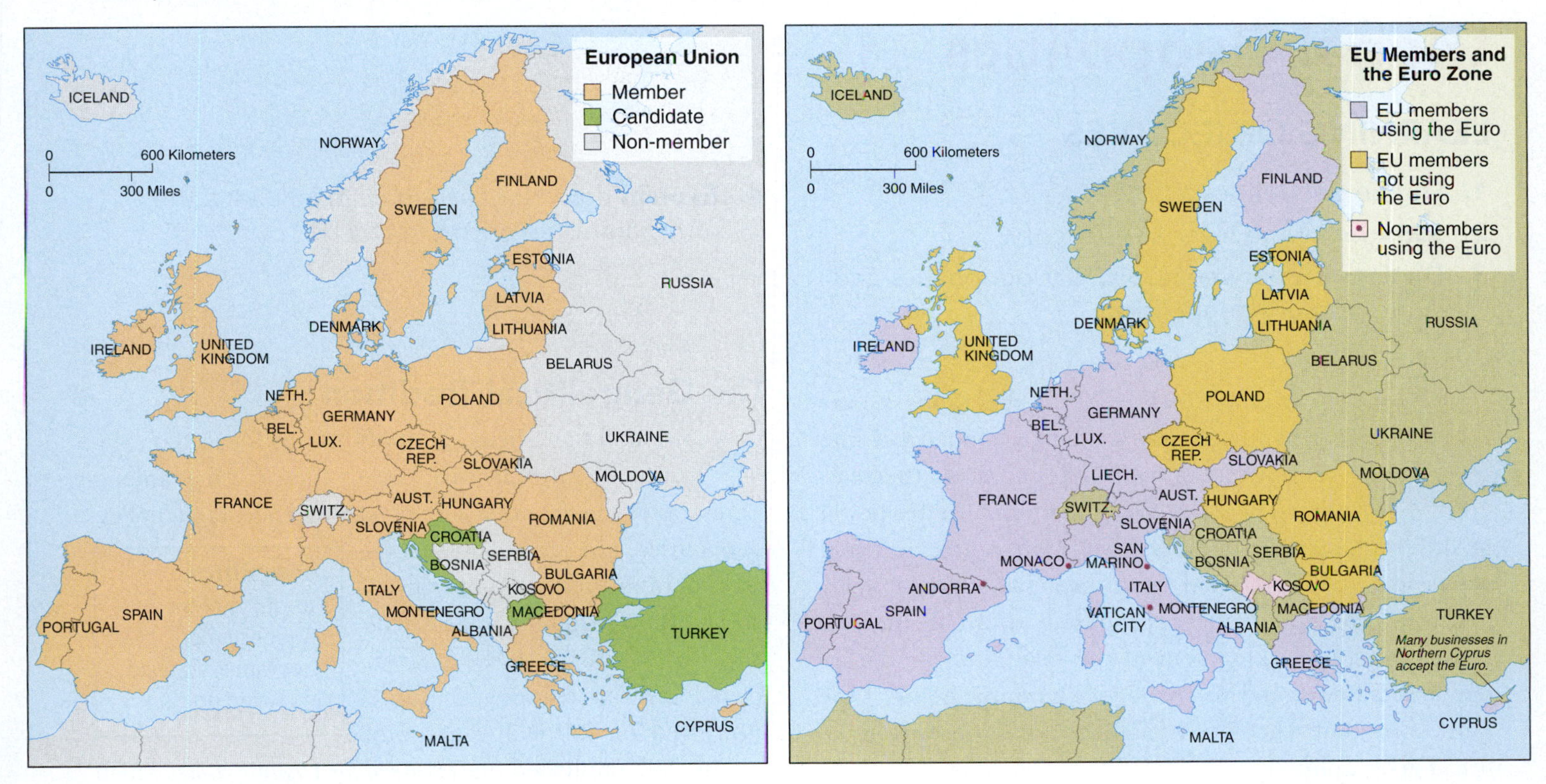

Political concerns include Turkey's stance on Cyprus (discussed previously in the chapter) and its human rights record, especially discrimination against Turkey's Kurdish minority.

The status of Turkey's application to the EU has become a divisive issue. Proponents of Turkey's membership highlight the value to be gained from additional market expansion, access to a large pool of labor, and the strategic importance of Turkey's geographic location, especially in terms of improved energy security for Europe. Proponents also maintain that admitting Turkey would send a powerful signal that the EU is not an elite, Christian-only club.

The EU is significant for many reasons, but two reasons deserve special attention. First, the EU is the best example of successful supranational economic cooperation and has become a model for other supranational economic organizations to follow. Second, it has pushed supranational cooperation to an unprecedented level such that the EU is beginning to take on some of the functions of a state in ways that challenge the traditional conception of the state. For example, the EU has a parliament, a central bank, a flag, and a national anthem. It has also developed a constitution, though it has not been ratified. These developments have led many scholars to ask whether the EU represents a new kind of supranational state.

CONCEPT CHECK STOP

1. **What** is a supranational organization, and what are some costs and benefits associated with membership in one?
2. **How** is the Security Council different from the General Assembly of the United Nations, and what role does the Security Council play?
3. **How** does the present-day EU differ from its forerunner supranational organizations Benelux, the ECSC, the EEC, and the EC?
4. **Why** is admitting Turkey into the EU a divisive issue?

Global Geopolitics

LEARNING OBJECTIVES

1. **Define** geopolitics.
2. **Summarize** the Heartland Theory.
3. **Distinguish** between Cold War geopolitics and contemporary geopolitics.
4. **Explain** how globalization can influence the diffusion of terrorism.

As we have seen, political geographers have spent much time trying to understand how states function—how they organize their territory, how the government distributes power to different territorial units within the state, and how states forge supranational organizations. Over the years political geographers have proposed different theories concerning the development of the state and the nature of political power. In this section we examine a few of the most important schools of thought or traditions within political geography.

The Geopolitical Tradition

As traditionally practiced, **geopolitics** has focused on the ways in which states acquire power, the relations among states, and the formulation of strategic foreign policy. Geopolitics, one branch of political geography, has its roots in the work of the German scholar Friedrich Ratzel (1844–1904). Ratzel, a zoologist by training, became interested in political geography and in 1897 published his *Theory of the Organic State*, which

geopolitics
The study of the relations among geography, states, and world power.

compared the growth of a state to the growth of an organism. Ratzel theorized that, like organisms, states needed sustenance in the form of resources and room to grow. Ratzel used the term *Lebensraum*—literally "living space"—to describe these needs.

The Theory of the Organic State draws a strong connection between the natural environment and the power of a state, as demonstrated by the idea that states develop and grow stronger through the addition of new territories. Ratzel's theory provides an example of the environmental determinism (see Chapter 1) that informed early geopolitical thought. It also shows the influence of Charles Darwin's ideas. Ratzel was familiar with Darwin's work, and the Theory of the Organic State placed importance on the concept of competition. According to Ratzel's theory, states competed with one another for resources and space, as do members of the animal world.

Ratzel never used his theory to guide foreign policy, but others did. Ratzel's ideas were adopted by Rudolf Kjellen (1864–1922), a Swedish professor and the person who actually coined the term *geopolitics*. Kjellen used Ratzel's ideas to argue that only large states would endure and that foreign policy should support the creation of a large state. Kjellen's work was translated into German and was used by the Nazis in the 1930s to support their goal of strengthening and enlarging the German state. As a result, German geopolitics was, for several decades, tainted by its connections with the Nazis.

The Heartland Theory

Halford Mackinder (1861–1947), a British geographer and member of parliament, contributed another geopolitical theory called the *Heartland Theory*. To understand this theory, it helps to know that Mackinder linked geopolitical stability with maintenance of a balance of power among states. Thus, if the balance of power was upset, a state or a combination of states could become *the* dominant world power. How might the balance of power be upset? It was not through control of the seas, Mackinder theorized, but rather through control of the large Eurasian landmass. In the interior of Eurasia was a region free from the danger of being attacked from the sea. Mackinder called this area the *geographical pivot*; later he referred to it as the **heartland** (**Figure 7.13**).

Mackinder's heartland • Figure 7.13

To Mackinder, the region between eastern Europe and Central Asia possessed the best combination of strategic geographic factors for world domination. Control of the heartland meant access to a sizable resource base and possession of a strategic, interior location that was safe from attack. In Mackinder's view, whoever dominated the heartland would be able to defeat any sea power.

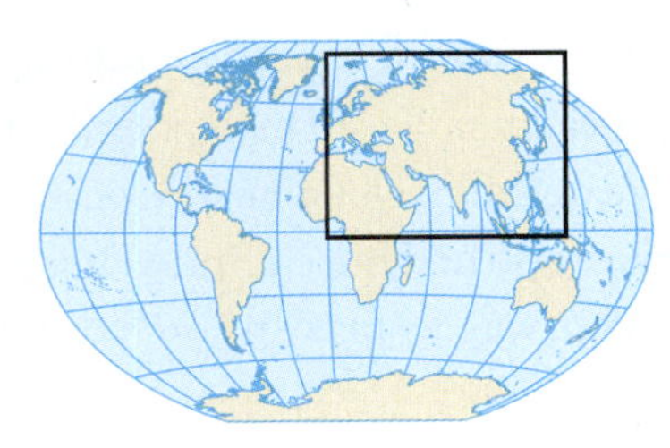

Cold War geopolitical configuration • Figure 7.14

The bipolar configuration during the Cold War pitted the capitalist West (the United States and its allies) against the Communist East (the Soviet Union and its allies). However, important military and ideological battles during the Cold War were waged in some unaligned states in an attempt to gain influence there. Note that the map shows 1980 political boundaries.

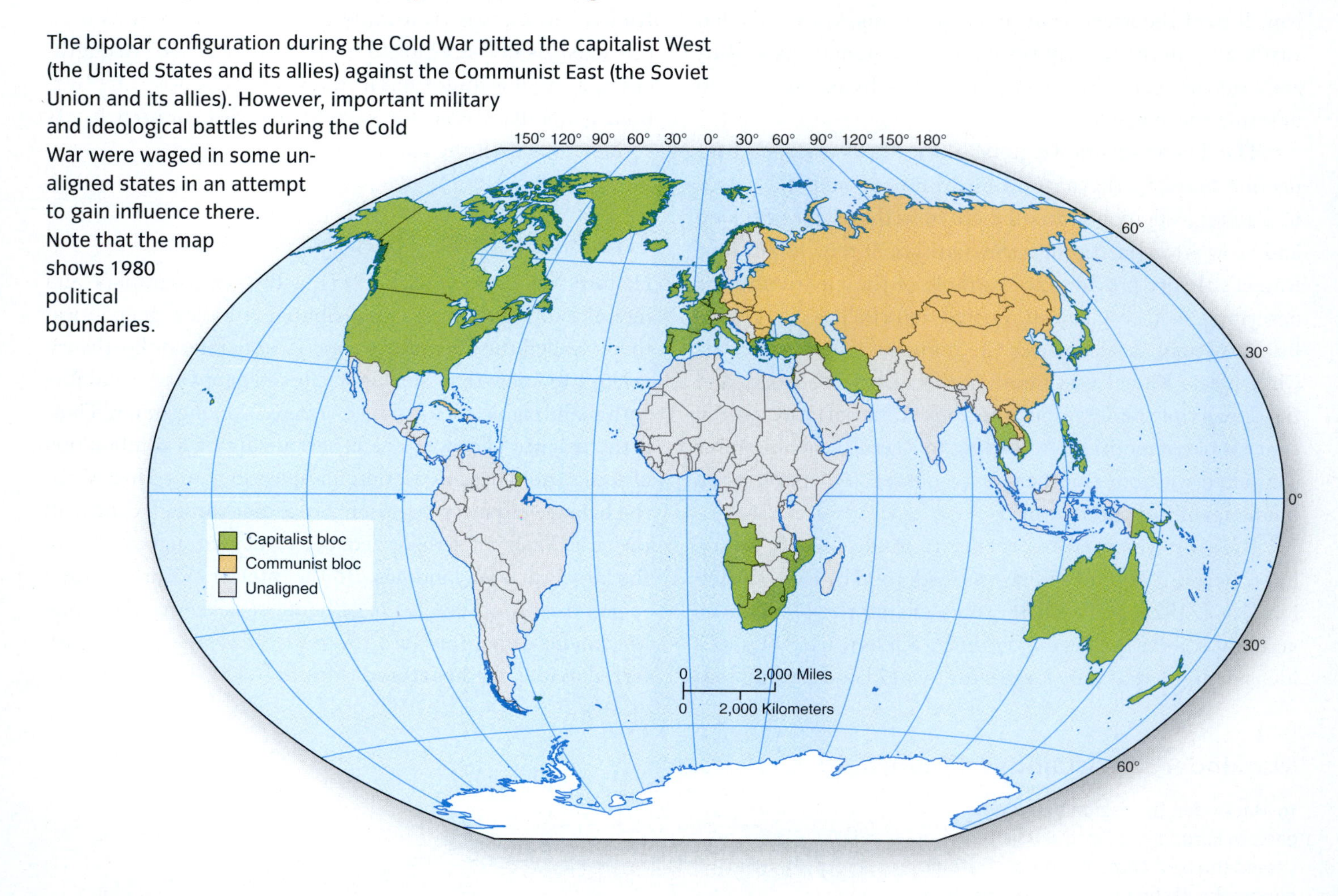

Mackinder's Heartland Theory also contains some elements of environmental determinism. Although he was aware of nonenvironmental factors, such as economic strength and transportation networks, Mackinder considered the territorial basis of states to be crucial to geopolitical power. In his words, "the [physical] geographical quantities in the calculation [of balance of power] are more measurable and more nearly constant than the human" (Mackinder 1942 [1919], p. 192). Thus, a major criticism of the Heartland Theory is that it oversimplifies the complexity of factors that shape global geopolitics.

Cold War Geopolitics

After World War II, relations between the United States and the Soviet Union cooled significantly. The term **Cold War** describes the hostility and rivalry that existed between the United States and the Soviet Union from the mid-1940s to the late 1980s. The race to build nuclear weapons was one expression of the rivalry between these superpowers. Geopolitically, the Cold War created a *bipolar world*—that is, a world divided into two opposing groups (**Figure 7.14**).

During the Cold War, the foreign policy of the United States was heavily influenced by the theory that if one country became Communist, other countries in the region would do the same and thus enable Communist domination of the world. Called the *domino theory*, this philosophy was used to justify the American policy of *containment*—the effort to limit the spread or influence of a hostile power or an ideology. One of the reasons the United States became involved in the Vietnam War was to prevent or "contain" the spread of communism in Asia.

Contemporary and Critical Geopolitics

The end of the Cold War in the early 1990s brought an end to the bipolar capitalist–Communist geopolitical configuration of the world. However, many political geographers maintain that there is still a bipolar configuration of the world that consists of a Global North and a Global South, separated on the basis of levels of development and wealth. We explore the basis for a North–South divide in greater detail in Chapter 9. Here, we focus on two other prevalent schools of thought. The first draws on the ideas of political scientist Samuel Huntington and his book, *Clash of Civilizations and the Remaking of the World Order*, published in 1996. In this book Huntington argues that instead of two opposing groups, there is a global configuration that is *multipolar* and consists of several groups or "civilizations." Using the terms *tribe, ethnic group, nation, civilization* as indicators of scale, Huntington argues that a civilization is the largest scale from which meaningful personal identity is derived. In his view, religion is the most important component that gives a civilization its identity, even more so than language or ancestry. Therefore, future conflict will result from the clash of these civilizations—in other words, cultural conflict, and the locations of this conflict will occur on fault lines. In this sense, a *fault line* is a place where civilizations meet, either within a country or along international boundaries (**Figure 7.15**).

This civilization-based view of geopolitics still privileges a geopolitical view of the world that sees strong bonds between people and territory. But a second school of thought sees a very different world order in which globalization enables *deterritorialization*, a loosening of ties between people and place. The modern state has its roots in the concepts of territoriality and sovereignty, but globalization—especially greater human mobility and technological integration—may facilitate deterritorialization. Members of a nation may be extremely geographically dispersed but can maintain close ties, even virtual communities, through

Huntington's "civilizations" • Figure 7.15

Huntington theorizes that geopolitics since the 1990s has been shaped more by factors affecting cultural identity—especially religion—than by the ideological differences that fueled the Cold War. Compare this map to Figure 7.14. His civilizations are, however, very broad categories that may give a false impression of unity when there is none or it is only weakly developed. Critics ask, for example, can we really speak of a single African or Islamic civilization today?

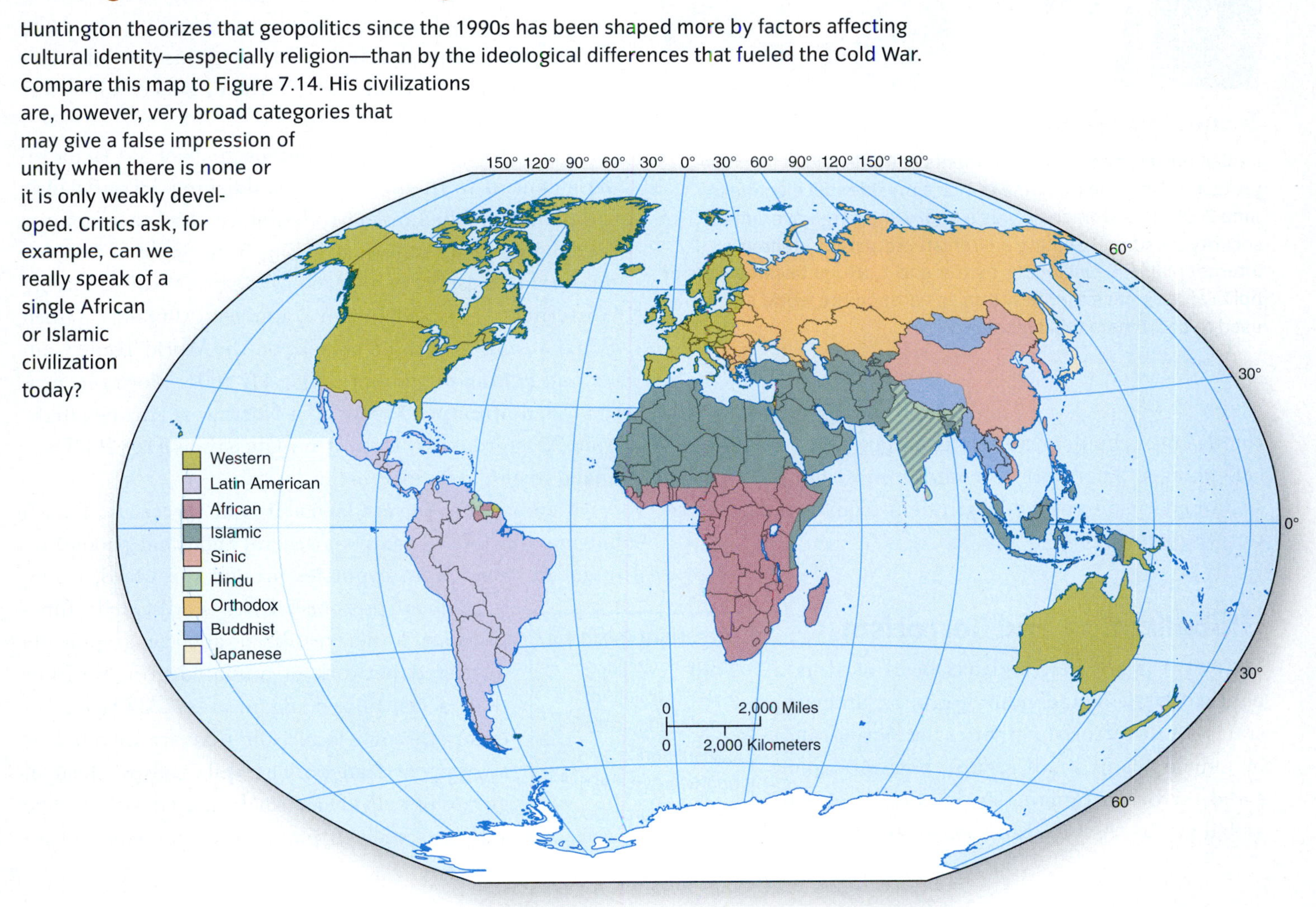

Technological integration • Figure 7.16

Iranian university exchange students in Rome join in a global protest against the outcome of Iran's presidential election in June 2009. This woman carries a photo showing government security personnel beating an Iranian protester in Tehran. Internet platforms including Twitter and YouTube became sites of political resistance for many Iranians around the world and were used to coordinate this global protest.

the use of technologies such as e-mail, the Internet, and cell phones. Therefore, a community's identity may be sustained in spite of its detachment from a specific territory (**Figure 7.16**).

Globalization and Terrorism

Terrorism is not new and has been used as a political tactic by individuals, groups, and even states. Most terrorist activities are perpetrated by individuals or small groups, but state-sponsored terrorism remains an important part of global affairs. A state can sponsor terrorism in several ways. It might provide a refuge for terrorists, help train them, provide them with weapons or equipment, share intelligence with them, or support them financially. A state might also be directly involved in designing terrorist activities. The United States Department of State currently recognizes four state sponsors of terrorism: Cuba, Iran, Sudan, and Syria.

> **terrorism** The threat or use of violence against civilians in order to inculcate fear, gain influence, and/or advance a specific cause or conviction.

There are four broad and often overlapping categories of terrorism: revolutionary, separatist, single-issue, and religious. Revolutionary terrorism seeks regime change. For example, the Algerian terrorist group Front de Libération Nationale (FLN) fought against French colonial rule from 1954 to 1962. Separatist terrorism may be perpetrated by groups seeking autonomy or independence, such as the Basque group ETA. Use of terrorism by individuals or groups to advance a specific cause, such as animal rights or environmental values, constitutes single-issue terrorism.

Al-Qaeda is a major terrorist organization whose motives are both revolutionary and religious. When al-Qaeda was formed by Osama bin Laden in 1988, its revolutionary cause was initially directed against the Soviet Union, whose troops had invaded Afghanistan and occupied the capital city. Gradually, bin Laden fused this revolutionary cause with a religious one as well: the waging of a holy war against the invaders.

The revolutionary and religious goals of al-Qaeda remain central to its mission of establishing a Pan-Islamic Caliphate—a Muslim-controlled state that encompasses the Islamic community extending from Spain to Indonesia. A related aspect of al-Qaeda's mission involves forcing Westerners to leave Muslim countries. Al-Qaeda gained global attention with its attacks on the World Trade Center and Pentagon on September 11, 2001. More recently, the failed attempt to blow up a Northwest Airlines flight from Amsterdam to Detroit on Christmas Day 2009 was linked to al-Qaeda in Yemen.

Membership in al-Qaeda is not precisely known but may include as many as several thousand people. Its members come from countries around the world, giving it a global presence. Even so, al-Qaeda functions as a decentralized and geographically dispersed network of affiliated groups rather than a centralized organization. Al-Qaeda operates through local cells that are directed by a group of leaders. The cells consist of small groups of individuals usually tasked with specific activities such as planning or carrying out an

Terrorism over time and by region, 1970–2007 • Figure 7.17

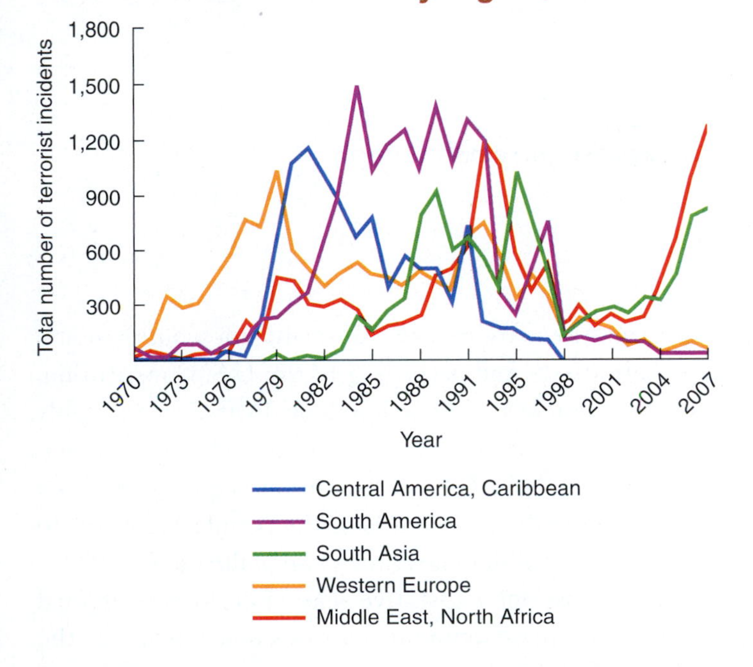

a. The graph shows the number of terrorist incidents by region. Terrorist activity related to the status of Northern Ireland, the Basque Country, and the French island of Corsica in the Mediterranean Sea helps explain the spike in western Europe in the 1970s.

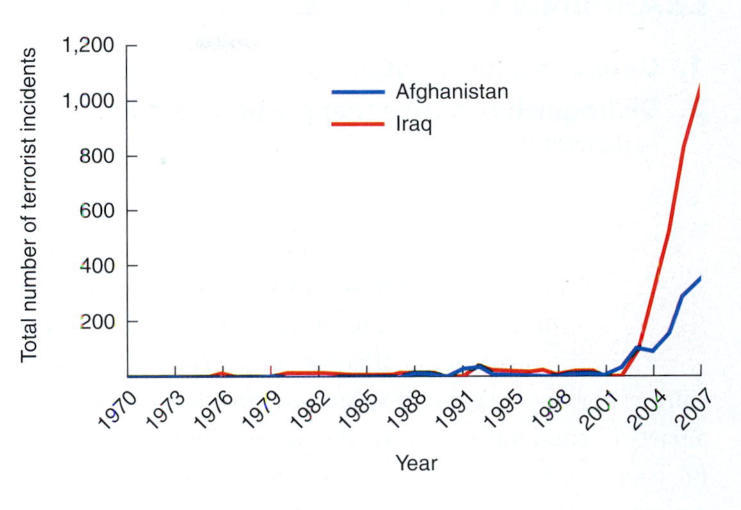

b. The Taliban has been responsible for most of the terrorist attacks in Afghanistan, whereas al-Qaeda and affiliated groups are leading perpetrators of terrorism in Iraq. Attacks often target government facilities and personnel, police, as well as the Coalition Forces and foreign workers in both places.

attack. Communications between cell members are carefully managed so that members of one cell do not know the nature of the work, identity, or location of members of another cell. Thus, effective counterterrorism against the al-Qaeda cells in one country is likely to have no effect on its cells in other countries.

Some experts consider the development of these decentralized networks to be a new strategy that has been facilitated by globalization. In other words, advances in communications and Internet technologies have made it easier, faster, and less expensive to share information. Moreover, the Internet has helped open up new spaces, including websites and chat rooms, that terrorist groups can use to spread threats or recruit members.

Other terrorist experts argue that terrorism today is an expression of resistance to globalization, and specifically the global diffusion of Western values associated with modernism (see Chapter 5). As evidence, these experts cite the rise in terrorist activity in places such as Afghanistan and Iraq since the U.S.-led invasions in 2001 and 2003, respectively. U.S. involvement in Afghanistan stems from the 9/11 al-Qaeda attacks in 2001 and the subsequent declaration of the "war on terror" by former President George W. Bush. The primary justification for the U.S.-led war on Iraq in 2003 was that Iraqi President Saddam Hussein was stockpiling weapons of mass destruction. Saddam Hussein was also thought to have links to al-Qaeda. Neither of these claims has been substantiated, however (**Figure 7.17**).

CONCEPT CHECK

STOP

1. **What** is geopolitics, and how has geopolitical thinking changed over time?
2. **How** does the environment factor in the Heartland Theory?
3. **What** political concerns shaped Cold War geopolitics?
4. **How** has the geography of terrorism changed over time?

Electoral Geography

LEARNING OBJECTIVES

1. **Define** electoral system.
2. **Distinguish** between reapportionment and redistricting.
3. **Explain** gerrymandering.

In a representative democracy, voters elect legislators whose duty is to develop and implement public policy on behalf of their constituents. The set of procedures used to convert the votes cast in an election into the seats won by a party or candidate is referred to as an **electoral system**. *Electoral geographers* study the spatial aspects of electoral systems, voting districts, and election results.

Several different electoral systems are used in the world, but they can be classified into two main systems: the majority-plurality system and the proportional system. With *majority-plurality representation* (also called geographic representation), the person who receives a majority or plurality of the votes is elected and represents all of the voters in an electoral district. Majority-plurality systems create single-member electoral districts that are territorially defined. In general, the majority-plurality system is commonly associated with countries that have two dominant political parties, as in the United States.

In contrast, with *proportional representation* (also called party-political representation) multiple representatives can be elected. When proportional representation is used, voters choose from among political parties rather than individual candidates. After the votes are tallied, legislative seats are divided on a proportional basis. For example, a party receiving 30% of the votes would receive 30% of the legislative seats. The proportional system is widely used in Europe.

Reapportionment and Redistricting

reapportionment The process of allocating legislative seats among voting districts so that each legislator represents approximately the same number of people.

For majority-plurality representation to be equitable, voting districts should have approximately the same number of people. **Reapportionment** becomes necessary because, over time, the population of a state can change. For example, in the U.S. House of Representatives, there are 435 seats for congressional representatives, and according to the U.S. Constitution, these seats must be apportioned or divided as equitably as possible among the 50 states according to their population. Indeed, the U.S. Constitution requires that the government conduct a census of the population every 10 years. Because of demographic change during the 1990s, Oklahoma lost one legislative seat while Arizona gained two in the reapportionment that took place following the 2000 U.S. Census.

redistricting Redrawing the boundaries of voting districts usually as a result of population change.

Reapportionment is ofen followed by **redistricting** (**Figure 7.18**). Three criteria, established by the Supreme Court, guide the redistricting process. Congressional districts: (1) are to have equal population; (2) to be contiguous and compact; and (3) are to respect the boundaries of other administrative units such as counties or parishes. Redistricting is the responsibility of each state and is usually carried out by the state legislature. As a result, redistricting often becomes a contentious exercise that is influenced by party politics.

Gerrymandering

Reapportionment and redistricting are intended to ensure equal representation on the basis of population in the House of Representatives. They are also supposed to treat political parties, as well as racial and ethnic minorities, equally. But legislators are well aware that how the boundaries of congressional districts are drawn can influence the outcome of elections. As a result, the redistricting process regularly raises concerns about **gerrymandering**.

gerrymandering The process of manipulating voting district boundaries to give an advantage to a particular political party or group.

Electoral geographers recognize two basic gerrymandering techniques: excess vote gerrymandering

Reapportionment and redistricting in the United States • Figure 7.18

THE PLANNER

PROCESS DIAGRAM

Every state is divided into congressional districts, with each represented by a single congressperson. To ensure equality among a state's districts, each representative is to speak for an equal number of people. If the population of a voting district changes, redistricting may become necessary to create districts of equal population. Arizona provides a good example of the reapportionment and redistricting processes.

1 In the 1990 census, Arizona had 3,665,228 people. The map shows the six congressional districts that were created following the 1990 census. They had equal populations of 610,871 people at the time. But the 2000 census (table) reveals how each district's population grew and how unequally distributed the state's population had become in only 10 years. (*Sources: Map*: Congressional Directory, 1997. *Data*: Arizona Independent Redistricting Commission.)

The six 1990 congressional districts	Their unequal populations in the 2000 census
1	829,492
2	773,824
3	997,565
4	735,344
5	793,256
6	1,001,151
Total	5,130,632

2 By the 2000 census, Arizona's population had sharply increased by 40%, to 5,130,632. As a result of this growth, the state gained two House seats through reapportionment. Because the state's population growth was spatially uneven, Arizona needed to redistrict. Each of the newly created congressional districts contained 641,329 people. Although each congressional district has the same number of people, it's important to recognize that each district's ethnic composition may be very different, as the pie charts reveal. (*Sources*: *Map*: Congressional Directory, 1997. *Chart data*: Arizona Independent Redistricting Commission.)

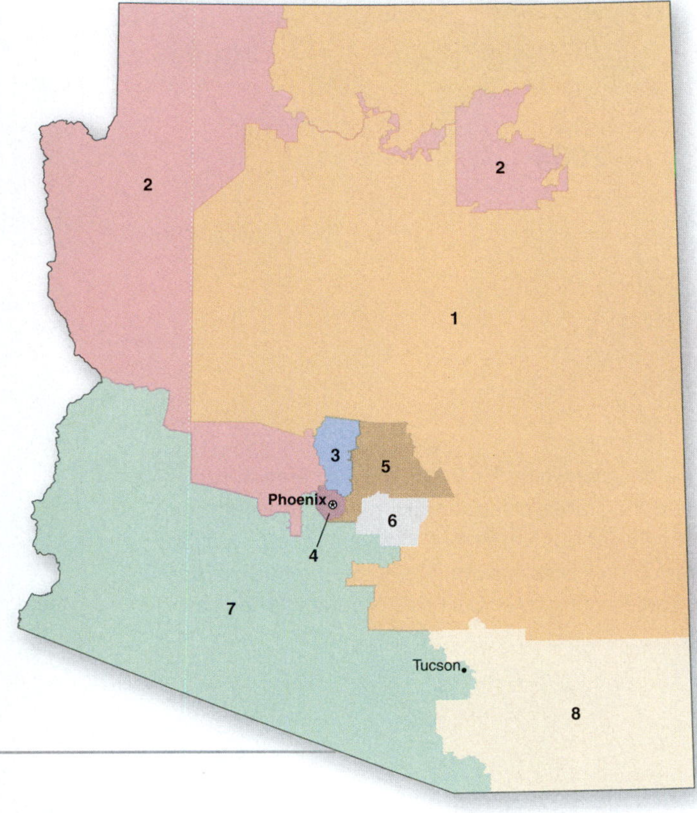

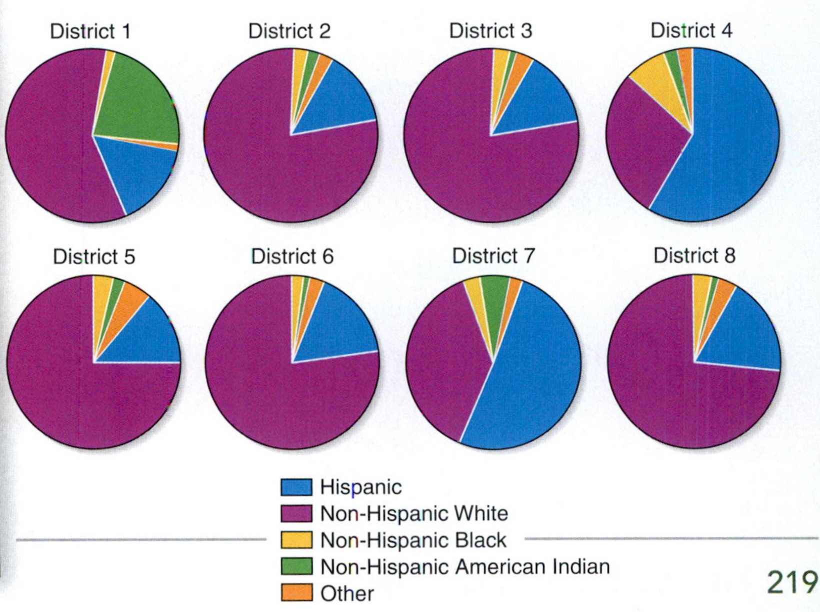

Gerrymandering in Texas • Figure 7.19

These maps show the Twenty-third Congressional District of Texas in 2002, 2004, and 2006.

a. The Twenty-third Congressional District was a majority-minority district (55% Hispanic, 41% White, 4% Other) in 2002.

b. Texas broke from the practice of redistricting once every 10 years and redistricted in 2003. By 2004, with redistricting completed, the Twenty-third district had been redrawn in such a way, shown here, that it excluded approximately 100,000 Hispanics in the vicinity of Laredo and created a district much more likely to elect a Republican. Lawsuits challenged the constitutionality of these actions.

c. In a decision issued in 2006, the Supreme Court upheld the state's calendar for redistricting, but found that the Twenty-third district was gerrymandered to dilute the power of the Hispanic vote. The Supreme Court ordered a further redistricting to correct this problem. What was the solution?

and wasted vote gerrymandering. The *excess vote technique* creates a few electoral districts in which support for the opposition forms a strong majority. In these districts, excess voting occurs because many more votes are cast than are needed to win the election. Although the opposition wins overwhelmingly in these few districts, it does not secure majority control and may lose seats in other districts. In contrast, the *wasted vote technique* disperses support for the opposition so that the opposition loses by a slim margin, say, 45–55%, or 40–60%. "Wasted votes" are the votes recorded for the losing candidate. When the support for the opposition draws heavily from racial or ethnic minorities, it is easy to see how these two gerrymandering techniques make it possible to create voting districts that diminish the effectiveness of the minority vote.

To try to prevent this from happening, the Voting Rights Act was amended in two important ways in the 1980s. First, it prohibited gerrymandering that dilutes minority voting power. Second, it stipulated that there may be some circumstances in which it is necessary to create voting districts that concentrate the strength of a specific minority group. This last change supported the creation of *majority-minority districts* (districts where minority group members form the majority) in order to improve minority representation. See **Figure 7.19** for an example of gerrymandering involving a majority-minority district.

CONCEPT CHECK STOP

1. **What** is the difference between majority-plurality representation and proportional representation?
2. **Why** are reapportionment and redistricting necessary?
3. **When** does redistricting become gerrymandering?

Political Landscapes

LEARNING OBJECTIVES

1. **Explain** what a landscape of central authority is.
2. **Distinguish** between security landscapes and landscapes of governance.
3. **Define** political iconography.

How do political affairs shape political landscapes? In what ways are cultural landscapes used to convey political power? How and why do certain landscapes, both cultural and natural, become the focus of intense political disputes? These are just some of the questions that help guide the study of political landscapes.

Landscapes of Central Authority

States exercise their political control through government. In turn, the policies, agencies, and laws of government affect the look of cities and towns as well as the countryside. When governments fund the design and construction of infrastructure including railroads, sewage, irrigation, or power facilities, they are creating landscapes of central authority. If you drive on a U.S. interstate to get to school or work, that interstate is part of a transportation network and landscape of central authority created by the federal government largely as a result of the 1956 Federal Highway Act. We can also see the stamp of central authority in the landscape of Egypt's Aswan High Dam, a major source of hydroelectric power for the entire country.

Landscapes of central authority are important because they contribute to the process of state-building. For example, they may help connect different parts of a country while reinforcing the power and significance of the central government.

Security landscapes in the West Bank • Figure 7.20

The separation barrier is a concrete wall, some 600 km (400 mi) long, that Israel has built in the West Bank. To some Israelis, the barrier provides security from Palestinian suicide bombings; to Palestinians, it is Israel's way of annexing territory. In an advisory opinion in 2004, the International Court of Justice found the barrier contrary to international law.

a. Here the barrier separates the Israeli town of Matan (foreground) from the Palestinian town of Hableh.

b. The barrier, 8 m (25 ft) high, divides the town of Abu Dis.

Security landscapes Boundaries establish the limits of a state's jurisdiction and, in effect, the limits of a state's political authority. Many demarcated boundaries are **security landscapes**—a specific type of political landscape created to protect the territory, people, facilities, and infrastructure of a state. Security cameras, metal detectors, and gated entrances are some measures intended to deter terrorist attacks and, in the event of one, minimize the damage and loss of life (**Figure 7.20**).

Landscapes of governance The imprint of central authority can also be revealed through an examination of legal policies. Laws can encourage or discourage certain human behaviors, which, in turn, may lead to the creation of distinctive landscapes of governance. Farm policy can influence agricultural practices. In Europe, where it is too cool to cultivate sugarcane, the sugar beet crop is an important source of sugar. Without sizable EU subsidies, however, sugar beet production would not be as widespread in Europe as it presently is, nor would the EU continue to be a net sugar exporter.

Landscapes of governance are also created when systems of land survey are commissioned by a government prior to the settlement of an area. Many other kinds of laws affect the landscape as well. Inheritance laws requiring that land be divided among all children, rather than passing to the oldest child, can result in fragmented landholdings. When governments change the laws and regulations in order to encourage particular businesses or industries to locate in an area, landscape change is likely to occur. The establishment of national parks and forests signals the impact of laws that protect, preserve, or regulate the use of the natural environment.

Political Iconography

Landscapes are not only affected by governmental laws and policies, but are also coded with political meaning. Examples of **political iconography** include flags, statues or images of political or military leaders, national anthems, war memorials, and symbols of political parties. These symbols come to represent certain ideals, such as freedom or democracy, and help build a shared sense of identity (*Where Geographers Click* and **Figure 7.21**).

political iconography An image, object, or symbol that conveys a political message.

Where Geographers CLICK

THE PLANNER

CAIN Web Service: Political Wall Murals in Northern Ireland

http://cain.ulst.ac.uk/murals/index.html

Learn more about political iconography and the conflict in Northern Ireland by exploring the murals that mark and demarcate the region's public spaces. There is a great deal to explore here. You might begin by following the link to **The Bogside Artists.**

Political iconography • Figure 7.21

U.S. Marines placed an American flag on a statue of Saddam Hussein in Baghdad in 2003. The Marines then helped topple the statue. The power of political iconography derives from the way it becomes part of the shared vocabulary of some groups but not others. How do you think Americans interpreted this image? Why is it provocative to many Middle Easterners?

CONCEPT CHECK STOP

1. **What** functions do landscapes of central authority serve?
2. **What** are landscapes of governance and why are they significant?
3. **How** is political iconography related to nationalism?

THE PLANNER

Summary

1 Key Concepts in Political Geography 194

- **Political geography** is a branch of human geography that focuses on the spatial aspects of political affairs and is heavily influenced by human **territoriality**, which can transform an open space into a divided or closed one, as shown here.
- An important turning point in the development of the **state** occurred during the middle of the 17th century, when ideas about **sovereignty** became strongly linked to specific units of territory.

A divided state • Figure 7.2

- When there is congruence between the boundaries of a **nation** and the boundaries of a state, a **nation-state** exists. Importantly, a **multinational state** can function as a nation-state if the different national groups are effectively integrated.
- Sizable portions of the globe have been shaped by **imperialism** and **colonialism**. These processes are especially visible on the contemporary map of Africa; they created an enduring framework for the political organization of African space even as Africans achieved **self-determination** and independence.

2 Geographical Characteristics of States 201

- The geographical characteristics of states include the **boundaries**, the character and configuration of a state's territory and **territorial seas**, the internal spatial organization of the state, and any **centripetal** or **centrifugal forces**. The shape of a state as well as the presence of **enclaves** and **exclaves** can influence a state's political affairs.
- For administrative purposes, states are divided internally into smaller units such as provinces or districts. In a federal system of government, the state distributes power to its territorial subdivisions; in a unitary system the power of the state is concentrated in the central government.
- Depending on the circumstances, nationalism can be a centripetal or centrifugal force. The map that is shown here, for example, can be seen as an expression of both nationalism and **separatism**.

A model of integration in multinational Spain • Figure 7.10

3 Internationalism and Supranational Organizations 209

- **Internationalism** can lead to the formation of **supranational organizations** such that multiple states agree to work together for common economic, military, political, or other purposes.
- The **United Nations** is a supranational political organization that is global in scale and promotes international peace and security. Concern for human well-being, as demonstrated through relief efforts shown here and other activities, is also a part of much UN-related work. The **European Union**, in contrast, began as an economic organization and operates on a regional scale.

UN peacekeeping operations • Figure 7.11

4 Global Geopolitics 212

- How states acquire and deploy their power are important dimensions of **geopolitics**. The Theory of the Organic State and the Heartland Theory were two early and influential geopolitical theories. According to the Heartland Theory, the Eurasian landmass, shown here, provided a territorial basis for a state or allied states to become the dominant world power.

Mackinder's heartland • Figure 7.13

- During the Cold War, the geopolitical configuration of the world was divided into two opposing groups: the capitalist West and the Communist East. The end of the Cold War has led some scholars to argue that the geopolitical configuration of the world has become multipolar.
- Most acts of **terrorism** can be classified into one of four categories, but it is often the case that terrorism is used for multiple purposes. An important geographical dimension of terrorist networks today is that they have a global reach but work through highly decentralized and localized cells.

5 Electoral Geography 218

- Electoral geography is the study of the spatial aspects of electoral systems, voting districts, and election results. Two main types of **electoral systems** exist: the majority-plurality system and the proportional system.
- In the United States, **reapportionment** and **redistricting** are electoral processes designed to ensure equal representation. Over time, shifts in population similar to those recorded here, create imbalances in the system of representation and underscore the necessity of redistricting. Redistricting often raises concerns about **gerrymandering**.

Reapportionment and redistricting in the United States • Figure 7.18

The six 1990 congressional districts	Their unequal populations in the 2000 census
1	829,492
2	773,824
3	997,565
4	735,344
5	793,256
6	1,001,151
Total	5,130,632

6 Political Landscapes 221

- Political landscapes can be thought of as a visual expression of political affairs. Landscapes of central authority are political landscapes that are the result of policies and programs implemented by the central or federal government.
- **Security landscapes** reflect a state's concern about its people, borders, and territory. But the construction of prominent security landscapes such as the one shown here can also be seen as a provocative political act.

Security landscapes in the West Bank • Figure 7.20

- Landscapes of governance reveal the impact of certain laws on the landscape. **Political iconography** refers to the use of symbols to convey political messages.

Key Terms

- boundary 202
- centrifugal force 206
- centripetal force 206
- Cold War 214
- colonialism 199
- devolution 207
- electoral system 218
- enclave 205
- European Union 210
- exclave 205
- geopolitics 212
- gerrymandering 218
- heartland 213
- imperialism 199
- internationalism 209
- multinational state 198
- nation 197
- nation-state 198
- political geography 194
- political iconography 222
- reapportionment 218
- redistricting 218
- security landscape 222
- self-determination 201
- separatism 207
- sovereignty 195
- state 196
- supranational organization 209
- territorial seas 202
- territoriality 194
- terrorism 216
- United Nations 209

Critical and Creative Thinking Questions

1. Will Internet voting ever replace the use of traditional polling places? What geographic and political conditions would be most conducive to such change?
2. List some advantages and disadvantages of majority-plurality and proportional representation systems.
3. Do some fieldwork in the area where you live and identify a relic boundary. What processes led to the creation of that boundary?
4. Not all ethnic groups are nations. Why?
5. Under what circumstances might devolution become a centrifugal force?
6. A political geographer might argue that the Berlin Conference was an exercise in gerrymandering. Explain what is meant by this statement and take a position on it.
7. What similarities and differences are there between the division of Germany after World War II and the division of Cyprus?
8. Review the geographic composition of the UN Security Council. Does it need reform? Why or why not?

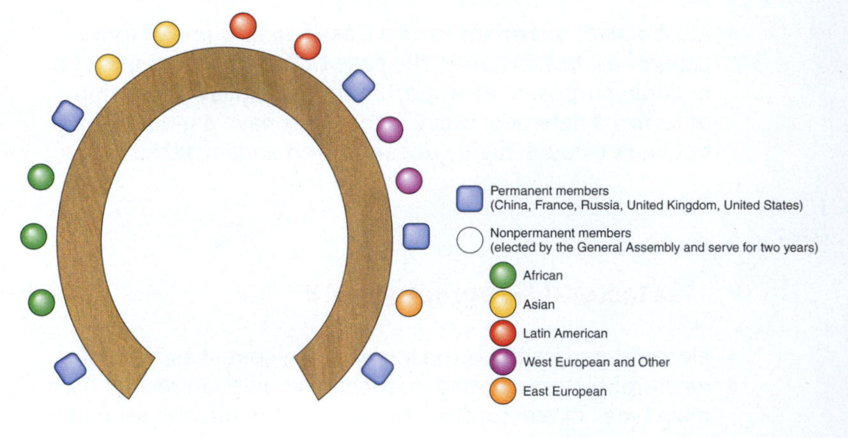

What is happening in this picture?

Kuwaiti women line up to vote. Kuwait's constitution, drafted in 1962, contained an electoral law that prevented women's political participation. In 2005, that law was amended to allow women the right to vote and run for election, but in accordance with Islamic law, polling places must be segregated.

Think Critically

1. Worldwide, what circumstances or developments have facilitated the spread of women's right to vote?
2. In any society, why has giving women the right to vote tended to be controversial?

Self-Test

(Check your answers in Appendix B.)

1. A _____ exists when the boundaries of a nation match the boundaries of the state.
   a. multinational state
   b. nation
   c. state
   d. nation-state
2. The Berlin Conference is associated with all of the following except _____.
   a. the granting of self-determination
   b. the creation of geometric boundaries
   c. the creation of multinational states
   d. the political map of Africa today
3. The areas where coastal states have the right to manage ocean resources are called _____.
   a. territorial seas
   b. boundary waters
   c. exclusive economic zones
   d. offshore waters
4. Turkey is not yet a member of the EU. Which is the least probable reason for this?
   a. Turkey has problems with inflation.
   b. Turkey has a predominantly Muslim population.
   c. Turkey's market is too small.
   d. Turkey has had problems with human rights violations.

5. A(n) _____ shape is most likely to be considered an efficient way of organizing territory.

   a. fragmented
   b. compact
   c. elongate
   d. prorupt

6. Use the accompanying map to describe the territorial configuration of this state.

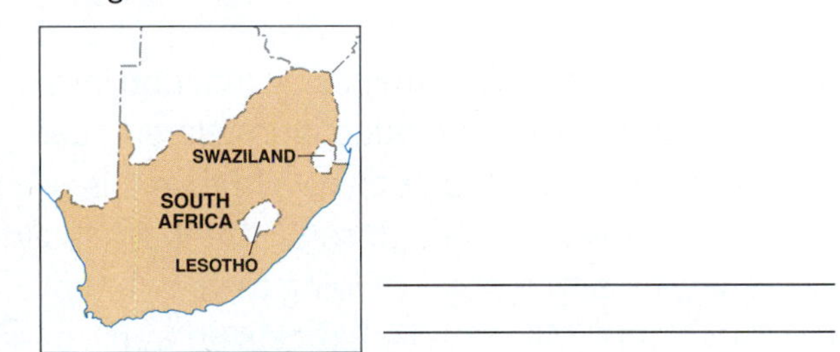

7. Which characteristic below is usually considered a centripetal force?

   a. a multinational population
   b. a raison d'être
   c. an exclave
   d. an elongate shape

8. The transfer of power from a state to a self-identified community within it is called _____.

   a. self-determination
   b. autonomy
   c. separatism
   d. devolution

9. In Europe, the first collaborative trading agreement and important forerunner of the EU was _____.

   a. the Treaty of Rome
   b. the Treaty of European Unity
   c. Benelux
   d. the European Coal and Steel Community

10. According to the Heartland Theory, world power was based on _____.

    a. land power
    b. sea power
    c. air power
    d. nuclear weapons

11. This diagram depicts what kind of geopolitical configuration of the world?

    a. unipolar
    b. bipolar
    c. multipolar
    d. deterritorialized

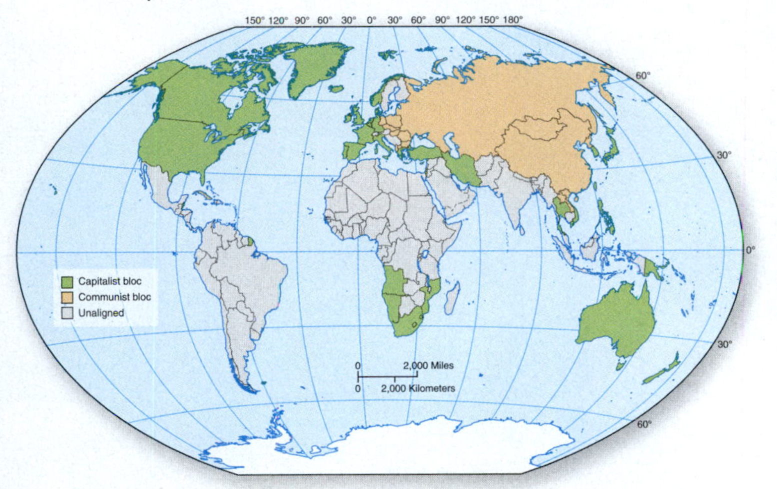

12. Countries that have more than two dominant political parties are most likely to have an electoral system based on _____.

    a. proportional representation
    b. majority-plurality representation
    c. federal representation
    d. unitary representation

13. The type of gerrymandering that creates voting districts where the opposition loses by a slim margin is called the _____.

    a. excess vote technique
    b. dilution technique
    c. wasted vote technique
    d. majority-minority technique

14. National parks provide landscape evidence of _____.

    a. centrifugal forces
    b. the imprint of central authority
    c. security landscapes
    d. devolution

15. This image provides a good example of _____.

    a. imperialism
    b. a relic boundary
    c. a geometric boundary
    d. political iconography

THE PLANNER

Review your Chapter Planner on the chapter opener and check off your completed work.

8 Urban Geographies

BUILDING A SUSTAINABLE CITY

Every city is a human creation that fulfills certain economic, political, and social functions. Until recently, however, few cities were built with the goal of environmental sustainability in mind. Masdar City, under construction in the United Arab Emirates since 2008, promises to change this.

Masdar City incorporates innovative traditional and futuristic design principles. Like many of the world's cities that took shape before the development of the automobile, Masdar City will be walled and car-free. City space will be organized mainly by function into residential, commercial, and education-related quarters. Approximately 40,000 people will live in this pedestrian-friendly city in the desert. Street-side buildings will shade the walkways, helping to keep them cool. Without cars, Masdar City expects no traffic or air quality problems. When traveling across town, people will use solar- powered personal rapid transit pods—a mode of transportation similar to a subway except that it uses individual car-like vehicles that are not limited to fixed routes between destinations.

Renewable energy will power Masdar City. Solar energy, wind energy, and the conversion of waste products into energy will supply the city's electricity. Approximately 80% of the water used in the city will be recycled. The emphasis on reusing and recycling should make it a zero-waste city and eventually one without a carbon footprint.

If we can create environmentally sustainable cities, what would it take to transform our cities into places that provide greater social equity—for example, in the form of adequate housing and jobs for all their residents?

Global Locator

UNITED ARAB EMIRATES

AFRICA

NATIONAL GEOGRAPHIC

CHAPTER OUTLINE

CHAPTER PLANNER

- ❑ Study the picture and read the opening story.
- ❑ Scan the Learning Objectives in each section: p. 230 ❑ p. 240 ❑ p. 248 ❑
- ❑ Read the text and study all visuals. Answer any questions.

Analyze key features

- ❑ Geography InSight, p. 238 ❑ p. 246 ❑
- ❑ Process Diagram, p. 243 ❑ p. 252 ❑
- ❑ What a Geographer Sees, p. 245
- ❑ Video Explorations, p. 250
- ❑ Stop: Answer the Concept Checks before you go on: p. 240 ❑ p. 247 ❑ p. 255 ❑

End of chapter

- ❑ Review the Summary and Key Terms.
- ❑ Answer the Critical and Creative Thinking Questions.
- ❑ Answer What is happening in this picture?
- ❑ Complete the Self-Test and check your answers.

Cities and Urbanization

LEARNING OBJECTIVES

1. **Explain** what is meant by functional complexity.
2. **Summarize** trends in global urbanization.
3. **Distinguish** between urban primacy and urban hierarchy.
4. **Explain** central place theory.

Love them or hate them, cities are a defining feature of our world and a driving force in the global economy. As hubs of activity, cities are constantly changing, and yet their role as important gateways endures. Globally, the pace of urbanization over the past 60 years has been swift, and more than half of the world's people now live in cities. But just what is a city, what does it mean to be urban, and how is the global urban network configured?

What Are Cities?

Someone from Tokyo, a city of 35 million people, probably has ideas about what a city is that are very different from those of someone from Stillwater, Oklahoma, a city of about 50,000 people. When asked about the differences between the two places, Japanese students from Tokyo who attend Oklahoma State University in Stillwater mention Stillwater's small population. To these students Stillwater feels empty, yet they find it difficult to get around because they do not have cars. They point out that even with the congestion in Tokyo it is still easier to go places because of the city's extensive subway system. Conversely, Stillwater natives who have visited Tokyo say that they were surprised by how densely packed Tokyo is—with people and high-rises—but that they enjoyed Tokyo's nightlife. These students also recall being shocked to learn that many families in Tokyo live in apartments no bigger than their dorm rooms. Clearly, Stillwater and Tokyo are both cities, but as these divergent views show, the cities function as different kinds of **central places** and they serve contrasting **hinterlands** (**Figure 8.1**).

In spite of their diversity, cities share six basic characteristics. (1) Cities possess dense concentrations of people. (2) Cities, unlike rural settlements, are distinguished by **functional complexity**. (3) Cities are centers of institutional power

central place A settlement that provides goods and services for its residents and its surrounding trade or market area.

hinterland The trade area served by a central place.

functional complexity The ability of a town or city to support sizable concentrations of people who earn their living from specialized, nonfarming activities.

Cities as central places • Figure 8.1

Tokyo, Japan, and Stillwater, Oklahoma, are very different kinds of central places not only because of their settlement histories but also because of the population and the specialized activities they support.

a. Tokyo, for example, is heavily involved in international trade, banking, and industry.

b. Stillwater, by contrast, is a small city whose largest employer is a university.

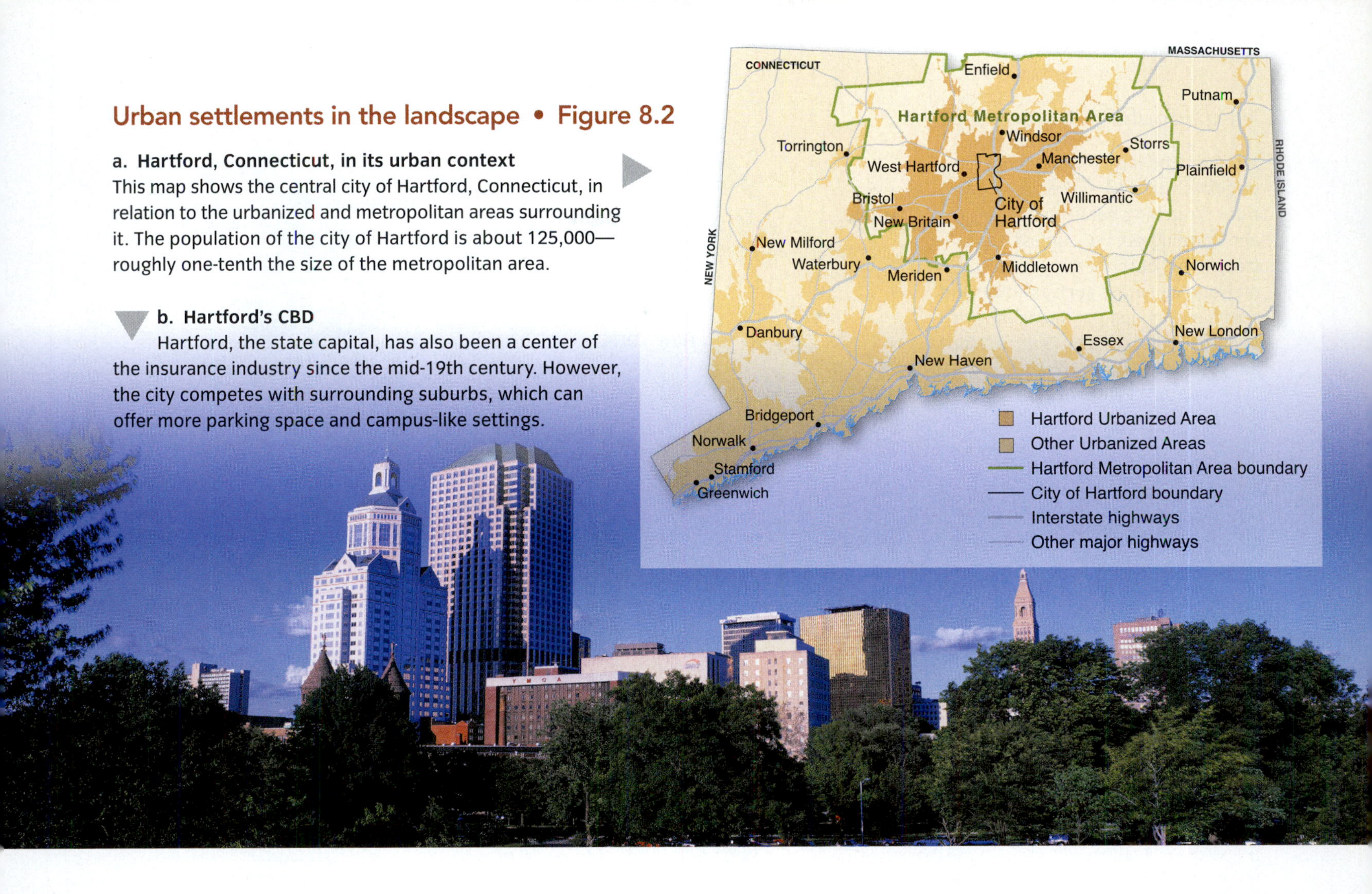

Urban settlements in the landscape • Figure 8.2

a. Hartford, Connecticut, in its urban context
This map shows the central city of Hartford, Connecticut, in relation to the urbanized and metropolitan areas surrounding it. The population of the city of Hartford is about 125,000—roughly one-tenth the size of the metropolitan area.

b. Hartford's CBD
Hartford, the state capital, has also been a center of the insurance industry since the mid-19th century. However, the city competes with surrounding suburbs, which can offer more parking space and campus-like settings.

associated with the business, governmental, or cultural activities of that place. (4) Cities are dynamic, human-created environments that possess complex patterns of specialized land use (for example, residential, industrial, public, and private). (5) Cities are linked, via trade, transportation, or communication, to other urban and rural places. (6) As places, cities are full of contradictions. They are hubs of creativity, opportunity, and hope but are simultaneously also places of poverty, deprivation, and despair.

Urban Settlements

What constitutes an urban settlement? There is no simple answer to this question in part because different countries use the term *urban* to mean different things. Some countries define urban on the basis of a minimum population size. For example, in Australia the minimum population for an urban settlement is 1,000. In Japan, however, to be designated urban a settlement must have 50,000 or more people. The U.S. Census Bureau considers a territory urban if it has a residential population of at least 2,500 and a population density of at least 1,000 people per square mile.

In other places the term *urban* refers to certain territories or administrative divisions. For example, New Caledonia (a French territory located in the southwest Pacific near Australia) designates its capital city as urban. Still other countries examine both the population size and the kinds of employment patterns, reserving the term *urban* for those places where the majority of the residents have nonagricultural jobs.

Most cities have boundaries (city limits) that are legally defined. The area enclosed by these boundaries is referred to as the **central city**. The part of the downtown where major office and retail businesses are clustered constitutes the **central business district** (CBD), whereas the **suburbs** are the built-up areas that surround the central city. In everyday usage the term **urbanized area** refers to land that has been developed for commercial, residential, or industrial purposes. The U.S. Census Bureau uses a more formal definition of urbanized area that refers to territory—usually the central city plus adjacent suburbs—that has at least 50,000 people and a population density of 1,000 people or more per square mile (**Figure 8.2**).

Megalopolis • Figure 8.3

Although the term *megalopolis* originally referred to the urban corridor stretching between Boston and Washington, D.C., unless capitalized, it now describes any coalescing metropolitan areas such as Tokyo and Yokohama in Japan or Rio de Janeiro and São Paulo in Brazil. (*Source*: Map adapted from Morril, 2006.)

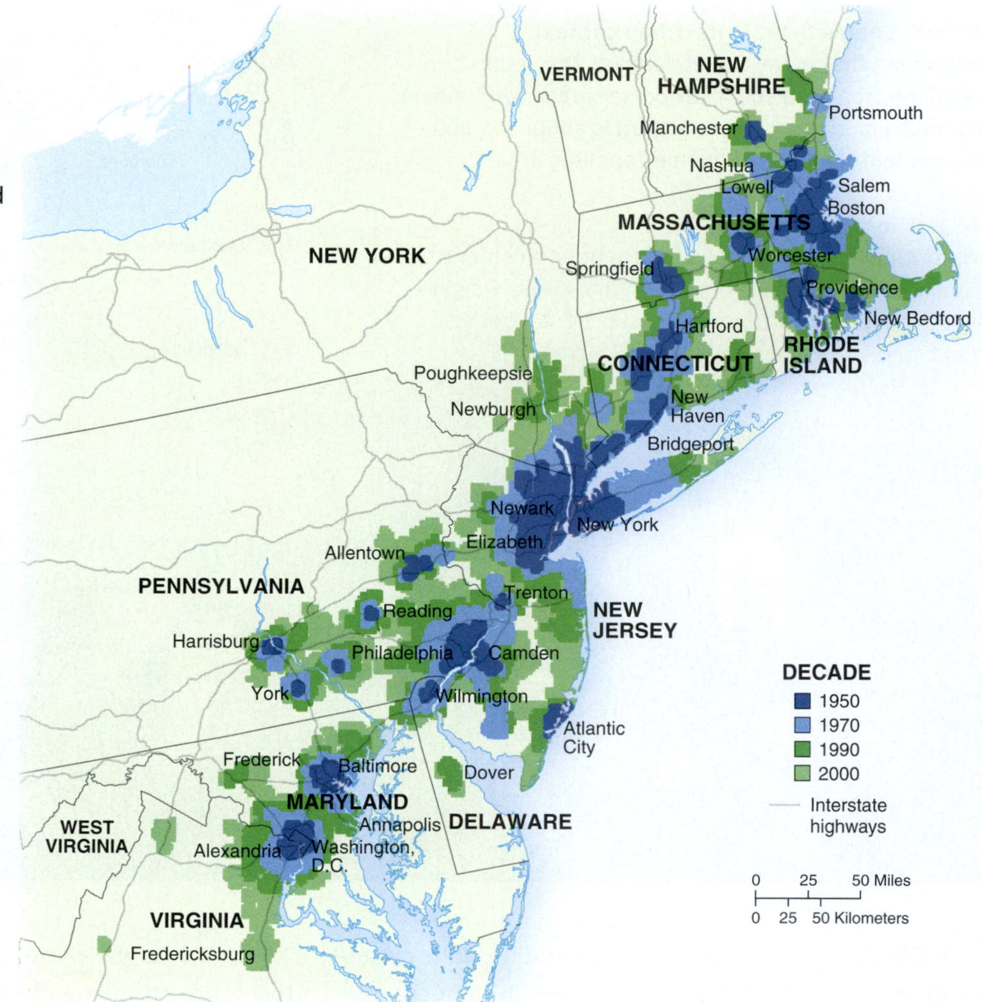

In contrast, a **metropolitan area** encompasses a large population center (50,000 people minimum) and the adjacent zones that are socially and economically connected to it, such as the places sending commuters into the city. The boundaries of urbanized areas follow those of U.S. Census blocks or block groups (see Chapter 6) since they are used to calculate population densities. The boundaries of metropolitan areas, however, follow local government units such as counties, or in the case of certain New England states, towns. Finally, a **megalopolis**, or conurbation, is a massive urban complex created by converging metropolitan areas (**Figure 8.3**).

Urbanization

Urbanization refers to the processes that concentrate people in urban places. When geographers measure urbanization, they can use one of two methods: the level of urbanization or the rate of urban growth. Urban geographers and other analysts use both measures to compare trends in urbanization in one country or region to another. The **level of urbanization** indicates the percentage of people living in urban places in some defined area (e.g., a county or a country). An area is considered urbanized if 50% or more of the population resides in urban places. The **rate of urban growth** refers to the annual percentage increase in an urban population. Geographers consider urban growth rates between 2% and 4% to be high.

Globally, the percentage of people living in urban places has increased dramatically over time: from 13% in 1900 to 29% in 1950. Today more than 50% of the world's people are urbanites. This level of urbanization represents a significant new moment in the geography and history of our world and underscores the importance of cities as systems of social organization. Current estimates indicate that by 2050, approximately 70% of the global population will be urban.

In general, developed countries tend to have higher levels of urbanization than developing countries. The level of urbanization is about 75% among developed countries and about 44% among developing countries (**Figure 8.4**).

Between 1975 and 2000, the urban growth rate of the world averaged 3.5% and the number of people living in cities increased from 1.5 to 2.8 billion. The rate of urban growth for the world has since slowed considerably and is now about 1.8%. Even so, the developing regions of the world are projected to have the highest rates of urban growth for the next two decades. Between 2010 and 2030, for example, the rate of urban growth in developing regions is estimated to be approximately 2.2%, with a population doubling time of 32 years. This compares to a projected urban growth rate of 0.5% in developed regions.

Levels of urbanization • Figure 8.4

a. The world reached a new milestone— an urban transition—in 2008 with half of the global population living in urban areas. Most of these urban residents live in towns and cities with fewer than 1 million people. (*Source*: Kaplan, 2004, with updates.)

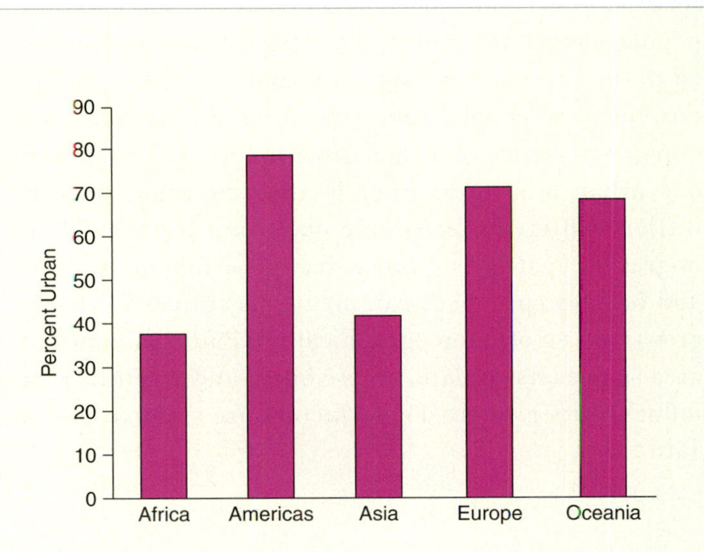

b. When levels of urbanization are mapped by country, we can see the unevenness of urbanization around the world. Although there is great variation from country to country, Africa and Asia are the least urbanized of the populated continents in the world. (*Source*: Data from Population Reference Bureau, 2009.)

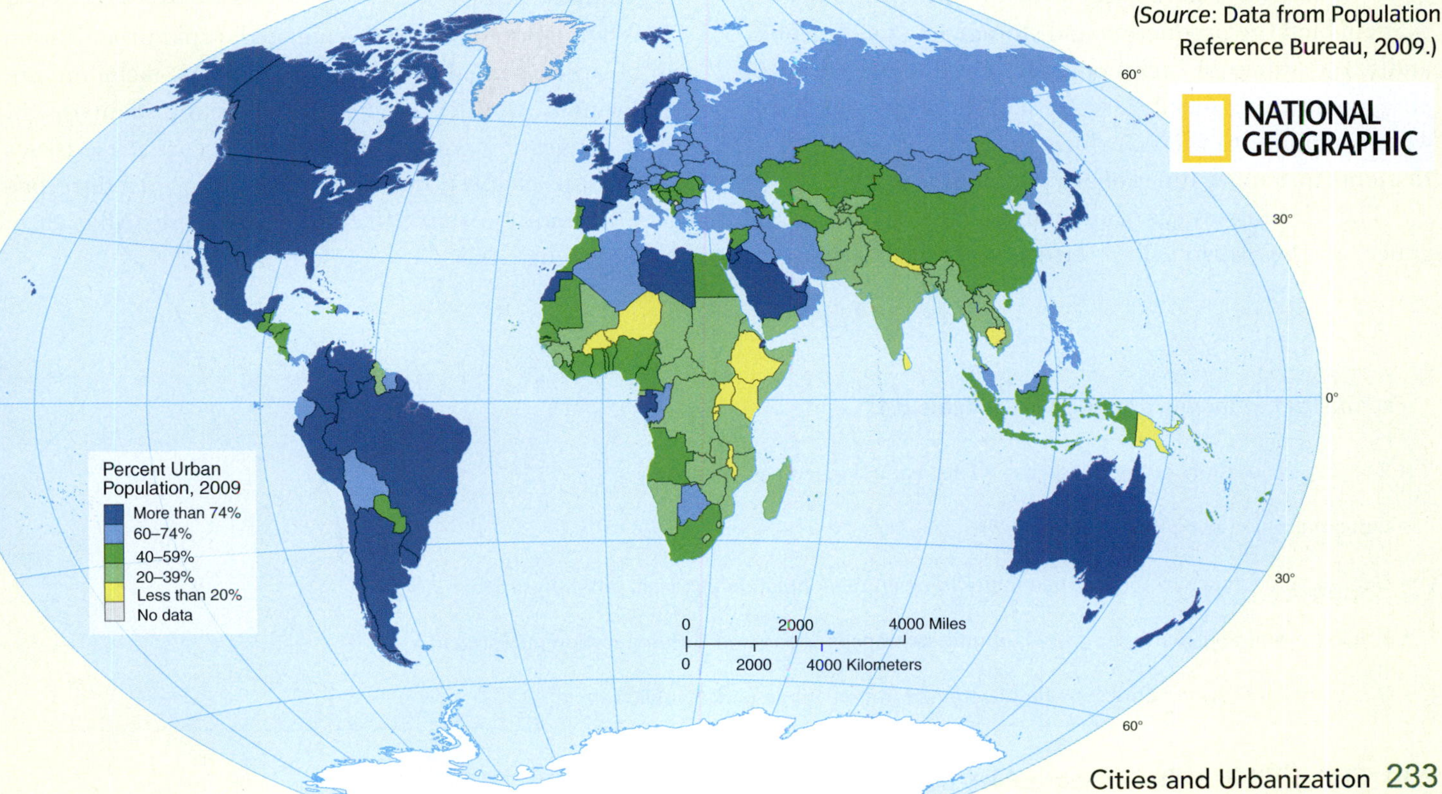

These rates of urban growth highlight three important trends in global patterns of urbanization. First, most of the population growth in the world is occurring in urban places. Second, most of this urban growth will take place in developing regions. Third, in general, rural populations in developed regions of the world are declining while the urban populations are growing slowly.

Urban growth usually is the result of numerous interrelated factors. For example, demographic characteristics of urban populations have a bearing on natural population growth. Similarly, urban growth is connected to the processes of migration and globalization. The movement of people from rural to urban areas has had a major influence on urban growth, but migration from one urban area to another is also important. Globalization can create economic opportunities that affect migration patterns. Likewise, government policies and business practices can create incentives for urban growth. Even other geographical attributes of an urban area such as its climate, accessibility, and amenities can influence its growth. These factors are summarized in **Table 8.1**.

Urbanization, development, and megacities

Cities have been a part of our world for about 6,000 years. By studying historic urban sites, scholars have identified two circumstances considered necessary for the emergence of cities: (1) development of an agricultural system that enabled the production and storage of a food surplus; and (2) a system of social organization dominated by an elite, nonfarming social class. Both of these circumstances were conducive to the transformation of agricultural settlements into more functionally complex places.

The circumstances that contributed to the emergence of the world's first cities are still relevant to discussions of contemporary cities, but additional considerations come into play. One way to recognize these other factors is to reflect on the geographic patterns of urbanization shown in **Figure 8.4b**. For example, why do more developed countries tend to have higher levels of urbanization and lower rates of urban growth? The answer to this question relates to social and technological changes associated with the Industrial Revolution that began in western Europe during the mid-18th century. As industrialization and mechanization spread, fewer people were needed to work the land. In turn, jobs at factories and industries in the growing cities drew rural residents from the countryside to the cities, fueling urbanization.

Since the Industrial Revolution, the regions of the world that were the first to industrialize, primarily the countries in western Europe and North America, also were the first to urbanize most extensively. Industrialization and urbanization contributed to the economic growth of these regions. The use of steel frames and electric elevators enabled the construction of skyscrapers and transformed the appearance of these cities.

For most of the 20th century, the largest cities in the world were geographically concentrated in the more developed regions of Europe and North America. After 1975, however, that pattern began to change as a result of rapid urbanization and the growth of **megacities** (**Figure 8.5**).

megacity A city with 10 million or more residents.

When ignored or poorly managed, rapid urbanization can be accompanied by a host of problems, including unemployment, the development of slums or shantytowns, traffic congestion, and pollution. We discuss these issues later in the chapter, but first we need to introduce primacy, a condition that affects the spatial distribution of people among cities.

Factors that influence urban growth Table 8.1
• Geographic location (accessibility to markets, natural resources, other cities)
• Industrialization and globalization
• Demographic trends (rural-to-urban migration, natural population increase, urban-to-urban migration)
• Policies promoting economic growth (improved transportation and communication infrastructure)
• Improved services and amenities (public transportation, urban parks, public safety)

Megacities • Figure 8.5

a. Megacities in 1950, 2007, and projected to 2025
As shown in the chart, the world had only two megacities in 1950. By 2007, there were 19, and most of them were in Asia. By 2025, 27 megacities are projected, many in developing countries, especially India and China. Despite their large populations, megacities house just 9% of the world's people. Population figures are for the urbanized area. (*Source*: United Nations Population Division, 2008.)

Asia
North America
South America
Africa
Europe

1950		2007		2025 (Projection)	
Megacity	Population (millions)	Megacity	Population (millions)	Megacity	Population (millions)
1. New York, USA	12.3	1. Tokyo, Japan	35.7	1. Tokyo, Japan	36.4
2. Tokyo, Japan	11.3	2. New York, USA	19.0	2. Mumbai, India	26.4
		3. Mexico City, Mexico	19.0	3. Delhi, India	22.5
		4. Mumbai, India	19.0	4. Dhaka, Bangladesh	22.0
		5. São Paulo, Brazil	18.8	5. São Paulo, Brazil	21.4
		6. Delhi, India	15.9	6. Mexico City, Mexico	21.0
		7. Shanghai, China	15.0	7. New York, USA	20.6
		8. Kolkata, India	14.8	8. Kolkata, India	20.6
		9. Dhaka, Bangladesh	13.5	9. Shanghai, China	19.4
		10. Buenos Aires, Argentina	12.8	10. Karachi, Pakistan	19.1
		11. Los Angeles, USA	12.5	11. Kinshasa, Democratic Republic of Congo	16.8
		12. Karachi, Pakistan	12.1	12. Lagos, Nigeria	15.8
		13. Cairo, Egypt	11.9	13. Cairo, Egypt	15.6
		14. Rio de Janeiro, Brazil	11.7	14. Manila, Philippines	14.8
		15. Osaka-Kobe, Japan	11.3	15. Beijing, China	14.5
		16. Beijing, China	11.1	16. Buenos Aires, Argentina	13.8
		17. Manila, Philippines	11.1	17. Los Angeles, USA	13.7
		18. Moscow, Russia	10.5	18. Rio de Janeiro, Brazil	13.4
		19. Istanbul, Turkey*	10.1	19. Jakarta, Indonesia	12.4
				20. Istanbul, Turkey*	12.1
				21. Guangzhou, China	11.8
				22. Osaka-Kobe, Japan	11.4
				23. Moscow, Russia	10.5
				24. Lahore, Pakistan	10.5
				25. Shenzhen, China	10.2
				26. Chennai, India	10.1
				27. Paris, France	10.0

*Istanbul spans both Europe and Asia.

b. Dhaka, Bangladesh
The world's fastest growing megacity is projected to gain more than 8 million residents within the next 18 years. By contrast, it took the metropolitan area of New York City more than a century to experience a similar population increase. Migration accounts for most of Dhaka's recent population growth.

Urban primacy Urban growth in the developing world is more likely to give rise to **primate cities** than urban growth elsewhere. For example, Mexico City has a population in excess of 19 million, whereas the second largest city in the country, Guadalajara, has a population of about 4 million. Thailand provides an even more extreme example. With a population of 6.9 million, Bangkok is more than 9 times the size of Samut Prakan (population 700,000), Thailand's next largest city.

primate city A city that has a population two or more times the population of the second largest city in the country.

Primate cities can become islands of growth and contribute to uneven development within a country. Like magnets, these cities attract factories and businesses, which in turn attract more people. Most primate cities are also

Urban primacy • Figure 8.6

Historical/ Political	Largest city is also the capital city Colonial or ex-colonial status Policy favoritism toward a specific city A unitary state with a free enterprise economy
Economic	Agricultural orientation Low level of economic development High concentration of wealth among elites and large gap between the rich and the poor Economic favoritism toward a specific city

a. Conditions associated with urban primacy
Most urban geographers agree that primacy is more likely to develop under certain conditions, including but not limited to those identified in the table. (*Source*: Mutlu, 1989.)

b. Buenos Aires, Argentina
With a population of 13 million, Buenos Aires is Argentina's capital city and primate city. Why is experience as a colony often associated with urban primacy?

capital cities; thus, they function as the main political and administrative center of the country. Similarly, primate cities tend to attract more cultural and educational resources. Urban primacy, then, involves more than the size of a city's population; it also includes the concentration of political, economic, and cultural functions within a city.

Some countries in the developing world, such as China and India, do not have primate cities. Moreover, primate cities are not unique to the developing world. Several European capitals, including London, Paris, Vienna, and Athens, are also primate cities. In Australia primacy exists at the state level rather than the national level. Within the state of New South Wales, for example, the primate city of Sydney is more than eight times the size of Newcastle. See **Figure 8.6** for an overview of conditions associated with primacy.

Urban Hierarchies and Globalization

As we have seen, a variety of different types of urban settlements or central places exists. In addition, the size of the hinterland is directly related to the size of the central place. For example, a village will have a much smaller hinterland than a city. The relationship between a central place and its hinterland is important because it indicates that a hierarchy of central places exists, and this, in turn, can affect their distribution. The first geographer to detect and analyze regularities in the system of central places was Walter Christaller. His ideas were formulated in the 1930s and provide the basis for central place theory—one of the fundamental theories in urban geography.

Central place theory Christaller developed two concepts—range and threshold—to help explain the emergence of a hierarchy of central places. Both concepts relate to the provision of goods or services. When purchasing a good or service, how far are you willing to travel to obtain it? The **range** expresses the maximum distance a consumer will travel for a particular good or service. Consumers are willing to travel longer distances to obtain luxury items or to make special purchases—to find that perfect wedding dress, for example, or to see their favorite band. However, consumers are unwilling to travel very far to buy a gallon of milk or mail a package.

If the range establishes the size of a market area, the threshold helps explain what goods and services are likely to be available in a given market area. In order to supply a particular good, a central place must have an adequate consumer base. The **threshold**, then, is the smallest number of consumers required to profitably supply a certain good or service. More specialized goods and services require larger thresholds. This helps explain why rare book and map stores, brain surgeons, and high-speed rail service are found in larger cities.

urban hierarchy A series of central places ranked on the basis of their threshold, range, and market area.

The **urban hierarchy**, therefore, consists of a ranked series of central places. At the top of the hierarchy are central places such as New York, Tokyo, and London that supply all of the basic necessities as well as the most specialized goods and services. Next, moving down the urban hierarchy, are the smaller cities, towns, and villages that, respectively, offer fewer goods and have smaller market areas. Hamlets, the central places characterized by the smallest thresholds and ranges, occupy the bottom of the hierarchy.

By focusing on the relationship between threshold and range, Christaller detected regularities in the distribution of central places and wondered whether it might be possible to predict the spatial arrangement of central places. Thus, **central place theory** posits that market forces account for the distribution of central places in an area, and that the optimal spatial arrangement of central places creates hexagonally shaped trade areas.

Christaller built his theory using three important assumptions:

1. The landscape consists of a uniform, flat surface.
2. The population is evenly distributed across this landscape.
3. As consumers, these people would always purchase the goods and services they need from the central place closest to them.

Christaller realized that these conditions would influence the shape of market areas. If there were no other central places in the area competing for consumers, the shape of the market area would be circular. But Christaller was not examining just one central place. Rather, he sought to understand the effects of multiple central places of different sizes in the urban hierarchy. Therefore, he theorized that the optimal shape of the market area would instead be hexagonal because market areas would be split between competing central places.

Christaller's work confirmed the interdependency of central places—that the size, retail functions, and location of one central place depend on the characteristics of other central places (**Figure 8.7**).

Spatial arrangement of central places • Figure 8.7

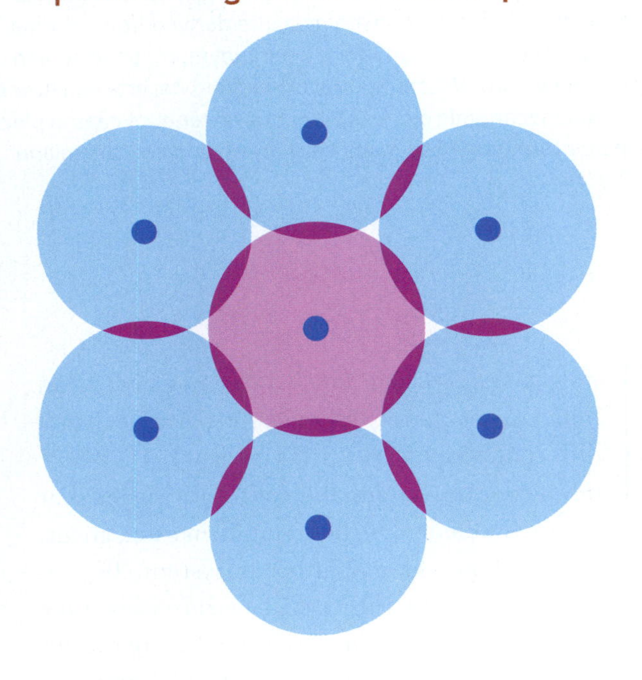

a. Optimal market areas
Following Christaller's assumption that people would travel to the nearest central place, market areas, usually circular, become hexagonal when there are other central places of the same rank in the area. (*Source*: Hartshorn, 1992.)

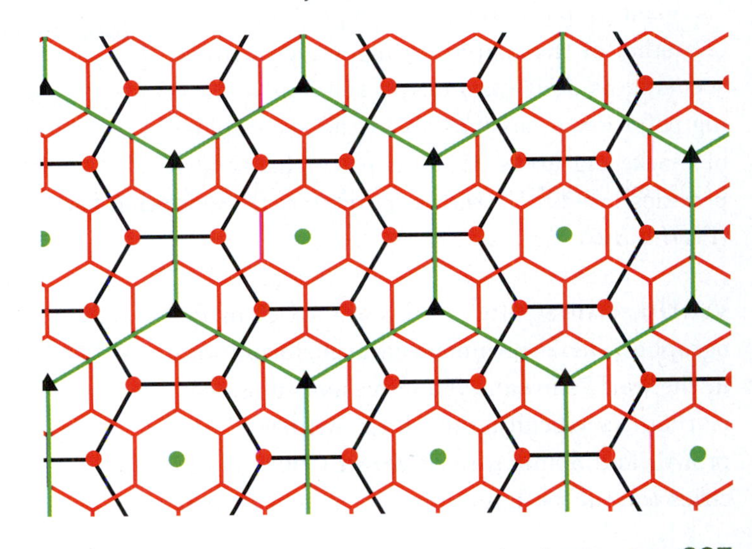

b. Nested hexagons
The idealized hierarchy of central places, as theorized by Christaller, yields a lattice of central places. Within this lattice, smaller central places occur most frequently and are closest together. In contrast, larger central places are more widely spaced. (*Source*: Hartshorn, 1992.)

Geography InSight Food deserts • Figure 8.8

a. Supermarket accessibility in London, Ontario, in 1961 and 2005
Service areas are the areas within 1000 m (3281 ft) of a supermarket that have street access. How did the geography of supermarkets change? (*Source*: Larsen and Gilliland, 2008.)

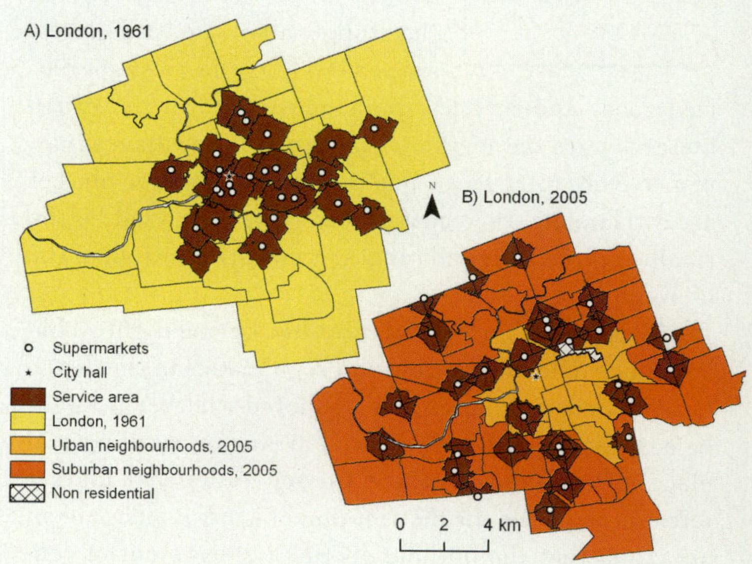

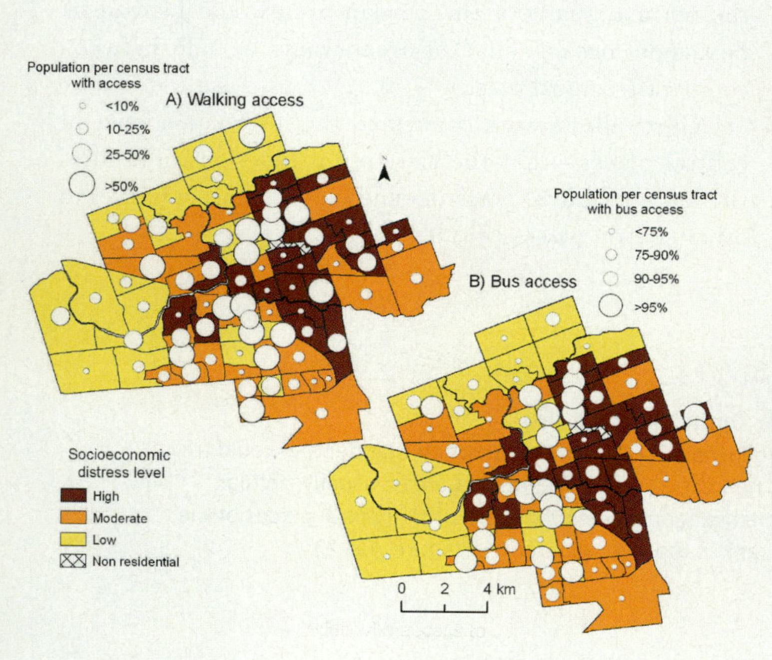

b. Neighborhood socioeconomic distress level and proportion of census tract population with access to supermarkets, 2005
Socioeconomic distress is a composite measure derived from information about high school graduation rates, unemployment, poverty, and single-parent households. Where are the urban food deserts and how do they relate to socioeconomic distress? What basic tenet of central place theory do the patterns on these maps challenge? (*Source*: Larsen and Gilliland, 2008.)

Central place theory prompted hundreds of studies of the spatial arrangement of urban settlements and continues to serve as a starting point for discussions about the geographies of marketing areas. Central place theory also provides insights on the geography of urban **food deserts** (**Figure 8.8**).

food desert An area characterized by a lack of affordable, fresh, and nutritious foods.

World cities and networks Urban geographers have documented transformations in the world's urban system because of globalization. Specifically, they have witnessed the growth and dominance of **world cities**. World cities tend to be large but are not necessarily megacities because city size is not always an indicator of global influence or power. What *is* different about world cities is the way they have developed into command centers or nodes that greatly influence the flow of information, goods, and capital throughout the global urban system. Two related factors help explain the rise of world cities. One factor has been the growth of multinational corporations and the concentration of their headquarters in certain cities. The second factor involves the increasing importance of advanced professional services such as banking, insurance, advertising, and legal services, and the concentration of these operations in certain cities.

world city A principal center of global economic power that significantly influences the world's business.

Some urban geographers think of world cities as the top tier in the hierarchy of urban places proposed by Christaller; other urban geographers, however, contend that world cities function so differently from other cities that traditional conceptualizations of the urban system need to be revised to emphasize global linkages. Therefore, the global urban system consists of hierarchies that can be understood as vertical relationships connecting central places *and* a global system of networked central places, linked through horizontal flows and connections. In other words, Christaller's model falls short when urban geographers use it to analyze or model the global urban system. Recall, for example, that in Christaller's model the market areas of central places in the same rank do not overlap. Today, however, overlapping market areas are common in part because people in the hinterland of one city can transact business—often electronically—with people in another city of the same urban rank on the other side of the world. The term **hinterworlds** conveys the idea that the area served by a world city potentially includes the entire globe, not just the territory adjacent to a specific city.

Most urban geographers agree that London, New York, and Tokyo are world cities. Identifying other world cities is challenging because experts can use many different criteria for evaluation and they disagree about which criteria are most important (**Figure 8.9**).

World cities • Figure 8.9

a. Selected indicators of world cities

How are primate cities and world cities alike? How are they different?

Key indicators of world cities

1. Recognized center of political power and influence, often because of the concentration of government or institutional functions, e.g.,United Nations, International Monetary Fund.
2. Major site of knowledge and information production from public and private sources, e.g., government documents, university studies, and leading business or financial companies.
3. Strong integration in the global economy.
4. High volume of interactions with other world cities.
5. Presence of a major international airport.
6. Major provider of advanced professional services such as accounting, financial services, insurance, and law.
7. State of the art telecommunications technologies and infrastructure, e.g., fiber optics, wireless networking.
8. Presence of a highly skilled, mobile, and multicultural labor pool.
9. Heavy reliance on a two-tiered structure of personnel in businesses and corporations that consists of an elite class associated with service sector jobs, e.g., financial managers, and an underclass associated with low or unskilled jobs, e.g., janitorial staff.
10. High-profile reputation as a center for arts and entertainment.

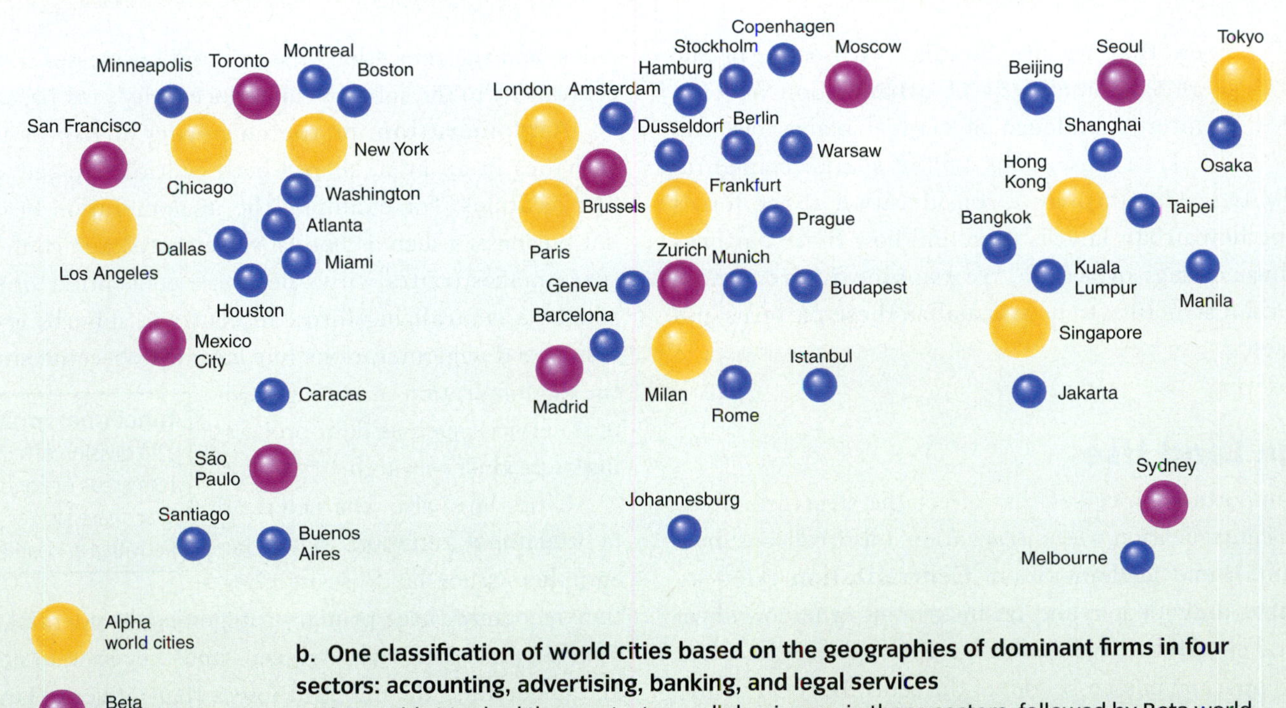

b. One classification of world cities based on the geographies of dominant firms in four sectors: accounting, advertising, banking, and legal services

Alpha world cities had the greatest overall dominance in those sectors, followed by Beta world cities and then Gamma world cities. If world cities reflect globalization, what message does this map convey? (*Source*: Beaverstock, Taylor, and Smith, 1999.)

Why are some large cities not world cities? The answer, according to urban geographer John Rennie Short, involves five factors: poverty, social collapse, avoidance of risk by managers of global firms, and resistance to globalization. These factors work to repel investment and exclude cities from the global economy. For example, large cities that have high poverty rates often cannot afford to build the high-tech infrastructure necessary to support advanced professional services and consequently do not attract corporate headquarters. Social tensions including ethnic conflict and political corruption can also adversely affect business development. Similarly, the publication by businesses and corporations of risk ratings for different countries also deters investment. Cities located in high-risk countries are likely to be overlooked by business managers trying to identify new locations for their firms. Lastly, exclusion from the global economy or resistance to globalization might reflect different political perspectives and help explain why cities such as Kolkata (Calcutta) or Tehran are not world cities.

CONCEPT CHECK STOP

1. **How** are the concepts of functional complexity and hinterlands related?
2. **What** three trends are shaping global urbanization?
3. **What** impact does a primate city have on the urban hierarchy?
4. **How** do world cities challenge central place theory?

Urban Structure

LEARNING OBJECTIVES

1. **Explain** what a bid-rent curve is.
2. **Identify** and explain four models of urban structure for North American cities.
3. **Account** for differences in the urban form of eastern and western European cities.
4. **Describe** the characteristics of a hybrid city.

Now that we are familiar with some of the global dimensions of urbanization and the interdependence of central places, we can take a closer look at the spatial configuration of cities. This involves a consideration of the forces affecting how urban land is used and how these patterns of land use change over time. We examine different models of urban structure to help visualize these patterns and processes.

Urban Land Use

Three important processes that affect the structure of cities are centralization, decentralization (also called suburbanization), and agglomeration. **Centralization** refers to forces that draw people and businesses into the downtown or central city. **Decentralization** has the opposite effect and draws people and businesses out of the central city, often into suburbs. Suburbanization can limit the territorial growth of cities, and the relocation of stores and businesses from the central city to the suburbs can reduce a city's tax base.

Agglomeration, or the clustering of like or unlike activities in an area, shapes both central city and suburban locations. For example, the agglomeration of different businesses such as hotels, restaurants, and conference space makes central cities desirable convention sites and acts as a centralizing force. In contrast, suburbs grew as residential agglomerations but have also become sites for the agglomeration of similar business services such as data processing or product research.

Cities are also characterized by **functional zonation**. When geographers study land use in cities, they recognize three primary categories—residential, commercial, and industrial land use—and they seek to identify and understand the different forces that influence land-use patterns.

functional zonation The division of a city into areas or zones that share similar activities and land use.

Land values and land use • Figure 8.10

a. Traditional bid-rent curves

Bid-rent curves depict changes in land value and land use across urban space and help to visualize functional zonation. Retailing activities that benefit from pedestrian flow or businesses such as mailing services that need to be close to their clients are willing to pay more for highly accessible, central locations. (*Source*: Hartshorn, 1992.)

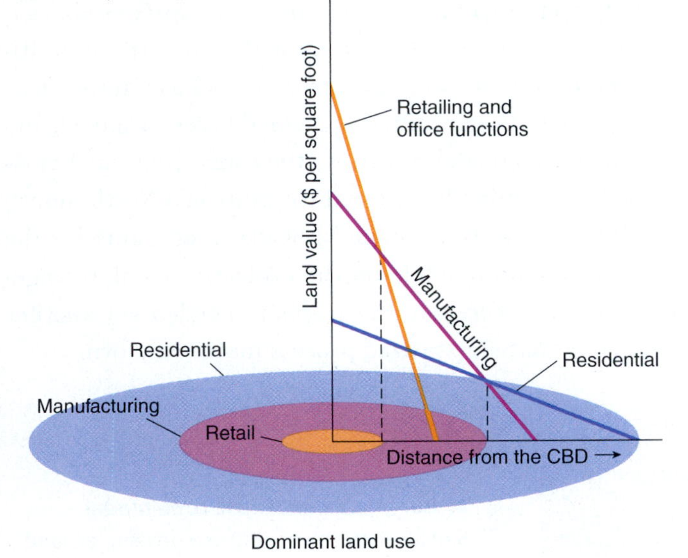

b. Bid-rent curves and varying accessibility

This graph hints at the complexity of urban land values and shows that suburban locations that are highly accessible tend to be associated with higher land values. What is the general relationship between land value and land-use density? (*Source*: Hartshorn, 1992.)

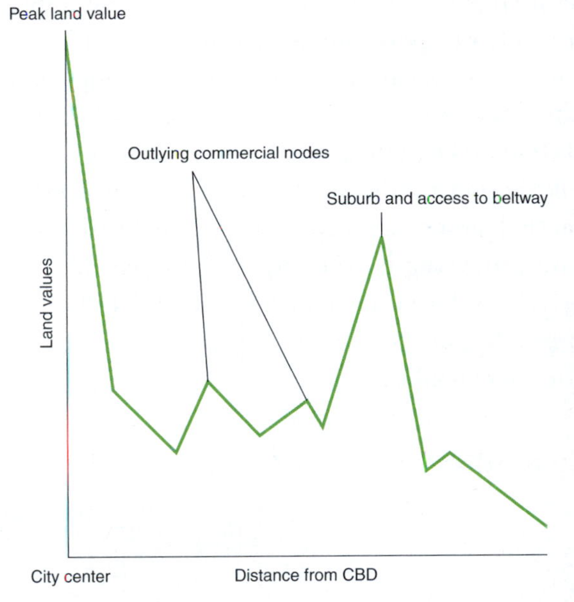

The value of land is one economic force that has a strong influence on land use in cities. Land values reflect, among other considerations, the accessibility and desirability of a particular site. Of course, these considerations vary from one enterprise to another, but as a rule, the more accessible or desirable a site is, the more expensive it is. Bid-rent curves help to visualize these economic forces. A **bid-rent curve** shows the amount a bidder (a business or individual, for example) is willing to pay for land relative to the distance from the central business district (CBD) (**Figure 8.10**).

Sometimes institutional forces have a major impact on urban land-use patterns. If left alone, market forces tend to inhibit certain activities. For example, high land-use values can exclude hospitals, schools, and parks. By intervening, city, state, or national governments can alter market forces and change land-use patterns. **Zoning** is another institutional force that directly affects urban land use. City councils, for example, can limit specific kinds of land use to certain parts of the city, concentrating factories or warehouses in particular areas.

> **zoning** Laws that regulate land use and development.

Urban geographers have long been interested in the spatial organization of cities and have developed models to describe their internal structure. Each of these models is a generalization and a simplification of the complexity of real cities. Nevertheless, these models help us understand the diverse forces that affect cities.

Urban Structure in North America

A number of models of urban structure were developed specifically through the study of U.S. cities. In this section we examine four of those models in chronological order and, in the process, show how the models have become more sophisticated.

Concentric zone and sector models In 1925, the sociologist Ernest Burgess produced the **concentric zone model**, one of the first models of urban spatial structure, based on his study of Chicago. He developed an ecological interpretation of urban growth and argued that in cities, as in the natural environment, groups competed for space and resources. These processes contributed to a sorting of social groups along economic and ethnic lines such that communities came to occupy distinctive niches or zones of urban space. Upward mobility, the arrival of new immigrants, or changes in land use or land value triggered the movement of people from one zone to another in a process he called *succession*. The concentric zone model shaped scholarly thinking about the structure of cities for many years, but critics point out that the model is overly simplistic and does not adequately account for the impact of transportation on urban form.

In 1939, the economist Homer Hoyt proposed the **sector model** to describe the land-use patterns and spatial structure of cities. This model places greater emphasis on the role of transportation and incorporates Hoyt's observation that the location of high-income groups influences the direction of a city's growth. New high-class neighborhoods tend to be built along the outer edges of cities. This sets in motion a process of succession called *filtering*. As Kenneth Jackson, an expert in urban history, concisely explains, **filtering** is "the sequential reuse of housing by progressively lower income households" (1985, p. 285). **Figure 8.11** compares and contrasts the sector and concentric zone models.

Multiple nuclei and urban realms models In 1945, two geographers, Chauncy Harris and Edward Ullman, proposed the **multiple nuclei model** as an alternative way of understanding the urban spatial structure of North American cities. Harris and Ullman observed that cities often had multiple cores, or nuclei, rather than a single core (the CBD). Specific nuclei varied from city to city but might include a harbor area, government district, university, or manufacturing areas. In turn, these multiple nuclei influenced the patterns of land use in the city.

Together, the concentric zone model, sector model, and multiple nuclei model constitute the three traditional models used to describe the internal structure of a North American city. None of these models, however, accounts for the impact of the automobile on the evolving spatial arrangement of cities. For that, geographers needed yet another model, one that emphasized process more than form.

Early models of North American cities • Figure 8.11

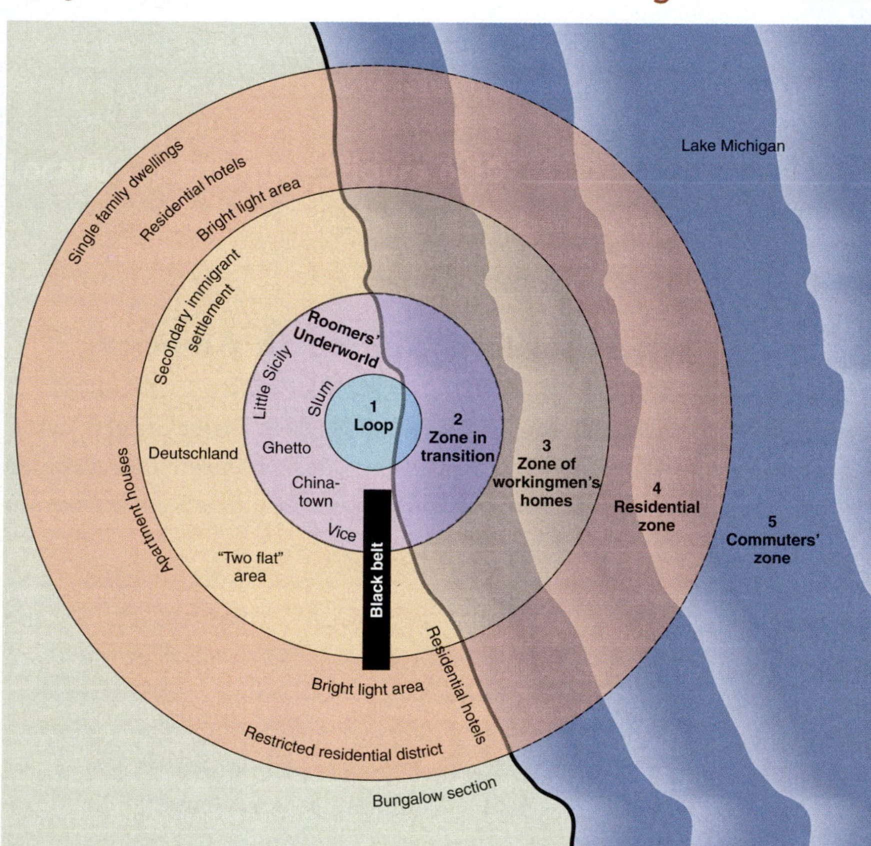

a. Burgess's concentric zone model Similar patterns of land use develop around the CBD in rings, or in the case of Chicago, concentric arcs because of the city's lakefront location. Nonresidential land use occurs primarily in the Loop, the local name for the CBD, and spreads into the zone in transition, which is also the destination of newly arrived immigrants. (*Source*: Harris and Ullman, 1945.)

1. CBD
2. Industrial
3. Low-class residential
4. Middle-class residential
5. High-class residential

b. Hoyt's sector model
Transportation influences the development of industrial sectors, which follow rail or river routes, for example, but high-income residential sectors also tend to grow in proximity to the fastest transportation routes. Filtering occurs as the well-to-do move into new homes and their vacated homes become available or "filter down" to middle-income households. Middle-income homes also filter down to low-income households. (*Source*: Burgess, 1925.)

PROCESS DIAGRAM

How changes in transportation influence urban form • Figure 8.12
THE PLANNER

In 1970, geographer John Adams related stages in the development of transportation to changes in the residential structure of cities. Although he concentrated on the Midwest, his stages are broadly applicable to other parts of the country. Because of changes in city shape since 1970, urban geographers have added a fifth stage. (*Source*: Adams, 1970; Hartshorn, 1992.)

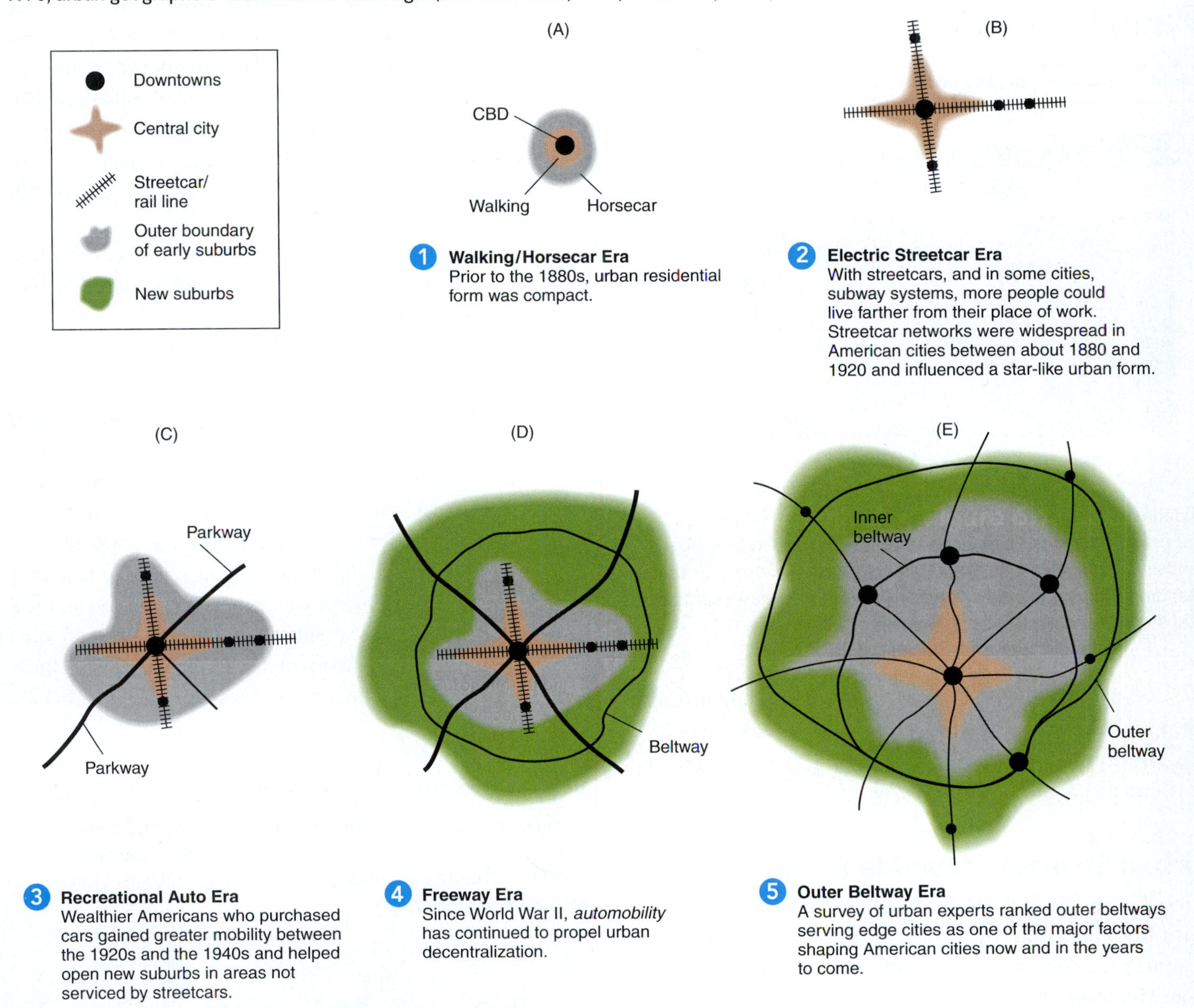

1 Walking/Horsecar Era Prior to the 1880s, urban residential form was compact.

2 Electric Streetcar Era With streetcars, and in some cities, subway systems, more people could live farther from their place of work. Streetcar networks were widespread in American cities between about 1880 and 1920 and influenced a star-like urban form.

3 Recreational Auto Era Wealthier Americans who purchased cars gained greater mobility between the 1920s and the 1940s and helped open new suburbs in areas not serviced by streetcars.

4 Freeway Era Since World War II, *automobility* has continued to propel urban decentralization.

5 Outer Beltway Era A survey of urban experts ranked outer beltways serving edge cities as one of the major factors shaping American cities now and in the years to come.

In 1964, the geographer James Vance developed the **urban realms model**. This model was based largely on his observations of San Francisco, where the suburbs had become self-sufficient centers. Each suburb was a new *urban realm*—an independent entity with its own downtown or commercial center. Vance's model recognizes the importance of the automobile in the evolution of the spatial form of the city. It also acknowledges that these new suburban downtowns, linked by beltways or ring roads, frequently rival the traditional CBD for commercial and retail trade. Because these suburban downtowns resemble cities and have grown on the edges of the traditional city, they are often referred to as **edge cities**.

> **edge cities** New downtowns consisting of clusters of business activity that develop in the suburbs surrounding a city.

The emergence of edge cities indicates that transportation has a major impact on the shape of cities. **Figure 8.12** depicts, in a general way,

Where Geographers CLICK

Library of Congress Panoramic Maps Collection

http://memory.loc.gov/ammem/pmhtml

A panoramic map present's a bird's-eye view of a place. For urban geographers, such maps allow us to view the urban landscape at a particular point in time, which is helpful for understanding urban change. Follow the link above to view panoramic maps of numerous U.S. and Canadian cities.

the relationship between changes in transportation, urban form, and the residential structure of North American cities.

For another method of visualizing urban street and settlement patterns, see *Where Geographers Click*.

Urban Structure Outside North America

Thus far the models we have discussed have been developed in conjunction with studies of North American cities. In Europe, cities have a very different look and configuration. Moreover, there are notable differences between cities in western and eastern Europe because of eastern Europe's long association with socialism. After examining European cities we discuss colonial cities and Islamic cities.

European cities Several of Europe's cities retain medieval characteristics, including remnants of a surrounding city wall, a historic core consisting of a church, marketplace, and dense concentration of buildings, an irregular street pattern, and a low central city skyline. Western European cities have retained a different spatial form because of the interplay of various kinds of economic, institutional, and sociocultural forces. More specifically, the following six centralizing tendencies have been especially influential:

1. Cities are pedestrian- and bicycle-friendly, and parts of the central city may even be closed to vehicular traffic.
2. Private transportation is expensive. Individuals pay more for vehicles (in the base cost of a vehicle, taxes, and registration) and for the gas to operate them.
3. Public transportation is widely accessible and affordable.
4. Private home ownership is not as widespread in part because house prices are higher and because 5- or 10-year mortgages are customary (as opposed to 30-year mortgages).
5. Public sentiment has favored the preservation of historic buildings and structures and resisted urban redevelopment.
6. Central cities have long been highly desirable and stable residential locations.

After World War II and until the fall of communism in 1989, cities across eastern Europe and the Soviet Union were influenced by political and economic systems different from those in western Europe. Such "socialist cities," as they were sometimes called, developed a distinctive urban form. One crucial difference is that since the land and economy were controlled by the state, there was no bidding process for land. Large, residential "housing estates" consisting of numerous apartment buildings were often built near factories and on the margins of the city (**Figure 8.13**).

Soviet-era housing estates • Figure 8.13

These apartment blocks on the outskirts of Tallinn, Estonia, stand in contrast to the city's medieval skyline. They illustrate that the attributes of socialist cities were sometimes grafted onto preexisting cities.

WHAT A GEOGRAPHER SEES

Spatial Imprints of Urban Consumption

The urban landscape of Warsaw has changed dramatically since 1989, following the shift in Poland from a socialist to a market economy.

a. Europlex

The first map shows that prior to 1989 this was the site of a public park with a nearby movie theater. A mall, multiplex theater, and adjacent office building have since been constructed. (*Source*: Kreja, 2006.)

Before 1989

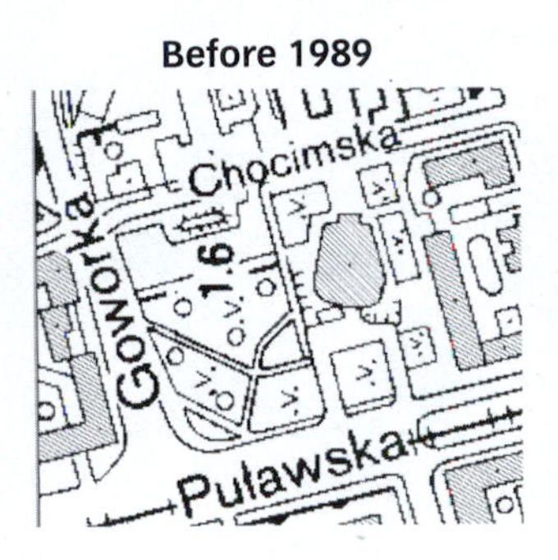

Today

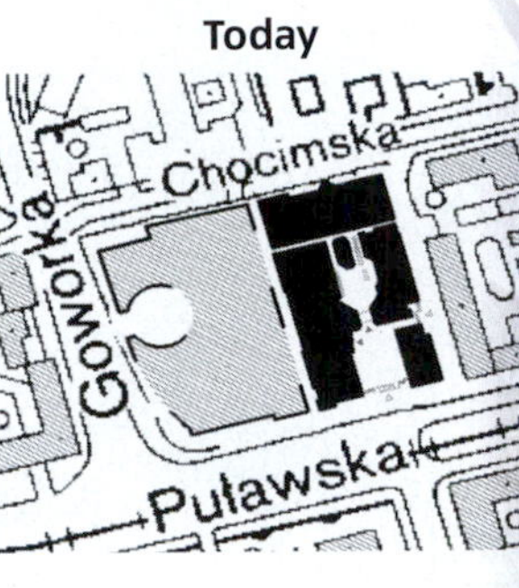

b. Galeria Mokotów

Compare the maps to see the transformation of this former industrial site into a huge commercial complex. When first opened in 2000, this was one of the largest shopping centers in Poland. (*Source*: Kreja, 2006.)

Before 1989

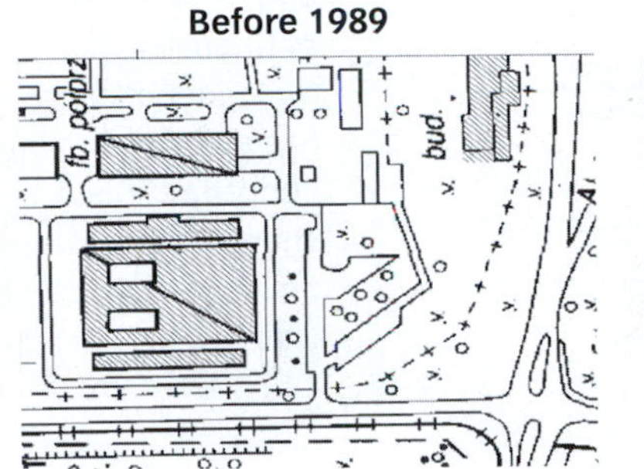

Today

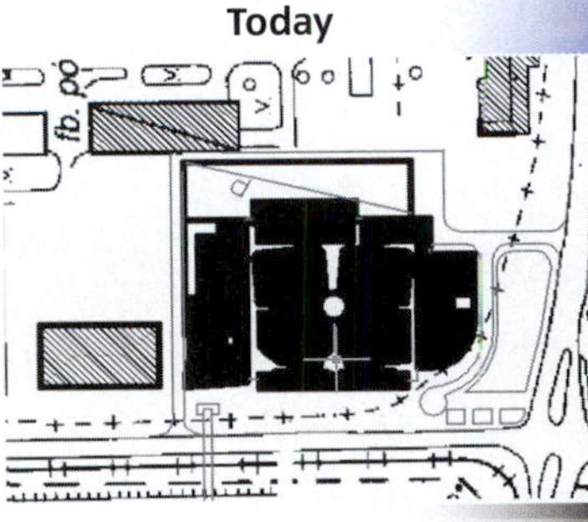

Think Critically

1. What happens to a bid-rent curve in a socialist city?
2. What are some implications of the changes captured in these photos and maps for public space?

Compared to cities in the West, retail and commercial land use in the socialist city was much more limited, but there were sizable plazas and parks for public use. Since the fall of communism, the urban landscapes in the former Soviet Union and eastern Europe have experienced major transformations, many of which are tied to globalization (see *What a Geographer Sees*).

Colonial cities The establishment of European colonies in Latin America, Southeast Asia, and Africa has had an enduring impact on the urban fabric of these regions. Depending on the needs of the colonizing power, colonial cities functioned as administrative, economic, and/or military centers. In some cases colonial cities were built near or even grafted onto an extant indigenous city or urban

Geography InSight

Hybrid city • **Figure 8.14**

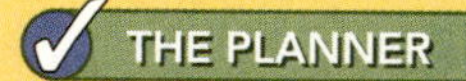

Accra, capital and largest city of Ghana, is a hybrid city. It has three central business districts (CBDs)—local, national, and global. Their development is linked to both historical and contemporary forces. (*Source*: Grant and Nijman, 2002.)

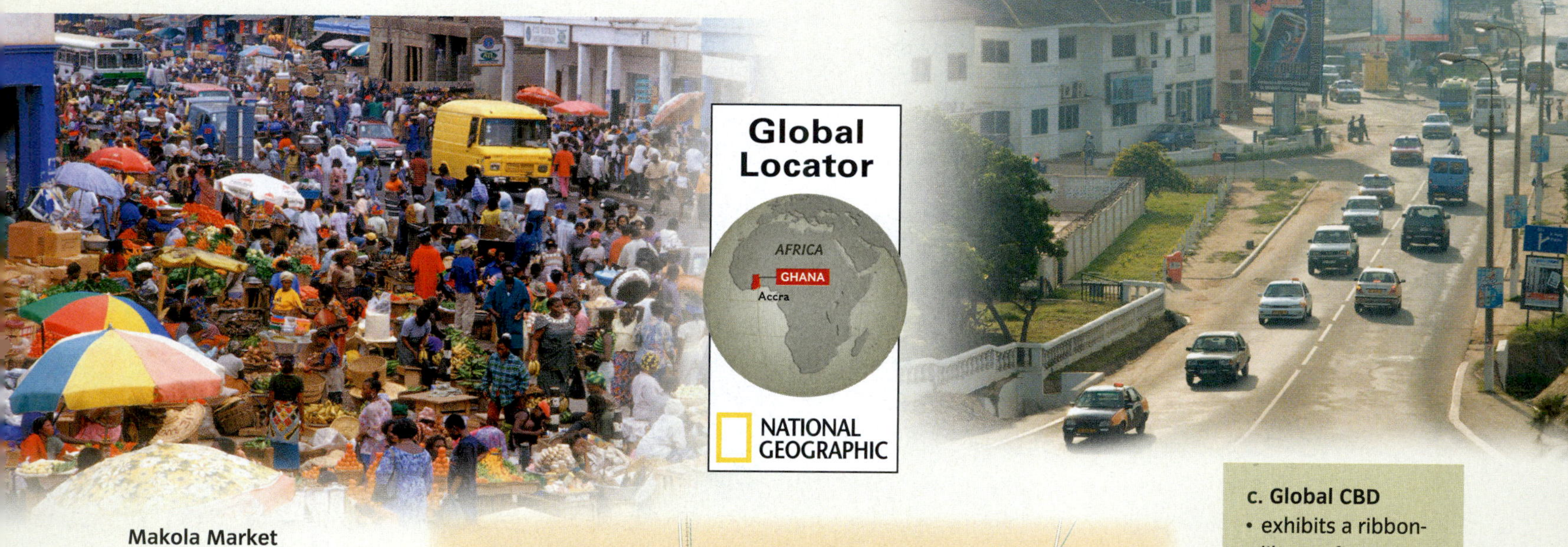

Near Cantonments Road

Makola Market

a. Local CBD

- includes indigenous markets and small-scale retailing that serves a local clientele and adjacent residential areas
- has been a chief destination of rural migrants

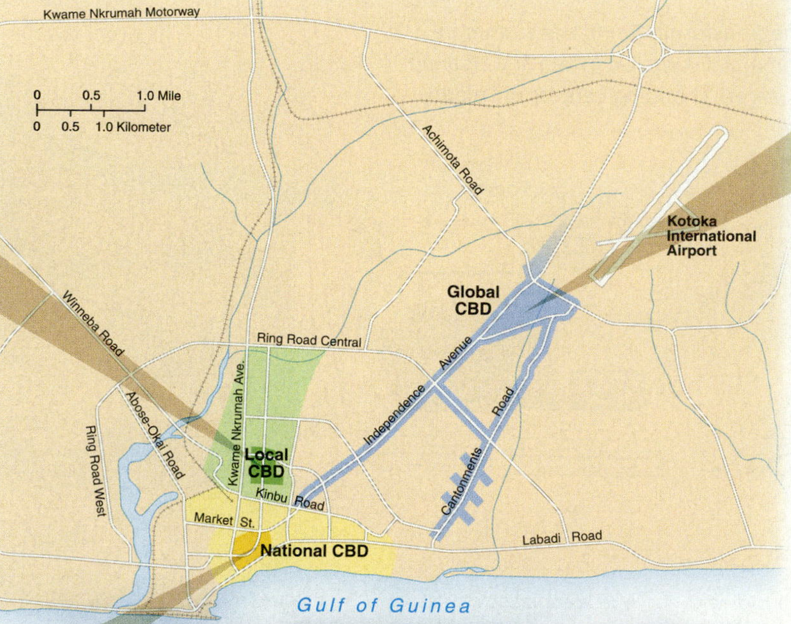

c. Global CBD

- exhibits a ribbon-like configuration along certain major thoroughfares
- includes subsidiaries of foreign firms and headquarters of Ghanaian multinational firms
- has a growing concentration of shops and amenities for the well-to-do

Supreme Court Building

b. National CBD

- corresponds to the CBD established when Ghana was a British colony
- includes a mix of domestic and some older foreign firms as well as government functions

settlement. In other cases entirely new cities were built. Significantly, colonial cities were defined by social segregation, in which European residential and commercial districts were spatially separated from African market areas and residential districts.

Like all cities, however, these former colonial cities are constantly changing under the pressures of urban growth and globalization. Consequently, some geographers now prefer the term **hybrid city**. A good example of a hybrid African city is Accra, Ghana (see **Figure 8.14**).

> **hybrid city**
> A city that exhibits a mixture of indigenous, colonial, and globalizing influences.

Islamic cities Is there such a thing as an Islamic city—one with an urban structure influenced by Islam? If we seek a specific type of city such that similar land-use patterns are replicated in all cities that were influenced by Islam, then the answer is no. In fact, many so-called Islamic cities have several features in common with the medieval cities of Europe: notably, a religious center (anchored by a cathedral in Europe and a mosque in Islamic cities), a central marketplace (called a *suq* in Islamic cities), residential quarters based on ethnicity or occupation, an irregular street pattern, and a surrounding defensive wall.

If, however, we shift our focus from form to process and acknowledge that Islam did (and still does) influence ideas about urban social relations, then we have a basis for identifying an Islamic city, though some urban experts prefer to speak of "traditional" Islamic cities. As an example, consider the *umma*, the global community of Muslims (see Chapter 5). An Islamic city is one that helps to link the local Muslim community with the international Muslim community. For example, geographer Michael Bonine has shown that some cities in Morocco are oriented according to the *qibla*, the sacred direction of Mecca. Another feature of some traditional Islamic cities that reveals the influence of Islam on social relations involves an emphasis on privacy and private space. This aspect has been expressed in the landscape through the construction of L-shaped entranceways, front doors that are offset rather than directly across from each other, and the absence of street-level windows (**Figure 8.15**).

Many of the cities in North Africa, the Middle East, Central Asia, and Indonesia bear evidence of the influence of Islam, but to identify an Islamic city requires that we consider how Islam shapes urban social relations. Thus, even though the design of the futuristic Masdar City, mentioned at the start of this chapter, will incorporate some elements of medieval urban form and is being built in a Muslim country, it would be premature to call it an Islamic city until we better understand how it works.

Islamic influence on the urban landscape • Figure 8.15

In Jeddah, Saudi Arabia, Islamic law places importance on maintaining visual privacy. Latticed balconies not only create private domestic spaces but also provide a kind of curtain such that women, who would be expected to be veiled when in public, can see out but cannot be seen by outsiders.

CONCEPT CHECK

STOP

1. **How** is urban land use related to land value?
2. **How** do the models of North American cities account for the impact of transportation?
3. **How** and why has urban land use in eastern European cities changed since 1989?
4. **Why** might hybrid cities be considered an effect of globalization?

Urban Dynamics

LEARNING OBJECTIVES

1. **Distinguish** between redlining and blockbusting.
2. **Define** sprawl and explain how it is measured.
3. **Summarize** the process of slum formation.
4. **Identify** the main goals of new urbanism.

Cities are dynamic entities that change in response to the decisions made by individuals, groups, institutions, as well as circumstances brought about by nature. Previously, we discussed such topics as the importance of transportation to urban form and structure, and touched on the influence of social relations on urban space. Here we explore in greater detail additional causes and consequences of urban transformation, including residential change, urban redevelopment, and slum formation.

Public Policy and Residential Change

Certain public policies adopted by U.S. government agencies such as the Federal Housing Administration (FHA) and the Veterans' Administration (VA), as well as actions by realtors, have contributed to the decentralization of American cities and the transformation of residential areas. The FHA was created to revitalize home construction and to expand home ownership during the Depression when many Americans had defaulted on their loans and lost their homes because of bank foreclosures. The FHA developed a new loan program in which the federal government guaranteed the full value of home loans (mortgages). Such FHA-insured loans significantly reduced the risk to private banks and lenders on home loans and lowered down payments, making it possible for more people to purchase homes. In the 1940s similar loan programs were implemented by the VA and helped thousands of returning soldiers finance homes.

The FHA and the VA did not agree to back every home loan, however. Instead, the agencies adopted a rating system and used it to identify areas that were considered to be insurance risks. High-risk areas—labeled "Fourth Grade" and described as "definitely declining" or "hazardous"—included those that appeared crowded, had older, deteriorating housing stock and less valuable housing, or were located near areas deemed problematic or inferior. Following the principles of succession and filtering, which suggest that older neighborhoods are prone to attract lower-income households, the loan insurance rating system also incorporated an evaluation of the social and economic character of urban areas. Low-income neighborhoods as well as those that were ethnically or racially mixed rarely received a rating other than high risk. The use of this discriminatory rating system became known as **redlining** (Figure 8.16).

redlining The biased practice of refusing to offer home loans on the basis of the characteristics of a neighborhood instead of the actual condition of the property being mortgaged.

The FHA and VA had a considerable effect on residential change because most of the loans these agencies approved were for new homes located in outlying residential subdivisions rather than older homes in the central city. Also, most FHA and VA loans supported the purchase of single-family homes, not duplexes or apartments. Therefore, FHA and VA policies tended to sustain the construction of single-family dwellings on city peripheries. Continued use of the biased rating system had the effect of stimulating the departure of whites from inner city areas to the suburbs. These public policies enforced a pattern of residential segregation that is often still visible in urban and suburban areas today.

Segregation was simultaneously enforced through the actions of realtors, especially those who encouraged **blockbusting**. For example, realtors, who stood to profit on the sale of the affected properties, drew on racial prejudices to promote the perception among white residents that a "black invasion" of their neighborhood was underway and that property values in the neighborhood would decline. In some instances, all of the properties on a block were put up for sale as whites sought to leave the area. Together, blockbusting and redlining contributed to *white flight*, the departure of whites from downtown neighborhoods to the suburbs.

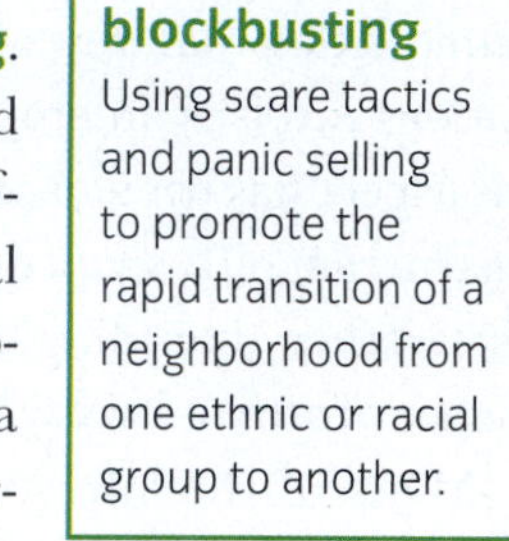
blockbusting Using scare tactics and panic selling to promote the rapid transition of a neighborhood from one ethnic or racial group to another.

Redlining St. Louis, 1937 • Figure 8.16

Properties that were likely to transition to lower-income households were considered higher risk and less eligible for loan insurance. Thus, the risk assessment of a property became linked to the race and ethnicity of its occupants. Fourth-grade, or high-risk, areas for loans are shown in red. (*Source*: Jackson, 1985.)

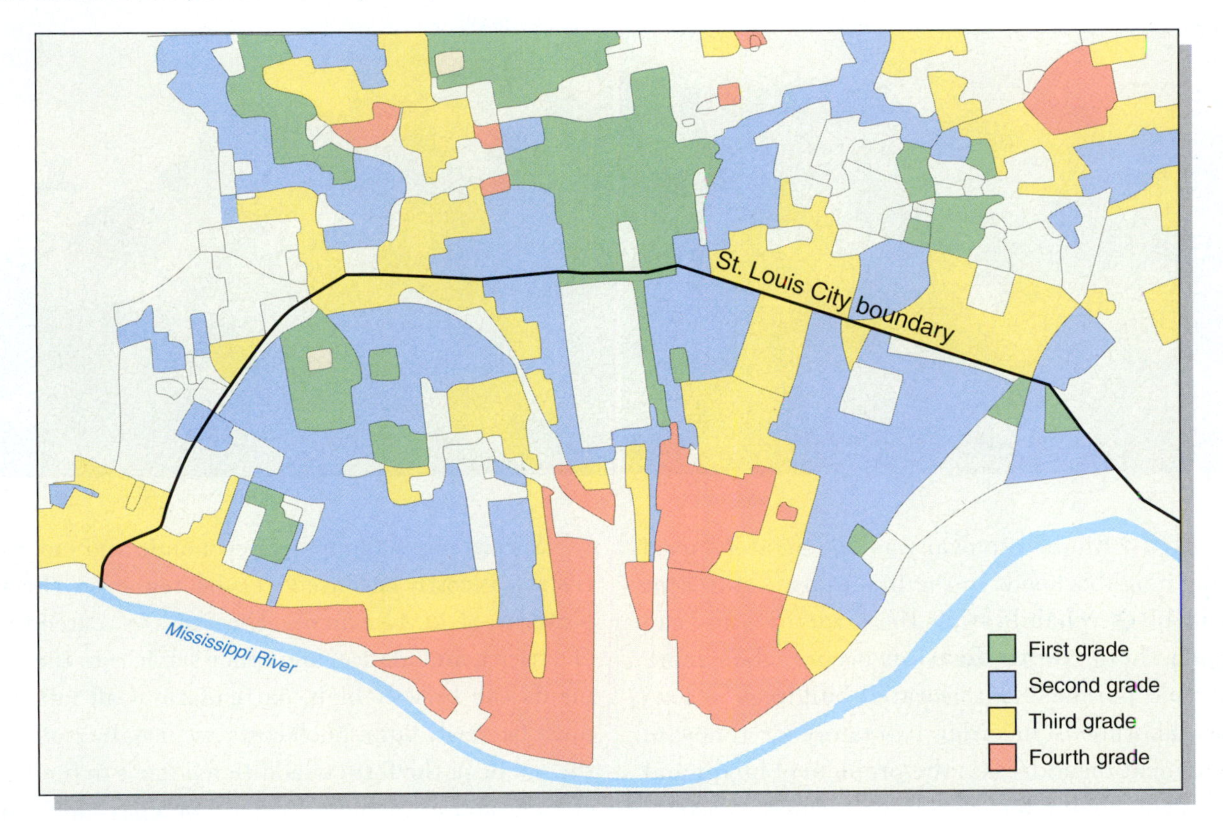

Urban Redevelopment

By the 1940s the tenements and apartments in many inner-city residential areas were aging and physically deteriorating. To local city officials, the presence of this impaired and dilapidated housing stock was a form of urban blight. Local officials sought federal assistance in attempting to solve the problem of blight, and this set in motion a new kind of change referred to as **urban redevelopment**—the process of renovating an area of a city, often by completely destroying dilapidated structures and rebuilding on the site.

Using provisions in the Federal Housing Act of 1949, the government can take possession of city property that has been classified as "blighted." The procedure typically involves the use of **eminent domain**. When eminent domain is invoked, the government must pay the private owner the market value of the property. Blight removal is considered to serve the public by improving the quality of the housing stock. After the government obtains the rights to the property, the buildings or structures on it are bulldozed. The property can then be sold or leased to developers for the eventual construction of hospitals, hotels, stadiums, convention centers, and other facilities.

eminent domain
The authority of a government to take private property when doing so serves the public's interests.

Urban redevelopment solved the problem of blight, but it created other problems in the process. Urban redevelopment displaced people and broke up long-standing neighborhoods. To counter these adverse effects, the government financed a number of public housing projects. In the 1950s and 1960s, inner-city public housing took the form of high-rise apartments. These apartments were initially celebrated as an innovative solution to the problem of blight but were soon heavily criticized. Elderly residents and families with young children have different needs, but these were not adequately considered in the design of this public housing, which reflected a "one size fits all" vision. Because of their shortcomings, many of these multistory public housing projects have since been torn down.

Before and after gentrification, Bricktown, Oklahoma City • Figure 8.17

Gentrification refashions the urban landscape in ways that are often visually and aesthetically appealing. But if we only focus on landscape change, such as that shown in these photos, we lose sight of the fact that gentrification can result in the displacement of lower-income individuals who can no longer afford to live in the gentrified districts.

Gentrification Blight removal was one strategy used to redevelop neighborhoods in the inner city. A different approach involves rehabilitating the structures instead of bulldozing them. **Gentrification** occurs when more affluent people purchase deteriorated buildings in low-income neighborhoods in order to restore or renovate them. Gentrification addresses the problem of blight and can act as a centralizing force by drawing middle-income residents, and retail establishments catering to them, into the inner city. Those who favor a lifestyle different from that of the typical suburbanite—artists, do-it-yourselfers, gays and lesbians, and others—often help propel gentrification. The results of gentrification are hotly contested, however. Supporters of gentrification maintain that it raises property values, but opponents argue that gentrification is a form of economic exploitation of the urban poor (**Figure 8.17**).

For another look at how gentrification transforms urban space and competing perspectives about those changes, see *Video Explorations*.

Sprawl Like gentrification, sprawl is a process that transforms urban landscapes. **Sprawl** occurs when the rate at which land is urbanized greatly exceeds the rate of population growth in a given period of time. To urbanize land means to develop it for residential, commercial, or industrial purposes. Thus, sprawl rapidly extends the footprint of urban space and leads to low-density land use in the form of strip mall development, office parks, and single-family residential subdivisions (**Figure 8.18**).

Sprawl has a number of economic and environmental costs. It is costly because it results in uneven, checkerboard patterns of development. The costs of extending water, electricity, ambulance, and other services to these developments can be very high, particularly if all infrastructures must be newly built. Such costs are usually passed along to the public in the form of additional taxes to fund the development and to the homeowners, who pay higher prices for their homes and often higher rates on basic utilities.

Video Explorations

Trastevere

WILEY PLUS NATIONAL GEOGRAPHIC Video

Trastevere, a neighborhood in Rome, Italy, has undergone much recent development and change. Residents and local business owners comment on the changes.

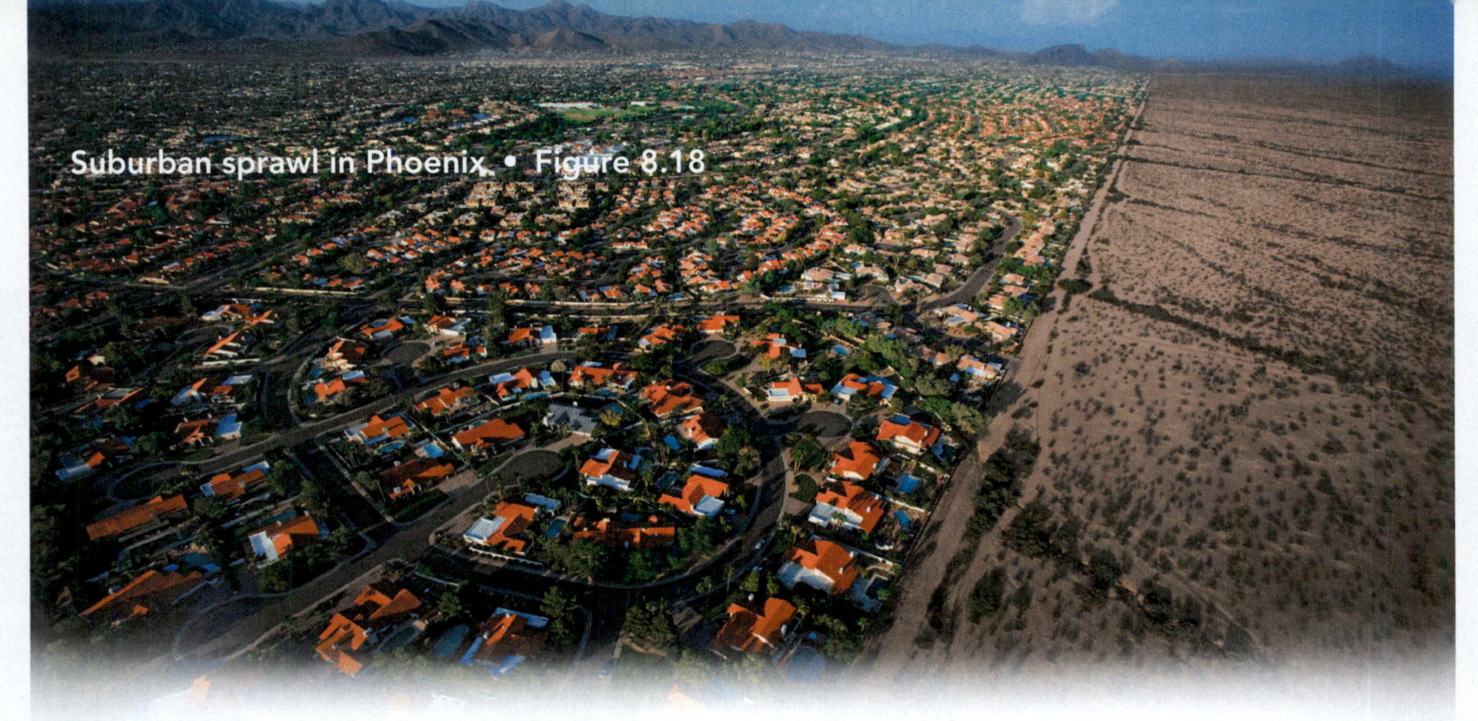

A wide variety of economic, political, social, and geographic forces contribute to sprawl. These might include such things as rising household incomes, municipal decisions on the construction of roads, population growth, and the presence of open, undeveloped land. Even divorce and the shift away from multigenerational households can affect sprawl.

slum An area of a city characterized by overcrowding, makeshift or dilapidated housing, and little or no access to basic infrastructure and services such as clean water and waste disposal.

Sprawl carries development beyond the areas that are served by public transportation. Because residential areas are often separated from commercial areas (refer again to Figure 8.18), sprawl contributes to a dependency on automobiles. Moreover, sprawl, and the spread of urbanized land in general, increase the area covered by impervious surfaces. More asphalt and concrete surfaces speed water runoff and alter drainage patterns. Sprawl has received a great deal of attention in the United States, but it can affect any city and raises important questions about how cities manage their growth. We will return to some of these issues in the section on urban planning, but first we need to examine the urban dynamics of poverty.

Urban Poverty and the Informal Sector

Urban poverty is a serious global problem. Broadly speaking, urban poverty is associated with three conditions: (1) increasing costs of basic necessities (food, water, electricity, transportation); (2) a widening gap between the money a person earns and the price of renting a dwelling; and (3) a failure to recognize the magnitude of urban poverty and/or the inability to allocate sufficient funds to address it.

The highest levels of urban poverty, as measured by the number and percentage of urban poor, exist in the developing world. There, the urban poor are highly visible in part because of the extensive **slums** or shantytowns that have grown up in and on the outskirts of the region's cities (**Figure 8.19**).

In a different sense, however, the urban poor are also highly invisible. High percentages of the urban poor, especially in the developing world, work in the informal sector or

Proportion of urban populations living in slums, by region • Figure 8.19

Approximately 1.2 billion people live in slums. Asia has the greatest number of urban poor, more than 500 million, but Sub-Saharan Africa has the highest percentage of urban residents who live in slums. In North America, slums often develop in inner-city residential areas, but elsewhere slums are associated with the urban periphery. (*Source*: United Nations Human Settlements Program, 2008.)

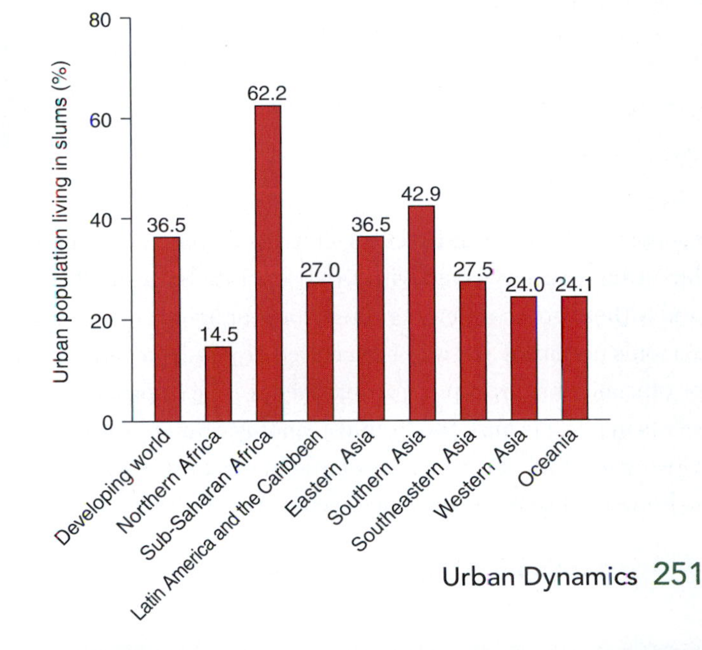

PROCESS DIAGRAM

Slum formation • Figure 8.20

THE PLANNER

Declining agricultural yields and/or environmental degradation
A Cambodian farmer

Conflict and human displacement
Conflict migrants in Sri Lanka receive aid.

1 Preconditions → Migration into cities and/or urban population growth → 2 Rapid urbanization → Increased demand for housing, jobs, urban infrastructure

Rapid urbanization remains closely associated with slum formation. Greater potential for slums to form exists when city governments have not planned for such rapid urbanization or lack the resources to do so. Slums provide a visual and spatial expression of differential access to resources and power. This diagram largely emphasizes slum formation in the developing world, but slums exist in developed countries as well.

Employment Prospects
Job seekers read about employment opportunities posted in Shenzhen, China.

economy. The **informal sector** consists of the retail, manufacturing, and service activities that operate on a small scale and without government regulation or oversight. The informal sector is not taxed and is not measured or monitored in formal or official statistics kept by governments. The informal sector exists in part because the formal economy cannot provide employment for all of the city residents. Informal sectors operate in both developed and developing countries, but they are more pronounced in developing countries. Countless different jobs exist in the informal sector, but two examples include working as street vendors selling food or beverages and washing windshields of vehicles in city traffic.

Slums demonstrate the existence of an informal housing sector. Indeed, slums are a kind of *informal settlement* in that they are unplanned and often develop as a result of transactions that do not follow procedures

Favela Morumbi,
São Paolo, Brazil

3 Other factors driving slum formation

Crowding; rundown or improvised housing on undesirable or hazardous land; lack of services (clean water, sanitation)

4 Slum formation

Inner-city slum, Camden, New Jersey

Makeshift dwellings in the Tanah Abang slum,
Jakarta, Indonesia

established by city or national governments. Not all slums are illegal, however. In fact, rental agreements and land tenure practices may be rooted in local customary law. Often different systems of land management coexist, and these different practices can complicate the ability to manage land use in and around a city, affecting slum development. When slums do develop illegally, as for example when people build shelters on property without the consent of the landowner, they are more accurately termed *squatter settlements*. Slums and informal settlements more generally are referred to as *favelas* in Brazil, *gecedondus* in Turkey, and *kampungs* in Malaysia.

No single cause triggers the formation of slums. Rather, slum formation is the result of myriad factors including economic, social, and institutional decisions. To better understand slum formation, see **Figure 8.20**.

Slum alleviation According to a recent estimate, it would cost between $60 and $70 billion to upgrade the slum dwellings of 100 million people. Although this sounds like a lot of money, government officials know all too well that the continued existence of slums costs a city and country far more because slums incur significant social, political, and economic costs as well.

A failed approach to slum alleviation is eviction. Forcing people out of slums and destroying their dwellings never solves the problem of slum development. Rather, it simply displaces it. Thus, there is no simple solution for alleviating slums. There is, however, one approach to slum alleviation that has a consistently strong record of success: community-driven projects that involve slum dwellers as agents of change. One of the most successful of these projects creates community-managed savings groups for **microfinancing**. Community members are encouraged to save some money—no matter how little—every day. The savings are pooled and used to improve the conditions of their dwellings or neighborhood, sometimes with additional financial assistance from governmental and nongovernmental organizations (**Figure 8.21**).

microfinancing Providing access to credit and other financial services for low-income individuals or groups.

Urban Planning

Slum alleviation and upgrading draw on the principles of urban planning, for urban planners seek to improve the physical and social conditions in towns and cities.

Women's savings group, Sri Lanka • Figure 8.21

Slum dwellers do not have access to the same lending and credit programs that higher- income individuals do. Obtaining credit or property rights can be additionally challenging for women because of discrimination. Nevertheless, microfinancing activities are often spearheaded by women and contribute significantly to slum improvement projects.

The scope of **urban planning** is broad and includes designing efficient systems of transportation, working with developers to coordinate housing construction in order to meet the needs of a city, providing facilities and open spaces for public use and recreation, and ensuring that different parts of a city are served by a safe water supply, schools, ambulance and fire service, and trash collection.

Many urban geographers pursue careers in planning, and those who do often work for local governments in city or county planning departments. Perhaps no other issue better illustrates the connection between planning and government functions than zoning. Planners use and recognize many different kinds of zoning, but three broad categories include zoning for residential, commercial, or industrial purposes. Zoning regulates the development of cities and this makes it an intensely political activity that sometimes pits homeowners against developers, or individual interests against the collective interests of corporations and local governments. How would you feel, for example, if a decision were made to change the zoning classification of a parcel of land near your home from residential to commercial so that a hotel or shopping center might be built? Planners are usually involved in both the development and enforcement of zoning codes.

The theory behind zoning today differs considerably from the zoning theory of just a few decades ago. In the past, zoning theory was guided by a greater emphasis on the creation of exclusive zones that permitted just one type of land use in a district. As a result, residential areas were completely separated from commercial and industrial areas. These areas were further partitioned so that, for example, in residential zones, apartment districts were separated from those with single-family houses. Critics have pointed out that exclusive zoning increased the need for and reliance on cars, and today zoning codes can more easily accommodate a mixture of land uses.

In the 1990s a new urban planning movement developed called neotraditional town planning or **new urbanism**. Proponents of new urbanism argue that too often urban planning has failed to consider the environmental impact of planning decisions, resulting in suburban sprawl, the loss of affordable housing, and the demise of cohesive urban neighborhoods. New urbanism has two main goals: to prevent sprawl and to create walkable neighborhoods.

A primary strategy used to achieve these goals involves **mixed-use development**—the combination of

New urbanist neighborhood, near Charleston, South Carolina • Figure 8.22

The principles of new urbanism apply across all scales, from the level of the city block to the metropolitan area. Three attributes are emphasized at the neighborhood scale: compactness, walkability, and mixed-use development that incorporates diverse house types and styles.

different types of land use within a particular neighborhood. For example, retail land use can be integrated with residential land use and transportation corridors by including shops and grocery stores that are within walking distance. Other strategies include using smaller lot sizes for residences and clustered development on them. Instead of building on the entire lot, clustered development encourages use of just part of a lot (Figure 8.22).

CONCEPT CHECK 

1. **How** has public policy in the United States affected residential change?
2. **Why** is gentrification contested?
3. **What** factors dive slum formation?
4. **What** approach to zoning does new urbanism promote?

THE PLANNER

Summary

1 Cities and Urbanization 230

- Cities are concentrations of large numbers of people who earn their living through nonagricultural pursuits. Cities function as **central places** serving their **hinterlands** in a system of commercial exchange. **Functional complexity** is a defining characteristic of cities.
- **Central cities** are legally defined and bounded, and possess an identifiable **central business district** (CBD). There is no standardized definition of the term *urban*; however, an **urbanized area** includes land converted to commercial, residential, or industrial uses. When **metropolitan areas** coalesce, a **megalopolis** can form.
- Two measures of **urbanization** include the **level of urbanization** and the **rate of urban growth**. Globally, higher levels of urbanization tend to be associated with higher levels of industrialization. The geography of **megacities**, however, illustrates the rapid rates of urbanization occurring in parts of the developing world. The development of **primate cities**, though not unknown in the developed world, is more commonly associated with developing countries.
- The concepts of **range** and **threshold** help explain the **urban hierarchy**. **Central place theory** advances the idea that market forces contribute to the development of an idealized lattice of central places, shown here.

Spatial arrangement of central places • Figure 8.7

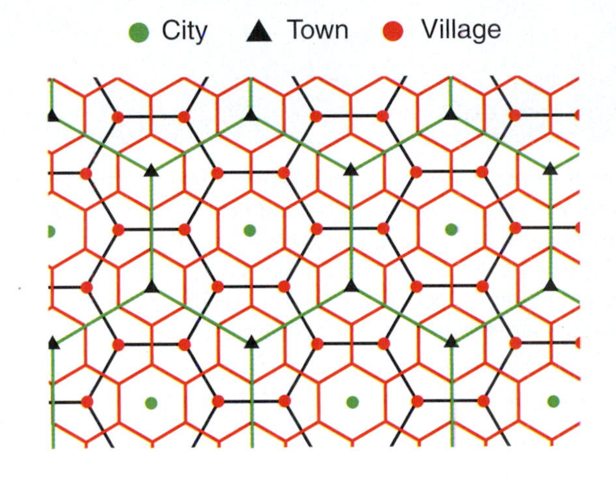

- The identification of **food deserts** and the emergence of **world cities** and **hinterworlds** highlight the complexities of the urban hierarchy.

2 Urban Structure 240

- The internal spatial organization of cities varies from one region of the world to another and is affected by **centralization, decentralization,** and **agglomeration**. Urban models and **bid-rent curves** help to visualize the **functional zonation** of cities.
- Several different models have been proposed to describe and explain the diverse forces that have helped structure the cities of North America. The **concentric zone model** and **sector model**, respectively, identify the processes of succession and **filtering** as key to understanding urban land-use patterns. Both models describe the land-use patterns in a city with a single CBD, in contrast to the **multiple nuclei model** and **urban realms model**. Changing transportation practices and the emergence of **edge cities** are two developments that have altered the spatial configuration of numerous cities.
- The geography of urban structure in cities outside of North America is just as complex. Western European cities have been shaped by a number of centralizing forces. Since 1989 the urban landscapes of eastern European cities have been dramatically transformed by capitalism and globalization, as shown here.

What a Geographer Sees

- Elsewhere, colonialism has influenced urban form, and many colonial cities are, in effect, **hybrid cities**. Throughout the Muslim world many cities bear the influence of Islam but in diverse ways. An orientation toward Mecca as well as design elements that reflect concern with private space and privacy are associated with traditional Islamic cities.

3 Urban Dynamics 248

- Cities are dynamic entities. Thus, understanding how and why they change remains essential to the work of urban geographers. Much recent change in American cities is related to policies that contributed to racially discriminatory practices including **redlining** and **blockbusting**.
- **Urban redevelopment** efforts target blight but also transform the look and socioeconomic makeup of cities. **Eminent domain** is a mechanism that facilitates urban redevelopment. **Gentrification** is a form of urban redevelopment that typically upgrades dilapidated buildings instead of razing them. Although gentrification may help counter the effects of **sprawl**, depicted here, gentrification has other social costs.

Suburban sprawl in Phoenix • Figure 8.18

- Urban poverty is a serious global problem made spatially visible through the existence of **slums** but economically invisible through the operation of the **informal sector**. Understanding slum formation constitutes an important first step toward slum alleviation. **Microfinancing** has become an increasingly popular way for slum dwellers to plan and implement slum upgrading projects.
- **Urban planning** involves gauging the future needs of a city and developing ways to prepare for and accommodate them. Making decisions about zoning is one important aspect of urban planning. **New urbanism** challenges conventional zoning regulations and promotes **mixed-use developments** instead.

Key Terms

- agglomeration 240
- bid-rent curve 241
- blockbusting 248
- central business district 231
- central city 231
- central place 230
- central place theory 237
- centralization 240
- concentric zone model 241
- decentralization 240
- edge cities 243
- eminent domain 249
- filtering 242
- food desert 238
- functional complexity 230
- functional zonation 240
- gentrification 250
- hinterland 230
- hinterworld 239
- hybrid city 247
- informal sector 252
- level of urbanization 232
- megacity 234
- megalopolis 232
- metropolitan area 232
- microfinancing 254
- mixed-use development 254
- multiple nuclei model 242
- new urbanism 254
- primate city 235
- range 236
- rate of urban growth 232
- redlining 248
- sector model 242
- slum 251
- sprawl 250
- suburb 231
- threshold 236
- urban hierarchy 237
- urban planning 254
- urban realms model 242
- urban redevelopment 249
- urbanization 232
- urbanized area 231
- world city 238
- zoning 241

Critical and Creative Thinking Questions

1. Select a city and do some fieldwork to evaluate the extent to which the concentric zone, sector, and multiple nuclei models apply. What other methods could you use to make this evaluation?
2. What informal economies exist in the United States?
3. Which cities have hosted the Olympics in the past five years? Ten years? Are cities that host the Olympics world cities? Explain and provide an assessment of the patterns you observe.
4. Watch the movie *Slumdog Millionaire* and evaluate its depiction of Mumbai's slums.
5. Should capitalism be considered a driver of slum formation? Explain your reasoning.

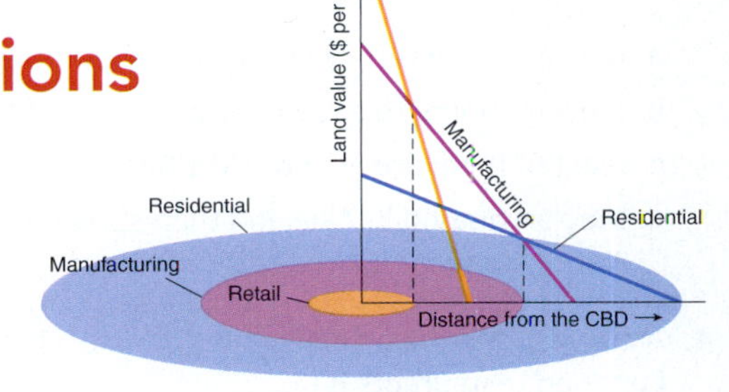

6. How can bid-rent theory, modeled here, help explain sprawl?
7. Watch *The End of Suburbia* (available on DVD). Evaluate its presentation of issues in urban geography.
8. What measures can be taken to reduce the social impact of gentrification?
9. Do you think that women and men experience cities differently, and can urban spaces be made to work well for all people? Why or why not?

What is happening in this picture?

A rickshaw puller in Kolkata, India, transports passengers against a heavy flow of traffic. Rickshaw pullers provide an affordable means of transportation around the clock as well as year-round. Unlike cars and buses, rickshaw pullers can still navigate inundated streets during the monsoon rains. But city officials consider rickshaw pullers a reason for traffic congestion and have proposed banning them.

Think Critically

1. What does this photo suggest are some other causes of traffic congestion?
2. Could rickshaws be considered a sustainable form of urban transportation? Why or why not?

NATIONAL GEOGRAPHIC

Self-Test

(Check your answers in Appendix B.)

1. Cities share all of the following characteristics *except* _______.

a. functional complexity

b. edge cities

c. hinterlands

d. accessibility

2. A city with 10 million or more inhabitants is a _______.

a. metropolitan area

b. megalopolis

c. metroplex

d. megacity

3. A significant aspect of the geography of megacities is that _______.

a. they are increasingly associated with the developing world

b. they are home to a majority of the world's people

c. most of them are also world cities

d. they originated in Asia and then spread to Europe and North America

4. A concept that helps explain why some small towns lack fast-food restaurants is _______.

a. threshold

b. range

c. decentralization

d. urban food deserts

5. Explain how this photo illustrates the concept of agglomeration.

____

____

____

6. The development of edge cities is most closely associated with the _______ model.

a. concentric zone

b. urban realms

c. sector

d. multiple nuclei

7. _______ occurs when financial institutions grade urban areas on the quality of their housing or ethnic composition and use the ratings to refuse home loans.

a. Blockbusting

b. Urban redevelopment

c. Redlining

d. Eminent domain

8. These photos illustrate the process of _______.

a. urban sprawl

b. gentrification

c. filtering

d. zoning

9. From the end of World War II until 1989, probably the most significant factor shaping urban form in East European and Soviet cities was _______.
   a. the development of large parks
   b. state control of the urban land markets
   c. the emphasis on manufacturing and industry
   d. the absence of slums

10. What do American suburbs and colonial cities have in common?
    a. large housing estates
    b. similar patterns of industrial land use
    c. a history of residential segregation
    d. low levels of population growth

11. Identify at least three urban processes that have contributed to this scene.

    ____

    ____

    ____

12. Which statement about primate cities is FALSE?
    a. Primate cities are unique to the developing world.
    b. Primate cities are associated with uneven economic development.
    c. Primate cities can exist at the national or state/provincial level within a country.
    d. Primate cities may result from historical forces.

13. Some large cities are not world cities because of their _______.
    a. language differences
    b. remote location
    c. political corruption
    d. bid-rent curves

14. The term _______ conveys the idea that a city's influence may extend well beyond its formal boundaries.
    a. urbanized area
    b. megalopolis
    c. megacity
    d. metropolitan area

15. Of the following items, which one best explains the reason for hexagonal hinterlands in central place theory?
    a. bid-rent curves
    b. nearby central places of the same rank
    c. land-use values
    d. the assumption of a uniform surface

THE PLANNER

Review your Chapter Planner on the chapter opener and check off your completed work.

9 Geographies of Development

BHUTAN'S QUEST FOR GROSS NATIONAL HAPPINESS

Imagine your own Shangri-la—that is, an idyllic place. What place on Earth, if any, comes closest to matching that? Did the country of Bhutan come to mind? Most likely it did not, although in recent years, this small mountainous state nestled between India and China has occasionally been described as a Shangri-la. This designation has less to do with Bhutan's striving to be a perfect place and more to do with its physical setting and its ideology of development.

Until the early 1970s, Bhutan was among the world's most impoverished countries. Then, King Jigme Singye Wangchuck conceived a development strategy that would balance economic growth with environmental protection, Bhutanese cultural traditions, and democratic governance. He envisioned an alternative path to development that, in his words, would bring "gross national happiness."

Bhutan has since invested heavily in education and health care. In the early 1980s, Bhutan had an adult literacy rate of 23% and an infant mortality rate of 163. Today, close to 60% of the adult population is literate, and the infant mortality rate has dropped to 40. In addition, Bhutan has set aside over 30% of its land area—more than any other country—as wildlife sanctuaries, national parks, and nature reserves.

Although Bhutan may not be a Shangri-la and still has room to improve the social well-being of its people, its example is instructive because it highlights a concerted effort to achieve development in a way that is environmentally sustainable and socially conscious.

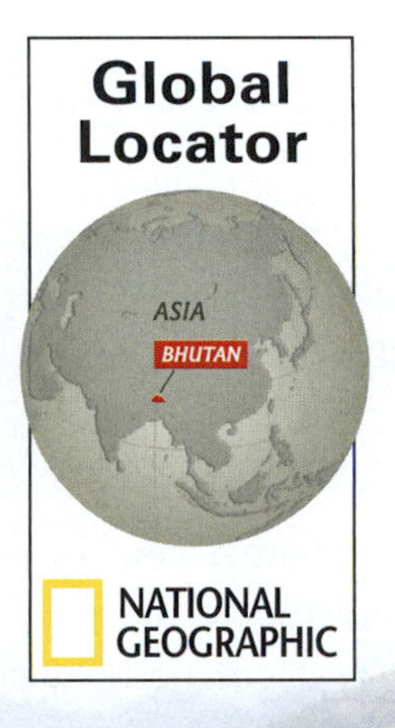

Bhutan's national highway zigzags across mountains.

CHAPTER OUTLINE

CHAPTER PLANNER

- ❑ Study the picture and read the opening story.
- ❑ Scan the Learning Objectives in each section: p. 262 ❑ p. 275 ❑ p. 280 ❑
- ❑ Read the text and study all figures and visuals. Answer any questions.

Analyze key features

- ❑ Geography InSight, p. 274
- ❑ Process Diagram, p. 281
- ❑ What a Geographer Sees, p. 286
- ❑ Video Explorations, p. 288
- ❑ Stop: Answer the Concept Checks before you go on: p. 275 ❑ p. 280 ❑ p. 288 ❑

End of chapter

- ❑ Review the Summary and Key Terms.
- ❑ Answer the Critical and Creative Thinking Questions.
- ❑ Answer What is happening in this picture?
- ❑ Complete the Self-Test and check your answers.

What Is Development?

LEARNING OBJECTIVES

1. **Explain** what development is.
2. **Distinguish** between development indicators and indexes.
3. **Contrast** the HDI, GDI, and GEM.
4. **Identify** geographic and institutional factors that can affect development.

One of the axioms of **development** is that, no matter how it is measured, it is geographically uneven. Human geographers study the differences in development from one place or region to another as well as the social and environmental consequences of development. We begin this chapter by considering what development is and how it is measured. We then examine global patterns of development and development theory.

development Processes that bring about changes in economic prosperity and the quality of life.

When comparing countries or regions on the basis of their levels of development, different terms and classifications are used including high-, middle-, and low-income countries, or more developed and less developed countries. The terms *First World* and *Third World* are problematic because they reinforce a view that less developed countries are intrinsically inferior, even backward. Today, most development practitioners use the terms *developed* and *developing*. These terms are not strictly defined, but general usage recognizes Australia, New Zealand, Japan, Europe, Canada, and the United States as developed. The Brandt Report, a study of international development issues published in 1980, drew attention to the global imbalance of development and observed

The global North-South divide • Figure 9.1

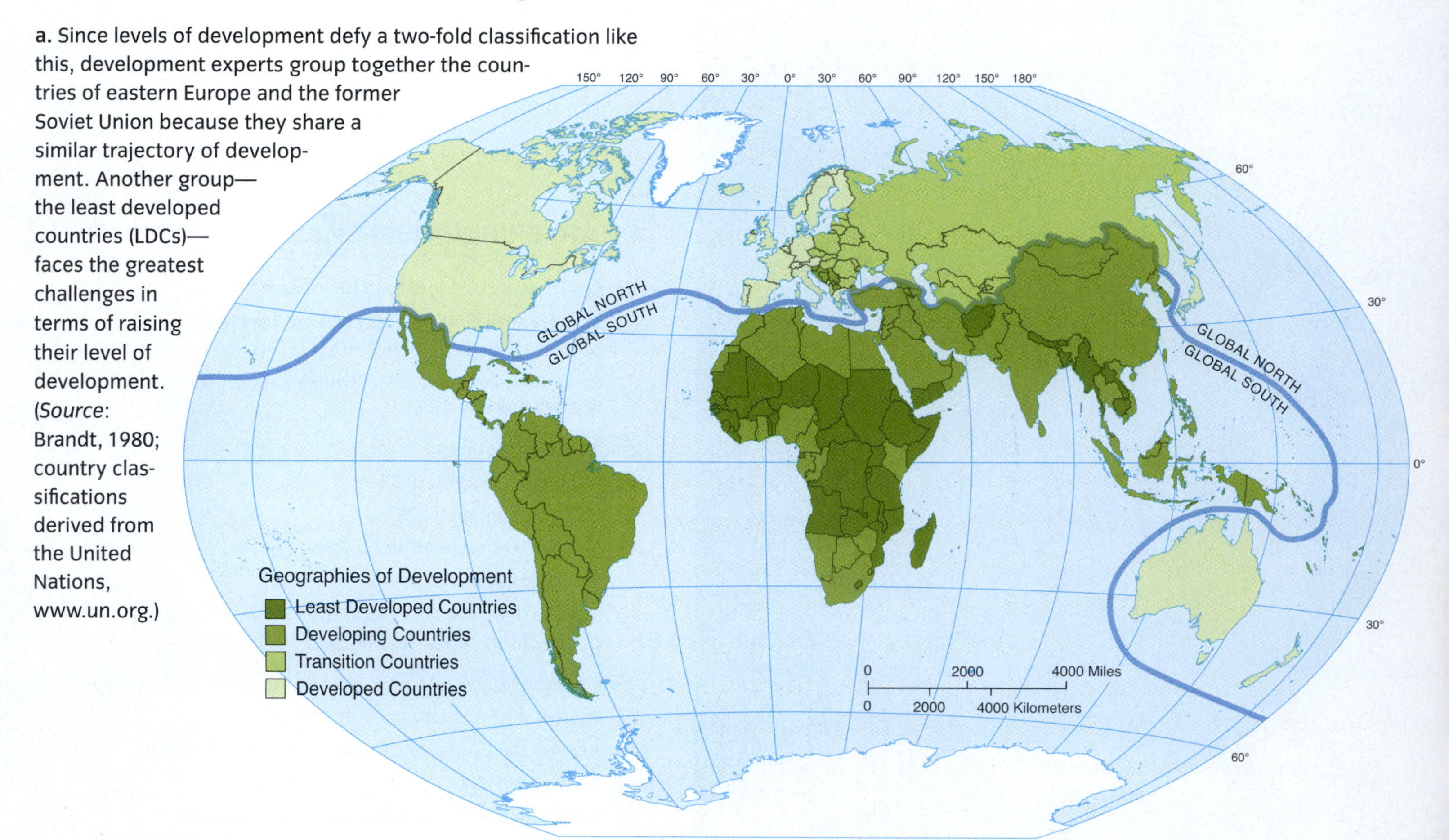

a. Since levels of development defy a two-fold classification like this, development experts group together the countries of eastern Europe and the former Soviet Union because they share a similar trajectory of development. Another group—the least developed countries (LDCs)—faces the greatest challenges in terms of raising their level of development. (*Source*: Brandt, 1980; country classifications derived from the United Nations, www.un.org.)

that a line drawn at about 30 degrees north latitude represented the global *North-South divide*. Accordingly, developed countries constitute the North whereas developing countries constitute the South (**Figure 9.1**).

The study of development is always a **normative** project. The term *normative* refers to the establishment of standards, or norms, to help measure the quality of life and economic prosperity of groups of people. Conventional views of development are strongly associated with normative ideas of progress, improvement, advancement, and social betterment. Consequently, *development* usually implies improvement in one or more of the following: a society's economic, social, or environmental conditions. Development experts recognize, however, that a gain in one area, such as economic growth, may have adverse consequences in another area, such as the environment. For this reason, whether development can be called an improvement remains hotly contested.

The global economic system depends not just on human and financial resources but on natural resources as well. This calls into question ways of thinking about the relationship among the economy, development, and the environment. Broadly speaking, there are two perspectives on this: conventional and sustainable. *Conventional development* favors economic and social gains but gives scant attention to the impact of these gains on resource use, consumption, or the state of the environment. In contrast, *sustainable development* favors economic and social gains achieved in ways that do not compromise natural resources or the state of the environment for future generations.

These viewpoints aside, what are the conditions or indicators that social scientists and others use to measure or gauge development? They recognize different types of indicators and group them into the following three categories: economic, sociodemographic, and environmental. When two or more indicators are combined, the result is called an *index*. Whereas indexes are most often used for country-level or international data, indicators can also describe very small areas.

b. Vehicular traffic on Cairo's streets sometimes includes donkey carts—compelling evidence of different development trajectories even within a single city.

Economic Indicators

We can use many different **economic indicators** to gauge development, from levels of debt to trade imbalances to the kinds of consumer goods that people purchase. In this section, we focus on three of the most basic economic indicators: gross national income, gross domestic product per capita, and poverty.

The most common measure of economic development historically has been the **gross national income (GNI)**, formerly known as the gross national product. The GNI expresses the total monetary value of goods and services produced in a year by a country, whether those operations are located within the country or abroad. For example, if a Canadian company operates a plant in India, the profits that plant earns are counted in Canada's GNI. Similarly, the income earnings of Canadians living abroad are counted in the GNI for Canada. In contrast, the **gross domestic product (GDP)** expresses the total monetary value of goods and services produced *within* a country's geographic borders. In the previous example, the profits of the Canadian-owned plant operating in India would contribute to India's GDP.

Dividing the GDP in a given year by the total population of the country in that same year yields the *GDP per capita*, which reflects the average output per person. To facilitate international comparisons, it is common to show values in purchasing power parity. In its most basic form, **purchasing power parity (PPP)** is an exchange rate used to compare output, income, or prices among

GDP per capita 2006 • Figure 9.2

This cartogram shows the size of a country in proportion to its GDP per capita and captures the global geographic disparities in economic production. Qatar and Luxembourg have per capita GDPs of about $80,000—among the highest in the world. Zimbabwe, in the throes of economic and political turmoil, has seen its GDP per capita slip in recent years.

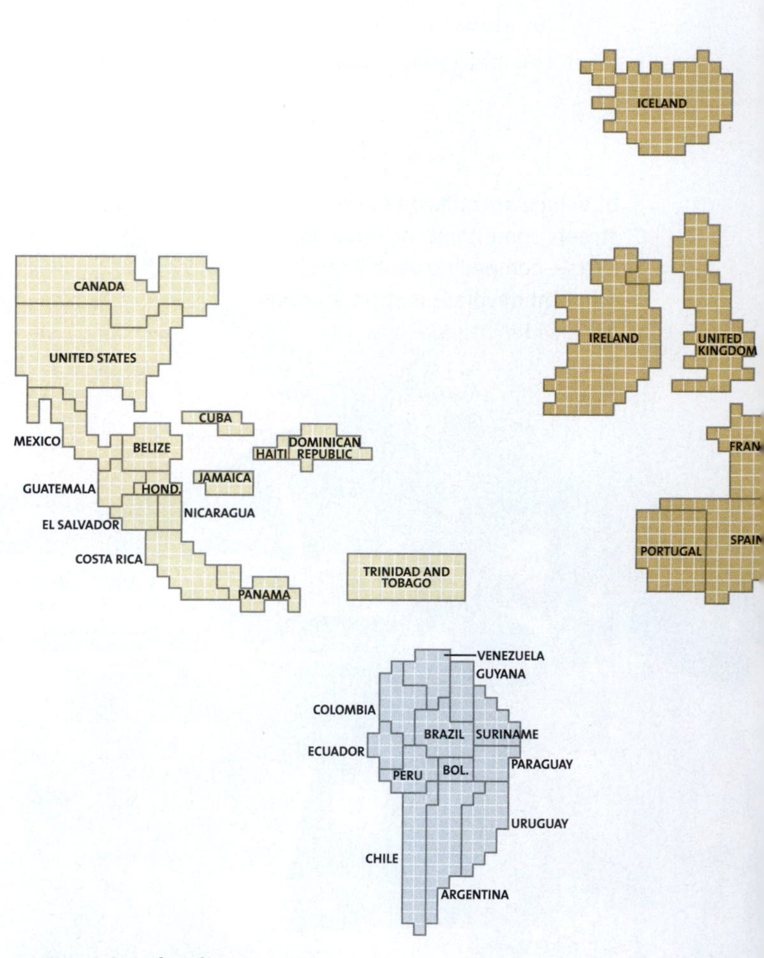

Gross Domestic Product (PPP) per capita, 2006

■ = U.S. $500
Not all countries or territories shown
Colors represent continental groupings
(*Source*: Central Intelligence Agency.)

countries with different currencies. PPP is based on the idea that the price of a good or service in one country should equal the price of that same good or service in another country when it is converted to a common currency. PPP is merely a unit indicating that a bottle of water in Cairo is equivalent to a bottle of water in Paris, even though the local prices for bottled water may be substantially different. If you are wondering why PPPs are used instead of currency exchange rates, the answer is that currency exchange rates can fluctuate tremendously from day to day. When a value is expressed in U.S. dollars (PPP), as in **Figure 9.2**, it shows how much a U.S. dollar can buy in any other country.

Economists frequently use the GNI and GDP as measures of economic development and, when examined over time, as measures of economic growth or decline. The GNI is regularly used to classify countries into low-, middle-, and high-income categories. However, these indicators have three important limitations. First, these indicators only reflect the monetary value of official receipts generated by the formal economy. They do not capture the value of goods and services produced through the informal economy, a key dimension of the economies of developing countries (see Chapter 8). Second, neither the GDP nor GNI provides information on the evenness or unevenness of distribution of wealth within a country. Third, the GDP and GNI do not take into consideration the social or environmental costs associated with the consumption of resources used in the production of the various goods and services.

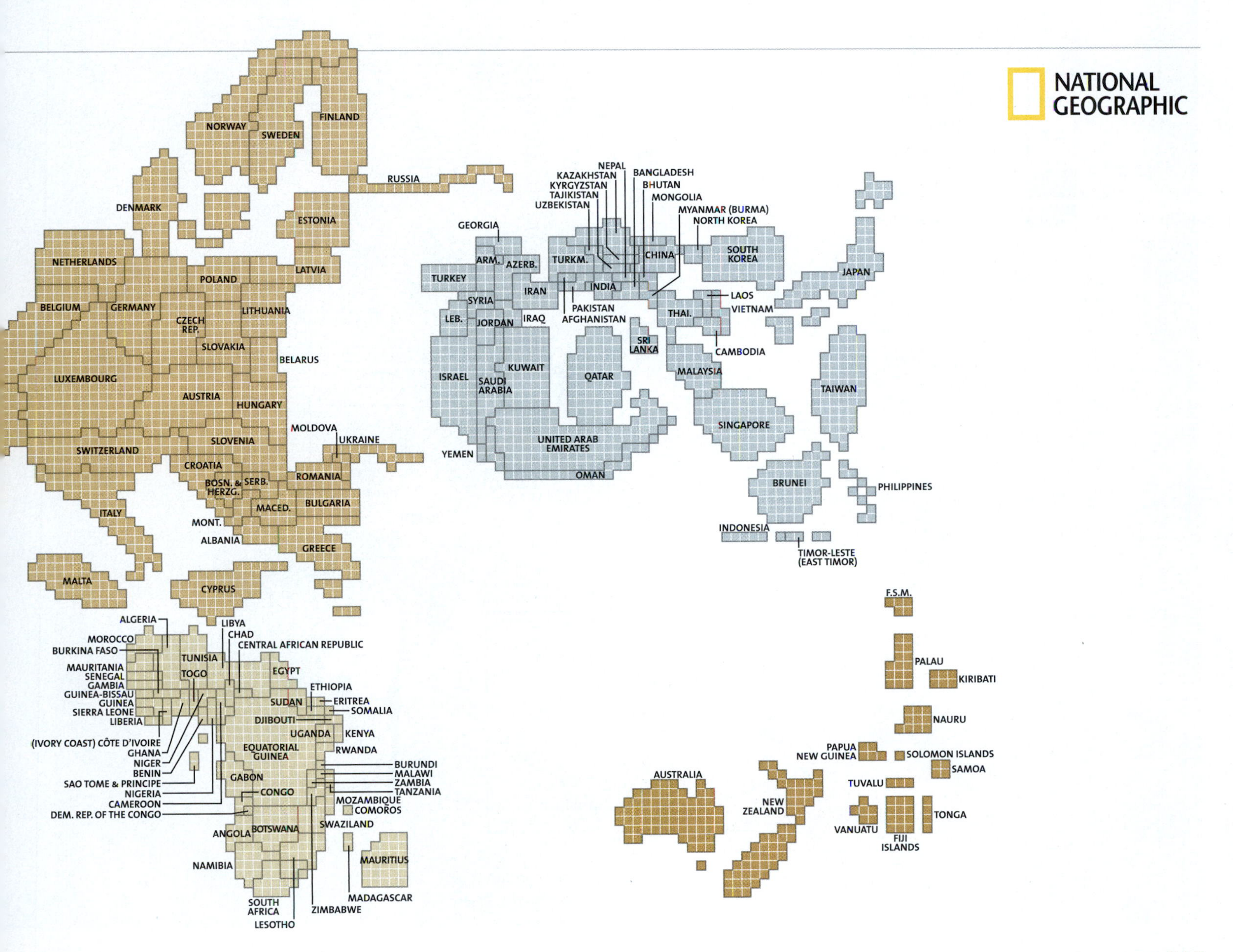

poverty Insufficient income to purchase the basic necessities of food, clothing, and shelter.

Another issue related to development is **poverty**—an extremely complex phenomenon that rarely has a single cause. Poverty is both an economic and social condition because it affects income as well as other facets of well-being, such as education and health. There are several economic indicators of poverty, including the poverty line and the poverty rate. In the United States, the government identifies **poverty lines** for different household units. For example, in 2009, the poverty line for a family of four with two children under the age of 18 was $22,050. The **poverty rate** is the most common measure used to express the occurrence or incidence of poverty. Approximately 40 million Americans live in poverty, giving the United States a poverty rate of about 13%.

poverty line The specific income amount social scientists and others use to separate the poor from the nonpoor.

poverty rate The percentage of the population below the poverty line.

Extreme poverty • Figure 9.3

More people in Sub-Saharan Africa live in extreme poverty than in any other region of the world, but extreme poverty is also a fact of life for many people living in parts of Asia as well as Central and South America. Haiti remains the poorest country in the Western Hemisphere.

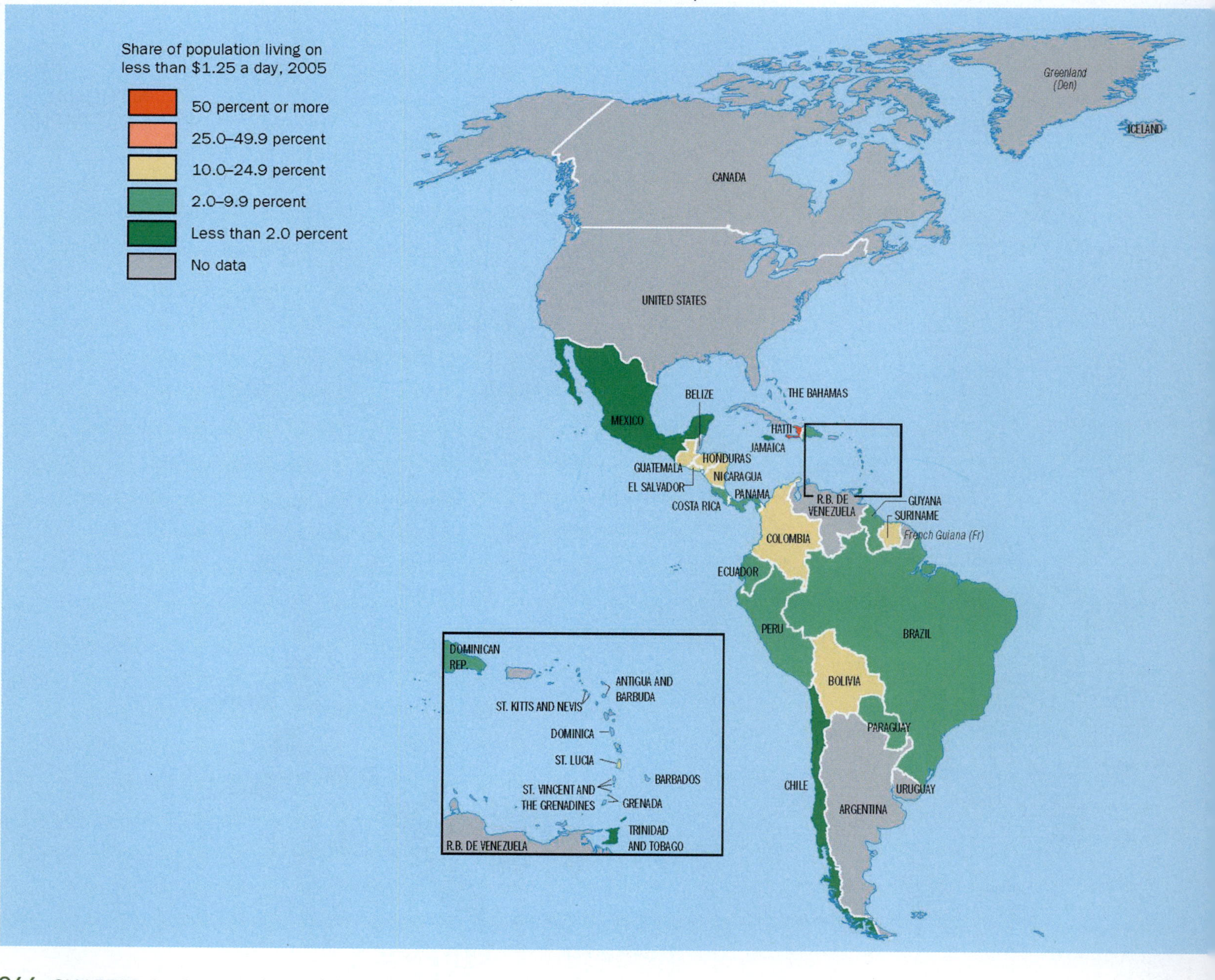

Because living standards and perceptions of poverty vary, individual countries often develop their own benchmarks for poverty. This practice makes comparisons of poverty among countries surprisingly difficult. In an attempt to solve this problem the World Bank, which provides financial and other forms of assistance to developing countries, has established two international poverty lines that distinguish between extreme poverty and moderate poverty: the $1.25/day (PPP) line and the $2/day (PPP) line, respectively.

Globally, 2.5 billion people live on less than $2 a day. Approximately 1.4 billion people live in extreme poverty. The geography of extreme poverty has changed since the 1980s. Significantly, the number of people living in extreme poverty dropped 26% between 1981 and 2005. Much of this success results from dramatic economic growth and rising prosperity in East Asia and the Pacific—specifically in China, where the poverty rate has dropped from 84% to 16%. For a depiction of the prevalence of extreme poverty, see **Figure 9.3**.

Infant mortality rates • Figure 9.4

a. Infectious diseases and malnourishment rank as leading causes of high infant mortality rates but medical care before, during, and after the delivery of a child is also crucial. How does this map compare to the one in Figure 9.1?

Infant mortality rate, 2005
(Deaths of infants under age 1 per 1,000 live births)
90.0-172.0
45.0-89.9
30.0-44.9
15.0-29.9
Less than 15.0

b. Children in Sudan receive the oral polio vaccine and midwives in Indonesia receive training on resuscitating infants. Studies suggest that small measures, such as immunization and improved access to skilled midwives, can significantly reduce infant mortality rates.

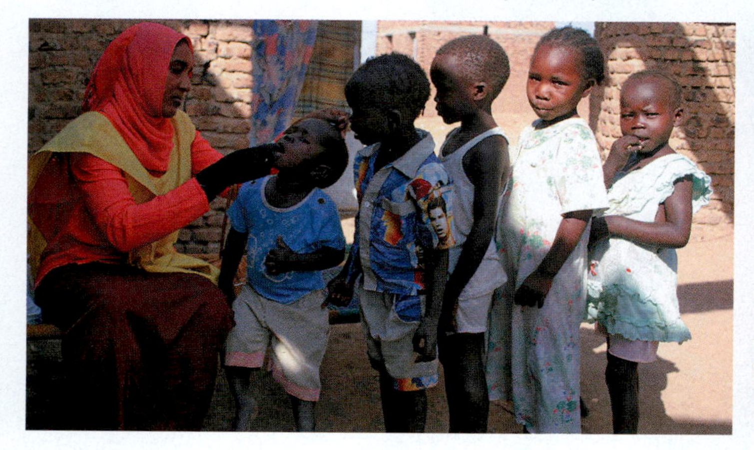

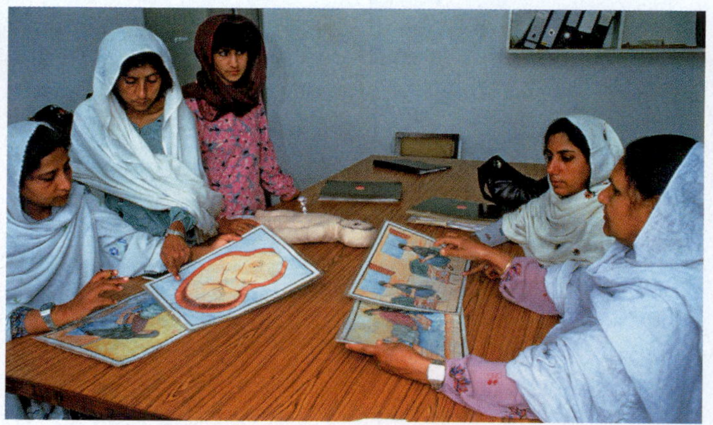

Sociodemographic Indicators

People are the most important resource a country has, as it is they who determine the use of the other resources—natural, manufactured, and creative—of the country. Thus, having a healthy, literate, and educated populace is an essential step toward successful development. **Sociodemographic indicators** provide information about the welfare of a population, including, for example, data on the prevalence of disease or the levels of literacy and education. The *literacy rate* is the percentage of a country's population over the age of 15 that can read and write. The literacy rate in developed regions exceeds 90%, but the rate in developing regions drops, on average, to about 60%. In five African countries and in Afghanistan the rate is less than 30%.

Sociodemographic indicators are complex and interrelated. For example, nutrition can impact health, which in turn, can affect the ability to work. Hunger and malnourishment are widespread issues that have the gravest consequences for children. Malnourishment affects nearly one-quarter of children under the age of five in developing countries. Other sociodemographic indicators include life expectancies and infant mortality rates (Chapter 3). Of the various sociodemographic indicators, infant mortality rates are considered the most telling because they so tragically highlight lost potential in a country's human resources (**Figure 9.4**).

Environmental Indicators

Compared to economic and sociodemographic indicators, the use of **environmental indicators** is a more recent phenomenon. Their development and use stem largely from the 1992 United Nations Conference on Environment and Development. Also known as the Earth Summit, this conference was held in Rio de Janeiro and grew out of widespread concern about the extent of global environmental problems, such as pollution and loss of biodiversity. The conference focused on making environmentally sustainable development a priority for all countries, rich and poor alike.

Agenda 21, the action plan that resulted from the Earth Summit, encouraged governments and other agencies to develop indicators that could be used to assess sustainable development. Since that time, hundreds of different indicators have been developed. Examples of environmental indicators include frequency of environmental hazards such as flooding, drought, and earthquakes; loss of biodiversity; and access to safe, potable water (**Figure 9.5**).

Development and Gender-Related Indexes

To what extent can the study of development be reduced to a consideration of one or more different indicators? Many development practitioners take the position that indicators alone are insufficient to gauge changes in development because development encompasses much more than just an increase in income, a rise in GDP per capita, or access to a safe water supply. Consequently, geographers and others are interested in combining a number of economic, sociodemographic, and environmental indicators in order to create development indexes. These indexes provide a broader, more inclusive

Access to clean water • Figure 9.5

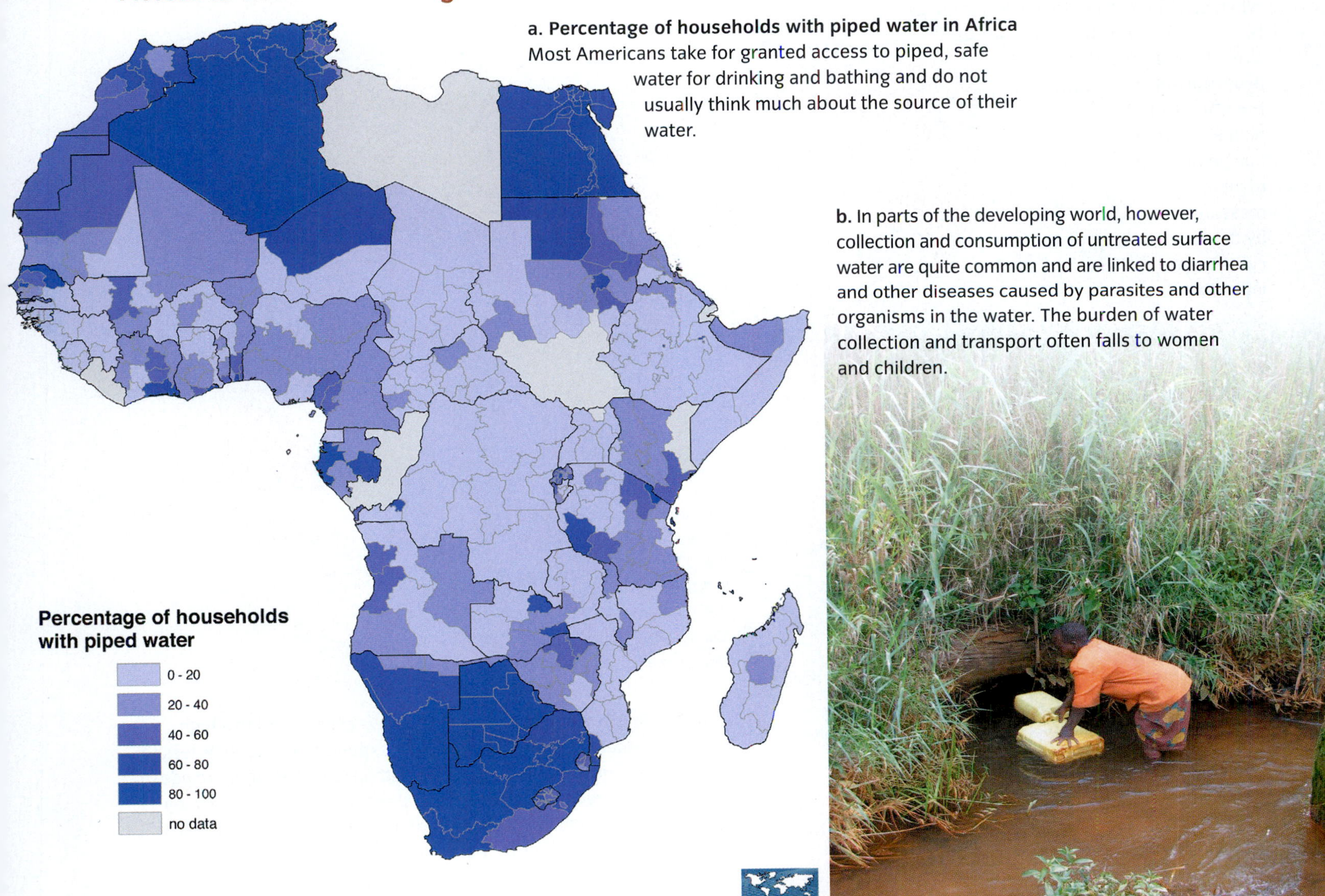

a. Percentage of households with piped water in Africa Most Americans take for granted access to piped, safe water for drinking and bathing and do not usually think much about the source of their water.

b. In parts of the developing world, however, collection and consumption of untreated surface water are quite common and are linked to diarrhea and other diseases caused by parasites and other organisms in the water. The burden of water collection and transport often falls to women and children.

assessment of a country's development. This section discusses three of these indexes: the human development index, the gender-related development index, and the gender empowerment index.

Created in 1990, the **human development index (HDI)** has since been adopted by the United Nations Development Programme (UNDP), the entity charged with measuring development worldwide and proposing strategies for improving it. Every year the UNDP produces the *Human Development Report*, a summary assessment of development in the world. In the words of Mahbub ul Haq, a Pakistani economist and one of the founders of UNDP:

> Human development is about . . . creating an environment in which people can develop their full potential and lead productive, creative lives in accord with their needs and interests. People are the real wealth of nations. Development is thus about expanding the choices people have to lead lives that they value. And it is thus about much more than economic growth, which is only a means—if a very important one—of enlarging people's choices. (UNDP, *The Human Development Concept*, http://hdr.undp.org/en/humandev/)

The HDI was the first development measure to incorporate information about the wealth, health, and education of a country's people in a single statistic. Four specific indicators are used to generate the HDI: the GDP (PPP) per capita, life expectancy, the adult literacy rate, and the gross enrollment ratio (the total enrollment in education as a percentage of the total school-age population) (**Figure 9.6**).

Human development index • Figure 9.6

When computed, the HDI yields a value between 0 and 1, which can be used as a relative measure of human development. The closer the HDI is to 1, the higher the level of human development as measured by the four component indicators.

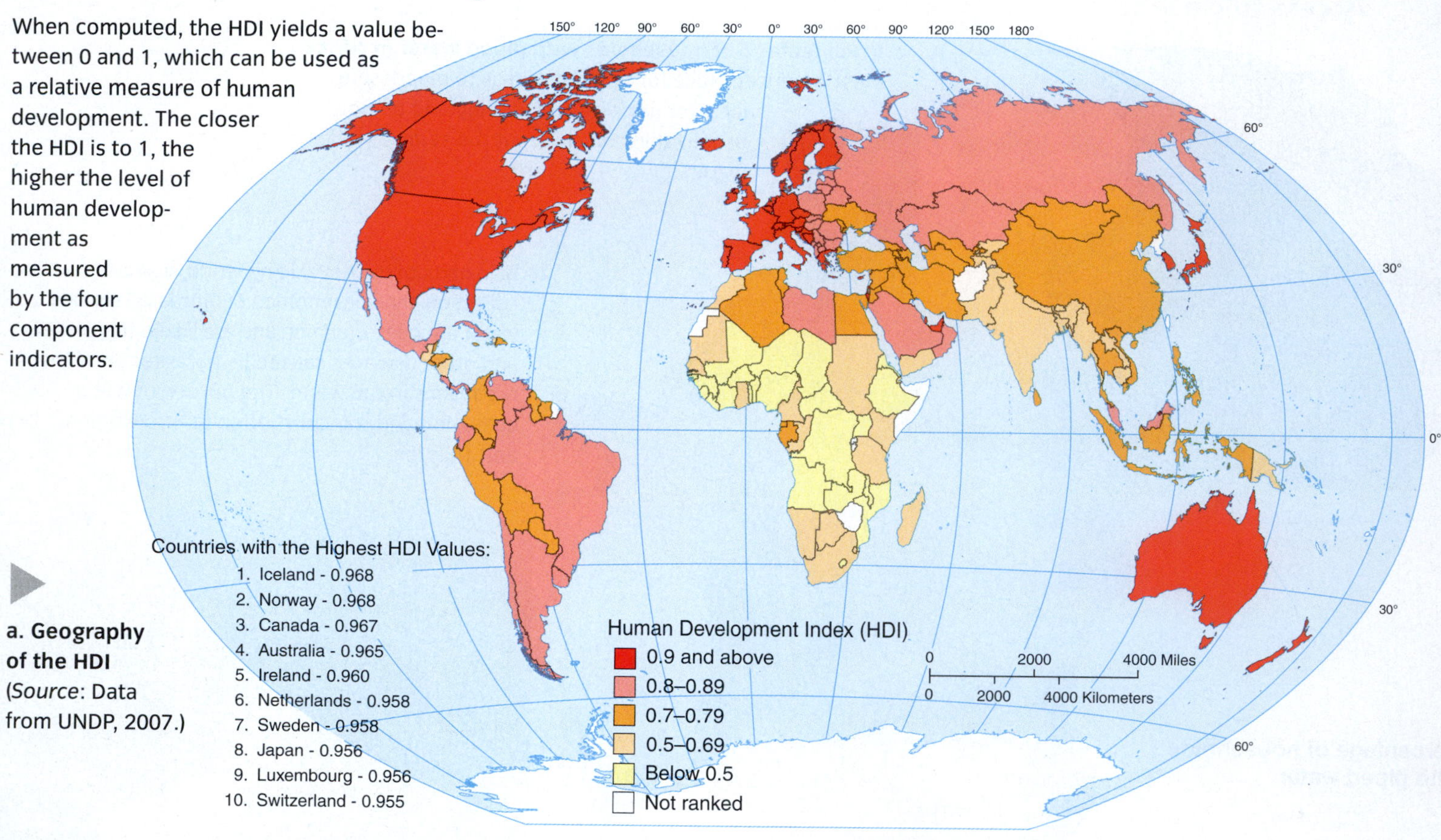

a. Geography of the HDI (*Source*: Data from UNDP, 2007.)

Country	Human Development Index (HDI)	GDP per capita (in PPP)
Cuba	0.855	$6,000
Botswana	0.664	$14,000

b. A high GDP per capita does not always correspond to a high HDI. In general, Cubans have high life expectancies, good access to education, and a government-financed health care system that provides coverage to all citizens. Despite its mineral resource wealth, Botswana has one of the lowest life expectancies of any country and low levels of literacy and education. (*Source*: Data from UNDP, 2007.)

Gender-related development index • Figure 9.7

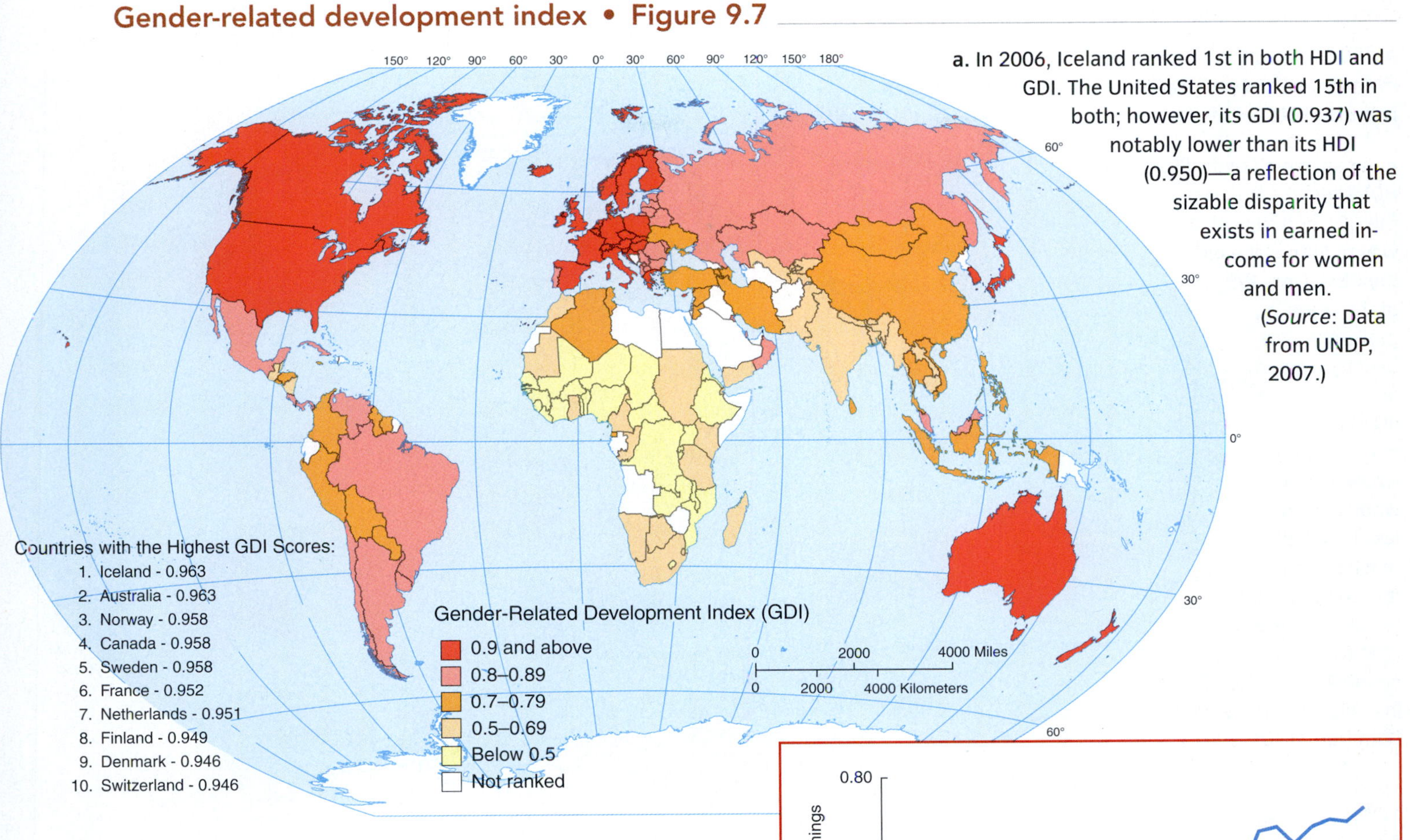

a. In 2006, Iceland ranked 1st in both HDI and GDI. The United States ranked 15th in both; however, its GDI (0.937) was notably lower than its HDI (0.950)—a reflection of the sizable disparity that exists in earned income for women and men. (*Source*: Data from UNDP, 2007.)

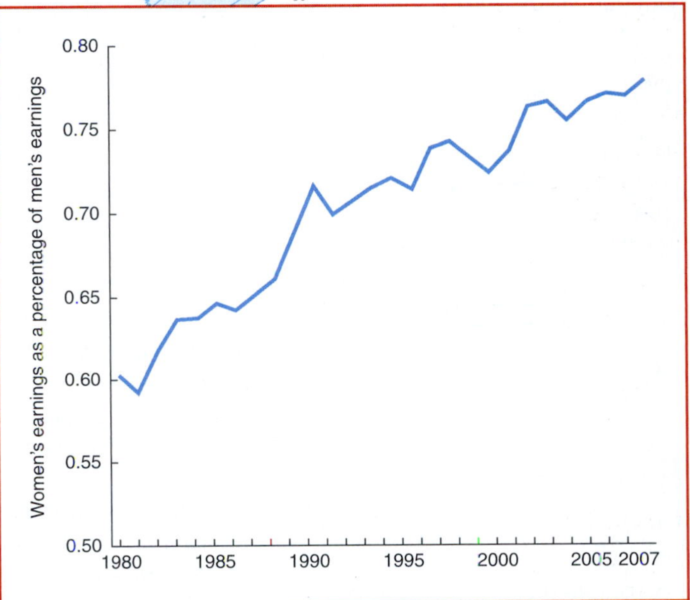

b. In the United States, women's earnings have improved; however, a woman earns just 77 cents for every dollar earned by a man. (*Source*: Data from U.S. Census Bureau.)

Prompted by the realization that development is not gender neutral—that the impacts of development affect women and men differently—the UNDP has called for the "engendering of development" since 1995. As stated in the *Human Development Report* for that year, "human development, if not engendered, is endangered." That is, if human development centers on improving and expanding people's choices, then the options, opportunities, and rights of women and men must be considered a crucial component of a society's overall development.

To examine gender disparities in development, the UNDP has produced and used two other development indexes: the gender-related development index and the gender empowerment measure. The **gender-related development index (GDI)** is the HDI adjusted to account for and, in effect, penalize countries that have wide disparities in achievement between women and men, or in which achievements for both women and men decline. To account for wealth, however, the GDI uses the estimated earned income for women and men instead of GDP per capita. A GDI value of 1 would indicate that a country has the highest possible attainments and gender equality in each of the four areas (life expectancy, adult literacy, gross enrollment ratio, and earned income). Countries rank consistently lower on the GDI than the HDI (**Figure 9.7**).

Gender empowerment measure • Figure 9.8

A GEM close to 1 indicates near parity between women and men in economic and political participation. Nine of the ten highest GEMs are in western Europe.

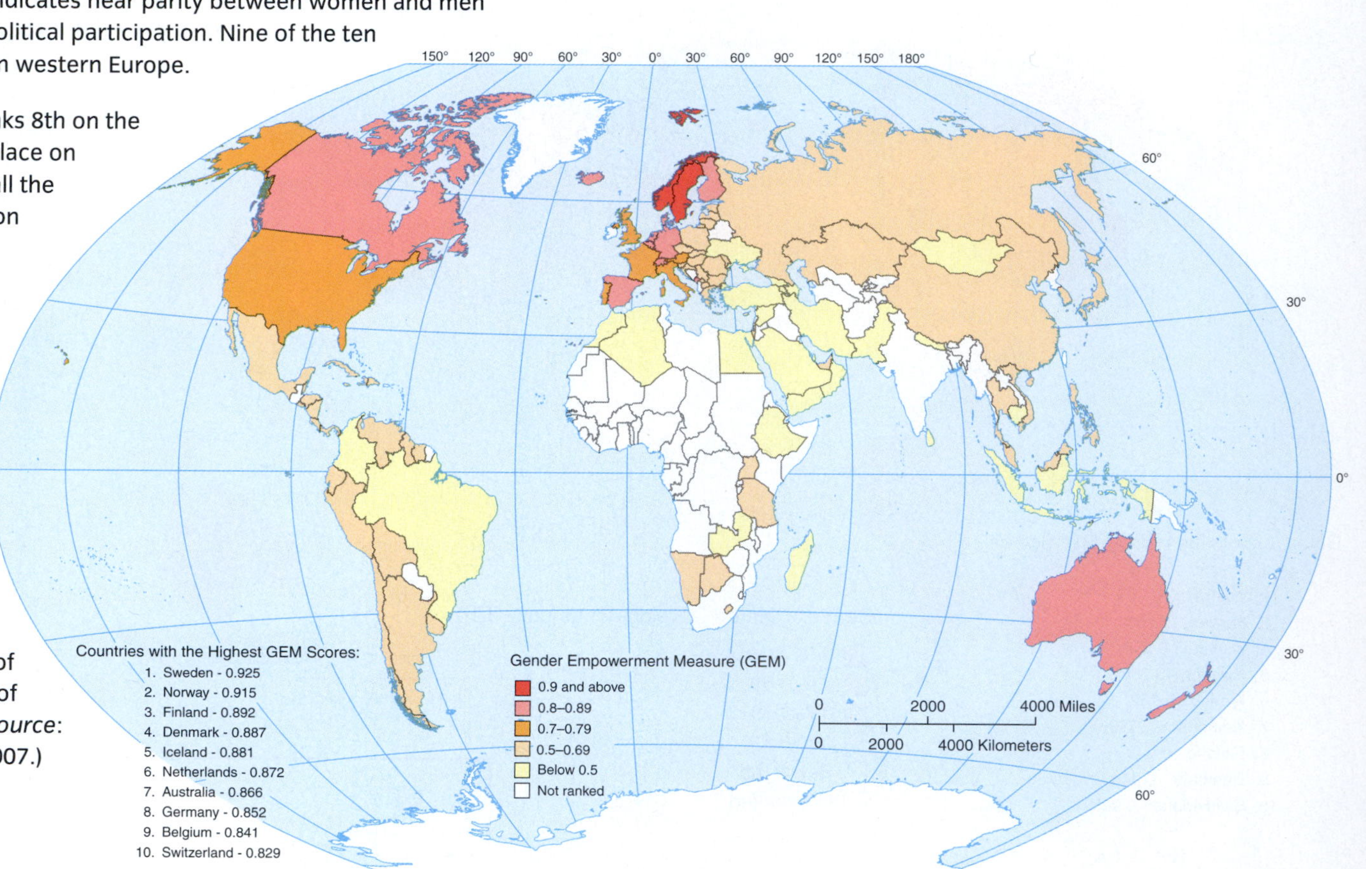

a. Japan, which ranks 8th on the HDI, slips to 12th place on the GDI, and slips all the way to 58th place on the GEM. Only 12% of the seats in parliament are held by women, only 10% of the legislators and managers are women, and women earn less than half of what men do. It is not possible to calculate the GEM for a number of countries because of insufficient data. (*Source*: Data from UNDP, 2007.)

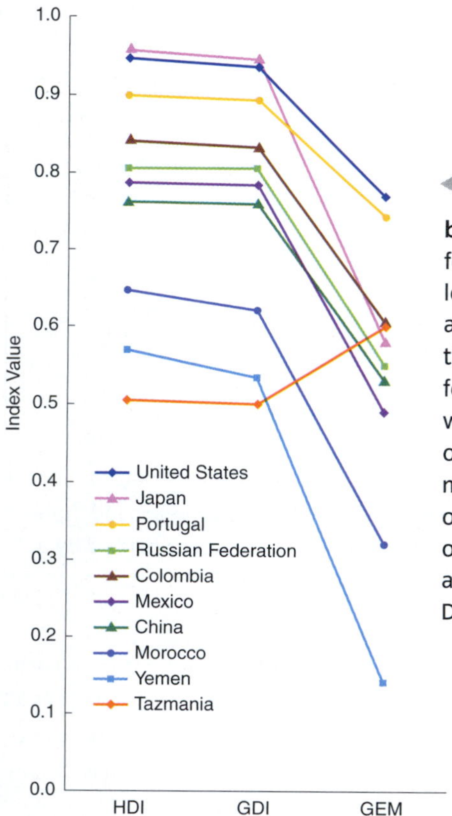

b. The value of the GEM for most countries is lower—often considerably lower—than either the GDI or HDI. Not so for Tanzania, however, where women hold 30% of the seats in parliament, and almost half of all legislators, senior officials, and managers are women. (*Source*: Data from UNDP, 2007.)

c. The use of gender quotas for political participation provides a partial explanation for the high GEMs in western Europe and Tanzania. Some form of voluntary or compulsory gender quotas exist in about half of the countries in the world. These women in New Delhi, India demonstrate in favor of legislation that would reserve 33% of seats in Parliament and state assemblies for women.

Unlike the GDI, the **gender empowerment measure (GEM)** assesses the extent to which women participate in the economic and political decision-making within a country. It, too, is derived from four indicators. Economic participation and empowerment are gauged by two indicators: estimated earned income for women and men and the proportion of women and men in professional and technical jobs. Political participation and empowerment are gauged by the proportion of women and men working as legislators and managers and the proportion of women and men who hold seats in parliament. One of the limitations of the GEM is that it does not reflect or consider empowerment at the household level (**Figure 9.8**). Also see *Where Geographers Click*.

Where Geographers CLICK

Human Development Reports

http://hdr.undp.org/en/statistics/data/hd_map/

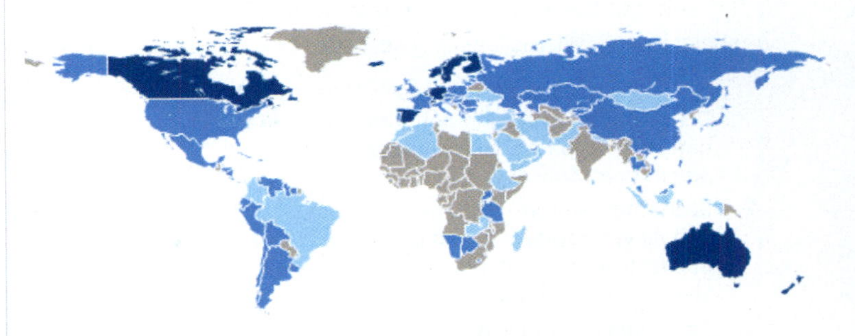

This site provides a very good place to learn more about measures of human development. From here it is possible to view additional maps and access data for specific countries.

Environmental Indexes, Vulnerability, and Development

Thus far we have examined a variety of different indicators and indexes that help to measure development, but what causes these different levels of development? The 18th-century economist Adam Smith suggested that the physical geography of an area influenced its economic growth and development. Similarly, during the 19th and early 20th centuries, geographers who held environmentally deterministic views argued that geographic factors explained why some countries were economically advanced and others were not.

Today, however, one explanation is that differences in development are the result of many diverse and interconnected conditions that fall into two broad categories: geographic conditions and institutional conditions. Thus, a more complete understanding of the development experience of a country or region calls for a consideration of its geographical endowment. Some significant components of a country's geographic endowment are shown in **Table 9.1**. In Sub-Saharan Africa, a combination of difficult or adverse conditions works to complicate the development process. For example, a number of countries are landlocked, have poor soils, and have a heavy disease burden associated with both malaria and HIV/AIDS.

Selected geographic conditions that may affect development Table 9.1

Situational	Landlocked state Small, isolated island state Limited natural resource base
Transport	Distant from major markets Population distant from coasts or navigable rivers Large population living in mountainous areas
Agroclimatic	Low and/or highly variable rainfall Poor or depleted soils
Health/Disease	High prevalence of tropical diseases High prevalence of HIV/AIDS
Disaster Vulnerability	High susceptibility to natural disasters (e.g., hurricanes, earthquakes, floods) Large population living in disaster-vulnerable regions

Geography InSight

The Environmental Vulnerability Index • Figure 9.9

The idea for the EVI was initially conceived in the late 1990s and grew out of discussions of sustainable development and impacts of climate change on small island developing countries—for example, Fiji. The index has much broader utility, however, and has been applied to most other countries.

a. The EVI incorporates data from 50 different indicators, although just 16 are shown here. Some of these indicators relate to the physical environment, but several involve institutional or social issues. Why is tracking sea temperatures especially important to a small island state?

1. Dry periods—average annual rainfall deficits over past 5 years in relation to 30-year monthly average
2. Sea temperatures—average annual temperature deviation in past 5 years in relation to 30-year monthly mean
3. Tsunamis—number with storm surge 2 m or more above high tide per 1000 km of coastline since 1900
4. Country dispersion—ratio of land and maritime borders to total land area
5. Borders—number of land and sea borders shared with other countries
6. Endemics—number of endemic species per million square km land area
7. Endangered species—number of endangered and vulnerable species per 1000 square km land area
8. Environmental openness—average annual freight imports over past 5 years per square km land area
9. Intensive farming—average tonnage of intensively farmed animal products over past 5 years per square km land area
10. Pesticides—average annual pesticide use over total land area over past 5 years
11. Mining—average annual mining production per square km land area over past 5 years
12. Spills—number of oil and hazardous substance spills of 1000 liters or more per million km maritime coast in past 5 years
13. Vehicles—number of vehicles per square km land area
14. Population growth—average annual growth rate over past 5 years
15. Environmental agreements—number of environmental treaties in force in a country
16. Conflicts—average number of conflict years per decade over past 50 years

Environmental Vulnerability Index
Resilient
At Risk
Vulnerable
Highly Vulnerable
Extremely Vulnerable

(*Source*: United Nations Environment Program and South Pacific Applied Geoscience Commission, 2005.)

b. Computing the EVI for a country yields a numeric score ranging between 174 (resilient) and 450 (extremely vulnerable). How does Bhutan, discussed in the chapter opener, fare? No index is perfect or provides a universal solution to the complex issue of development. Rather, indexes serve as tools to help make decisions about human and environmental resource management.

c. Drilling down into the EVI for the United States shows that environmental openness is one area the United States fares poorly in because of its high freight imports. The recent introduction of zebra mussels, shown clinging to a pier from Lake Erie, illustrates this. Zebra mussels were introduced via the ballast water of transatlantic ships and now threaten the ecosystem of the upper Mississippi River and Great Lakes.

Think Critically What criteria might be used to measure the resilience of a country or population?

Geographic conditions, however, are just one part of the development picture. Institutional conditions also have a bearing on development. For example, authoritarian rule can stifle debate about human and environmental well-being, and corruption can result in mismanagement of financial or natural resources. Similarly, discrimination against ethnic or minority groups can marginalize or entirely exclude them from access to public services, such as education or health care. We should realize that linkages also exist between geographic and institutional conditions. In some cases, the development process itself can lead to increased vulnerability to natural disasters. Rapid urbanization, for example, can lead to the construction of housing developments on unstable slopes or in floodplains.

Today, another facet of sustainable development involves examining vulnerability and resilience. *Vulnerability* refers to how prone a country or group is to suffering damage from economic, environmental, or other shocks, whereas *resilience* refers to the ability to withstand or resist those shocks. Because vulnerability and resilience are multifaceted conditions, indexes are helpful in evaluating them. One such index is the environmental vulnerability index (EVI) (**Figure 9.9**).

CONCEPT CHECK

STOP

1. **Why** is development a normative project?
2. **How** is a development indicator different from a development index?
3. **How** is it possible for a country to have a high HDI but also a low GDI or GEM?
4. **How** can geographic and institutional conditions affect development?

Development and Income Inequality

LEARNING OBJECTIVES

1. **Distinguish** between income distribution and income inequality.
2. **Describe** techniques for measuring and mapping income inequality.
3. **Identify** factors that can affect income distribution.
4. **Summarize** opposing views on the relationship between globalization and income inequality.

Have you ever thought about how income is distributed among the employees of a company or between men and women? In our earlier discussion of the gender-related development index, we touched briefly on the difference between what women and men earn in the United States. Unless income is equally distributed, there will be a gap between those who earn more and those who earn less.

We have already explored some dimensions of the geography of poverty, but tracking **income distribution** and **income inequality** is also important. Although it is possible to use the average income of a country to get an overall picture of the presence or absence of poverty, information about income distribution and income inequality is more revealing because this kind of data shows the share of the income held by the richest and poorest groups.

income distribution How income is divided among different groups or individuals.

income inequality A ratio of the earnings of the richest to the earnings of the poorest.

Development geographers examine income distribution and income inequality at different scales and between different clusters of countries. The **Organisation for Economic Co-operation and Development (OECD)** is one such cluster. Established in 1961, the OECD was created to enhance development by promoting economic growth and improving the standard of living of its member

OECD members • Figure 9.10

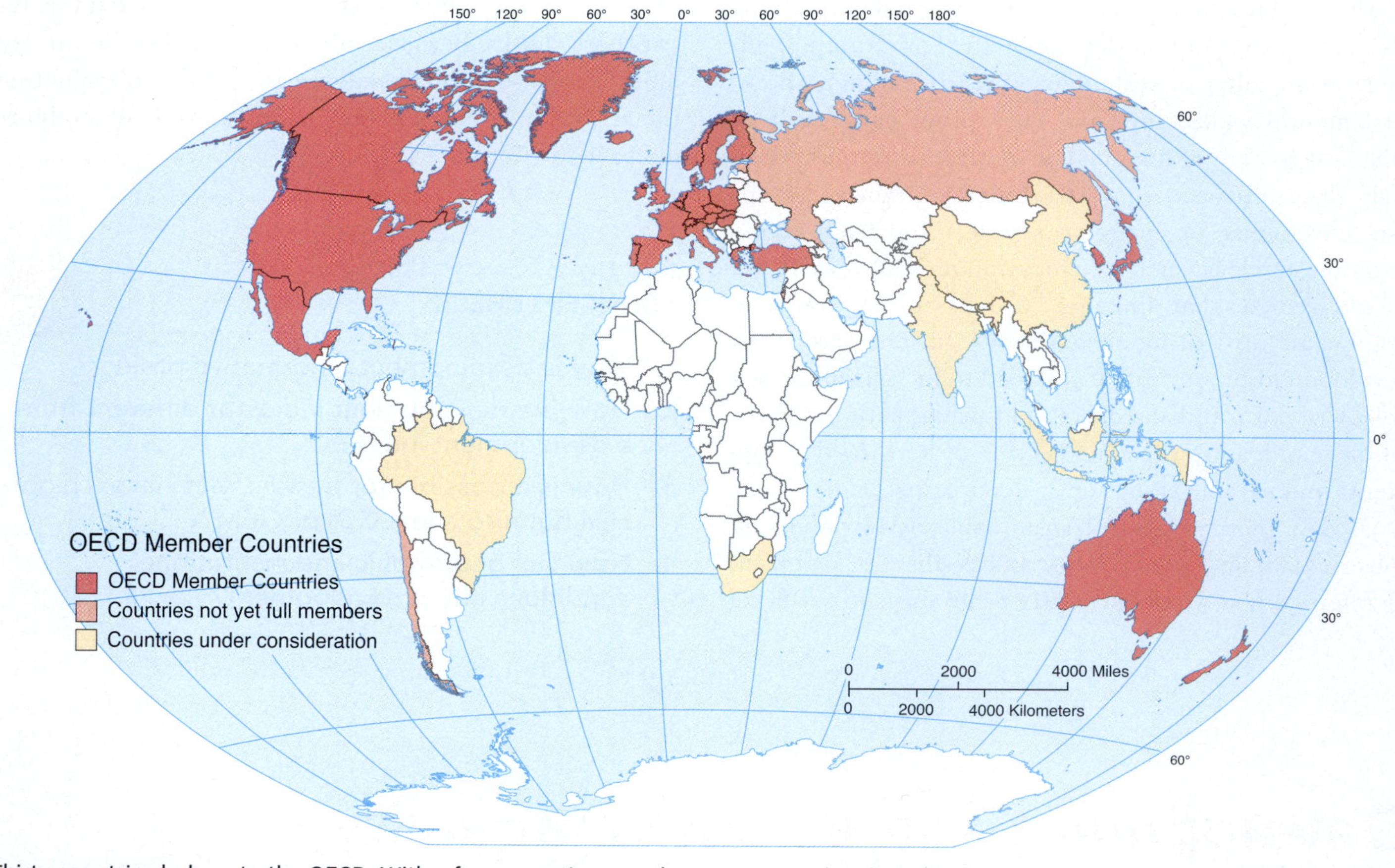

Thirty countries belong to the OECD. With a few exceptions, such as Mexico and Turkey, OECD members are very wealthy, industrialized countries. Thus, the OECD is sometimes called the "rich country club of the world." How does the planned expansion of OECD membership challenge the global North-South divide illustrated in Figure 9.1?

countries. Since that time, its mission has expanded to include financial and other forms of assistance for developing countries that are not OECD members (**Figure 9.10**).

The Gap Between the Rich and the Poor

At the global scale, income inequality is very high. One of the most comprehensive studies of income distribution to date shows the concentration of large shares of income among the elites who form a minority of the population. This phenomenon has a *champagne-glass effect* on the pattern of income distribution (**Figure 9.11**).

Is income inequality increasing or decreasing? No clear consensus exists within the research community about these trends, in part because there is considerable variation in the quality and availability of the data to measure this and also because it depends on the period of time as well as the countries being examined. Studies of eastern and central Europe show that income inequality increased by more than 40% between 1970 and 2000. The transition from communism to capitalism helps explain this increase.

In the United States there is also evidence that income inequality has been rising. At the start of 2007, the Federal Reserve Chairman, Ben S. Bernanke, drew attention to the gap between the rich and the poor, noting that

> . . . the share of income received by households in the top fifth of the income distribution . . . rose from 42 percent in 1979 to 50 percent in 2004, while the share of income received by those in the bottom fifth of the distribution declined from 7 percent to 5 percent. (Bernanke, 2007)

Global income inequality • Figure 9.11

This series of graphs examines income distribution and inequality first at the global scale and then at the regional scale. Household surveys conducted by the various countries, usually as a part of their population census, provide the data on income.

a. This chart depicts the distribution of income in five classes, from the richest to the poorest, of the world's people. In 2000, the most recent year for which comparable data exist, the richest 20% of the world's population held 74% of the income in the world. Conversely, the poorest 20% held a mere 1.5% of global income, creating the champagne-glass effect. (*Source*: UNDP, 2005.)

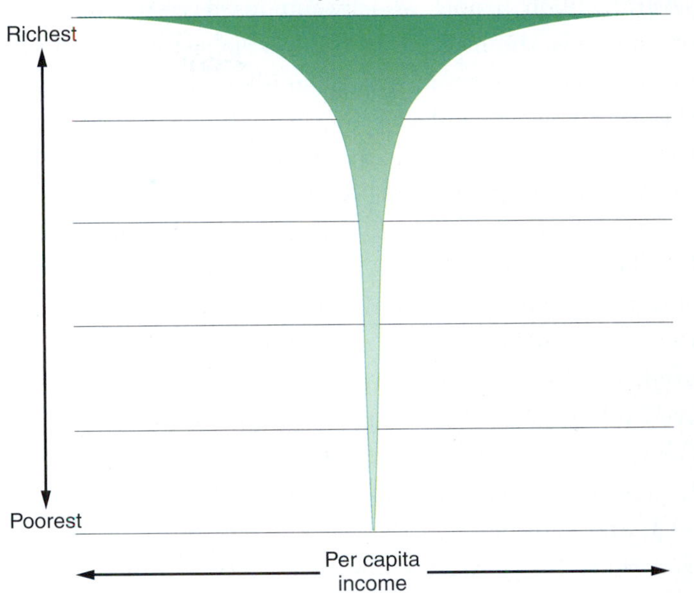

b. The widest gap exists in Africa, but sizable divides separate the richest and poorest groups in the various world regions and the OECD. Income inequality is powerfully captured through juxtapositions, such as that shown in this photo of a homeless man, his belongings in tow, in front of the U.S. Treasury Department.

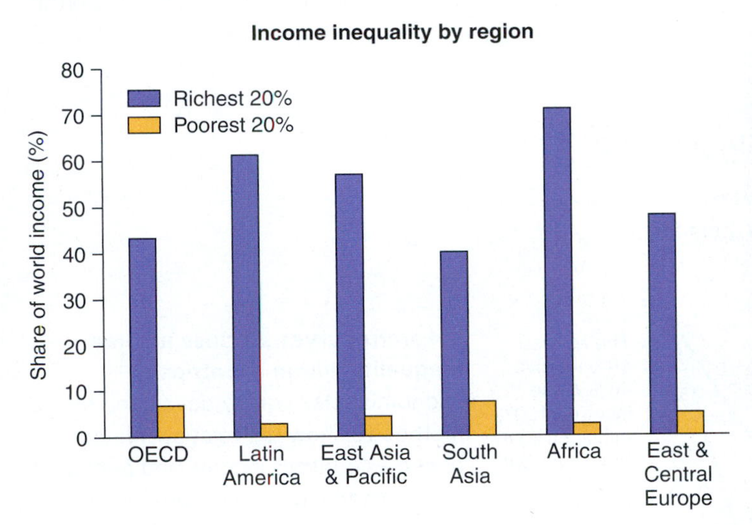

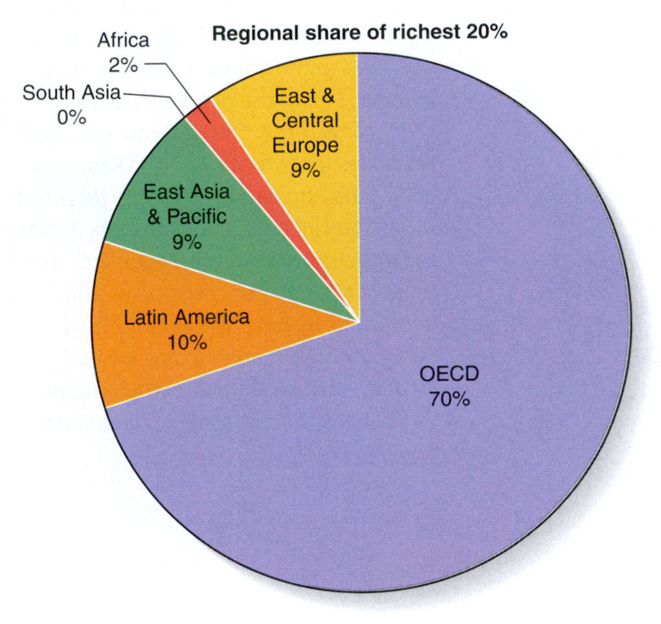

c. No other group of countries in the world has such a large share of the global income as the OECD countries do. For every $100 held by the richest 20% of the world's population, $70 of those are in OECD countries; $28 are split among the regions of Latin America, East Asia and the Pacific, and eastern and central Europe; and $2 are in Africa. (*Source*: Data from Dikhanov, 2005.)

Globally, one fact about income distribution remains clear: there are a few very, very wealthy individuals but several billion people who live in poverty. According to one analysis, the incomes of the 500 richest people in the world exceed the incomes of 400 million others. For some, this is evidence enough that we should be concerned about the patterns of income inequality.

When geographers measure income inequality, one statistic they often use is the Gini coefficient. Values for the Gini coefficient range between 0 and 100; the closer to 0, the more equally income is distributed, and the closer to 100, the more unequally income is distributed. The lowest Gini coefficients recorded in the world are 24.7 and 24.9, for Denmark and Japan, respectively. For the world as a whole, the Gini coefficient is 67. The United States has a Gini coefficient of 40.8, Brazil's is 59.3, and Namibia's is 70.7. This latter figure indicates that there is greater income inequality within Namibia than there is in the world. In fact, the richest 20% in Namibia hold nearly 80% of the country's income. On this basis, Namibia proves the exception rather than the rule because globally the trend is that greater income inequality exists between countries rather than within them. **Figure 9.12** explains how Gini coefficients are derived and shows their global variation.

Factors Affecting Income Distribution

There is no consensus on the causes of income inequality. What analysts do agree on is that multiple, often interrelated factors affect the equality or inequality of income distribution. The factors affecting income distribution can be grouped into four categories: individual, social, policy-related, and historical. Broadly speaking, what all of these factors have in common is that they contribute to circumstances that affect the ability of people to earn or to save money.

Individual factors that affect income distribution include the skills and abilities each one of us possesses as well as the attitudes we have about work and leisure. *Social factors*

Measuring and mapping income inequality • Figure 9.12

The Gini coefficient helps compare income inequality among countries. Since the Gini coefficient is derived from Lorenz curves, we discuss them first.

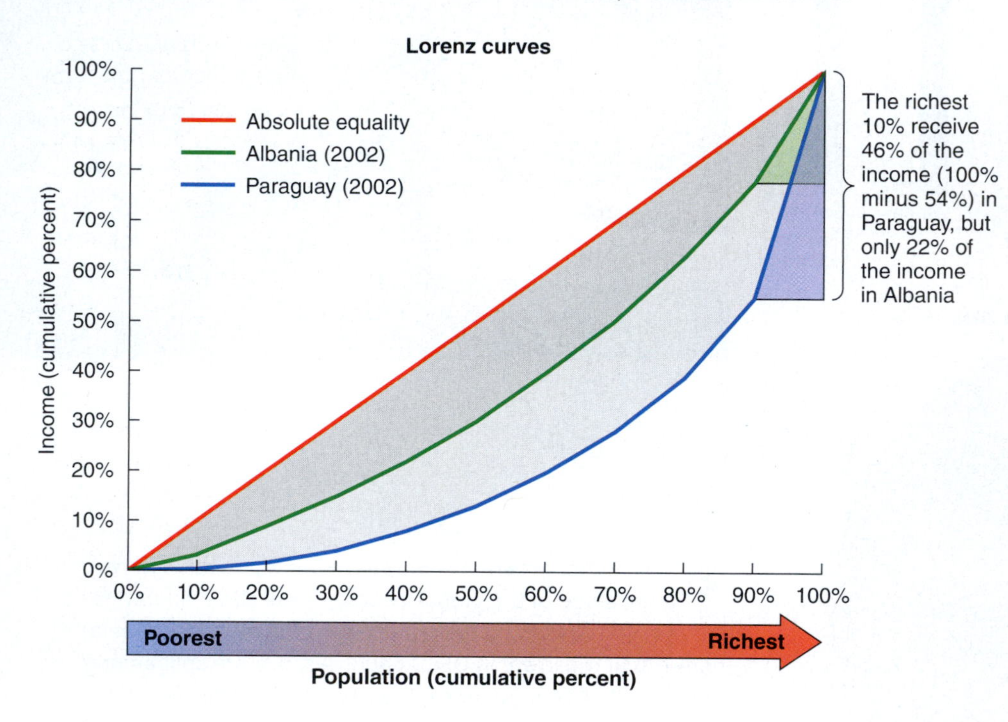

a. Lorenz curves disclose income inequality within countries
Economist Max Lorenz developed this method to show the relationship between a country's income and population, broken into deciles (groups of 10%). The red curve is a baseline showing an idealized equal-income distribution (10% of the people have 10% of the income, 50% of the people have 50% of the income, etc.). The other curves graph examples for two countries. In Paraguay, 40% of the population holds *less than 10%* of the total income, but in Albania, 40% of the population holds *more than 20%*. The "lower" a country's Lorenz curve, the more unequal its income distribution. Thus, Paraguay has greater income inequality than Albania. (*Source*: Data from the World Bank, http://iresearch.worldbank.org/PovcalNet/.)

refer to conditions or circumstances within society at large that introduce different kinds of pressures. For example, the social pressure to be successful can affect the distribution of income. Even within different kinds of employment, the social expectations related to work can vary tremendously. The age of a country's work force is also an important social factor.

Policy-related factors are numerous and often reveal the priorities of a particular country or society. These include policies affecting taxation, international trade, labor, immigration, and education. Some scholars point to the decline in unionism as one factor associated with an increase in income inequality in the United States. *Historical factors* are often closely connected with social and policy-related factors and include events that have had a significant impact on the structure of a country's society and economy. Slavery, colonization, warfare, and internal unrest are some examples of historical factors that can affect income distribution. There is considerable income inequality in Latin America, a region long affected by both slavery and colonization. In Brazil, for example, the richest 10% of the population holds almost half of the income while the poorest 10% holds less than 1% of the income.

Globalization and Income Distribution

Many geographers and others are asking questions about the impact of globalization on the distribution of income. To what extent has globalization been the world's great equalizer? There are two very different schools of thought on this: a trickle-down theory and the theory of the widening gap between the rich and the poor.

Proponents of the trickle-down theory of globalization argue that income convergence or equality follows trade. Historically, they point to the convergence in incomes between the United States and Europe that occurred in the late 19th century as trade expanded. In other words, trade is essential because it leads to specialization, increased competition, and rising prosperity. For these trickle-down effects to occur, barriers to trade need to be removed.

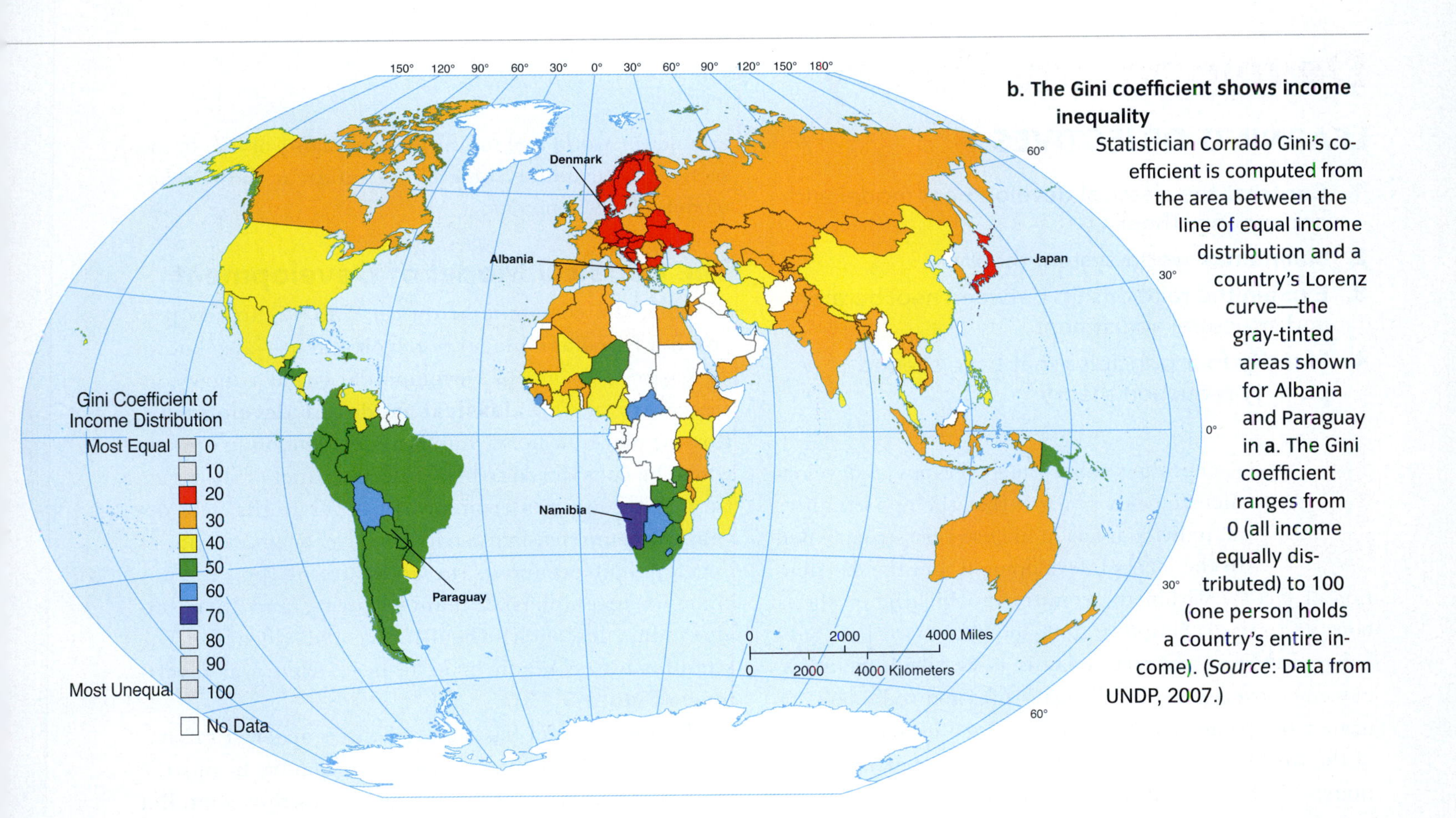

b. The Gini coefficient shows income inequality
Statistician Corrado Gini's coefficient is computed from the area between the line of equal income distribution and a country's Lorenz curve—the gray-tinted areas shown for Albania and Paraguay in **a**. The Gini coefficient ranges from 0 (all income equally distributed) to 100 (one person holds a country's entire income). (*Source*: Data from UNDP, 2007.)

In contrast, proponents of the theory of the widening gap between the rich and the poor argue that globalization works against a level playing field. Instead, it creates a higher demand for skilled workers, placing a premium on those with a college education or graduate degree. The more skilled the labor force, the higher the earning potential of that labor force. Those who lack the skills and education are more likely to find themselves left behind, unable to gain access to or compete for jobs. In some circumstances, globalization can create unemployment, which in turn, affects the distribution of income.

No matter which school of thought you subscribe to, the degree of income inequality in a country can have a number of serious consequences for development. First, income inequality is associated with a higher incidence of poverty, which in turn, can deter investment and development. Second, income inequality can exacerbate tensions between the rich and the poor, upsetting the political and social stability of a country. This unrest can derail economic growth and development. Third, and perhaps most importantly, when high income inequality coincides with unemployment, countries squander valuable human resources that could be used for the betterment of the society and economy.

CONCEPT CHECK STOP

1. **How** are income distribution and income inequality related?
2. **What** does a Lorenz curve show?
3. **What** are some factors that can affect income distribution?
4. **What** are two perspectives on the relationship between globalization and income inequality? Explain them.

Development Theory

LEARNING OBJECTIVES

1. **Contrast** the classical development model and dependency theory.
2. **Summarize** world-system theory.
3. **Explain** the relationship between neoliberalism and structural adjustment.
4. **Identify** four principles that have shaped poverty-reduction theory.

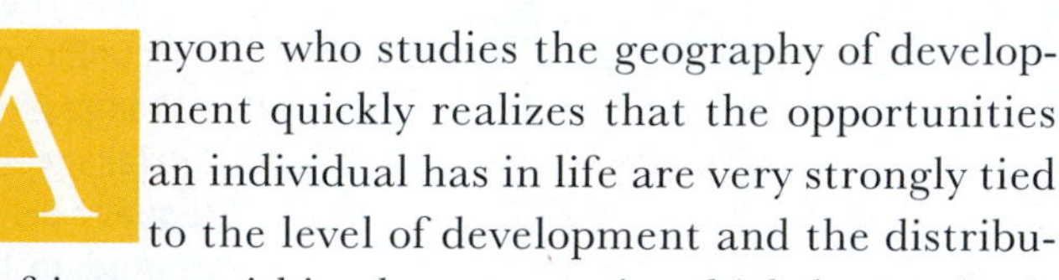

Anyone who studies the geography of development quickly realizes that the opportunities an individual has in life are very strongly tied to the level of development and the distribution of income within the country in which he or she is born and raised. Simply stated, someone born in a Sub-Saharan African country today is more likely to live in extreme poverty, to live a shorter life, and to be less educated than someone born in a country in another part of the world. This reality challenges development practitioners to think about ways of raising standards of living around the world and improving the quality of life. In this section, we examine some of the different theories and models of development.

The Classical Model of Development

In 1960, the economic historian Walt W. Rostow proposed a five-stage model of development—also known as Rostow's Stages of Development—that is now considered to be the **classical model of development**. Rostow placed a heavy emphasis on economic growth, which he considered to be a direct product of the structure of a country's economy. In other words, less developed countries had agricultural economies and, as development occurred, the structure of the economy changed to emphasize manufacturing and then service industries, such as health care and education. The stimulus for economic growth, in his view, was investment (**Figure 9.13**).

Rostow's model has suffered three major criticisms. First, the model assumes that every country begins the process of development from the same starting point. But

PROCESS DIAGRAM

THE PLANNER

Classical model of development • Figure 9.13

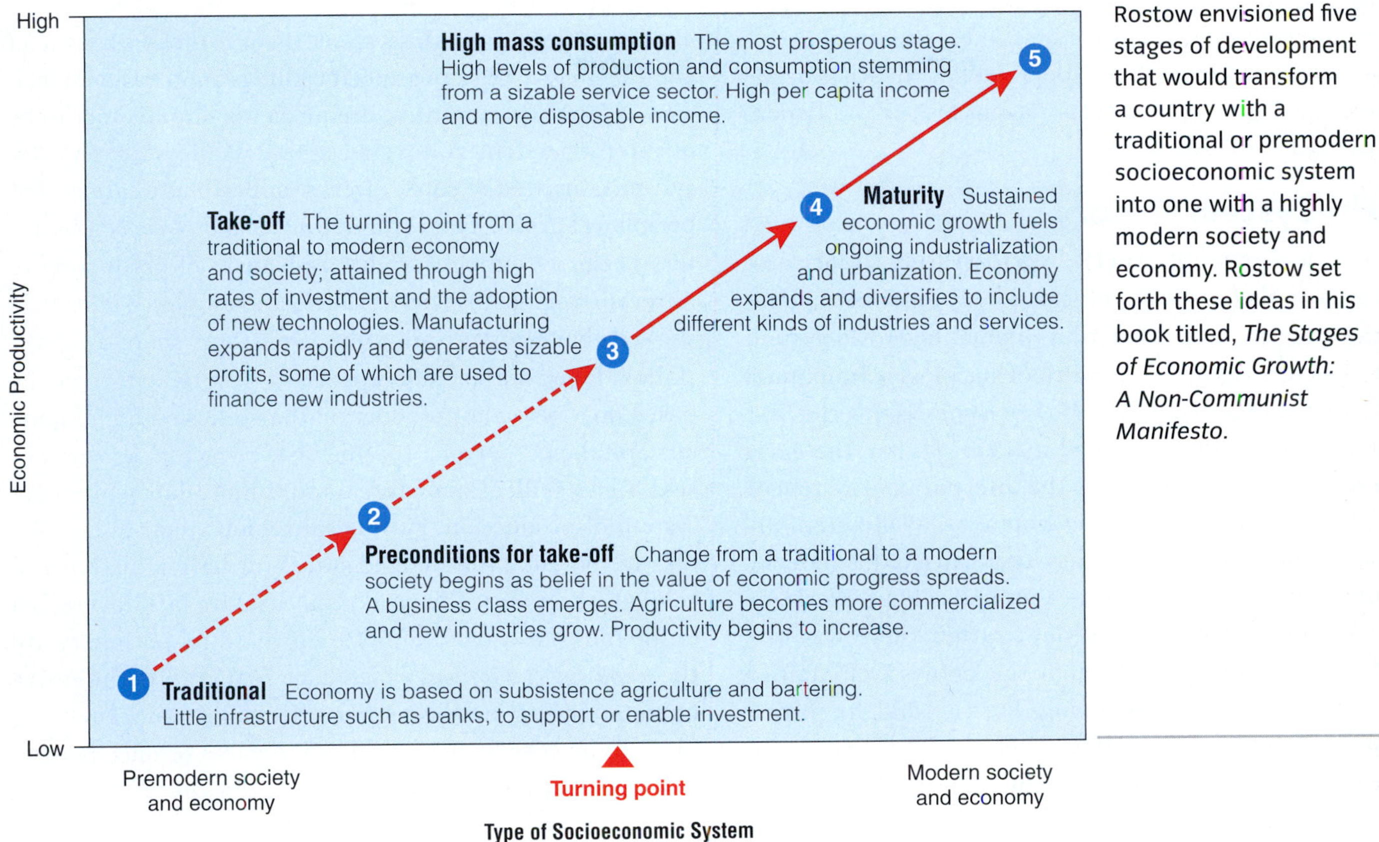

High
Low
Economic Productivity
High mass consumption The most prosperous stage. High levels of production and consumption stemming from a sizable service sector. High per capita income and more disposable income.
5
Maturity Sustained economic growth fuels ongoing industrialization and urbanization. Economy expands and diversifies to include different kinds of industries and services.
4
Take-off The turning point from a traditional to modern economy and society; attained through high rates of investment and the adoption of new technologies. Manufacturing expands rapidly and generates sizable profits, some of which are used to finance new industries.
3
2
Preconditions for take-off Change from a traditional to a modern society begins as belief in the value of economic progress spreads. A business class emerges. Agriculture becomes more commercialized and new industries grow. Productivity begins to increase.
1
Traditional Economy is based on subsistence agriculture and bartering. Little infrastructure such as banks, to support or enable investment.
Premodern society and economy
Turning point
Modern society and economy
Type of Socioeconomic System

Rostow envisioned five stages of development that would transform a country with a traditional or premodern socioeconomic system into one with a highly modern society and economy. Rostow set forth these ideas in his book titled, *The Stages of Economic Growth: A Non-Communist Manifesto.*

has Peru, for example, faced the same conditions as the United Kingdom? Critics note that Rostow's model does not effectively account for such economic differences. A second criticism of Rostow's model is that it works from a very narrow understanding of development with a singular focus on a pattern of linear economic growth. The model fails to consider that receipt of monetary aid from another country might stimulate economic growth in the short term but can also result in high levels of debt that stifle economic growth over the long term. Finally, the third major criticism of the model is that it is strongly Eurocentric in the way it envisions that development will yield a technologically advanced and modernized Western society. Moreover, Rostow's model assumes that what worked for the West will necessarily work for non-Western countries as well.

Dependency Theory

Rostow's Stages of Development were so controversial that several alternative theories of development were proposed. In the 1960s and 1970s, a school of thought known as **dependency theory** gained prominence. Dependency theorists argued that development might be better understood as a relational process rather than a series of stages, and that this process was linked to international trade.

Studying the system of international trade revealed the existence of two kinds of states: dominant and dependent. *Dominant states* are the most developed countries—the industrialized states of Europe, North America, and Japan—that command the economic resources and power to shape the policies and practices of international trade. *Dependent states* lack these economic resources and power. They represent the developing countries.

Dependency is, therefore, a condition that stems from patterns of international trade and results in underdevelopment, or low levels of development. According to dependency theorists, for example, the development of Europe resulted in the dependency and underdevelopment of Africa and Latin America. Contrary to Rostow's model, as Europe grew more developed and richer, development in Africa and Latin America was stymied. Dependency theorists argue that imperialism played a crucial role in this process because it helped shape a pattern of international relations built on a system of domination and dependency. Some dependency theorists argue that today's multinational corporations also create dependency (**Figure 9.14**).

World-System Theory

For some development experts, **world-system theory** provides a more theoretically informed body of ideas explaining dependency and underdevelopment. World-system theory has its origins in the work of sociologist Immanuel Wallerstein and his book, *The Modern World-System*, the first volume of which was published in 1974. Unlike the early dependency theorists, who saw the international system of trade as the cause of underdevelopment, Wallerstein argued that the capitalist world-system caused dependency and underdevelopment. There was not a First World or Third World in Wallerstein's view; rather there was one world connected by and through the network of capitalism—in effect, a world economy. This is what he meant by the term *world-system*. Wallerstein traced the emergence of the capitalist world-system to 16th century Europe and argued that during the 20th century, the system became fully global, with capitalist markets reaching and integrating all parts of the world.

According to world-system theory, the operation of capitalism gives rise to a specific kind of international division of labor that, in turn, creates a geographic hierarchy of interdependent states or regions. Wallerstein's world-system consisted of core states, semiperipheral areas, and peripheral areas. The term *international division of labor* refers to the assignment of different tasks of production to different regions of the world. For example, Wallerstein showed that *core states* are militarily strong, share a highly-skilled labor force, and possess a diversified economy based on a system of production that relies on high inputs of capital per person. In contrast, *peripheral regions* possess a less-skilled labor force and a more labor-intensive system of production. Peripheral regions tend to be colonies rather than full-fledged states (or have a history as a colony), are politically weak, and usually lack diversified economies. Positioned in between core and periphery are the *semiperipheral regions*, which have some capital-intensive manufacturing and economic diversification. They are likely to be core states that are in decline or once peripheral regions that are now on the rise (**Figure 9.15**).

Dependency theory • Figure 9.14

Dependency can develop through imperial dominance or financial-technological dominance. Dependency theory has also been heavily criticized for encouraging a simplistic view of international relations, for neglecting to consider the role of local politics and social classes in shaping development, and for treating dependency as a natural outcome of international relations.

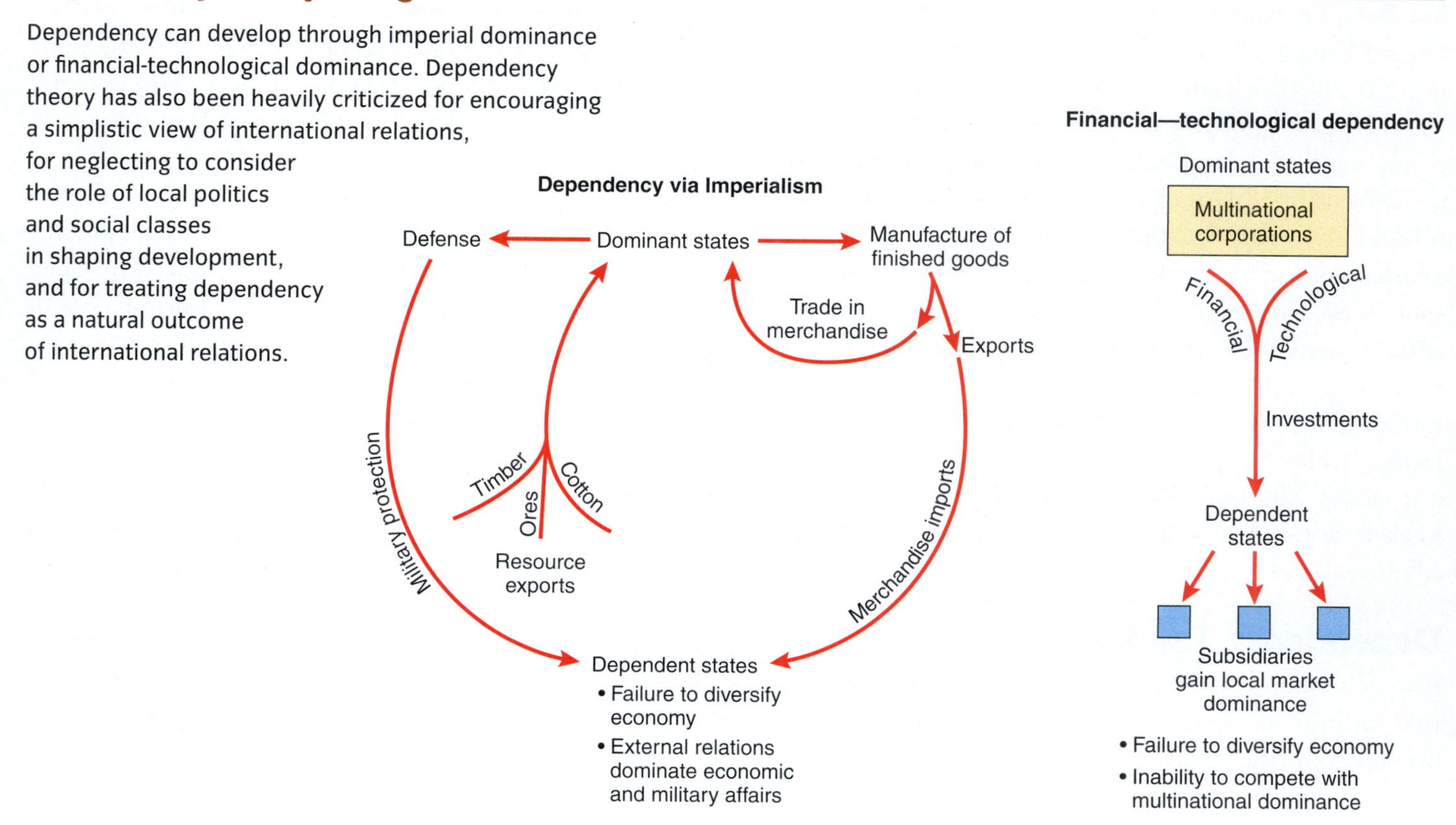

The world-system then and now • Figure 9.15

a. Wallerstein's world-system consisted of a core, semiperiphery, periphery, and external arenas. The external arenas represented the most isolated places, detached from the workings of capitalism but likely to become part of the periphery as capitalism expanded. This map shows the world-system in 1900, following his descriptions.

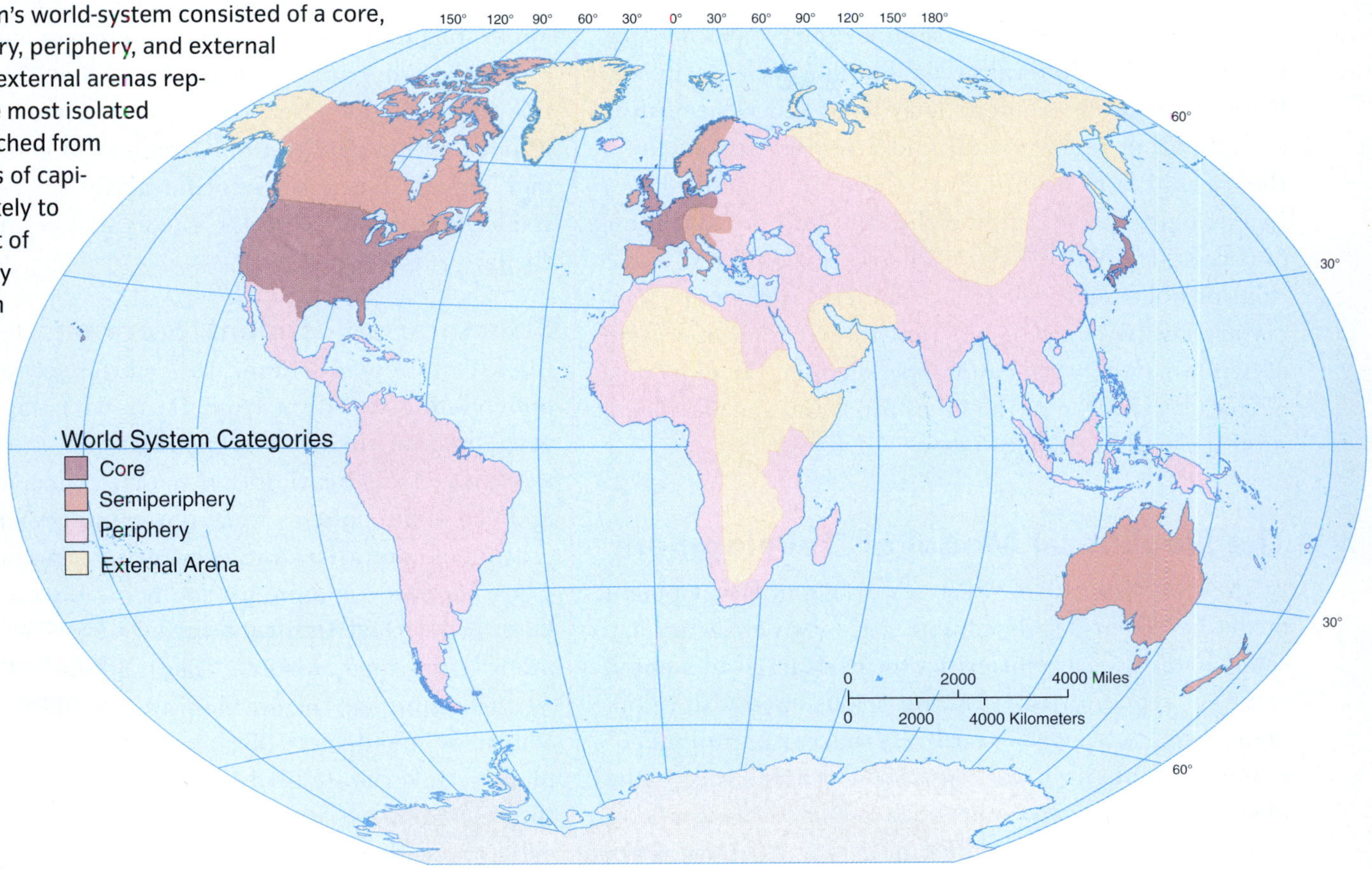

b. Applying the concepts of core, semiperiphery, and periphery to the world-system today yields this map. How has Africa's position in the world-system changed?

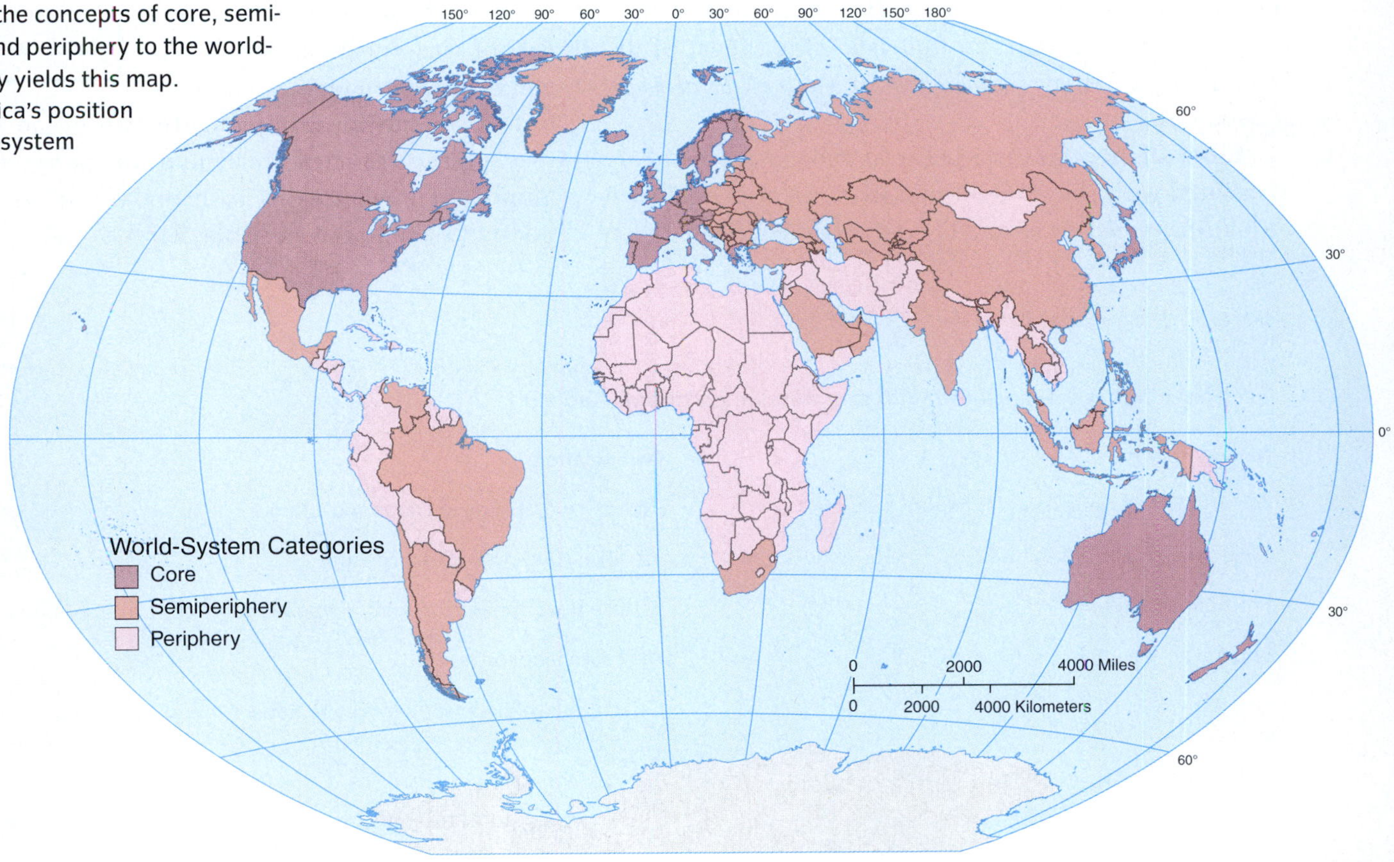

Wallerstein argues that the operation of capitalism creates a system of unequal exchange in which core states dominate the semiperiphery and periphery, and semiperipheral areas dominate the periphery. Core states profit tremendously from this relationship, accumulating more capital and more wealth. Some of this capital is used to finance the development of new technologies that reinforce the competitive advantage of core states. World-system theory recognizes that the relations between the core, periphery, and semiperiphery are always dynamic but that the efficient operation of capitalism demands and depends upon the existence of spatial variation in the labor force and specifically an international division of labor.

The Neoliberal Model of Development

By the end of the 1970s, Rostow's Stages of Development model had been dismissed and world-system theory had come under fire by some critics for its Marxist-influenced critique of capitalism. In the 1980s, many development experts turned to an alternative paradigm, or school of thought, that was strongly anchored to the idea that capitalism could in fact drive development, rather than underdevelopment, as long as appropriate reforms were implemented to allow economic competition and free markets to flourish. This theory of development is called *neoliberalism* or the *neoliberal model of development*.

Liberalism refers to a political and economic theory associated with the thought and writings of people such as Jean-Jacques Rousseau, Thomas Jefferson, and Adam Smith. Liberalism promotes political equality through the development of laws and rights that are universally applicable. It emphasizes the protection of human rights, property rights, and individual freedom. Economically, liberalism favors a free, unregulated market and the removal of barriers to the movement of goods, services, and capital. The term **neoliberalism** refers to the revival and application of the theory of liberalism during the late 20th century.

Structural adjustment programs From a neoliberal standpoint then, the causes of underdevelopment did not stem from flaws with capitalism, as world-system theorists argued. Rather, underdevelopment was an indication that poorly conceived political and economic policies were impeding the efficient operation of capitalism and preventing economic growth. Thus, underdevelopment could be solved with **structural adjustment programs (SAPs)**. That is, the economies of developing countries needed to be completely overhauled—to have their economic structure adjusted—following the principles of neoliberalism. The actual process of structural adjustment involves many economic changes, but these can be grouped into two broad categories: market reforms and deregulation. Some specific examples of mechanisms to bring about structural adjustment are listed in **Table 9.2**.

structural adjustment program (SAP) A country-specific economic policy based on neoliberal principles intended to promote economic growth and development.

Economic reforms associated with structural adjustment Table 9.2

Market Reforms	Deregulation
Reduce budget deficits and inflation	Reduce role of state in economic affairs
Meet debt-payment schedule	Privatize state-owned enterprises
Promote exports	Reduce government spending on public services
Reduce tariffs	Liberalize labor laws
Devalue currency	Liberalize foreign investment regulations

Structural adjustment programs became the cornerstone of the neoliberal development model during the 1980s and 1990s, and they influenced the policies of the International Monetary Fund (IMF) and the World Bank. These international financial institutions are independent agencies within the United Nations system. IMF programs help countries avoid financial crises and develop sufficient exports to pay for imported goods. A main function of the World Bank is to foster long-term development by providing loans or other technical assistance to developing countries.

By the end of the 1980s, SAPs accounted for nearly one-quarter of loans provided by the World Bank. Within another decade, SAPs had been developed or implemented in almost every developing country. Many observers agree that SAPs helped to curb the high rates of inflation and reduce the budget deficits in a number of countries. Others take the position that neoliberalism in general and structural adjustment in particular caused more harm than good because they have adversely affected the poorest individuals.

Criticisms of structural adjustment Since the early 1980s, both neoliberalism and structural adjustment have been heavily criticized. Much of this criticism has centered on the following five points. First, because SAPs call for reduced government spending and cuts in public services, health care systems are often negatively impacted. Clinics that receive government funding may be forced to reduce their hours, lay off personnel, or stop providing certain services. Alternatively, without government funding these facilities may be forced to charge user fees for certain services, making it even more difficult for the poorest individuals to afford health care. Second, and related to the previous point, SAPs encourage the removal of agricultural subsidies as another way of reducing government spending. Usually the removal of subsidies raises the price of foodstuffs, with the most severe consequences for those who are most impoverished.

A third criticism of SAPs is that they often recommend the devaluation of the local currency. One of the effects of currency devaluation is to raise the price of all imported goods, whether it is medicine, equipment, or consumer goods. Fourth, SAPs promote the development of exports. Thus, developing countries often fall back on the troublesome pattern of exporting an agricultural or mineral commodity instead of diversifying their economies. Fifth, SAPs are a form of external involvement by the World Bank and IMF in the internal affairs of states. Specifically, the World Bank and IMF used various criteria (e.g., currency devaluation) as conditions that had to be met before either institution would provide aid or loans. Joseph Stiglitz, the former chief economist at the World Bank, has become one of the leading critics of structural adjustment.

Poverty-Reduction Theory and Millennium Development

For many development experts, structural adjustment and neoliberalism were too narrowly focused on economic growth and did not give enough consideration to improvements in the quality of people's lives. Would the 21st century be another century marked by extensive poverty? The sheer number of people living in extreme poverty worldwide and the prospect that most of the population growth in coming years will be in developing countries prompted yet another paradigm shift in development theory. Since 1999, **poverty-reduction theory** has been the focal point of present development theory, particularly as advanced by the World Bank and IMF. Small area mapping has become an important tool geographers use for reducing poverty (see *What a Geographer Sees* on the next page).

poverty-reduction theory A development theory focused specifically on lowering the incidence of poverty in a developing country.

Poverty-reduction strategies have often been developed in concert with the Millennium Development Goals. The *Millennium Development Goals* (MDGs) were conceived in September 2000 when the leaders of the countries that are members of the United Nations convened the Millennium Summit in New York City. At this meeting, they recognized that every country shares the responsibility of helping attain economic and social development, and that the United Nations must play a fundamental role in this process. As a step in that direction, the United Nations adopted eight MDGs and

WHAT A GEOGRAPHER SEES

Poverty Mapping

Poverty is etched in both landscapes and statistics. It is also highly variable from place to place, so understanding its spatial distribution at various scales helps governments and aid organizations to manage it. Madagascar is a good example, as Earth's fourth-largest island (see locator map) and one of its poorest countries.

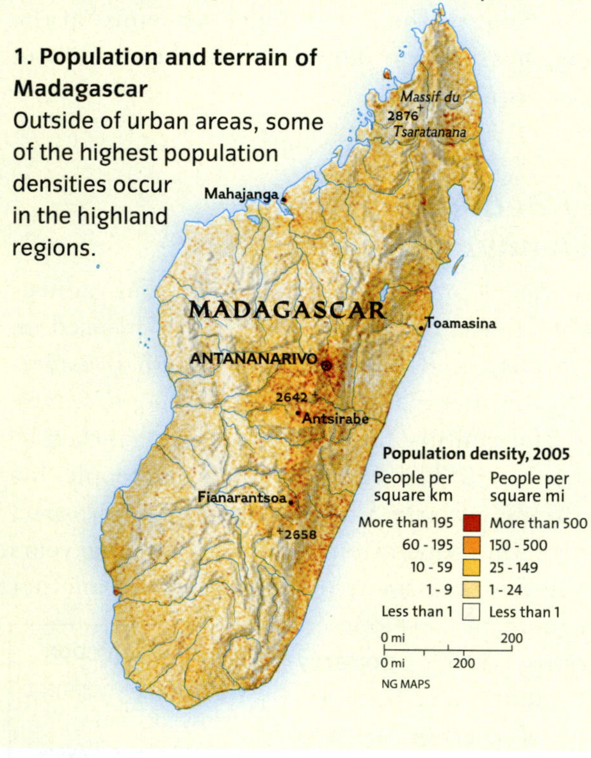

1. Population and terrain of Madagascar
Outside of urban areas, some of the highest population densities occur in the highland regions.

2. Distribution of Madagascar's poverty
Poverty can be expressed in consumption patterns, such as spending. This lets geographers use census data and household-expenditure surveys to create *small area estimation maps* that reveal poverty at administrative or municipal scales. The maps here show Madagascar's poverty at three scales—provincial, district, and commune.

2a. Madagascar's poverty mapped at the provincial level
Poverty ranges from 33% to 82%, and in every province, the highest poverty rates are in rural areas. (*Source*: Mistiaen, J. A., et al., 2002.)

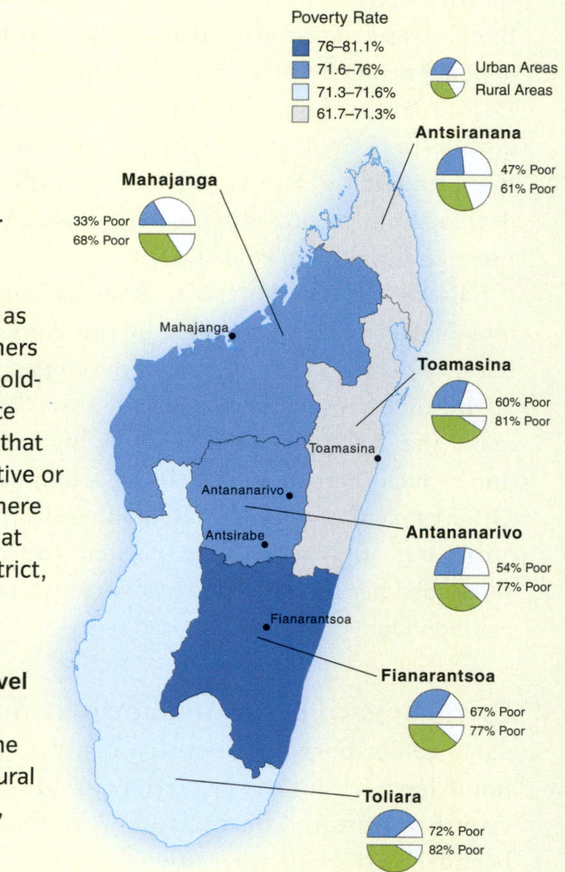

3. Rural poverty is compounded by water availability, which varies widely across Madagascar.

MADAGASCAR

Water availability, 2005
(in millimeters per-person per-year)
More than 750
251 - 750
26 - 250
Less than 26

0 mi 200
0 mi 200
NG MAPS

2b. Madagascar's poverty mapped at the district level
The island's poverty rate is 70%, and districts are colored to show whether they are above or below that rate. Compare the patterns on this map to the patterns on map **2a.** (*Source*: Mistiaen, J. A., et al., 2002.)

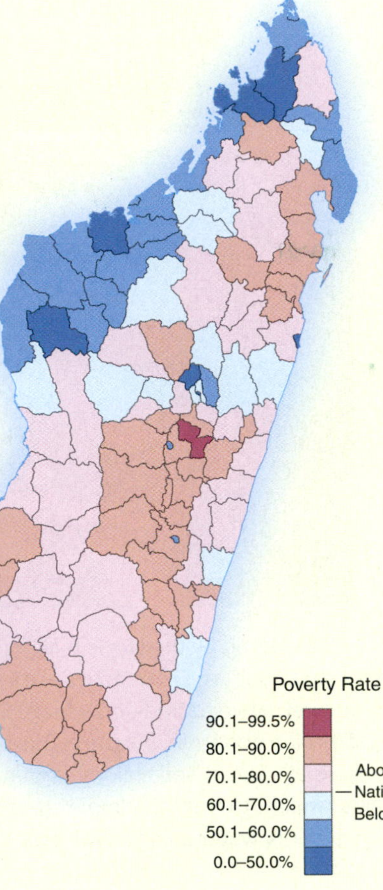

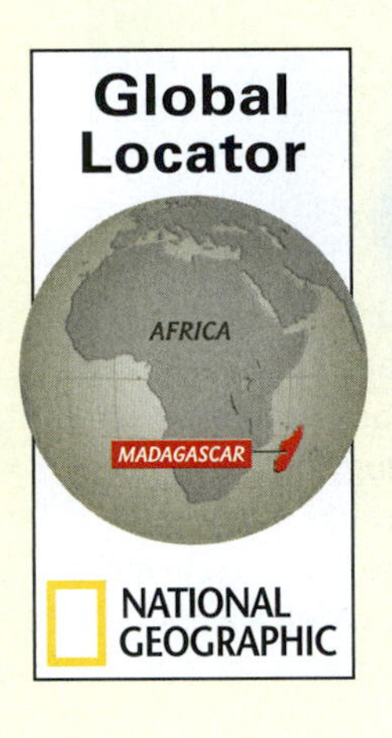

2c. Madagascar's poverty mapped at the commune level
This map details the spatial variation of poverty that is not visible at larger scales. The circled area shows that a high-poverty commune can be nearly surrounded by communes having lower poverty rates—a pocket of poverty that could be missed if relief administrators rely on data mapped at provincial or district scales. District boundaries are shown for comparative purposes. (*Source*: Mistiaen, J. A., et al., 2002.)

Poverty Rate
90.1–99.5%
80.1–90.0%
70.1–80.0%
60.1–70.0%
50.1–60.0%
0.0–50.0%
Above
National Average (70%)
Below

3a. Most rural Madagascar households cultivate wet rice, requiring a reliable water supply.

3b. Deforestation and soil erosion, in addition to long distances to markets and a poor transportation network, create additional economic challenges.

Think Critically

1. What different geographies of poverty do the maps in **2** present?
2. How does the geography of poverty vary by scale?
3. Compare map **2c** to the maps in **1** and **3**. How might topography, population density, or water availability contribute to poverty?

identified the year 2015 as the target date for achieving them. They aim to:

1. Halve extreme poverty and hunger;
2. Achieve universal primary education;
3. Empower women and promote equality between women and men;
4. Reduce mortality of children under five by two-thirds;
5. Reduce maternal mortality by three-quarters;
6. Reverse the spread of diseases, specifically HIV/AIDS and malaria;
7. Ensure environmental sustainability;
8. Create a global partnership for development, with targets for aid, trade, and debt relief.

Poverty reduction is fundamental to achieving the MDGs. The theory of poverty reduction grows out of four key principles of development. The first principle is that because poverty is a complex, multifaceted problem, reducing poverty requires a comprehensive approach—one that balances concern for economic growth with improvements in social and environmental well-being. The second principle is that both the action plan for poverty reduction and the desired goals need to come from the developing country instead of being imposed by external institutions. The third principle follows from the first two in that successful strategies for development require effective partnerships between domestic and external agencies. Lastly, the fourth principle is that in order to be sustainable, development and poverty reduction demand a long-term perspective.

In 2003, Albania became one of the first countries to link its poverty-reduction strategy to the MDGs. For example, it has set a target of halving its extreme poverty by 2015. We should note that Albanian development experts have also been tracking personal perceptions of poverty as a way of more fully understanding the nature of the poverty problem. Compared to other countries, Albanians are more likely to avoid paying income tax, which adversely impacts government revenue. However, simply raising the tax rate or increasing the penalties on tax evasion may not be the best approach to building such confidence. Recognizing and understanding unique local circumstances such as this are important parts of poverty-reduction theory. Of course, large institutions such as the World Bank and IMF are not the only ones involved in poverty reduction and human development. Indeed, many nonprofit organizations play an important role too, especially through their efforts at the local level. As an example, see *Video Explorations*.

Solar Cooking

A stove powered by the sun is making a big difference in impoverished communities. Solar stoves are a great energy saver, providing an alternative to traditional fuels. Solar Cookers International, a Californian nonprofit, supports this technology that could save the lives of women and children around the world. The group trains families to use this simple stove and use it as a sustainable way to purify water using a wax-based gauge called a *wabi*.

CONCEPT CHECK

1. **How** is development understood according to the classical model of development and dependency theory?
2. **What** makes world-system theory a geographic theory?
3. **What** kinds of reforms are associated with structural adjustment programs?
4. **How** is poverty-reduction theory different from neoliberalism?

Summary

THE PLANNER

1 What Is Development? 262

- **Development** refers to processes that bring about changes in economic and social well-being. Development is **normative** because it involves setting standards and measuring achievements compared to those standards.
- Many different indicators are used to measure development and these indicators are commonly grouped into three categories: economic, sociodemographic, and environmental. Common **economic indicators** include the **gross national income (GNI)**, **gross domestic product (GDP)**, **poverty lines**, and **poverty rates**. **Sociodemographic indicators** provide information about the welfare of a population. **Environmental indicators** provide a means to assess sustainable development.
- Development indexes are created by combining several indicators into a single measure. The **human development index (HDI)**, an example of which is shown here, incorporates information on the wealth, health, and education of a population.

Human development index • Figure 9.6

Countries with the Highest HDI Values:
1. Iceland - 0.968
2. Norway - 0.968
3. Canada - 0.967
4. Australia - 0.965
5. Ireland - 0.960
6. Netherlands - 0.958
7. Sweden - 0.958
8. Japan - 0.956
9. Luxembourg - 0.956
10. Switzerland - 0.955

Human Development Index (HDI)
- 0.9 and above
- 0.8–0.89
- 0.7–0.79
- 0.5–0.69
- Below 0.5
- Not ranked

0 2000 4000 Miles
0 2000 4000 Kilometers

- The **gender-related development index (GDI)** and the **gender empowerment measure (GEM)** help to identify countries with notable development-related disparities between men and women. One important difference between an indicator and an index is that indicators can be used for very small areas but indexes are typically used with country-level or international data. Indicators and indexes also point to the ways in which geographic and institutional conditions can impact development.

2 Development and Income Inequality 275

- The study of development necessarily involves examining the spatial variation in poverty with techniques such as poverty mapping. It also raises important questions about **income distribution** within a population and **income inequality**, or the extent of the gap between the rich and the poor.
- Geographers frequently use Lorenz curves, shown here, and Gini coefficients to assess income distribution and income inequality. Individual, social, policy-related, and historical factors can affect the distribution of income.

Measuring income inequality inside countries • Figure 9.12

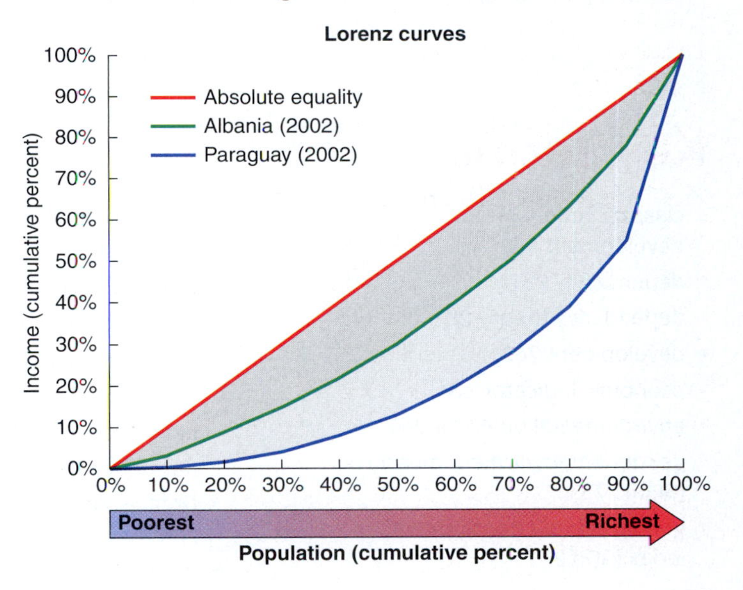

- Views on the effect that globalization has on income distribution fall mainly into two camps: the trickle-down theory of globalization and the theory of the widening gap between the rich and the poor.

3 Development Theory 280

- Development theory has been strongly shaped by five main models. Rostow's Stages of Development has become the **classical model of development**. **Dependency theory** focuses on development as a relational process rather than a series of stages. Wallerstein's **world-system theory** highlights the influence capitalism has on the emergence of global core-periphery patterns, and uneven development (see map).
- The neoliberal model of development emphasizes economic growth through structural adjustment. The intellectual roots of **neoliberalism** derive from the theory of **liberalism**, which promotes free trade and unregulated markets. Today, a dominant approach to development applies **poverty-reduction theory** to achieve the Millennium Development Goals.

The world-system then and now • Figure 9.15

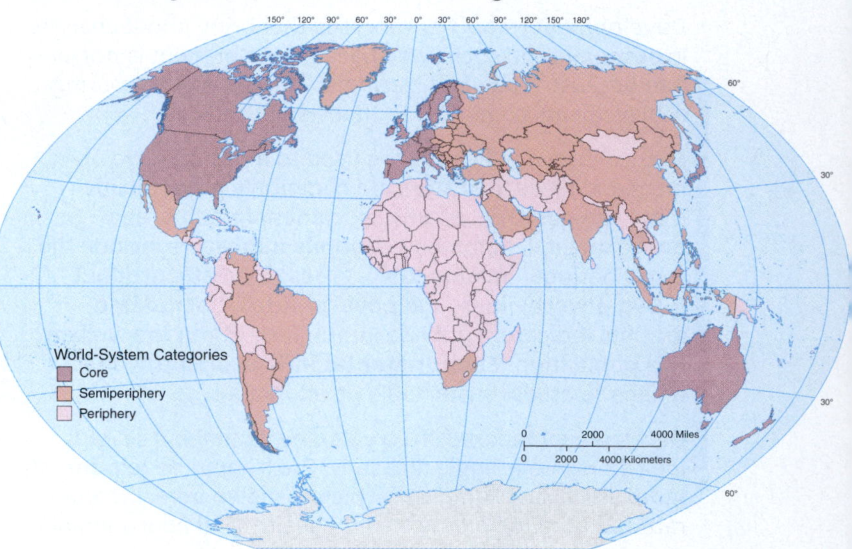

Key Terms

- classical model of development 280
- dependency 281
- dependency theory 281
- development 262
- economic indicator 264
- environmental indicator 269
- gender empowerment measure (GEM) 273
- gender-related development index (GDI) 271
- gross domestic product (GDP) 264
- gross national income (GNI) 264
- human development index (HDI) 270
- income distribution 275
- income inequality 275
- liberalism 284
- neoliberalism 284
- normative 263
- Organisation for Economic Co-operation and Development (OECD) 275
- poverty 266
- poverty line 266
- poverty rate 266
- poverty-reduction theory 285
- purchasing power parity (PPP) 264
- sociodemographic indicator 268
- structural adjustment program (SAP) 284
- world-system theory 282

Critical and Creative Thinking Questions

1. Do the people and governments of developed countries have a moral and financial obligation to improve the life chances of people in developing countries? Explain your thinking.
2. In 1969, the Commission on International Development recommended that developed countries should commit 0.7% of their GNI to foreign aid by 1975. As of 2004, five countries had achieved this. In that same year, the United States contributed about 0.2% of its GNI to foreign aid. The United States contributes more money to foreign aid than any other country, but as a proportion of its GNI, the aid disbursed is still very low. Should the United States raise its contribution to meet the goal of 0.7% of its GNI? Defend your position.
3. What are some advantages and disadvantages of a country providing foreign aid to another?
4. A nongovernmental organization (NGO) is a nonprofit organization providing humanitarian assistance and development work around the world. CARE and Oxfam are two examples of NGOs. Do some research to establish how they are similar and different from one another.
5. Review the theories about development. To what extent do they reflect either "top-down" or "bottom-up" approaches to development and why is that significant?

6. Pick one of the least developed countries and one of the developing countries from the map in Figure 9.1 and contrast them using the criteria in Table 9.1 (below). Provide specific data to illustrate the contrasts.

Selected geographic conditions that may affect development	Table 9.1
Situational	Landlocked state Small, isolated island state Limited natural resource base
Transport	Distant from major markets Population distant from coasts or navigable rivers Large population living in mountainous areas
Agroclimatic	Low and/or highly variable rainfall Poor or depleted soils
Health/Disease	High prevalence of tropical diseases High prevalence of HIV/AIDS
Disaster Vulnerability	High susceptibility to natural disasters (e.g., hurricanes, earthquakes, floods) Large population living in disaster-vulnerable regions

7. Using data from the U. S. Census Bureau website, work with a classmate to construct a Lorenz curve for your state and, then interpret the curve.

8. Working individually and using data from the U.S. Census Bureau or your state's website, find or compute the poverty rate for your state. How does it compare to the national poverty rate? Using what you learned about location quotients in Chapter 6, can you calculate a location quotient for your state using the poverty data?

What is happening in this picture?

(PRODUCT) RED PROMOTION

U2 singer and activist Bono and Bobby Shriver, the chairman of DATA (Debt, AIDS, Trade, Africa), appear with Jane Skinner to promote (PRODUCT) RED, an initiative to link consumer purchasing power with human development. Certain corporations have created specific RED-branded, red-colored products and donate a portion of the profits from sales of these products to the Global Fund to Fight AIDS, Tuberculosis, and Malaria in Africa.

Think Critically

1. Should corporations be obligated to play a greater role in development efforts?
2. Is development a right or a need?

Self-Test

(Check your answers in Appendix B.)

1. The global North-South divide refers to _______.
   a. the separation between conventional versus sustainable development perspectives
   b. the separation between liberal and neoliberal development initiatives
   c. the separation between developed and developing countries and regions
   d. the separation between regions of poverty and extreme poverty

2. The difference between the GNI and GDP is that _______.
   a. the GNI cannot be calculated per capita
   b. the GNI includes the value of goods produced abroad
   c. the GNI is limited to the value of goods produced within a country's borders
   d. the GNI provides information about the value of goods produced in the informal sector

3. An example of a sociodemographic indicator is the _______.
   a. infant mortality rate
   b. poverty rate
   c. vulnerability to drought
   d. poverty line

4. Which of the following statements about poverty is *false*?
   a. Every country may have a slightly different poverty line.
   b. The poverty rate tells us how far below the poverty line a country's poor are.
   c. Extreme poverty has been reduced in East Asia and the Pacific since the 1980s.
   d. There are different economic indicators of poverty.

5. An important reason for creating and using the GDI and GEM was _______.
   a. to be able to penalize countries for poor performance on them
   b. the realization that men often earn more than women
   c. to be able to reward countries for strong performance on them
   d. the realization that development does not always proceed in gender-neutral ways

6. A basic difference between the GDI and GEM is that _______.
   a. the GEM includes an indicator for political participation
   b. the GEM includes only indicators about women
   c. the GDI emphasizes gender development at the household level
   d. the GDI emphasizes differences in extreme poverty between men and women

7. Environmental indicators _______.
   a. were part of the first development index, the HDI
   b. can be used to measure pollution but probably not access to clean water
   c. reflect the growing concern for assessing sustainable development
   d. always include data on two or more environmental problems or concerns

8. As discussed in the chapter, factors affecting development can be grouped into two broad categories. What are they?
   a. vulnerability and resilience
   b. conventional and sustainable
   c. historical and individual
   d. geographic and institutional

9. The OECD _______.
   a. is a branch of the World Bank
   b. still reflects the global North-South divide through its membership
   c. is an international organization that aids developing countries
   d. has no notable income inequality on the part of its members

10. This chart shows that _______.

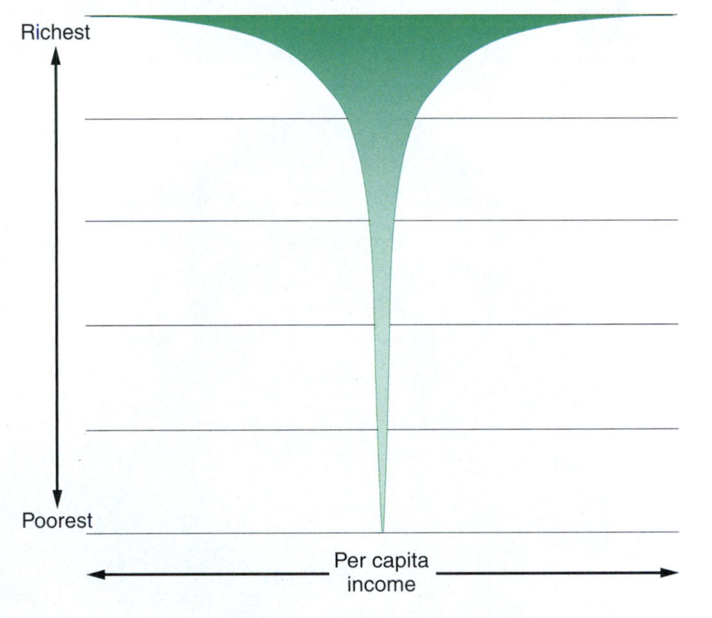

a. the richest 20% of the world's people hold most of the world's per capita income
b. the share of per capita income held by the poorest 20% of the world's people is increasing
c. globally, per capita income levels are rising rapidly
d. none of the above

11. A _______ curve, shown here, helps visualize the relationship between a country's income and population.

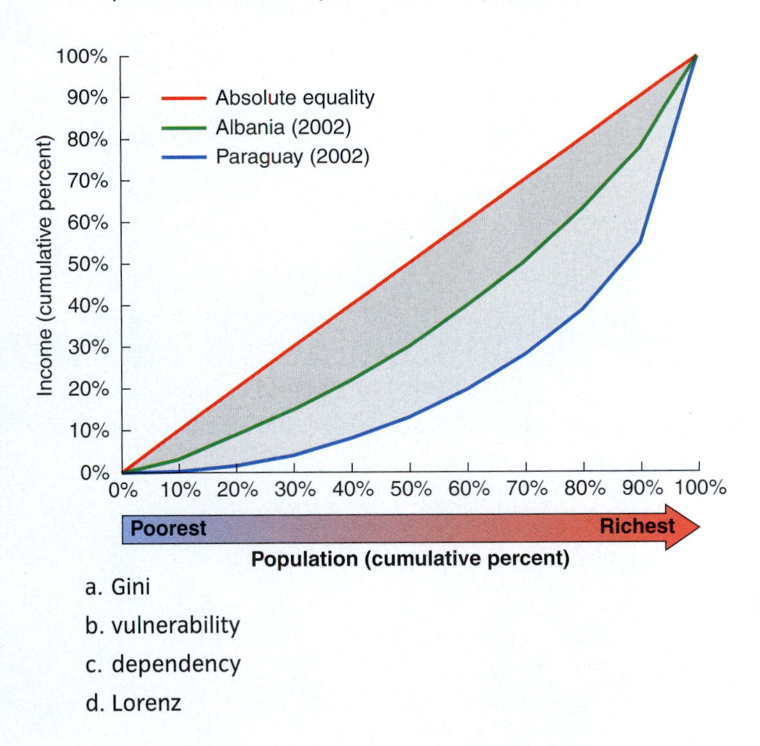

a. Gini
b. vulnerability
c. dependency
d. Lorenz

12. Which of the following is *least* likely to affect income distribution?
a. peer pressure to be successful
b. individual credit ratings
c. the age of a workforce
d. tax laws

13. According to the classical model of development, the turning point coincides with which stage?

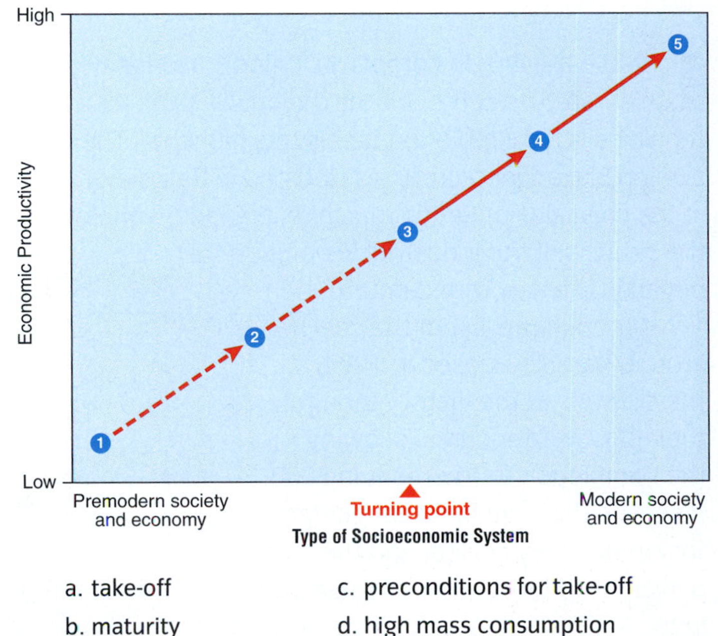

a. take-off
b. maturity
c. preconditions for take-off
d. high mass consumption

14. Which statement best summarizes world-system theory?
a. Development involves investment that leads to mass consumption.
b. Development involves core dominance of the periphery.
c. Development creates conditions of dependency before mass consumption.
d. Development requires belief in the value of economic progress.

15. Which of the following pairs is *incorrectly* matched?
a. poverty-reduction theory – Millennium Development Goals
b. neoliberalism – structural adjustment
c. dependency theory – currency devaluation
d. world-system theory – international division of labor

THE PLANNER

Review your Chapter Planner on the chapter opener and check off your completed work.

10

Changing Geographies of Industry and Services

GEOGRAPHY AND THE MANUFACTURE OF TENNIS BALLS

In 2002, the British company Dunlop Slazenger celebrated its centennial as the official supplier of tennis balls to Wimbledon tennis tournaments. The Dunlop Slazenger factory was located in Barnsley, a city in the coalfields of South Yorkshire, England—one of the hearths of the Industrial Revolution that began in the late 18th century.

Although the last of the coal mines around Barnsley closed in 1994, the clientele at the factory thought their jobs were secure. Following the Wimbledon tournament in 2002, however, the Dunlop Slazenger factory in Barnsley closed, and the company moved its operations to Bataan in the Philippines. Bataan is designated an export-processing zone—that is, it offers incentives to industries producing goods for export. Factories that locate in Bataan are not charged tax, do not pay duties on imported resources or exported goods, and have access to a pool of low-wage workers who are not permitted to strike.

Whether we consider the production of tennis balls, clothing, electronics, or the staffing of call centers, the geography of industry increasingly relies on parts, labor, and knowledge from countries around the world. In this chapter we examine the geographies of industry and services. In the process we give consideration to two fundamental topics: the spatial patterns of industry and services, and how and why these patterns have changed.

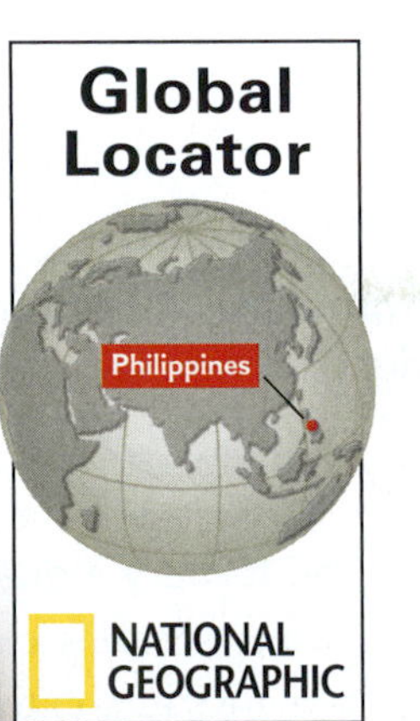

CHAPTER OUTLINE

CHAPTER PLANNER

- ❑ Study the picture and read the opening story.
- ❑ Scan the Learning Objectives in each section:
  p. 296 ❑ p. 301 ❑ p. 307 ❑ p. 314 ❑
- ❑ Read the text and study all visuals.
  Answer any questions.

Analyze key features

- ❑ What a Geographer Sees, p. 304
- ❑ Process Diagram, p. 313
- ❑ Video Explorations p. 315
- ❑ Geography InSight, p. 316 ❑
- ❑ Stop: Answer the Concept Checks before you go on:
  p. 301 ❑ p. 307 ❑ p. 312 ❑ p. 319 ❑

End of chapter

- ❑ Review the Summary and Key Terms.
- ❑ Answer the Critical and Creative Thinking Questions.
- ❑ Answer What is happening in this picture?
- ❑ Complete the Self-Test and check your answers.

Types of Industry

LEARNING OBJECTIVES

1. **Identify** primary, secondary, and tertiary industries.
2. **Explain** staple theory.
3. **Summarize** the origins and diffusion of the Industrial Revolution.

What is the difference between industry and manufacturing? Although the terms are broadly synonymous, we should recognize the distinctions between them. For human geographers, economists, and others, *industry* refers to distinct groups of economic activities (e.g., the electronics industry or the automotive industry), whereas *manufacturing* is one kind of industrial activity that involves the physical or chemical transformation of materials into new products.

Industrial activities can be grouped into one of three broad categories: **primary**, **secondary**, and **tertiary industries**. **Figure 10.1** and the following sections explore these types of industrial activities in greater detail.

primary industries Industries that extract natural resources from the Earth.

secondary industries Industries that assemble, process, or convert raw or semiprocessed materials into fuels or finished goods.

tertiary industries Industries that provide services, usually in the form of nontangible goods, to other businesses and/or consumers.

Primary Industry

All primary industries extract natural resources. When people assign economic value to these resources and trade them, they become *commodities*. This is another dimension of the process of commodification that we discussed in Chapter 2. Because resources are unevenly distributed, trade in commodities has become an enormously important part of the global

Types of industry • Figure 10.1

NATIONAL GEOGRAPHIC

We can group industries according to broadly similar economic activities, as shown here.

a. Aerial view of a swath of forest cleared by logging in British Columbia, Canada
Primary industries are extractive and, in addition to logging, include fishing and mining, among other activities.

b. A worker in a poultry processing plant in Arkansas
Secondary industries process raw materials or assemble component parts into finished goods or products.

c. A waiter at an Indonesian hotel
Tertiary industries provide useful services as opposed to manufactured goods, for consumers, other businesses, and the public.

Share of primary products in merchandise exports of developing regions, 1970–2001, by percent • Figure 10.2

Commodity dependency has lessened over time but is still a common characteristic of developing regions. Whereas Latin America and Asia have significantly reduced their reliance on primary products and have expanded their production and export of manufactured goods, Africa and the Middle East still rely heavily on primary product exports, particularly fuels. (*Source*: WTO, 2002.)

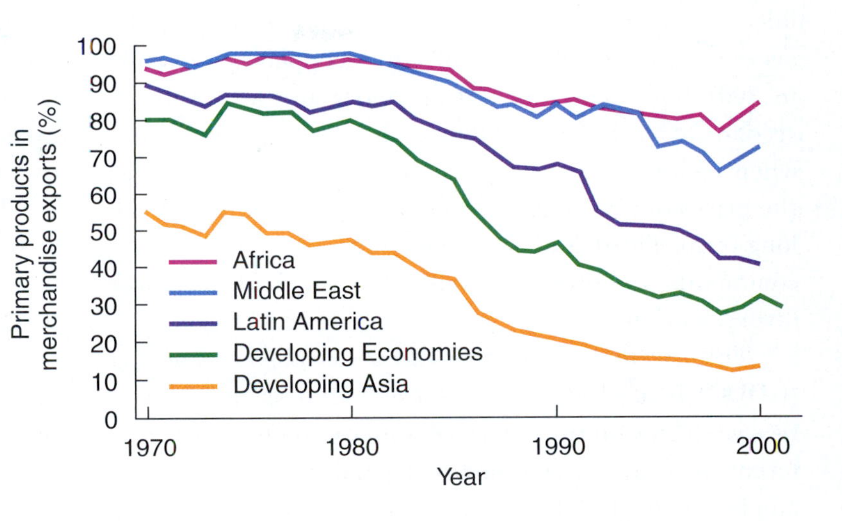

economy. In addition, reliance on commodities has consequences for the kinds of industries established in a place.

Staple theory How do commodities affect the economic development of a country or region? This question was independently explored by two Canadian scholars, W. A. Mackintosh and Harold Innis, beginning in the 1920s. Their work generated a collection of ideas referred to as staple theory. A *staple* is a primary resource that dominates the exports of an economy. **Staple theory** posits that the resource geography of an area shapes its economic system through linkages. Broadly speaking, **linkages** refer to the other economic activities that emerge in conjunction with a specific primary industry.

There are three types of linkages: forward, backward, and demand. *Forward linkages* process the staple resource. Sawmills are forward linkages associated with logging. *Backward linkages*, by contrast, consist of the economic activities that help access or extract the staple. These could include saw blade manufacturers or engineering firms that develop road or other transportation routes. *Demand linkages* refer to the demand for and purchase of consumer goods, especially by workers as the production of a staple commodity stimulates the accumulation of income.

As Mackintosh and Innis showed, the economic consequences of staple production vary at different geographic scales. At a local scale, the economic system of the staple-producing area is shaped by the natural resources. At a regional or even global scale, however, the staple-producing area functions as a hinterland economy that supplies commodities to more powerful economic centers. Interestingly, Mackintosh and Innis came to different conclusions about the impact of staples on long-term economic development. Mackintosh held the view that the development of a staple-based economy would enable subsequent industrial development, whereas Innis took the position that reliance on staples inhibited economic growth and contributed to staple or commodity dependency.

There are many reasons for these different interpretations. One that is particularly significant involves the geographic impact of forward and backward linkages. For example, Innis believed that linkages frequently fail to stimulate growth in the local economy. Australia during the late 19th and early 20th centuries provides a good example. One of Australia's staples was wool, but forward linkages to textile processing did not develop there. Instead, mills in England processed the wool. Similarly, backward linkages in Australia were also weak since equipment to shear the animals was imported from England and not supplied locally.

A number of scholars argue that a high degree of commodity dependency—a characteristic of many developing countries—lends support to Innis's interpretation. Until the mid-1980s most of the world's developing regions were heavily dependent on exports of primary goods (**Figure 10.2**).

Commodity dependency From an economic standpoint heavy reliance on commodities is problematic for three reasons. First, commodity prices are volatile and fluctuate a great deal over time. As an example, the average price of an ounce of gold dropped below $300 in 2001 but topped $1,100 an ounce in early 2010. A second and related problem with commodities is that when compared to the prices of manufactured goods, the prices of commodities do not rise as rapidly over the long term. Third, and as we have seen, heavy reliance on commodities is often associated with lack of economic diversification.

The term *commodity-dependent developing countries* (CDDC) is used to refer to those countries that have a heavy reliance on the export of primary commodities. Different measures of commodity dependency exist. For example, commodity dependency could be expressed as the ratio of the value of commodity exports to a country's gross domestic product (GDP; see Chapter 9). Alternatively, commodity dependency could be expressed as the percentage of a country's total exports from the four leading commodities exported as shown in (**Figure 10.3**).

Secondary Industry

Secondary industries assemble, process, or manufacture raw or semiprocessed materials into useful products, fuels, or finished goods. Sometimes a distinction is made between heavy and light manufacturing. *Heavy manufacturing* refers to the fabrication of items such as steel, nuclear fuel, chemical products, or petroleum as well as durable goods such as motor vehicles, refrigerators, and military equipment. *Light manufacturing* includes activities related to the assembly of clothing or small appliances such as irons or light fixtures as well as the manufacture of food products, beverages, or medical instruments.

The Industrial Revolution The geography of secondary industry has been dramatically shaped by technological innovations—especially since the Industrial Revolution. The term **Industrial Revolution** refers to the fundamental changes in technology and systems of production that began in England in the late 18th century. The phrase *system of production* refers to the dominant ways of organizing and coordinating the manufacture of goods. Prior to the Industrial Revolution, manufacturing was largely characterized by small-scale craft production of ceramics, cloth, and metal goods. The labor was provided by members of a household or community, giving rise to the term *cottage industries*. With the onset of the Industrial Revolution, cottage industries in England gave way to factories, which were then an innovative way of organizing labor.

Two major developments helped spur the Industrial Revolution. The first involved greater access to capital, much of which was generated by England's commanding position in the system of global trade and its control over resources in its colonies. The second major development involved not one but a series of technological innovations that worked to raise output. Some of these innovations improved agricultural production, whereas others improved the processing of raw materials such as cotton. For example, the spinning jenny—a device that twists cotton fibers into thread—was developed in 1764. Another crucial innovation was the development of the steam engine by James Watt in 1769.

Resources such as coal and iron ore influenced the geography of industrialization in England because factories were initially located near energy sources, especially coalfields. Workers moved to live near the factories in urban places, and transportation networks that funneled raw materials to the factories and finished goods to markets grew denser. Thus, industrialization has been strongly associated with urbanization.

Diffusion of the Industrial Revolution The global diffusion of the system of production associated with the Industrial Revolution has occurred slowly and in three general phases. During the first phase, which ran from roughly 1760 to 1880, the Industrial Revolution diffused to Belgium, the Netherlands, France, Germany, and the United States. The second phase, spanning the years 1880 to 1950, carried the Industrial Revolution to Russia, Japan, and Canada among other primarily Western places, especially British dominions. During this time, some industrial centers developed in places that were still predominantly agricultural, including Shanghai, China; Bombay, India; Monterrey, Mexico; and São Paolo, Brazil. The third phase, which began in

Primary commodity dependence among developing countries • Figure 10.3

Countries where the four major commodities account for 60% or more of the exports have high commodity dependence. Commodity fuel dependency is high in the Middle East and Africa. What do many of the countries with high dependency on agricultural products also have in common? (*Source*: UNCTAD, 2008.)

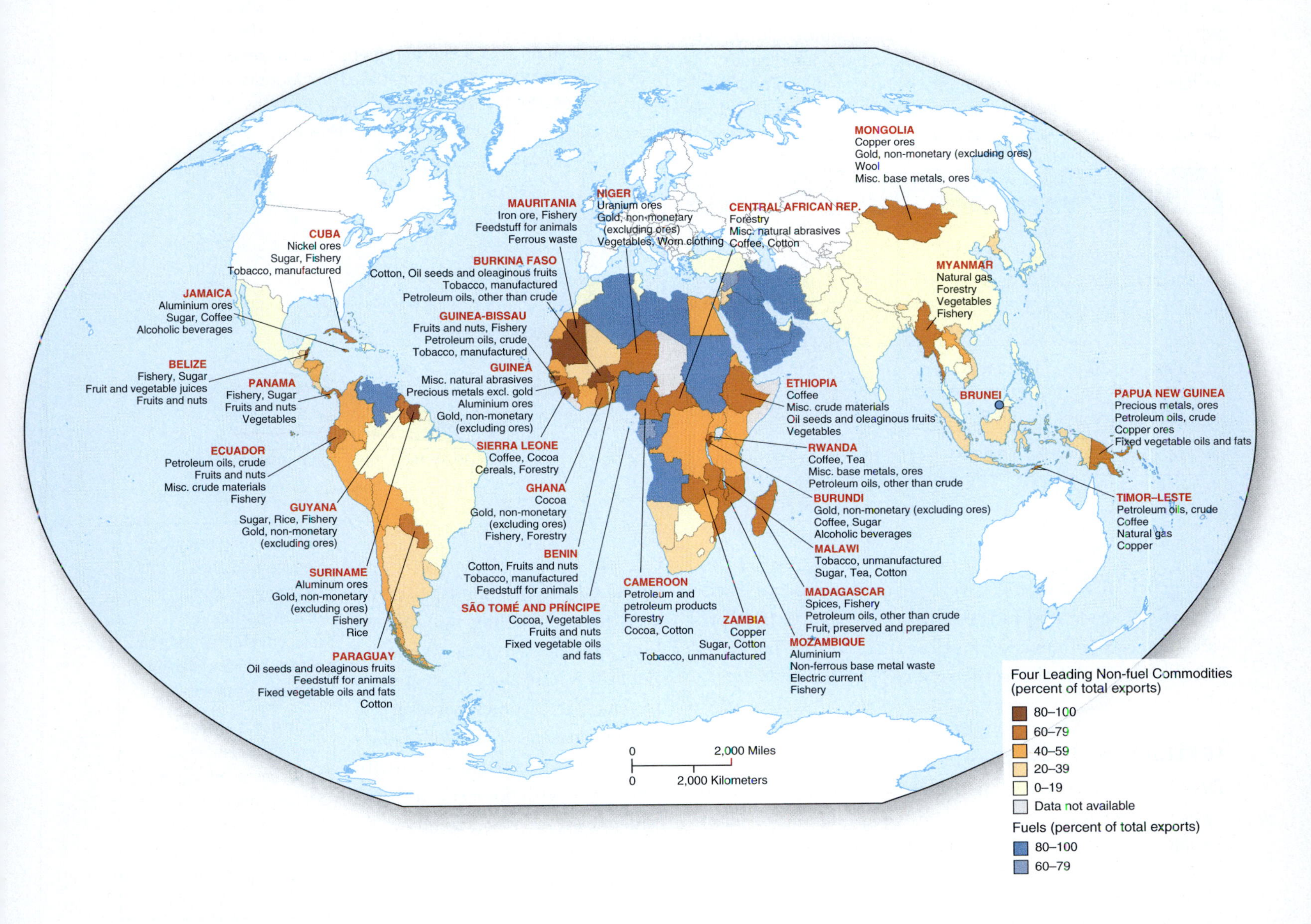

the 1950s and is still underway, has seen the continued industrialization of countries affected by phase two and the industrialization of Israel and several Pacific Rim countries.

Diffusion of the Industrial Revolution has been highly uneven not only at the global scale but also within individual countries. As a result, *core-periphery patterns* of industrialization and development are often discernible. Within the United States, for example, manufacturing initially concentrated within New England, making it a core area of secondary industry focused on textile mills, while the South, a cotton-supplying region, functioned as the commodity-supplying periphery. As we discussed in Chapter 9 in conjunction with world-system theory, the

Industrialization in historical perspective • Figure 10.4

The Industrial Revolution first affected the textile industry and then the metal-working industries.

a. Percentage of global manufacturing production, by major world region

When the Industrial Revolution began to unfold, household cottage and craft industries generated most of the manufactured goods. As mechanization transformed the scale and system of production, new centers of manufacturing emerged in areas that would become the developed world. (*Source*: Bairoch, 1982, p. 275.)

Year	Manufacturing production by developed countries (%)	Manufacturing production by developing countries (%)
1750	27.0	73.0
1800	32.3	67.7
1860	63.4	36.6
1900	89.0	11.0
1953	93.5	6.5
1980	88.0	12.0

b. Selected annual estimates of GDP per capita, by region

GDP per capita is often used to gauge the level of industrialization in a country, and the international dollar is a unit used to make comparisons between different countries and at different times. Western offshoots include the United States, Canada, Australia, and New Zealand. This graph shows the divide that emerged among major world regions as a consequence of uneven industrialization. (*Source*: Data from Maddison, 2003.)

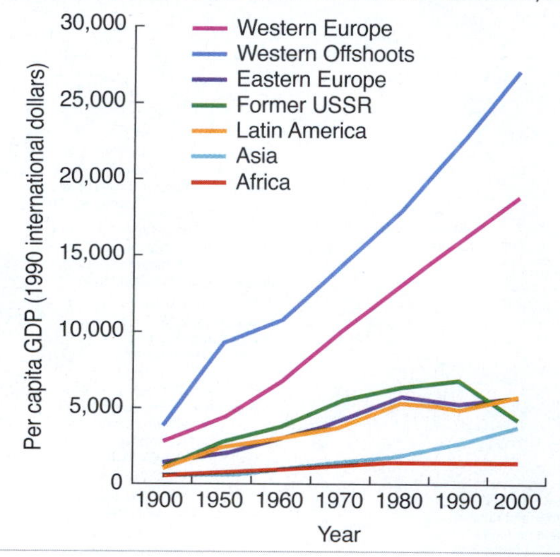

emergence of the global core is strongly associated with patterns of trade and industrialization. **Figure 10.4** provides two ways of envisioning the impact of the Industrial Revolution on global core-periphery patterns.

Tertiary Industry

Thus far we have taken an expansive view of tertiary or service industries. However, many geographers and other scholars point out that important distinctions can be made between different kinds of service activities, including **quaternary** and **quinary services**. When we identify quaternary and quinary services, it becomes necessary to adopt a narrower view of tertiary activities that associates them with the provision of domestic and quasi-domestic services, as shown in **Figure 10.5**.

quaternary services Service industries including or related to transportation, telecommunications, real estate, insurance, finance, and management.

quinary services Service industries such as research, education, and engineering that facilitate the creation of innovations through the production of new knowledge and skills.

Categories of service activities • Figure 10.5

The diversity of tertiary industries has prompted the identification of different categories of service activities. Advanced professional or producer services are a subset of quaternary services that require specialized expertise. Quinary services are those that enhance human and societal capacities for development.

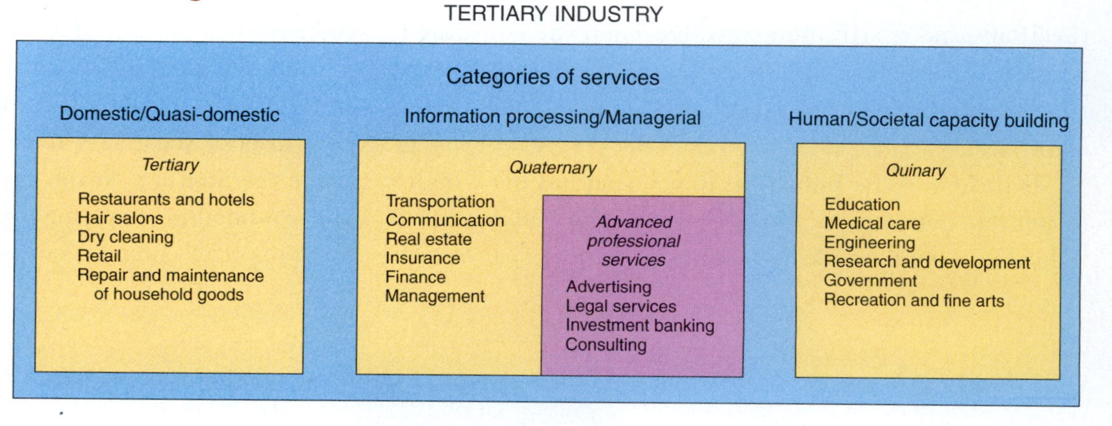

For a neat collection of maps depicting different primary, secondary, and tertiary activities, see *Where Geographers Click*.

Where Geographers CLICK

Worldmapper

http://www.worldmapper.org/index.html

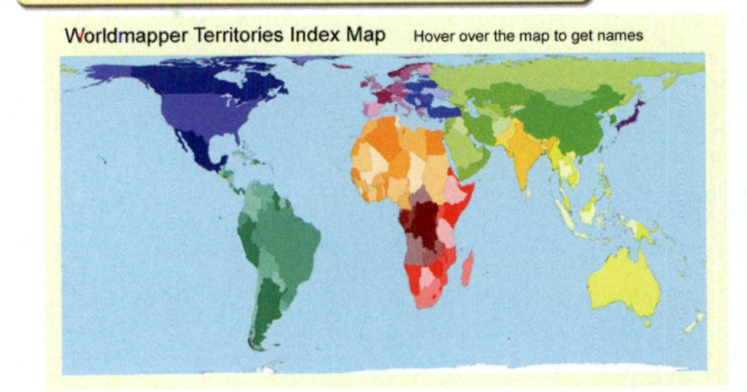

This website has many cartograms available for viewing. Click on the categories tab to see a list of different maps.

CONCEPT CHECK 

1. **Why** are tertiary industries distinctively different from primary and secondary industries?
2. **How** does staple theory address economic development?
3. **Why** is a core-periphery pattern often associated with the diffusion of industrialization?

Evolution of Manufacturing in the Core

LEARNING OBJECTIVES

1. **Identify** two groups of factors that can influence the location of manufacturing.
2. **Define** Fordism and explain its development.
3. **Contrast** Fordist and flexible systems of production.
4. **Distinguish** between outsourcing and offshoring.

Since the Industrial Revolution there has been considerable change in the geography of manufacturing. This section explores the broad trends that have characterized manufacturing in the core—those regions that constitute the developed world.

Factors Affecting the Location of Manufacturing

What is the optimal location of a particular factory? Usually the answer involves a consideration of the costs and benefits of different sites and situations. See **Table 10.1** for a description of some of these factors.

Site and situation factors in industrial location   Table 10.1

Site Factors	Situation Factors
Labor costs and availability	Proximity to the market
Land availability, accessibility, value, and taxes	Transportation options and costs
Energy costs (e.g., oil, electricity)	Agglomeration effects
Local environmental regulations	Political policies (e.g., right-to-work laws)
Local amenities or tax advantages	Other (e.g., proximity to schools for families of employees)

Major urban-industrial agglomerations prior to the 1970s • Figure 10.6

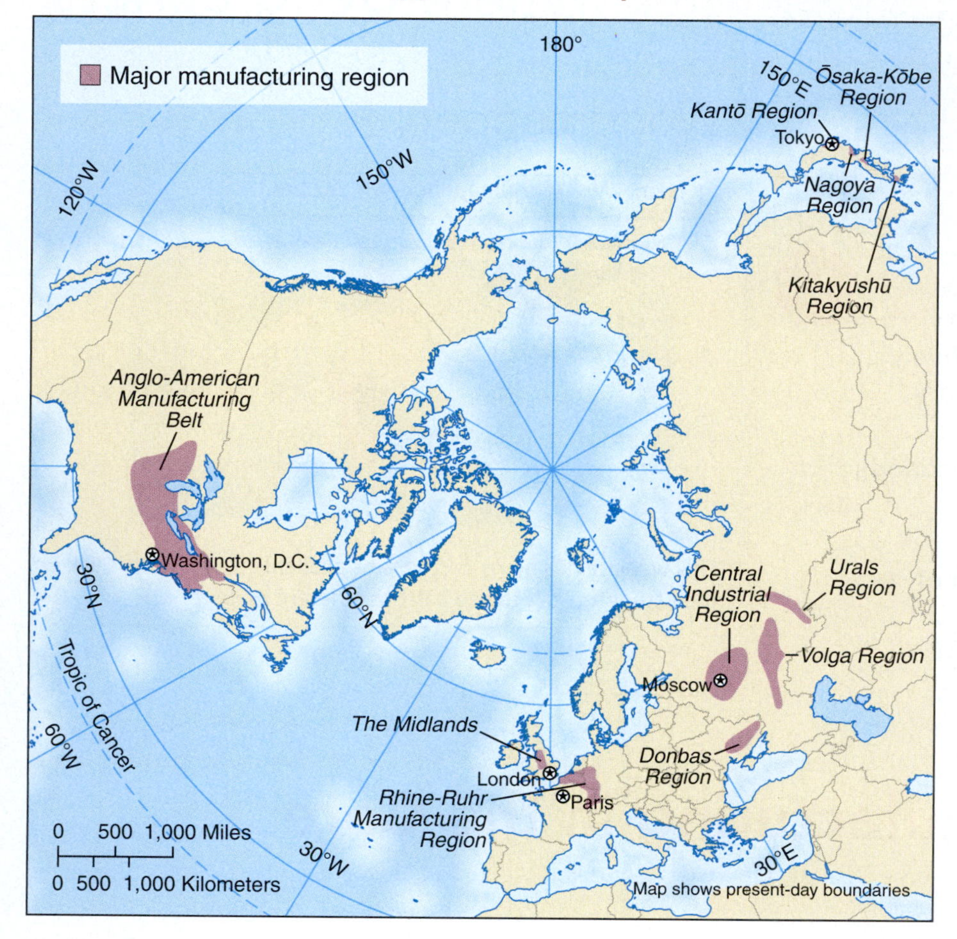

Historically speaking, proximity to raw materials, a good transportation network, and dense concentrations of people help explain the emergence of these major centers of manufacturing. Because of its limited resource base, Japan has been significantly more dependent on imported raw materials. In the planned economy of the former Soviet Union, decisions about industrial development and production were made in Moscow.

Through the first and second phases of the Industrial Revolution, decisions about where to locate factories were strongly influenced by the cost of transporting raw materials. To minimize these costs, iron and steel mills were located close to the coalfields and iron ore deposits. For many firms today, having access to the market so that finished goods can readily be distributed and sold has taken on greater significance in decisions about factory location. Manufactured goods typically cost more to transport, mainly because such goods are often bulk-gaining. That is, they are heavier than the component parts that compose them, and they take up larger volumes (e.g., televisions). Thus, access to the market can help minimize transportation costs.

agglomeration The spatial clustering of people and economic activities, especially industries that are related or interdependent, in a place.

Another important factor that can affect a firm's profitability involves **agglomeration**. In addition to labor, firms need inputs such as equipment, which they acquire from other enterprises called suppliers. Firms can achieve cost savings by locating in urban areas that have a sizable pool of skilled labor and also house needed suppliers. Such savings are called *agglomeration economies*. Sometimes, however, urban growth can result in increased taxes or increased transportation costs because of congestion, creating *agglomeration diseconomies*.

Deciding where to locate an industry is not simply a matter of listing the site and situational factors. Rather, it also involves analyzing and predicting how the contingencies of a place may change (Chapter 1). See **Figure 10.6** for a map of the major manufacturing centers that developed in the core.

Fordism

To varying degrees, firms in each of the urban-industrial agglomerations depicted in Figure 10.6 used systems of factory production that were influenced by the ideas of F. W. Taylor and Henry Ford. Taylor was a mechanical engineer whose book *Principles of Scientific Management* (1911) promoted the division of labor into the most elemental tasks. Scientific management, now known as **Taylorism**, involved studying the tasks performed by workers, timing the workers, and, if necessary, altering their movements in order to minimize wasted effort. Taylor did not want workers to think about what they were doing because he considered that a distraction and potential waste of time.

Ford, who was familiar with Taylor's ideas, desired to mass produce automobiles and is credited with developing a moving assembly line at his Highland Park factory in 1913. Using a production process based on interchangeable parts, clear divisions of labor, and product standardization, the assembly line cut in half the time to produce a single car. The word **Fordism**, then, refers to a system of industrial production designed for mass production and influenced by the principles of scientific management.

The impact of Fordism on the evolution of manufacturing can hardly be overstated. The implementation of Fordist principles had three major consequences. First, Fordism contributed significantly to the de-skilling of labor. Whereas early factories employed craftspeople who were experts in their particular specialty such as metal fabrication, the fragmentation of production into a series of tasks for the assembly line meant that factory workers did not need to be craft experts any longer. Therefore, Fordism depended on and contributed to the de-skilling of labor.

A second and related consequence of Fordism is that it reinforced the existence of a rigid social hierarchy between the workers and the managers. In fact, the spread of Fordism, especially in the United States, is closely associated with the unionization of the labor force. Third, Fordism contributed to the rise of multinational corporations (MNCs). The Ford Motor Company grew to become an MNC over time, establishing factories and assembly plants in the United Kingdom, France, Denmark, and other countries. Today it has operations in more than 100 countries. As we discussed in Chapter 2, tens of thousands of MNCs exist today.

Fordist Production

The moving assembly line is not only efficient, it is also very productive. There are, however, three major weaknesses. One is that it requires a regular and steady supply of inputs at the scale of the assembly line and also at the scale of the company. For example, if a machine on the assembly line malfunctions, the entire operation has to be halted. Likewise, should the company fail to receive adequate quantities of steel, assembly-line production will be affected. A second weakness is that, in order to be successful, Fordism relies on a mass market that can consume the goods that are produced. In terms of the history of manufacturing, Fordism helped to connect mass production with mass consumption. This relationship is summed up by the saying, "Everything we produce has already been sold." A third weakness of Fordism is that assembly-line work can be extremely boring for the employees.

If you were a manager of a company using a Fordist system of production, what would you do to address these weaknesses? Most likely you would find ways to stabilize the system. Although the specific strategies that were used varied from industry to industry, some common practices emerged. These include purchasing large quantities of stock and storing it in warehouses so that the manufacturer always has a supply of necessary inputs. Another strategy to stabilize the system of Fordist production was to vigilantly maintain and service the equipment on the assembly lines in order to prevent breakdowns. Since labor is also a major input, having a reliable workforce was considered essential. High wages and long-term labor contracts were used to minimize employee turnover, as were *collective agreements*—contracts between unions and management stipulating employment practices and employee benefits.

For many business managers, vertically integrating the company was another solution. **Vertical integration** occurs when a company controls two or more stages in the production or distribution of a commodity either directly or through contractual arrangements. Thus, vertical integration is a strategy of extending a company's ownership and control "up" the supply stream and/or "down" the distribution stream of a good or service in order to lessen the company's vulnerability. The Ford Motor Company became vertically integrated early in the 20th century, especially in an upstream direction. The facilities Ford constructed at the River Rouge plant in Michigan provide an excellent example of vertical integration (**Figure 10.7**).

A portion of Ford's River Rouge Plant in the 1940s • Figure 10.7

By this time, Ford owned mines, quarries, and forests in different states and built facilities at the Rouge to manufacture iron, steel, tires, glass, and cardboard. The massive coal and ore bins are a testament to the vast scale of Ford's operations made possible by vertical integration.

As Fordist systems of production developed, they became closely associated with producer-driven commodity chains, which also have important limitations. A **commodity chain** (also called a production chain or value-added chain) is the linked sequence of operations from the design to the production and distribution of a good. *Producer-driven commodity chains* are associated with large, vertically integrated MNCs, and they influence decisions about production. These decisions are made months in advance of the actual production, and they are communicated in a top-down fashion through the MNC to the manufacturers as well as the distributors and retailers. Because these decisions have to be made so far in advance, they may not reflect actual consumer demand when a product is finally manufactured. To better understand the geographical dimensions of commodity chains, see *What a Geographer Sees*.

By the 1950s, Fordist systems prevailed in the industrial regions of the core but were most extensively

WHAT A GEOGRAPHER SEES

A Commodity Chain

A commodity chain can be thought of as a network that connects the different steps in the production of a good, from information and resource gathering to the manufacture, distribution, and marketing of it. Every manufacturing operation makes use of one or more commodity chains, and such networks are increasingly globalized. Here we present a commodity chain for fur garments, which is strongly anchored to the northern hemisphere.

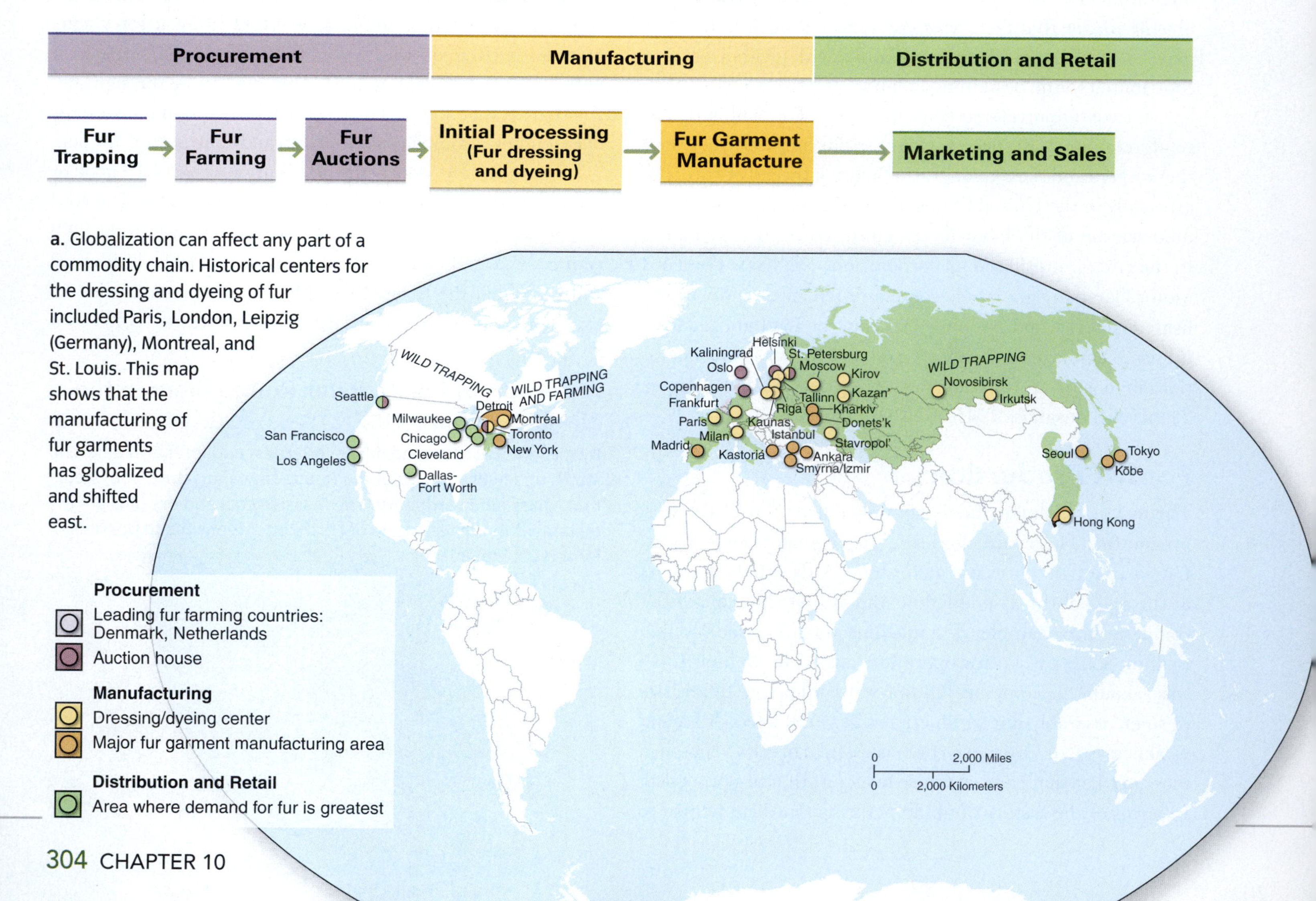

a. Globalization can affect any part of a commodity chain. Historical centers for the dressing and dyeing of fur included Paris, London, Leipzig (Germany), Montreal, and St. Louis. This map shows that the manufacturing of fur garments has globalized and shifted east.

developed in North America and Western Europe. Firms and employees benefited from rising profits and wages during a boom that lasted nearly 20 years. The *crisis of Fordism* marks the end of that boom and specifically refers to the declining productivity and competitiveness of firms. Two major developments contributed to the crisis of Fordism: (1) the energy crisis of the 1970s, which increased manufacturing and transportation costs, and (2) improvements in computers and electronics, which began to transform industrial practices. Together, these two developments showed that Fordism is a highly inflexible system of production that does not easily or rapidly adjust to changing economic or market conditions. Indeed, some of the financial troubles of the U.S. auto industry today stem from its inability to shift quickly to the production of more fuel-efficient vehicles and hybrid technologies as the price of oil increased in the mid-2000s.

THE PLANNER

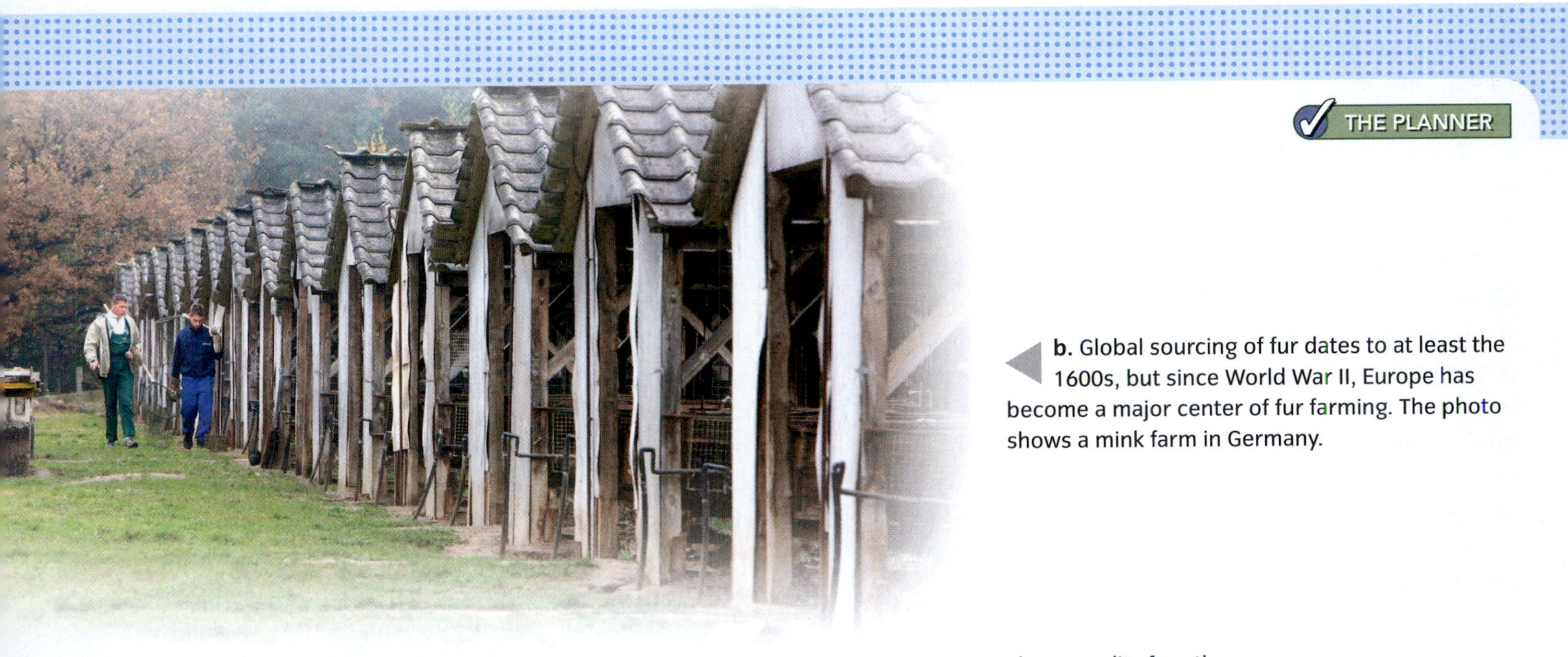

b. Global sourcing of fur dates to at least the 1600s, but since World War II, Europe has become a major center of fur farming. The photo shows a mink farm in Germany.

c. Fur garments are a status symbol for some and a contested commodity for others. The fur industry describes itself as a "responsible industry." Some experts consider the eastward shift of fur manufacturing to be partially the result of animal rights campaigns.

Think Critically

1. How and in what way is responsibility a component of the global commodity chain?
2. How would a commodity chain for an automobile manufacturer differ from this one?

In Japan, however, Fordism developed differently. Unlike the United States with its large domestic market, manufacturing and production in Japan had to be adapted to a much smaller market. Many of these adaptations originated in the automobile industry and specifically within the Japanese company Toyota, where **flexible production** was pioneered. Flexible, or lean, production uses information technologies such as computer networking, high-tech sensors, and automation technology to make the production of goods more responsive to market conditions and therefore more efficient. In contrast to Fordism, flexible production relies on *consumer-driven commodity chains* in which consumer demand shapes the amount and kind of products produced.

Table 10.2 highlights some of the major differences between Fordism and flexible production. Even the assembly-line system had its limitations. In the United States, each model of car required a completely different assembly line. This was not the case in Japan, however, where a single assembly line could produce different models, depending on specific market conditions. Flexible production is also based on the idea that workers should be empowered to think, troubleshoot, and perform multiple work responsibilities. One of the mottoes of flexible production is to promote the continuous improvement of the product. Thus, workers and managers need to communicate and collaborate on a frequent basis.

Two strategies that have been crucial to the success of flexible production are just-in-time delivery and outsourcing. **Just-in-time delivery** refers to how a company manages its inventory and obtains the materials, components, or supplies it needs. Supplies are ordered in smaller quantities on an as-needed basis. Just-in-time delivery enables a company to match production quantities to actual customer demand without the need for or costs of maintaining large warehouses and inventory stockpiles. By **outsourcing,** a company subcontracts a business activity that was previously performed in-house (such as the manufacture of a part, packaging, or customer support) to another firm. Outsourcing can be thought of as a kind of vertical disintegration. Many business processes (sometimes called *back-office functions*) such as data entry, bookkeeping, and other administrative, legal, or information technology (IT) operations are outsourced. Except for strategic planning, virtually all internal company operations are subject to outsourcing (Figure 10.8).

It is important to note that outsourcing always involves another firm, or subcontractor, and this firm may be located in the same country as the company or abroad (i.e., offshore, as shown in Figure 10.8). **Offshoring** is the transfer of an internal or outsourced business activity from a domestic to an international location. Nike, for example, offshores the manufacture of its footwear to factories located in Southeast Asia and China, while Dell offshores the manufacture of its desktop computers to Taiwan. Offshoring has shaped the globalization of industry and is so important to the geography of manufacturing today that we will return to it in the next section.

Major contrasts between Fordism and flexible production Table 10.2

Fordism	Flexible Production
Maximize inventory held in warehouses	Minimize inventory; no need for warehouses
Vertical integration	Outsourcing
Producer-driven commodity chain	Consumer-driven commodity chain
Highly standardized product design	Made-to-order product design
Strongly hierarchical management style	Flat management style
Minimally skilled labor	Multiskilled labor
Large labor force	Smaller and more efficient labor force

Business process outsourcing • Figure 10.8

These employees work in Hewlett-Packard's business process outsourcing center in Bangalore, India. Reasons for outsourcing vary depending on the type of business, but generally companies outsource in order to streamline their business operations. This includes achieving reductions in the cost of utilities, labor, rent, or personnel and resource management. Are there other reasons, not listed here, for outsourcing?

CONCEPT CHECK STOP

1. **Where** did the leading urban-industrial agglomerations of the core develop and why?
2. **Why** is Fordism associated with vertical integration?
3. **What** are two types of commodity chains, and how are they different?
4. **What** are examples of activities that might be outsourced and when does outsourcing become a form of offshoring?

Evolution of Manufacturing Beyond the Core

LEARNING OBJECTIVES

1. **Discuss** the rise of the Asian NIEs.
2. **Explain** what an export-processing zone is.
3. **Distinguish** between a maquiladora and a special economic zone.

Manufacturing transforms a product and adds value to it. One way to measure industrial output is to use the **manufacturing value-added (MVA)**. We can calculate this by taking the cost of the finished product and subtracting from it the cost of purchased inputs necessary to produce it such as fuel, electricity, and the cost of other parts or materials.

By the 1970s, an important two-stage shift in the geography of manufacturing was beginning to unfold. The first stage involved a shift *within* the core as Japan experienced rapid growth in its MVA and began to rival in output centers of manufacturing in Europe and the United States. The second stage involved a rise in importance of manufacturing centers in certain semiperipheral areas in Asia in conjunction with the third phase of the diffusion of the Industrial Revolution.

Newly Industrialized Economies

Japan's manufacturing success was paralleled by rapid economic growth, improved living standards, and reductions in poverty. The country became an example for others to emulate. Indeed, by the 1970s four other East Asian centers were showing signs of substantial growth and productivity in manufacturing. These **newly industrialized economies (NIEs)**, sometimes referred to as the Four Asian Tigers, are Hong Kong, Singapore, South Korea, and Taiwan. A second tier of Asian NIEs began to emerge as important centers of manufacturing in the 1980s. This tier includes Indonesia, Malaysia, the Philippines, and Thailand.

The economic transformation of the NIEs is related to three major factors: (1) government-supported initiatives to increase manufacturing productivity and improve trade; (2) a gradual shift from low-skill, labor-intensive industries to higher value-added technology-intensive industries such as the manufacture of computer components and scientific instruments; and (3) the presence of a skilled labor force. In most of the NIEs, growth in the textile/apparel industry helped fuel the initial production

NIEs and change in manufacturing • Figure 10.9

The growth of manufacturing in the Asian NIEs is related to the diffusion of the Industrial Revolution and illustrates how that diffusion has played out differently among regions.

a. First- and second-tier Asian NIEs
Between the 1960s and 1980s, high rates of growth in manufacturing output—often in excess of 10%—became common. In addition, manufacturing shifted to the production of higher-technology goods.

of manufactured goods for export, though its importance has diminished over the years (**Figure 10.9**).

The NIEs increasingly compete against one another to attract manufacturing. The production of hard disk drives for computers provides a good example. During the 1990s the leading manufacturers of hard disk drives—multinational corporations such as Seagate, Western Digital, Maxtor, and Hitachi—had opened manufacturing facilities in Singapore. However, by 1996 Seagate had established two major plants for the production of hard disk drives and their components in Thailand. Shortly thereafter, Western Digital relocated its manufacturing operations to Thailand and Malaysia. Singapore remains an important center for the design and development of hard disk drives, but the actual manufacture of these devices has become more diffuse. Although not classified as an NIE, China has experienced rapid growth in the production of high-technology products as well and appears to be following a similar path in its efforts to upgrade industrially.

Since the rise of the NIEs, the geographic unevenness of manufacturing within the developing world has taken on a different pattern. We can visualize this by comparing the value added in manufacturing among regions (**Figure 10.10**).

Export-Processing Zones

Some of the manufacturing growth in the semiperiphery and periphery is attributed to the spread of export-processing zones such as the one in the Philippines described at the start of this chapter. An **export-processing zone** (EPZ), also

	Apparel as a percentage of total exports		
Region/Country	1980	1990	2000
Hong Kong	25.4	18.7	12.0
South Korea	17.0	12.4	2.9
Taiwan	12.3	5.8	2.0
Indonesia	2.4	10.3	7.8
Malaysia	1.2	4.5	2.3
Thailand	4.2	12.2	5.5
Philippines	4.9	8.4	6.9

b. Apparel as a share of total exports in selected Asian NIEs, 1980, 1990, and 2000
Notice the geographic pattern that distinguishes first-tier NIEs, which experienced significant declines in the 1980s, from the second-tier NIEs, whose declines are more recent. (*Source*: Gereffi and Memedovic, 2003.)

c. High-technology exports as a percent of manufactured exports for selected Asian NIEs, 1990 and 2005
In recent years, high-technology goods have constituted more than half of the share of exports from Singapore and Malaysia, and in the Philippines they have risen above 70%. The Philippines has expanded rapidly into electronics, including the manufacture of cell phone chips. (*Source*: Data from UNDP, 2007.)

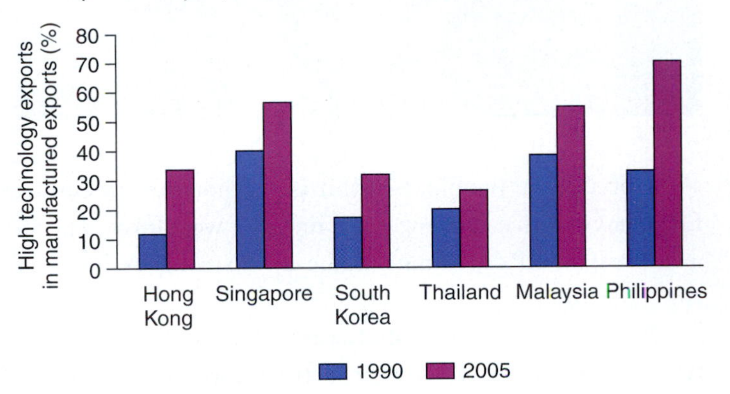

d. Funan Digitalife Mall
Also known as Singapore's IT Mall, Funan Digitalife Mall is not only a place to find the latest high-tech gadgets but also reflects the country's reorientation toward the manufacture and consumption of higher value-added goods.

called a free-trade zone, is an industrial area that operates according to different policies than the rest of the country in which it is located in order to attract and support export-oriented production. For example, an EPZ may be a duty-free zone or may have simplified customs and/or tax regulations. EPZs also provide access to transportation and communication networks. As an additional incentive to firms, trade union activity in EPZs is often prohibited or closely monitored.

EPZs vary in size, but most usually cover about half a square mile (1.3 km²) and are fenced. EPZs are not unique to the developing world, however; the geography of manufacturing in the periphery and semiperiphery has been strongly influenced by the explosive growth in the number of EPZs. Between 1975 and 2006, for example, the number of EPZs increased from 79 to 3,500, and EPZs now employ

Percentage of global manufacturing value added (MVA), 2005 • Figure 10.10

Industrialized economies account for nearly 70% of the global MVA, whereas industrializing economies account for just over 30%. Together, the Asian NIEs and China help explain the importance of the East Asia and Pacific region to global MVA. (*Source*: Data from UNIDO, 2009.)

South Asia, 1.8%
Sub-Saharan Africa, 0.7%
Middle East and North Africa, 2.2%
Latin America and the Caribbean, 6.4%
East Asia and the Pacific, 17.5%
Least developed countries, 0.3%
Countries with economies in transition, 1.7%
Industrialized economies, 69.4%

Maquiladoras and manufacturing • Figure 10.11

a. A maquiladora in Ciudad Juarez, Mexico

These women make wire harnesses for automobiles. Work at other U.S.-owned maquiladoras includes assembling garments, furniture, car seats, and televisions, among other products.

b. Percentage difference between female and male wages in manufacturing in selected countries and areas, 1995 and 2004

Whether they work in maquiladoras or other manufacturing plants, women tend to be paid less than men. Sometimes this pay gap stems from the fact that women have not had as much job experience or as many years of schooling as men, but gender discrimination remains a persistent cause as well. (*Source*: ILO, 2007.)

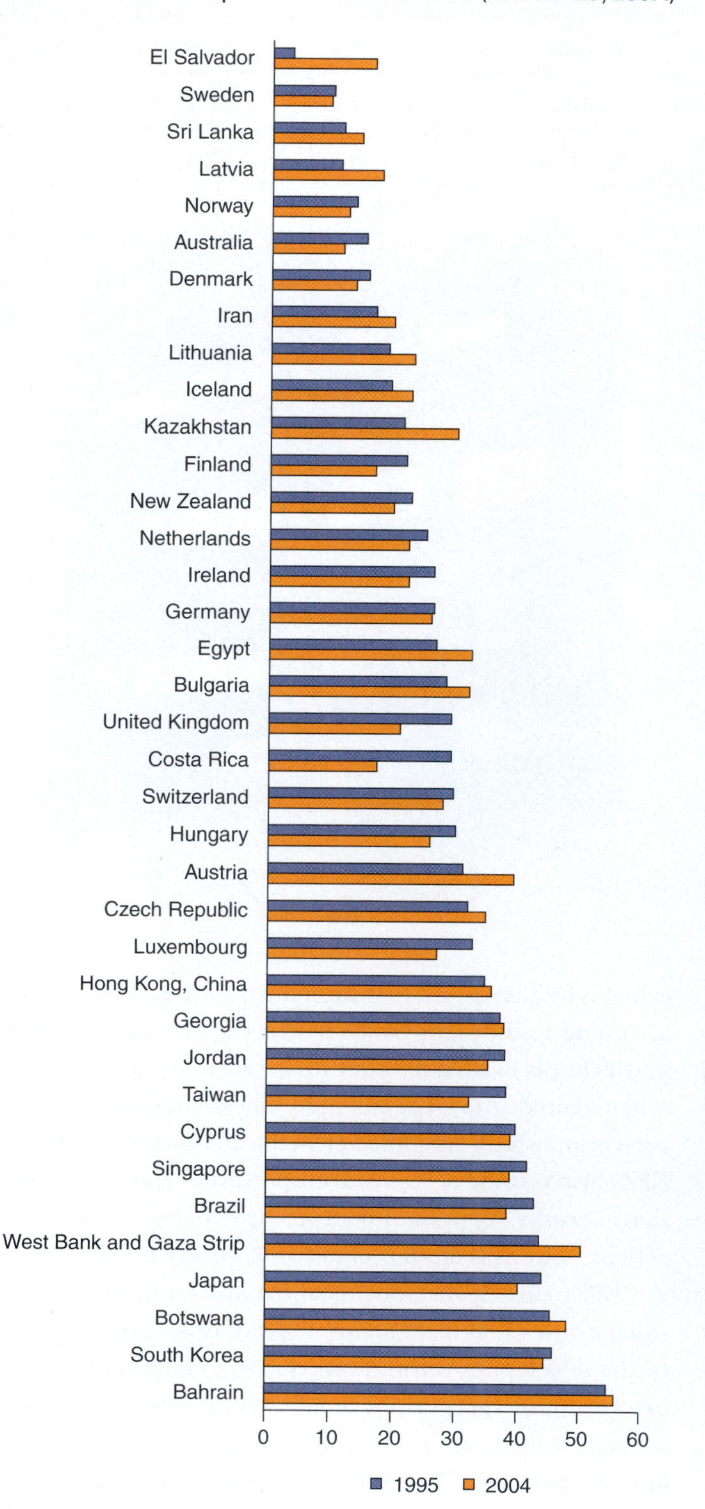

some 66 million people, two-thirds of whom are in China. EPZs have drawn more women into the workforce, and in a number of EPZs, women make up a strong majority of the labor force.

EPZs have been promoted as a strategy for helping countries industrialize. They help attract foreign investment and trade, can enable the production of new, nontraditional exports, and can generate jobs. EPZs, however, have also been criticized. From a geographic standpoint, they can exacerbate uneven development by concentrating resources and infrastructure in them, to the neglect of other regions. From a labor standpoint, EPZs vary considerably in their practices and treatment of workers. On the one hand, studies indicate that workers benefit from the ability to earn bonuses and overtime. Similarly, worker pay, though it is low by Western standards, is often higher than that provided by other jobs outside of the EPZ. On the other hand, some factories in EPZs have been likened to sweatshops where labor is severely exploited. For example, low wages, long hours, and the failure of firms to pay workers for their overtime contributed to the worker protests in an EPZ in Bangladesh in May 2006.

Maquiladoras A *maquiladora,* or *maquila,* is a manufacturing plant, often foreign-owned, that receives duty-free imported materials, assembles or processes them, and then exports them. A maquiladora can be thought of as an EPZ that consists of a single factory. Historically, maquiladoras have been associated with Mexico, but today the term *maquiladora* is sometimes used to describe similar assembly-for-export plants in other parts of Latin America and the Caribbean.

In Mexico, maquiladoras were part of a government-based strategy to alleviate unemployment in the states along the border with the United States and to disperse some of the industry away from the region around Mexico City. Following the implementation of the North American Free Trade Agreement (NAFTA) by the United States, Canada, and Mexico in 1994, the number of maquiladoras in Mexico surged. The reasons for this surge include geographic proximity to the American market, low wages in Mexico, and a growing U.S. economy. At their peak, maquiladoras employed more than 1 million laborers in 3,500 different plants located mainly along the U.S.-Mexico border and in central Mexico (**Figure 10.11**).

Since 2000, however, the maquiladora industry in Mexico has struggled, as a result of several factors including a slump in the U.S. economy and new regulations stipulating that maquiladoras have to pay taxes on goods imported from non-NAFTA countries. Within just the electronics sector—an industry relying on many components from Asia—employment in the maquiladoras dropped 30% by 2002. This example points to the geographically "footloose" nature of assembly plants in general, which tend to migrate to the places that can provide the lowest costs.

Special economic zones Special economic zones (SEZs) are export-processing zones established in China as part of a national policy to create a more open, market-oriented economy. When this policy was first implemented in 1979, the creation of such "islands" of capitalism within communist China represented a massive change in the economic functioning of the country. Therefore, SEZs were developed on an experimental basis. To restrict the spread and influence of capitalism, SEZs were initially created in just four cities (**Figure 10.12**).

Like other export-processing zones, China's SEZs were conceived as a tool to attract foreign investment with a variety of incentives, including tax holidays, exemptions from customs duties on imported and exported goods, and reduced rates on the lease of land or buildings.

Two characteristics of SEZs differentiate them from other export-processing zones. The first is their size. As a whole, SEZs tend to be larger than EPZs. For example, the Shenzhen SEZ covers about 126 square miles (328 km^2)—nearly twice the size of Washington, D.C. The second characteristic is that they tend to be more comprehensively conceived. That is, in addition to the production of goods for export, other economic facets of the SEZs are promoted such as research and development as well as tourism. Arguably, China has made a concerted effort to use SEZs to generate both forward and backward linkages.

Few scholars dispute the overall success of China's SEZs. Between 1980 and 1988, for example, Shenzhen's exports as a percentage of all exports from Guangdong Province grew from less than 1% to nearly 25%. As a result of this and similar success in other SEZs, China has continued to transform its economy in a variety of ways. In 1984 the country "opened" 14 cities to foreign investment, enabling the creation of multiple SEZs within them. These open areas have since been extended geographically to include major delta, peninsular, and island zones, as well as interior locations (refer again to Figure 10.12). Numerous other countries in Asia and Latin America have recently made plans to create their own SEZs.

China's special economic zones • Figure 10.12

Zhuhai, Shenzhen, Shantou, and Xiamen were the first SEZs. Hainan Island was then added as a fifth SEZ. Until recently, the geography of China's SEZs and open areas has been predominantly coastal.

How manufacturing value added changes across the commodity chain • Figure 10.13

Value added in manufacturing and the required skill level of the workers increase as the work performed changes from assembly to brand-name designs and research and development. Places that have export-processing zones and large pools of low-skilled labor tend to be especially attractive to low value-added types of manufacturing.

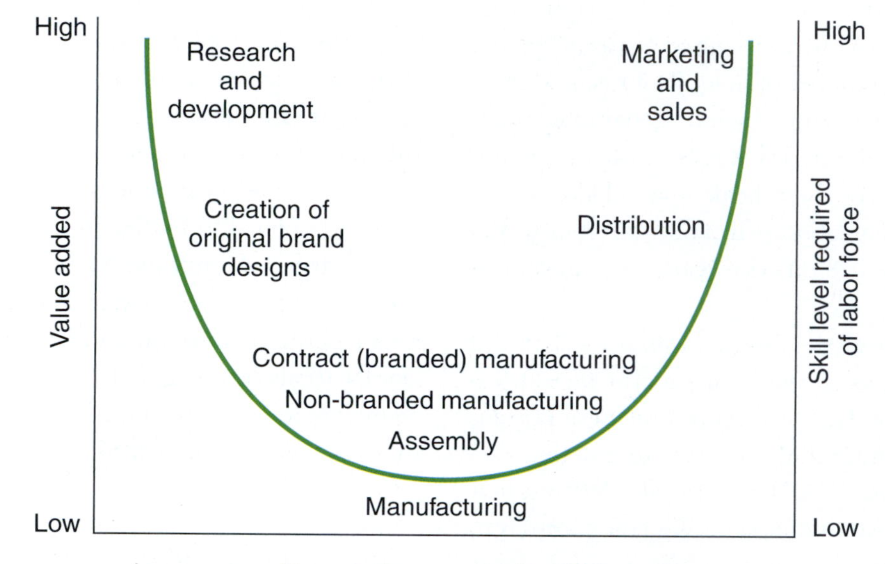

Offshoring

The establishment of export-processing zones is linked to the offshoring of certain aspects of manufacturing to developing regions. We can better comprehend why semiperipheral and peripheral areas have become popular locations for offshore manufacturing when we consider the relationship between different types of manufacturing and the value they add to the finished product (**Figure 10.13**).

It is easier to comprehend the extent and impact of offshoring when we realize that companies such as Nike, Reebok, Ikea, and The Gap do not own any manufacturing plants. Rather, these companies create original brand designs, and they offshore and subcontract the production of their goods mainly to manufacturing plants in developing countries in a process called contract or branded manufacturing. By some estimates, Nike subcontracts 61% of its product manufacturing to Southeast Asian countries and 38% to China.

Propelled by multinational corporations, offshoring has had three important geographic consequences. First, it has given manufacturing a much more global character because different operations—from the supply of materials and components to the assembly, production, and sale of final products—now take place in countries other than the one in which a firm is based. Second, it has contributed to a *new* international division of labor such that certain kinds of manufacturing and product assembly are increasingly handled by countries in peripheral and semiperipheral regions. The term *new* is used to draw a contrast with the old or historical division of labor associated with commodity extraction in the semiperiphery and periphery and manufacturing in the core (see Chapter 9). Third, offshoring has impacted the geography of profit captured from manufacturing (**Figure 10.14**).

CONCEPT CHECK 

1. **Why** have the Asian NIEs been able to increase their MVA?
2. **Where** have export-processing zones tended to concentrate, and why?
3. **How** has the creation of globalized commodity chains affected the maquiladora industry in Mexico?

Manufacturing value added and profit captured in a video iPod • Figure 10.14

Production of the Apple video iPod demonstrates the geography of manufacturing value added (MVA), but there is also a geography to the value captured (profit) realized from sales.

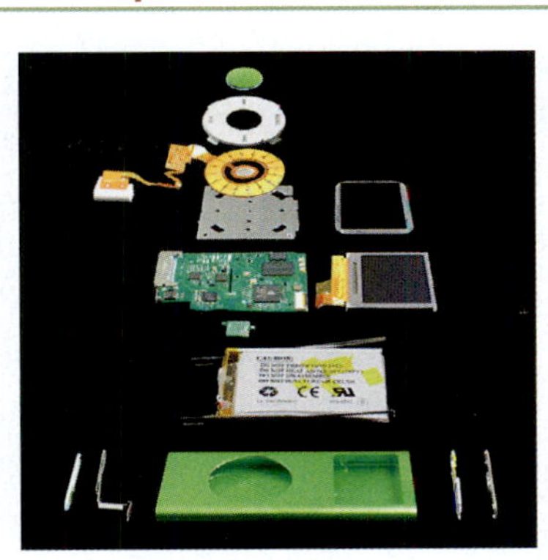

1 Manufacturing an Apple 30 GB video iPod
This iPod contains both some high-cost components, such as the hard drive and display module, and many low-cost components including the video processor, memory, and display driver.

2 Geography of countries supplying the video iPod's most expensive inputs
All purchased inputs for a video iPod total $144. Subtracting that amount from the cost of the finished product, $299, generates a MVA of $155. As shown below, Japan creates most of the MVA in a video iPod. (*Source*: Data from Linden, Kraemer, and Dedrick, 2007.)

Country	Component or Process	Estimated Cost ($)
Japan	Hard drive display module and driver	97
USA	Video processor and CPU	13
Taiwan	Insert, test, assembly	4
Korea	SDRAM, mobile memory	2
Subtotal of most expensive inputs		116
Other purchased inputs		28
	Total	144

3 Geography and flow of profit captured and role in the commodity chain
Apple outsources the manufacture of certain components to companies including Toshiba (Japan) and Samsung (Korea). (*Source:* Data from Linden, Kraemer, and Dedrick, 2007.)

Country	Role	Profit ($)*
USA	Brand design, distribution, retail	163
Japan	Original part supplier	25
Korea	Original part supplier	1
Total		189

*From sale in the USA

4 Breakdown of profit for the United States from a $299 video iPod
Note that sale of a video iPod in another country, such as Japan, shifts the geography of profit, because Japanese companies are likely to handle more of the distribution and retail. (*Source*: Data from Linden, Kraemer, and Dedrick, 2007.)

a. Apple, Inc. branding captures $80, or 49% of U.S. profit.

b. Microchip manufacture for the video iPod may occur in the U.S. or offshore, possibly in Taiwan. This manufacturing process captures $8, or 5% of U.S. profit.

c. Retail and distribution capture $75, or 46% of U.S. profit.

Services

LEARNING OBJECTIVES

1. **Define** deindustrialization.
2. **Characterize** a postindustrial society.
3. **Explain** the development of technopoles.

When human geographers and economists speak of the structural makeup of an economy, they are referring to the comparative importance of the primary, secondary, and tertiary sectors as contributors to GDP and/or as sources of employment. Among developed countries a common trend in the structural change of an economy is for the primary sector to be the most important sector initially, followed by growth in industry and expansion of the service sector. This section explores the spatial patterns of structural or sectoral change.

Deindustrialization and Globalization

The crisis of Fordism marked a period of structural change within the core industrialized countries that involved declines in manufacturing employment. But these declines do not mean that the core industrialized countries have lost their global dominance in manufacturing, whether that dominance is measured by the value added in dollars or by a percentage of global manufacturing value added. As we have seen, industrialized countries still generate nearly 70% of global MVA. The geography of *growth* in MVA, however, has shifted significantly in recent decades (**Figure 10.15**).

Job losses in manufacturing in core countries raise questions about the causes and consequences of **deindustrialization**. Three broad explanations help account for deindustrialization: (1) greater productivity gains from manufacturing than from services; (2) changing resource endowments; and (3) economic globalization. We will discuss each of these in turn.

deindustrialization The long-term decline in industrial employment.

According to the first explanation, productivity gains in manufacturing outpace productivity gains in services, and this triggers deindustrialization. Think of the gains in productivity of the systems of flexible production we discussed earlier in the chapter in which one assembly line can manufacture several

Trends in manufacturing • Figure 10.15

a. Manufacturing as a percent of total employment for selected countries, 1970–2003

The declines recorded here point to noteworthy changes in employment patterns and signal change in the manufacturing sector and economic structure of these countries. (*Source*: Pilat, Cimper, Olsen, and Webb, 2006.)

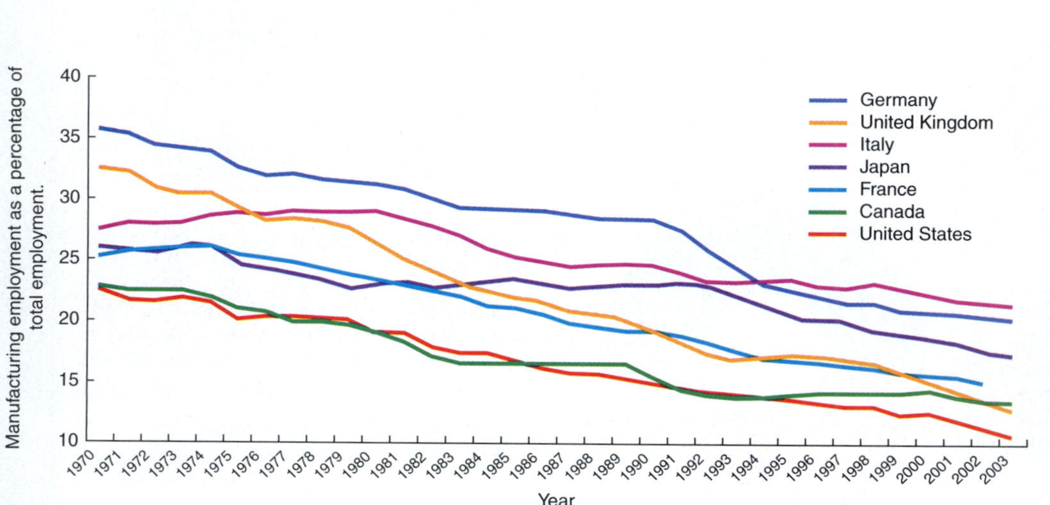

b. Growth rates in manufacturing value added for the world and different regions, 1995–2000 and 2000–2005

What do these trends suggest about the globalization of manufacturing? (*Source*: Data from UNIDO, 2009.)

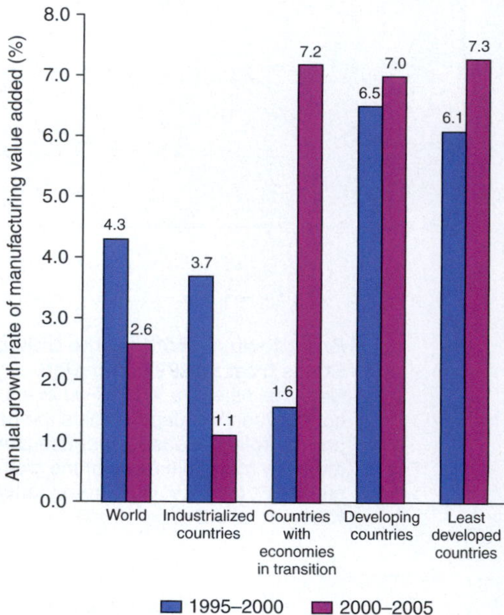

Deindustrialization and service sector growth • Figure 10.16

a. The former Bethlehem Steel mill in Pennsylvania. The U.S. steel industry experienced substantial declines between the 1960s and 1980s and lost its comparative advantage, in part because it was unable to compete with more efficient steel manufacturers in Japan and Korea. Steel plants from Pennsylvania to Illinois closed, resulting in a 65% drop in employment. Dubbed the "Rust Belt," the region's abandoned mills became icons of its deindustrialization.

b. In 2009, a casino opened at the site of the former steel mill, promising to revitalize the city of Bethlehem and highlighting the process of structural change in the local economy.

different models of vehicles. Now contrast this example with one of a lawyer who has to learn the details of each specific case and client.

These differences in productivity are also related to the adoption of technology to perform automated tasks. In contrast to manufacturing, fewer job tasks in the service sector can be mechanized. Thus, productivity gains in manufacturing not only enhance industrialization, they also trigger deindustrialization.

A second explanation of deindustrialization draws on the significance of changing resource endowments. When resource endowments change, a firm in another country may be able to take advantage of lower costs of production and, as a consequence, gain a comparative advantage in the manufacture of a good (**Figure 10.16**).

A third and related explanation of deindustrialization involves economic globalization and the new international division of labor that has shifted manufacturing jobs to developing regions. Globalization and the increased trade that it has fostered have helped to disperse manufacturing, with consequences for deindustrialization. For a different example of the nature of sectoral change, see *Video Explorations*.

Video Explorations

Essaouira, Morocco

WILEY PLUS NATIONAL GEOGRAPHIC Video

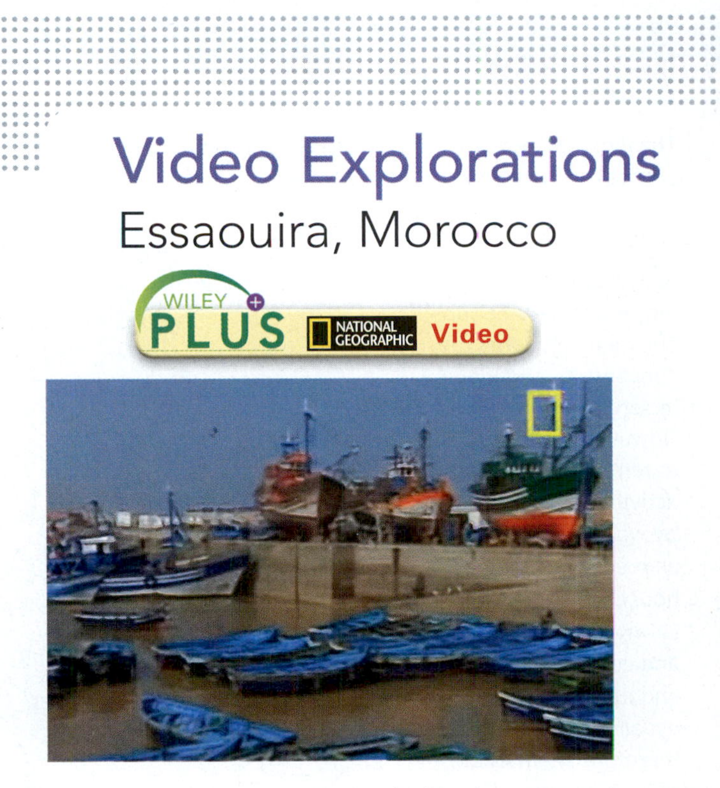

As fishing opportunities in Essaouira decline, its economy is increasingly focused on tourism. http://www.natgeoeducation-video.com/film/421/essaouira

Global share in employment, by sector, 1970 and 2008 • Figure 10.17

These pie charts show the proportional growth in the service sector. The geography of employment in the service sector, however, remains highly uneven. Today, nearly three-quarters of the labor force in developed countries is employed in services, yet in developing countries service jobs employ just over one-third of the labor force. (*Source*: Data from USAID, 2000; ILO, 2009.)

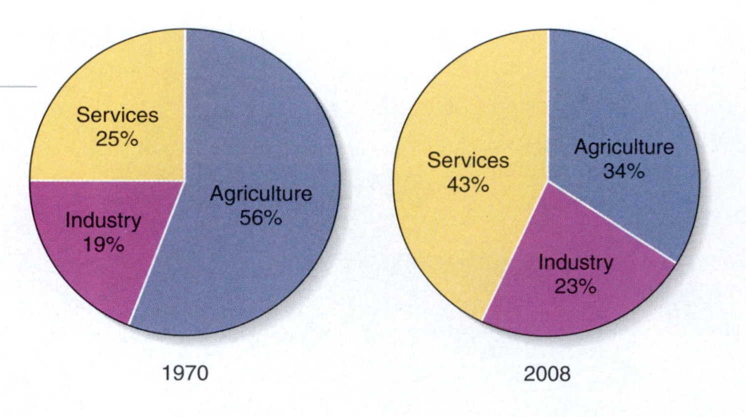

Services, Gender, and Postindustrial Society

The rapid growth of services has brought about a significant change in the global economic structure. The percentage of people employed in services worldwide has nearly doubled since 1970, and the service sector now employs a higher percentage of the world's people than agriculture (**Figure 10.17**).

When the world's countries are ranked by the share of their GDP derived from services, the United States comes out at the top. Remarkably, services account for 77% of the U.S. GDP. In general, services constitute a higher percentage of the GDP (approximately 71%) in developed countries compared to less than 50% of the GDP in developing countries.

Geography InSight — Gender mainstreaming in the service sector • Figure 10.18

The aim of gender mainstreaming is to promote the equitable participation of women and men in all types of service jobs.

a. Percentage of women in service jobs The map highlights the concentration of women in services worldwide. However, many essential service activities performed by women—such as preparing meals, housekeeping, caring for children, and collecting water and fuelwood—are usually not counted, and therefore are invisible, in official statistics. (*Source:* UNDP, 2007.)

Service sector growth has also had a major impact on the employment of women. Even though women form a significant part of the labor force in EPZs and maquiladoras, on a global scale neither agriculture nor industry has as feminized a labor force as the service sector does. Between 1995 and 2005, women accounted for 75% or more of the employment in services in more than 50 countries. Data for Saudi Arabia indicates that employment of women in services is 98%. This is a result of practices that have channeled women into careers in teaching and health care, while presenting few opportunities for them to obtain degrees in fields such as engineering and architecture.

It is not surprising, then, that when we examine specific occupations more closely—both at the country level and worldwide—women are strongly concentrated in health, education, and social services. This creates a kind of gendered occupational segregation (**Figure 10.18**).

Growth in the service sector not only points to important changes in employment patterns, but is also associated with the emergence of **postindustrial societies**. There are five main characteristics of postindustrial societies: (1) high levels of urbanization; (2) dominance of the service sector, especially in total employment; (3) prevalence of "white-collar" workers (professionals and highly skilled specialists) in the labor force; (4) an infrastructure heavily based on information and communication technology (ICT) such as computer software, networks, and satellites; and (5) a knowledge-based economy. In a *knowledge-based economy* expertise, know-how, and resourceful ideas drive innovation and create value. Knowledge is a productive asset that rivals traditional productive assets such as land or labor. If Fordist mass production was the dominant system of production in developed countries from World War II until the 1980s, then the ICT system prevails in postindustrial societies. Fordist production systems based on linear processes have been supplanted by networked systems driven by high-tech innovations.

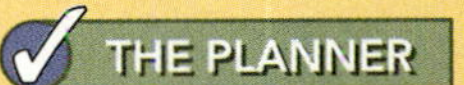

b. A male teacher with students
Teaching at the preschool and primary school levels has long been feminized. What impact do you think gender mainstreaming will have on gender roles?

c. A female manager consults with some workers
Worldwide, women are severely underrepresented in senior management positions, holding fewer than 10% of them.

Technopoles • Figure 10.19

Factors conducive to the establishment of technopoles

1. Highly educated labor force (e.g., college graduates and postgraduates)
2. Presence of research labs and facilities
3. Venture capital to support new enterprise
4. Infrastructure that supports high connectivity (e.g., fiber optic networks)
5. Attractive location (e.g., corporate campuses, residential amenities)

a. Technopoles may grow in urban, rural, or suburban settings. What impacts do technopoles and knowledge-based economies have on those who are less educated?

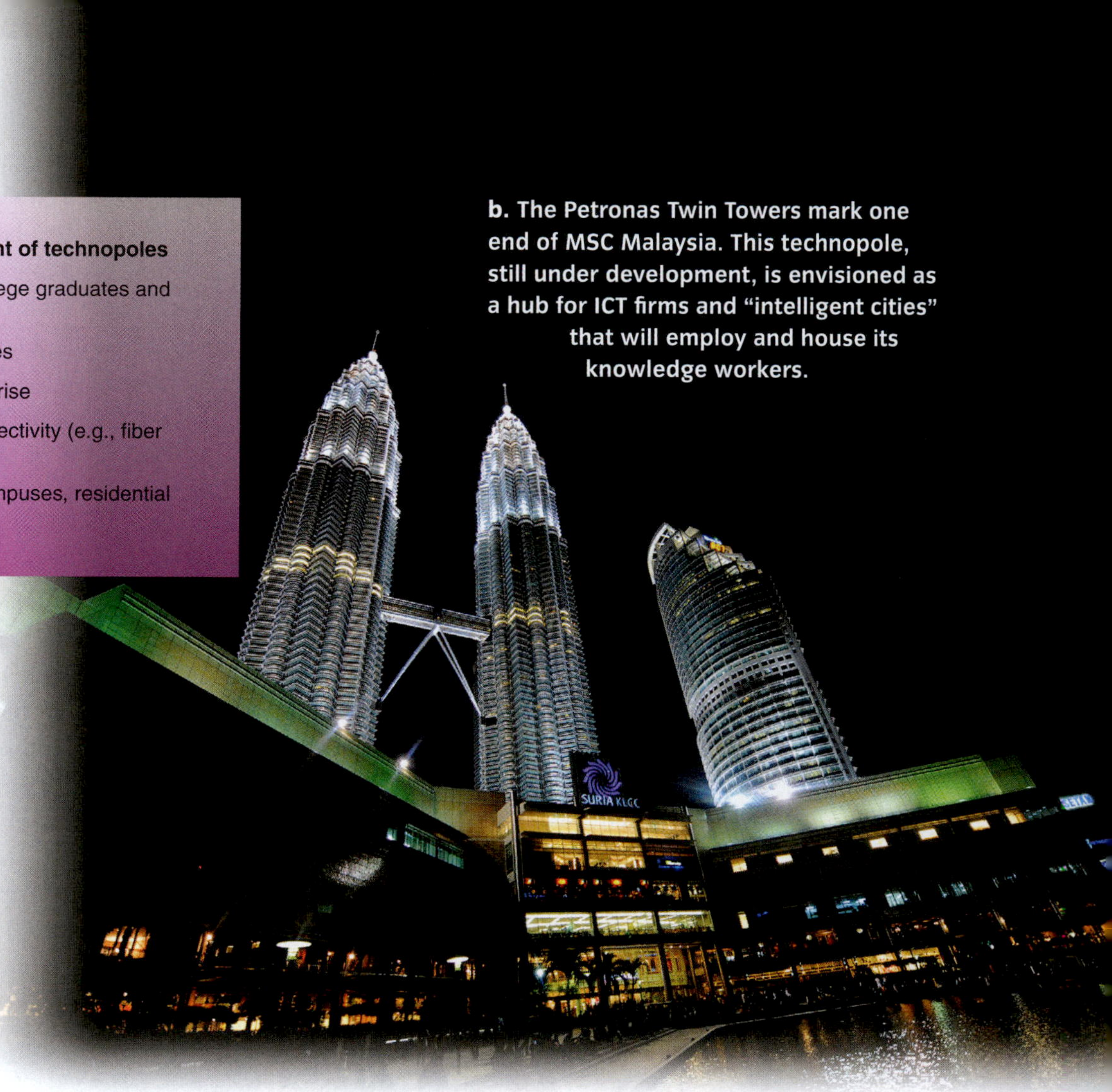

b. The Petronas Twin Towers mark one end of MSC Malaysia. This technopole, still under development, is envisioned as a hub for ICT firms and "intelligent cities" that will employ and house its knowledge workers.

One of the best indicators of a knowledge-based economy is the amount spent on research and development (R&D). High-income countries spend three or more times as much money on R&D as do middle- and low-income countries. Funds for R&D are provided by a mix of private firms, nonprofit organizations, and government agencies. One of the landscape expressions of R&D is the **technopole**—an area with a cluster of firms conducting research, design, development, and/or manufacturing in high-tech industries such as wireless communications, integrated circuitry, or software development. California's Silicon Valley, which began to develop as a technopole in the 1950s, is the recognized prototype. Google, Intel, Hewlett-Packard, and numerous other ICT firms have their headquarters in the Silicon Valley.

Technopoles provide a good example of agglomeration that is associated with high-tech industries. Manuel Castells, a sociologist who has written several books on the information revolution, gives three reasons for the development of technopoles: reindustrialization, regional development, and synergy. The first two reasons are closely related. In an area that has experienced job loss, creation of a technopole can provide a source of employment. Similarly, as nodes of economic activity, technopoles can rejuvenate regional economies. In some cases, technopoles are specifically used as a strategy for urban and economic development. By synergy, Castells refers to the benefits achieved through cross-fertilization. Technopoles can be thought of as "incubators of innovation" that capitalize on agglomeration and the reduced transaction costs that agglomeration brings. As shown in **Figure 10.19**, several factors are conducive to the establishment of technopoles.

Spatially, technopoles usually begin as a collection of nodes that, over time, grow along transportation routes. The Research Triangle in North Carolina has coalesced around the cities of Raleigh, Durham, Chapel Hill, and the several highways and interstates that serve the area. Silicon Fen has developed around Cambridge, England; Sweden has its Wireless Valley outside Stockholm; and the

Sophia Antipolis technopole has grown between Nice and Cannes in France.

Although technopoles are not unique to the developed world, their presence in semiperipheral and peripheral countries is often the result of planned government initiatives. Since 1999 the government of the State of Karnatka, India, has promoted the development of an IT Corridor on the outskirts of Bangalore; it is now called the Silicon Valley of India. Similarly, the city of Hyderabad, also in India, has supported the creation of Hi-Tec City (Hyderabad Information Technology and Engineering Consultancy), sometimes referred to as "Cyberabad." Other technopoles include the MSC (Multimedia Super Corridor) in Malaysia, and the emerging Smart City in Cairo.

CONCEPT CHECK STOP

1. **How** are industrialization and deindustrialization interconnected?
2. **What** is the dominant system of production in postindustrial societies?
3. **Where** are technopoles likely to develop, and why?

Summary

THE PLANNER

1 Types of Industry 296

- Human geographers recognize the existence of three types of industry. **Primary industry,** shown here, is extractive; **secondary industry** includes manufacturing and is transformative; and **tertiary industry** provides services to consumers or other businesses.

Types of industry • Figure 10.1

- **Staple theory** explores the relationship between primary industry, the creation of **linkages,** the economic development of a region or country, and commodity dependency. In the developing world, commodity dependency has weakened since the 1980s but remains a serious problem for many countries.
- The **Industrial Revolution** refers to interrelated changes in technology and changes in the system of production of goods that began in England in the late 18th century. Greater access to capital combined with technological innovations helped unleash the Industrial Revolution. Geographically, the diffusion of the Industrial Revolution has been highly uneven at local, national, and global scales.
- Expansion of the service sector since the Industrial Revolution makes it helpful to distinguish among **tertiary, quaternary**, and **quinary services**.

2 Evolution of Manufacturing in the Core 301

- A key aspect of the geography of manufacturing includes identifying the optimal location for firms. Consequently, geographers examine the costs and benefits of diverse site and situation factors. They also try to understand how such factors may change over time and what the consequences for their firm might be.
- The diffusion of the Industrial Revolution and **agglomeration** help explain the emergence of the major centers of manufacturing in the core prior to the 1970s. Within the core, **Fordism** and **Taylorism** strongly influenced the systems of production in North America and Western Europe. Between the 1930s and the 1970s, **vertical integration** developed in concert with producer-driven **commodity chains** geared toward the mass production of standardized goods for mass consumption, sometimes involving vast scales of production (see photo).

A portion of Ford's River Rouge Plant in the 1940s • Figure 10.7

- **Flexible production**, a system pioneered in Japan, contributed to the crisis of Fordism in the 1970s and played a major role in reshaping the geography of manufacturing with the use of **just-in-time delivery** and **outsourcing**.
- Outsourcing may or may not involve **offshoring,** but the occurrence of offshoring is significant for the globalization of manufacturing.

3 Evolution of Manufacturing Beyond the Core 307

- By the 1980s, remarkable gains in industrial output had been achieved by the **newly industrialized economies (NIEs)** in Asia.
- The practice of offshoring and the existence of **export-processing zones**, including maquiladoras such as this one, have helped diffuse manufacturing into semiperipheral and peripheral areas to a greater extent than ever before. This diffusion has contributed to the growth of transnational supply chains and has resulted in a new international division of labor.

Maquiladoras and manufacturing • Figure 10.11

4 Services 314

- Recent structural change in the core industrialized countries involves **deindustrialization** and the expansion of the service sector. Industrialization and deindustrialization are complicated and interconnected processes.
- Globally, the percentage of people employed in the service sector has nearly doubled since the 1970s. Tertiary industries have high concentration of women, more so than either primary or secondary industries. Gender mainstreaming constitutes one effort to address these occupational disparities.
- The emergence of a knowledge-based economy built around information and communication technologies and **technopoles,** such as the one shown here, indicates the ongoing importance of technological advances to structural or sectoral change, and specifically the emergence of **postindustrial societies**.

Technopoles • Figure 10.19

Key Terms

- agglomeration 302
- commodity chain 304
- deindustrialization 314
- export-processing zone 308
- flexible production 306
- Fordism 303
- Industrial Revolution 298
- just-in-time delivery 306
- linkages 297
- manufacturing value added (MVA) 307
- newly industrialized economies (NIEs) 307
- offshoring 306
- outsourcing 306
- postindustrial societies 317
- primary industries 296
- quaternary services 300
- quinary services 300
- secondary industries 296
- staple theory 297
- Taylorism 302
- technopole 318
- tertiary industries 296
- vertical integration 303

Critical and Creative Thinking Questions

1. Staple theory has been criticized for being environmentally and technologically deterministic. Explain the rationale for these criticisms.
2. Review the two different methods for measuring commodity dependency on page 298. Discuss the advantages and disadvantages of each.
3. Identify an industrial firm in your area and do some research to explain what it produces and why it located where it did.
4. In 1942, the economist Joseph Schumpeter used the term *creative destruction* to explain that innovation drives competition and economic growth, but ironically, also unravels the status quo as old business practices and technology give way to newer ones. Why is the concept of creative destruction significant, and how is it relevant to the geography of industry and services?
5. What is a "right-to-work" state, and do you live in one? Has right-to-work legislation affected industry in your region?
6. To what extent does specialization in the provision of services resemble Fordism? Justify your response.
7. In addition to research and development expenditures, what other national data might be used to identify a knowledge-based economy?
8. The greater participation of women in the workforce might be considered both advantageous and problematic for women. Why?

What is happening in this picture?

These women break up and carry rocks at a construction site in Udaipur, Rajasthan, India. In parts of South and Southeast Asia, women and men work alongside one another in construction, though most laborers are male.

Think Critically

1. What does this photo suggest about the geography of gender mainstreaming?
2. What factors contribute to the gendering of certain occupations?

Self-Test

(Check your answers in Appendix B.)

1. Identify two service activities associated with each category and write them in the appropriate boxes

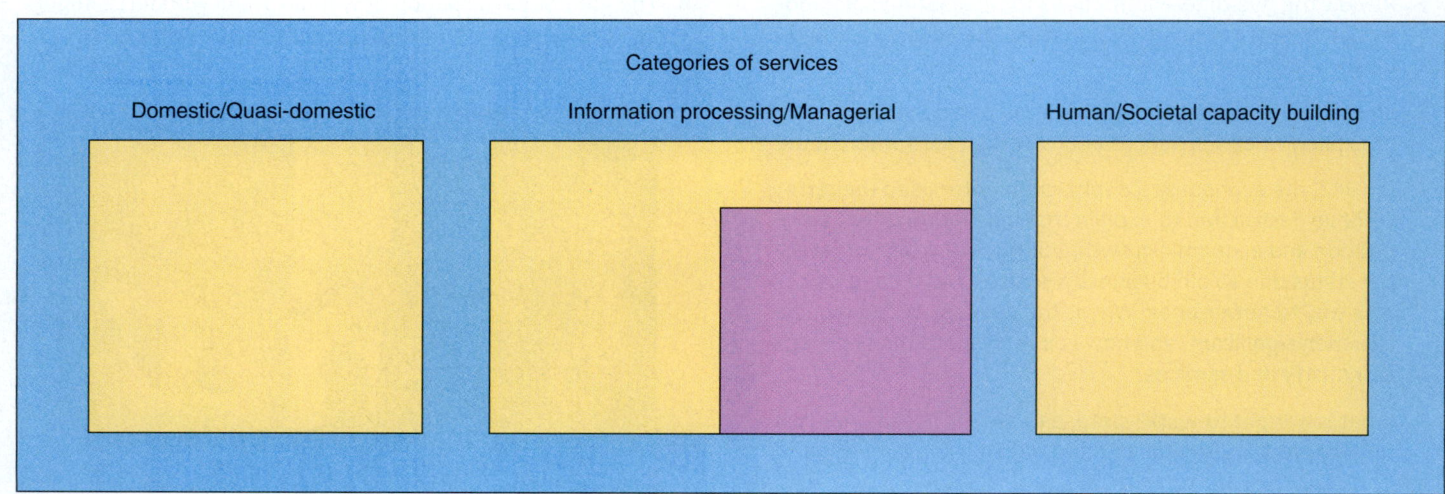

2. What kind of activity is advanced professional services?
   a. primary
   b. secondary
   c. quaternary
   d. quinary

3. The construction of railroad lines to supply a steel mill with iron ore and coal provides an example of _______ linkages.
   a. forward
   b. backward
   c. demand
   d. industrial

4. Staple theory _______:
   a. explains the growth of export-processing zones
   b. uses natural resources to explain an area's economic development
   c. relates secondary industry to Fordism and Taylorism
   d. accounts for the growth of services

5. _______ is considered to be the hearth of the Industrial Revolution.
   a. England
   b. Germany
   c. The United States
   d. Belgium

6. Identify the characteristic below that is associated with a Fordist system of production.
   a. multiskilled labor
   b. little inventory
   c. producer-driven commodity chain
   d. flat management style

7. Manufacturing value added (MVA) is defined as _______:
   a. the cost of the finished product
   b. the sum of all inputs used to make a product
   c. the value gained by putting a brand name on a manufactured good
   d. the cost of the finished product minus the cost of all inputs used to make it

8. Using your knowledge of MVA, place these labels on the diagram: research and development; distribution; contract manufacturing; marketing and sales; assembly; creation of original brand designs.

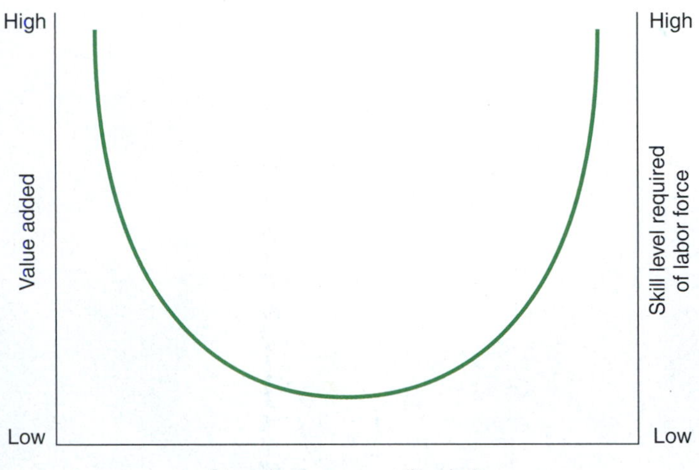

9. _______ is a second-tier newly industrialized economy.
   a. Thailand
   b. Hong Kong
   c. Singapore
   d. Taiwan

10. Location in the global periphery and simplified tax regulations are characteristics of _______.
    a. outsourcing
    b. export-processing zones
    c. multinational corporations
    d. commodity dependency

11. Briefly identify three factors that help explain the rise of the Asian NIEs as important centers of manufacturing.

____

____

____

12. Explain the difference between outsourcing and offshoring.

____

____

____

____

13. Use these images to explain what geographers mean by structural or sectoral change.

____

____

____

14. Which statement about offshoring is *false*?
    a. Offshoring originated with China's special economic zones.
    b. Offshoring has affected the global character of manufacturing.
    c. Offshoring has affected the geography of profit from the sale of manufactured goods.
    d. Offshoring is associated with a new international division of labor.

15. As discussed in the chapter, which of the following is not one of the reasons given to explain the development of technopoles?
    a. synergy among local businesses
    b. a need to reindustrialize
    c. a desire to promote regional development
    d. structural change associated with deindustrialization

THE PLANNER

Review your Chapter Planner on the chapter opener and check off your completed work.

11 Agricultural Geographies

THE RISE OF FAIR TRADE

Over the past two decades, the price of coffee has fluctuated erratically, falling to record lows in some years with devastating effects on coffee producers, most of whom are small farmers. In fact, about 25 million small farmers grow and harvest nearly three-fourths of the world's coffee. Even though it is not unusual for Americans to pay as much as $25 for a pound of specialty Ethiopian coffee, most of the small coffee farmers in Ethiopia remain impoverished and earn less than $10 a week. Simultaneously, coffee purchasers such as Kraft and Nestlé earn large profits and control much of the revenue generated from coffee sales.

Enter Fair Trade—a social movement and trading partnership committed to improving the plight of farmers whose situation has worsened rather than improved as a result of global trade. In these partnerships, companies agree to pay higher, stable prices to farmers along with an additional premium for investing in social projects. In return, farmers organize cooperatives, provide safe working conditions, reinvest in community development, and adopt sustainable practices. Fair Trade products must meet other criteria and be certified.

The popularity of certified Fair Trade goods has never been greater. As a social movement, it has generated Fair Trade towns, schools, churches, and universities. In 2008 the University of Wisconsin at Oshkosh became the first Fair Trade university in the United States. The university has committed to selling Fair Trade certified products in its campus stores and dining facilities.

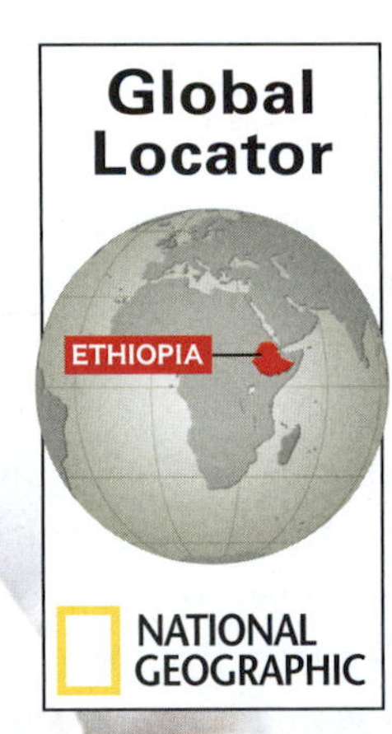

CHAPTER OUTLINE

CHAPTER PLANNER

- ❑ Study the picture and read the opening story.
- ❑ Scan the Learning Objectives in each section: p. 326 ❑ p. 333 ❑ p. 344 ❑
- ❑ Read the text and study all visuals. Answer any questions.

Analyze key features

- ❑ Video Explorations, p. 327
- ❑ Process Diagram, p. 328 ❑ p. 334 ❑
- ❑ Geography InSight, p. 330 ❑ p. 341 ❑
- ❑ What a Geographer Sees, p. 344
- ❑ Stop: Answer the Concept Checks before you go on: p. 332 ❑ p. 343 ❑ p. 348 ❑

End of chapter

- ❑ Review the Summary and Key Terms.
- ❑ Answer the Critical and Creative Thinking Questions.
- ❑ Answer What is happening in this picture?
- ❑ Complete the Self-Test and check your answers.

Agriculture: Origins and Revolutions

LEARNING OBJECTIVES

1. **Define** agriculture and identify its hearths.
2. **Distinguish** among the first, second, and third agricultural revolutions.
3. **Contrast** the Green Revolution and the Gene Revolution.

lthough we seldom think about it, our lifestyle is intimately connected with and highly dependent on **agriculture**, especially the ability to produce, process, and transport agricultural commodities. Agriculture involves the ongoing process of **domestication**—selecting plants or animals for specific characteristics and influencing their reproduction. Domestication not only makes plants and animals visibly or behaviorally distinct from their wild ancestors but also increases the interdependency between people and the domesticate. Domestication reflects human agency.

> **agriculture** Activities centered on cultivating domesticated crops and livestock in order to procure food and fiber for human use or consumption.

Until very recently, agriculture employed the highest percentage of people worldwide. As we saw in Chapter 10, the service sector now employs the highest percentage of people—about 43% of the world's workforce. In comparison, the agricultural sector employs approximately 35% of the world's workforce.

This decline in the share of agricultural employment is a testament to the ongoing urbanization of our world as well as the mechanization and industrialization of agriculture. Even so, employment in the agricultural sector differs vastly from one region and country to another. Although women have been involved in agriculture since it began, certain trends increasingly point to a new feminization of agriculture (**Figure 11.1**).

Origins of Agriculture

Hunting and gathering is the oldest method of obtaining food, and historically all people obtained their food this way. Most hunters and gatherers moved frequently in pursuit of game and seasonally available plants, although some groups that relied heavily on fishing might settle permanently in one location. Strictly speaking, however, hunters and gatherers are not classified as agriculturalists because they use wild rather than domesticated plants and animals.

> **hunting and gathering** Hunting wild animals, fishing, and gathering wild plants for food.

A geography of employment in agriculture • Figure 11.1

a. Employment in agriculture by region

Agriculture employs a higher percentage of people in Sub-Saharan Africa than in any other world region. For example, in Tanzania, Rwanda, and Ethiopia the share of the labor force employed in agriculture exceeds 80%. (*Source*: Data from ILO, 2008.)

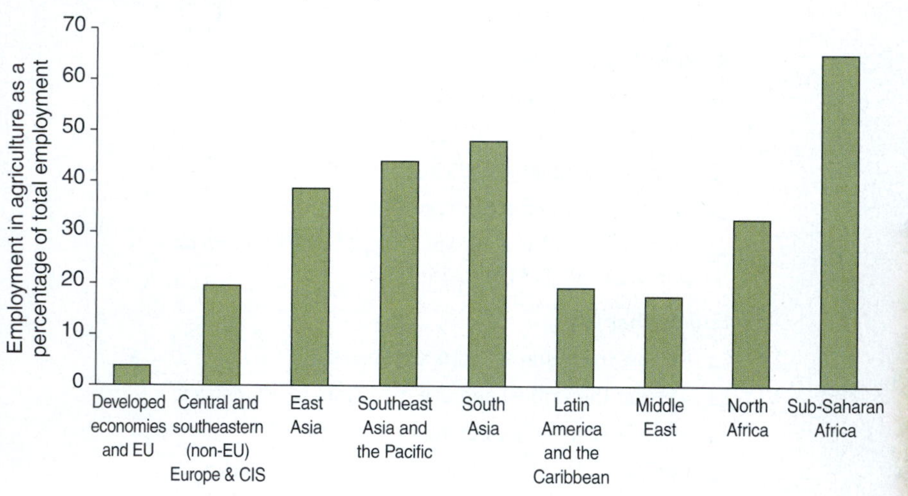

b. Women threshing millet in Niger

Women produce a majority of the food in developing countries. The new feminization of agriculture in these regions stems in part from the migration of men into cities for work, the increase in households headed by women, and the willingness of women to perform tasks once assigned to their children so that their children can attend school.

As a way of life, hunting and gathering is in decline because of the dominance of settled agriculture and its close association with the modern state and global economy. Today hunting and gathering is confined to peripheral areas, where it is practiced by small numbers of people including, among others, the San of southern Africa, some Aboriginals in the interior of Australia, and the Moken of Myanmar (see *Video Explorations*). The notion that hunters and gatherers live on the brink of starvation is a popular misconception; numerous studies have shown that hunters and gatherers are generally well nourished. Historically, the transition from hunting and gathering to farming marks the first of three sweeping revolutions that have transformed the world.

The First and Second Agricultural Revolutions

The development of agriculture constitutes the **first agricultural revolution,** which began with the domestication of plants and animals some 11,000 years ago. Most geographers agree that agriculture was independently invented at different locations and at different times (Figure 11.2).

Video Explorations

Moken

WILEY PLUS NATIONAL GEOGRAPHIC Video

This video provides an introduction to the Moken people of Myanmar (Burma), sometimes called hunters and gatherers of the sea.

Hearths of agriculture • Figure 11.2

The map shows the five hearths of agricultural innovation and three secondary centers. They are secondary centers because it is not clear if diffusion of crops and food production practices contributed to the origins of agriculture in these locations or if people independently developed agriculture there.

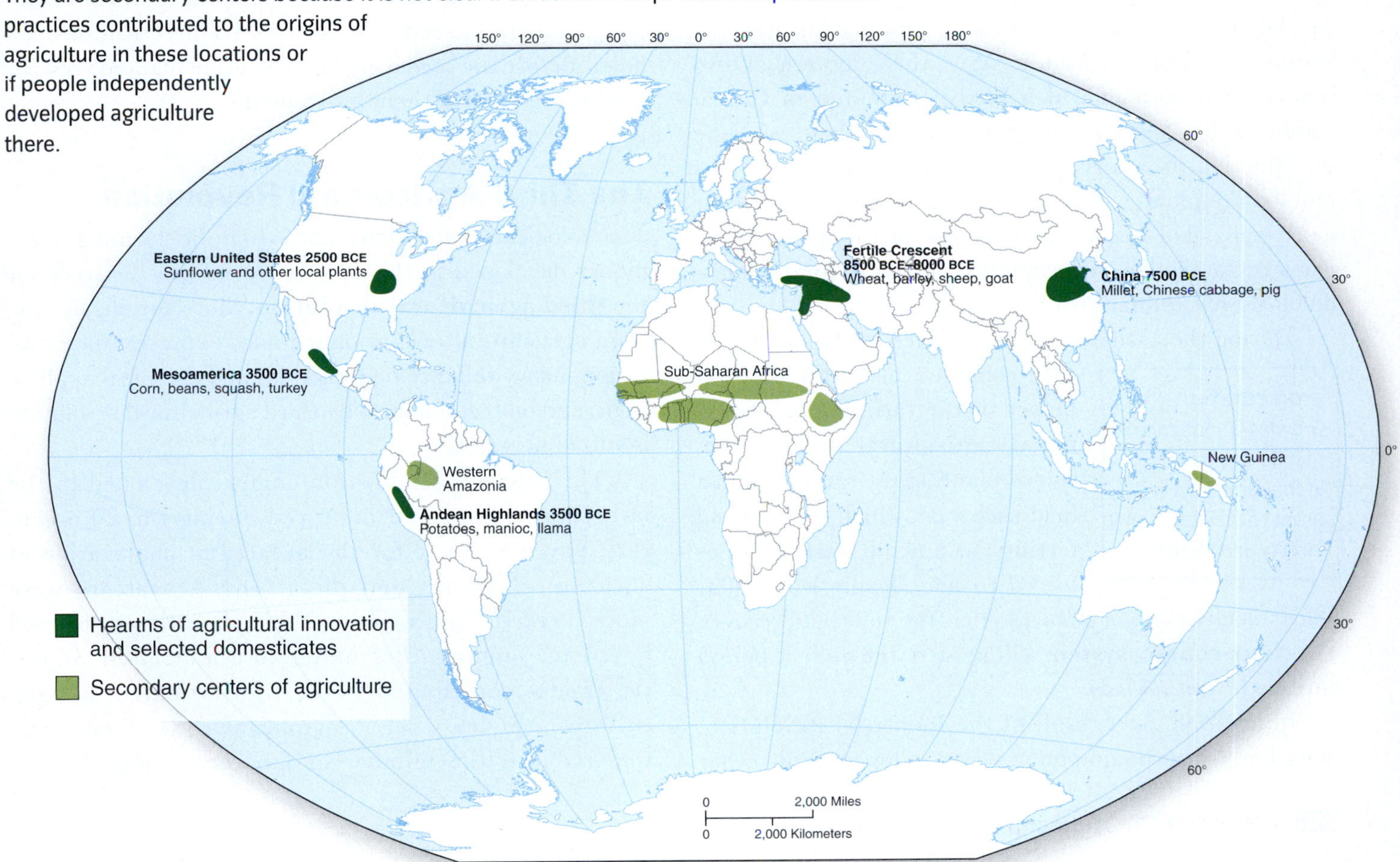

PROCESS DIAGRAM

THE PLANNER

Four-course crop rotation • Figure 11.3

This system, introduced to England from Holland in the 18th century, is based on a four-year planting regime. This crop cycle balances the planting of food crops with feed crops and incorporates legumes that enrich the soil. By removing the need for a fallow period, this practice also enables higher agricultural yields. Many variations on this system exist today.

Course 1 Small Grain Crop

A small grain, such as wheat, barley, rye, or oats is planted and provides a marketable crop.

Course 2 Root Crop

Fields planted to small grain crops are susceptible to weeds. Thus, root crops, such as turnips, are planted in rows that can be hoed to remove weeds. Turnips also provide feed for livestock.

The roots of the **second agricultural revolution** can be traced to new agricultural practices in western Europe. During the Middle Ages the adoption of two innovations, both of which likely originated in China, significantly raised farming yields. The first was the introduction of a curved metal plate used to make the moldboard plow, which enabled farmers to turn over heavy soils. The second was the use of the horse collar. Prior to the collar's introduction, farmers used oxen—much slower animals.

During the 17th and 18th centuries a modification in the technique of **crop rotation** helped boost farm yields. Traditional agricultural practice involved planting the same crop in a field each year, which reduced soil fertility; as a result, farmers were forced to periodically leave their fields uncultivated, or *fallow*, so that the soil could recover. The **four-course system** eliminated the fallow period entirely (Figure 11.3).

crop rotation Growing a sequence of different crops in the same field in order to maintain soil fertility and health.

In addition, as a result of the Industrial Revolution, more horse-drawn equipment was developed, while other tools improved the efficiency of farm practices. For example, Jethro Tull's seed drill placed seed directly into small holes. Before the seed drill, farmers often planted seeds by tossing handfuls of them into a field.

The Third Agricultural Revolution

Technological innovations and scientific farming techniques developed in the 20th century form the basis for the **third agricultural revolution.** More specifically, the third agricultural revolution includes extensive mechanization, heavy reliance on irrigation and chemical applications, and biotechnology. The third agricultural revolution is still in progress.

The internal combustion engine, developed in the late 19th century and improved during the 20th century, paved the way for the greater mechanization of agriculture, as gas and diesel engine tractors were more powerful and maneuverable than those powered by steam engines. Beginning in the United States, the tractor contributed to the transformation of agriculture in at least three significant ways. First, tractors reduced the number of laborers required for a

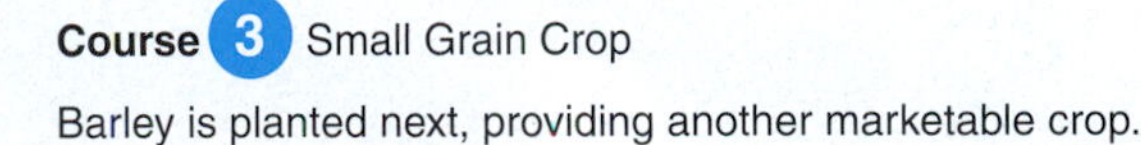

Course 3 Small Grain Crop

Barley is planted next, providing another marketable crop.

Course 4 Legume

A forage legume, such as clover, is planted. Legumes boost soil fertility by converting atmospheric nitrogen into a form more useful for plants. Livestock are grazed on the clover and their manure, also nitrogen rich, helps fertilize the soil.

particular task and simultaneously improved the efficiency and productivity of farming. Second, the tractor helped bring more land into cultivation. Third, tractors facilitated the shift to **monoculture**, leading to a substantial alteration of environments and landscapes. Similar impacts have occurred elsewhere as adoption and use of tractors continue.

monoculture
Planting a single crop in a field, often over a large area.

Scientific farming, another hallmark of the third agricultural revolution, relies on technology and synthetic chemicals to promote crop growth, deter crop disease, prevent weeds, or solve other agricultural challenges. Although irrigation has long been practiced, improved irrigation technologies have facilitated the spread of crops into areas once considered too dry for them. Since the 1960s, for example, the amount of irrigated land in the world has more than doubled. Although use of chemical fertilizers and pesticides has increased yields, our dependence on them has significant ecological costs—including pollution and greater reliance on petroleum, which is used to apply and manufacture many of them.

Agricultural biotechnology, or agro-biotech, is an additional facet of the third agricultural revolution. Broadly speaking, **agro-biotech** seeks to improve the quality and yield of crops and livestock through the use of such techniques as cross-breeding, hybridization, and, more recently, genetic engineering. The impact of agro-biotech developments on agriculture can best be understood by distinguishing between the **Green Revolution** and the **Gene Revolution**. In this context, *green* refers to the expansion of productive agriculture, not to the adoption of organic or eco-friendly practices in the way we popularly use the term today.

Green Revolution
The dramatic increase in grain production between 1965 and 1985 in Asia and Latin America from high-yielding, fertilizer- and irrigation-dependent varieties of wheat, rice, and corn.

Gene Revolution
The shift, since the 1980s, to greater private and corporate involvement in and control of the research, development, intellectual property rights, and genetic engineering of highly specialized agricultural products, especially crop varieties.

There are two fundamental differences between the Green Revolution and the Gene Revolution. The first is that innovations associated with the Green Revolution,

Geography InSight The Green Revolution • Figure 11.4

The Green Revolution grew out of an effort to alleviate world hunger. In the 1950s, scientists in Mexico developed a high-yield strain of wheat responsive to fertilizer and irrigation. High-yielding seed varieties were exported to India and Pakistan in the 1960s; in less than a decade, wheat production nearly doubled in both countries.

b. This farmer in the Punjab region of India examines his wheat crop.

a. The Green Revolution transplanted a new system of agriculture dependent on irrigation, heavy inputs of synthetic fertilizers, greater mechanization, and the monoculture of wheat or rice. (*Source*: Adapted from Borlaug and Dowswell, 2004.)

Year	Area irrigated in million hectares (acres)	Fertilizer nutrient use (million tons)	Tractors (millions)
1961	87 (215)	2	0.2
1970	106 (262)	10	0.5
1980	129 (319)	29	2.0
1990	158 (390)	54	3.4
2000	175 (432)	70	4.8

c. Government policies encouraged farmers to adopt this new system. These policies were implemented in India.

Selected government policies implemented in India
• A minimum support price for wheat and rice: If the market price of the grain fell below this price, the government reimbursed the farmer the difference. This policy lessened the financial risk of adopting high-yielding strains. • Subsidies for synthetic fertilizer • Subsidies on electricity, which is necessary to power well pumps for irrigatiion • Increased opportunities for agricultural loans and credit-based purchases of various supplies (seeds, fertilizer, pesticides) and equipment

such as high-yielding varieties, were shared with governments and agencies in developing countries, whereas genetically engineered crops produced during the Gene Revolution have been protected by patents. The second difference is that the Gene Revolution is more closely associated with multinational corporations and the spread of global capitalism. See **Figure 11.4** for a visual depiction of other facets of the Green Revolution.

The Green Revolution staved off famine in Asia and enabled India to become self-sufficient in grain production. However, yields have begun to level off, raising concerns about future food security. Will advances associated with the Gene Revolution help extend the gains of the Green Revolution? This is a contentious issue, in part because it leads many critics to question the motives for and consequences of the Gene Revolution. For

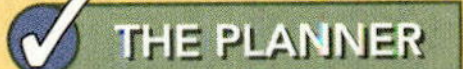

d. The diffusion of the Green Revolution has been highly uneven. Within Asia and Latin America, regions with more reliable rainfall and regions where irrigation is possible have benefited the most. Africa, which depends more heavily on other crops, such as sorghum, millet, and cassava, remains largely untouched by the Green Revolution.

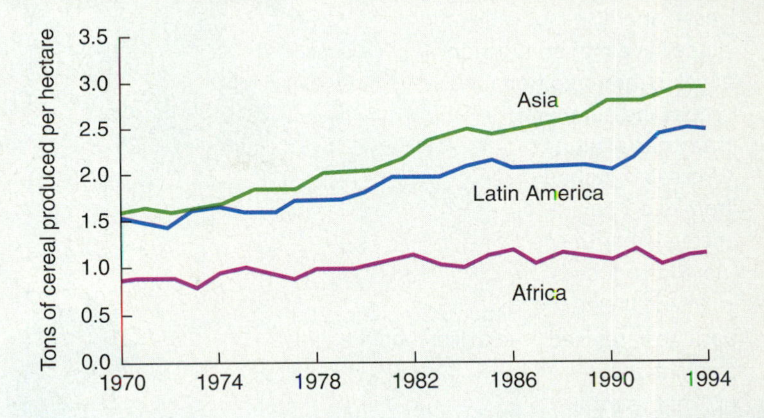

e. The Green Revolution has also brought about some problems. Farmers' debt has risen, groundwater has been overexploited, soil fertility has declined, and fertilizer and pesticide residues have built up in the environment. Ironically, pumps for irrigation are electric, but many households lack electricity and rely on dung, shown here being shaped into patties by a Punjabi woman, as a source of fuel.

example, whose interests are served when the genetic makeup of crops is protected by patents? Monsanto, an agro-biotech multinational, acquired the company that developed *terminator seeds*—seeds that produce sterile plants so that farmers have to purchase new seeds from Monsanto each year. Monsanto has not marketed these seeds because of public outcry, but the example is instructive nonetheless.

As we have seen, genetic engineering is an important part of the Gene Revolution. More specifically, genes with the code for certain traits, such as drought tolerance or stronger stalks, are transferred from one organism to another. Part of the controversy of this kind of genetic engineering is that genes from animals or even viruses can be transferred to crops in ways that are at odds with natural reproductive processes. Plants and animals changed as

GM cropping • Figure 11.5

a. Genetically modified (GM) crops can be made more resistant to weeds, pests, and disease, and sometimes have higher nutritional value than traditional crops. The number of countries planting GM crops climbed from 6 in 1996, the first year they were introduced, to 25 in 2008. GM varieties are used to plant about 40% of the total acreage sown to soybeans, maize (corn), cotton, and canola. (*Source*: Data from ISAAA, 2008.)

Biotech Cropland, 2008 (in million hectares)
- More than 10
- 2.5–10
- 0.5–2.49
- 0.1–0.49
- Less than 0.1
- Data not available

UNITED STATES alfalfa, canola, cotton, maize, papaya, soybean, squash, sugarbeet
CANADA canola, maize, soybean, sugarbeet
MEXICO cotton, soybean
HONDURAS maize
COLOMBIA carnation, cotton
BOLIVIA soybean
PARAGUAY soybean
BRAZIL cotton, maize, soybean
URUGUAY maize, soybean
CHILE canola, maize, soybean
ARGENTINA cotton, maize, soybean
GERMANY maize
SPAIN maize
PORTUGAL maize
POLAND maize
CZECH REPUBLIC maize
ROMANIA maize
SLOVAKIA maize
EGYPT maize
BURKINA FASO cotton
SOUTH AFRICA cotton, maize, soybean
INDIA cotton
CHINA cotton, papaya, petunia, poplar, sweet pepper, tomato
PHILIPPINES maize
AUSTRALIA canola, carnation, cotton

0 2,000 Miles
0 2,000 Kilometers

NATIONAL GEOGRAPHIC

Keep our fields GM free

b. GM crops remain controversial because they mix genes from different organisms. Also, some GM foods were originally marketed without labeling to indicate that they had been genetically modified. This protest is in London.

a result of biotechnology are referred to as *transgenic* or *genetically modified organisms* (GMOs). Proponents of this kind of genetic engineering consider it a viable means of overcoming environmental problems and generating more reliable yields, though the long-term ecological consequences of genetic engineering are not fully understood (**Figure 11.5**).

CONCEPT CHECK

1. **Why** do geographers distinguish between hearths of agriculture and secondary centers of agricultural innovation?
2. **What** agricultural advances led to the movement from the first to the second to the third agricultural revolutions?
3. **What** are the similarities and differences between the Green Revolution and the Gene Revolution?

Agricultural Systems

LEARNING OBJECTIVES

1. **Distinguish** between subsistence and commercial systems of agriculture.
2. **Identify** four types of subsistence agriculture and discuss the distribution of each.
3. **Provide** examples of specialization in different types of commercial agriculture.
4. **Summarize** the von Thünen model.

We might think of agriculture as a system of food production and thus a strategy for human survival. A *system* is a set of interacting components that functions as a unit. Food-producing systems include the land, the inputs (e.g., labor, machinery, fertilizer), the outputs or commodities that are produced, the consumers, and the various flows among the different components (e.g., migrant farm laborers to available jobs, of seeds to farmers, or of grain to food processors).

There are different ways of categorizing agricultural systems. Some experts prefer to distinguish between **subsistence agriculture** and **commercial agriculture**. **Figure 11.6** describes some of the differences between these systems and maps several of the world's major agricultural systems that are discussed in greater detail in the following sections.

subsistence agriculture A farming system that is largely independent of purchased inputs and in which outputs are typically used or consumed by farmers and their family or extended family.

commercial agriculture A farming system that relies heavily on purchased inputs and in which products are sold for use or consumption away from the farm.

The world's major agricultural systems • Figure 11.6

a. The table distinguishes between subsistence and commercial systems of agriculture. This categorization is useful as long as we remember that the different types of agriculture form a continuum marked by a great many variations.

The continuum of agricultural systems		
	Subsistence	**Commercial**
Farm size	small	large
Agricultural activity	diverse	specialized
Scale of consumption	household, local	national, international
Land tenure	communal, private	private, corporate
Purchased inputs	low	high
Contract farming	infrequent	frequent
Vertical integration	low	high
Proportion of output sold	minority	majority or all

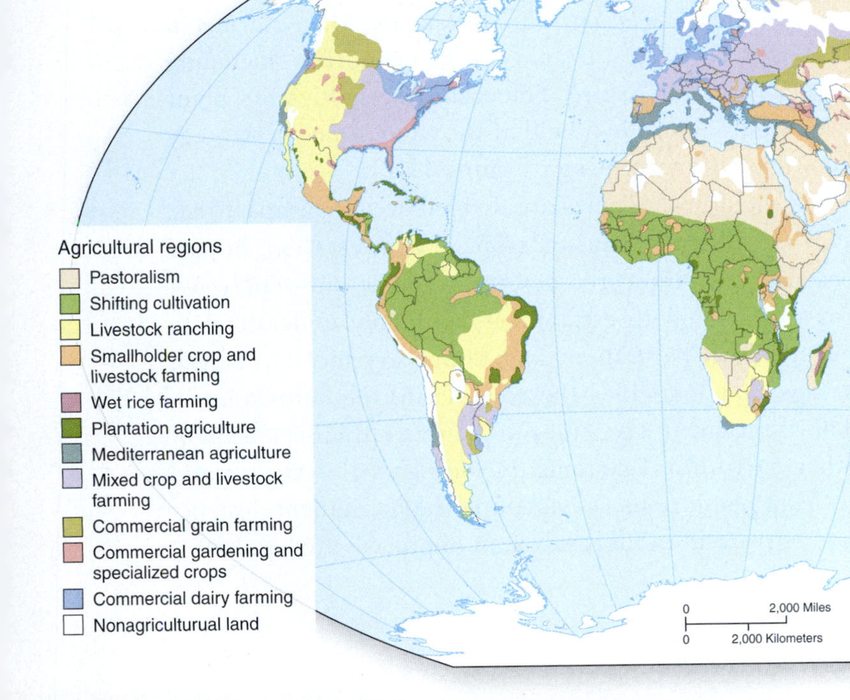

b. Hundreds of different agricultural systems exist around the world, but this map depicts just a fraction of them. For instance, urban agriculture is not included on the map. Keep in mind that even though some areas share similar agricultural systems, local agricultural practices may vary considerably because of different policies, climate patterns, or traditions.

PROCESS DIAGRAM

Shifting cultivation • Figure 11.7

THE PLANNER

Shifting cultivation is associated with rainforest zones, in both lowland and upland regions of the humid tropics. The cultivated land is usually owned or controlled by local families or by an entire village. Although the total area of land controlled by a village can be sizable, individual fields are small, perhaps 5 acres (2 ha).

1 Selecting a site
Farmers consider family needs, ecological conditions, past successes or failures in the area, and other factors when choosing a site.

2 Clearing a field
Using a "slash-and-burn" technique, farmers kill the trees by cutting into them and removing a ring of bark from each. They then burn the trees and undergrowth, sometimes selectively retaining those that provide a resource or serve an important function, such as preventing soil erosion. The residual ash adds nutrients and improves the soil.

A farmer plants cassava in a cleared field in Borneo.

3 Planting
Fields may include a single crop, have patches planted to different crops, or may be intercropped. If intercropped, tall, sturdy, or broadleaved crops provide support, shade, and even protection from heavy rains for lower growing crops and reduce the need for weeding.

Sorghum grows amid diverse trees in Uganda.

Subsistence Agriculture

Worldwide, millions of people earn their living through subsistence agriculture. This system is especially prevalent in Africa, much of Asia, and parts of Central and South America. There is a stunning array of subsistence agricultural practices around the world. Here, we will consider a few of them so that you may better understand some of the constraints and opportunities that subsistence agriculturalists face. From the many different of types of subsistence agriculture, we will focus on four, each of which is suited to a different climate, environment, and land type.

shifting cultivation An agricultural system that uses fire to clear vegetation in order to create fields for crops; it is based on a cycle of land rotation that includes fallow periods.

Shifting cultivation

Also known as *swidden* or *slash-and-burn agriculture*, **shifting cultivation** has been practiced for thousands of years in the tropical and subtropical regions of Southeast Asia, Central and South America, and Africa. Shifting cultivation has different local names: *milpa* in Central America, *ladang* in Indonesia and Malaysia, *roca* in Brazil, and *chitimene* in some African countries, including Zambia and Zimbabwe. Some shifting cultivators plant two or more crops in a field at the same time—a strategy known as **intercropping.** (Figure 11.7).

Agricultural experts are divided over the impact that shifting cultivation has on tropical deforestation, in part because they interpret deforestation in different ways. For example, deforestation is commonly understood to include both permanent and temporary forest removal; a view that magnifies the role of shifting cultivation in deforestation. In the Amazon, however, the expansion of other agricultural systems, including cattle ranches and soybean plantations, is hastening the permanent destruction of tropical rainforests on an unprecedented scale.

4 Harvesting

The harvest cycle depends on the crops that are planted. Intercropping may extend the harvest as different crops reach maturity at different times and may lessen the impact if one crop fails. As soil fertility declines, usually in about two to four years, yields begin to diminish.

Swidden rice field after harvest in central Laos.

5 Fallowing the land

The field is fallowed in order to regain fertility. The length of the fallow period—5, 7, 10, or more years—depends on local conditions. A single family will have several fields in different phases: Some ready to be fallowed, others about mid-way through the fallow cycle, and still others soon to be cleared for cultivation.

In Thailand, previously cleared forest regrows in the background.

Shifting cultivation can be sustainably practiced, but under certain conditions it can adversely affect the environment, or fail. Conditions that shorten the fallow period present a serious problem because they inhibit the ability of the soils to regain their fertility. An increase in the number of households engaged in shifting cultivation or loss of land to urbanization or highways can shorten the fallow period and affect the sustainability of the system.

When faced with such pressures, one strategy used by shifting cultivators to maintain adequate yields and improve the soil involves **agroforestry**, or the purposeful integration of trees with crops and/or livestock in the same field simultaneously or sequentially, one after the other. Many shifting cultivators now increasingly "manage the fallows" by planting species that help restore the soil's fertility or provide another resource, such as a fruit crop.

Pastoralism Domesticated livestock form the centerpiece of **pastoralism**. Pastoralists favor reindeer in the cold lands, and camels, cattle, goats, or sheep in arid regions. Because of their importance as a resource, the livestock are rarely killed and consumed for their meat. Consequently, in varying degrees pastoralists trade with, and rely on, settled farmers for cereal crops, fruits, and vegetables. Pastoralism is well adapted to arid and semiarid regions.

pastoralism An agricultural system in which animal husbandry based on open grazing of herd animals is the sole or dominant farming activity.

Mobility is an important dimension of pastoralism, since pastures cannot support livestock herds year-round. **Transhumance**—moving herds on a seasonal basis to new pastures or water sources—is a common practice, but the nature and frequency of mobility varies among pastoralists.

Often the women and children may not move with the animals; instead, they will settle and farm small plots of land in areas where rainfall or access to water is more reliable (**Figure 11.8**).

Is pastoral life incompatible with modernization? Some government officials claim that the mobility of pastoralists interferes with government programs such as population censuses, the provision of schools or basic medical care, and the establishment of protected areas and reserves. Increasingly, pastoral groups are pressured to take up sedentary agriculture in permanent settlements or seek nonagricultural occupations. For example, since the breakup of the Soviet Union, land privatization in Kyrgyzstan has limited pastoralists' access to pastures.

Wet rice farming Where rice is the primary crop and staple food, **wet rice farming** constitutes a prime example of **intensive agriculture**. In the world's most densely populated regions, the amount of land owned or worked by a family may be only 3 or 4 acres (1–2 ha), sometimes even smaller. In these circumstances, wet rice farming also constitutes a form of **smallholder agriculture**. Farmers cultivate wet rice in coastal lowlands, deltas, and river valleys (**Figure 11.9**).

In order to produce yields sufficient to provide for a family, the land must be intensively worked year-round. Following the harvest, the paddy may be prepared for planting a second time. The technique of **double cropping**—completing the cycle from planting to harvesting on the same field twice in one year—is common. Where a humid winter aids the cultivation cycle, as in southeastern China, rice is double cropped. In drier areas, farmers double crop by growing rice in the summer and wheat or barley during the winter.

Asia produces and consumes most of the rice in the world, and wet rice production benefited from the Green Revolution. Leading rice exporters include Thailand, Vietnam, India, and the United States, but rice production in the United States differs vastly from that in these other countries because it is not a smallholder system. In addition, rice production is highly mechanized in all stages in Japan, Korea, Taiwan, and the United States.

Smallholder crop and livestock farming In those places in Asia where conditions are not conducive to wet rice farming, **smallholder crop and livestock farming** prevails. This system also occurs in other parts of the developing world, but the specific combination of crop types and livestock varies significantly from one place to another because of different socioeconomic,

wet rice farming Rice cultivation in a flooded field.

intensive agriculture An agricultural system characterized by high inputs, such as labor, capital, or equipment, per unit area of land.

smallholder agriculture A farming system characterized by small farms in which the household is the main scale of agricultural production and consumption.

Pastoralism • Figure 11.8

a. Yak herders in India practice vertical transhumance—moving their herds into mountain pastures in the summer and into lowland pastures in the winter. In Nigeria, the Fulani practice horizontal transhumance, moving their cattle north to avoid the brunt of the wet season. They return south in the dry season.

b. Governmental policies can unleash changes that cause pastoralism to break down. After the Ngorongoro Conservation Area was created in Tanzania, Maasai herders like the one shown here have had their rangelands reduced and have been banned from growing crops, even though crop cultivation improves their overall food security.

Wet rice farming • Figure 11.9

In smallholder wet rice systems, most rice is consumed by the members of the household but any surplus rice is sold. Households also typically keep some pigs or poultry and cultivate small plots for vegetables.

a. Population pressures and lack of flat land have led to the terracing of hillsides and mountains to cultivate rice in Longsheng, China. Wet rice cultivation begins as seeds are sown in planting beds. In one to two months the seedlings will be ready for transplanting, and the wet field where the rice seedlings will grow to maturity—the paddy—is plowed.

b. Women in Vietnam transplant rice seedlings into a field that has been flooded with water. Workers apply large amounts of fertilizer, manually weed and harvest the rice, and after harvesting, thresh it. Paddy preparation and seeding are activities usually performed by men, but women contribute half or more of the labor through their work transplanting, weeding, harvesting, and threshing.

climatic, and soil conditions. Crop cultivation generally revolves around a grain crop, a tuber or root crop, legumes, and some vegetable crops. Many households also keep different kinds of livestock but in small numbers—for example, a single cow and a few pigs or chickens.

Across the drier parts of Asia corn and wheat are common grains, sweet potatoes a key root crop, and cattle and pigs the main livestock. In the Middle East and North Africa, farmers cultivate wheat as the staple grain, barley primarily as animal feed, and legumes such as lentils and chickpeas. Sheep, goats, and cattle are important livestock. Corn, millet, and sorghum are common grains cultivated in Africa south of the Sahara. Cassava, a tuber, is also widely cultivated there. In contrast to wet rice cultivation, smallholder crop and livestock farmers apply fewer inputs of fertilizer and irrigated water, and do not double crop (**Figure 11.10**).

Smallholder crop and livestock farming • Figure 11.10

a. A smallholder farm in the Andes Mountains of Peru. Smallholders are more likely to depend on rainfall than on irrigation.

b. A Nigerian woman carries her baby and cassava home from the field. The root crop is a staple starch across Sub-Saharan Africa.

Commercial Agriculture

Commercial farmers and their families are not the primary users or consumers of the agricultural goods they produce. Rather, they sell their farm products to food-processing companies. Commercial agriculture is one part of the large industry of food production often referred to as **agribusiness**. One of the hallmarks of agribusiness is *vertical integration*—when a company controls two or more stages in the production or distribution of a commodity directly or through contractual arrangements (see Chapter 10).

agribusiness The interconnected industry of food production involving farmers, processors, distributors, and retailers.

A number of food processors, such as Tyson, Kraft, and Kellogg's, have become household names. These companies consider commercial farmers to be their suppliers and negotiate contracts with them in order to secure the beef, poultry, wheat, corn, or other products they need for processing into the packaged meats, soups, cereals, and other items for consumers. In commercial agriculture, food-processing companies serve as an intermediary between producers and consumers.

As you read about some of the different types of commercial agriculture in this section, think about the impact they have had on landscape change, how they have altered social relations, and to what extent conventional ideas about the importance of physical proximity to market matter today.

Plantation agriculture

Plantations have long been associated with the production of high-value **cash crops**, such as coffee, tea, palm oil, and sugar, that are sold on the international market. As we discussed in Chapter 10, many developing countries have become highly dependent on the export of staple commodities, including cash crops, because they contribute significantly to the national economy. Cotton, for example, ranks as the most valuable fiber crop. Plantations are not the sole source of the world's so-called plantation crops, however. Smallholder farmers also cultivate a number of these crops, including cacao, coffee, and coconuts among others (**Figure 11.11**).

plantation A large estate in tropical or subtropical areas that specializes in the production of a cash crop.

The Portuguese established the first plantations in Africa in the 15th century. These plantations used slave labor to produce sugarcane. Relying on unskilled or semiskilled labor to plant, harvest, and process the

Plantation agriculture • Figure 11.11

a. A sugar plantation near Durban, South Africa
Historically, plantations were established near the coast or serviced by rail lines in order to facilitate the export of plantation commodities. It is presently more cost-effective to burn the vegetation that remains once the sugarcane is harvested, though such burning is contested by environmentalists.

b. Modest housing for tea plantation workers in Malaysia
Work conditions at plantations are demanding, sometimes dangerous, and the pay is low. Why are the houses virtually identical?

farm commodities remains a defining feature of plantation agriculture today. Indeed, the plantation system perpetuates a **dual society** consisting of two distinct social classes—the upper-class plantation managers and the lower-class laborers.

In addition, the plantations are frequently owned by multinational corporations, a number of which are headquartered in Europe or North America. One example is the fruit company Dole. It is headquartered in California, operates plantations in the Philippines and Costa Rica, owns and runs a cannery in Thailand, and has cold-storage facilities in Chile. These pockets of commercial agriculture in developing countries also foster a **dual economy**, with large-scale, export-oriented agriculture operating alongside smallholder agriculture.

Commercial gardening, specialized crops, and Mediterranean agriculture Geographically, **commercial gardening** zones developed just beyond the built-up areas of towns and cities and supplied urban residents with fresh produce. Historically, farmers located near the markets to minimize the problem of spoilage, and local products were destined for local consumption. However, well-developed transportation networks and long-distance trucking industries mean that fresh produce can now be shipped from farm to market over hundreds of miles in a matter of hours.

commercial gardening The intensive production of nontropical fruits, vegetables, and flowers for sale off the farm.

Since World War II, a form of commercial gardening known as **truck farming** has emerged in the United States. Important crops produced on truck farms include tomatoes, lettuce, melons, broccoli, onions, and strawberries. Most of these farms are large, specialize in the production of one commodity, are distant from the markets they serve, and rely on migrant farm laborers during the harvest season. Although product shipment involves large-capacity trucks, this is not the source of the term *truck farming*. Rather, one of the meanings of the word *truck* is "vegetables grown for market." Truck farming is heavily concentrated in the southeastern United States, but numerous other zones of specialized crop production exist, including parts of Maine and Idaho where potatoes are produced, and the Caucasus (located between Europe and Asia) where such crops as cabbage, onions, and eggplant are farmed.

Mediterranean agriculture • Figure 11.12

Sheep graze in between the grape vines, an illustration of one facet of the integration of livestock and vine crops in Portugal. The sheep keep the vineyard weeded, provide manure, and can be trained to avoid eating the grape vines.

The lands surrounding the Mediterranean Sea constitute the hearth of **Mediterranean agriculture**. In its traditional form, Mediterranean agriculture was a kind of agroforestry centered on the integrated cultivation of livestock, a grain crop, and a tree or vine crop. Olives, grapes, and citrus fruits are strongly associated with regions of Mediterranean agriculture, but, as with commercial gardening, Mediterranean agriculture has been affected by specialization. This is especially the case in the Central Valley of California and in the region surrounding Valparaiso, Chile, which increasingly focus on the production of specialized crops. Around the Mediterranean Sea, wheat remains a principal grain crop, and some farms still manage livestock; however, tree and vine crops, especially olives and grapes, provide the most valuable commodities (**Figure 11.12**). Seasonal demand for work on these farms in Europe draws many farm laborers from Romania, Bulgaria, and Albania as well as North Africa. Migrant labor has long been crucial to agricultural harvesting in the Central Valley of California as well.

Commercial dairy farming

Commercial dairy farming is an intensive and heavily mechanized form of agriculture. Fresh milk production relies on automatic milking machines, vacuum systems, and pipelines to move the milk into refrigerated storage tanks before it is transferred to tank trucks for shipment to milk-processing plants. In spite of the mechanization, dairy farming requires constant vigilance. The cows need to be milked twice a day and have their nutrition closely monitored, or they will not produce the desired quality or volume of milk.

commercial dairy farming The management of cattle for producing and marketing milk, butter, cheese, or other milk by-products.

The geography of fluid milk production has evolved in connection with the rise of cities. Because milk is perishable, it initially had to be consumed on the farm or made into another, less perishable dairy product such as cheese. Dairy farming areas on the outskirts of cities that supply fluid milk constitute the *milkshed*. Improvements in transportation and refrigeration now enable dairy farms producing milk to locate farther from cities, expanding the milkshed. California has been the largest fluid milk–producing state in the United States since the early 1990s, a development closely associated with the emergence of *dry-lot dairies* that facilitate high-volume production (**Figure 11.13**).

Mixed crop and livestock farming

As historically practiced, **mixed crop and livestock farming** was an integrated system that involved raising crops to feed livestock. The animal products were then sold off the farm, generating most of the farm's revenue. This type of farming once defined an extensive part of Europe, stretching from France across central Europe and into Russia, where corn, barley, and oats were grown as feed crops for beef cattle and hogs. Across the Corn Belt of the United States (from central Ohio to eastern Nebraska), corn and soybeans were raised to feed cattle and hogs.

Agricultural specialization continues to transform these practices, however. In Europe, some regions once associated with mixed crop and livestock farming in countries including Germany, France, and Poland now concentrate on producing high-value oilseed crops such as canola. Specialization in the Corn Belt has involved two main trends. One trend emphasizes cash-grain farming of corn and soybeans in rotation, with corn planted one year and soybeans the next year. There are different ways to define *cash-grain farms* (also called commercial grain farms) but the distinction is usually based on revenue, with grains sales accounting for 50% or more of farm products sold. The second trend involves specialized hog production on **factory farms**, also known as concentrated animal feeding operations (CAFOs). **Feedlots** have become emblematic of factory farms. The dry-lot dairies mentioned in the previous section are a type of feedlot developed for dairy cattle. **Figure 11.14** illustrates some of the changes affecting the Corn Belt.

factory farm A farm that houses huge quantities of livestock or poultry in buildings, dry-lot dairies, or feedlots.

feedlot Confined space used for the controlled feeding of animals.

Dairy farming and milk production • Figure 11.13

a. A California dry-lot dairy, like this one, contains no pasture and holds some 600 dairy cows on open lots, often with sun shades. In contrast, dairy farms in the Upper Midwest typically maintain about 70 pasture-fed dairy cows. Globally, dry-lot dairying is the exception rather than the rule.

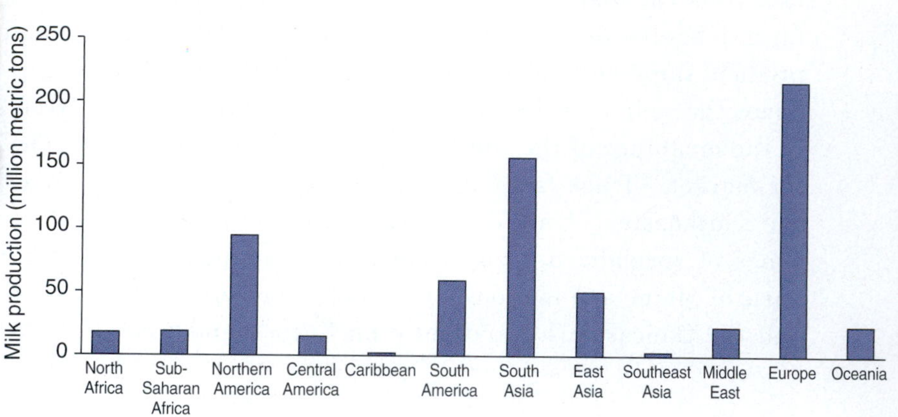

b. Regionally, the three main centers of milk production are in Europe, South Asia, and Northern America. More than half of the milk produced in South Asia comes from water buffalo. (*Source*: Data from FAOSTAT, 2010.)

Geography InSight

Change in the Corn Belt • Figure 11.14

THE PLANNER

Many Corn Belt farms do still integrate the production of corn, soybeans, and hogs. However, changes in farming practices in the region are underway.

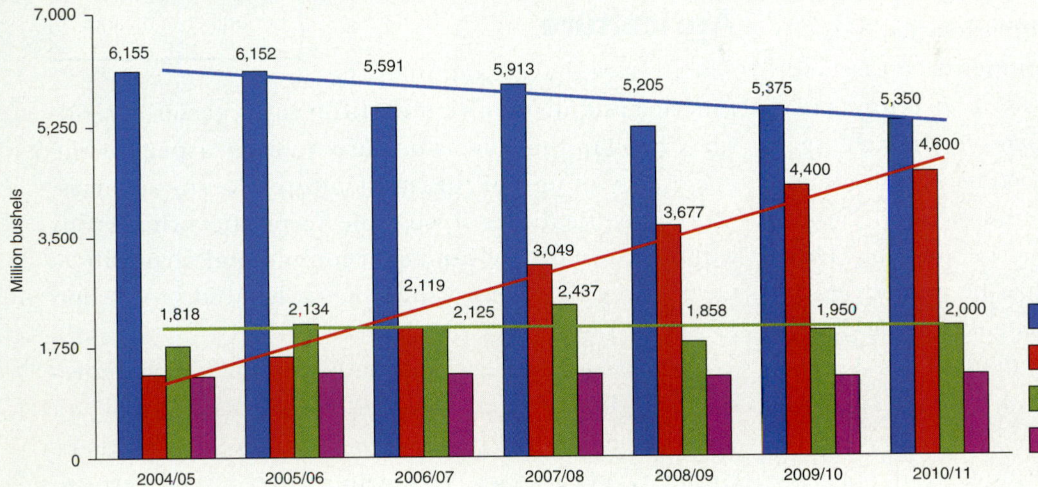

a. Trends in U.S. corn use
This graph underscores some important domestic developments, namely, that the amount of corn used for feed is declining, whereas the amount of corn used to make ethanol is rising sharply. The renewable fuel standards, part of the country's attempt to reduce its reliance on foreign oil, are expected to sustain this trend. (*Source*: O'Brien, 2010.)

Corn Production (Bushels)
Not Estimated
< 1,000,000
1,000,000 - 4,999,999
5,000,000 - 9,999,999
10,000,000 - 14,999,999
15,000,000 - 19,999,999
20,000,000 +

Ethanol Plants
Construction
Producing
Not producing

U.S. Department of Agriculture, National Agricultural Statistics Service

USDA
AGRICULTURE COUNTS NASS

b. Corn produced for grain, 2008, and location of ethanol plants as of January 2009
Strong market prices for corn help to increase the acreage planted to the crop and can tempt farmers to plant corn two years consecutively before rotating in soybeans, even though corn yields diminish significantly in the second year.

lagoon

lagoon

c. Hog CAFO near Milan, Missouri
Hog farms are distinguishable by long, rectangular buildings that are grouped together and adjacent to a lagoon, which holds the waste from the hog houses. Since the 1990s, the Corn Belt has lost hog farms to North Carolina and Oklahoma, in part the result of lower labor and production costs and economies of scale. Some of the largest hog farms can produce in excess of 200,000 hogs per year.

Commercial grain farming and livestock ranching The staple item of most people's diets is the grain of a cereal grass. Common grains include wheat, rice, corn, barley, oats, millet, and sorghum. These grains not only feed people and animals, they frequently have industrial uses as well. **Commercial grain farming** is closely associated with temperate grassland environments. Monoculture prevails, with farms covering large areas of flat to gently rolling land that is planted to a single crop. Some of these farms can top 2,000 acres (800 ha). This type of large-scale grain farming has been made possible because of mechanization. Indeed, commercial grain farming remains heavily dependent on fossil fuels used in the production of fertilizers and in the gas consumed in working the fields.

commercial grain farming Agriculture involving the large-scale, highly mechanized cultivation of grain.

Lands that are unsuitable for more valuable agricultural uses—for example, in arid and semiarid regions—tend to be used for **livestock ranching**. Ranchers have fixed places of residence and graze their livestock on the open range or on fenced land. The rangelands used cover sizable areas. Historically, these regions have been distant from the centers of demand—the cities and towns. Getting the animals to slaughterhouses required that they be herded long distances overland to railroads. Both commercial grain farming and livestock ranching are considered examples of **extensive agriculture** (Figure 11.15).

livestock ranching A form of agriculture devoted to raising large numbers of cattle or sheep for sale to meat processors.

extensive agriculture An agricultural system characterized by low inputs of labor, capital, or equipment per unit area of land.

Spatial Variations in Agriculture

Thus far we have examined the practice and distribution of different types of agriculture, but land use decisions are another aspect of the geography of agriculture that interest many scholars. More specifically, is it possible to predict what crops will be grown or how land on a commercial farm will be used if we know where the farm is located in relation to the market?

One of the people who initiated the study of this question was Johann Heinrich von Thünen (1783–1850), a farmer, scholar, and estate owner who lived in northern Germany. Over years of traveling from his property to towns in the region, he observed that agricultural practices and crops changed as he got farther away from the marketplace. He used these observations to devise a model, the **von Thünen model**, to account for spatial variations in commercial agriculture.

To simplify the complexity of real-world conditions, von Thünen assumed that the quality of the land is the same everywhere. Following the economic principle that land-use decisions are profit-maximizing decisions, he reasoned that transportation costs determine how farmers can make the most profitable use of their land (Figure 11.16).

Extensive agriculture • Figure 11.15

a. Wheat being harvested in Saskatchewan, Canada

In the Southern Plains wheat is harvested in early summer. In the Northern Plains wheat is harvested at the end of the summer. Crews of laborers and "custom cutters" with combines move across the wheat belt to take advantage of these seasonal differences.

b. Sheep grazing in Wanaka, New Zealand

In New Zealand, more than 40% of the farmland in the country is devoted to sheep raising, and the country remains the world's leading exporter of mutton.

The von Thünen model • Figure 11.16

Farmers located near the market or city have low transportation costs and can afford to engage in more intensive agriculture than farmers farther away, creating rings of agricultural land use. The forestry ring provides one exception to this pattern. Timber, still needed for fuel and building in von Thünen's time, is a less intensive, low-value good that would not be profitable if transported long distances. What accounts for our ability to profitably transport timber over longer distances today? What does the diagram suggest about how terrain might influence land use? (*Source*: Adapted from Chisholm, 1968.)

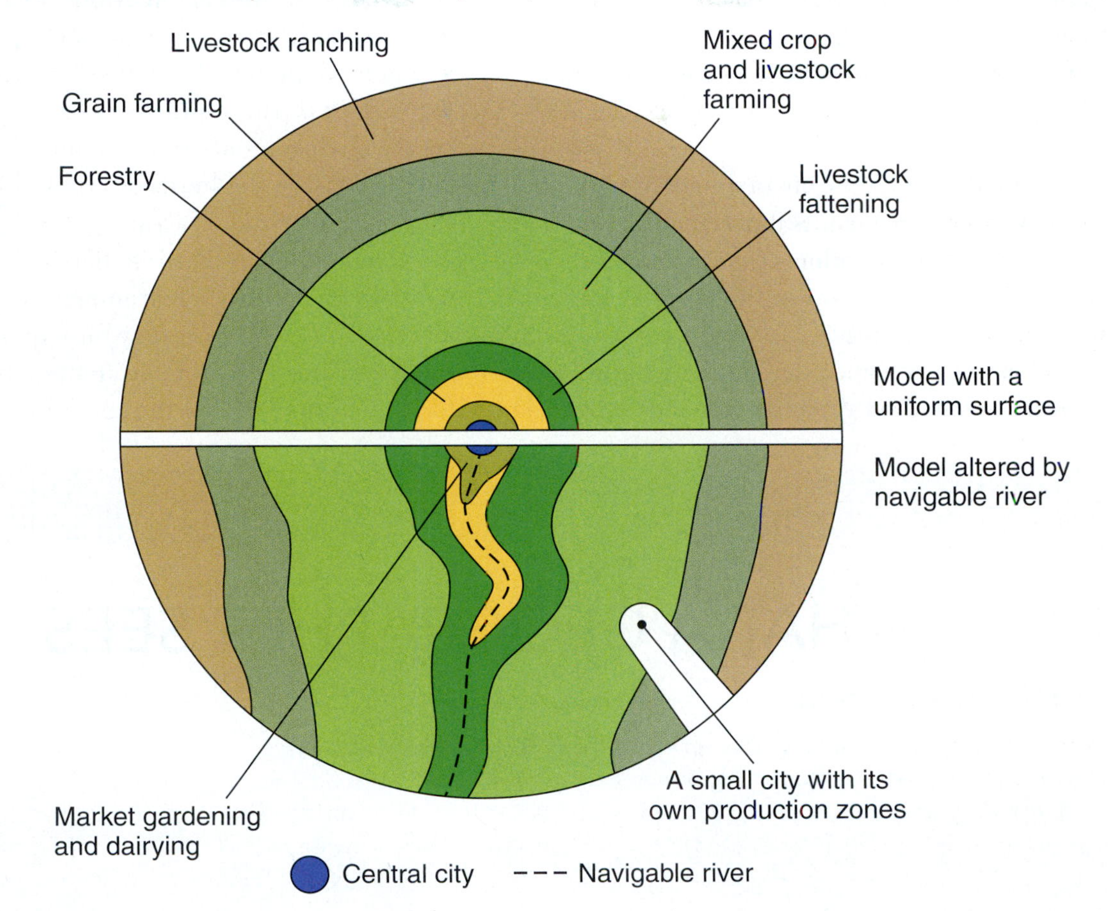

Although von Thünen's model may seem oversimplified, its principles still have relevance and help us explore such questions as what economic forces make it possible for cut flowers from Colombia to be shipped to Miami and sold in other U.S. cities, and how does globalization affect the geography of intensive agriculture? Taking a different approach, Arild Angelsen, a scientist with the Center for International Forestry Research, has used von Thünen's model to examine the relationship between profitable land use and tropical deforestation. Angelsen shows that land used for agriculture—for example, beef cattle production—generates more profit than land that is forested. Thus, economic considerations can drive tropical deforestation and influence decisions about land use.

CONCEPT CHECK

1. **What** is meant by the continuum of agricultural systems?
2. **How** is mobility a factor in shifting cultivation and nomadic pastoralism, and why are these systems of production changing?
3. **Where** have plantations tended to locate and why?
4. **How** does von Thünen's model account for spatial variations in intensive and extensive agriculture?

Agriculture, the Environment, and Globalization

LEARNING OBJECTIVES

1. **Define** desertification and salinization.
2. **Distinguish** between sustainable agriculture and organic agriculture.
3. **Explain** how agriculture has been affected by globalization.
4. **Summarize** the causes of the recent global food crisis.

Agriculture and the environment are intimately interconnected. Soil or climatic conditions in an area can influence decisions about what to grow or how to use the land. At the same time, the practice of agriculture can have a significant impact on the environment. Since the first agricultural revolution, clearing forests and draining wetlands have been common strategies for increasing the acreage devoted to cropland.

Human actions as well as changes in climate can contribute to **desertification**. Overgrazing damages vegetation, while poor crop management depletes the soil's fertility. Both of these practices can create environments that are unable to sustain the herds or crops that they once did.

desertification The creation of desertlike conditions in nondesert areas through human and/or environmental causes.

Irrigation can have a detrimental effect on the environment, even though it is usually thought of as a strategy for expanding agriculture. If irrigation draws on groundwater aquifers, water usage has to be monitored so that the aquifer is not depleted. In soils

WHAT A GEOGRAPHER SEES

The Shrinking Aral Sea

For more than 40 years, water has been diverted from the rivers that feed the Aral Sea in Central Asia to irrigate land for cotton and rice production. As a result, the lake has lost more than 60% of its water and has shrunk from over 65,000 sq km (25,000 sq mi) to less than half that size. In the process, the sea has been transformed from a fresh to saltwater environment.

Global Locator

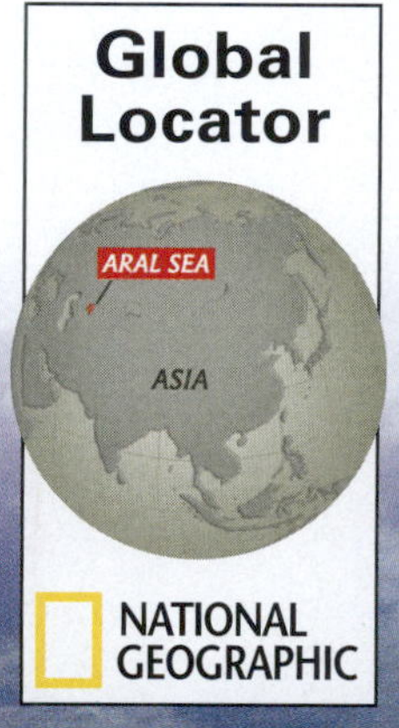

a. A rusted boat is left high and dry. An increase in the lake's salt concentration from 10% to more than 23% has devastated fish populations. Camel breeding has replaced fishing as a primary source of local income. How might the sea's decline affect the region's climate?

that drain poorly, irrigation can lead to waterlogging and crop death. When irrigation is used in arid and semiarid regions where evaporation rates are high, **salinization** becomes an issue and can result in decreased productivity (see *What a Geographer Sees*).

Applications of chemical fertilizers, herbicides, pesticides, and fungicides also have an impact on the environment. The production of fertilizers such as ammonia uses large amounts of energy. In addition, runoff from fields treated with other chemicals used in commercial farming can pollute surface water and groundwater supplies.

salinization The accumulation of salts on or in the soil.

sustainable agriculture Farming practices that carefully manage natural resources and minimize adverse effects on the environment while maintaining farm profits.

Sustainable Agriculture

The environmental impacts of agriculture, especially large-scale commercial agriculture, have prompted experts to question its ability to provide for future generations. As a result, there is growing interest in **sustainable agriculture**. Examples of sustainable farming practices include measures taken to conserve soil and water resources, such as contour plowing, strip cropping, and the establishment of filter or buffer strips. *Contour plowing* follows the slopes in a field, rather than cutting across them. *Strip cropping* alternates the planting of row crops, such as cotton, with bands of sod crops, such as alfalfa or soybeans. *Filter strips* or *buffer strips* are belts of vegetation that surround fields and act to prevent runoff.

No-till farming also encourages sustainable land use. *No-till farming* avoids agitating the soil with tractor-drawn implements that remove weeds, mix in fertilizers, or shape the soil for seeding—all of which can lead to erosion. Crop rotations that help prevent disease or pest problems, and actions taken to reduce reliance on fossil fuels, especially petroleum, are also sustainable practices.

b. Satellite images illustrate the lake's reduction from 2000 to 2009 and relative to the shoreline in 1960. Soil from the exposed lake beds feeds salt particles and pesticide residues into dust storms. The resulting pollution has been linked to respiratory problems for those who live in the region. The lake's increased salinity reduces crop yields in the fields that are irrigated with the diverted water.

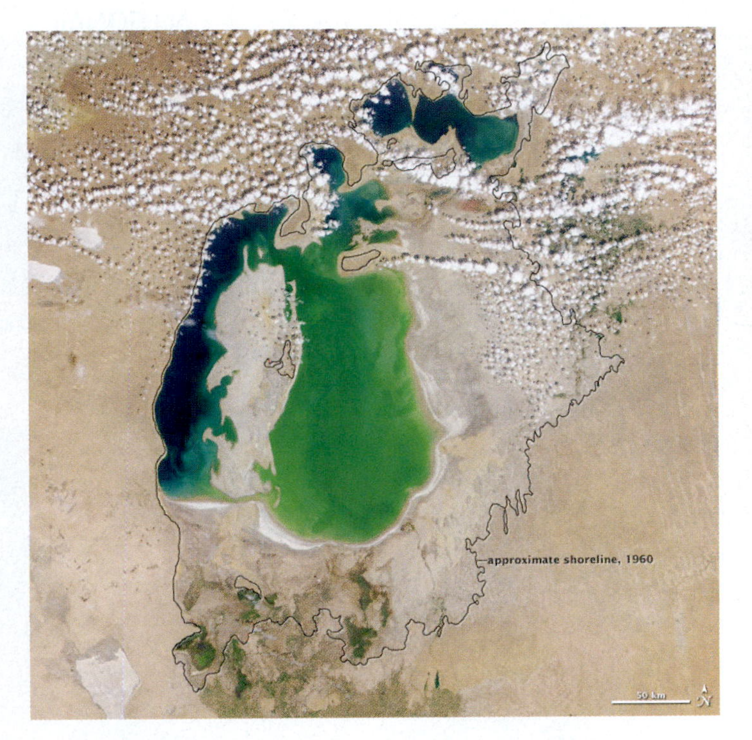

Precision agriculture employs technologies such as the global positioning system (GPS) and aerial imagery to measure and map the spatial variation in environmental conditions within a field. Soil fertility is rarely uniform across an agricultural field, for example. Mapping the site-specific soil conditions reveals the geography of soil nutrients in a field. This information can then be used to calibrate farm machinery to apply fertilizer at variable rates, releasing more in those areas where the soil is deficient in nutrients. Precision agriculture can also be used to manage pesticide applications, determine the best sowing density, and more accurately predict crop yields. In that they are closely tied to effective soil and field management, certain precision agriculture techniques can support sustainable practices. It should be noted, however, that some experts contest the association of precision agriculture with sustainability, in part because precision agriculture often uses synthetic chemicals that require large amounts of energy to manufacture.

Another expression of concern about the sustainability of agriculture involves the growing demand for **organic agriculture**. Organic agriculture accounts for a very small share of all agricultural products sold; however it is the fastest growing sector of agriculture today. Globally, Australia, Argentina, and Brazil have the largest areas under organic management, but the highest percentages of organic land are consistently found in Europe. This is partly the result of agricultural policies in Europe that have subsidized organic farming. At present, most of the organic farm products from Africa and Latin America are exported (**Figure 11.17**).

organic agriculture A farming system that promotes sustainable and biodiverse ecosystems and relies on natural ecological processes and cycles, as opposed to synthetic inputs such as pesticides.

Sustainable agricultural practices • Figure 11.17

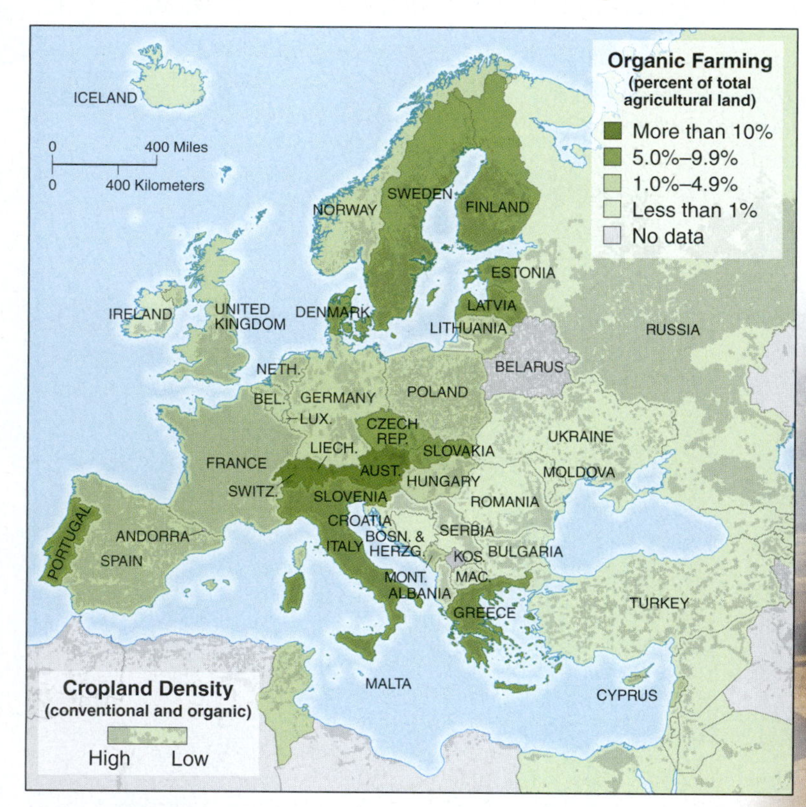

a. Organic farming in Europe
Nearly 25% of all organically managed land is in Europe. Tiny Liechtenstein boasts the highest proportion—29%—of its agricultural land used for organic farming, followed by Austria and Switzerland with 13% and 11%, respectively. What other regional patterns are evident on the map?

b. Soil conservation techniques
Contour plowing and strip cropping in the Palouse region of eastern Oregon and Washington create a distinctive agricultural landscape. Dark fallow strips ready to be sown alternate with strips of wheat stubble. The wheat stubble aids water infiltration and prevents the soil from freezing.

Globalization and Agriculture

Our ability to purchase grapes from Chile, tea from Sri Lanka, or apples from New Zealand at grocery stores here in the United States is certainly one expression of the globalization of agriculture. Although globalization brings increased trade and access to a greater variety of agricultural products, it is also clear that the globalization of agriculture creates significant challenges, especially for poorer countries. We can glimpse this problem through the workings of the World Trade Organization (WTO), discussed briefly in Chapter 2.

The WTO seeks to make trade freer through the removal of tariffs and other policies that distort the market. Although the least developed countries have been given longer time frames to dismantle trade barriers, it is still reasonable to ask how a smaller and much poorer country such as Jamaica can compete with a country like the United States in terms of producing and selling its agricultural goods. One particular issue that creates an unlevel playing field and that the WTO has been slow to address involves government subsidies to farmers. Poorer countries cannot provide such subsidies. Thus, their farmers bear a higher share of the production costs, and this translates into higher prices for their agricultural products. Many trade experts have argued that domestic subsidies create severe market distortions and prevent free trade in agricultural goods.

The globalization of agriculture also affects diets. Patterns of food consumption are changing as the availability of processed foods increases. Asian diets, specifically those of the urban and middle classes, are becoming westernized and are contributing to a **nutrition transition**—a shift characterized by a decline in the consumption of rice and an increase in meat, wheat-based food products, and convenience foods. Although Asian diets now include a greater variety of foodstuffs, many of these items also have more fats and refined sugars, with the potential for adverse health consequences, including obesity and diabetes. To find a photo gallery of other examples of the diffusion of Western, and specifically American, foods and beverages, see *Where Geographers Click*.

Over the past decade, developing countries in East and Southeast Asia, Latin America, and Africa have also witnessed the rapid spread of supermarkets. This *supermarket revolution* affects the way that fruits and vegetables are grown and sold throughout world. Although supermarkets can lower food prices for consumers, small retailers often cannot compete with them. In addition, supermarket chains are often just as likely to source their products from distant rather than local suppliers, with negative consequences for the local agricultural sector.

Where Geographers CLICK

Cultural Landscapes from Around the World

http://www2.geog.okstate.edu/users/lightfoot/lightfoot.html

Use the photographs on this website to track the diffusion of products such as Frosted Flakes and Coca-Cola.

Agriculture and resource use • Figure 11.18

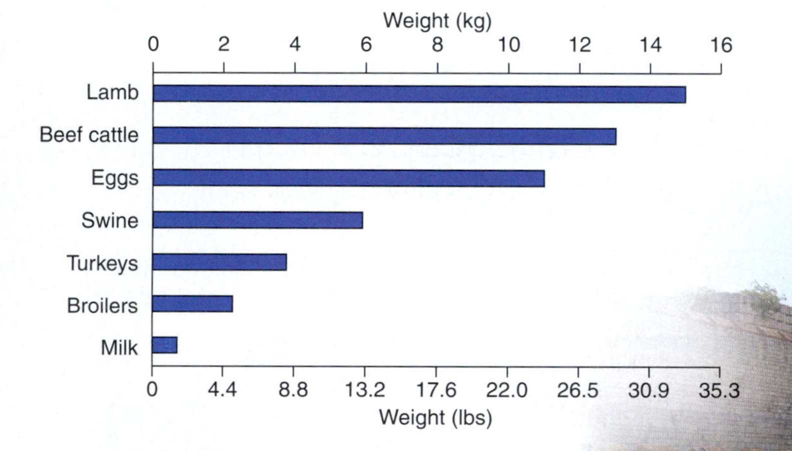

a. Amount of grain needed to produce 1 kilogram (2.2 pounds) of meat or other animal product
Livestock are inefficient producers of protein and, as shown by this graph, must consume inordinate amounts of grain to generate adequate weight gain to produce meat. Should our agricultural practices change in times of food crisis, so that grain that is processed into feed for animals could be used instead to feed people? (*Source*: Data from USDA, 2001.)

b. Urban agriculture
In Istanbul, Turkey, spaces devoted to urban agriculture are called *bostans*. These bostans are next to the old city walls. Urban agriculture exists in cities around the world but in recent years has been heralded as an important mechanism for improving nutrition and access to fresh produce among the urban poor.

The Global Food Crisis

Food prices worldwide increased, on average, by 43% in the year from March 2007 to March 2008, setting in motion a **global food crisis**. How can this happen when enough food is produced to feed each person? The answer is complicated and involves a mix of diverse factors.

Late 2006 saw unseasonable droughts in grain-producing countries. Rising oil prices also contributed to increasing costs of fertilizers and fuel. A related and contributing factor involved increased production of **biofuels**, especially the use of corn to produce ethanol. Although biofuels are renewable and reduce reliance on fossil fuels, farmlands and crops must be diverted away from food production to generate biofuels. Rising consumer demand also increased pressure on the grain market. These developments, coupled with declines in food stockpiles, all contributed to a dramatic worldwide rise in food prices. When food prices rise, the poor—who spend a disproportionate share of their income on food—suffer the most. Such food crises raise questions not only about the efficiency of certain agricultural practices but also about strategies that might help people reduce their vulnerability to them. One practice that could help improve future food security at the household level is **urban agriculture** the use of vacant lots, rooftops, balconies, or other spaces to raise food for the household or neighborhood (**Figure 11.18**).

biofuel Fuel derived from renewable biological material, such as plant matter.

CONCEPT CHECK

STOP

1. **How** are desertification and salinization related to changes in the Aral Sea?
2. **What** are some specific techniques associated with sustainable agriculture, and why are they beneficial to the environment?
3. **Why** might the globalization of agriculture be considered a mixed blessing?
4. **What** triggered the recent global food crisis?

Summary

1 Agriculture: Origins and Revolutions 326

- **Agriculture** is an economic activity centered on the purposeful tending of crops and livestock in order to procure food and fiber for human use or consumption. Prior to the development of agriculture, people subsisted by **hunting and gathering**, which still forms the basis of some livelihoods today.
- Three agricultural revolutions have transformed human geographies, including both social and environmental dynamics. The rise of farming about 11,000 years ago, made possible by the domestication of plants and animals in at least five hearths and numerous secondary centers, shown here, marked the **first agricultural revolution**.

Hearths of agriculture • Figure 11.2

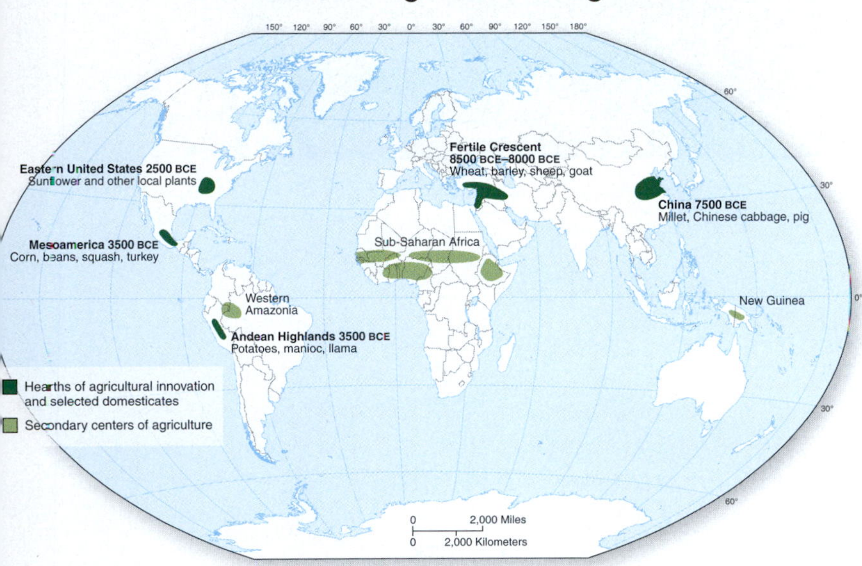

- The **second agricultural revolution** began in Europe during the Middle Ages and was prolonged because of developments during the Industrial Revolution. Innovations that made the second agricultural revolution possible include the development of the moldboard plow, the horse collar, and the **four-course system** of **crop rotation**.
- The **third agricultural revolution** began in the 20th century and is still underway. It is associated with the development of the internal combustion engine and greater reliance on chemical applications and **agro-biotech** practices.
- The **Green Revolution** and the **Gene Revolution** are both associated with the third agricultural revolution. The Green Revolution involved increased grain production in certain developing regions as a result of high-yielding, fertilizer- and irrigation-dependent varieties of wheat, rice, and corn. The **Gene Revolution** is marked by the shift toward greater control of the research, development, intellectual property rights, and genetic engineering of highly specialized agricultural products.

2 Agricultural Systems 333

- Many different types of agriculture are practiced in the world, and they can be placed along a continuum from **subsistence agriculture** to **commercial agriculture**.
- There are four major types of subsistence agriculture. **Shifting cultivation** is practiced in the humid tropics, whereas **pastoralism** remains confined to arid and semiarid regions. Both of these agricultural systems support low population densities. **Wet rice farming**, which prevails in more humid parts of Asia, and **smallholder crop and livestock farming**, which takes place in regions too dry to support wet farming, are both forms of **intensive agriculture**.
- Today, commercial agriculture is one component in an interconnected system of food production that involves farmers, processors, distributors, and retailers. Vertical integration has become a defining feature of **agribusiness** and has led to greater involvement of corporations in farming. Corporations increasingly participate in the technical aspects of crop and stock management, as well as processing agricultural products and moving them to market.
- Specialization has had a strong impact on commercial agriculture. **Plantation agriculture** is practiced in tropical and subtropical areas and has a strong presence in developing regions. **Commercial gardening** is a kind of **truck farming** and is increasingly associated with large, specialized farms. **Mediterranean agriculture** is closely associated with the production of tree and vine crops. Within the United States, dry-lot dairies have altered the geography of **commercial dairy farming. Factory farms** and **feedlots** are changing farming practices in **mixed crop and livestock farming** regions. **Commercial grain farming** occurs in the temperate grassland regions of North and South America, Australia, and eastern Europe and Russia. The practice of **livestock ranching**, a type of extensive agriculture, tends to be spatially associated with regions that are drier and/or more remote from major markets, as in New Zealand (see photo).

Extensive agriculture • Figure 11.15

- The **von Thünen model** helps to depict the relationship between location, or nearness to the market, and how land is used for commercial agriculture.

3 Agriculture, the Environment, and Globalization 344

- All types of agriculture transform the environment. The nature and extent of the impact on the environment differs from place to place and from farmer to farmer. Soil degradation and the impacts of climate change are serious issues that all farmers confront. **Desertification**, whether caused by human or climatic factors, can prohibit the practice of agriculture. Irrigation brings another set of impacts.
- **Sustainable agriculture** and **organic agriculture** have developed in response to concerns about the adverse impacts that commercial farming can have on the environment. Although **precision agriculture** was not developed strictly for reasons of sustainability, some aspects of it support the careful management of resources.
- Globalization has had an impact on food consumption and agricultural practices around the world. Many cities in the developing world are experiencing a **nutrition transition** as Western, high-fat foods are gaining popularity. Westernized diets are becoming increasingly popular in Asia, placing greater demands on wheat production. Developing countries in Asia, Latin America, and Africa have also seen a rapid rise in supermarkets over the past decade, which has had important consequences for local farmers and retailers.
- Unseasonable droughts in grain-producing nations, rising oil prices, the conversion to **biofuels** and changing dietary patterns around the globe contributed to the **global food crisis** of 2008. That year an increase in food prices had a significant impact on producers and consumers around the world.
- **Urban agriculture**, glimpsed here, could help improve future food security at the scale of the household.

Agriculture and resource use • Figure 11.18

Key Terms

- agribusiness 338
- agriculture 326
- agro-biotech 329
- agroforestry 335
- biofuel 348
- cash crops 338
- commercial agriculture 333
- commercial dairy farming 340
- commercial gardening 339
- commercial grain farming 342
- crop rotation 328
- desertification 344
- domestication 326
- double cropping 336
- dual economy 339
- dual society 339
- extensive agriculture 342
- factory farm 340
- feedlot 340
- first agricultural revolution 327
- four-course system 328
- Gene Revolution 329
- global food crisis 348
- Green Revolution 329
- hunting and gathering 326
- intensive agriculture 336
- intercropping 334
- livestock ranching 342
- Mediterranean agriculture 339
- mixed crop and livestock farming 340
- monoculture 329
- nutrition transition 347
- organic agriculture 346
- pastoralism 335
- plantation 338
- precision agriculture 346
- salinization 345
- second agricultural revolution 328
- shifting cultivation 334
- smallholder agriculture 336
- smallholder crop and livestock farming 336
- subsistence agriculture 333
- sustainable agriculture 345
- third agricultural revolution 328
- transhumance 335
- truck farming 339
- urban agriculture 348
- von Thünen model 342
- wet rice farming 336

Critical and Creative Thinking Questions

1. What is slow food? It might be said that slow food is fundamentally about the relationship between people and place. Do you agree? Does slow food provide a viable mechanism for eliminating global hunger?
2. How does the Fair Trade movement reflect a concern with moral geographies?
3. How has the diffusion and popularity of pizza affected the geography of dairy products?
4. What kind of evidence would a diffusionist use to support the theory that agriculture developed only once and spread from that location around the world?
5. How did European colonization affect global patterns of agriculture?
6. What is aquaculture, and should we consider it a type of agriculture? Explain your answer.
7. Will there be a fourth agricultural revolution, and if so, what might it involve?

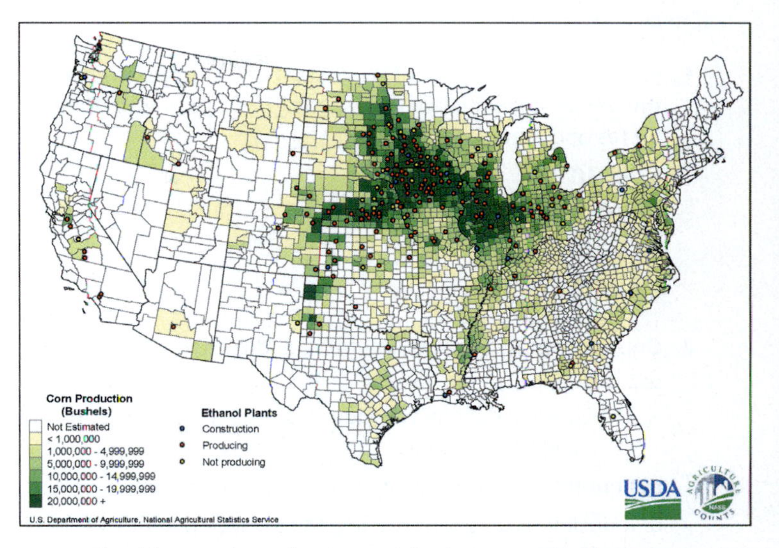

8. Using this map, how do you explain the presence of producing ethanol plants (refineries) in counties where the production of corn is low?

What is happening in this picture?

The photograph shows a tea picker in Munnar, India. Pluckers take new growth from the top, wedging leaves between thumb and forefinger before placing them in a basket.

Think Critically

1. What type of agriculture is this woman involved in?
2. In what ways might increased mechanization help her? In what ways might it have a negative impact on her life?
3. What impact might the use of biotechnology have on the harvest, the workers, and the environment?

Self-Test

(Check your answers in Appendix B.)

1. In most of Africa south of the Sahara and in parts of Asia, the percentage of farmers exceeds _______%. In the United States and Canada it is about _______%.
   a. 90; 10
   b. 35; 15
   c. 70; 2
   d. 60; 12

2. One misconception about hunting and gathering is that _______.
   a. its historical decline was due to a reduction in wild animals and plant resources
   b. hunters and gatherers were close to starvation
   c. it declined as agriculture began to develop about 11,000 years ago
   d. it is still practiced in some parts of the world

3. _______ was introduced during the first agricultural revolution. The second agricultural revolution included the development of _______. The third agricultural revolution involves _______.
   a. Domestication; four-course crop rotation; biotechnology
   b. Seed propagation; fertilizers; scientific farming
   c. The horse collar; the seed drill; mechanization
   d. Crop rotation; the internal combustion engine; organic agriculture

4. List three characteristics of subsistence agriculture.

   ____

   ____

   ____

5. Identify the type of agriculture and mobility pattern depicted in this photograph.

   ____

6. Swidden is another name for _______.
   a. shifting cultivation
   b. intercropping
   c. crop rotation
   d. subsistence farming

7. Intensive wet rice farming often employs a system of _______.
   a. buffer strips
   b. agroforestry
   c. monoculture
   d. double cropping

8. _______ perpetuates a dual economy.
   a. Truck farming
   b. Plantation agriculture
   c. Livestock ranching
   d. Mediterranean agriculture

9. The image shows _______.
   a. double cropping
   b. intercropping
   c. precision agriculture
   d. strip cropping

10. Place the following labels on this diagram of the von Thünen model.
   a. Dairy farming
   b. Grain farming
   c. Mixed crop and livestock farming
   d. Livestock ranching

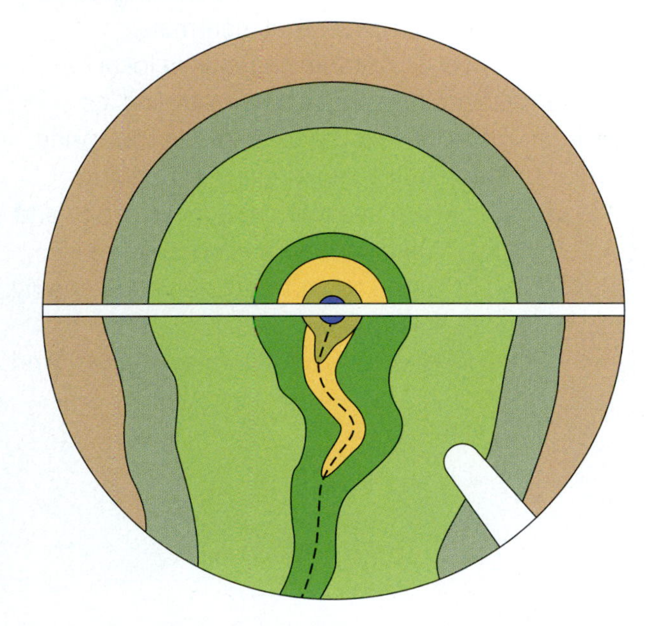

11. Mediterranean agriculture has had an important association with _______.
   a. dry-lot dairies
   b. plantations
   c. agroforestry
   d. organic agriculture

12. This photograph shows _______.
   a. a desalinization plant in Central Asia
   b. a dry-lot dairy in California
   c. specialized hog production in the Corn Belt
   d. none of the above

13. _______ caused the Aral Sea to _______ and its salinity to _______.
   a. Drought; shrink in size; fluctuate
   b. Irrigation; shrink in size; increase
   c. A series of wet years; increase in size; decrease
   d. Altered drainage patterns; increase in size; stay about the same.

14. Which is not a major milk-producing region?
   a. East Asia
   b. South Asia
   c. Europe
   d. Northern America

15. Find the false statement about the globalization of agriculture.
   a. The globalization of agriculture is linked with a nutrition transition.
   b. The globalization of agriculture is resisted by the WTO.
   c. The globalization of agriculture is associated with specialization.
   d. The globalization of agriculture is related to the supermarket revolution.

THE PLANNER

Review your Chapter Planner on the chapter opener and check off your completed work.

12 Environmental Challenges

THE INUIT AND ARCTIC ENVIRONMENTAL CHANGE

The Inuit are one of several indigenous peoples who inhabit the Arctic and sub-Arctic regions of the globe. They have lived in the circumpolar region for several thousand years and have adapted to its extremely cold conditions. These adaptations are closely related to their detailed environmental observations, which have enabled them to understand local conditions such as the thickness of the ice, where cracks are likely to form in the ice, and the migration patterns of aquatic and terrestrial mammals.

Recently, however, notable changes in local environmental conditions have occurred. The ice freezes later in the fall and thaws earlier in the spring. The earlier breakup of ice can affect the breeding grounds of seals, which use the ice to give birth to and nurse their young. The disappearance of the sea ice has exposed areas of the coast to erosion and flooding. For the Inuit these environmental changes mean that they must travel farther to reach hunting grounds. And,

NATIONAL GEOGRAPHIC

lately, the poor quality of the snow has complicated their ability to build their hunting shelters. The historical record, which shows that over the past century temperatures in the Arctic have risen at nearly twice the rate for the world, supports Inuit observations.

Environmental change and the relationships between the Inuit and the environment are complex and interconnected. Environmental challenges, the title of this chapter, captures this complexity, as we study in greater detail the human–environment interactions we introduced in Chapter 1.

CHAPTER OUTLINE

CHAPTER PLANNER

- ❑ Study the picture and read the opening story.
- ❑ Scan the Learning Objectives in each section: p. 356 ❑ p. 360 ❑ p. 368 ❑ p. 374 ❑
- ❑ Read the text and study all visuals. Answer any questions.

Analyze key features

- ❑ Geography InSight, p. 360
- ❑ Process Diagram, p. 365
- ❑ Video Explorations, p. 373 ❑ p. 380 ❑
- ❑ What a Geographer Sees, p. 377
- ❑ Stop: Answer the Concept Checks before you go on: p. 359 ❑ p. 368 ❑ p. 373 ❑ p. 380 ❑

End of chapter

- ❑ Review the Summary and Key Terms.
- ❑ Answer the Critical and Creative Thinking Questions.
- ❑ Answer What is happening in this picture?
- ❑ Complete the Self-Test and check your answers.

Ecosystems

LEARNING OBJECTIVES

1. **Define** ecosystem.
2. **Distinguish** between human and natural causes of environmental degradation.
3. **Identify** examples of common property and open-access resources.

When you use the word *environment*, do you mean just the natural world? Limiting the word in this way is a common practice in popular usage. As we discussed in Chapter 1, human geographers and environmental scientists, however, usually take a broader view. For them, the **environment** refers to one's surroundings—that is, all of the biotic (living) and abiotic (nonliving) factors with which people, animals, and other organisms coexist.

ecosystem The living organisms, their physical environment, and the flows of energy and nutrients cycling through them.

Since the 1930s scholars have used the ecosystem concept to study the interactions between different components of the environment. **Ecosystems** can be defined at a variety of scales. The Earth, for example, is an ecosystem; so are oceans, deserts, tropical rainforests, estuaries, grasslands, and even the neighborhood pond (**Figure 12.1**). Among academics, the general consensus is that the complexity of an ecosystem derives from its **biodiversity,** or the variety of species contained within it. On this basis, tropical ecosystems are considered some of the most complex. Although scientists may attempt to study a single ecosystem as though it could be isolated, the fact is that all ecosystems are interconnected. Together, these interconnected ecosystems constitute the **biosphere**—the zone of the Earth, extending from the soil and waters to the lower parts of the atmosphere, that supports and includes living organisms.

Ecological Concepts

To capture the interconnectedness of people and the environment, the biologist Garrett Hardin (1985, p. 471) used the maxim "we can never do merely one thing." With this simple expression he helped codify what is sometimes called the **First Law of Ecology**. He hoped that drawing attention to this fundamental principle would lead not just to additional study of the environment but also to greater concern for and awareness of the consequences of human actions.

Since the late 1980s another expression of the interdependence and interconnectedness of nature and society has emerged. It involves the concept of natural capital. The term *capital* refers to assets derived from human creativity (e.g., financial assets or knowledge). In contrast, **natural capital** refers to the goods and services provided by nature. Natural capital includes four component parts: (1) the renewable

NATIONAL GEOGRAPHIC

Ecosystems • Figure 12.1

Ecosystems include visible and nonvisible components such as heat energy, oxygen, and carbon dioxide. People are part of ecosystems. This estuary, an aquatic ecosystem, is in Santa Barbara County, California.

Unsustainable and sustainable harvests • Figure 12.2

a. A large net traps tuna off the coast of southern Spain. In the Mediterranean Sea, recent catches of the endangered Bluefin tuna—a prized fish for sushi—have been more than three times the established sustainable yield. Nearly one-third of the world's harvested fish stocks are overfished.

b. To promote responsible forest management, the international Forest Stewardship Council (FSC) certifies wood harvested from forests that are sustainably managed. Compliance with guidelines in order to obtain FSC certification is voluntary.

resources, (2) the nonrenewable resources, (3) the Earth's biodiversity, and (4) the ecosystems. The first three of these components constitute the goods or stocks of natural resources, whereas the fourth component illustrates the importance of services or processes provided by the natural world. These include the cycling of nutrients through ecosystems and processes such as photosynthesis. Without natural capital there would be no life on Earth. Moreover, natural capital makes possible the functioning of the economy, which is based on human use of natural resources.

There are two different types of natural resources, nonrenewable and renewable, and both types can be depleted. **Nonrenewable resources** are considered finite because they are not self-replenishing or they take very long amounts of time to do so. **Renewable resources** are replenished naturally or through human intervention, such as planting trees. The quantities of nonrenewable resources are fixed—once they are used they are gone forever. In theory, human use could entirely exhaust these nonrenewable resources; however, economic depletion usually precedes and in effect prevents total resource depletion. *Economic depletion* occurs when the cost of extracting the resource exceeds the economic value of it. As a rule, economic depletion exists when 80% of the resource has been extracted. The concept of economic depletion is related to expected future revenue and is an important business concern.

The concept of economic depletion can be applied to renewable resources as well, but scholars usually apply the concept of sustainable yield instead. *Sustainable yield* refers to the maximum quantity of a resource that can be harvested or used without impairing its ability to renew or replenish itself. One problem with this concept, however, is that it is often applied on a species-specific basis without considering the impacts that harvesting a particular species may have on the ecosystem. That is, a yield of a particular species that is determined to be sustainable may still have significant consequences for the functioning of its ecosystem. For this reason, some scholars prefer the term *ecologically sustainable yield*, which takes a systems view of the impact of extracting or harvesting renewable resources. See **Figure 12.2** for contrasting approaches to resource extraction.

Environmental Degradation

To degrade something is to impair one or more of its physical properties. The leaching, or draining, of minerals or nutrients from soil—for example, by rain—is a form of natural environmental degradation. In popular usage, however, environmental degradation is understood to be *anthropogenic*, or caused by human activities. Defining environmental degradation is no simple task, largely because the perception of what constitutes environmental degradation varies from one group to another and even from one person to another.

It is also helpful to note that human activities can directly or indirectly trigger environmental degradation. When crude oil is spilled on land or in water, toxins in it pose a direct risk to people and wildlife. Road construction in mountainous regions of the tropics can cause unstable slopes. Government policies that support road construction but do not consider its environmental impacts can indirectly trigger environmental degradation.

If we think about degradation from the standpoint of sustainability and ecologically sustainable yields, it is possible to offer a broad definition of the term that draws on three major conditions. **Environmental degradation** occurs when one or more of these conditions are met: (1) when a resource is used at a rate faster than its rate of replenishment, (2) when human activities impair the long-term productivity or biodiversity of a location, or (3) when concentrations of pollutants exceed recognized standards for maximum allowable levels.

One limitation of this definition of environmental degradation is that it fails to acknowledge that some human activities are beneficial for the environment. Therefore, geographers Piers Blaikie and Harold Brookfield (1987) have proposed another way of conceptualizing degradation. In their view a more complete assessment of environmental degradation should add together all degrading processes, both natural and human, and then subtract from this total all natural replenishment and all of the ways in which human activities have contributed to environmental restoration. See **Figure 12.3** for one example of environmental restoration.

Common Property Resources

Common property resources (CPRs) include community forests, pastures, and fishing grounds. Worldwide, many of the landless—those people who do not own individual units of land—depend on common property resources to obtain necessities such as firewood, food items, and pasture grasses for livestock (**Figure 12.4**). Common property resources differ from **open-access resources**. The air we breathe, the open seas, solar energy, national parks, and outer space are open-access resources. There may or may not be any rules controlling the use of open-access resources.

common property resources Natural resources, equipment, or facilities that are shared by a well-defined community of users.

open-access resources Goods that no single person can claim exclusive right to and that are available to everyone.

An issue that has long fascinated researchers involves the relationship between common property resources and environmental degradation. The most famous essay on this subject remains "The Tragedy of the Commons," written by

Riverine wetland restoration • Figure 12.3

a. Intrusive woody vegetation along the Central Platte River in Nebraska degraded the wetland habitat and clogged the river channel.

b. Restoration work removed the intrusive vegetation to re-create a wetland environment. Migratory waterfowl, some of which are endangered, use the islands and sandbars in the river to roost.

Common property resources in Africa • Figure 12.4

a. Women in Cameroon collect nontimber forest products
Common property in forests includes timber and nontimber forest resources such as berries, seeds, leaves, and oils.

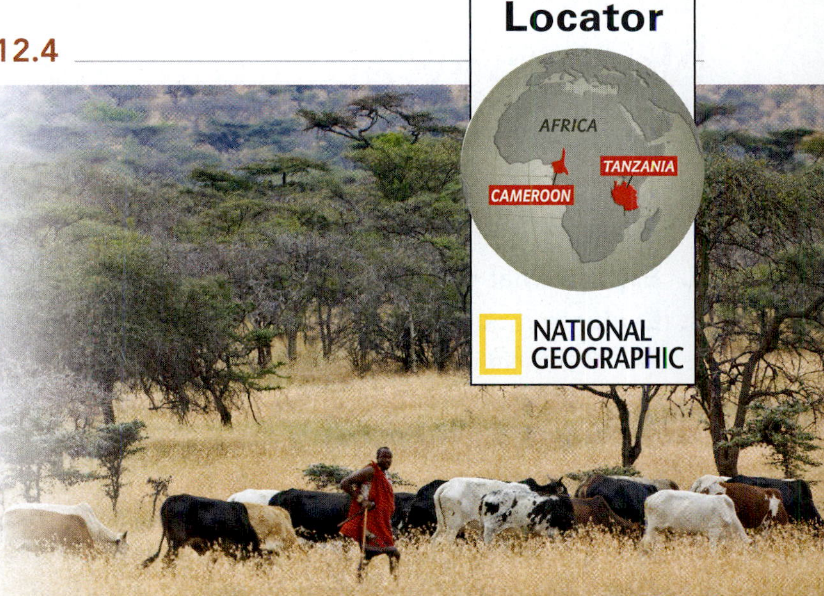

b. Maasai pastoralist in Tanzania
Historically, the Maasai have managed their pasturelands as common property resources. Since the 1980s, however, their pastures have been subdivided into individual ranches—in effect enclosing the commons. What consequences might such change bring?

Garrett Hardin and published in 1968. Hardin focused his essay on this question: Does pursuit of self-interest contribute to the public good? To answer this question, Hardin used the example of a community pasture open to all. In this scenario, the costs of keeping a large herd on the commons are shared by all of the herders, but the profits are not. Instead, profits from the sale of an animal accrue to the individual herder. Thus, it is clearly in every herder's self-interest to maximize her or his use of the commons, but if every herder does this the commons will be destroyed by overgrazing. The tragedy, according to Hardin, is that "Freedom in a commons brings ruin to all" (1968, p. 1244).

Hardin argued that private property provided only a partial solution to the problem of the tragedy of the commons. He felt that individual ownership of land would prevent environmental degradation of it but would do nothing to solve the problem of air pollution since the atmosphere cannot be divided into individually owned units. For Hardin, this meant that government policies, including taxes and regulations, were also necessary to prevent the tragedy of the commons.

The most significant flaw in Hardin's work was his mistaken assumption that common property resources lacked rules governing their use and that they are the same as open-access resources. As we have discussed, common property resources are used and managed in accordance with the established laws or customary practices within a particular community. Moreover, with common property resources, the community of users is clearly defined and is exclusive. That is, not just anyone can use common property resources. It is also the case that the usage rights granted to different members of the community vary. For example, rights of use can vary depending on one's age, sex, and/or social status.

Scholars now realize that traditional or indigenous knowledge (see Chapter 5) plays a major role in the management of common property resources. Fishers in the Torres Strait north of Australia, for example, stop fishing when they notice the size of their catches declining. In Canada, the Cree do not hunt for waterfowl during the breeding season. These and other practices serve as checks to prevent destruction of the commons. Not every system of common property resource management is successful, but successful examples raise important questions about some of the premises of Hardin's argument.

CONCEPT CHECK STOP

1. **Do** ecosystems have boundaries?
2. **How** is economic depletion related to environmental degradation?
3. **Why** is it important to distinguish between common property and open-access resources?

Nonrenewable Energy Resources

LEARNING OBJECTIVES

1. **Identify** important regional variations in the distribution of global oil reserves.
2. **Explain** what OPEC is and how it operates.
3. **Evaluate** mountaintop removal and its consequences.
4. **Cite** specific factors that have shaped the geography of nuclear energy use.

Nonrenewable energy resources include fossil fuels and uranium. **Fossil fuels** derive from the buried remains of plants and animals that lived millions of years ago. Over time, sand and other sediments covered these deposits while heat and pressure gradually transformed them into coal, oil, or natural gas. Sources of renewable energy include solar, wind, water, geothermal, and biomass (from wood products or waste products). As the pie chart in **Figure 12.5** shows, the world depends very heavily on nonrenewable energy sources.

Global energy consumption by source • Figure 12.5

The chart shows only major commercially traded energy sources, not firewood, wind, or solar power. Note the dependence on fossil fuels, which provide some 88% of the commercial energy consumed. (*Source*: BP, 2008.)

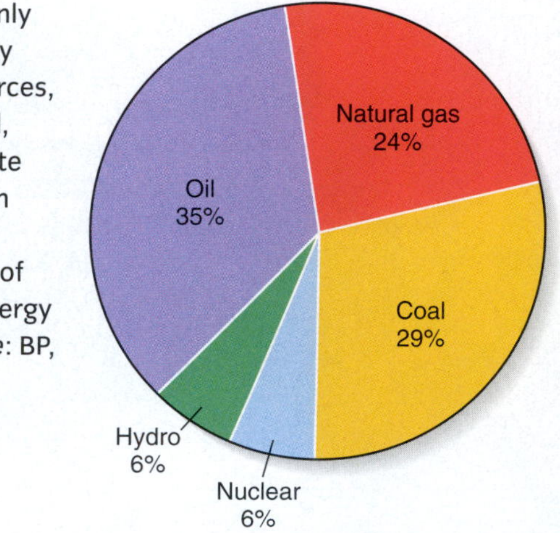

Geography InSight Oil reserves • Figure 12.6

The size of each country in the cartogram is proportional to its proved oil reserves in billions of barrels (bbs). The dominance of the Middle East is pronounced. (*Source*: Data from BP, 2008; Energy Information Administration, 2009.)

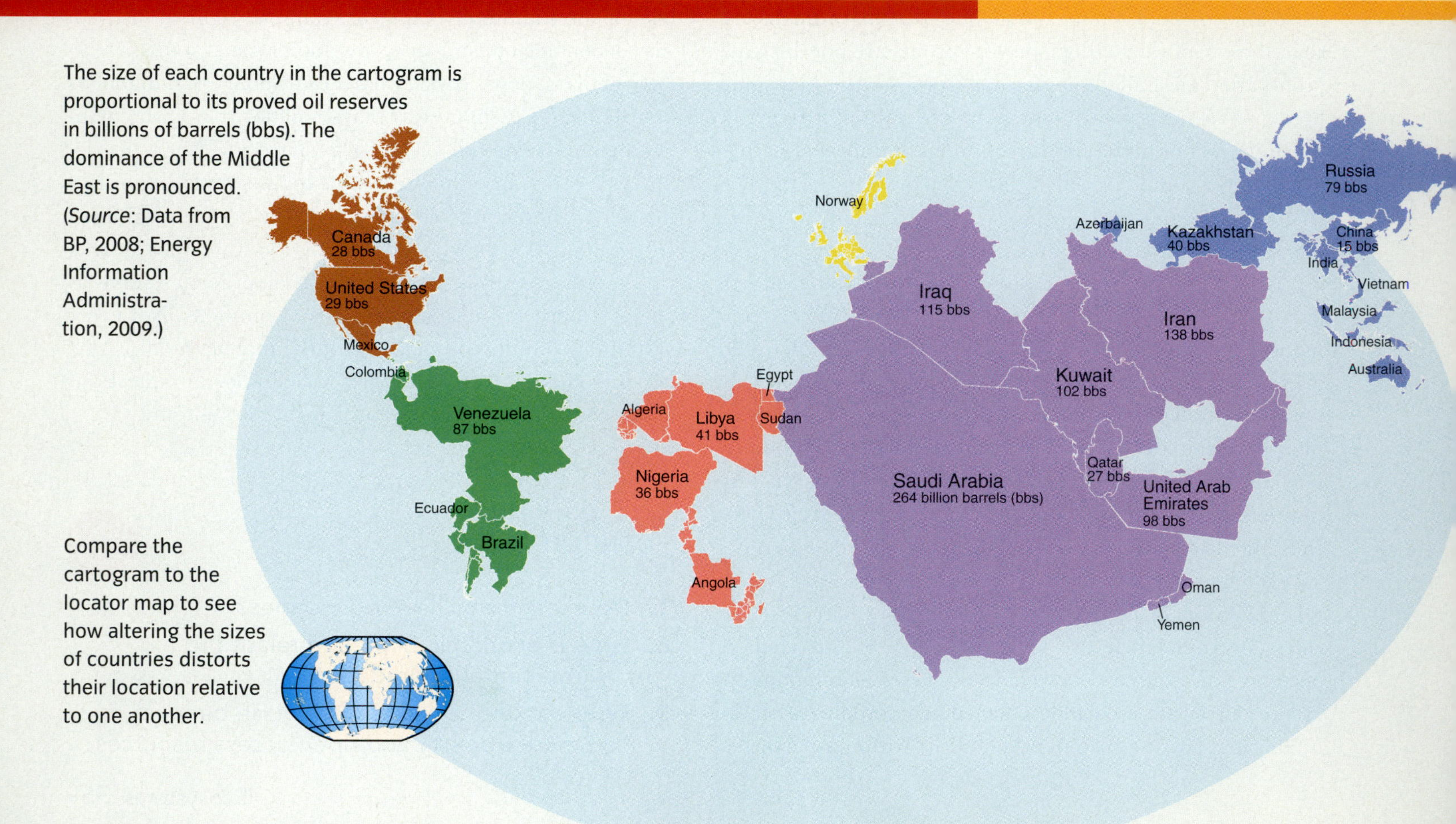

Compare the cartogram to the locator map to see how altering the sizes of countries distorts their location relative to one another.

Oil

Though nonrenewable, oil is a versatile energy resource for industrialized countries that have the necessary infrastructure to store, refine, and transport it. Oil can be burned as a fuel to heat buildings and to generate electricity, or it can be refined into gasoline, jet, or diesel fuels. From water bottles to cell phones, most of our plastic products are made from oil, as are cosmetics and even some pharmaceuticals.

With nonrenewable resources the term *proved* (or *proven*) *reserves* expresses the estimated quantity of a resource that could be extracted in the future given present financial, technological, and geological conditions. Proved reserves are not fixed; they change as a result of consumption, discoveries of additional resources, or advances in the process of recovering resources. Since 1998, for example, global proved oil reserves have increased 14%, but over the same time period U.S. proved oil reserves have dropped almost 4%.

We should bear in mind that, in spite of their name, proved reserves are always estimated amounts. Despite technological advances, there is no way to know for certain how much oil the Earth contains. Even if we did know how much oil exists, there is no assurance that we would be able to recover it (see **Figure 12.6**).

The future of oil How long will the world's current oil reserves last? The *reserves-to-production ratio* (or R/P ratio) provides one estimate. Dividing the total remaining global reserves by the annual rate of oil production gives the R/P ratio in years. The R/P ratio for the world is 41.6. Thus, at the present rate of production and barring substantial new discoveries, the world's oil will last a little more than four decades.

The countries with the largest proved oil reserves
If Canada's oil sands are counted, the country moves into second place. Strictly speaking, the oil from oil sands is not directly comparable to conventionally drilled crude oils. Oil sands are mined in open pits. The product first recovered is bitumen, a heavier and much lower grade of crude oil that must first be upgraded and prepared for refining. (*Source*: Data from BP, 2008; Energy Information Administration, 2009.)

Country	Proved Oil Reserves (billions of barrels)	Percent of World Reserves
Saudia Arabia	264	21.3
Iran	138	11.2
Iraq	115	9.3
Kuwait	102	8.2
United Arab Emirates	98	7.9
Venezuela	87	7.0
Russia	79	6.4
Libya	41	3.3
Kazakhstan	40	3.2
Nigeria	36	2.9
United States	29	2.4
China	15	1.3
Canada	28	2.2
Qatar	27	2.2

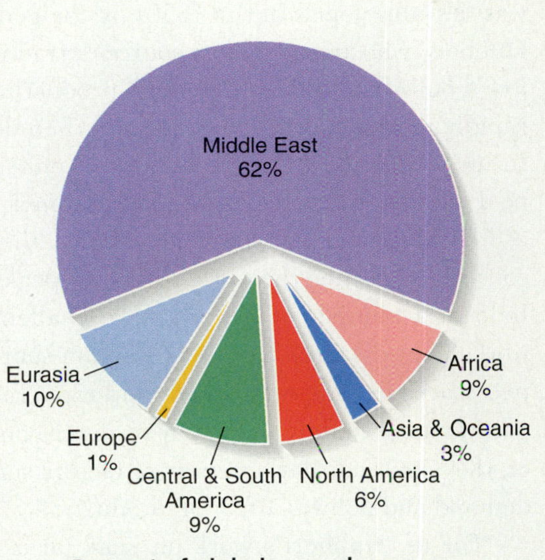

Percent of global proved reserves, by region
Russia is included in Eurasia and has the largest proved reserves in that region. North America includes Mexico. Venezuela and Brazil have the largest proved oil reserves in South America. (*Source*: Data from BP, 2008.)

OPEC and patterns of oil production • Figure 12.7

a. OPEC members
Venezuela advanced the idea of a petroleum organization and played a leading role in the establishment of OPEC. What did Venezuela stand to gain by doing this?

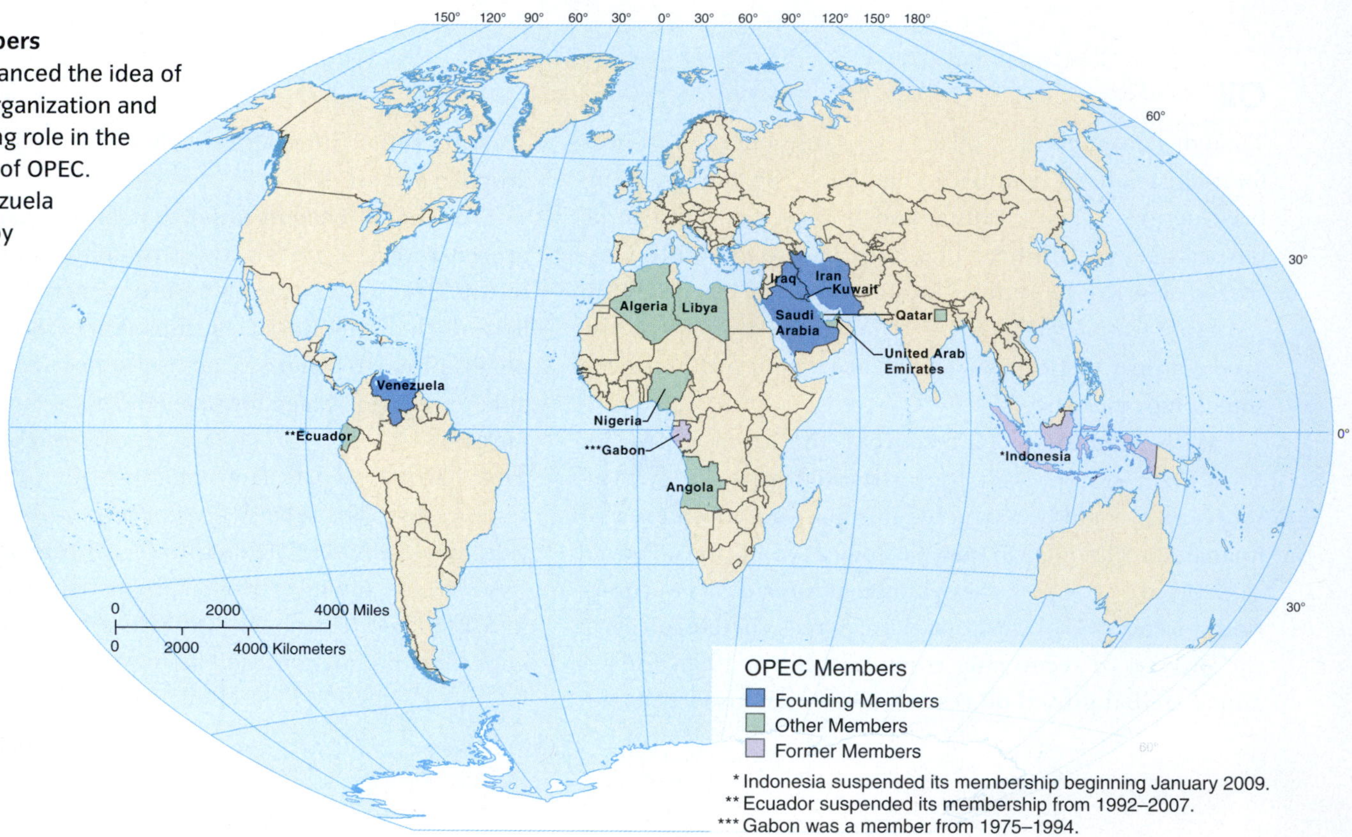

According to some oil industry experts, another important concern is the time at which the world will pass the point of peak oil production. The concept of *peak oil* was first developed in the 1950s by the geologist M. King Hubbert, who argued that resource extraction tends to follow a bell-shaped curve in which production tends to rise rapidly at first, reaches a peak, and then declines rapidly. In his view, the rate of production declines following the peak because resource extraction becomes more costly.

As with most complex questions, there is no clear consensus on when the world will reach peak oil production. Indeed, a number of experts have challenged Hubbert's work. They disagree with his assertion that oil production necessarily declines at a steady and rapid rate, and that oil production is mainly shaped by resource constraints. Rather, they contend that a variety of factors such as global oil demand and politics affect oil production.

For us, Hubbert's work on peak oil is significant because he drew attention to a foreseeable *energy transition*. In other words, he recognized that oil production would decline and that this would compel people to use a different energy source. He also thought that an energy transition would have serious consequences for the global economy unless people anticipated it and were prepared for it. Does the increasing use of electric buses and gas-electric hybrid vehicles mark the onset of an energy transition?

Oil production and consumption The Persian Gulf countries, with their massive oil reserves, are major oil producers. The Organization of the Petroleum Exporting Countries (OPEC) also influences oil production. OPEC was formed in 1960 in an attempt to counter the dominance of the oil market by a few Western (mainly American and British) oil companies. OPEC seeks to coordinate oil production among its members. It does this by functioning as a *cartel*, an organization that controls the supply of a commodity and therefore its demand and price (see Chapter 2). The first sign of OPEC's international influence came in 1973 during the Arab Oil Embargo when it restricted the movement of oil tankers to ports in Europe and the United States, triggering the first major spike in oil prices. See **Figure 12.7** to learn more about OPEC and oil production.

Much of the success of a cartel depends on the ability of its members to coordinate their production. OPEC has long used oil production quotas to regulate the supply of crude oil but has had a difficult time enforcing them. When the price of crude oil is high, for example, greater temptation exists for a country to cheat and exceed its quota in order to

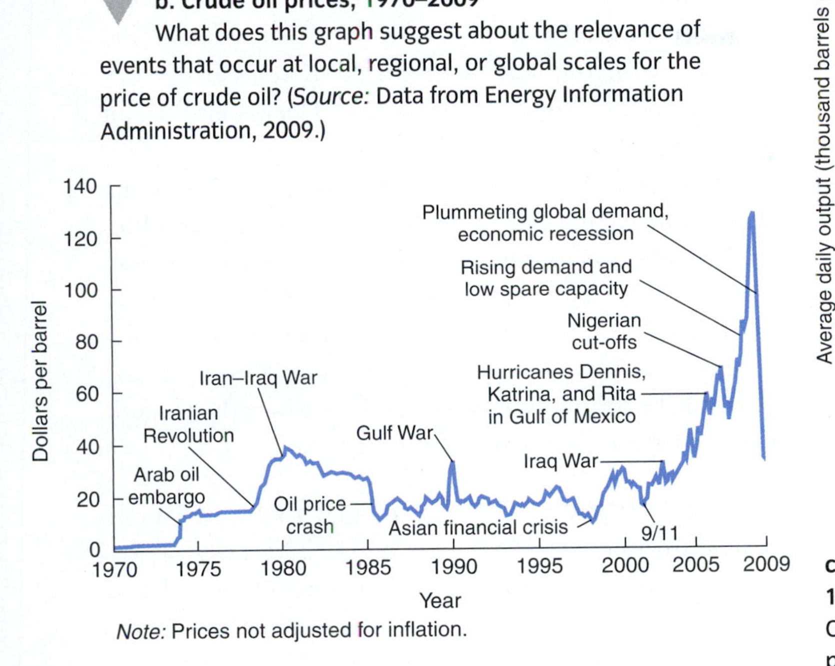

b. Crude oil prices, 1970–2009

What does this graph suggest about the relevance of events that occur at local, regional, or global scales for the price of crude oil? (*Source:* Data from Energy Information Administration, 2009.)

Note: Prices not adjusted for inflation.

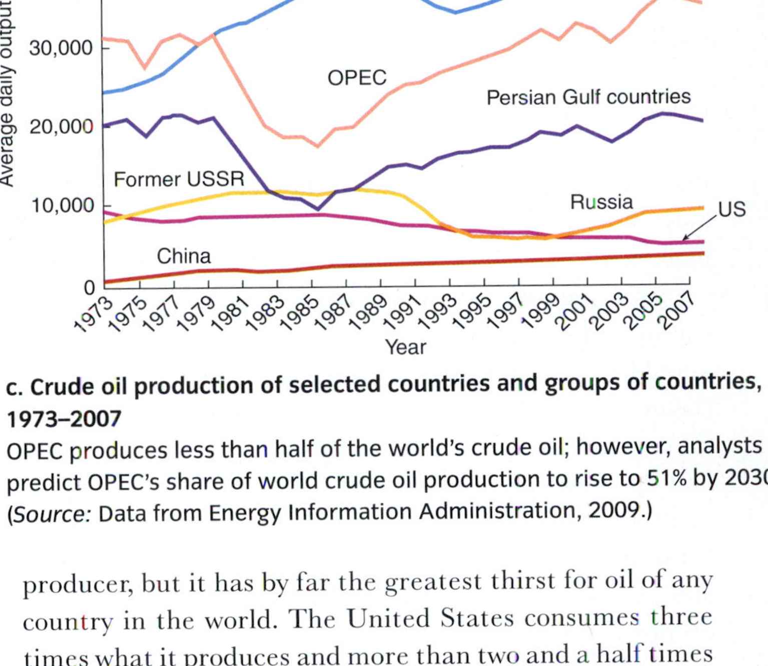

c. Crude oil production of selected countries and groups of countries, 1973–2007

OPEC produces less than half of the world's crude oil; however, analysts predict OPEC's share of world crude oil production to rise to 51% by 2030. (*Source:* Data from Energy Information Administration, 2009.)

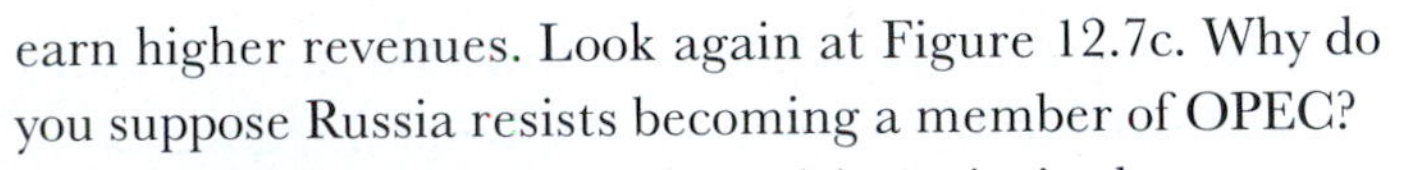

earn higher revenues. Look again at Figure 12.7c. Why do you suppose Russia resists becoming a member of OPEC?

Significant geographic disparities exist in the patterns of production and consumption of oil (**Figure 12.8**). For example, the United States ranks third as a global oil producer, but it has by far the greatest thirst for oil of any country in the world. The United States consumes three times what it produces and more than two and a half times what China, the second leading consumer, uses. On a daily basis, the United States uses nearly 21 million barrels of

Oil production and consumption • Figure 12.8

The Middle East is the greatest contributor to global oil flow, followed by Russia and other parts of the former Soviet Union, West Africa, and South and Central America.

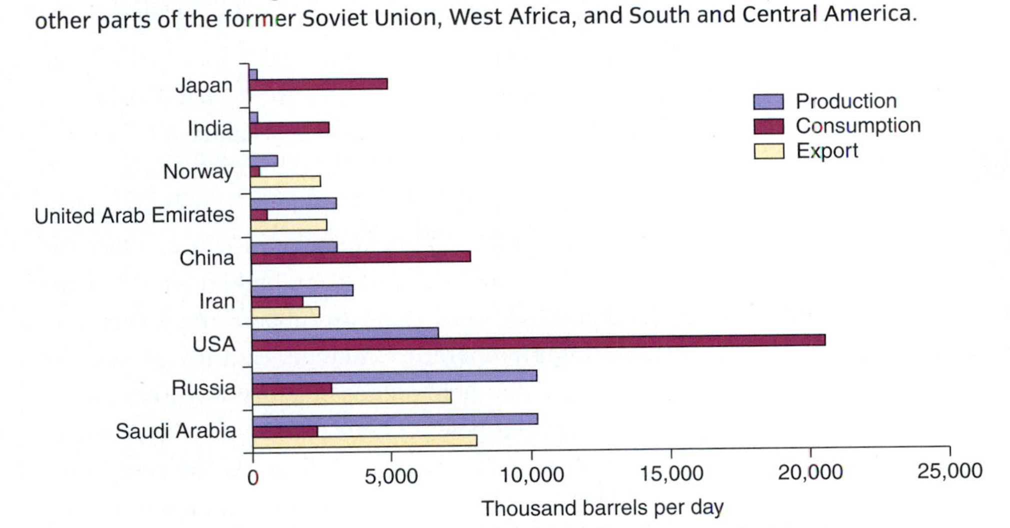

a. National variation among major oil producers, exporters, and consumers

These average daily volumes show that five leading consumer nations account for 46% of global oil consumption. The United States consumes about three times more oil than it produces and exports very little. (*Source*: Data from BP, 2008; Energy Information Administration, 2009.)

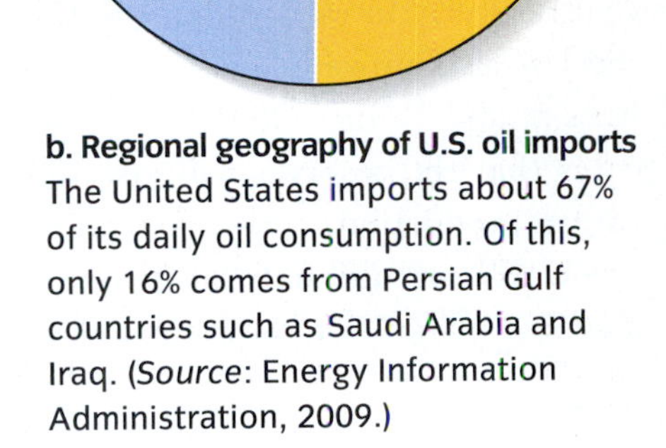

b. Regional geography of U.S. oil imports

The United States imports about 67% of its daily oil consumption. Of this, only 16% comes from Persian Gulf countries such as Saudi Arabia and Iraq. (*Source*: Energy Information Administration, 2009.)

oil (one barrel equals 42 gallons). In contrast, Saudi Arabia, the world's leading oil producer, drops out of the list of the top five consumers of oil. Japan, the third leading consumer of oil, has only limited oil reserves and must import almost all of its oil.

The industrialized countries account for a disproportionate share of the daily global consumption of oil. In one day the world consumes in excess of 85 million barrels of oil. Nearly 60% of that total is consumed by the 30 wealthiest and most industrialized countries of the world each day.

For decades the United States and Japan were the two leading consumers of oil in the world. In about 2003, however, China overtook Japan and moved into second place and in 2007 India moved into fourth place. What we are witnessing is a major transformation in the global use of and demand for oil. By some estimates, the demand for oil is expected to increase as much as 60% over the next three decades. Most striking is the rate at which oil use in China and India has increased. Between 1997 and 2007 oil consumption grew 88% in China—almost doubling—and in India it grew 50%. By comparison, oil consumption increased 16% for the world as a whole and 11% in the United States. In Japan, however, oil use dropped nearly 6% between 2005 and 2007, in part because of new government policies designed to reduce the country's dependence on oil.

Coal

The coal mined and extracted today derives from the woody, partially decomposed remains of plants and trees that accumulated in swampy environments some 300 to 400 million years ago. Coal seams or deposits formed over millions of years as this plant matter was covered by other sediment and then compacted under very high pressure.

Coal distribution and consumption Coal is the most abundant and the most geographically widespread of the fossil fuels. More than 70 countries have workable coal reserves. In rank order, the three countries with the most sizable proved coal reserves are the United States, Russia, and China. Globally, the R/P ratio for coal indicates that the reserves will last 133 years at the current rate of production.

Unlike oil and natural gas, people have used coal as a source of fuel for thousands of years. Coal, which was burned to heat water and generate steam power, also literally fueled the Industrial Revolution. After oil, coal is the second most widely consumed fossil fuel in the world today. Whereas oil is used mainly for transportation and heating, coal is the leading fuel used in generating electricity (by coal-burning power plants). Coal is also one of the key ingredients needed to produce steel.

China is heavily dependent on coal, and is by far the number one coal producer and consumer in the world today. In 2007, for example, China consumed more than twice as much coal as the United States, the second largest coal consumer in the world. The amount of coal China consumes each day is equivalent to about 26 million barrels of oil. This figure is more than three times the amount of oil China presently uses in one day.

The challenges of coal Although coal is the most abundant of the fossil fuels, extracting and using the resource present a number of serious environmental and social challenges. The most controversial method of coal mining involves mountaintop removal (MTR); the controversy stems from the scale and impact of the operations involved in MTR.

The use of MTR as a coal extraction method dates to the 1970s. In the United States, federal regulation of MTR officially began in 1977 with passage of the Surface Mining Control and Reclamation Act. This law requires that mining companies restore land that has been surface mined unless they have been granted an exception. Not all countries have such laws, however, and it is usually much cheaper for a mining company if it does not have to restore the site. **Figure 12.9** depicts the steps in MTR.

Not only does the extraction of coal present serious environmental challenges, so does the use of coal. Coal burns far less cleanly than the other fossil fuels and contributes to air pollution and smog. In addition, the combustion of coal releases mercury into the air. Atmospheric processes such as precipitation deposit mercury on land or in water. Once the mercury is in lakes and streams it can build up in the tissue of fish and enter the food chain. Scrubbers and special procedures for high-temperature mercury capture are some of the technologies that help reduce mercury and other pollutants from coal-fired power plants. When the Environmental Protection Agency issued the Clean Air Mercury Rule in 2005, the United States took an important step toward becoming the first country to regulate the emissions from utility companies.

THE PLANNER

PROCESS DIAGRAM

Understanding mountaintop removal • Figure 12.9

Mountaintop removal is a controversial mining method. Although it profitably produces vast tonnages of coal, it also contributes to massive landscape change and alters local and regional watersheds.

1 Removing mountaintops
First, all vegetation is removed from the area to be mined. Then explosives loosen thick rock above the coal (called overburden), and mammoth draglines scoop the overburden into large trucks.

2 Filling valleys
The trucks haul the overburden and dump it in nearby valleys. These "valley fills" are unstable, slide-prone, and can leach toxic metals into streams and alter drainage patterns. These changes not only affect the environment but also can have serious consequences for people who live in the area.

3 Mining the coal
Large front loaders scoop coal from the exposed seam, pouring it into trucks that haul it to loading docks, where the coal is transferred to railroad cars or river barges.

4 Rehabilitating the site
Former mining site being hydroseeded (sown with a watery mixture of seeds, fertilizer, and mulch) to establish vegetation that will retain the soil and reduce erosion. The plants that are sown are not necessarily native to the site or region and sometimes have difficulty becoming established.

5 Reshaping the land
The law allows mining companies to reshape land into a gently rolling surface instead of restoring mountaintops to their original contour. Sometimes entirely new uses of the mined site are created. Here a golf course was built on a mine site where the mountaintop was removed.

Acidity of precipitation, 2005–2007 • Figure 12.10

The pH expresses the measured acidity of water or another solution. On the pH scale, values below 7 are acidic, and the lower the value, the greater the acidity. Unpolluted rainwater has a pH of about 5.6.

Although the map does not show Canada, the problem of acid rain does not stop at the border. What is the main cause of acid rain in the eastern United States? What is the main cause of acid rain in the western United States? (*Source:* Data from National Atmospheric Deposition Program; Moran, 2007.)

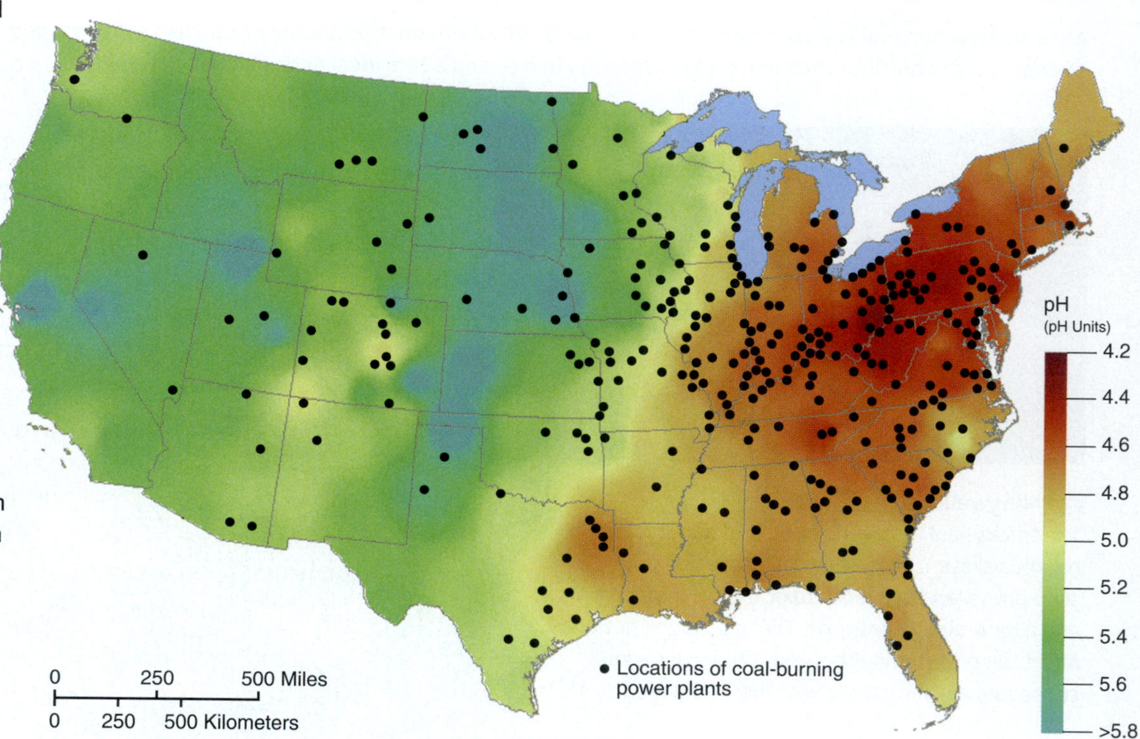

Acid rain Fossil-fuel combustion constitutes a major source of air pollution worldwide and contributes to **acid rain**. In addition to releasing mercury, coal combustion emits sulfur dioxide and nitrogen oxide. Motor vehicle exhaust also releases sulfur and nitrogen oxides into the air. These compounds interact with water, oxygen, and other chemicals in the atmosphere to form acidic substances that fall to the Earth in rain and snow. You may have noticed that coal-burning power plants and other factories often have tall smokestacks. These smokestacks release pollutants higher in the atmosphere so that they are carried by the wind. Thus, acid rain becomes a *transboundary* problem. That is, the effects of acid rain may be exported to other places, sometimes to different countries (**Figure 12.10**).

> **acid rain**
> Precipitation that, primarily because of human activities, is significantly more acidic than normal and can harm aquatic and terrestrial ecosystems.

Uranium

Uranium is a naturally occurring radioactive element that is found in certain mineral ores. It is not a fossil fuel but is a nonrenewable resource and the basic fuel supply for nuclear energy and nuclear weapons. At the present rate of consumption, uranium reserves are expected to last about another 100 years.

Uranium is used primarily to generate nuclear energy, which heats water to produce steam that turns turbines and generates electricity. The ability to harness nuclear energy dates to the 1940s—a time when much nuclear energy research focused on producing the atomic bomb. In 1957 the International Atomic Energy Association, an organization promoting peaceful uses of nuclear energy, was formed. Commercial development of nuclear energy followed rapidly in the 1950s and 1960s.

Nuclear energy constitutes a small fraction—about 6%—of the energy consumed worldwide (refer again to Figure 12.5). Globally, the geography of nuclear power plants is highly uneven and strongly associated with the most industrialized countries. There are three main reasons for this pattern. First, the ability to harness and control the production of nuclear energy as well as its waste materials requires specialized knowledge and expertise. Second, there are very high costs (of up to several billion dollars) involved in building nuclear reactors. Third, nuclear power plants also require a substantial supporting infrastructure including power generators, appropriate sites for waste storage or management, and other facilities. **Figure 12.11** depicts different aspects of the geography of nuclear energy.

Uranium and nuclear energy • Figure 12.11

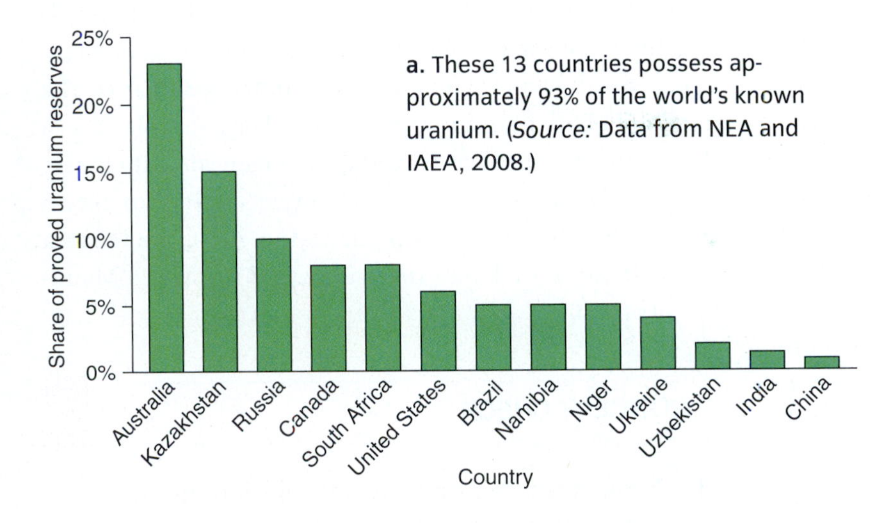

a. These 13 countries possess approximately 93% of the world's known uranium. (*Source:* Data from NEA and IAEA, 2008.)

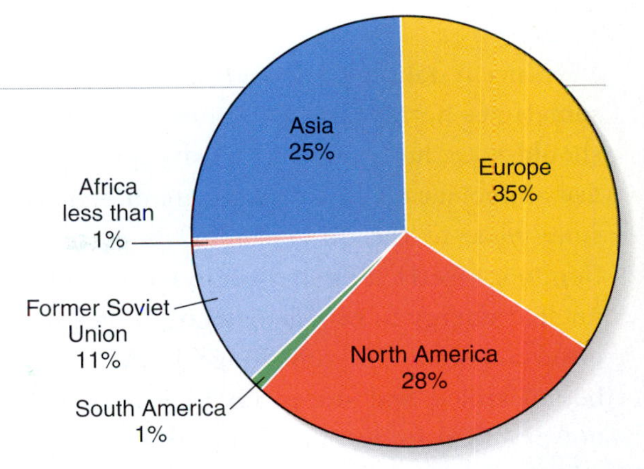

b. Operating nuclear reactors, by region, 2007 There are fewer than 450 operating nuclear power plants in the world today. The United States has more operating nuclear power plants than any other country and has more than three-quarters of the nuclear reactors in North America. Most of Asia's nuclear reactors are in Japan. (*Source:* Data from IAEA, 2008.)

c. Share of electricity from nuclear power, 2007 The United States produces nearly one-third of all the nuclear generated electricity in the world, but less than 20% of the electricity generated in the United States comes from nuclear energy. Contrast this with the situation in France or Sweeden. What factors help explain this circumstance? (*Source*: Data from NEA and IAEA, 2008.)

Electricity from Nuclear Power
50% or more
40–49%
20–39%
10–19%
Less than 10%
Not reported or no data

d. Kashiwazaki-Kariwa nuclear power plant, Japan Nuclear power plants in Japan are designed to withstand earthquake damage and shut down in the event of one. Even so, an earthquake damaged the road near this nuclear plant and caused a radioactive leak. With no known uranium reserves and minimal petroleum reserves, energy imports are crucial to Japan.

Since it does not rely on fossil fuels, nuclear energy is sometimes heralded as an alternative energy source. We should note, however, that most experts do not consider use of nuclear energy a substitute or replacement for oil since most of the oil consumed is used for transportation, not electricity generation. Nevertheless, the use of nuclear energy to produce electricity presents a number of advantages. One advantage is that nuclear fuel can be stockpiled and stored for long periods of time. It is not practical to do this with coal, oil, or natural gas. In addition, on a per-weight basis, uranium generates more electricity than do the fossil fuels. If less uranium is needed, some experts argue, the disturbance to the landscape caused by mining it will be significantly less than the disturbance caused by mining coal. Other advantages of nuclear power include low air pollution and very low emissions of carbon dioxide.

One disadvantage of nuclear power is that it is expensive to develop. Another major concern is that an accident at or sabotage against a nuclear power plant could cause its nuclear reactor to melt down and trigger a radioactive catastrophe comparable to or worse than the disaster at the Chernobyl nuclear plant in Ukraine in 1986. Also, the use of nuclear fuel yields different kinds of radioactive waste. The safe storage of these waste materials has become a major issue and is central to the ongoing debate about nuclear power. For example, Yucca Mountain, in southern Nevada, has been identified as a long-term storage site for radioactive waste, but Nevadans strongly oppose that plan, and in 2010, the Obama administration took measures to cancel the Yucca Mountain program.

CONCEPT CHECK STOP

1. **How** does the distribution of oil reserves compare and contrast to geographical patterns in the production and consumption of oil?
2. **Why** was OPEC formed, and what challenges does it face?
3. **Why** is mountaintop removal controversial?
4. **What** factors help explain the geography of nuclear energy use?

Renewable Energy Resources

LEARNING OBJECTIVES

1. **Explain** how biogas digesters contribute to sustainable practices.
2. **Identify** some of the environmental problems associated with large hydropower facilities.
3. **Identify** factors affecting the adoption of solar and wind power.
4. **Distinguish** between direct and indirect uses of geothermal energy.

When we think about different systems of energy production around the world, it is useful to distinguish between commercial and noncommercial energy. Commercial energy has historically been produced largely from fossil fuels, nuclear fuel, or large-scale hydropower facilities. Consumers access and purchase commercial energy through a networked infrastructure such as an electric grid. Commercial energy is often used or consumed a long way from its original source or where it was refined, whereas noncommercial energy is produced and consumed locally or regionally—for instance, at the village or household level. Thus, noncommercial energy can be thought of as "off-grid" energy.

On a daily basis, noncommercial energy satisfies the energy needs of hundreds of millions of people across rural areas throughout much of the developing world. Although considerable data exist on the supply and consumption of commercial energy, much less information is available about the regional geographies of noncommercial energy. One measure that suggests the extent of noncommercial energy use is the population without electricity. About one-quarter of the world's people live without electricity.

Renewable energy, also called *alternative energy*, has, until very recently, been largely a component of noncommercial energy. In fact, the only renewable fuel type that is a major source of commercial energy is water—that is, hydropower (refer again to Figure 12.5). It is important to note that most of the energy that is generated from hydropower comes from large-scale facilities (with dams that are more than 15 meters or 50 feet high) such as the Three Gorges Dam in China or the Grand Coulee Dam in Washington State. Today, however, these massive facilities are considered to have such substantial environmental impacts that they are excluded from the category of renewable energy. **New renewables,** therefore, include biomass, tidal, solar, wind, and geothermal energy as well as small-scale hydropower.

Biomass Energy

Biomass, the organic matter in an ecosystem, is an important source of energy worldwide. More specifically, **biomass energy** is obtained from plant matter and/or animal wastes. Common sources of biomass energy include wood, charcoal, crop residues such as plant stalks or coconut shells, and cattle manure. Even though fossil fuels ultimately derive from the buried remains of plants, they are not counted as a source of biomass energy because of the chemical changes in their makeup that have occurred over lengthy periods of time. Unlike fossil fuels, biomass energy is renewable as long as the resource that supplies it is sustainably managed.

There are two methods for harnessing the energy from biomass: direct and indirect. The direct method is to burn unprocessed biomass and use the energy released for cooking or heating. The indirect method involves converting the biomass into a gas (biogas) or liquid fuel (biofuel), with the help of naturally occurring microbes. Methane, a biogas, can be used for cooking, heating, or lighting. Liquid ethanol, a biofuel that can be obtained from the residues of certain crops, notably corn and sugarcane, can be used to power certain vehicles. Biodiesel can be manufactured from vegetable oils, including used cooking oils (**Figure 12.12**).

Biomass is the leading type of renewable energy used worldwide. A large part of the global demand for biomass stems from its use as fuel for cooking. Indeed, more than 3 billion people prepare and cook their foods with the heat generated by burning fuelwood, dung, or various crop residues. Of these, fuelwood remains the principal source of biomass energy in many developing countries. Recent estimates indicate that fuelwood is the dominant source of energy in more than 90% of rural households in Sub-Saharan Africa.

Biodiesel at a gas station near Los Angeles • Figure 12.12

Whereas most developed countries including the United States have generally been slow to adopt biofuels, Brazil has been using sugarcane to generate ethanol as a fuel for cars and light trucks since the 1970s.

A serious problem associated with reliance on fuelwood, however, is that it can intensify pressures on local forest resources. From a socioeconomic standpoint, the time spent collecting fuelwood, a task most often relegated to women and girls, can interfere with their education or opportunities to earn additional income. Also, burning fuelwood is a major source of indoor air pollution and has been linked to tuberculosis, lung disease, cataracts, and other ailments. Across the developing world, indoor air pollution poses the greatest threat to women and children because they are most likely to spend long hours working inside and in poorly ventilated rooms. To address this problem, both governmental and nongovernmental organizations are increasingly promoting improved ventilation systems.

In the Caucasus, a Georgian villager tends to her biogas digester • Figure 12.13

Biogas digesters are specially designed tanks that are usually placed in the ground and have an above-ground opening. People simply add biomass such as crop residues or animal waste to the digester, which facilitates the conversion of the biomass into a biogas that can be used as a household fuel supply for cooking and lighting. Collecting livestock and poultry waste in biogas digesters instead of discharging them into streams helps prevent water pollution.

Another type of technology that enables people to make use of biomass—most commonly dung and vegetable matter—is a *biogas* or *methane digester*. Biogas digesters are fairly easy to build, can be used by rural and urban dwellers alike, and can be built to supply energy at the industrial or household scale (**Figure 12.13**).

Hydropower

Hydropower is a nonpolluting, renewable source of energy. Less than one-third of the world's economically viable hydropower potential has been tapped. The countries or regions with the greatest hydropower potential include China, Russia, Sub-Saharan Africa, Southeast Asia, and Central and South America. On a per-person basis, Nepal has one of the highest hydropower potentials in the world. Hydropower ranks third after coal and natural gas as a source of electricity. Large hydropower facilities produce about 15% of the world's electricity. As we have seen, in most industrialized countries, use of hydropower to generate electricity constitutes a small part of the overall energy picture. However, a few countries, including Norway, Brazil, and Argentina, obtain most of their electricity from hydropower.

Globally, the heyday of large dam construction occurred between the 1930s and 1970s. At the time, many development experts believed that these massive facilities would help solve the problem of uneven economic development by improving food security, lessening import dependency, and providing jobs. More dams were approved for construction during the 1970s than in any other decade. Although they have provided numerous benefits, such as a year-round water supply for irrigation, large dams have not solved the problem of uneven development and have contributed to a host of environmental problems.

Large dams fragment rivers and alter the ecosystems of river basins. According to a report of the World Commission on Dams, "Large dams generally have a range of extensive impacts on rivers, watersheds and aquatic ecosystems—these impacts are more negative than positive and, in many cases, have led to irreversible loss of species and ecosystems" (WCD, 2000, p. xxxi). Beyond the impact on the physical environment, large dam construction also has major consequences for people, including forced resettlement or loss of livelihood (**Figure 12.14**).

Hydropower production • Figure 12.14

a. Leading hydropower producers
The leading producers account for nearly half of the world's hydroelectricity production. (*Source*: Data from Energy Information Administration, 2009.)

Producer	Percent of World Total
China	14.4%
Canada	12.0%
Brazil	11.5%
United States	9.7%
Russia	5.8%
Norway	3.9%
India	3.8%
Japan	2.8%
Venezuela	2.7%
Sweden	2.0%

b. The world's largest dam
The massive Three Gorges Dam on the Chang Jiang (Yangtze) River in China provides flood control and has boosted China's production of hydropower, but construction of the dam also forced the relocation of more than 1 million people and has altered the river's water quality and ecosystems.

Today, small hydropower (SHP) is favored as a more sustainable alternative. By definition, SHP includes those facilities that generate fewer than 10 megawatts of electricity. Depending on specific river and site characteristics, the electricity generated from SHP facilities can serve a local community or a single household. Like the use of biomass, the development of SHP holds promise as another mechanism that can spread rural electrification in developing areas.

Solar and Wind Energy

Energy from the sun can be harnessed in two ways: passively and actively. *Passive solar collection* uses the design of a building and its materials to capture sunlight. In the northern hemisphere, for example, south-facing windows are advantageous for passive solar collection. *Active solar collection* uses different devices including solar panels, mirrors, or photovoltaic cells to capture, store, or use the sun's energy. Photovoltaic (PV) cells enable the conversion of sunlight directly into electricity. In other systems the sun's energy warms a liquid such as water. Solar hot water heaters work on this principle. Prompted by electricity shortages, Cape Town, South Africa, has implemented a city by-law that encourages households to install solar water heaters.

Technological and financial barriers have limited our ability to maximize use of this form of renewable energy. Solar energy systems tend to be expensive to install, and at present it can take from 5 to 15 years to recoup these costs. Nevertheless, growth in the solar energy sector has been rapid in recent years, with most of the installed capacity making use of PV cells. Government subsidies have helped to offset these costs, and in 2009 California enacted legislation to provide incentives for solar installations. Like hydropower, one of the major advantages of solar and wind power is that they are emission free.

The sun is the ultimate source of wind power. Winds are created through the uneven heating of the Earth's surface by the sun. Wind turbines harness the energy from moving air and convert it to electricity. The precursor to the wind turbine was the windmill, a device used in Europe from the 1200s to grind grain and pump water. Electricity can be generated from a single turbine or multiple turbines. Most commercial uses of wind power now involve the construction of *wind farms*, clusters of turbines in an area.

At present, wind power contributes only minimally to the world's energy supply. Denmark provides one important exception to this trend—in recent years wind power has supplied approximately 20% of the country's electricity. No other country comes close to this figure. For example, in the United States wind power produces a little more than 1% of the country's electricity. Nevertheless, the wind energy industry is growing very rapidly. New turbine installations in 2007 increased the world's wind power capacity by nearly 30%. Two factors contributing to the popularity of wind energy are concerns about energy security associated with reliance on oil imports and concerns about climate change (**Figure 12.15**).

Geothermal Energy

Geothermal energy comes from the interior of the Earth. High pressures combined with the slow radioactive decay of elements in the Earth's core produce vast amounts of heat. The surrounding rock materials absorb this heat such that the materials closest to the Earth's core are the hottest. In some places, such as Iceland, which is situated along a rift in the Earth's surface, it is possible to find extremely hot rock material and water near the Earth's surface. At some locations, for example, underground waters about 3 km (2 mi) deep have temperatures in excess of 150 degrees Celsius (300 degrees Fahrenheit).

Geothermal energy is harnessed by drilling deep wells in order to tap into reservoirs of heated groundwater. This hot water can then be used as a direct source of heat for

Installed wind power capacity • Figure 12.15

The geographic centers of the wind energy industry have traditionally included western Europe and the United States. Today, however, the wind energy industry is becoming more globalized. In rank order, the five countries with the highest installed wind power capacity are Germany, the United States, Spain, India, and China.

NATIONAL GEOGRAPHIC

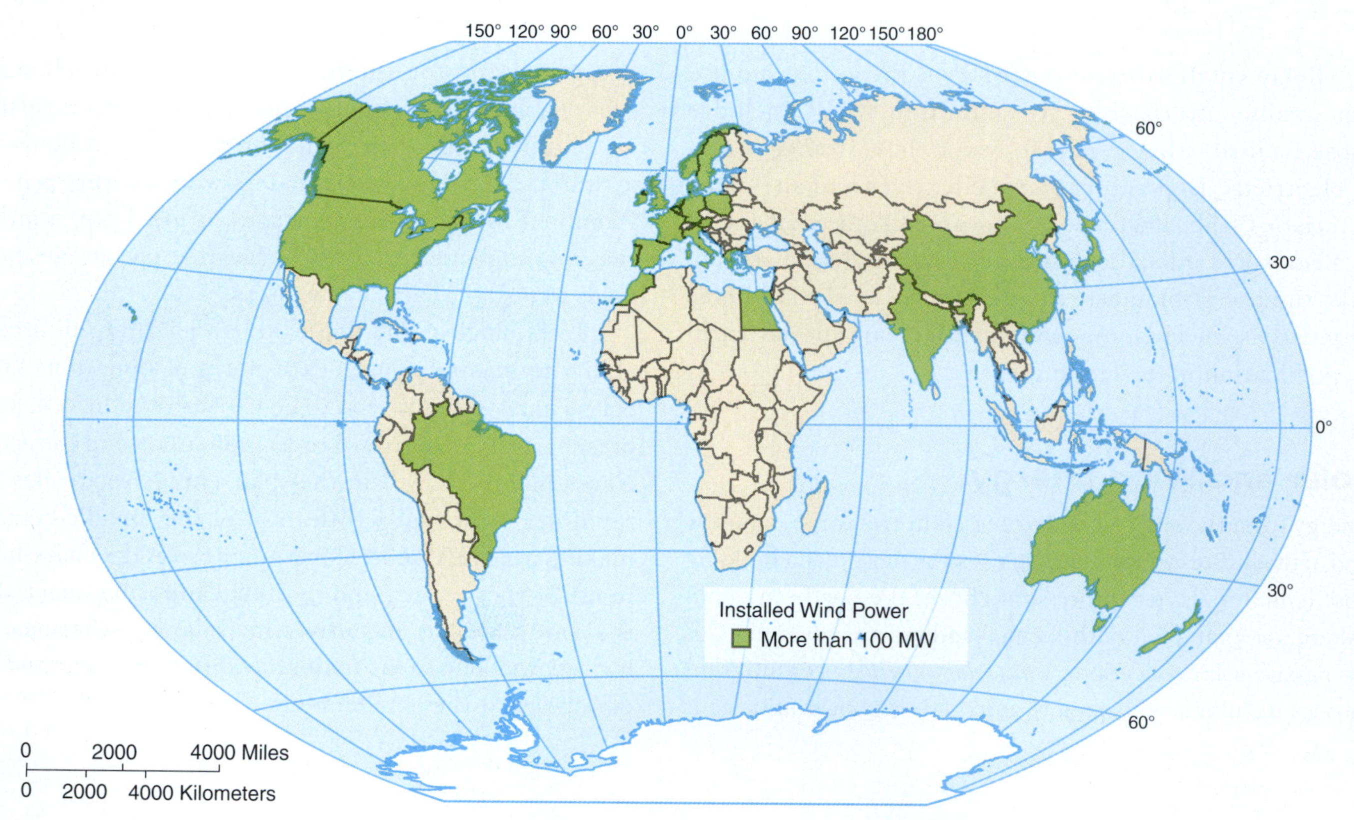

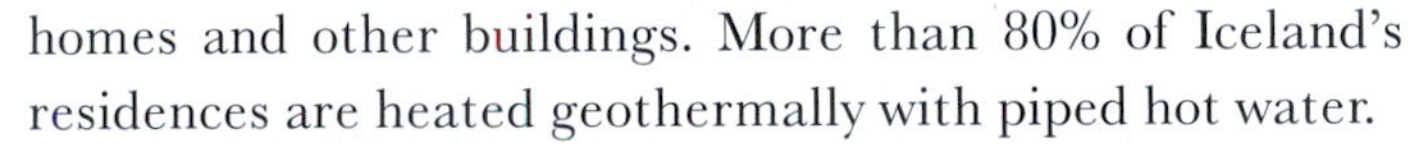

Geothermal energy, Iceland • Figure 12.16

Geothermal energy warms this outdoor spa and powers the energy plant in the background.

NATIONAL GEOGRAPHIC

homes and other buildings. More than 80% of Iceland's residences are heated geothermally with piped hot water.

If the hot water is converted to steam, it can be used to rotate turbines and generate electricity. Geothermal energy supplies 25% of Iceland's power—far more than any other country. At present, most of the existing capacity for geothermal power generation is confined to a handful of countries, including the United States (especially the western states), the Philippines, Mexico, Italy, Indonesia, Japan, Iceland, and New Zealand (**Figure 12.16**).

Many types of renewable energy exist, but different barriers have slowed adoption of them. For another perspective on this topic, see *Video Explorations*.

Video Explorations

Alternative Energy

This video clip reviews several different forms of alternative energy; specifically wind power, solar energy, and biomass fermentation. It explores the current uses of each alternative energy form in the United States and interviews several alternative energy advocates. It also discusses why the uses of alternative energy forms are much rarer in the United States.

CONCEPT CHECK STOP

1. **What** are some environmental and social issues related to the use of fuelwood?
2. **Why** are large-scale hydropower facilities not considered renewable sources of energy?
3. **What** developments have aided the adoption of solar and wind power?
4. **How** is geothermal energy produced and harnessed and where is most of the existing capacity for it?

Human-Environment Interactions

LEARNING OBJECTIVES

1. **Distinguish** between the greenhouse effect and global warming.
2. **Relate** land-use and land-cover change to sustainability.
3. **Explain** the significance of the Kyoto Protocol and the Copenhagen Accord.

As we saw in the previous section, important strides are being made in the adoption and use of renewable forms of energy. Collectively, however, we still have a long way to go before we can end our addiction to fossil fuels. In this final section we examine the relationship between human activities and global environmental change.

The greenhouse effect • Figure 12.17

Solar radiation passes through the atmosphere and warms the Earth. As the Earth's surface warms, it emits heat, some of which passes into space and some of which is "trapped" or absorbed by molecules of water vapor, carbon dioxide, methane, and other greenhouse gases. The heat that these gases emit warms the atmosphere.

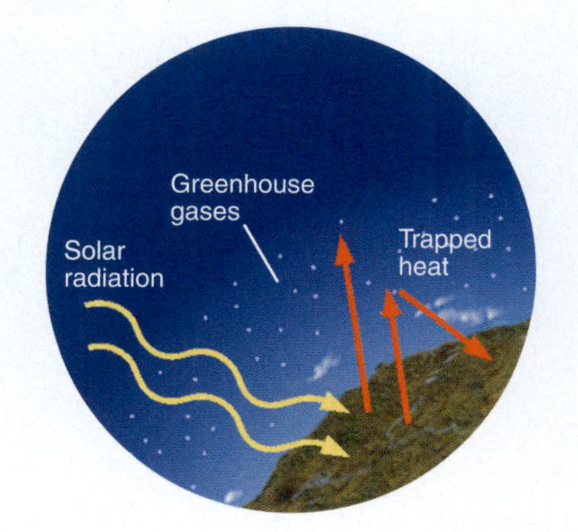

The Greenhouse Effect and Global Warming

Academics and nonacademics alike regularly use the term *greenhouse effect* to explain how the Earth's atmosphere works. Unfortunately, the term is slightly misleading. A greenhouse receives incoming solar radiation that warms the inside air. This warm air is then trapped within the greenhouse by the glass windows and doors that enclose it. The atmosphere—that thin gaseous layer that surrounds the Earth—works in a different manner. Recall that the atmosphere consists of nitrogen (78%), followed by oxygen (21%) and a combination of other gases. It is some of these other gases, including water vapor, carbon dioxide, methane, nitrous oxide, and hydrochlorofluorocarbons (HCFCs), that are *greenhouse gases*. **Figure 12.17** provides a simplified depiction of the **greenhouse effect**.

> **greenhouse effect** A natural process, involving solar radiation and atmospheric gases, that helps to warm the Earth.

The greenhouse effect is a naturally occurring process that helps enable life on Earth. For example, we know that if the Earth did not have its atmosphere, temperatures would be significantly colder. The concern about the greenhouse effect, however, centers on the increasing concentrations of greenhouse gases in the atmosphere because of certain human activities. Let's consider carbon dioxide (CO_2). Burning fossil fuels and plant matter releases CO_2 into the atmosphere.

Considerable historical data confirm that the atmospheric concentrations of CO_2 have increased sharply since the Industrial Revolution. Since there is no known natural process or source that could account for this increase in CO_2, scientists attribute the rise in CO_2 concentrations to human activities, primarily the burning of fossil fuels (**Figure 12.18**).

Like CO_2, atmospheric concentrations of methane and nitrous oxide have also increased significantly since the Industrial Revolution. These increases, too, are tied to human activities. Worldwide, the human activity responsible for releasing the greatest amounts of methane into the atmosphere is livestock production. Ruminant livestock such as cattle, sheep, and goats produce methane during digestion. Nearly 30% of the methane from human activities comes from livestock. Wet rice farming (see Chapter 11) is another important source of methane because the processes of anaerobic (without oxygen) decay that occur in the flooded fields produce methane. In addition, methane can leak into the atmosphere when we drill gas wells, transport natural gas through pipelines, or use gas dryers. Even though methane is a clean-burning fuel, there is presently no practical way to harness the methane produced by livestock or agricultural processes.

Atmospheric concentrations of carbon dioxide • Figure 12.18

Global emissions of CO_2 nearly doubled between 1970 and 2004. Land-use change—especially deforestation—also contributes to increased concentrations of CO_2. Forests help remove CO_2 from the atmosphere; without forests, that process is significantly diminished. (*Source*: IPCC, 2007.)

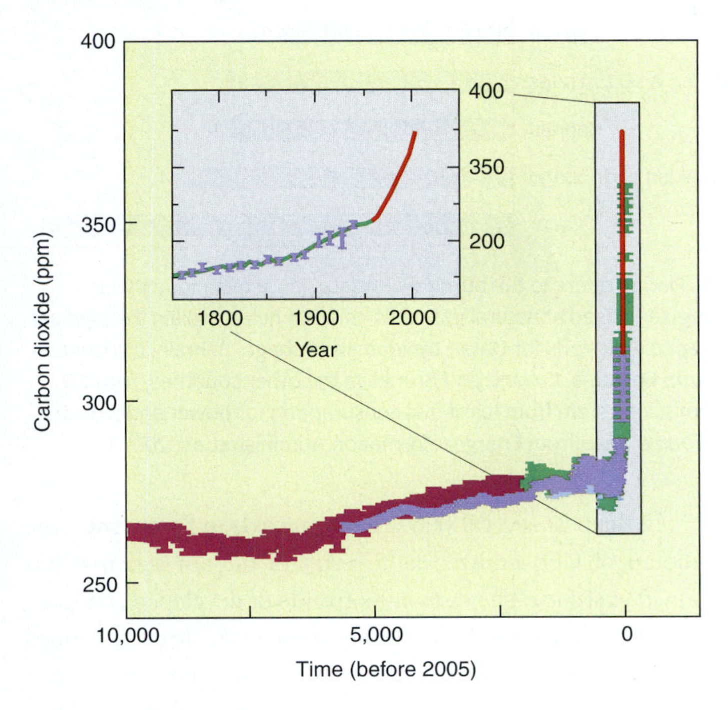

Anthropogenic sources of nitrous oxide include agriculture—specifically, the application of nitrogen fertilizers. Under certain soil and moisture conditions, the natural bacterial action that occurs in the presence of these fertilizers can result in the formation of nitrous oxide. Energy use is another major source of nitrous oxide. Motor vehicles emit nitrous oxide; this is the leading anthropogenic source of it in the United States. Coal-fired power plants are also an important source of nitrous oxide.

Unlike the other greenhouse gases, HCFCs do not occur naturally. Rather, they are synthesized specifically for human use. HCFCs are used as coolants in refrigerators and air conditioners. The manufacture of HCFCs is now being phased out around the world.

global warming A rise in global temperatures primarily attributed to human activities that have increased concentrations of greenhouse gases in the atmosphere.

The consensus among many scientists is that our activities have amplified the greenhouse effect and have contributed to **global warming**. According to the Intergovernmental Panel on Climate Change, "Warming of the climate system is unequivocal, as is now evident from observations of increases in global average air and ocean temperatures, widespread melting of snow and ice and rising global average sea level" (IPCC 2007, p. 30). Based on temperature records that date to the 1850s, scientists recorded the warmest surface temperatures between 1995 and 2006. Over an even longer perspective, the data convey a similar message about global warming: the past 50 years were warmer than any other 50-year period in 1,300 years (**Figure 12.19**).

As noted at the start of the chapter, changes to the Arctic sea ice are having consequences for local people such as the Inuit. However, global warming has the potential to affect the continental ice sheets in Greenland and West Antarctica. These ice sheets are so substantial that if they melt they will cause sea level to rise. Even a 1-meter (3-foot) rise in sea level means that low-lying coastal areas around the globe will be flooded, or put at risk of hazards caused by storm surges or erosion, with additional costs to maintain infrastructure and protect people or property in these areas.

Global annual surface temperatures • Figure 12.19

Fluctuations in global temperatures are normal and in any single year are the result of a variety of natural and human factors including solar radiation and greenhouse gases. Climate scientists focus on long-term trends, such as the upward trend in temperatures since the 1960s, shown here.

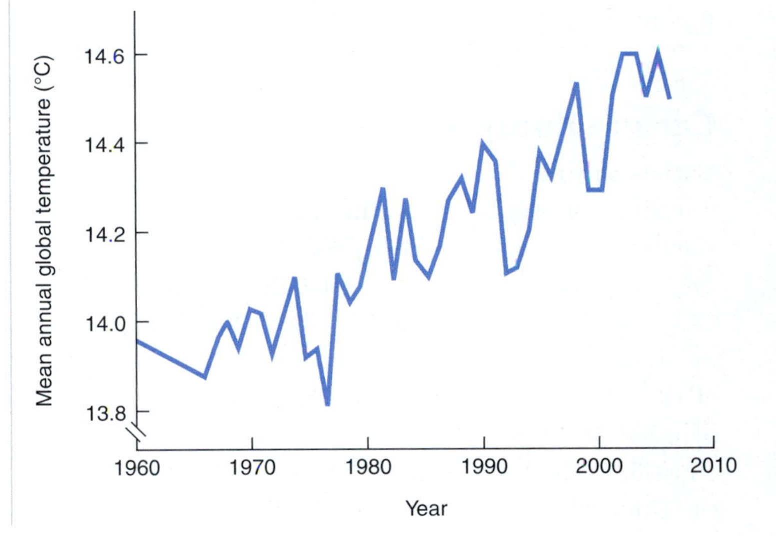

Leading carbon dioxide emitters • Figure 12.20

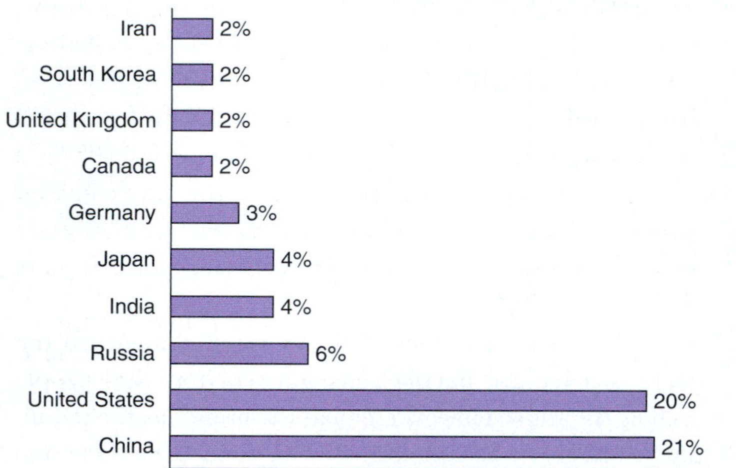

a. China is now the world's leading emitter of CO_2. Collectively, China and the United States are responsible for 41% of the world's CO_2 emissions from the consumption and flaring of fossil fuels. (*Source*: Data from Energy Information Administration, 2009.)

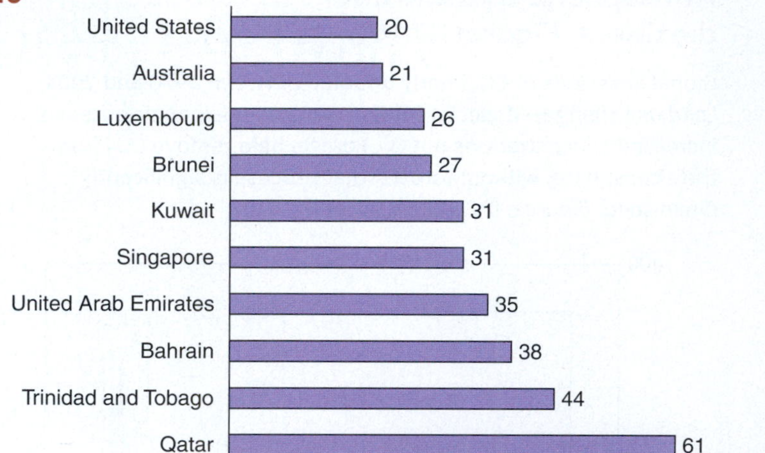

b. Flaring refers to the burning of waste gas, a technique that is regularly used at natural gas wells and that helps explain the high per capita emissions for Qatar, Trinidad and Tobago, Bahrain, the United Arab Emirates, Kuwait, and Brunei. In the other countries, high CO_2 emissions stem from fossil-fuel consumption for power and industry. (*Source*: Data from Energy Information Administration, 2009.)

Rising global temperatures also have consequences for ecosystems. Some species of plants, trees, and amphibians among others, may be endangered by significantly warmer temperatures. Climate models indicate that precipitation patterns would also be affected by global warming, raising questions about the impact on agriculture. In arid and semiarid regions, for example, problems of water scarcity may worsen, and droughts may become more frequent. For human geographers and others, global warming raises numerous environmental and ethical questions. As an example, Pacific Island countries contribute less than 1% of the world's carbon dioxide emissions, yet will be among the first places affected by climate change, especially because many of the islands face inundation and erosion from rising sea levels.

Carbon Footprints

Carbon dioxide factors prominently in studies of global warming not only because atmospheric concentrations of it have increased so markedly, but also because it stays in the atmosphere a long time. Globally, the geography of CO_2 emission is highly uneven. The richest and most industrialized countries in the world account for nearly half of the CO_2 emissions. In contrast, the countries of Sub-Saharan Africa generate less than 3% of the world's CO_2 emissions. The two greatest CO_2 emitters are China and the United States (**Figure 12.20**).

In the past several years the term **carbon footprint**—the amount of CO_2 emitted as a result of human activity—has gained currency. The carbon footprints of developed countries as well as China and India swamp those of the least developed countries. To promote environmental awareness and clean energy, Carbonfund.org created *carbon offsets*—a way for a businesses or individuals to make a financial contribution to support a renewable energy, energy efficiency, or reforestation project.

Land-use and land-cover change Scientists recognize that another key factor affecting global climate patterns is **land-use and land-cover change** (LULCC). Some examples of LULCC include the conversion of woodlands to agricultural fields, the drainage of wetlands to construct shopping centers, the expansion of paved surfaces as cities grow, desertification caused by overgrazing, or water withdrawals that drastically alter a river's flow. Not all LULCC is human-induced. Drought or other natural stresses can affect the ability of vegetation to regenerate and alter local or regional biodiversity. Under certain conditions, human activity and natural fluctuations in weather or climate patterns may coincide and intensify LULCC. Changes in the Louisiana coastal zone provide an instructive example (see *What a Geographer Sees*).

> **land-use and land-cover change** An interdisciplinary approach to studying human–environment interactions based on an awareness of the linkages between ecosystem processes and social conditions.

WHAT A GEOGRAPHER SEES

Environmental Change

About half of the coastal wetlands in the United States are located along the shores the Gulf of Mexico. These wetlands not only provide habitats for a wide variety of aquatic and terrestrial species, they also help buffer storm surges, protect against erosion, improve water quality, and offer numerous opportunities for tourism and recreation. But these coastal wetlands continually face impacts that are both natural and human-induced. One result is the significant loss of land along the Louisiana coast. What has happened to cause this land-cover change?

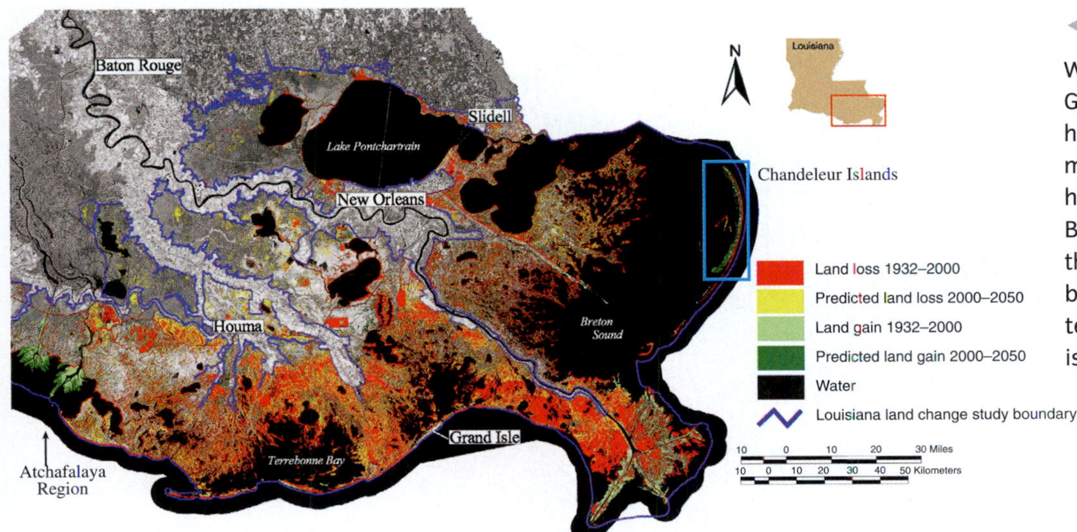

a. 100+ years of land change for southeast coastal Louisiana With the help of remote sensing and GIS, geographers and other scientists have identified where as well as how much of Louisiana's coastal wetlands have been lost to the Gulf of Mexico. Between 1932 and 2000, an area about the size of Delaware washed away, and between 2000 and 2050 an area about ten times the size of Washington, D.C. is expected to be lost.

b. Chandeleur Islands, 2001–2005 These satellite images show the effects that hurricanes Ivan and Katrina had on the northern half of these barrier islands. Tropical storms can drastically reconfigure coastal landscapes.

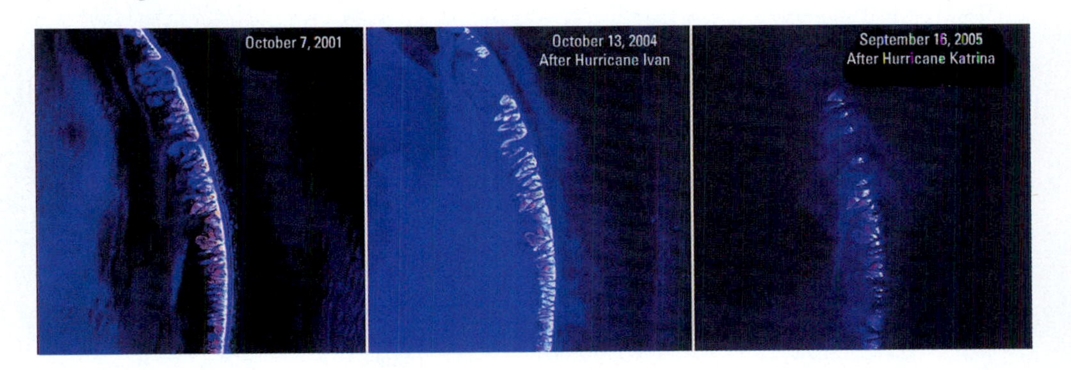

c. A portion of the Atchafalaya region, with locations of oil and gas wells Oil and gas production from both on- and offshore wells is a mainstay of Louisiana's economy. Many well sites in the coastal zones are not accessible by roads; consequently, numerous channels have been built through the wetlands. In this image, the channels are the straight or nearly straight waterways. Note the spatial association between the channels and the locations of well sites (in purple). (*Source*: Lyles and Namwamba, 2005.)

Think Critically

1. Why is channel construction in this coastal zone linked to environmental degradation? What would environmental restoration here involve?
2. What impacts on land use, land cover, and livelihoods has the BP oil spill of 2010 had?

LULCC has implications not only for conditions at the local, national, and global scales, but also for sustainability. For geographers, LULCC fits squarely within the tradition, discussed in Chapter 1, that emphasizes the view of the Earth as a dynamic, integrated system.

In the tropical rainforests of the Brazilian Amazon, for example, deforestation has been a major aspect of LULCC, especially since the 1980s. Rainforest clearance often results in warmer temperatures and drier conditions because less water evaporates or cycles from the

Land-use and land-cover change in the Brazilian Amazon • Figure 12.21

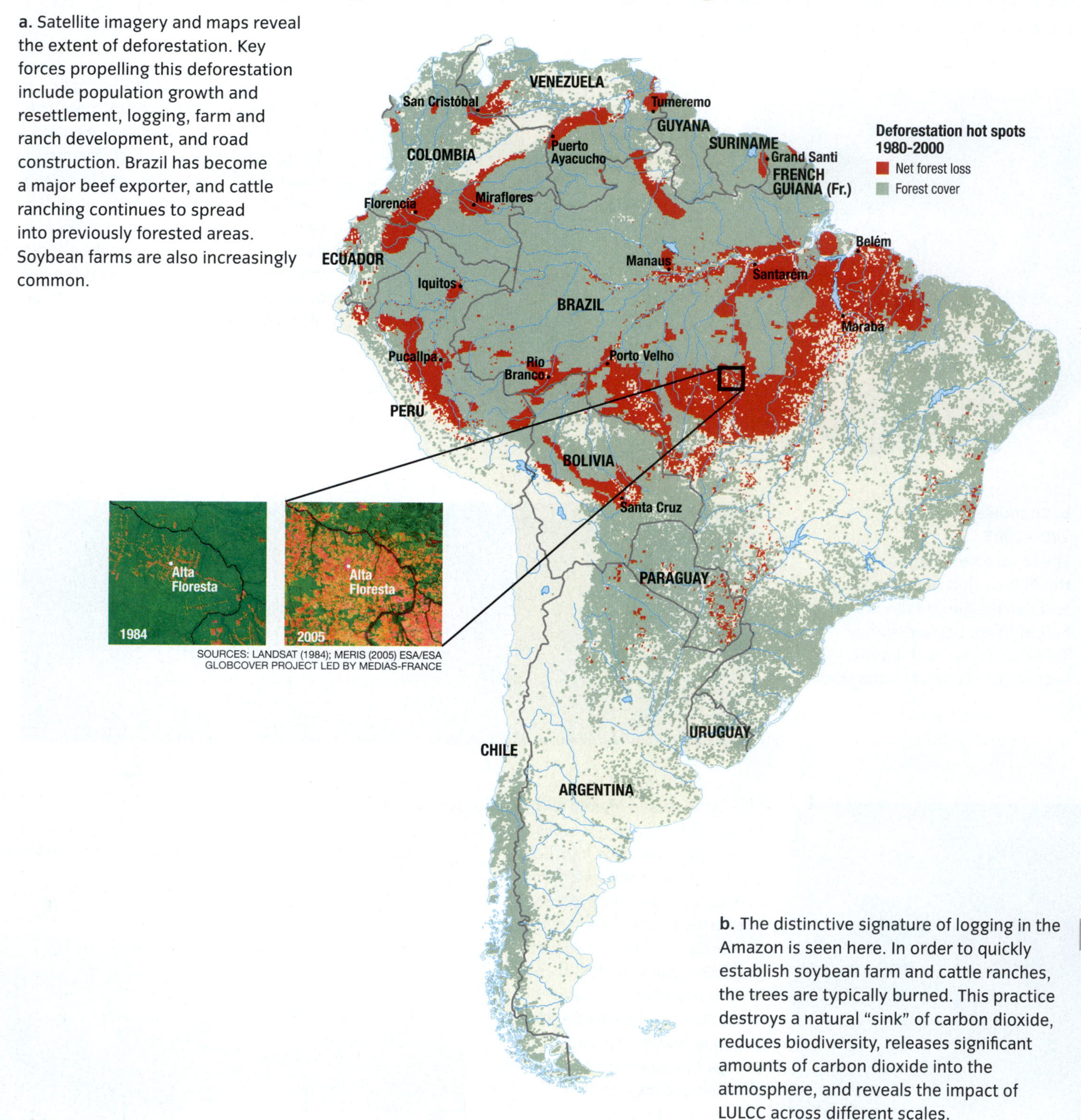

a. Satellite imagery and maps reveal the extent of deforestation. Key forces propelling this deforestation include population growth and resettlement, logging, farm and ranch development, and road construction. Brazil has become a major beef exporter, and cattle ranching continues to spread into previously forested areas. Soybean farms are also increasingly common.

b. The distinctive signature of logging in the Amazon is seen here. In order to quickly establish soybean farm and cattle ranches, the trees are typically burned. This practice destroys a natural "sink" of carbon dioxide, reduces biodiversity, releases significant amounts of carbon dioxide into the atmosphere, and reveals the impact of LULCC across different scales.

vegetation into the atmosphere. These conditions have consequences for cloud formation and, in turn, precipitation patterns. When these changes are coupled with an awareness of economic and demographic factors influencing human behavior, such as job availability and migration, we can more fully understand how and why changes in land management occur. **Figure 12.21** depicts some of these changes.

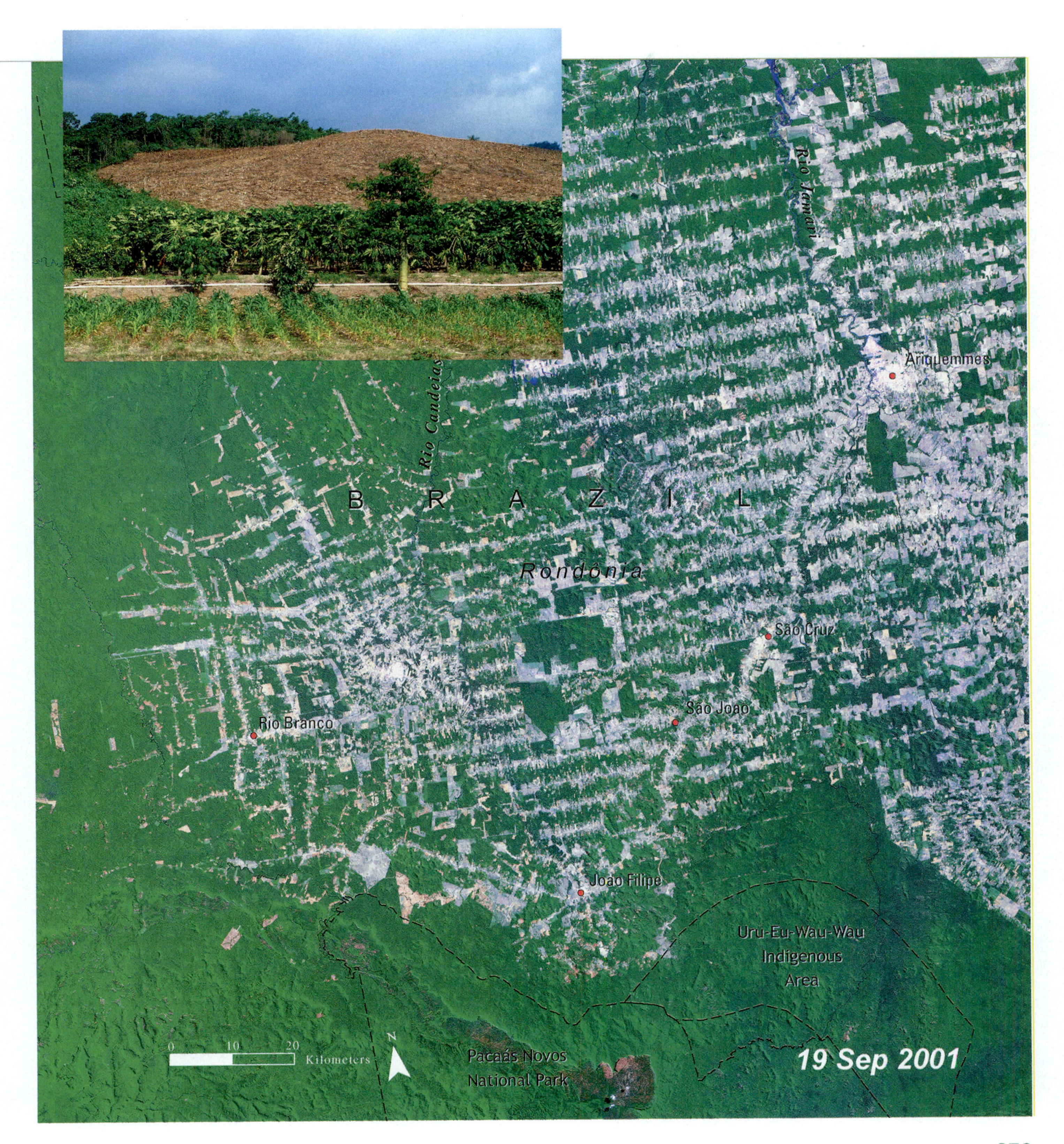

Where Geographers CLICK

Earth Trends

http://earthtrends.wri.org

The World Resources Institute provides an online database with information on a vast array of environmental topics—from different kinds of ecosystems to carbon emissions and other environmental news. Go to http://earthtrends.wri.org.

Targeting Greenhouse Gas Reduction

For more than 150 years the developed countries have contributed disproportionately to the atmospheric concentrations of carbon dioxide and many other anthropogenic greenhouse gases. Consequently, many experts argue that these same countries are obliged to acknowledge the long historical trajectory of their contribution to greenhouse gases and to implement measures to stabilize or reduce their greenhouse gas emissions. See *Where Geographers Click* for access to a website that tracks different environmental trends.

One international agreement that marks an important step toward improved management of greenhouse gases is the **Kyoto Protocol**. A key aspect of the Kyoto Protocol is the commitment of a large number of developed countries to reducing their greenhouse gas emissions 5% below their 1990 levels by 2012. The treaty was adopted in 1997 and went into effect in 2005. It has been ratified by more than 30 developed countries. This number, however, does not include the United States. Former President George W. Bush opposed the Kyoto Protocol and argued that it would harm U.S. economic interests.

The first phase of the Kyoto Protocol expires in 2012, and many observers expected the Copenhagen Climate Change Summit in 2009 to generate the successor to the Kyoto Protocol. It did not; the result instead was the Copenhagen Accord. The Accord is an important achievement that is now supported by about 120 countries, many of which have pledged to reduce their emissions. Nevertheless, three weakness of the Accord deserve mentioning. First, the Accord is not a legally binding global agreement. Second, no specific emission-reduction targets were established for developed countries. Third, no specific target for the reductions in global emissions was identified (e.g., cutting global emissions in half by 2050).

With the Copenhagen Accord, the global community has acknowledged the importance of regulating activities that affect global warming. The challenge is to achieve a consensus on specific targets and to find ways to actually meet those targets. To learn how carbon farming can help manage greenhouse gases, see *Video Explorations*.

Video Explorations

Carbon Farming

Each year farmers release huge amounts of carbon dioxide into the atmosphere when they plow their fields. This video describes new plowing techniques that help farmers keep the carbon in the soil.

CONCEPT CHECK

1. **How** is the greenhouse effect related to global warming?
2. **How** can human activities and natural processes affect local and global climate patterns?
3. **What** prompted the Kyoto Protocol and Copenhagen Summit, and what have they achieved?

Summary

1 Ecosystems 356

- An **ecosystem** includes the living organisms, their physical **environment**, and the flows of energy and nutrients cycling through them. Ecosystems exist at a variety of scales, from a local estuary such as the one shown here, to a desert that spans several countries. The Earth's interconnected ecosystems constitute the **biosphere**.

Ecosystems • Figure 12.1

- The **First Law of Ecology** expresses the principle that the environment is an interconnected web that includes people, and that human actions have environmental consequences.
- The environment is also a form of **natural capital**, which includes its **nonrenewable** and **renewable resources**. When assessing **environmental degradation**, a more complete picture considers both beneficial and detrimental processes.
- Another way of classifying resources distinguishes between **common property** and **open-access resources**. The idea of the tragedy of the commons, as presented by Garrett Hardin, was based on some incorrect assumptions about common property resources but raised provocative questions about resource management and environmental degradation.

2 Nonrenewable Energy Resources 360

- **Fossil fuels** are **nonrenewable resources** and include oil, natural gas, and coal.
- The world is heavily dependent on fossil fuels, especially oil, a versatile and valuable energy resource. How long the oil will last and when the world will reach peak oil production are issues that continue to stir controversy.
- Most of the proved oil reserves are concentrated in the Middle East. The single greatest consumer of the world's oil is the United States. Worldwide, demand for oil is expected to grow by 60% in the next 30 years.
- Coal, the most abundant of fossil fuels, presents a number of environmental challenges when extracted and used. Mountaintop removal, depicted here, remains a controversial technique, and burning coal is linked to mercury pollution and **acid rain**.

Understanding mountaintop removal • Figure 12.9

- Though not a fossil fuel, uranium is a nonrenewable resource used to produce nuclear energy. Worldwide, nuclear energy constitutes a minor part of the energy picture; however, some countries rely very heavily on nuclear energy.

3 Renewable Energy Resources 368

- **Renewable energy resources** have long been used noncommercially but are an increasingly important component of commercial energy as well. **New renewables** are energy resources, such as biomass, small-scale hydropower, solar, wind, and geothermal energy.
- Globally, the most-used source of renewable energy is **biomass energy**, specifically fuelwood for cooking. Biogas digesters built for household, village, or industrial operations provide an ecologically sustainable energy alternative and can help relieve pressure on forests for fuelwood.

- Most of the world's hydropower resources remain untapped. Because of the way large dams such as the Three Gorges Dam (see photo) fragment habitat and ecosystems, their sustainability has been questioned.

Hydropower production • Figure 12.14

- The sun provides solar energy and through the uneven heating of the Earth's surface also creates wind energy. At present, both forms of energy contribute minimally to the world's energy supply but are gaining popularity as emission-free sources of energy. The Earth's core generates **geothermal energy** that heats surrounding rock and water.

4 Human–Environment Interactions 374

- The **greenhouse effect,** depicted here, enables life on Earth but also is associated with global climate change.

The greenhouse effect • Figure 12.17

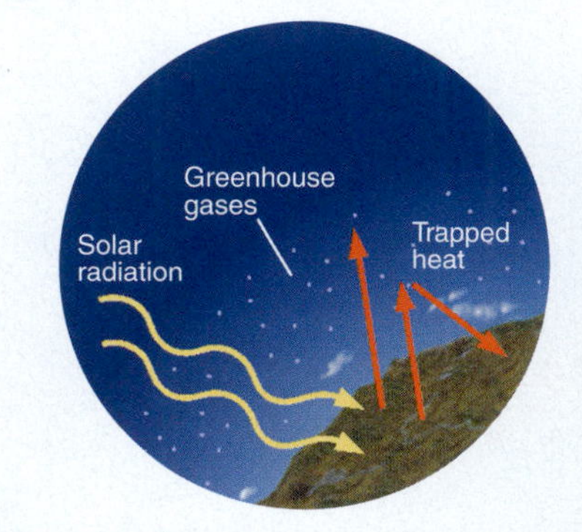

- Many climate scientists agree that human activities have increased atmospheric concentrations of greenhouse gases, thereby contributing to **global warming**.
- The concept of a **carbon footprint** provides one means of expressing the amount of carbon dioxide emissions from human activities. **Land-use and land-cover change** also plays an important role in global environmental change.
- The **Kyoto Protocol** and the Copenhagen Accord are international agreements aimed at achieving reductions in emissions of greenhouse gases.

Key Terms

- acid rain 366
- biodiversity 356
- biomass energy 369
- biosphere 356
- carbon footprint 376
- common property resources 358
- ecosystem 356
- environment 356
- environmental degradation 358
- First Law of Ecology 356
- fossil fuels 360
- geothermal energy 372
- global warming 375
- greenhouse effect 374
- Kyoto Protocol 380
- land-use and land-cover change (LULCC) 376
- natural capital 356
- new renewables 369
- nonrenewable resources 357
- open-access resources 358
- renewable resources 357

Critical and Creative Thinking Questions

1. Some ecologists have compared the Earth to an island ecosystem. What are the strengths and weaknesses of this analogy?
2. In what ways are the concepts of regions and ecosystems similar to and different from one another?
3. Do you agree with the statement that national parks are open-access resources? Is the Internet a common property or an open-access resource? Defend your answers.
4. The coal industry has been promoting the use of "clean coal." Do some research on this, and assess the extent to which clean coal lives up to its name.
5. Use this photo to develop three or four questions that a land-use land-cover change scientist might ask about mountaintop removal.

6. Use a carbon footprint calculator, such as the one hosted by the University of California at Berkeley, to calculate your personal carbon footprint (http://bie.berkeley.edu/calculator.swf). What could you do to reduce the size of your footprint? Are you willing to change your lifestyle to do so?
7. What is carbon sequestration? Do some research on it and evaluate its viability as a solution to the challenge of carbon emissions.
8. What are some major environmental consequences of globalization?
9. Give some reasons to explain the slow development and adoption of alternative energy uses for transportation.

What is happening in this picture?

NATIONAL GEOGRAPHIC

This is the Puente Hills Landfill (PHL) near Los Angeles, the largest operating landfill in the United States. A state-of-the-art facility, PHL collects methane from the buried, decomposing waste. Most of the methane is burned to produce steam and generate electricity, enough to power approximately 70,000 homes.

Think Critically

1. How different would this look to passing motorists, and why is that significant?
2. What similarities and differences exist between landfill management and mountaintop removal?

Self-Test

(Check your answers in Appendix B.)

1. The term _______ refers to the zone of life on Earth.
   a. ecosystem
   b. biosphere
   c. environment
   d. biomass

2. Natural capital includes all of the following except _______.
   a. the greenhouse effect
   b. photosynthesis
   c. fossil fuels
   d. hydrochlorofluorocarbons (HCFCs)

3. Economic depletion of a mineral means that _______.
   a. the cost of extracting the mineral exceeds the economic value of it
   b. the value of a mineral has peaked and is declining rapidly
   c. the mineral has lost all value as a commodity
   d. the mineral has been completely exhausted

4. _______ are natural resources that are shared by a well-defined community of users.
   a. Open-access resources
   b. Degraded resources
   c. Common property resources
   d. Renewable resources

5. A flaw in Hardin's argument on "The Tragedy of the Commons" is its failure to _______.
   a. acknowledge the existence of customary rules
   b. see the commons as a scarce resource
   c. incorporate the First Law of Ecology
   d. specify where the commons was located

6. Geologist M. King Hubbert used a bell-shaped graph of oil production to demonstrate _______.
   a. a need for improved oil shale technology
   b. the concept of sustainable yield
   c. the reserves-to-production ratio over time
   d. the concept of peak oil

7. As discussed in the chapter, this sequence of before and after photos relates to environmental degradation through the idea of _______ .

   a. natural degrading processes
   b. human interference
   c. natural reproduction
   d. environmental restoration

8. The leading consumer of oil in the world is _______, but the leading carbon dioxide emitter is _______.
   a. Japan; Singapore
   b. Saudi Arabia; Russia
   c. the United States; China
   d. China; Saudi Arabia

9. Of the fossil fuels, the one that burns the least cleanly is _______.
   a. oil
   b. natural gas
   c. coal
   d. uranium

10. Using the photo of mountaintop removal, identify the step shown here and explain why it is controversial.

11. List three problems associated with large-scale hydropower facilities:
    _______, _______, and _______.

12. What is the leading type of renewable energy used worldwide?
    a. oil
    b. coal
    c. biomass
    d. hydropower

13. The rise in atmospheric concentrations of carbon dioxide since the Industrial Revolution is primarily attributed to _______.
    a. the widespread application of fertilizers
    b. natural causes
    c. wet rice farming
    d. the burning of fossil fuels

14. The term _______ refers to the amount of carbon dioxide emitted as a result of human activity.
    a. carbon footprint
    b. greenhouse effect
    c. global warming
    d. emissions credit

15. What has happened here and why?

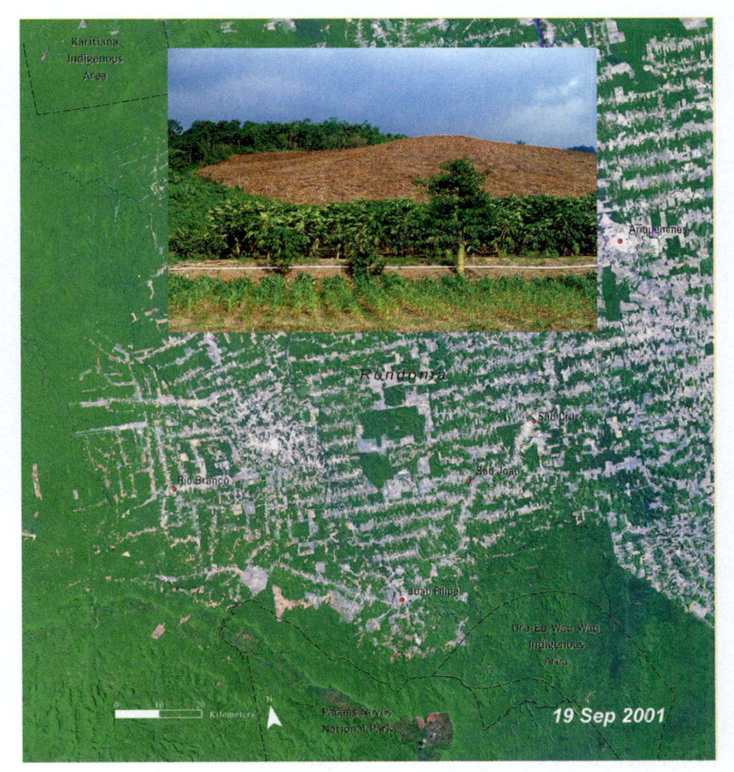

THE PLANNER

Review your Chapter Planner on the chapter opener and check off your completed work.

Appendix A Understanding Map Projections

When we want to see what the Earth looks like, we use a globe or a digital version of a globe as found in GoogleEarth, for example. Globes are scale models of the Earth that help us see places, regions, or other features in geographic context. Much of the usefulness of globes stems from their faithful depiction of the relative shapes and areas of landmasses, as well as distances and directions from one point to another. The small scale of a globe limits the amount of detail that can be shown, although digital globes more easily allow us to change scale by zooming in and out. Even as digital globes have helped make globes less bulky, maps, whether in digital or paper form, provide a practical alternative to depicting the Earth's geography.

The process of converting the Earth's spherical properties into flat map or planimetric form is called *map projection*. Imagine trying to flatten the peel of an orange. Flattening it will cause it to tear, deforming the spherical surface in the process. In analogous fashion, projecting a map introduces the problem of *distortion*, or the misrepresentation of the spatial and geometric properties of the Earth. Another way to think about distortion on map projections is to consider what happens to the *graticule*, the imaginary grid consisting of lines of latitude that parallel the Equator and meridians of longitude that converge at the poles (**Figure A.1**).

The graticule or geographical grid • Figure A.1

On a sphere, lines of latitude (parallels) and lines of longitude (meridians) intersect at right angles, and lines of latitude become shorter closer to the poles. Even an inexperienced map user can make observations about map distortion by thinking about how closely the graticule on a particular map resembles the graticule on a globe.

Understanding the process of map projections becomes a little easier if we visualize the kind of surface that is used when transforming the spherical Earth into a map. For example, if you wrap a transparent globe with a piece of paper so that it forms a cylinder around the globe and then place a light source nearby, the graticule and outlines of landmasses will be projected onto the paper. You can use a similar approach and project the features of the globe onto a cone-shaped piece of paper or a flat piece of paper. Use of these different surfaces, respectively, gives rise to three broad families of map projections: *cylindrical*, *conic*, and *planar* (**Figure A.2**).

Map projection families • Figure A.2

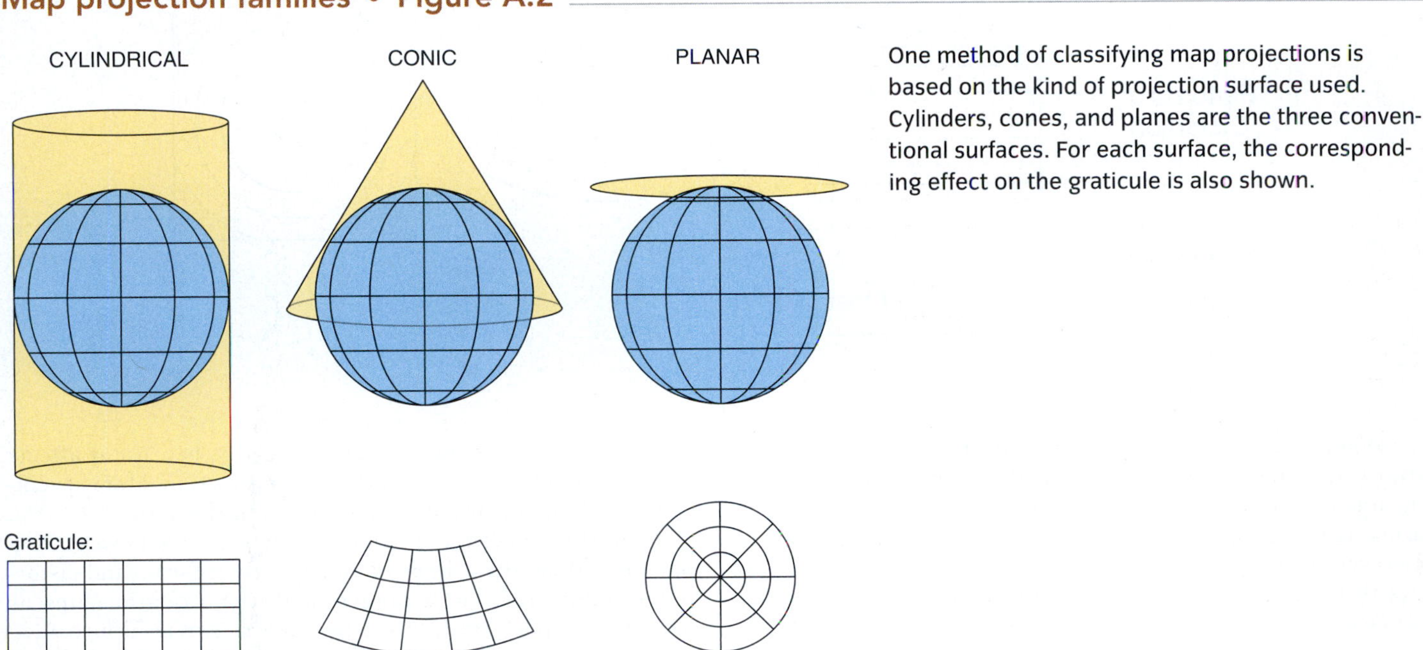

One method of classifying map projections is based on the kind of projection surface used. Cylinders, cones, and planes are the three conventional surfaces. For each surface, the corresponding effect on the graticule is also shown.

Projections and patterns of distortion • Figure A.3

A standard line represents a line of contact between the projection surface and the globe. Tangent relationships use a single standard line, whereas secant relationships use two standard lines. The darker the color in this illustration, the greater the distortion.

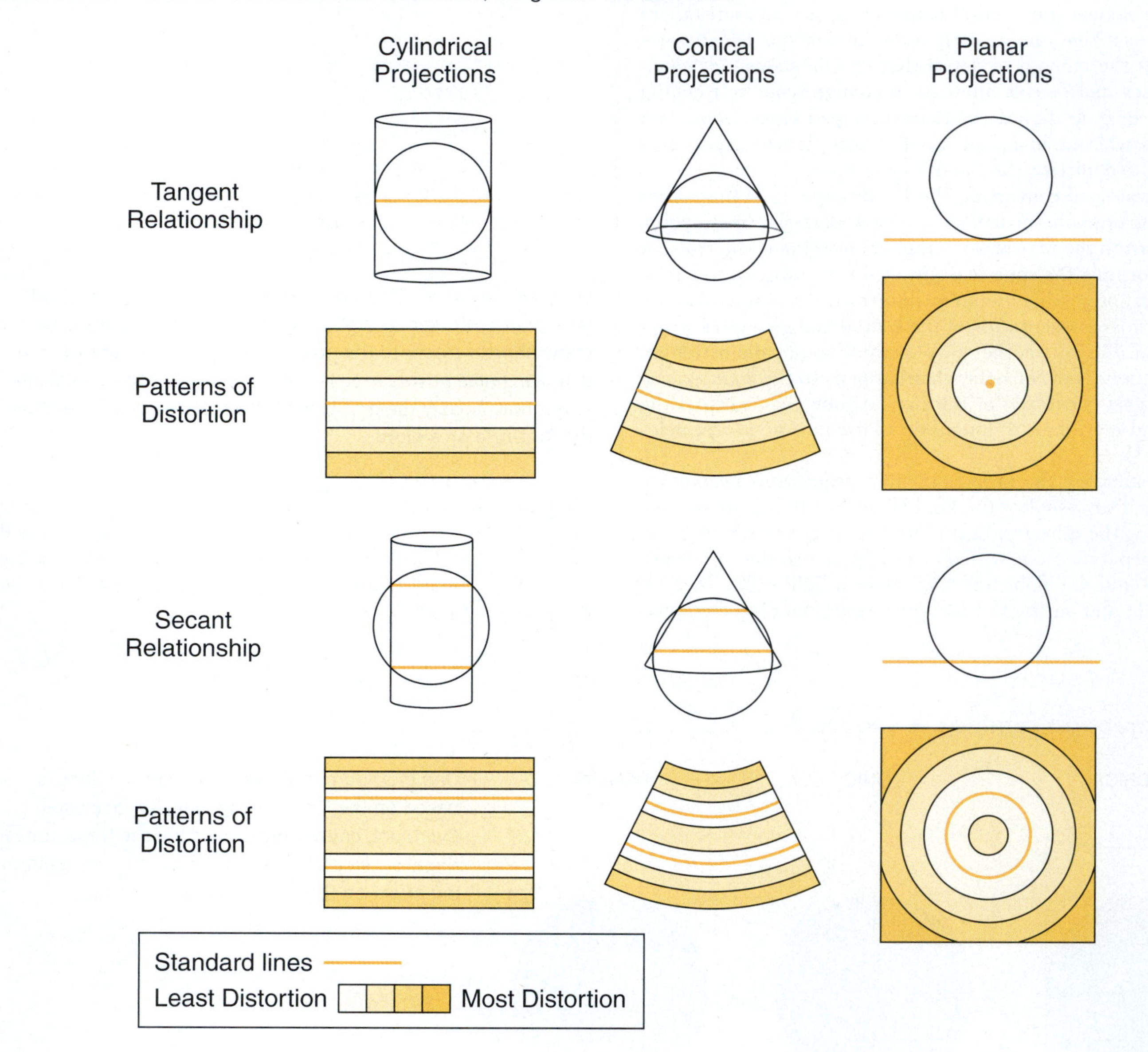

When making cylindrical or conic projections, we can wrap the paper so that it touches the globe along a single line and creates a *tangent* relationship. In the case of a planar projection, the paper is tangent to the globe at a single point. However, we can also envision situations in which the paper cuts through the globe along two lines, creating a *secant* relationship of the projection surface to the globe. The way in which the paper touches the globe matters because it affects the patterns of distortion on the map projection. More specifically, the line or lines of contact between the surface and the globe in the tangent and secant examples are known as *standard lines;* along these lines the map is true to scale and distortion is minimized. These tangent and secant relationships, as well as their patterns of distortion, are shown in **Figure A.3**.

Of course, today it is much more efficient to use computers and projection programs to make map projections instead of transparent globes and a light source. Even with the use of more sophisticated technologies, it is still impossible to represent the spherical Earth on a flat surface without some distortion. Moreover, no single map projection is perfect for all mapping needs. When deciding which map projection to use, cartographers (people who design and create maps) and geographers consider where distortion on a map is minimized and more specifically, the effect a map projection has on geometric and spatial properties such as area, shape, scale, and direction. It is practical, then, to know about a slightly different classification system that groups map projections according to similar properties. Here we focus on a fourfold classification consisting of *equal-area*, *conformal*, *azimuthal*, and a broad category of *other* map projections.

Conformal Projections

In the context of map projections, *conformal* means to show small areas in their correct form or shape, and conformal projections are sometimes described as shape-preserving. Strictly speaking, however, the property of conformality (being conformal) applies to points and the angles about them. This property makes conformal projections useful for showing the relief or topography of an area. A projection that maintains angles and shapes and that is well suited to mapping regions in the middle latitudes with a predominantly east-to-west extent is the Lambert conformal conic projection with two standard parallels (**Figure A.4**).

In order to maintain proper angles and shapes, however, conformal projections sacrifice the property of size. This is because maintaining certain angular relationships when converting a spherical to a flat form requires that some distances be stretched or compressed, and such alterations affect the sizes of regions. Thus, conformal projections cannot also show large areas in their proper relative sizes. This characteristic is especially noticeable on the Mercator projection, a projection that is both conformal and cylindrical.

Lambert conformal conic projection • Figure A.4

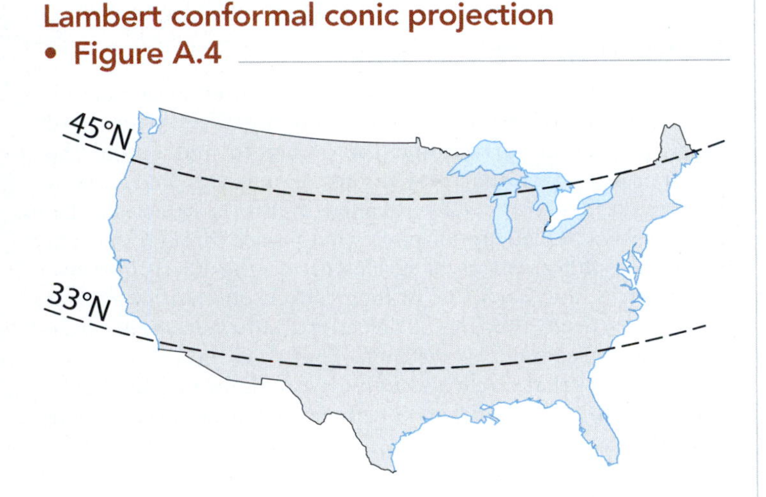

This projection is frequently used for representing the coterminous United States. The standard lines or more specifically here, the standard parallels, are at 33° and 45° north latitude.

In 1569, the Flemish cartographer Gerardus Mercator (1512–1594) created the map projection that now bears his name. He specifically devised this projection as a navigational aid for mariners. The value of Mercator's projection stems from the fact that any straight line drawn between two places on the map is a line of constant compass bearing, or a *rhumb line*. On a Mercator projection, the rhumb line always crosses lines of longitude at the same or constant angle.

The Mercator projection is a cylindrical projection that is often developed with a standard parallel along the Equator. Cartographers needing a conformal map of equatorial regions often select the Mercator projection because distortion is minimized there. However, since the Mercator projection is conformal, the correct representation of sizes of regions is sacrificed for the correct representation of angular relationships. This is most visible in the high-latitude zones where the meridians, instead of converging toward the poles, remain far apart (**Figure A.5**).

Mercator projection • Figure A.5

The Mercator projection is a cylindrical and conformal map projection. Rhumb lines always appear as straight lines on these maps, so navigators measure the angle between their planned route and a meridian to identify the compass bearing or direction of travel. Relative sizes of areas especially in the higher latitudes, however, are exaggerated. Despite what the Mercator projection shows, Greenland is just one-eighth the size of South America.

Examples of equal-area world map projections • Figure A.6

a. Molleweide projection
Developed in the early 19th century, this projection represents the globe as an oval or, technically, an ellipse.

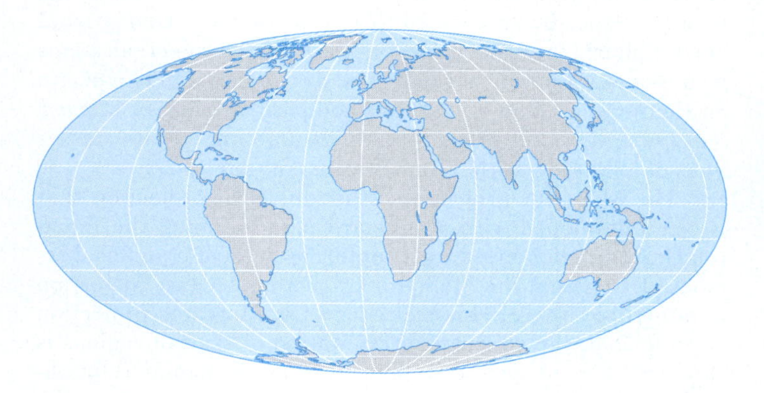

b. Eckert IV projection
To preserve area on this and the Molleweide projection, the parallels are not equally spaced. As a result, shapes in the low latitudes appear stretched.

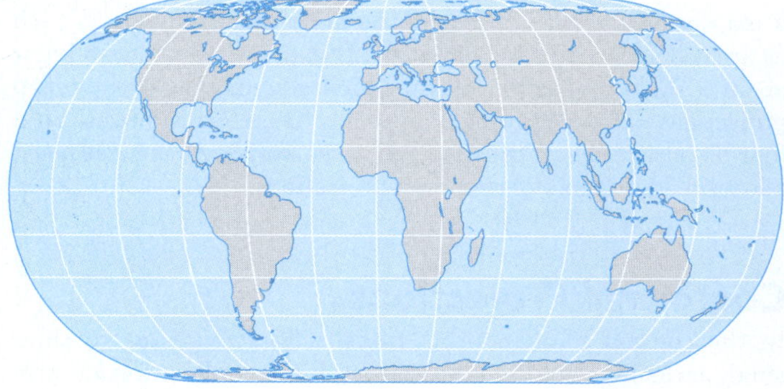

Equal-Area Projections

As its name suggests, an equal-area projection (sometimes also called an equivalent projection) is one that shows mapped regions in their true relative sizes. As we have discussed, no map can be both conformal and equal-area. Therefore, if you need a map of the world that shows landmasses and oceans in their correct proportions, it is advisable to select an equal-area projection. However, for a long time and mainly out of complacency, the Mercator projection was widely used as a reference map of the world in classrooms, in the media, and in books and atlases, even though it was not designed to be used for such purposes.

Repeated use of the Mercator and similar cylindrical projections in popular books and media inadvertently reinforces the notion that such maps provide a reliable representation of what the Earth looks like, when, in fact, this is far from the case. Look again at the Mercator projection in Figure A.5. Where are the North and South Poles? What shape does the Mercator projection give to the Earth?

In the late 1980s, the Map Projections Committee of the American Cartographic Association issued a formal resolution urging book and map publishers to stop using the Mercator projection and other rectangular world maps as general-purpose or basic reference maps because of their considerable distortion. Two equal-area projections recommended, among other suitable alternatives, were the Molleweide projection and the Eckert IV projection (**Figure A.6**).

Briefly, then, equal-area projections facilitate areal comparisons, enabling us to visually compare the extent of countries, continents, or bodies of water, for example. Similarly, we should choose an equal-area map projection when mapping densities (e.g., population density or livestock density) or if we have other area-based data, such as the percentage of land area in a country planted to a certain kind of crop. To varying degrees, conformal projections distort size dimensions, and equal-area maps distort shapes. Compare and contrast the projections in Figures A.5 and A.6.

Azimuthal Projections

Azimuthal projections are planar projections created from either a tangent or secant relationship between the plane used as the projection surface and the globe. To understand the properties of azimuthal projections, it helps to know about great circles. A *great circle* is formed by the intersection of a sphere with a flat surface or plane that passes through the center of the sphere. All great circles divide the Earth into two hemispheres, and each line of longitude is one-half of a great circle. An *azimuth* is a line of direction, and every great circle forms what is called a *true azimuth*. That is, if you begin at one point along a great circle and follow its path around the Earth and return to your starting point, then you have followed a true azimuth.

Examples of azimuthal projections • Figure A.7

a. Gnomonic projection
Great circles show as straight lines on these projections. Since rhumb lines on Mercator projections are not great circle routes and do not identify the shortest distances, navigators consult both of these projections when determining their routes.

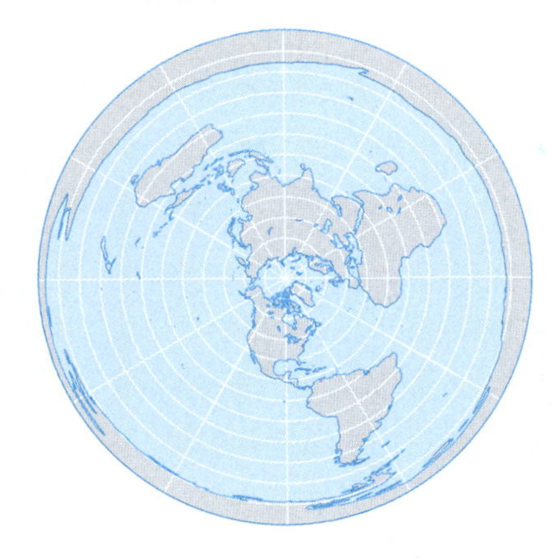

b. Azimuthal equidistant projection
Both distances and directions are true from the center point to any other point on these projections. Thus, they are favored when mapping seismic waves, the vibrations caused by earthquakes.

c. Orthographic projection
These projections provide a hemispheric perspective of the globe. Distortion occurs along the edges. The global locators used in this book provide examples of orthographic projections.

Great circles have significance for distances as well. More specifically, the shortest distance over the Earth between any two locations follows the arc of a great circle between them. On an azimuthal map projection, great circles that pass through the center point of the projection show as straight lines. This makes sense if we think about what happens to the graticule when we create an azimuthal projection with the plane tangent to the Earth at either the North or South Pole: Lines of longitude (parts of great circles) are represented as straight lines extending from the center (refer again to Figure A.2).

There are numerous different kinds of azimuthal projections. Three of them are the gnomonic projection, the azimuthal equidistant projection, and the orthographic projection. You can think of a gnomonic projection as one created by placing a light source at the center of a transparent globe and projecting the sphere onto a tangent plane. Gnomonic projections are particularly useful for navigation because they show great circles as straight lines anywhere on the map, not just from the center point.

When projecting a map it is not possible to preserve scale or map distances properly everywhere, but distances can be preserved between certain points or along specific lines. An important property of azimuthal equidistant map projections is that distances are correct from the center point to any other point on the map. A third kind of azimuthal projection, called an orthographic projection, gives a hemispheric view of the Earth as if observed from a location far above it. Illustrations of these different azimuthal projections are shown in **Figure A.7**.

Other Projections

There are so many other kinds of projections that, for simplicity, we have grouped them into this very broad and inclusive category. We discuss three examples as a way of highlighting some of the diversity within this grouping. Two of these examples illustrate different types of interrupted projections, and the third illustrates a compromise projection.

Technically speaking, all maps are interrupted because they show the Earth as having edges instead of showing it as the continuous surface that it is. When cartographers and map projection experts speak of interrupted map projections, however, they are referring to the practice of breaking up either the oceans or the landmasses in order to highlight the other.

Examples of interrupted projections • Figure A.8

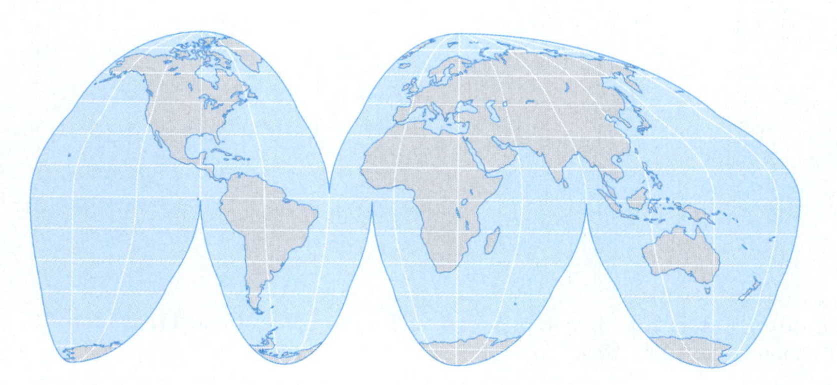

a. Interrupted Goode Homolosine projection
This equal-area projection, with its elliptical polar zones, is partly based on the Molleweide projection. The continuity of the oceans is sacrificed to highlight land areas.

b. Dymaxion map projection
R. Buckminster Fuller was an inventor. His unique map projection defies the conventional practice of orienting maps so that north is at the top and emphasizes the near connectedness of the Earth's land areas. This projection modestly distorts land areas and shapes.

Winkel Tripel projection • Figure A.9

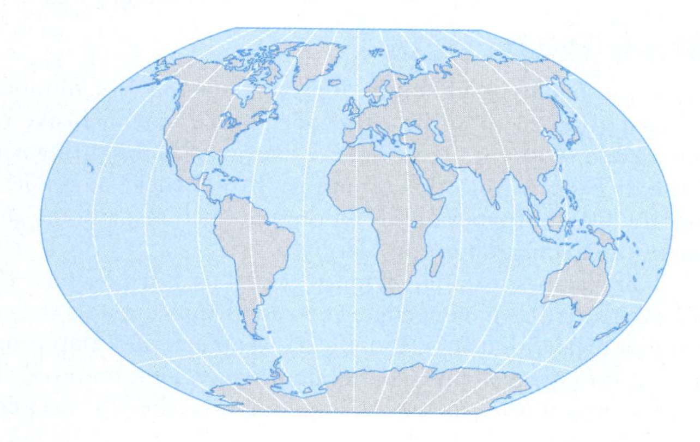

The shapes of the landmasses on this compromise projection very closely resemble how they look on a globe, while the distortion of areas is kept to a minimum. This projection is used on most of the world maps in this book.

The Interrupted Goode Homolosine projection is an equal-area projection that interrupts the oceans so that land areas can be emphasized. The Dymaxion map projection created by R. Buckminster Fuller (1895–1983) also interrupts the oceans (**Figure A.8**).

The Winkel Tripel projection, a projection used throughout this book, provides a third example of yet another kind of projection. In contrast to the Dymaxion map and the Interrupted Goode Homolosine projections, the Winkel Tripel projection is not classified as an interrupted projection. It is, however, usually described as a *compromise projection*. This is to say that the map projection strikes a balance between aesthetics and distortion (**Figure A.9**).

Appendix B Answers to Self-Tests

Chapter 1

1. d; **2.** b; **3.** d; **4.** d; **5.** a; **6.** b; **7.** c; **8.** b; **9.** d; **10.** c; **11.** d; **12.** a; **13.** c; **14.** a; **15.** c

Chapter 2

1. c; **2.** a; **3.** a; **4.** d; **5.** c; **6.** a; **7.** It contributed to the view that members of a folk culture were less civilized and usually excluded certain social classes; **8.** d; **9.** a; **10.** d; **11.** c; **12.** Acupuncture diffused with Buddhism; via trade; within the medical profession and via Chinese immigrants; **13.** b; **14.** d; **15.** b

Chapter 3

1. b; **2.** d; **3.** c; **4.** Geographic dimensions include attention to spatial scale, especially via strict management at the local level, and varying implementation spatially, especially between rural and urban areas; **5.** a; **6.** a; **7.** c; **8.** d; **9.** c; **10.** See Figure 3.9.; **11.** Fiji has a high birth rate, low death rate, and high rate of natural increase; **12.** c; **13.** See Figure 3.13; **14.** a; **15.** c

Chapter 4

1. a; **2.** c; **3.** d; **4.** a; **5.** b; **6.** See Figure 4.5c; **7.** Answers may vary. Some possible answers: omission of "l" between vowels; omission of "r" between vowels; use of "be" to indicate a regular activity; **8.** b; **9.** c; **10.** d; **11.** a; **12.** a; **13.** Their low LDIs are a legacy of European colonization, which caused the extinction of many indigenous peoples and their languages, and led to the diffusion of Spanish and Portuguese; **14.** b; **15.** See Figure 4.15

Chapter 5

1. c; **2.** b; **3.** c; **4.** a; **5.** It illustrates a reaction to the caste system, a legacy of Hinduism; **6.** d; **7.** a; **8.** informal consecration; 9. Answers may vary and might include Christianity's spread from large cities to smaller settlements and rural areas; **10.** d; **11.** b; **12.** b; **13.** c; **14.** b; **15.** d

Chapter 6

1. d; **2.** b; **3.** d; **4.** a; **5.** b; **6.** There is no biological basis for exclusive racial categories; **7.** c; **8.** c; **9.** b; **10.** c; **11.** d; **12.** a; **13.** b; **14.** c; **15.** d

Chapter 7

1. d; **2.** a; **3.** c; **4.** c; **5.** b; **6.** It is prorupt and perforated; **7.** b; **8.** d; **9.** c; **10.** a; **11.** b; **12.** a; **13.** c; **14.** b; **15.** d

Chapter 8

1. b; **2.** d; **3.** a; **4.** a; **5.** Answers will vary depending on what part of the image the student focuses on but should include mention of the clustering of activities (residential and religious). More thorough answers will incorporate reference to clusters of like and unlike activities; **6.** b; **7.** c; **8.** b; **9.** b; **10.** c; **11.** Possible answers include sprawl, decentralization, single-use zoning; **12.** a; **13.** c; **14.** d; **15.** b

Chapter 9

1. c; **2.** b; **3.** a; **4.** b; **5.** d; **6.** a; **7.** c; **8.** d; **9.** c; **10.** a; **11.** d; **12.** b; **13.** a; **14.** b; **15.** c

Chapter 10

1. See Figure 10.5; **2.** c; **3.** b; **4.** b; **5.** a; **6.** c; **7.** d; **8.** See Figure 10.13; **9.** a; **10.** b; **11.** Answers should include mention of government initiatives to increase manufacturing productivity, the shift from labor-intensive to technology-intensive industries, and presence of a skilled labor force; **12.** Outsourcing subcontracts an in-house business activity to another firm, whereas offshoring moves a business activity to a foreign country; **13.** Answers should include a discussion of the shift from secondary to tertiary industries (service sector jobs) as measured by employment in them; **14.** a; **15.** d

Chapter 11

1. c; **2.** b; **3.** a; **4.** See Figure 11.6a; **5.** nomadic herding and vertical transhumance; **6.** a; **7.** d; **8.** b; **9.** d; **10.** See Figure 11.16; **11.** c; **12.** c; **13.** b; **14.** a; **15.** b

Chapter 12

1. b; **2.** d; **3.** a; **4.** c; **5.** a; **6.** d; **7.** d; **8.** c; **9.** c; **10.** Step 2, (valley fill) and see Figure 12.9. Controversy relates to scale and impact of operations; **11.** habitat fragmentation, species loss, forced resettlement; **12.** c; **13.** d; **14.** a; **15.** Image shows a signature of logging in the Amazon to clear land for soybean farms or cattle ranches.

accent Differences in pronunciation among speakers of a language.

acid rain Precipitation that, primarily because of human activities, is significantly more acidic than normal and can harm aquatic and terrestrial ecosystems.

actor-network theory A body of thought that emphasizes that humans and nonhumans are linked together in a dynamic set of relations that, in turn, influence human behavior.

acupuncture An ancient form of traditional Chinese medicine that promotes healing through the insertion of needles into the body at specific points.

age-dependency ratio The number of people under the age of 15 and over the age of 65 as a proportion of the working-age population.

agglomeration The spatial clustering of people and economic activities, especially industries that are related or interdependent, in a place.

agribusiness The interconnected industry of food production involving farmers, processors, distributors, and retailers.

agriculture Activities centered on cultivating domesticated crops and livestock in order to procure food and fiber for human use or consumption.

agro-biotech (short for agricultural biotechnology) One facet of the third agricultural revolution that seeks to improve the quality and yield of crops and livestock using techniques such as cross-breeding, hybridization, and, more recently, genetic engineering.

agroforestry The purposeful integration of trees with crops and/or livestock in the same field simultaneously or sequentially.

allopathic medicine Medical practice that seeks to cure or prevent ailments with procedures and medicines that have typically been evaluated in clinical trials.

Americanization The diffusion of American brands, values, and attitudes throughout the world.

animistic religion A system of beliefs and practices that incorporates veneration of spirits or deities associated with natural features such as mountains, trees, or rivers.

apartheid In South Africa, the government-sponsored policy of racial segregation and discrimination that regulated the social relations and opportunities of its population from 1948 to 1991.

arithmetic density The number of people per unit area of land.

ascription A process that occurs when people assign a certain quality or identity to others, or to themselves.

assimilation A model of ethnic interaction that describes the gradual loss of the cultural traits, beliefs, and practices that distinguish immigrant ethnic groups and their members.

asylum Protection from persecution granted by one country to a refugee from another country.

atheism The belief that there is no deity.

authorized immigrants In the United States, legal permanent residents who are sometimes also called green-card holders.

bid-rent curve Graph that shows the amount a bidder (e.g., a business or individual) is willing to pay for land relative to the distance of that land from the central business district.

biodiversity The variety of species contained within an ecosystem.

biofuel Fuel derived from renewable biological material, such as plant matter.

biomass energy Energy that is obtained from plant matter and/or animal wastes.

biosphere The zone of the Earth, extending from the soil and waters to the lower parts of the atmosphere, that supports and includes living organisms.

blockbusting Using scare tactics and panic selling to promote the rapid transition of a neighborhood from one ethnic or racial group to another.

boundary A vertical plane, usually represented as a line on a map, that fixes the territory of a state.

brain drain The migration of skilled professionals (e.g., teachers, engineers, doctors) to another country, where they can obtain a higher paying job and better quality of life.

capital Financial, social, intellectual, or other assets that are derived from human creativity and are used to create goods and services.

carbon footprint The amount of carbon dioxide emitted as a result of human activity.

carrying capacity The number of people the Earth can support at a comfortable standard of living given current technology and habits of resource use.

cartel Entity consisting of individuals or businesses that control the production or sale of a commodity or group of commodities, often worldwide.

cash crop Broadly, any high-value crop that is sold for profit; historically, traditional cash crops are coffee, tea, tobacco, cocoa, rubber, and sugar—nonfood crops that have been closely associated with plantation agriculture.

caste system A hierarchical form of social stratification historically associated with Hinduism.

central business district (CBD) The part of the downtown where major office and retail businesses are clustered.

central city The area enclosed by the legal boundaries of a city.

central place A settlement that provides goods and services for its residents and its surrounding trade or market area.

central place theory A theory developed by Walter Christaller that posits that market forces account for the distribution of central places in an area, and that the optimal spatial arrangement of central places creates hexagonally shaped trade areas.

centralization Forces that draw people and businesses into the downtown or central city.

centrifugal force An event or a circumstance that weakens a state's social and political fabric.

centripetal force An event or a circumstance that helps bind together the social and political fabric of a state.

circulation The temporary, often cyclical, relocation of an individual or a group from one place to another.

civil religion A set of beliefs and practices that takes shape when religious notions, symbols, and rituals infuse the political culture of an area, as for example when people share the collective belief that a country's constitution is sacred.

classical model of development An accounting of economic development that was formulated by Walt W. Rostow as a series of five stages through which countries pass as they are transformed, by economic investment, from a traditional to a modern society.

Cold War The hostility and rivalry that existed between the United States and the Soviet Union from the mid-1940s to the late 1980s.

colonialism A form of imperialism in which a state takes possession of a foreign territory, occupies it, and governs it.

commercial agriculture A farming system relying heavily on purchased inputs and in which products are sold for use or consumption away from the farm.

commercial dairy farming The management of cattle for producing and marketing milk, butter, cheese, or other milk by-products.

commercial gardening The intensive production of nontropical fruits, vegetables, and flowers for sale off the farm.

commercial grain farming Agriculture involving the large-scale, highly mechanized cultivation of grain.

commodification The conversion of an object, a concept, or a procedure once not available for purchase into a good or service that can be bought or sold.

commodity chain (also a production chain or value-added chain) The linked sequence of operations from the design to the production and distribution of a good.

common property resources Natural resources, equipment, or facilities that are shared by a well-defined community of users.

complementarity A situation in which one place or region can supply the demand for resources or goods in another place or region.

concentric zone model A description of urban structure that was created by Ernest Burgess and emphasizes the development of circular rings of similar land use around the central city, each occupied by different ethnic and socioeconomic groups.

conflict diamonds (also blood diamonds) Diamonds sold to finance wars or terrorist activities.

consumption Broadly defined as the use of goods to satisfy human needs and desires.

cornucopian theory A theory positing that human ingenuity will result in innovations that make it possible to expand the food supply.

cosmogony A theory that provides an explanation of the beginning of the world.

creole language A language that develops from a pidgin language and is taught as a first language.

crop rotation Growing a sequence of different crops in the same field in order to maintain soil fertility and health.

crude birth rate (CBR) The annual number of births per 1,000 people.

crude death rate (CDR) The annual number of deaths per 1,000 people.

cultural ecology A subfield within human geography that studies the relationship between people and the natural environment.

cultural geography A branch of human geography that emphasizes human beliefs and activities and how they vary spatially, utilize the environment, and change the landscape.

cultural landscape The collection of structures, fields, or other features that result from human transformation of the natural environment; any landscape created or modified by people.

culture A social creation consisting of shared beliefs and practices that are dynamic rather than fixed, and a complex system that is shaped by people and, in turn, influences them.

decentralization In urban geography, forces that draw people and businesses out of the central city, often into suburbs. In political geography, a process whereby a state transfers functions or authority from the central government to lower-level internal subdivisions.

deindustrialization The long-term decline in industrial employment.

demographic equation A technique for measuring population change in a region over a specified period of time by adding population growth through natural increase and net migration to the population at the start of the time period being examined.

demographic transition model A simplified representation of a common demographic shift from high birth and death rates to low birth and death rates over time and in conjunction with urbanization and industrialization.

dependency In development studies, a condition that stems from patterns of international trade and results in underdevelopment, or low levels of development among states that lack the resources to command or control the system of trade.

dependency theory A theory that relates disparities in levels of development to relations between dominant and dependent states in the system of international trade.

desertification The creation of desert-like conditions in nondesert areas through human and/or environmental causes.

development A process that brings about changes in economic prosperity and the quality of life.

devolution A kind of decentralization whereby a state transfers some power to a self-defined community, such as one of its national groups.

dialect A particular variety of a language characterized by distinctive vocabulary, grammar, and/or pronunciation.

dialect geography The study of the spatial patterns of dialect usage.

diaspora The scattering of a people through forced migration.

diffusionism A belief, associated with the rationality doctrine and cultural superiority, that the spread of Western science, technology, and practices to nonWesterners (deemed inferior) would enable them to advance socially and economically.

discourse Communication that provides insight into social values, attitudes, priorities, and ways of understanding the world.

dissonance The quality of being inconsistent; within heritage studies the idea that the meaning and value of heritage vary from group to group.

distance decay The tapering off of a process, a pattern, or an event over distance.

distribution The arrangement of phenomena on or near the Earth's surface.

domestication An ongoing process of selecting plants or animals for specific characteristics and influencing their reproduction in ways that make them visibly or behaviorally distinct from their wild ancestors.

double cropping Completing the cycle from planting to harvesting on the same field twice in one year.

dual economy An economy in which two production systems operate virtually independently from one another, as for example when large-scale export-oriented agriculture exists alongside smallholder agriculture.

dual society A society that is sharply divided into two social classes such as upper-class plantation managers and lower-class plantation laborers.

economic indicator A value or measure that can be used to gauge development, such as gross national income, gross domestic product per capita, or incidence of poverty.

ecosystem The living organisms, their physical environment, and the flows of energy and nutrients cycling through them.

edge cities New downtowns consisting of clusters of business activity that develop in the suburbs surrounding a city.

electoral system The set of procedures used to convert the votes cast in an election into the seats won by a party or candidate.

emigration The out-migration or departure of people from a location.

eminent domain The authority of a government to take private property when doing so serves the public's interests.

enclave Territory completely surrounded by another state but not controlled by it.

endangered language A language that is no longer taught to children by their parents and is not used for everyday conversation.

environment All of the biotic (living) and abiotic (nonliving) factors with which people, animals, and other organisms coexist.

environmental degradation The result of human or natural processes that impair the Earth's physical properties or ecosystems.

environmental determinism A theory maintaining that natural factors control the development of human physiological and mental qualities.

environmental indicator A value or measure that provides information about the state of the environment such as levels of pollution or loss of biodiversity.

environmental justice A social movement and an approach to public policy that is concerned with analyzing and managing the impacts of environmental hazards so that no people are disproportionately burdened by exposure to such hazards because of their race, ethnicity, national origin, income, gender, or disability.

environmental stewardship The view that people should be responsible managers of the Earth and its resources.

epidemiological transition A shift from infectious to chronic diseases as lifestyle changes associated with urbanization and industrialization occur.

ethnic cleansing The forced removal of an ethnic group from an area.

ethnic geography A subfield of human geography that studies the migration and spatial distribution of ethnic groups, ethnic interaction and networks, and the various expressions or imprints of ethnicity in the landscape.

ethnic group People who share a collective identity that may derive from common ancestry, history, language, or religion, and who have a conscious sense of belonging to that group.

ethnic island A pattern of settlement, quite variable in size and population, formed by some ethnic groups in rural areas.

ethnic neighborhood A pattern of settlement formed by some ethnic groups in urban areas.

ethnic religion A belief system largely confined to the members of a single ethnic or cultural group.

ethnicity The personal and behavioral basis of an individual's identity that generates a sense of social belonging.

ethnoburb A multiethnic residential, commercial, or mixed suburban cluster in which a single ethnic group is unlikely to form a majority of the population.

ethnoscape A cultural landscape that reveals or expresses aspects of the identity of an ethnic group.

European Union (EU) A supranational organization that has enlarged considerably since its establishment in western Europe and is characterized by a significant degree of both economic and political integration among its members.

exclave Territory that is separated from the state to which it belongs by the intervening territory of another state.

export-processing zone (EPZ) (also free-trade zone) An industrial area that operates according to different policies than the rest of the country in which it is located in order to attract and support export-oriented production.

extensive agriculture An agricultural system characterized by low inputs of labor, capital, or equipment per unit area of land.

extinct language (also dead language) A language that has no living speakers.

factory farm A farm that houses huge quantities of livestock or poultry in buildings, dry-lot dairies, or feedlots.

feedlot Confined space used for controlled feeding of animals.

feng shui The Chinese art and science of situating settlements or designing cultural landscapes in order to harmonize the cosmic forces of nature with the built environment.

filtering In urban geography, the process whereby home ownership in a neighborhood gradually transitions from high- to middle- to lower-income households over time.

first agricultural revolution The rise of agriculture, which began with the domestication of plants and animals some 11,000 years ago.

First Law of Ecology The axiom that people and the environment are intimately interconnected such that any single human action has multiple consequences.

flexible production (also lean production) A system of industrial production that uses information technologies such as computer networking, high-tech sensors, and automation technology to make the production of goods more responsive to market conditions and therefore more efficient.

food desert An area characterized by a lack of affordable, fresh, and nutritious foods.

food insecurity A situation in which people do not have physical or financial access to basic foodstuffs.

forced migration A situation in which a person, group, government, or other entity insists that another individual or group must relocate. The people being forced to move have no say about where they are moving, when, or other conditions of the move.

Fordism A system of industrial production designed for mass production and influenced by the principles of scientific management.

foreign direct investment (FDI) The transfer of monetary resources by a multinational corporation from its home country abroad in order to finance its overseas business activities.

formal region An area that possesses one or more unifying physical or cultural traits.

fossil fuels Nonrenewable energy resources that derive from the buried remains of plants and animals that lived millions of years ago.

four-course system A system of crop rotation that is based on a four-year planting regime that removes a fallow period, balances the planting of food crops with feed crops, and incorporates legumes that enrich the soil.

functional complexity The ability of a town or city to support sizable concentrations of people who earn their living from specialized, nonfarming activities.

functional region An area that is unified by a specific economic, political, or social activity and possesses at least one node.

functional zonation The division of a city into areas or zones that share similar activities and land use.

gender The cultural or social characteristics society associates with being female or male.

gender empowerment measure (GEM) A statistic for measuring development that assesses the extent to which women participate in the economic and political decision making within a country. It is derived from estimated earned income for women and men, the proportion of women and men in professional and technical jobs, the proportion of women and men working as legislators and managers, and the proportion of women and men who hold seats in parliament.

gender gap A disparity between men and women in their opportunities, rights, benefits, behavior, or attitudes.

gender-related development index (GDI) A statistic for measuring development that is sensitive to wide disparities in achievement between women and men. The indicators on which the GDI is based are estimated earned income for women and men, life expectancy, the adult literacy rate, and the gross enrollment ratio (the total enrollment in education as a percentage of the total school-age population).

gender role The idea that there are certain social expectations, responsibilities, or rights associated with femininity and masculinity.

Gene Revolution The shift, since the 1980s, to greater private and corporate involvement in and control of the research, development, intellectual property rights, and genetic engineering of highly specialized agricultural products, especially crop varieties.

gentrification A process of urban residential change that occurs when more affluent people purchase deteriorated buildings in low-income neighborhoods in order to restore or renovate them.

geodemography The study of the spatial variations among human populations.

geographic information systems (GIS) A combination of hardware and software that enables the input, management, analysis, and visualization of georeferenced (location-based) data.

geographic scale A way of depicting, in reduced form, all or part of the world, or a level of analysis used in a specific project or study.

geopiety The religious-like reverence that people may develop for the Earth.

geopolitics The study of the relations among geography, states, and world power.

gerrymandering The process of manipulating voting district boundaries to give an advantage to a particular political party or group.

global food crisis A protracted condition of food insecurity worldwide in scope or significance.

global positioning system (GPS) A constellation of artificial satellites, radio signals, and receivers used to determine the absolute location of people, places, or features on Earth.

global warming A rise in global temperatures primarily attributed to human activities that have increased concentrations of greenhouse gases in the atmosphere.

globalization The greater interconnectedness and interdependence of people and places around the world.

glocalization The idea that global and local forces interact and that both are changed in the process.

Green Revolution The dramatic increase in grain production between 1965 and 1985 in Asia and Latin America from high-yielding, fertilizer-, and irrigation-dependent varieties of wheat, rice, and corn.

greenhouse effect A natural process involving solar radiation and atmospheric gases that helps to warm the Earth.

gross domestic product (GDP) The total monetary value of goods and services produced within a country's geographic borders in a year.

gross national income (GNI) The total monetary value of goods and services produced in a year by a country, whether those operations are located within the country or abroad.

haka A collective, ritual dance of the New Zealand Maori and other Polynesian peoples.

hearth A place or region where an innovation, idea, belief, or cultural practice begins.

heartland Generally, any area of vital interest to a state; according to Halford Mackinder's heartland theory, a region possessing the best combination of strategic geographic factors for world domination, specifically the interior of the Eurasian landmass.

heritage Property transmitted to an heir or, more broadly, any contemporary use of the past.

heritage industry Enterprises, such as museums, monuments, and historical and archaeological sites, that manage or market the past.

heterolocalism A reflection of the impact of globalization on ethnic interaction such that members of an ethnic group maintain their sense of shared identity even though they are residentially dispersed.

heterosexual norm The conventional, binary division of the sexes based on clearly defined masculine and feminine gender roles.

hinterland The trade area served by a central place.

hinterworld The idea that the trade area served by a world city potentially includes the entire globe because of increased opportunities for spatial interaction.

holistic approach Within traditional medicine, a manner of understanding health such that it encompasses all aspects—physical, mental, social, and spiritual—of a person's life.

horizontal expansion In the context of globalization, the increase in international connections among places via rapid flows of goods, people, and ideas.

human development index (HDI) The first development measure to incorporate information about the wealth, health, and education of a country's people in a single statistic. The HDI is based on the GDP (PPP) per capita, life expectancy, the adult literacy rate, and the gross enrollment ratio (the total enrollment in education as a percentage of the total school-age population).

human geography A branch of geography centered on the study of people, places, spatial variation in human activities, and the relationship between people and the environment.

human trafficking The forcible and/or fraudulent recruitment of people for work in exploitative conditions, for example, as child soldiers or prostitutes.

hunting and gathering Hunting wild animals, fishing, and gathering wild plants for food.

hybrid city A city that exhibits a mixture of indigenous, colonial, and globalizing influences.

ideology A system of ideas, beliefs, and values that justify the views, practices, or orientation of a group.

immigration The in-migration or arrival of people at a location.

imperialism One state's exercise of direct or indirect control over the affairs of another political society.

income distribution How income is divided among different groups or individuals.

income inequality A ratio of the earnings of the richest to the earnings of the poorest.

Industrial Revolution Fundamental changes in technology and systems of production that began in England in the late 18th century and transformed manufacturing from small-scale craft to factory-based production.

infant mortality rate The number of deaths of infants under one year of age per 1,000 live births.

informal sector The retail, manufacturing, and service activities that operate on a small scale, without government regulation or oversight, and that are not measured or recorded in formal or official statistics.

institutional discrimination A situation in which the policies, practices, or laws of an organization or government disadvantage people because of their cultural differences.

intensive agriculture An agricultural system characterized by high inputs such as labor, capital, or equipment per unit area of land.

intercropping The strategy of planting two or more crops in a field at the same time.

internally displaced persons People forcibly driven from their homes into a different part of their country.

international migration A situation in which people cross international boundaries and take up long-term or permanent residence in another country.

internationalism The development of close political and economic relations among states.

intervening opportunity A situation in which a different location can provide a desired good more economically than another location.

Islamic traditionalism (or Islamism) A movement that favors a return to or preservation of traditional, premodern Islam and resists Westernization and globalization.

jihad A term that is popularly understood to mean "holy war" but is preferably translated as "utmost struggle" and refers to a personal struggle to uphold the tenets of Islam.

Kyoto Protocol An international agreement since 1997 with the goal of achieving significant reductions in greenhouse gas emissions, particularly among the industrialized countries that have ratified it.

land-use and land-cover change (LULCC) An approach to studying human–environment interactions based on an awareness of ecosystem processes and human activities.

language A system of communication based on symbols that have agreed-upon meanings.

language family A collection of languages that share a common but distant ancestor.

level of urbanization The percentage of people living in urban places in some defined area (e.g., a county or a country).

liberalism An 18th-century political and economic theory that emphasizes the protection of human rights, property rights, and individual freedom. Economically, liberalism favors a free, unregulated market and the removal of barriers to the movement of goods, services, and capital.

life expectancy The average length of time from birth that a person is expected to live given current death rates.

lingua franca A language that is used to facilitate trade or business between people who speak different languages.

linguistic diversity The assortment of languages in a particular area.

linguistic diversity index (LDI) A measure that expresses the likelihood that two randomly selected individuals in a country speak different first languages.

linguistic dominance A situation in which one language becomes comparatively more powerful than another language.

linkage Narrowly, an economic activity that emerges in conjunction with a specific primary industry, such as blade manufacturers that supply the lumber industry. More broadly, the interconnections that develop among businesses.

livestock ranching A form of agriculture devoted to raising large numbers of cattle or sheep for sale to meat processors.

loanword A word that originates in one language and is incorporated into the vocabulary of another language.

local culture The practices, attitudes, and preferences held in common by the members of a community in a particular place.

local knowledge The collective wisdom of a community that derives from the everyday activities of its members.

location quotient A measure that can be used to show how the proportional presence of an ethnic group in a region compares to the proportional presence of that same ethnic group in the country.

manufacturing value-added (MVA) A measure of industrial output calculated by taking the cost of the finished product and subtracting from it the cost of purchased inputs necessary to produce it, such as fuel, electricity, and the cost of other parts or materials.

material culture The tangible and visible artifacts, implements, and structures created by people such as dwellings, musical instruments, and tools.

Mediterranean agriculture As historically practiced, a form of agroforestry that integrated cultivation of livestock, a grain crop, and a tree or vine crop, and that is today increasingly affected by specialization.

megachurch A church with 2,000 or more members that follows mainline or Renewalist Christian theologies.

megacity A city with 10 million or more residents.

megalopolis (also conurbation) A massive urban complex created as a result of converging metropolitan areas.

metropolitan area A large population center (50,000 people minimum) and the adjacent zones that are socially and economically connected to it, such as the places from which people commute.

microfinancing Lending practices that provide access to credit and other financial services for low-income individuals or groups.

migration The long-term or permanent relocation of an individual or group from one place to another.

mixed crop and livestock farming A farming system in various stages of evolution worldwide from an integrated system based on raising crops to feed livestock and selling the animal products off the farm to a more specialized emphasis on cash grain production.

mixed-use development An approach to urban design that combines different types of land use within a particular neighborhood or district.

modernism An intellectual movement that has roots in the European Enlightenment of the 1700s and encourages scientific thought, the expansion of knowledge, and belief in the inevitability of progress.

monoculture The planting of a single crop in a field, often over a large area.

monotheism The belief in or devotion to a single deity.

multinational corporation (MNC) (also a transnational corporation, or TNC) A company that owns offices or production facilities in one or more other countries.

multinational state A state whose population consists of two or more nations.

multiple nuclei model An accounting of urban spatial structure that was developed by Chauncy Harris and Edward Ullman and that relates land-use patterns to the influence of two or more cores, as opposed to a single core represented by the central business district.

mutual intelligibility The ability of speakers of different but related languages to understand one another.

mystical ecology The interrelationship between an awareness of cosmic forces and human use of the environment.

nation A sizable group of people with shared political aspirations whose collective identity is rooted in a common history, heritage, and attachment to a specific territory.

nation-state In a narrow sense, a political entity in which the boundaries of a nation coincide with the boundaries of the state and the people share a sense of political unity. More broadly, a state whose population possesses a shared political identity that sees the nation and the state as the same.

natural capital The goods and services provided by nature, including renewable resources, nonrenewable resources, Earth's biodiversity, and its ecosystems.

nature In one sense, the physical environment that is external to people, but also a social construction derived from ideas that people have about the physical environment.

nature–culture dualism A conceptual framework that separates nature from culture (nature is not culture, and vice versa) and is rejected by many scholars today.

neoliberalism The revival and application of the theory of liberalism, especially since the late 20th century.

neolocalism A renewed interest in sustaining and promoting the uniqueness of a place.

neo-Malthusians People who share the same general views of Thomas Malthus and argue that, since the world's resources are limited, there is also a natural limit to the number of people the Earth can support at a comfortable standard of living.

net migration A measure of migration-based population change in a place; calculated as the number of immigrants minus the number of emigrants.

new renewables Energy sources that include biomass, tidal, solar, wind, and geothermal energy as well as small-scale hydropower facilities but exclude large hydropower facilities because of their significant environmental impacts.

new urbanism (also neotraditional town planning) An urban planning movement that developed in the 1990s around the two main goals of preventing sprawl and creating walkable neighborhoods.

newly industrialized economies (NIEs) Economies in which rapid growth, improved living standards, and reductions in poverty have been achieved in East Asia first by Hong Kong, Singapore, South Korea, and Taiwan, and more recently by Indonesia, Malaysia, the Philippines, and Thailand.

nonmaterial culture The oral traditions, behavioral practices, and other nontangible components of a cultural group's way of life, including recipes, songs, or philosophies.

nonrenewable resource A resource that is considered finite because it is not self-replenishing or takes a very long period of time to do so.

normative In reference to development, the establishment of standards, or norms, to help measure the quality of life and economic prosperity of groups of people.

nutrition transition A change in patterns of food consumption toward an increasingly Westernized diet consisting of more meat, wheat-based food products, and convenience foods.

official language A language that a country formally designates, usually in its constitution, for use in political, legal, and administrative affairs.

offshoring A kind of outsourcing that moves a business activity to another country.

open-access resources Goods that no single person can claim exclusive right to and that are available to everyone.

organic agriculture A farming system that promotes sustainable and biodiverse ecosystems and relies on natural ecological processes and cycles, as opposed to synthetic inputs such as pesticides.

Organization for Economic Cooperation and Development (OECD) An organization founded in 1961 to enhance development by promoting economic growth among its member countries. Most members have historically been the very wealthy, industrialized countries.

othering The act of differentiating between individuals and groups such that distinctions are made between "me" and "you," and between "us" and "them."

outsourcing A business practice whereby a company subcontracts an activity that was previously performed in-house (such as the manufacture of a part, packaging, or customer support) to another firm.

pastoralism An agricultural system in which animal husbandry based on open-grazing of herd animals is the sole or dominant farming activity.

perceptual region An area that people perceive to exist because they identify with it, have an attachment to it, or imagine it in a certain way.

personal approach Within traditional medicine, a manner of understanding health in which it is possible for two people to have the same symptoms but receive different treatments because of their different individual circumstances.

physiological density The number of people per unit area of arable land.

pidgin language A language that combines vocabulary and/or grammatical practices from two or more languages that have come in contact.

piety (or piousness) A state of deep devotion to a religion.

pilgrimage A journey to a sacred place or site for religious reasons.

place A locality distinguished by specific physical and social characteristics.

placelessness The loss of the unique character of different places and the increasing standardization of places and cultural landscapes that is sometimes associated with the diffusion of popular culture.

plantation A large estate in tropical or subtropical areas that specializes in the production of a cash crop.

pluralism A model of ethnic interaction that characterizes members of immigrant ethnic groups as resisting pressures to assimilate and retaining those traits, beliefs, and practices that make them distinctive.

political ecology An offshoot of cultural ecology that studies how economic forces and competition for power influence human behavior, especially decisions and attitudes involving the environment.

political geography The study of the spatial aspects of political affairs.

political iconography An image, object, or symbol that conveys a political message.

polytheism The belief in or devotion to multiple deities.

popular culture The practices, attitudes, and preferences held in common by large numbers of people and considered to be mainstream.

population doubling time The number of years it takes a population to double; calculated by dividing the number 72 by the rate of natural increase.

population ecology The study of the impacts populations have on their environments as well as the ways in which environmental conditions affect people and their livelihoods.

population pyramid A bar graph that shows the age and gender composition of a population.

possibilism A theory that people use their creativity to decide how to respond to the conditions or constraints of a particular natural environment.

postindustrial society A society characterized by high levels of urbanization, dominance of the service sector especially in total employment, prevalence of skilled professionals in the labor force, infrastructure that is heavily based on information and communication technology, and a knowledge-based economy.

poverty Insufficient income to purchase the basic necessities of food, clothing, and shelter.

poverty line The specific income amount social scientists and others use to separate the poor from the nonpoor.

poverty rate The percentage of the population below the poverty line.

poverty-reduction theory A development theory focused specifically on lowering the incidence of poverty in a developing country.

precision agriculture The application of technologies such as the global positioning system (GPS) and aerial imagery to measure and map the spatial variation in environmental conditions within a field, and the related use of this information to calibrate machinery for site-specific applications of fertilizers or pesticides, for example.

primary industry An industry that extracts natural resources from the Earth.

primate city A city that has a population two or more times the population of the second largest city in the country.

public space A kind of commons; a space intended to be open and accessible to anyone.

pull factors Favorable conditions or attributes of a place that attract migrants.

purchasing power parity (PPP) An exchange rate that is used to compare output, income, or prices among countries with different currencies and that is based on the idea that the price of a good or service in one country should equal the price of that same good or service in another country when it is converted to a common currency.

push factors Unfavorable conditions or attributes of a place that encourage migration.

qanat A system of water supply that uses shaft and tunnel technology to tap underground water resources.

quaternary services Service industries including or related to transportation, telecommunications, real estate, insurance, finance, and management.

quinary services Service industries such as research, education, and engineering that facilitate the creation of innovations through the production of new knowledge and skills.

race The mistaken idea that one or more genetic traits can be used to identify distinctive and exclusive categories of people; hence, race is today understood to be a social construction.

racism Intolerance of people perceived to be inherently or genetically inferior.

range The maximum distance a consumer will travel for a particular good or service.

rate of natural increase The percentage of annual growth in a population excluding migration.

rate of urban growth The annual percentage increase in an urban population.

rationality doctrine The attitude and belief that Europeans were rational and that non-Europeans, especially colonized peoples, were irrational; closely associated with diffusionism.

reapportionment The process of allocating legislative seats among voting districts so that each legislator represents approximately the same number of people.

redistricting Redrawing the boundaries of voting districts, usually as a result of population change.

redlining The biased practice of refusing to offer home loans on the basis of the characteristics of a neighborhood instead of the actual condition of the property being mortgaged.

refugee One who flees to another country out of concern for personal safety or to avoid persecution.

regional analysis The study of the cultural, economic, political, physical, or other factors that contribute to the distinctiveness of geographical areas.

religion A system of beliefs and practices that help people make sense of the universe and their place in it.

religious ecology The interdependency between people, their religious beliefs and practices, and nature.

religious fundamentalism An interpretation of the principles of one's faith in such a way that they come to shape all aspects of one's private and public life.

remittance The money, goods, or services sent by immigrants to family members or relatives in their home countries.

remote sensing A means of acquiring information about something that is located at a distance from you or the sensing device, such as a satellite.

renewable resource A resource that is replenished naturally or through human intervention.

Renewalism A branch of Christianity that includes the Pentecostal and Charismatic movements.

replacement level The fertility rate necessary for a population to replace itself.

ritual Behavior, often regularly practiced, that has personal and symbolic meaning.

sacred space An area that has special religious significance or meaning that makes it worthy of reverence or devotion.

salinization The accumulation of salts on or in the soil.

sanctification The process by which people come to associate a place or site with sacredness; the making of a sacred site.

second agricultural revolution A fundamental change involving the adoption of new agricultural practices in western Europe, such as the moldboard plow and the horse collar, beginning in the Middle Ages.

secondary industry An industry that assembles, processes, or converts raw or semiprocessed materials into fuels or finished goods.

sector model An accounting of urban spatial structure that was developed by Homer Hoyt and that relates the formation of sectors or wedges of similar land use to transportation factors and the influence of high-income groups.

secularization A process that reduces the scope or influence of religion.

security landscape A type of political landscape created to protect the territory, people, facilities, and infrastructure of a state.

self-determination The ability of people in a territory to choose their own political status.

separatism The desire of a nation or other group to break apart from its state.

sex ratio The proportion of males to females in a population.

sexuality A basis for personal and social identity that stems from sexual orientation, attitudes, desires, and practices.

sharia Islamic law derived from the Qur'an, the teachings of Muhammad, and other sources.

shifting cultivation An agricultural system that uses fire to clear vegetation in order to create fields for crops and is based on a cycle of land rotation that includes fallow periods.

site The physical characteristics of a place, such as its topography, vegetation, and water resources.

situation The geographic context of a place, including its political, economic, social, or other characteristics.

slum An area of a city characterized by overcrowding, makeshift or dilapidated housing, and little or no access to basic infrastructure and services such as clean water and waste disposal.

smallholder agriculture A farming system characterized by small farms in which the household is the main scale of agricultural production and consumption.

smallholder crop and livestock farming An agricultural system that is based on the management of a combination of plants and livestock that varies significantly from one region to another.

social capital The social ties, networks, institutions, and trust that members of a group use to achieve mutual benefits.

social construction An idea or a phenomenon that does not exist in nature but is created and given meaning by people.

sociodemographic indicator A value or measure that provides information about the welfare of a population, such as data on the prevalence of disease or the levels of literacy and education.

sovereignty Supreme authority of a state over its own affairs and freedom from control by outside forces.

space A bounded (absolute) or unbounded (relative) area. Absolute space can be precisely measured; relative space is shaped by contingency.

spatial association The degree to which two or more phenomena share similar distributions.

spatial diffusion The movement of a phenomenon, such as an innovation, information, or an epidemic, across space and over time.

spatial interaction The connections and relations that develop among places and regions as a result of the movement or flow of people, goods, or information.

spatial variation Change in the distribution of a phenomenon from one place or area to another.

sprawl A process that occurs when the rate at which land is urbanized greatly exceeds the rate of population growth in a given period of time, leading to the spread of low-density land use.

standard dialect The designation of a specific dialect as the norm or authoritative model of language usage.

staple theory A system of ideas developed by W. A. Mackintosh and Harold Innis that posits that the commodities of an area shape its economic system by triggering the formation of related industries.

state An internationally recognized political unit with a permanently populated territory, defined boundaries, and a government with sovereignty over its domestic and international affairs.

structural adjustment program (SAP) Country-specific economic policy favored by the International Monetary Fund and World Bank during the 1980s and 1990s, and based on neoliberal principles intended to promote economic growth and development.

subsistence agriculture A farming system that is largely independent of purchased inputs and in which outputs are typically used or consumed by farmers and their family or extended family.

suburb The built-up area that surrounds the central city.

supranational organization A political entity created when multiple states agree to work together for a common economic, military, cultural, or political purpose, or a combination of several of these.

sustainable agriculture Farming practices that carefully manage natural resources and minimize adverse effects on the environment while maintaining farm profits.

sustainable development An approach to resource use and management that meets economic and social needs without compromising the resources for future generations.

symbolic ethnicity The way in which a collection of symbols (e.g., flags, music, dress styles) imparts meaning and identity to members of an ethnic group.

syncretic A term often used to describe religions that demonstrate a notable blending of beliefs and practices, usually as a result of contact between people who practice different religions.

Taylorism (also scientific management) A philosophy about industrial production that was developed by F. W. Taylor and that promotes the division of labor into the most elemental tasks for greatest efficiency.

technopole An area with a cluster of firms conducting research, design, development, and/or production in high-tech industries.

territoriality Strong attachment to or defensive control of a place or an area.

territorial seas The waters that are enclosed by the boundaries of a coastal state and that are considered part of the territory of that state.

terrorism The threat or use of violence against civilians in order to inculcate fear, gain influence, and/or advance a specific cause or conviction.

tertiary industry An industry that provides services, usually in the form of nontangible goods, to other businesses and/or consumers.

third agricultural revolution A fundamental change in agriculture associated with technological innovations and scientific farming techniques developed in the 20th century including extensive mechanization, heavy reliance on irrigation and chemical applications, and biotechnology.

threshold The smallest number of consumers required to profitably supply a certain good or service.

time-space convergence The process by which places seem to become closer together in both time and space as a result of innovations in transportation and communication that weaken the barrier or friction of distance.

toponym A place-name.

total fertility rate (TFR) The average number of children a woman is expected to have during her childbearing years (between the ages of 15 and 49), given current birth rates.

traditional medicine Medical practices, derived from a society's long-established health-related knowledge and beliefs, that are used to maintain or restore well-being.

transferability The cost of moving a good and the ability of the good to withstand that cost.

transgender People who do not identify with the gender assigned them at birth.

transhumance Moving herds on a seasonal basis to new pastures or water sources.

transnationalism In migration studies, the process by which immigrants develop and cultivate ties to more than one country.

truck farming A form of commercial gardening centered on the specialized production of fruits and vegetables for market.

unauthorized immigrants Also called undocumented or illegal immigrants in the United States; people who enter a country on a temporary visa but remain in the country after their visa expires, or who cross the border without being detected.

United Nations (UN) A supranational organization founded in 1945 for the purpose of promoting international peace and security.

universalizing religion A belief system that is worldwide in scope, welcomes all people as potential adherents, and may also work actively to acquire converts.

urban agriculture The use of vacant lots, rooftops, balconies, or other urban spaces to raise food for metropolitan households or neighborhoods.

urban hierarchy A series of central places ranked on the basis of their threshold, range, and market area.

urban planning A field of study broadly concerned with improving the physical and social conditions in towns and cities through the wise use and management of urban space.

urban realms model An accounting of urban structure that was developed by James Vance and that emphasizes the influential effects of suburbanization on the evolution of urban form.

urban redevelopment The process of renovating an area of a city, often by completely destroying dilapidated structures and rebuilding on the site.

urbanization Processes that concentrate people in urban places.

urbanized area Less formally, land that has been developed for commercial, residential, or industrial purposes. According to the U.S. Census Bureau, the territory, usually the central city plus adjacent suburbs, that has at least 50,000 people and a population density of 1,000 people or more per square mile.

vernacular architecture The common structures—dwellings, buildings, barns, churches, and so on—associated with a particular place, time, and community.

vertical expansion In the context of globalization, the deepening of connections between places through the development of policies, such as trade agreements that formalize and strengthen those linkages.

vertical integration A strategy of extending a company's ownership and control "up" the supply stream and/or "down" the distribution stream of a good or service in order to lessen the company's vulnerability to disruptions and stabilize the system of production.

voluntary migration A long-term or permanent move that stems from choice.

von Thünen's model A model that relates transportation costs to agricultural land-use decisions and yields a concentric ring pattern showing progressively more extensive forms of agriculture practiced at greater distances from the city or market.

wet rice farming Rice cultivation in a flooded field.

world city A principal center of global economic power that significantly influences the world's business.

world heritage (also global heritage) Sites perceived to have outstanding universal value for all of humanity.

world-system theory A body of ideas that was developed by Immanuel Wallerstein and that links dependency and underdevelopment to capitalism and its role in creating an international division of labor that shapes relations between core, semiperipheral, and peripheral regions of the world.

zoning Laws that regulate land use and development.

Chapter 1

Adams, B. J., Huyck, C. K., Mansouri, B., Eguchi, R. T., and Shinozuka, M. 2004. Application of High-Resolution Optical Satellite Imagery for Post-Earthquake Damage Assessment: The 2003 Boumerdes (Algeria) and Bam (Iran) Earthquakes. *MCEER Research Progress and Accomplishments 2003–2004*. Buffalo, NY: Multidisciplinary Center for Earthquake Engineering Research, University at Buffalo.

Agnew, J., Livingstone, D. N., and Rogers, A., eds. 1996. *Human Geography: An Essential Anthology*. Oxford: Blackwell.

Bauman, K. J., and Graf, N. L. 2003. Educational Attainment: 2000. Census 2000 Brief. Washington, DC: U.S. Census Bureau. Available online at: http://www.census.gov/prod/2003pubs/c2kbr-24.pdf

Blaut, J. M. 1961. Space and Process. *Professional Geographer* 13(4): 1–7.

Block, R., and Bernasco, W. 2008. Finding a Serial Burglar's Home Using Distance Decay and Conditional Origin-Destination Patterns: A Test of Empirical Bayes Journey-to-Crime Estimation in The Hague. Unpublished manuscript available online at: http://www.aic.gov.au/en/events/seminars/2008/rblock.aspx.

Castree, N. 2001. Socializing Nature: Theory, Practice, and Politics. In *Social Nature: Theory, Practice and Politics*, eds. N. Castree and B. Braun, 1–21. Malden, MA: Blackwell.

Cronon, W. 1995. The Trouble with Wilderness; or, Getting Back to the Wrong Nature . In *Uncommon Ground: Rethinking the Human Place in Nature,* ed. W. Cronon, 69–90. New York: W. W. Norton & Co.

Cutter, S. L., Holm, D., and Clark, L. 1996. The Role of Geographic Scale in Monitoring Environmental Justice. *Risk Analysis* 16(4): 517–526.

DeFries, R. H., Hansen, M., Townshend, J. R. G., Janetos, A. C., and Loveland, T. R. 2000. A New Global 1 km Data Set of Percent Tree Cover Derived from Remote Sensing. *Global Change Biology* 6: 247–254.

DeGroote, J. P., Sugumaran, R., Brend, S. M., Tucker, B. J., and Bartholomay, L. C. 2008. Landscape, Demographic, Entomological, and Climatic Associations with Human Disease Incidence of West Nile Virus in the State of Iowa, USA. *International Journal of Health Geographics* 7: 19. Available online at: http://www.ij-healthgeographics.com/content/7/1/19.

Dobson, J. E., and Fisher, P. F. 2003. Geoslavery. *IEEE Technology and Society Magazine* (Spring): 47–52.

Downs, R. M. 1997. The Geographic Eye: Seeing Through GIS? *Transactions in GIS* 2(2): 111–121.

ESRI. 2008. *GIS Best Practices: Essays on Geography and GIS*. Redlands, CA: ESRI.

Fenneman, N. M. 1919. The Circumference of Geography. *Annals of the Association of American Geographers* 9: 3–11.

Foucault, M. 1977. *Discipline & Punish: The Birth of the Prison*. Trans. by Alan Sheridan. New York: Vintage Books.

Gaile, G. L., and Willmott, C. J., eds. 2003. *Geography in America at the Dawn of the 21st Century*. New York: Oxford University Press.

Goodchild, M. F. 2004. The Validity and Usefulness of Laws in Geographic Information Science and Geography. *Annals of the Association of American Geographers* 94(2): 300–303.

Greiner, A. L., Wikle, T. A., and Spencer, J. 2000. *Where Geography Can Take You: An Interactive CD-ROM*. Stillwater, OK: Department of Geography.

Grossner, K. E., Goodchild, M., and Clarke, K. 2008. Defining a Digital Earth System. *Transactions in GIS* 12(1): 145–160.

Holdsworth, D. W. 1997. Landscape and Archives as Text. In *Understanding Ordinary Landscapes*, eds. P. Groth and T. W. Bressi, 44–55. New Haven, CT: Yale University Press.

Hubbard, P., Kitchin, R., Bartley, B., and Fuller, D. 2002. *Thinking Geographically: Space, Theory and Contemporary Human Geography.* London: Continuum.

Jackson, P. 1989. *Maps of Meaning: An Introduction to Cultural Geography*. London: Unwin Hyman.

Johnston, R. J., Gregory, D., Pratt, G., and Watts, M., eds. 2000. *The Dictionary of Human Geography*. 4th ed. Malden, MA: Blackwell.

Jordan, T. G. 1992. The Concept and Method. In *Regional Studies: The Interplay of Land and People*, ed. G. E. Lich, 9–24. College Station: Texas A&M University Press.

Kwan, M.-P. 1999. Gender, the Home-Work Link, and Space-Time Patterns of Nonemployment Activities. *Economic Geography* 75(4): 370–394.

Kwan, M.-P. 2000a. Interactive Geovisualization of Activity-Travel Patterns Using Three-Dimensional Geographical Information Systems: A Methodological Exploration with a Large Data Set. *Transport Research Part C* 8: 185–203.

Kwan, M.-P. 2000b. Gender Differences in Space-Time Constraints. *Area* 32(2): 45–56.

Kwan, M.-P. 2002. Feminist Visualization: Re-envisioning GIS as a Method in Feminist Geographic Research. *Annals of the Association of American Geographers* 94(2): 645–661.

Kwan, M.-P. 2008. From Oral Histories to Visual Narratives: Re-presenting the Post-September 11 Experiences of the Muslim Women in the USA. *Social & Cultural Geography* 9(6): 653–669.

Marston, S. A. 2000. The Social Construction of Scale. *Progress in Human Geography* 24(2): 219–242.

Minot, N., Baulch, B., and Epprecht, M. 2003. *Poverty and Inequality in Vietnam: Spatial Patterns and Geographical Determinants*. Washington, DC and Brighton, UK: International Food Policy Research Institute (IFPRI) and Institute of Development Studies.

Mitchell, D. 1995. There's No Such Thing as Culture: Towards a Reconceptualization of the Idea of Culture in Geography. *Transactions of the Institute of British Geographers* NS, 20(1): 102–116.

Mitchell, D. 2000. *Cultural Geography: A Critical Introduction*. Malden, MA: Blackwell.

Plumwood, V. 2006. The Concept of a Cultural Landscape: Nature, Culture, and Agency in the Land. *Ethics & Environment* 11(2): 115–150.

Robbins, P. 2004. *Political Ecology: A Critical Introduction*. Malden, MA: Blackwell.

Robbins, P. 2007. *Lawn People: How Grasses, Weeds, and Chemicals Make Us Who We Are*. Philadelphia: Temple University Press.

Sauer, C. O. 1963 [1925]. The Morphology of Landscape. In *Land and Life: A Selection from the Writings of Carl Ortwin Sauer*, ed. J. Leighley, 315–350. Berkeley: University of California Press.

Shuurman, N. 2000. Trouble in the Heartland: GIS and Its Critics in the 1990s. *Progress in Human Geography* 24(4): 569–590.

Sui, D. Z. 2004. Tobler's First Law of Geography: A Big Idea for a Small World? *Annals of the Association of American Geographers* 94(2): 269–277.

Sunderlin, W., Dewi, S., and Puntodewo, A. 2008. Poverty and Forests: Multi-Country Analysis of Spatial Association and

Proposed Policy Solutions. *CIFOR Occasional Paper 47*, rev. ed. Bogor Barat, Indonesia: Center for International Forestry Research.

Taaffe, E. J. 1997. Spatial Organization and Interdependence. In *Ten Geographic Ideas that Changed the World*, ed. S. Hanson, 145–162. Piscataway, NJ: Rutgers University Press.

Thrift, N. 2003. Space: The Fundamental Stuff of Human Geography. In *Key Concepts in Geography*, eds. S. Holloway, S. P. Rice, and G. Valentine, 95–108. London: Sage Publications.

Tobler, W. 1970. A Computer Movie Simulating Urban Growth in the Detroit Region. *Economic Geography* 46(June supplement): 234–240.

Tobler, W. 2004. On the First Law of Geography: A Reply. *Annals of the Association of American Geographers* 94(2): 304–310.

Ullman, E. L. 1954. Geography as Spatial Interaction. *Annals of the Association of American Geographers* 43: 54–69.

Wiehe, S. E., Carroll, A. E., Liu, G. C., Haberkorn, K. L., Hoch, S. C., Wilson, J. S., and Fortenberry, J. D. 2008. Using GPS-Enabled Cell Phones to Track the Travel Patterns of Adolescents. *International Journal of Health Geographics* 7 (22). Available online at: http://www.ij-healthgeographics.com/content/7/1/22.

Chapter 2

Alanen, A. R. 1988. Architecture and Landscapes in Colombia: The Viability of the Vernacular. *Journal of Popular Culture* 22(1): 99–119.

Alvarez, M. 2008. Striking a Global Pose: Considerations for Working with Folk and Traditional Cultures in the 21st Century. Issues in Folk Arts and Traditional Culture Working Paper Series, The Fund for Folk Culture. Available online at: http://www.folkculture.org/AboutOurWork/ResearchandPublications/tabid/67/Default.aspx.

Appadurai, A. 1997. *Modernity at Large: Cultural Dimensions of Globalization*. Minneapolis: University of Minnesota Press.

Basser, S. 1999. Acupuncture: A History. *The Scientific Review of Alternative Medicine* 3(1): 34–41.

Blaut, J. 1987. Diffusionism: A Uniformitarian Critique. *Annals of the Association of American Geographers* 77: 30–47.

Bowers, J. Z. 1973. Acupuncture. *Proceedings of the American Philosophical Society* 117(3): 143–151.

Brenner, N. 1999. Beyond State-Centrism? Space, Territoriality, and Geographical Scale in Globalization Studies. *Theory and Society* 28(1): 39–78.

Castells, M. 1996. *The Information Age: Economy, Society and Culture*. Vol. 1, *The Rise of the Network Society*. Cambridge, MA: Blackwell.

Cressy, G. B. 1958. Qanats, Karez, and Foggaras. *Geographical Review* 48(1): 27–44.

Foster, G. M. 1953. What Is Folk Culture? *American Anthropologist* 55: 159–173.

Giddens, A. 2000. *Runaway World: How Globalization Is Reshaping Our Lives*. New York: Routledge.

Graff, T. O., and Ashton, D. 1993. Spatial Diffusion of Wal-Mart: Contagious and Reverse Hierarchical Diffusion. *Professional Geographer* 46(1): 19–29.

Graham, B., Ashworth, G. J., and Tunbridge, J. E. 2000. *A Geography of Heritage: Power, Culture & Economy*. London: Arnold.

Hinrichs, T. H. 1998. New Geographies of Chinese Medicine. *Osiris* 2nd Series 13: 287–325.

Holson, L. M. 2005. The Feng Shui Kingdom. *New York Times*, April 25.

Holton, R. 2000. Globalization's Cultural Consequences. *Annals of the American Academy of Political and Social Science* 570: 140–152.

Jackson, P. 1999. Commodity Cultures: The Traffic in Things. *Transactions of the Institute of British Geographers* NS 24(1): 95–108.

Jordan, T. G., and Kaups, M. 1987. Folk Architecture in Cultural and Ecological Context. *Geographical Review* 77: 52–75.

Knapp, R. G. 1999. *China's Living Houses: Folk Beliefs, Symbols, and Household Ornamentation*. Honolulu: University of Hawaii Press.

KPMG. 2006. *The Global Gems and Jewellery Industry. Vision 2015: Transforming for Growth*. Mumbai: GJEPC-KPMG.

Lightfoot, D. L. 2000. The Origin and Diffusion of Qanats in Arabia: New Evidence from the Northern and Southern Peninsula. *The Geographical Journal* 166(3): 215–226.

Lowenthal, D. 1975. Past Time, Present Place: Landscape and Memory. *Geographical Review* 65: 1–36.

Mikesell, M. W. 1978. Tradition and Innovation in Cultural Geography. *Annals of the Association of American Geographers* 68: 1–16.

Mintz, S. W. 1953. The Folk-Urban Continuum and the Rural Proletarian Community. *The American Journal of Sociology* 59: 136–143.

Mitchell, D. 1995. There's No Such Thing as Culture: Towards a Reconceptualization of the Idea of Culture in Geography. *Transactions of the Institute of British Geographers* NS 20: 102–116.

Oliver, P., ed. 1997. *Encyclopedia of Vernacular Architecture of the World*. 3 vols. Cambridge: Cambridge University Press.

Parsons, J. J. 1991. Giant American Bamboo in the Vernacular Architecture of Colombia and Ecuador. *Geographical Review* 81(2): 131–152.

Pieterse, J. N. 2009. *Globalization and Culture: Global Mélange*. 2nd ed. Lanham, MD: Rowman & Littlefield.

Pocock, D. 1997. Some Reflections on World Heritage. *Area* 29(3): 260–268.

Redfield, R. 1947. The Folk Society. *The American Journal of Sociology* 52: 293–308.

Revenge of Geography, The. 2003. *The Economist*. March 15.

Salih, A. 2006. Qanats a Unique Groundwater Management Tool in Arid Regions: The Case of Bam Region in Iran. Paper presented at the International Symposium on Groundwater Sustainability (ISGWAS), January. Available online at: http://aguas.igme.es/igme/ISGWAS/Ponencias%20ISGWAS/6-Salih.pdf.

Sauer, C. O. 1962 [1931]. Cultural Geography. In *Readings in Cultural Geography*, eds. P. L. Wagner and M. W. Mikesell, 30–34. Chicago: University of Chicago Press.

Teather, E. K., and Chow, C. S. 2000. The Geographer and the Fengshui Practitioner: So Close and Yet So Far Apart? *Australian Geographer* 31(3): 309–332.

Tunbridge, J. E., and Ashworth, G. J. 1996. *Dissonant Heritage: The Management of the Past as a Resource in Conflict*. New York: John Wiley.

United Nations Conference on Trade and Development (UNCTAD). 2008. *World Investment Report: Transnational Corporations and the Infrastructure Challenge*. New York: United Nations.

United Nations Conference on Trade and Development (UNCTAD). 2009. *World Investment Report: Transnational Corporations, Agricultural Production and Development*. New York: United Nations.

United Nations Educational, Scientific and Cultural Organization (UNESCO) 2009. World Heritage List. Available online at: http://whc.unesco.org/en/list.

Van Elteren, M. 2003. U.S. Cultural Imperialism Today: Only a Chimera? *SAIS Review* 23(2): 169–188.

Wagner, P. L. 1994. Foreword: Culture and Geography: Thirty Years of Advance. In *Re-reading Cultural Geography*, eds. K. E. Foote, P. J. Hugill, K. Mathewson, and J. M. Smith, 3–8. Austin: University of Texas Press.

Wayland, C. 2001. Gendering Local Knowledge: Medicinal Plant Use and Primary Health Care in the Amazon. *Medical Anthropology Quarterly*, NS 15(2): 171–188.

World Health Organization (WHO). 2002. *WHO Traditional Medicine Strategy 2002–2005*. Geneva: WHO.

Wu, J.-N. 1996. A Short History of Acupuncture. *Journal of Alternative and Complementary Medicine* 2(1): 19–21.

Chapter 3

Boserup, E. 1965. *The Conditions of Agricultural Growth*. London: G. Allen & Unwin.

Brun, C. 2001. Reterritorializing the Relationship Between People and Place in Refugee Studies. *Geografiska Annaler. Series B, Human Geography* 83(1): 15–25.

Carling, J. 2005. Gender Dimensions of International Migration. *Global Migration Perspectives No. 35*. Geneva: Global Commission on International Migration.

Castles, S. 2002. Migration and Community Formation Under Conditions of Globalization. *International Migration Review* 36(4): 1143–1168.

Central Intelligence Agency (CIA). 2009. *The World Factbook 2009*. Washington, DC: Central Intelligence Agency. Available online at: https://www.cia.gov/library/publications/the-world-factbook/index.html.

Citizenship and Immigration Canada. 2009. *Facts and Figures: Immigration Overview Permanent and Temporary Residents 2008*. Ottawa: Citizenship and Immigration Canada Research and Evaluation Branch. Available online at: http://www.cic.gc.ca/english/resources/statistics/menu-fact.asp.

Democratic People's Republic of Korea (DPRK). 2005. *DPRK 2004 Nutrition Assessment Report of Survey Results*. N.p.: DPRK Central Bureau of Statistics. Available online at: http://www.unicef.org/dprk/dprk_national_nutrition_assessment_2004_final_report_07_03_05.pdf.

Ehrlich, P. R. 1978. *The Population Bomb*. Rev. ed. New York: Ballantine Books.

Ehrlich, P. R., and Ehrlich, A. 2009. The Population Bomb Revisited. *The Electronic Journal of Sustainable Development* 1(3): 63–71.

Hinrichsen, D. 1999. The Coastal Population Explosion. In *Trends and Future Challenges for U.S. National Ocean and Coastal Policy, Proceedings of a Workshop*, eds. B. Cicin-Sain, R. W. Knecht, and N. Foster, 27–29. Silver Spring, MD: NOAA, National Ocean Service.

Hoefer, M., Rytina, N., and Baker, B. C. 2009. Estimates of the Unauthorized Immigrant Population Residing in the United States: January 2008. Washington, DC: U.S. Department of Homeland Security. Available online at: http://www.dhs.gov/xlibrary/assets/statistics/publications/ois_ill_pe_2008.pdf.

International Organization for Migration (IOM). 2005. *World Migration 2005: Costs and Benefits of International Migration*. Geneva: International Organization for Migration.

International Organization for Migration (IOM). 2008. *World Migration Report 2008: Managing Migration Mobility in the Evolving Global Economy*. Geneva: International Organization for Migration.

Juchno, P. 2007. Asylum Applications in the European Union. *Statistics in Focus: Population and Social Conditions. Luxembourg: Eurostat*. Available online at http://epp.eurostat.ec.europa.eu/cache/ITY_OFFPUB/KS-SF-07-110/EN/KS-SF-07-110-EN.PDF.

Kates, R. W. 1995. Labnotes from the Jeremiah Experiment: Hope for a Sustainable Transition. *Annals of the Association of American Geographers* 85(4): 623–640.

Kenya National AIDS Control Council. 2009. *Kenya: HIV Prevention Response and Modes of Transmission Analysis*. Nairobi: Kenya National AIDS Control Council.

Lee, E. S. 1966. A Theory of Migration. *Demography* 3(1): 47–57.

Lee, R. 2003. The Demographic Transition: Three Centuries of Fundamental Change. *Journal of Economic Perspectives* 17(4): 167–190.

Lopez, A., Mathers, C. D., Ezzati, M., Jamison, D., and Murray, C. J. L. 2006. Global and Regional Burden of Disease and Risk Factors, 2001. *The Lancet* 367: 1747–1757.

Malthus, T. 1998 [1798]. *An Essay on the Principle of Population*. Available from the Electronic Scholarly Publishing Project online at: http://www.esp.org.

Marcus, A. P. 2009. (Re)creating Places and Spaces in Two Countries: Brazilian Transnational Migration Processes. *Journal of Cultural Geography* 26(2): 173–198.

Méda, D., and Pailhé, A. 2008. Fertility: Is there a French Model? *The Japanese Journal of Social Security Policy* 7(2): 31–40.

Population Reference Bureau (PRB). 2009. *World Population Data Sheet: 2009*. Washington, DC: PRB.

Poulain, M., Pes, G. M., Grasland, C., Carru, C., Ferrucci, L., Baggio, G., Franceschi, C., and Deiana, L. 2004. Identification of a Geographic Area Characterized by Extreme Longevity in the Sardinia Island: The AKEA Study. *Experimental Gerontology* 39: 1423–1429.

Ratha, D., and Xu, Z., comps. 2008. *Migration and Remittances Factbook 2008*. Washington, DC: World Bank. Available online at: http://econ.worldbank.org.

Ravenstein, E. G. 1885. The Laws of Migration. *Journal of the Statistical Society of London* 48(2): 167–235.

Ravenstein, E. G. 1889. The Laws of Migration. *Journal of the Royal Statistical Society* 52(2): 241–305.

Roseman, C. C. 1971. Migration as a Spatial and Temporal Process. *Annals of the Association of American Geographers* 61(3): 589–598.

Simon, J. L. 1981. *The Ultimate Resource*. Princeton, NJ: Princeton University Press.

Simon, J. 1996. *The Ultimate Resource 2*. Rev. ed. Princeton, NJ: Princeton University Press.

Skinner, G. W., Henderson, M., and Jianhua, Y. 2000. China's Fertility Transition Through Regional Space: Using GIS and Census Data for a Spatial Analysis of Historical Demography. *Social Science History* 24(3): 613–652.

Suzuki, M., Willcox, C., and Willcox, B. 2007. The Historical Context of Okinawan Longevity: Influence of the United States and Mainland Japan. *The Okinawan Journal of American Studies* 4: 46-61

Tanner, A. 2005. Brain Drain and Beyond: Returns and Remittances of Highly Skilled Migrants. *Global Migration Perspectives No. 24*. Geneva: Global Commission on International Migration.

Torres, A. B. 2006. Colombian Migration to Europe: Political Transnationalism in the Middle of Conflict. *COMPAS Working Paper No. 39*. Oxford: Center on Migration, Policy and Society.

UC Atlas of Global Inequality. 2009. Available online at: http://ucatlas.ucsc.edu/cause.php.

United Nations Department of Economic and Social Affairs (UNDESA). 2009. World Population Prospects: The 2008 revision. Available online at: http://esa.un.org/unpp/index.asp.

United Nations High Commissioner for Refugees (UNHCR). 2009. Asylum Levels and Trends in Industrialized Countries 2008. Available online at: http://www.unhcr.org/49c796572.html.

United Nations Population Fund (UNFPA). 2008. *State of World Population 2008*. New York: UNFPA.

United Nations Program on HIV/AIDS (UNAIDS). 2008. *2008 Report on the Global AIDS Epidemic*. Geneva: UNAIDS.

United States Census Bureau, International Data Base. 2009. Available online at: http://www.census.gov/ipc/www/idb/index.php.

United States Department of Homeland Security 2009. *Yearbook of Immigration Statistics: 2008*. Washington, DC: U.S. Department of Homeland Security, Office of Immigration Statistics.

Watts, M. J. 1983. *Silent Violence: Food, Famine and Peasantry in Northern Nigeria*. Berkeley: University of California Press.

Watts, M., and Bohle, H. 1993. Hunger, Famine, and the Space of Vulnerability. *Geojournal* 30(2): 117–126.

World Bank. 1986. *Poverty and Hunger: Issues and Options for Food Security in Developing Countries*. A World Bank Policy Study. Washington, DC: World Bank.

World Bank. 2009. Migration and Remittances Data. Available online at: http://econ.worldbank.org.

Zhu, W. X., and Lu, L. 2009. China's Excess Males, Sex Selective Abortion, and One Child Policy: Analysis of Data from 2005 Intercensus Survey. *British Medical Journal* 338(18 April): 920–923.

Chapter 4

Ardila, A. Spanglish: An Anglicized Spanish Dialect. *Hispanic Journal of Behavioral Sciences* 27(1): 60–81.

Crystal, D. 1987. *The Cambridge Encyclopedia of Language*. Cambridge: Cambridge University Press.

Crystal, D. 1997. *English as a Global Language*. Cambridge: Cambridge University Press.

Delahunty, J. L. 2009. Yew Tree Toponyms and Their Connection to the Irish Landscape. *Focus on Geography* 51(4): 1–6.

de Swaan, A. 2001. *Words of the World: The Global Language System*. Cambridge: Polity Press.

Diamond, J., and Bellwood, P. 2003. Farmers and Their Languages: The First Expansions. *Science*, New Series 300(5619): 597–603.

Frawley, W. J., ed. 2003. *International Encyclopedia of Linguistics*. 2nd ed. Oxford: Oxford University Press.

Gamkrelidze, T. V., and Ivanov, V. V. 1990. The Early History of Indo-European Languages. *Scientific American* (March): 110–116.

Gordon, R. G., ed. 2005. *Ethnologue*. 15th ed. Dallas, TX: SIL International.

Graddol, D. 2004. The Future of Language. *Science* 303(5662): 1329–1331.

Greenberg, J. H. 1987. *Language in the Americas*. Stanford, CA: Stanford University Press.

Grounds, R. A. 2007. English Only, Native-Language Revitalization and Foreign Languages. *Anthropology News* (November): 6–7.

Hale, K., Krauss, M., Watahomigie, L. J., Yamamoto, A. Y., Craig, C., Jeanne, L. M., and England, N. C. 1992. Endangered Languages. *Languages* 68(1): 1–42.

Hinton, L., and Hale, K., eds. *The Green Book of Language Revitalization in Practice*. New York: Academic Press.

Jackson, J. B., and Linn, M. S. 2000. Calling in the Members: Linguistic Form and Cultural Context in a Yuchi Ritual Speech Genre. *Anthropological Linguistics* 42(1): 61–80.

Kurath, H. 1970 [1949]. *A Word Geography of the Eastern United States*. Ann Arbor: University of Michigan Press.

Language Hotspots Project. Living Tongues Institute for Endangered Languages. http://www.livingtongues.org/hotspots.html.

Laponce, J. A. 1987. *Languages and Their Territories*. Translated by A. Martin-Sperry. Toronto: University of Toronto Press.

Leutwyler, K. 2000. Preserving the Yuchi Language. *Scientific American*. December 12. http://www.scientificamerican.com/article.cfm?id=preserving-the-yuchi-lang.

Lewis, M. P., ed. 2009. *Ethnologue: Languages of the World*. 16th ed. Dallas, TX: SIL International.

Lieberson, S. 1981. *Language Diversity and Language Contact*. Compiled by A. S. Dil. Stanford, CA: Stanford University Press.

Linn, M. S. 2000. A Grammar of Euchee (Yuchi). Ph.D. diss., University of Kansas.

Mackey, W. F. 1991. Language Diversity, Language Policy and the Sovereign State. *History of European Ideas* 13(1–2): 51–61.

Maurais, J., and Morris, M. A., eds. 2003. *Languages in a Globalizing World*. Cambridge: Cambridge University Press.

Mazrui, A. A., and Mazrui, A. M. 1998. *The Power of Babel: Language and Governance in the African Experience*. Oxford: James Curry.

McArthur, T. 1998. *The English Languages*. Cambridge: Cambridge University Press.

Mufwene, S. S., Rickford, J. R., Bailey, G., and Baugh, J., eds. 1998. *African-American English: Structure, History and Use*. London: Routledge.

Nettle, D. 1998. Explaining Global Patterns of Language Diversity. *Journal of Anthropological Archeology* 17: 354–374.

Peterson, C. B. 1977. The Nature of Soviet Place-Names. *Names* 25: 15–24.

Renfrew, C. 1989. The Origins of Indo-European Languages. *Scientific American* 261(4): 106–114.

Ruhlen, M. 1987. *A Guide to the World's Languages*. Vol. 1, *Classification*. Stanford, CA: Stanford University Press.

Sappenfield, M., and Joshi, S. 2006. Tear Up the Maps: India's Cities Shed Colonial Names. *Christian Science Monitor*, September 7.

Schmemann, S. 1991. Leningrad, Petersburg and the Great Name Debate. *New York Times*, June 13.

Tonkin, H., and Reagan, T., eds. 2003. *Language in the Twenty-First Century: Selected Papers of the Millennial Conferences of the Center for Research and Documentation on World Language Problems*. Amsterdam: John Benjamins Publishing Company.

Web Atlas of Oklahoma. 2005. Available online at: http://www.okatlas.org/.
Williams, C. H., ed. 1988. *Language in Geographic Context*. Philadelphia: Multilingual Matters.
Wolfram, W. 1991. *Dialects and American English*. Englewood Cliffs, NJ: Prentice Hall.

Chapter 5

Barrett, D. B., and Johnson, T. M. 2001. *World Christian Trends, AD 30–AD 2200: Interpreting the Annual Christian Megacensus*. Pasadena, CA: William Carey Library.
Barrett, D. B., Johnson, T. M., and Crossing, P. F. 2005. Missiometrics 2005: A Global Survey of World Mission. *International Bulletin of Missionary Research* 29(1): 27–30.
Bellah, R. N. 2005 [1967]. Civil Religion in America. *Daedalus* (Fall): 40–55.
Bhardwaj, S. M. 1998. Non-Hajj Pilgrimage in Islam: A Neglected Dimension of Religious Circulation. *Journal of Cultural Geography* 17(2): 69–87.
Bogan, J. 2009. America's Biggest Megachurches. *Forbes*, June 26.
Breuilly, E., O'Brien, J., and Palmer, M. 1997. *Religions of the World: The Illustrated Guide to Origins, Beliefs, Traditions & Festivals*. New York: Facts on File, Inc.
Chatwin, B. 1987. *The Songlines*. New York: Penguin Books.
Encyclopaedia Britannica, Inc., and NetLibrary, Inc. 2008. *Britannica Book of the Year 2008*. Chicago: Encyclopædia Britannica.
Foote, K. E. 1992. Stigmata of National Identity: Exploring the Cosmography of America's Civil Religion. In *Person, Place, and Thing: Interpretive and Empirical Essays in Cultural Geography*. Geoscience and Man, vol. 31, ed. S. T. Wong, 379–402. Baton Rouge: Department of Geography and Anthropology, Louisiana State University.
Foote, K. E. 1997. *Shadowed Ground: America's Landscapes of Violence and Tragedy*. Austin: University of Texas Press.
Francaviglia, R. V. 1971. The Cemetery as an Evolving Cultural Landscape. *Annals of the Association of American Geographers* 61(3): 501–509.
Griswold, E. 2008. God's Country. *The Atlantic Monthly* (March): 40–55.
Jenkins, P. 2002. The Next Christianity. *The Atlantic Monthly* (October): 53–68.
Johnson, T. M., and Chung, S. Y. 2004. Tracking Global Christianity's Statistical Centre of Gravity, AD 33–AD 2100. *International Review of Mission* 93(369): 166–181.
Kong, L. 2001. Mapping "New" Geographies of Religion: Politics and Poetics in Modernity. *Progress in Human Geography* 25(2): 211–233.
Lehren, A., and Ericson, M. 2007. Where Megachurches Are Concentrated. *New York Times*, October 23.
Park, C. C. 1994. *Sacred Worlds: An Introduction to Geography and Religion*. London: Routledge.
Pew Research Center. 2006. *Spirit and Power: A 10-Country Survey of Pentecostals*. Washington, DC: Pew Forum on Religion & Public Life. Available online at: http://religions.pewforum.org.
Pew Research Center. 2008. *U.S. Religious Landscape Survey*. Washington, DC: Pew Forum on Religion & Public Life. Available online at: http://religions.pewforum.org.
Pew Research Center. 2009. Religious Groups' Views on Global Warming. April 16. Washington, DC: Pew Forum on Religion & Public Life. Available online at: http://religions.pewforum.org.
Rinehart, R., ed. 2004. *Contemporary Hinduism: Ritual, Culture, and Practice*. Santa Barbara, CA: ABC-CLIO.
Rose, D. 1996. *Nourishing Terrains: Australian Aboriginal Views of Landscape and Wilderness*. Canberra: Australian Heritage Commission.
Smart, N., ed. 1999. *Atlas of the World's Religions*. Oxford: Oxford University Press.
Sopher, D. E. 1967. *Geography of Religions*. Englewood Cliffs, NJ: Prentice-Hall.
Stahnke, T., and Blitt, R. C. 2005. *The Religion-State Relationship and the Right to Freedom of Religion or Belief: A Comparative Textual Analysis of the Constitutions of Predominantly Muslim Countries*. Washington, DC: United States Commission on International Religious Freedom. Available online at: http://uscirf.gov.
Stoddard, R. H., and Morinis, A., eds. 1997. *Sacred Places, Sacred Spaces: The Geography of Pilgrimages*. Geoscience and Man, vol. 34. Baton Rouge: Department of Geography and Anthropology, Louisiana State University.
Stump, R. W. 2008. *The Geography of Religion: Faith, Place, and Space*. Lanham, MD: Rowman & Littlefield.
Teather, E. K. 1998. Themes from Complex Landscapes: Chinese Cemeteries and Columbaria in Urban Hong Kong. *Australian Geographical Studies* 36(1): 21–36.
Teather, E. K. 1999. High-Rise Homes for the Ancestors: Cremation in Hong Kong. *Geographical Review* 89(3): 409–430.
Whyte, L. 1971 [1967]. The Historical Roots of Our Ecological Crisis. In *Man's Impact on Environment*, ed. T. R. Detwyler, 27–35. New York: McGraw-Hill Book Co.
Zelinsky, W. 1994. Gathering Places for America's Dead: How Many, Where, and Why? *Professional Geographer* 46(1): 29–38.
Zelinsky, W. 2001. The Uniqueness of the American Religious Landscape. *Geographical Review* 91(3): 565–585.

Chapter 6

Airriess, C. A. 2007. Conflict Migrants from Mainland Southeast Asia. In *Contemporary Ethnic Geographies in America*, eds. I. M. Miyares and C. A. Airriess, 291–312. Lanham, MD: Rowman & Littlefield.
Anderson, K. J. 1987. The Idea of Chinatown: The Power of Place and Institutional Practice in the Making of a Racial Category. *Annals of the Association of American Geographers* 77(4): 580–597.
Australian Bureau of Statistics. 2009. Experimental Estimates of Aboriginal and Torres Strait Islander Australians, June 2006. Available online at: http://www.abs.gov.au.
Bonnet, C. 1779–1783. *Oeuvres d'histoire naturelle et de philosophie de Charles Bonnet*. 18 vols. Neuchatel: S. Fauche.
Bullard, R. D. 2000. *Dumping in Dixie: Race, Class, and Environmental Quality*. 3rd ed. Boulder, CO: Westview Press.
Christopher, A. J. 1994. *The Atlas of Apartheid*. London: Routledge.
Central Intelligence Agency (CIA) 1979. *Racial Concentrations and Homelands* [map]. Scale not given. N.p.: CIA. Available online at http://www.lib.utexas.edu/maps/africa/south_africa_racial_1979.jpg
Eltis, D., and Richardson, D. 2009. *An Atlas of the Transatlantic Slave Trade*. New Haven, CT: Yale University Press.
Gans, H. 1979. Symbolic Ethnicity: The Future of Ethnic Groups and Cultures in America. *Ethnic and Racial Studies* 2(1): 1–20.
Hattam, V. 2005. Ethnicity & the Boundaries of Race: Rereading Directive 15. *Daedalus* (Winter): 61–69.

Human Rights Campaign (HRC). 2010a. Marriage Equality & Other Relationship Recognition Laws. Available online at: http://www.hrc.org/state_laws.
Human Rights Campaign (HRC). 2010b. Statewide Marriage Prohibition Laws. Available online at: http://www.hrc.org/state_laws.
International Labor Organization (ILO). 2005. *A Global Alliance against Forced Labour*. Geneva: ILO.
International Labor Organization (ILO). 2008. *Global Employment Trends*. Geneva: ILO.
Levi, J., and Maybury-Lewis, B. 2010. Becoming Indigenous: Identity and Heterogeneity in a Global Movement. In *Indigenous Peoples, Poverty and Development*, eds. G. Hall and H. Patrinos, 1–44. Available online at: http://siteresources.worldbank.org/EXTINDPEOPLE/Resources/407801-1271860301656/Chapter_2_Becoming_Indigenous.pdf.
Li, W. 1998. Anatomy of a New Ethnic Settlement: The Chinese Ethnoburb in Los Angeles. *Urban Studies* 35(3): 479–501.
Longhurst, R. 2000. Geography and Gender: Masculinities, Male Identity and Men. *Progress in Human Geography* 24(3): 439–444.
Mahgoub, E-T. M. 2004. Inside Darfur: Ethnic Genocide by a Governance Crisis. *Comparative Studies of South Asia, Africa and the Middle East* 24(2): 3–17.
Massey, D. 1994. *Space, Place, and Gender*. Minneapolis: University of Minnesota Press.
McDowell, L., and Sharp, J. P., eds. 1997. *Space, Gender, Knowledge: Feminist Readings*. London: Hodder Arnold.
Merchant, C. 2003. Shades of Darkness: Race and Environmental History. *Environmental History* 8(July): 380–394.
Miyares, I. M., and Airriess, C. A., eds. 2007. *Contemporary Ethnic Geographies in America*. Lanham, MD: Rowman & Littlefield.
Morning, A. 2008. Ethnic Classification in Global Perspective: A Cross-National Survey of the 2000 Census Round. *Population Research and Policy Review* 27(2): 239–272.
Pastor, M., Sadd, J., and Morello-Frosch, R. 2007. *Still Toxic after All These Years: Air Quality and Environmental Justice in the San Francisco Bay Area*. Santa Cruz: Center for Justice, Tolerance & Community, University of California, Santa Cruz.
Prewitt, K. 2005. Racial Classification in America: Where Do We Go from Here? *Daedalus* (Winter): 5–17.
Pulido, L. 2000. Rethinking Environmental Racism: White Privilege and Urban Development in Southern California. *Annals of the Association of American Geographers* 90(1): 12–40.
Reddy, G. 2005. *With Respect to Sex: Negotiating Hijra Identity in South India*. Chicago: University of Chicago Press.
Sciorra, J. 1989. Yard Shrines and Sidewalk Altars of New York's Italian-Americans. *Perspectives in Vernacular Architecture* 3: 185–198.
Silvey, R. 2004. Transnational Domestication: State Power and Indonesian Migrant Women in Saudi Arabia. *Political Geography* 23: 245–264.
Statistics South Africa 2008. *Labor Force Survey, Historical Revision, March Series*. Pretoria: Statistics South Africa.
Strauss, S. 2005. Darfur and the Genocide Debate. *Foreign Affairs* 84(1): 123–133.
Suchan, T. A., Perry, M. J., Fitzsimmons, J. D., Juhn, A. E., Tait, A. M., and Brewer, C. A. 2007. *Census Atlas of the United States*. Series CENSR-29. Washington, DC: U.S. Census Bureau.
Townsend, J. 1991. Towards a Regional Geography of Gender. *The Geographical Journal* 157(1): 25–35.
U.S. Census Bureau. 2000a. American FactFinder, Census Tract 2132.01, Los Angeles County, California. Available online at: http://factfinder.census.gov.
U.S. Census Bureau. 2000b. American FactFinder, Census 2000 Redistricting Data Summary File, Census Tract 2132.01, Los Angeles County, California. Available online at: http://factfinder.census.gov.
U.S. Census Bureau. 2000c. *Ancestry 2000 Census Brief*. Available online at: http://www.census.gov.
U.S. Census Bureau. 2000d. Census Tract Outline Map: Los Angeles County, California. Available online at: http://www.census.gov.
U.S. Census Bureau 2010. *The 2010 Statistical Abstract: The National Data Book*. Table 685, Median Income of People with Income in Constant (2007) Dollars by Sex, Race, and Hispanic Origin: 1990 to 2007. Available online at: http://www.census.gov/compendia/statab.
Vestal, C. 2009. Gay Marriage Legal in Six States. Updated June 4. Available online at: http://www.stateline.org.
Wood, J. 1997. Vietnamese American Place Making in Northern Virginia. *Geographical Review* 87(1): 58–72.
World Economic Forum. 2007. *The Global Gender Gap Report 2007*. Geneva: World Economic Forum.
Zelinsky, W., and Lee, B. 1998. Heterolocalism: An Alternative Model of the Sociospatial Behavior of Immigrant Ethnic Communities. *International Journal of Population Geography* 4: 281–298.

Chapter 7

Agnew, J. 1998. *Geopolitics: Re-visioning World Politics*. 2nd ed. New York: Routledge.
Chan, P. C. W. 2009. The Legal Status of Taiwan and the Legality of the Use of Force in a Cross-Taiwan Strait Conflict. *Chinese Journal of International Law* 8(2): 455–492.
Child, J. 2005. The Politics and Semiotics of the Smallest Icons of Popular Culture: Latin American Postage Stamps. *Latin American Research Review* 40(1): 108–137.
Council of the European Union. 2009. *Independent International Fact-Finding Mission on the Conflict in Georgia Report*. 3 vols. Brussels: Council of the European Union. Available online at: http://www.ceiig.ch.
Farrell, D. M. 2001. *Electoral Systems: A Comparative Introduction*. New York: Palgrave.
Flint, C., and Radil, S. M. 2009. Terrorism and Counter-Terrorism: Situating al-Qaeda and the Global War on Terror within Geopolitical Trends and Structures. *Eurasian Geography and Economics* 50(2): 150–171.
Guibernau, M. 2000. Spain: Catalonia and the Basque Country. *Parliamentary Affairs* 53(1): 55–68.
Hartshorne, R. 1950. The Functional Approach in Political Geography. *Annals of the Association of American Geographers* 40(2): 95–130.
Huntington, S. P. 1997. *The Clash of Civilizations and the Remaking of the World Order*. New York: Simon & Schuster, 1996; reprint, New York: Touchstone.
Jackson, R. 1999. Sovereignty in World Politics: A Glance at the Conceptual and Historical Landscape. *Political Studies* 47: 431–456.
Mackinder, H. J. 1942 [1919]. *Democratic Ideals and Reality: A Study in the Politics of Reconstruction*. Reprint, with a new introduction by S. Mladineo, Washington, DC: National Defense University Press Publications, 1996.

McRoberts, K. 2001. Canada and the Multinational State. *Canadian Journal of Political Science / Revue canadienne de science politique* 34(4): 683–713.

Morrelli, V., and Migdalovitz, C. 2009. European Union Enlargement: A Status Report on Turkey's Accession Negotiations. *Congressional Research Service* Report RS22517. Available online at: http://www.crs.gov.

Muir, R. 1997. *Political Geography: A New Introduction*. New York: John Wiley & Sons.

Philpott, D. 1995. Sovereignty: An Introduction and Brief History. *Journal of International Affairs* 48(2): 353–368.

Philpott, D. 1999. Westphalia, Authority, and International Society. *Political Studies* 47: 566–589.

Rosendorff, B. P., and Sandler, T. 2005. The Political Economy of Transnational Terrorism. *Journal of Conflict Resolution* 49(2): 171–182.

Shelley, F. M., Archer, J. C., Davidson, F. M., and Brunn, S. D. 1996. *Political Geography of the United States*. New York: Guilford.

Shughart, W. F. 2006. An Analytical History of Terrorism, 1945–2000. *Public Choice* 128(1–2): 7–39.

Smith, A. D. 1987. *The Ethnic Origins of Nations*. New York: Blackwell.

Symonds, P., Alcock, M., and French, C. 2009. Setting Australia's Limits. *AUSGEO News* 93(March): 1–8.

Chapter 8

Abu-Lughod, J. 1987. The Islamic City—Historic Myth, Islamic Essence, and Contemporary Relevance. *International Journal of Middle East Studies* 19(2): 155–176.

Acioly, C. 2007. The Challenge of Slum Formation in the Developing World. *Land Lines* 19(2): 1–6. Available online at: http://www.lincolninst.edu.

Adams, J. S. 1970. Residential Structure of Midwestern Cities. *Annals of the Association of American Geographers* 60(1): 37–62.

Beaverstock, J. V., and Taylor, P. J. 1999. A Roster of World Cities. *Cities* 16(6): 445–458.

Bonine, M. E. 1990. The Sacred Direction and City Structure: A Preliminary Analysis of the Islamic Cities of Morocco. *Muqarnas, An Annual on Islamic Art and Architecture* 7: 50–72.

Burgess, E. W. 1925. The Growth of the City. In *The City*. University of Chicago Studies in Urban Sociology, eds. R. E. Park, E. W. Burgess, R. D. McKenzie, and L. Wirth, 47–62. Chicago: University of Chicago Press.

Congress for the New Urbanism. 1996. Charter for the New Urbanism. Available online at: http://www.cnu.org/charter.

Ford, L. 1996. A New and Improved Model of Latin American City Structure. *Geographical Review* 86(3): 437–440.

Ford, L., and Griffin, E. 1979. The Ghettoization of Paradise. *Geographical Review* 69(2): 140–158.

Getis, A., and Getis, J. 1966. Christaller's Central Place Theory. *Journal of Geography* May: 220–226.

Grant, R., and Nijman, J. 2002. Globalization and the Corporate Geography of Cities in the Less-Developed World. *Annals of the Association of American Geographers* 92(2): 320–340.

Griffin, E., and Ford, L. 1980. A Model of Latin American City Structure. *Geographical Review* 70(4): 397–422.

Harris, C. D., and Ullman, E. L. 1945. The Nature of Cities. *Annals of the American Academy of Political and Social Science* 242(November): 7–17.

Hartshorn, T. 1992. *Interpreting the City: An Urban Geography*. 2nd ed. New York: John Wiley and Sons.

Hoyt, H. 1939. *The Structure and Growth of Residential Neighborhoods in American Cities*. Washington, DC: Federal Housing Administration.

Jackson, K. T. 1985. *Crabgrass Frontier: The Suburbanization of the United States*. New York: Oxford University Press.

Kaplan, D. H., Wheeler, J. O., and Holloway, S. R. 2004. *Urban Geography*. New York: John Wiley and Sons.

Kreja, K. 2006. Spatial Imprints of Urban Consumption: Large-Scale Retail Development in Warsaw. In *The Urban Mosaic of Post-Socialist Europe: Space, Institutions and Policy*, eds. S. Tsenkova and Z. Nedovic-Budic, 253–272. New York: Physica-Verlag.

Larsen, K., and Gilliland, J. 2008. Mapping the Evolution of "Food Deserts" in a Canadian City: Supermarket Accessibility in London, Ontario, 1961–2005. *International Journal of Health Geographics* 7: 16. Available online at: http://www.ij-healthgeographics.com/content/7/1/16.

Morrill, R. 2006. Classic Map Revisited: The Growth of Megalopolis. *Professional Geographer* 58(2): 155-161.

Mutlu, S. 1989. Urban Concentration and Primacy Revisited: An Analysis and Some Policy Conclusions. *Economic Development and Cultural Change* 37(3): 611–639.

Palca, J. 2008. Abu Dhabi Aims to Build First Carbon-Neutral City. National Public Radio, May 6. Available online at: http://www.npr.org/templates/story/story.php?storyId=90042092.

Park, R. E., Burgess, E. W., McKenzie, R. D., and Wirth, L., eds. 1925. *The City*. University of Chicago Studies in Urban Sociology. Chicago: University of Chicago Press.

Short, J. R. 2004. *Global Metropolitan: Globalizing Cities in a Capitalist World*. London: Routledge.

Taylor, P. J. 2001. Urban Hinterworlds: Geographies of Corporate Service Provision Under Conditions of Contemporary Globalization. *Geography* 86(1): 51–60.

United Nations Human Settlements Program 2003. *The Challenge of Slums: Global Report on Human Settlements 2003*. London: Earthscan.

United Nations Human Settlements Program 2008. *State of the World's Cities 2008/2009: Harmonious Cities*. London: Earthscan.

United Nations Population Division. 2008. *World Urbanization Prospects: The 2007 Revision, Highlights*. New York: United Nations.

Vance, J. E. 1964. *Geography and Urban Evolution in the San Francisco Bay Area*. Berkeley: Institute of Government Studies, University of California.

Chapter 9

Bernanke, B. S. 2007. Chairman of the Board of Governors of the Federal Reserve System. *The Level and Distribution of Economic Well-Being*. Address, Greater Omaha Chamber of Commerce, Omaha, NE, February 6.

Blaikie, P. 2000. Development, Post-, Anti-, and Populist: A Critical Review. *Environment and Planning A* 32: 1033–1050.

Brandt, W. 1980. *North-South: A Programme for Survival: Report of the Independent Commission on International Development Issues*. Cambridge, MA: MIT Press.

Dicken, P. 1998. *Global Shift: Transforming the World Economy*. 3rd ed. New York: Guilford Press.

Dickenson, J., Gould, B., Clarke, C., Mather, S., Prothero, M., Siddle, D., Smith, C., and Thomas-Hope, E. 1996. *A Geography of the Third World*. 2nd ed. London: Routledge.
Dikhanov, Y. 2005. *Trends in Global Income Distribution, 1970–2000, and Scenarios for 2015*. Occasional Paper, Human Development Report, 2005. New York: United Nations Development Program.
International Monetary Fund (IMF). 2006. *Albania: Poverty Reduction Strategy Paper Annual Progress Report*. IMF Country Report 06/23. Washington, DC: IMF.
Mistiaen, J. A., Özler, B., Razafimanantena, T., and Razafindravonoma, J. 2002. Putting Welfare on the Map in Madagascar. Africa Region Working Paper Series, No. 34. Washington, DC: World Bank.
Paternostro, S., Razafindravonona, J., and Stifel, D. 2001. Changes in Poverty in Madagascar 1993–1999. Africa Region Working Paper Series, No. 19. Washington, DC: World Bank.
Porter, P. W., and Sheppard, E. S. 1998. *A World of Difference: Society, Nature, Development*. New York: Guilford Press.
Revkin, A. C. 2005. A New Measure of Well-Being from a Happy Little Kingdom. *New York Times*, October 4.
Rostow, W. W. 1971. *The Stages of Economic Growth: A Non-Communist Manifesto*. 2nd ed. Cambridge, MA: Cambridge University Press.
Sachs, J. D., Mellinger, A. D., and Gallup, J. L. 2001. The Geography of Poverty and Wealth. *Scientific American* 284(3): 70–76.
Slater, D. 1997. Geopolitical Imaginations across the North-South Divide: Issues of Difference, Development, and Power. *Political Geography* 16(8): 631–653.
Smith, N. 1984. *Uneven Development: Nature, Capital, and the Production of Space*. New York: Blackwell.
Taylor, P. J. 1992. Understanding Global Inequalities: A World-Systems Approach. *Geography* 77: 10–21.
Therien, J. 1999. Beyond the North-South Divide: The Two Tales of World Poverty. *Third World Quarterly* 20(4): 723–742.
Organization for Economic Cooperation and Development (OECD). 2008. *OECD Annual Report*. Paris: OECD.
United Nations Development Program (UNDP). 2005. *Human Development Report: International Cooperation at a Crossroads: Aid, Trade and Security in an Unequal World*. New York: UNDP.
United Nations Development Program (UNDP). 2007. *Human Development Report 2007/2008: Fighting Climate Change: Human Solidarity in a Divided World*. New York: Palgrave Macmillan.
United Nations Environment Program and South Pacific Applied Geoscience Commission 2005. *Building Resilience in SIDS: The Environmental Vulnerability Index*. N.p.: UNEP and SOPAC. Available online at: http://islands.unep.ch/EVI%20Final%20Report%202005.pdf.
Wallerstein, I. 1974. *The Modern World-System*. 3 vols. New York: Academic Press.
World Bank. 2008. *Poverty Data: A Supplement to World Development Indicators 2008*. Washington, DC: World Bank.
World Bank. 2009. Online Atlas of the Millennium Development Goals. http://devdata.worldbank.org/atlas-mdg/.
Zurick, D. 2006. Gross National Happiness and Environmental Status in Bhutan. *Geographical Review* 96(4): 657–681.

Chapter 10

Bairoch, P. 1982. International Industrialization Levels from 1750 to 1980. *Journal of European Economic History* 11: 269–310.
Benko, G. 2000. Technopoles, High-Tech Industries and Regional Development: A Critical Review. *GeoJournal* 51(3): 157–167.
Berry, B. J. L., Conkling, E. C., and Ray, D. M. 1997. *The Global Economy in Transition*. 2nd ed. Upper Saddle River, NJ: Prentice Hall.
Dicken, P. 1998. *Global Shift: Transforming the World Economy*. 3rd ed. New York: Guilford Press.
Engman, M., Onodera, O., and Pinali, E. 2007. Export Processing Zones: Past and Future Role in Trade and Development, *OECD Trade Policy Working Papers*, No. 53. Paris: OECD Publishing.
Gereffi, G., and Memedovic, O. 2003. *The Global Apparel Value Chain: What Prospects for Upgrading by Developing Countries?* Vienna: UNIDO Strategic Research and Economics Branch.
Hayter, R. 1997. *The Dynamics of Industrial Location: The Factory, the Firm and the Production System*. New York: John Wiley & Sons.
Ho, A. L. 2007. Breaking into the Tennis World a Ball at a Time. *Philippines Daily Inquirer*, May 20. Available online at: http://business.inquirer.net/money/features/view/20070520-67042/Breaking_into_the_tennis_world_a_ball_at_a_time.
Innis, H. A. 1933. *Problems of Staple Production in Canada*. Toronto: University of Toronto Press.
International Fur Trade Federation (IFTF). n.d. *The Socio-Economic Impact of International Fur Farming*. Available online at: http://www.iftf.com.
International Labor Organization (ILO). 2007. Equality at Work: Tackling the Challenges. Report of the Director General, International Labor Conference. 96th Session 2007. Geneva: ILO. Available online at: http://www.ilo.org.
International Labor Organization (ILO). 2009. *Global Employment Trends for Women*. Geneva: ILO.
Linden, G., Kraemer, K. L., and Dedrick, J. 2007. *Who Captures Value in a Global Innovation System? The Case of Apple's iPod*. Irvine, CA: Personal Computing Industry Center.
Mackintosh, W. A. 1923. Economic Factors in Canadian History. *Canadian Historical Review* 4: 12–25.
Maddison, A. 2001. *The World Economy: A Millennial Perspective*. Paris: OECD Development Center Studies.
Maddison, A. 2003. *The World Economy: Historical Statistics*. Paris: OECD Development Center Studies.
Pilat, D., Cimper, A., Olsen, K., and Webb, C. 2006. The Changing Nature of Manufacturing in OECD Economies. *STI Working Paper 2006/9*. Paris: OECD.
Rowthorn, R., and Ramaswamy, R. 1999. Growth, Trade, and Deindustrialization. *IMF Staff Papers* 46(1): 18–41.
Skov, L. 2005. The Return of the Fur Coat: A Commodity Chain Perspective. *Current Sociology* 53(1): 9–32.
Stearns, P. N. 2007. *The Industrial Revolution in World History*. 3rd ed. Boulder, CO: Westview Press.
United Nations Conference on Trade and Development (UNCTAD). 2008. *Development and Globalization: Facts and Figures*. New York: UNCTAD.
United Nations Development Program (UNDP). 2007. *Human Development Report 2007/2008: Fighting Climate Change: Human Solidarity in a Divided World*. New York: Palgrave Macmillan.
United Nations Industrial Development Organization (UNIDO). 2005. Capability Building for Catching-Up: Historical, Empirical and Policy Dimensions. Vienna: UNIDO.
United Nations Industrial Development Organization (UNIDO). 2009. *Breaking In and Moving Up: New Industrial Challenges for the Bottom Billion and the Middle-Income Countries*. Vienna: UNIDO.

United States Agency for International Development (USAID) 2000. *Latin America and the Caribbean: Selected Economic and Social Data, 1999*. Retrieved from LexisNexis Statistical.
World Trade Organization (WTO). 2002. *Annual Report 2002*. Geneva: WTO.
Yeung, Y., Lee, J., and Kee, G. 2009. China's Special Economic Zones at 30. *Eurasian Geography and Economics* 50(2): 222–240.

Chapter 11

Angelsen, A. 2007. Forest Cover Change in Space and Time: Combining the von Thunen and Forest Transition Theories. World Bank Policy Research Working Paper 4117. Washington, DC: World Bank. Available online at: http://econ.worldbank.org.
Bauman, P. R. 2008. Grand Prairie of Illinois: Cash Grain Farming. *Geocarto International* 23(3): 235-244.
Borlaug, N. E., and Dowswell, C. 2004. The Green Revolution: An Unfinished Agenda. CFS Distinguished Lecture Series. Committee on World Food Security, Rome, 20-23 September. Available online at: http://www.fao.org/docrep/meeting/008/J3205e/j3205e00.htm.
Chisholm, M. 1968. *Rural Settlement and Land Use: An Essay in Location*. 2nd rev. ed. London: Hutchinson University Library.
Clement, C. R. 1989. A Center of Crop Genetic Diversity in Western Amazonia. *BioScience* 39(9): 624-631.
Cowan, C. W., and Watson, P. J., eds. 1992. *The Origins of Agriculture: An International Perspective*. Washington, DC: Smithsonian Institution Press.
Evenson, R. E., and Gollin, D. 2003. Assessing the Impact of the Green Revolution, 1960 to 2000. *Science* 300(5620): 758–762.
FAOSTAT 2010. Milk Production, 2008. Available online at: http://faostat.fao.org/site/339/default.aspx. Last accessed May 5, 2010.
Goodman, M. K. 2004. Reading Fair Trade: Political Ecological Imaginary and the Moral Economy of Fair Trade Foods. *Political Geography* 23: 891–915.
Grimes, K. M. 2005. Changing the Rules of Trade with Global Partnerships: The Fair Trade Movement. In *Social Movements: An Anthropological Reader*, ed. J. Nash. Malden, MA: Blackwell Publishing, 237–248.
Hart, J. F. 1998. *The Rural Landscape*. Baltimore, MD: Johns Hopkins University Press.
Hart, J. F. 2003. *The Changing Scale of American Agriculture*. Charlottesville, VA: University of Virginia Press.
Hawkes, C. 2006. Uneven Dietary Development: Linking the Policies and Processes of Globalization with the Nutrition Transition, Obesity and Diet-Related Chronic Diseases. *Globalization and Health*. Available online at: http://www.globalizationandhealth.com/content/2/1/4.
International Food Policy Research Institute (IFPRI). 2002. Green Revolution: Curse or Blessing? Washington, DC: IFPRI. Available online at: http://www.ifpri.org.
International Labor Organization (ILO). 2008. *Global Employment Trends: January 2008*. Geneva: ILO.
International Service for the Acquisition of Agri-Biotech Applications (ISAAA). 2008. *Global Status of Commercialized Biotech/GM Crops: 2008*. ISAAA Brief 39-2008. Available online at: http://www.isaaa.org/default.asp.
Malcolm, S. A., Aillery, M., and Weinberg, M. 2009. *Ethanol and a Changing Agricultural Landscape*. Economic Research Report 86. U.S. Department of Agriculture, Economic Research Service. Available online at: http://www.ers.usda.gov.
Mannion, A. M. 1999. Domestication and the Origins of Agriculture: An Appraisal. *Progress in Physical Geography* 23(1): 37–56.
Mitchell, D. 2008. A Note on Rising Food Prices. Policy Research Working Paper 4682. Washington, DC: World Bank, Development Prospects Group. Available online at: http://econ.worldbank.org.
Nair, P. K. R. 1993. *An Introduction to Agroforestry*. Dordrecht: Kluwer Academic Publishers in cooperation with International Center for Research in Agroforestry.
O'Brien, D. 2010. U.S. Grain Supply-Demand Projections for 2010 from the USDA Agricultural Outlook Forum. Manhattan, KS: Kansas State University Research and Extension. Available online at: http://www.agmanager.info/marketing/outlook/newletters/archives/
Parayil, G. 2003. Mapping Technological Trajectories of the Green Revolution and the Gene Revolution from Modernization to Globalization. *Research Policy* 32: 971–990.
Pingali, P. 2006. Westernization of Asian Diets and the Transformation of Food Systems: Implications for Research and Policy. *Food Policy* 32(2006): 281–298.
Piperno, D. R., and Pearsall, D. M. 1998. *The Origins of Agriculture in the Lowland Neotropics*. San Diego, CA: Academic Press.
Robbins, P. 2004. *Political Ecology: A Critical Introduction*. Malden, MA: Blackwell Publishing.
Smith, B. D. 1994. Origins of Agriculture in the Americas. *Evolutionary Anthropology* 3(5): 174–184.
Spencer, J. E., and Stewart, N. R. 1973. The Nature of Agricultural Systems. *Annals of the Association of American Geographers* 63(4): 529–544.
Sunderlin, W. D., and Resosudarmo, I. A. P. 1996. *Rates and Causes of Deforestation in Indonesia: Towards a Resolution of the Ambiguities*. CIFOR Occasional Paper No. 9. Jakarta: CIFOR.
Timmer, P. 1969. The Turnip, the New Husbandry, and the English Agricultural Revolution. *The Quarterly Journal of Economics* 83(3): 375–395.
Turner II, B. L., and Brush, S. B., eds. 1987. *Comparative Farming Systems*. New York: Guilford Press.
United States Agency for International Development (USAID). 2009. Global Food Insecurity and Price Increase Situation Report #1. Available online at: http://www.usaid.gov/our_work/humanitarian_assistance/foodcrisis/.
USDA, National Agricultural Statistics Service. 2001. *Agricultural Statistics, 2001*. Washington, D.C.: U.S. GPO. Available online at: http://www.nass.usda.gov/Publications/Ag_Statistics/index.asp.
Willer, H., and Klicher, L., eds. 2009. *The World of Organic Agriculture: Statistics and Emerging Trends 2009*. Bonn, Germany and Frick, Switzerland: International Federation of Organic Agriculture Movements and Research Institute of Organic Agriculture. Available online at: http://www.organic-world.net.

Chapter 12

Blaikie, P., and Brookfield, H. 1987. *Land Degradation and Society*. London: Methuen & Co.
BP 2008. *BP Statistical Review of World Energy June 2008*. Available online at: http://www.bp.com/statisticalreview.
Cavallo, A. J. 2002. Predicting the Peak in World Oil Production. *Natural Resources Research* 11(3): 187–195.
Cavallo, A. J. 2004. Hubbert's Petroleum Production Model: An Evaluation and Implications for World Oil Production Forecasts. *Natural Resources Research* 13(4): 211–221.

Ecological Society of America. *Ecological Principles for Managing Land Use*. Available online at: http://www.esa.org/science_resources/publications/landUse.php.

Ecological Society of America. 1997. Ecosystem Services: Benefits Supplied to Human Societies by Natural Ecosystems. *Issues in Ecology* 2: 1–16.

Ellis, E. 2007. Land-Use and Land-Cover Change. In *Encyclopedia of the Earth*, ed. C. J. Cleveland. Washington, DC: Environmental Information Coalition, National Council for Science and the Environment. Available online at: http://www.eoearth.org/article/Land-use_and_land-cover_change.

Foley, J. A., DeFries, R., Asner, G. P., Barford, C., Bonan, G., Carpenter, S. R., Chapin, F. S., Coe, M. T., Daily, G. C., Gibbs, H. K., Helkowski, J. H., Holloway, T., Howard, E. A., Kucharik, C. J., Monfreda, C., Patz, J. A., Prentice, I. C., Ramankutty, N., and Snyder, P. K. 2005. Global Consequences of Land Use. *Science* 309: 570–574.

Gearheard, S. 2008. What Changes are Indigenous Peoples Observing in the State of Sea Ice? National Oceanic and Atmospheric Administration (NOAA) Arctic Theme Page. Available online at: http://www.arctic.noaa.gov/essay_gearheard.html.

Global Wind Energy Council (GWEC). 2008. *Global Wind 2007 Report*. Brussels: GWEC.

Greene, D. L, Hopson, J. L., and Li, J. 2006. Have We Run Out of Oil Yet? Oil Peaking Analysis from an Optimist's Perspective. *Energy Policy* 34: 515–531.

Hardin, G. 1968. The Tragedy of the Commons. *Science* New Series 162(3859): 1243–1248.

Hardin, G. 1985. Human Ecology: The Subversive, Conservative Science. *American Zoologist* 25(2): 469–476.

Hardin, G. 1998. Extensions of "The Tragedy of the Commons." *Science*. New Series 280 (5364): 682–683.

Intergovernmental Panel on Climate Change (IPCC) 2007. *Climate Change 2007: Synthesis Report*. Contribution of Working Groups I, II and III to the Fourth Assessment Report of the Intergovernmental Panel on Climate Change, eds. R.K Pachauri and A. Reisinger. Geneva: IPCC.

International Atomic Energy Agency (IAEA). 2008. *Energy, Electricity and Nuclear Power Estimates for the Period up to 2030*. Vienna: IAEA.

International Energy Agency (IEA). 2006. *World Energy Outlook 2006*. Paris: IEA.

International Energy Agency (IEA). 2007. *Key World Energy Statistics*. Paris: IEA.

Laidler, G. 2006. Inuit and Scientific Perspectives on the Relationship between Sea Ice and Climate Change: The Ideal Complement? *Climatic Change* 78: 407–444.

Le Treut, H., Somerville, R., Cubasch, U., Ding, Y., Mauritzen, C., Mokssit, A., Peterson, T., and Prather, M. 2007. Historical Overview of Climate Change. In: *Climate Change 2007: The Physical Science Basis. Contribution of Working Group I to the Fourth Assessment Report of the Intergovernmental Panel on Climate Change,* eds. S. Solomon, D. Qin, M. Manning, Z. Chen, M. Marquis, K. B. Avery, M. Tignor and H. L. Miller. Cambridge: Cambridge University Press

Lyles, L. D., and F. Namwamba. 2005. Louisiana Coastal Zone Erosion: 100+ Years of Landuse and Land Loss Using GIS and Remote Sensing. *Proceedings of 5th Annual ESRI Education User Conference*. Available online at: http://proceedings.esri.com/library/userconf/educ05/papers/pap1222.pdf

Mann, M. E., and Kump, L. R. 2009. *Dire Predictions: Understanding Global Warming*. New York: DK Publishing.

Meyer, W. B., and Turner II, B. L. 1994. *Changes in Land Use and Land Cover: A Global Perspective*. Cambridge: Cambridge University Press.

Moran, S. The Energy Challenge: Strangers as Allies. *New York Times*, October 20, 2007.

Mwangi, E., and Ostrom, E. 2009. Top-Down Solutions: Looking Up from East Africa's Rangelands. *Environment* 51(1): 34–44.

NEA and IAEA 2008. *Uranium 2007: Resources, Production and Demand*. Joint Publication of the OECD Nuclear Energy Agency (NEA) and the International Atomic Energy Agency (IAEA). Paris: OECD Publications.

Ostrom, E. 2008. The Challenge of Common-Pool Resources. *Environment Science and Policy for Sustainable Development* (July/August). Available online at: http://www.environmentmagazine.org/Archives/Back%20Issues/July-August%202008/ostrom-full.html.

Robbins, P. 2004. *Political Ecology: A Critical Introduction*. Malden, MA: Blackwell.

Soublière, M. 2006. Meltdown: Climate Change Hits Home. *Inuktitut* (Winter): 23–31.

Sutton, M. Q., and Anderson, E. N. 2004. *An Introduction to Cultural Ecology*. Walnut Creek, CA: AltaMira Press.

Turner II, B. L., Clark, W. C., Kates, R. W., Richards, J. F., Matthews, J. T., and Meyer, W. B. 1990. *The Earth as Transformed by Human Action: Global and Regional Changes in the Biosphere over the Past 300 Years*. Cambridge: Cambridge University Press.

Turner II, B. L., and Robbins, P. 2008. Land-Change Science and Political Ecology: Similarities, Differences, and Implications for Sustainability Science. *Annual Review of Environment and Resources* 33(8): 295–316.

United Nations 2003. *Integrated Economic and Environmental Accounting*. Handbook of National Accounting, Studies in Methods, Series F, no. 61. New York: United Nations.

United Nations Development Program (UNDP). 2007. *Human Development Report 2007/2008: Fighting Climate Change: Human Solidarity in a Divided World*. New York: Palgrave Macmillan.

United Nations Environmental Program (UNEP). 2007. *Dams and Development: Relevant Practices for Improved Decision-Making*. Nairobi: UNEP.

United Nations Framework Convention on Climate Change (UNFCC). Local Coping Strategies Database. Available online at: http://maindb.unfccc.int/public/adaptation/.

United Nations Framework Convention on Climate Change (UNFCC). Kyoto Protocol. Available online at: http://unfccc.int/kyoto_protocol/items/2830.php.

World Commission on Dams (WCD). 2000. *Dams and Development: A New Framework for Decision-Making*. London: Earthscan Publications.

Text, Table, and Line Art Credits

Chapter 1

Figure 1.3: *Lawn People: How Grasses, Weeds, and Chemicals Make Us Who We Are* by Paul Robbins. Adapted by permission of Temple University Press. © by Temple University. All Rights Reserved. Figure 1.5b: Adapted from *The Primary Market Area.* (2006). Media One Utah. Available online at http://www.nacorp.com/NAC2/adv/mktg/PDFS/primarymarketarea.pdf. There are instances where we have been unable to trace or contact the copyright holder. If notified, the publisher will be pleased to rectify any errors or omissions at the earliest opportunity. Figure 1.7: Courtesy of the University Libraries, The University of Texas at Austin. Figure 1.8: Adapted with permission granted by the Monitor Institute. Figure 1.9: From Sunderland, W., Dewi, S., and Puntodewo, A. Poverty and Forests: Multi-Country Analysis of Spatial Association and Proposed Policy Solutions. *CIFOR Occasional Paper No. 47.* Center for International Forestry Research, 2007, p. 25. Permission granted by CIFOR. Figure 1.13: Block, R. and Bernasco, W., 2009. Finding a Serial Burglar's Home Using Distance Decay and Condition Origin-Destination Patterns: A Test of Empirical Bayes Journey to Crime Estimation in the Hague. *Journal of Investigative Psychology and Offender Profiling* 6. Copyright © 2009. Reprinted with permission of John Wiley & Sons, Ltd.. Figure 1.16b: From Wiehe, S. E., Carroll, A. E., Liu, G. C., Haberkorn, K. L., Hoch, S. C., Wilson, J. S., and Fortenberry, J. D., 2008. Using GPS-enabled cell phones to track the travel patterns of adolescents. *International Journal of Health* Geographics 7 (22). Figure 1.19: From DeGroote, J. P., Sugumaran, R., Bend, S. M., Tucker, B. J., and Bartholomay, L. C. 2008. Landscape, demographic, entomological, and climatic associations with human disease incidence of West Nile virus in the state of Iowa, USA. *International Journal of Health Geographics* 7(19).

Chapter 2

Figure 2.4a, map: Adapted from Graff, T. O. and Ashton, D. (1994). Spatial Diffusion of Wal-Mart: Contagious and Hierarchical Elements. *Professional Geographer*, 1994, 46(1), p. 23. Taylor & Francis Group, http://www.informaworld.com. Figure 2.13b: *WHO Traditional Medicine Strategy 2002-2005*, Geneva, World Health Organization, 2002, p. 11. What a Geographer Sees Part a (page 56): Adapted from Salih, A. (2006). Qanats a Unique Groundwater Management Tool in Arid Regions: The Case of Bam Region in Iran. International Symposium on Groundwater Sustainability (ISGWAS). UNESCO. What a Geographer Sees Part b (page 56): Adapted from Lightfoot, D. L. (2000). The Origin and Diffusion of Qanats in Arabia: New Evidence from the Northern and Southern Peninsula. *The Geographical Journal* 166(3): 215-226. Copyright © 2000 Wiley-Blackwell. Reprinted with permission of John Wiley & Sons Ltd.

Chapter 3

Figure 3.4b: Adapted from Riley, N. E., 2004. China's Population: New Trends and Challenges. *Population Bulletin* 59(2), p. 11, a publication of the Population Reference Bureau, Washington, DC. Figure 3.10: Adapted from Infectious diseases kill 1/3 worldwide; AIDS is top cause of death in developing region. *UC Atlas of Global Inequality*. Available online at http://ucatlas.ucsc.edu/cause.php. Figure 3.12a: Adapted from Estimated food security conditions, 2nd Quarter 2010 (April-June). USAID/Famine Early Warning Systems Network. Available online at http://www.fews.net/Pages/imageryhome.aspx?map=0. Figure 3.12b: Adapted from Lee, E. S. (1966). A Theory of Migration. *Demography* 3(1), p. 50. Population Association of America. Figure 3.17, map: Adapted from 2010 MTM Map on Irregular and Mixed Migration Routes. West, North and East Africa, Europe, Mediterranean, and Middle East. © ICMPD, EUROPOL, FRONTEX, Interpol, UNHCR, UNODC, and Odysseus Academic Network. January 2010. Available online at https://www.imap-migration.org/index2.html.

Chapter 4

Figure 4.4a: Used by permission, © SIL International, Adapted from Gordon, Jr., R. G. (Ed.). (2005). *Ethnologue: Languages of the world, fifteenth edition*. Dallas, TX: SIL International. Figure 4.7a: Adapted from *Human Geography: People, Place, and Culture, Ninth Edition* by Erin H. Fouberg, Alexander B. Murphy, and H.J. de Blij, John Wiley & Sons, Inc., Copyright © 2010. Reprinted with permission of John Wiley & Sons, Inc. Original from Gamkrelidze, T. V. and Ivanov, V. V. The Early History of Indo-European Languages, *Scientific American*, March 1990. Copyright © 1990 Scientific American, a division of Nature America, Inc. *All rights reserved*. Figure 4.9b: Adapted from U.S. English, Inc. www.usenglish.org. Figure 4.12: Adapted from Crystal, D. 2003. *The Cambridge Encyclopedia of the English Language*. 2nd ed. Cambridge: Cambridge University Press. Copyright Cambridge University Press 1995, 2003. Reprinted with the permission of Cambridge University Press. Figure 4.16a-b: Adapted from Kurath, H. (1949). *A Word Geography of the Eastern United States*. Ann Arbor: The University of Michigan Press, 1950. Where Geographers Click (page 118): Dictionary of American Regional English Project. Figure 4.17a: © Bert Vaux. Figure 4.17b: Gregory A. Plumb. Map derived from *Web Atlas of Okalhoma*, www.okatlas.org.

Chapter 5

Figure 5.3a (chart): Excerpted with permission from 2008 Britannica Book of the Year, © 2008 by Encyclopaedia Britannica, Inc. Figure 5.3c: Adapted from Johnson, T. M. and Chung, S. Y. (2004). Tracking Global Christianity's Statistical Centre of Gravity, AD 33-AD 2100. *International Review of Mission* 93(369): p. 167. Wiley-Blackwell Copyright © 2004. Reprinted with permission of John Wiley & Sons Ltd. Where Geographers Click (page 140): From the U.S. Religious Landscape Survey, Pew Research Center's Forum on Religion & Public Life, © 2009, Pew Research Center. http://religions.pewforum.org/. Figure 5.17b: From the Pew Research Center's Forum on Religion & Public Life, © 2009, Pew Research Center. http://pewforum.org/. Full question wording: From what you've read and heard, is there solid evidence that the average temperature on Earth has been getting warmer over the past few decades, or not? [If "yes," ask]: Do you believe that the earth is getting warmer? 1 - Mostly because of human activity, such as burning fossil fuels, OR 2 - Mostly because of natural patterns in the earth's environment? [options rotated]. What a Geographer Sees (page 154): Adapted from Zelinsky, W. (1994). Gathering Places for America's Dead: How Many, Where, and Why? *Professional Geographer* 46(1): 29-38. Reprinted by permission of the publisher, Taylor & Francis Group, http://www.informaworld.com.

Chapter 6

Figure 6.8a: Adapted from Airriess, C. A. (2007). Conflict migrants from mainland Southeast Asia. In Miyares, I. M. and Airriess, C. A. (Eds.), *Contemporary Ethnic Geographies in America* (pp. 291-312). Lanham, MD: Rowman & Littlefield. U.S. Census Bureau, American Fact Finder, Summary File 2 (SF2). Figure 6.13: Pastor, M., Sadd, J., and Morello-Frosch, R. 2007. *Still Toxic After All These Years: Air Quality and Environmental Justice in the San Francisco Bay Area.* Center for Justice, Tolerance & Community, University of California Santa Cruz. Figure 6.15a: 2007 Gender Gap Heatmap. *The Global Gender Gap Report.* © World Economic Forum. Available online at http://www.weforum.org/pdf/gendergap/2007_heatmap.pdf.

Chapter 7

Figure 7.4b: Adapted from English, P. W. and Miller, J. A. (1989). *World Regional Geography: A Question of Place, 3rd ed.* New York: John Wiley & Sons, Inc. There are instances where we have been unable to trace or contact the copyright holder. If notified, the publisher will be pleased to rectify any errors or omissions at the earliest opportunity. Figure 7.5: © Commonwealth of Australia (Geoscience Australia) 2009. Based upon an original drawing by Geoscience Australia 2009. Figure 7.13: Extract adapted from Mackinder, H. J. (1996 [1942]). *Democratic ideals and reality: a study in the politics of reconstruction.* Mladineo, S. V., ed. Washington, DC: National Defense University Press. By kind permission of Constable & Robinson Ltd. There are instances where we have been unable to trace or contact the copyright holder. If notified, the publisher will be pleased to rectify any errors or omissions at the earliest opportunity. Figures 7.14-7.15: Adapted with the permission of Simon & Schuster, Inc. from THE CLASH OF CIVILIZATIONS AND THE REMAKING OF WORLD ORDER by Samuel P. Huntington. Copyright © 1996 by Samuel P. Huntington. Base map: © Hammond World Atlas Corp. Figure 7.17: Adapted from National Consortium for the Study of Terrorism and Responses to Terrorism (START). Where Geographers Click (page 223): © CAIN (cain.ulster.ac.uk).

Chapter 8

Figure 8.3: Adapted from Morrill, R. (2006). Classic map revisited: The growth of megalopolis. *The Professional Geographer, 58: 2*, 155-160. Taylor & Francis. Reprinted by permission of the publisher (Taylor & Francis Group, http://www.informaworld.com). Figure 8.4a: Adapted from *Urban Geography, Second Edition* by David Kaplan, James O. Wheeler, and Steven Holloway, John Wiley & Sons, Inc., Copyright © 2008. Reprinted with permission of John Wiley & Sons, Inc. Figure 8.5a: Adapted from *World Urbanization Prospects: The 2007 Revision.* (2007). United Nations (available online at http://www.un.org/esa/population/meetings/EGM_PopDist/Heilig.pdf) and *State of the World's Cities 2008/2009.* (2008). United Nations Human Settlements Programme. © 2008 United Nations. Reprinted with the permission of the United Nations. Figure 8.7: Adapted from *Interpreting the City: An Urban Geography, Second Edition* by Truman Asa Hartshorn, John Wiley & Sons, Inc., Copyright © 1992. Reprinted with permission of John Wiley & Sons, Inc. Figure 8.8: Larsen, K. and Gilliland, J. (2008). Mapping the evolution of 'food deserts' in a Canadian city: Supermarket accessibility in London, Ontario, 1961-2005. *International Journal of Health Geographics.* BioMed Central. Figure 8.9b: Adapted from *Cities* 16(6), Beaverstock, J. V., Taylor, P. J., and Smith, R. G. A Roster of World Cities, pages 445-458, Copyright 1999, with permission from Elsevier. Figure 8.10: Adapted from *Interpreting the City: An Urban Geography, Second Edition* by Truman Asa Hartshorn, John Wiley & Sons, Inc., Copyright © 1992. Reprinted with permission of John Wiley & Sons, Inc. Figure 8.11a: Adapted from Burgess, E. (1929). Urban Areas. In Smith, T. V., and White, L. D., eds., *Chicago: An Experiment in Social Science Research.* Chicago: University of Chicago Press, 115. Copyright 1929 by The University of Chicago. All rights reserved. Published December 1929. Figure 8.11b: Adapted from Harris, C. and Pullman, E. (1945). The Nature of Cities. *Annals of the Association of the American Academy of Political and Social Science* 242(Nov), p. 13. Copyright © 1945 by Sage Publications. Reprinted by Permission of SAGE Publications. Figure 8.12: Adapted from *Interpreting the City: An Urban Geography, Second Edition* by Truman Asa Hartshorn, John Wiley & Sons, Inc., Copyright © 1992. Reprinted with permission of John Wiley & Sons, Inc. Figure 8.14 center map: Adapted from Grant, R. and Nijman, J. (2002). Globalization and the corporate geography of cities in the less-developed world. *Annals of the Association of American Geographers, 92:2*, 320-340. Taylor & Francis. Reprinted by permission of the publisher (Taylor & Francis Group, http://www.informaworld.com). Figure 8.19: *State of the World's Cities 2008/2009.* (2008). United Nations Human Settlements Programme. © 2008 United Nations. Reprinted with the permission of the United Nations.

Chapter 9

Figure 9.1a: Courtesy of Dr. Arno Peters, University of Bremen. Adapted cover image of NORTH-SOUTH: A PROGRAMME FOR SURVIVAL, edited by Willy Brandt, published by The MIT Press. Figure 9.3: From 2008 World Development Indicators. Poverty Data. A Supplement to World Development Indicators 2008. Washington, DC: The World Bank, pp. 12-13. Permission granted by World Bank Publications. Figure 9.5a: CIESIN Columbia University, available online at http://sedac.ciesin.columbia.edu/wdc/map_gallery.jsp. © 2006. The Trustees of Columbia University in the City of New York. Figure 9.6a: From United Nations Development Programme, Human Development Report 2007/08, published 2007, Palgrave Macmillan, reproduced with permission of Palgrave Macmillan. Figure 9.7a: From United Nations Development Programme, Human Development Report 2007/08, published 2007, Palgrave Macmillan, reproduced with permission of Palgrave Macmillan. Figure 9.8a: From United Nations Development Programme, Human Development Report 2007/08, published 2007, Palgrave Macmillan, reproduced with permission of Palgrave Macmillan. Where Geographers Click (page 273): From United Nations Development Programme, Human Development Report 2007/08, published 2007, Palgrave Macmillan, reproduced with permission of Palgrave Macmillan. Figure 9.9b: CIESIN Columbia University, available online at http://sedac.ciesin.columbia.edu/wdc/map_gallery.jsp. © 2008 The Trustees of Columbia University in the City of New York. Figure 9.10: Adapted from OECD Annual Report 2008, p. 9, www.oecd.org. Figure 9.11a: From *Human Development Report 2005: International Cooperation at a Crossroads, Aid, Trade and Security in an Unequal World.* NY: United Nations Development Program, p. 37. Permission granted by Oxford University Press. What a Geographer Sees Part 2a (page 286): From *Changes in poverty in Madagascar 1993–1999.* Joint project of The World Bank, NSTAT, and Cornell University, 2001. Permission granted by World Bank Publications. What a Geographer Sees Parts 2b-2c (page 287): From Mistiaen, J. A., Özler, B., Razafimanantena, T., and Razafindravonona, J. Putting Welfare on the Map in Madagascar. *Africa Region Working Paper Series No. 34.* The World Bank, 2002. Permission granted by World Bank Publications.

Chapter 10

Figure 10.2: Adapted from the World Trade Organization, Annual Report 2002, Geneva: WTO, 2002. http://www.wto.org/english/res_e/booksp_e/anrep_e/anrep02_e.pdf. Figure 10.3: *Development and Globalization: Facts and Figures 2008.* (2008). New York, NY: United Nations Conference on Trade and Development. Available online at: http://www.unctad.org/en/docs/gdscsir20071_en.pdf. Figure 10.4a: Adapted from Bairoch, P. (1982). International Industrialization Levels from 1750 to 1980. *Journal of European Economic History*, Volume 11, p. 275. Banca di Roma. Where Geographers Click (page 301): © Copyright 2006 SASI Group (University of Sheffield) and Mark Newman (University of Michigan). Figure 10.9b: Adapted from Gereffi, G. and Memedovic, O. (2003). *The Global Apparel Value Chain: What Prospects for Upgrading by Developing Countries?* Vienna: UNIDO Strategic Research and Economics Branch. Figure 10.11b: Adapted from *Equality at Work: Tackling the Challenges.* Report of the Director General, International Labor Conference. 96th Session 2007. Geneva: ILK. Copyright © International Labour Organization, 2007. Figure 10.12: Adapted from Dicken, P. (2003). *Global Shift, Fourth Edition.* Sage Publications. Original from Phillips, D. R. and Yeh, A. G. O., (1990). Foreign investment and trade: impact on spatial structure of the economy. In Cannon, T. and Jenkins, A. (Eds.), *The Geography of Contemporary China: The Impact of Deng Xiaoping's Decade.* London: Routledge. Figure 10.13: Adapted from Stan Shih's Smile Curve from Dedrick, J. and Kraemer, K. (1998). *Asia's Computer Challenge: Threat or Opportunity for the U.S. and the World.* By permission of Oxford University Press, Inc. Figure 10.15a: Adapted from Pilat, D., Cimper, A., Olsen, K., and Webb, C. (2006). The Changing Nature of Manufacturing in OECD Economies, *OECD Science, Technology and Industry Working Papers*, p. 6, http://dx.doi.org/10.1787/308452426871

Chapter 11

Figure 11.4a (table): Adapted from the Committee on World Food Security, Thirtieth Session, Rome, 20-23 September 2004, CFS Distinguished Lecture Series: The Green Revolution: An Unfinished Agenda. Food and Agriculture Organization of the United Nations, 2004. Figure 11.4d: Adapted from Fritschel, H., Pandya-Lorch, R., and Rose, B. 1996. Key trends in feeding the world, 1970-95. A guide to the 2020 vision video 2020 *Hindsight: Successes, failures, and lessons learned in feeding the world,* 1970-95. Washington, DC: International Food Policy Research Institute. Due acknowledgement shall be made to the Food and Agriculture Organization of the United Nations. "FAO Stat-PC, Production Domain, 1995" (computer disk), version 3.0 (Rome, 1995). Figure 11.6b: Adapted from Hammond. Comparative World Atlas, 2004. Copyright Hammond World Atlas - Used with permission. Figure 11.14a: Adapted from O'Brien, D. (2010). *U.S. Grain Supply-Demand Projections for 2010 from the USDA Agricultural Outlook Forum.* K-State Research and Extension. Data from the U.S. Department of Agriculture. Figure 11.16: Adapted from Chisholm, M. (1968). *Rural Settlement and Land Use: An Essay in Location.* London: Hutchinson University Library.

Chapter 12

Figure 12.6 (map): Adapted from Drilling Deeper Into Dependence. Environmental Action, www.environmental-action.org. Figure 12.18: Adapted from Climate Change 2007: Synthesis Report. Contribution of Working Groups I, II and III to the Fourth Assessment Report of the Intergovernmental Panel on Climate Change, Figure 2.3. IPCC, Geneva, Switzerland. Figure 12.19: From *Visualizing Environmental Science, Second Edition* by Linda R. Berg and Mary Catherine Hager, John Wiley & Sons, Inc., Copyright © 2009. Reprinted with permission of John Wiley & Sons, Inc.

Appendix A

Figure A.1: Adapted from *Visualizing Physical Geography* by Alan H. Strahler, John Wiley & Sons, Inc., Copyright © 2008. Reprinted with permission of John Wiley & Sons, Inc. Figures A.2-A.3: Adapted from *Elements of Cartography. Fifth edition* by Arthur H. Robinson, Randall D. Sale, Joel L. Morrison, and Phillip C. Muehrcke. John Wiley & Sons, Inc., Copyright © 1984. Reprinted with permission of John Wiley & Sons, Inc

Photo Credits

Chapter 1

Pages 2-3: Data courtesy Marc Imhoff, NASA GSFC, and Christopher Elvidge of NOAA NGDC. Image by Craid Mayhew and Robert Simmon, NASA GSFC; page 4: iStockphoto; Pages 6-7: NG Maps; page 9: Mattias Klum/NG Image Collection; page 11: Courtesy Alyson Greiner; page 12: iStockphoto; page 13: (top) Panoramic Images/NG Image Collection; page 13: (bottom) Alex Webb/NG Image Collection; page 15: Used with permission from CIFOR, www.cifor.cgiar.org. William Sunderland, Sonya Dewi, and Atie Puntodewo. Poverty and Forests: Multi-Country Analysis of Spatial Association and Proposed Policy Solutions. CIFOR Occasional Paper No. 47. Center for International Foresty Research, 2007, p.25; page 18: NG Maps; page 19: Justin Guariglia/NG Image Collection; page 20: Richard Block and Wim Bernasco "Finding a Serial Burglar's Home Using Distance Decay and Conditional Origin-Destination Patterns: A Test of Empirical Bayes Journey to Crime Estimation in The Hague." Journal of Investigative Psychology and Offender Profiling. Wiley, November 12, 2009; page 21: NG Maps; page 22: (left) Bill Freeman/PhotoEdit; page 22: (right) Moses Harris/Detroit Free Press; page 23: National Geographic; page 24: (top left and bottom) Beverley J. Adams, Charles K. Huyck, B. Mansouri, R. T. Eguchi, M. Shinozuka. (2004). Application of High-Resolution Optical Satellite Imagery for Post-Earthquake Damage Assessment: The 2003 Boumerdes (Algeria) and Bam (Iran) Earthquakes. In Research Progress & Accomplishments: 2003-2004, MCEER-04-SP01, MCEER, University of Buffalo, pp. 173-186; page 24: (top right) Claude Paris/©AP/Wide World Photos; page 25: Beverley J. Adams, Charles K. Huyck, B. Mansouri, R. T. Eguchi, M. Shinozuka. (2004). Application of High-Resolution Optical Satellite Imagery for Post-Earthquake Damage Assessment: The 2003 Boumerdes (Algeria) and Bam (Iran) Earthquakes. In Research Progress & Accomplishments: 2003-2004, MCEER-04-SP01, MCEER, University of Buffalo, pp. 173-186; page 26: (left) NG Maps; page 26: (right) ©2008 S.E. Wiehe et al; licensee BioMed Central Ltd.; page 27: (bottom left) NG Maps; page 28: NG Maps; page 29: (top left) From J. P. DeGroote, R. Sugumaran, S. M. Bend, B. J. Tucker, and L. C. Bartholomay. 2008. Landscape, demographic, entomological, and climatic associations with human disease incidence of West Nile virus in the state of Iowa, USA. International Journal of Health Geographics 7(19). Original publisher: BioMed Central; page 29: (top right) From J. P. DeGroote, R. Sugumaran, S. M. Bend, B. J. Tucker, and L. C. Bartholomay. 2008. Landscape, demographic, entomological, and climatic associations with human disease incidence of West Nile virus in the state of Iowa, USA. International Journal of Health Geographics 7(19). Original publisher: BioMed Central; page 29: (bottom) From J. P. DeGroote, R. Sugumaran, S. M. Bend, B. J. Tucker, and L. C. Bartholomay. 2008. Landscape, demographic, entomological, and climatic associations with human disease incidence of West Nile virus in the state of Iowa, USA. International Journal of Health Geographics 7(19). Original publisher: BioMed Central; page 30: (center left) Alex Webb/NG Image Collection; page 31: (top) NG Maps; page 32: (right) ©2008 S.E. Wiehe et al; licensee BioMed Central Ltd.; page 31: (bottom right) ©2008 S.E. Wiehe et al; licensee BioMed Central Ltd.; page 32: (center) Steve Parsons/©AP/Wide World Photos; page 33: Courtesy Alyson Greiner.

Chapter 2

Pages 34-35: Courtesy Samoa News Archives; page 37: Danita Delimont/Alamy; page 40: REUTERS/Mohamed Nureldin Abdallh/©Corbis; page 41: Courtsy Wal-Mart Stores, Inc.; page 42: Wikipedia. From: http:/www.flickr.com/photos/35034362831@N01/2870473171/; page 43: (left) REUTERS/ Robert Sorbo/Landov LLC; page 43: (right) Scott Olson/Getty Images, Inc.; page 44: David Pearson/Alamy; page 46: (top) Jonathan Torgovnik/Getty Images, Inc.; page 46: (center) Reuters/Francois Lenoir/Landov LLC; page 46: (just above bottom left) Candace Feit/The New York Times/Redux Pictures; page 46: (bottom left) Schalk van Zuydam/©AP/Wide World Photos; page 46: (bottom right) Itsuo Inouye/©AP/Wide World Photos; page 47: (right inset) NG Maps; page 47: Peter Morrison/©AP/ Wide World Photos; page 48: (left) Michael McNutt/©AP/Wide World Photos; page 48: (right) Argus/©AP/Wide World Photos; page 49: (center left) Robert W. Nicholson/NG Image Collection; page 49: (center right) Travel Ink/Getty Images, Inc.; page 49: (bottom) Stuart Franklin/NG Image Collection; page 50: Guy Vanderelst/Getty Images, Inc.; page 52: National Geographic; page 53: (center left) iStockphoto; page 53: (bottom left) Tetra Images/Alamy; page 53: (bottom right) Vincent Du/ Reuters/Landov LLC; page 54: Richard Levine/Alamy; page 56: Photo by Dale R. Lightfoot; page 55: (top) Georg Gerster/Photo Researchers, Inc.; page 55: (bottom) Photo by Dale R. Lightfoot; page 58: (top) Alison Wright/NG Image Collection; page 58: (center left) The New York TImes Rights and Permissions; page 58: (bottom right) CONBAM/Dipl.-Ing. Christoph Tonges; page 59: (left) REUTERS/Mohamed Nureldin Abdallh/©Corbis; page 59: (right) David Pearson/Alamy; page 60: (left) Peter Morrison/©AP/Wide World Photos; page 60: (right) Tetra Images/Alamy; page 61: (top) Alison Wright/NG Image Collection; page 61: (bottom) Apichart Weerrawong/©AP/Wide World Photos; page 62: David Pearson/Alamy; page 63: Georg Gerster/ Photo Researchers, Inc.

Chapter 3

Pages 64-65: Ed Kashi/NG Image Collection; page 66: NG Maps; pages 68-69: NG Maps; page 68: (bottom left) Kim Kyung-Hoon/Landov LLC; page 69: (bottom right) Kenneth Garrett/ NG Image Collection; page 70: Lynn Johnson/NG Image Collection; page 71: (left) Bruce Dale/NG Image Collection; page 71: (right) Randy Olson/NG Image Collection; page 72: (bottom left) NG Maps; page 72: (bottom right) David McLain/Aurora Photos, Inc.; page 73: National Geographic; page 74: Luis Liwanag/NewsCom; page 75: (top right) U.S. Census Bureau; page 75: (bottom left) Kamran Jebreili/©AP/Wide World Photos; page 79: Lori Adamski Peek/Getty Images, Inc.; page 80: ©WFP/Gerald Bourke. Reproduced with permission; page 84: Guillermo Arias/©AP/Wide World Photos; page 87: (top) Photo by Alan P. Marcus, Towson University, Marietta, May 2007. From Marcus, Alan P. 2009a. (Re)Creating Places and Spaces in Two

Countries: Brazilian Transnational Migration Processes. Journal of Cultural Geography (26) 2: 173-198; page 87: (center) Photo by geographer, Alan P. Marcus (taken during fieldwork June, 2007); page 88: Katja Hoffmann/laif/Redux Pictures; page 89: Darrin Zammit Lupi/REUTERS/Landov LLC; page 90: Uriel Sinai/Getty Images, Inc.; page 91: Kim Kyung-Hoon/Landov LLC; page 92: (left) Lori Adamski Peek/Getty Images, Inc.; page 92: (right) Darrin Zammit Lupi/REUTERS/Landov LLC; page 93: David McLain/Aurora Photos, Inc.; page 94: Martin Gray/NG Image Collection.

Chapter 4

Pages 96-97: Photographer, Brian Mott, Southern California Collegiate Ski & Snowboard Conference, Inc (www.sccsc.com); page 98: (bottom left inset) NG Maps; page 98: (bottom) Chris Johns/NG Image Collection; page 99: (top left) Justin Guariglia/NG Image Collection; page 99: (top right) arabianEye/Getty Images, Inc.; page 100: (top right) Steve Raymer/NG Image Collection; page 100: (center left) DOMINIQUE FAGET/AFP/Getty Images, Inc.; page 103: NG Maps; page 106: (top left) Jody Amiet/AFP/Getty Images, Inc.; page 106: (bottom left) Michael S. Lewis/NG Image Collection; page 107: Peter Jordan/Alamy; page 109: (center) Courtesy Alyson Greiner; page 109: (bottom) Todd Gipstein/NG Image Collection; page 110: David Austen/NG Image Collection; page 112: NG Maps; page 113: (top left) NG Maps; page 113: (top right) Photo by Chris Rainier, Swarthmore College Linguistics; page 116: Photo by sAh@IA Renee Grounds; page 117: National Geographic; page 119: (bottom) Adapted from Geography of Everyday Life, National Geographic Magazine, December 2005/NG Maps; page 120: (bottom) Courtesy Gregory Plumb, East Central University, OK, Department of Cartography and Geography; page 121: Brad Barket/PictureGroup/©AP/Wide World Photos; page 122: Chris Hammond/Alamy; page 123: (top) AFP PHOTO/Sebastian D'SOUZA/NewsCom; page 123: (center) Ric Feld/©AP/Wide World Photos; page 124: Courtesy Alyson Greiner; page 125: (top right) AFP PHOTO/Sebastian D'SOUZA/NewsCom; page 125: (bottom left) DOMINIQUE FAGET/AFP/Getty Images, Inc.; page 126: Anand Giridharadas/IHT/Redux Pictures; page 127: NG Maps.

Chapter 5

Pages 128-129: Sam Abell/NG Image Collection; page 130: (left) Stephanie Maze/NG Image Collection; page 130: (right) National Geographic; page 131: NG Maps; page 132: Ed Kashi/NG Image Collection; page 134: (top left) Carlos Sanchez Pereyra/Getty Images, Inc.; page 134: (top right) Craig Lassig/©AP/Wide World Photos; page 134: (center right) Zinhua/Landov LLC; page 134: (bottom) Branky Baker/The Detroit News; page 134: (center left) iStockphoto; page 135: DIBYANGSHU SARKAR/AFP/Getty Images, Inc.; page 136: (left) Martin Gray/NG Image Collection; page 136: (right) Alison Wright/NG Image Collection; page 137: George F. Mobley/NG Image Collection; page 139: (top right) NG Maps; page 141: Courtesy National Khalsa (Khalistan) Day Parade Organization; page 142: NG Maps; page 143: (center left) Michael Melford/NG Image Collection; page 143: (bottom right) Martin Gray/NG Image Collection; page 144: Steve McCurry/NG Image Collection; page 145: (left) Robert Nickelsberg/Getty Images, Inc.; page 145: (right) Alankar Chandra/NG Image Collection; page 147: (top right) ©AP/Wide World Photos; page 147: (top left) ©AP/Wide World Photos; page 147: (center left) Jeff Haynes/AFP/Getty Images, Inc.; page 147: (center right) J. Pat Carter/©AP/Wide World Photos; page 147: (bottom left) Jerry Laizure/©AP/Wide World Photos; page 147: (bottom right) Doug Hoke/The Oklahoman; page 149: Rajesh Nirgude/©AP/Wide World Photos; page 150: (bottom right inset) NG Maps; page 150: (bottom) National Geographic; page 151: (left) REUTERS/Jo Yong-Hak /Landov LLC; page 151: (center) ©New York Times Graphics; page 152: Reuters/Dadang Tri/Landov LLC; page 153: Michele Falzone/Getty Images, Inc.; page 155: (top left) Photo courtesy of Amaury Laporte (pix.alaporte.net); page 155: (top right) Courtesy Alyson Greiner; page 155: (bottom left) Courtesy Aberdeen Chinese Permanent Cemetary, Hong Kong; page 155: (bottom right) Courtesy Alyson Greiner; page 156: (left) Stephanie Maze/NG Image Collection; page 156: (right) Courtesy National Khalsa (Khalistan) Day Parade Organization; page 157: (left) REUTERS/Jo Yong-Hak /Landov LLC; page 157: (right) Courtesy Alyson Greiner; page 157: (bottom) Courtesy National Khalsa (Khalistan) Day Parade Organization; page 158: (top) Steve McCurry/NG Image Collection; page 158: (bottom) Rajesh Nirgude/©AP/Wide World Photos; page 159: (center left) ©AP/Wide World Photos; page 159: (center right) Photo courtesy of Amaury Laporte (pix.alaporte.net).

Chapter 6

Pages 160-161: Katie Clementson/NBAE/Getty Images, Inc.; page 163: (top) AUDE GUERRUCCI/UPI/Landov LLC; page 163: (bottom) Charles Bonnet, uvres d'histoire naturelle et de philosophie, 1779-83, Neuchatel, vol. IV, p. 1; page 164: From David Eltis and David Richardson, An Atlas of the Translatlantic Slave Trade (New Haven, 2009), reproduced with permission of Yale University Press; page 165: Kay Chernush for the U.S. State Department; page 166: from Kay J. Anderson. 1987 The Idea of Chinatown: The Power of Place and Institutional Practice in the Making of a Racial Category. Annals of the Association of American Geographers, 77(4):580-597;1987, p. 588; page 166: (bottom inset) NG Maps; page 169: (top left) Tomasz Tomaszewski/NG Image Collection; page 169: (center) David Turnley/©Corbis; page 169: (top right) Dennis Farrell/©AP/Wide World Photos; page 171: (right) Dado Galdieri/©AP/Wide World Photos; page 171: (left) Catherine Karnow/NG Image Collection; page 173: (top) Courtesy Kyonghwan Park, Chonnam National University, South Korea; page 173: (bottom) S. Meltzer/PhotoLink/Photodisc/Getty Images, Inc.; page 177: Compliments of Eden Center; page 179: Richard B. Levine/NewsCom; page 180: (inset) NG Maps; page 180: Scott Nelson/Getty Images, Inc.; page 182: National Geographic; page 185: (left) Pavel Rahman/©AP/Wide World Photos; page 185: (right) Hassan Ammar/©AP/Wide World Photos; page 186: (left) Charles Bonnet, Œuvres d'histoire naturelle et de philosophie, 1779-83, Neuchatel, vol. IV, p. 1; page 186: (right) David Turnley/©Corbis; page 187: Catherine Karnow/NG Image Collection; page 188: Pavel Rahman/©AP/Wide World Photos; page 189: Tim Laman/NG Image Collection; page 190: Dado Galdieri/©AP/Wide World Photos.

Chapter 7

Pages 192-193: Kyodo News/©AP/Wide World Photos; page 193: (inset) NG Maps; pages 194-195: NG Maps; page 197: (center left) Petros Karadjias/©AP/Wide World Photos; page 197: (center right) STEFANOS KOURATZIS/AFP/Getty Images, Inc.; page 198: NG Maps; page 199: (right) Mindaugas Kulbis/©AP/Wide World Photos; page 199: (left) Remy de la Mauviniere/©AP/Wide World Photos; page 200: (top left) NG Maps; page 200: (bottom left) UNHCR/P.Moumtzis; page 200: (bottom right) UNHCR/C. Sattleberger; page 203: (top) Image courtesy of photographer Scott Cutler; page 203: (bottom) Dean Conger/NG Image Collection; page 204: (bottom right) NG Maps; page 204: (top left) NG Maps; page 204: (top right) Courtesy Dr. Jack Child, American University; page 204: (bottom) NASA/Goddard Space Flight Center Scientific Visualization Studio; page 206: National Geographic; page 207: (left) AFP/Getty Images, Inc.; page 207: (right) Bela Szandelszky/©AP/Wide World Photos; page 208: (center left) Alvaro Barrientos/©AP/Wide World Photos; page 208: (center right) Manu Fernandez/©AP/Wide World Photos; page 210: Marco Dormino/United Nations Photo; page 216: REUTERS/Dario Pignatelli/Landov LLC; page 222: (left) MENAHEM KAHANA/AFP/Getty Images, Inc.; page 222: (right) Lefteris Pitarakis/©AP/Wide World Photos; page 223: (top right) RAMZI HAIDAR/AFP/Getty Images, Inc.; page 223: (top left) ©CAIN (cain.ulster.ac.uk); page 223: (bottom) Petros Karadjias/©AP/Wide World Photos; page 224: (center left) Alvaro Barrientos/©AP/Wide World Photos; page 224: (top right) Marco Dormino/Courtesy United Nations; page 225: MENAHEM KAHANA/AFP/Getty Images, Inc.; page 226: YASSER AL-ZAYYAT/AFP/Getty Images, Inc.; page 227: Courtesy Dr. Jack Child, American University.

Chapter 8

Pages 228-229: Courtesy Foster and Partners; page 229: (inset) NG Maps; page 230: (left) Justin Guariglia/NG Image Collection; page 230: (right) ©2009 Journal Communications, Inc; page 231: (left) Panoramic Stock Images/NG Image Collection; page 235: Justin Guariglia/NG Image Collection; page 236: James P. Blair/NG Image Collection; page 238: Kristian Larsen and Jason Gilliland, Mapping the evolution of 'food deserts' in a Canadian city: Supermarket accessibility in London, Ontario 1961-2005, International Journal of Health Geographics 7:16-xx, 2008; page 244: (top left) Library of Congress Prints and Photographs Division; page 244: (bottom right) Cotton Coulson/NG Image Collection; page 245: Courtesy Karina Kreja; page 246: (top center) NG Maps; page 246: (top left) Thomas Cockrem/Alamy; page 246: (bottom) Eitan Simanor/Alamy; page 246: (top right) Olivier Asselin/Alamy; page 247: Julian Nieman/Alamy; page 250: (top left) Courtesy Greater Oklahoma City Chamber; page 250: (top right) Courtesy Greater Oklahoma City Chamber; page 250: (bottom) National Geographic; page 251: Vincent Laforet/NG Image Collection; page 252: (top right) Pritt Veslind/NG Image Collection; page 252: (top left) Thomas J. Abercrombie/NG Image Collection; page 252: (bottom) Michael S. Yamashita/NG Image Collection; page 253: (top) Alexandre Meneghini/©AP/Wide World Photos; page 253: (center) Michael Nichols/NG Image Collection; page 253: (right) Maria Stenzel/NG Image Collection; page 254: Manish Swarup/©AP/Wide World Photos; page 255: Sarah Leen/NG Image Collection; page 256: (left) Courtesy Karina Kreja; page 256: (right) Vincent Laforet/NG Image Collection; page 257: Amy Vitale/NG Image Collection; page 258: (bottom left) Cotton Coulson/NG Image Collection; page 258: (center and bottom right) Courtesy Greater Oklahoma City Chamber; page 259: Vincent Laforet/NG Image Collection.

Chapter 9

Pages 260-261: John Scofield/NG Image Collection; page 261: (inset) NG Maps; page 263: AMR Nabil/AFP/Getty Images, Inc.; pages 264-265: NG Maps; page 266: Courtesy Development Data Group, The World Bank, www.worldbank.org/data; page 267: Courtesy Development Data Group, The World Bank, www.worldbank.org/data; page 268: (top) NG Maps; page 268: (center left) Saim Al-Hajj/AFP/Getty Images, Inc.; page 268: (center right) Neil Cooper/Alamy; page 269: (left) This work, including access to the data and technical assistance, is provided by CIESIN, with funding from the National Aeronautics and Space Administration under Contract NAS5-03117 for the Continued Operation of the Socioeconomic Data and Applications Center (SEDAC)/NASA; page 269: (right) Karyn D. Ellis, courtesy Solar Cookers International, www.solarcookers.org; page 272: Sebastian John/©AP/Wide World Photos; page 273: Source: Human Development Reports website (http:/hdr.undp.org). Data from: United Nations Development Programme (UNDP). 2009. Overcoming barriers: Human mobility and development. Palgrave Macmillan: Hampshire and New York. Courtesy Human Development Report Office Outreach and Advocacy Unit, United Nations Development Programme; page 274: (top) From CIESIN, with funding from the National Aeronautics and Space Administration under Contract NAS5-03117 for the Continued Operation of the Socioeconomic Data and Applications Center (SEDAC); page 274: (bottom) Jim West/Alamy; page 277: Marcy Nighswander/©AP/Wide World Photos; page 287: (center left) NG Maps; page 286: NG Maps; page 287: (top) NG Maps; page 287: (top) NG Maps; page 287: (bottom left) IFAD Photo by Robert Grossman; page 287: (bottom right) IFAD Photo by Horst Wagner; page 291: REUTERS/Michael Simon/FOX News /Landov LLC; page 288: National Geographic.

Chapter 10

Pages 294-295: Romeo Ranoco/Reuters/Landov LLC; page 295: (inset) NG Maps page 296: (center) Stephen Sharnoff/NG Image Collection; page 296: (bottom left) Jim Richardson/NG Image Collection; page 296: (bottom right) Alison Wright/NG Image Collection; page 303: From the Collections of The Henry Ford; page 305: (center) PETER FOERSTER/dpa /Landov LLC; page 305: (bottom left) Paul Chesley/NG Image Collection; page 305: (bottom right) Mike Theiler/Getty Images, Inc.; page 307: Aijaz Rahi/©AP/Wide World Photos; page 309: Image provided courtesy of Sengkang on Wikipedia; page 310: Joe Raedle/Newsmakers/Getty Images, Inc.; page 313: Courtesy iSuppli Corporation; page 315: (top left) Matt Rourke/©AP/Wide World Photos; page 315: (top right) Rick Smith/©AP/Wide World Photos; page 315: (bottom) National Geographic; page 317: (left) Blend Images/

Getty Images, Inc.; page 317: (right) image100/Alamy; page 318: AFP/Saeed Khan/Getty Images, Inc.; page 319: (top left) Stephen Sharnoff/NG Image Collection; page 320: (top left) From the Collections of The Henry Ford; page 320: (top right) Joe Raedle/Newsmakers/Getty Images, Inc.; page 320: (bottom right) AFP/Saeed Khan/Getty Images, Inc.; page 321: (top) image100/Alamy; page 321: (bottom) Blaine Harrington III/Alamy; page 323: (top) Matt Rourke/©AP/Wide World Photos; page 323: (center) Rick Smith/©AP/Wide World Photos.

Chapter 11

Pages 324-325:; page 325: (inset) NG Maps page 326: Finbarr O Reilly/Landov LLC; page 327: (top) National Geographic; page 328: (left) dem10/iStockphoto; page 328: (right) Elena Elisseeva/iStockphoto; page 329: (left) Mlenny/iStockphoto; page 329: (right) Rachel Dunn/iStockphoto; page 330: Narinder Nanu/AFP/Getty Images, Inc.; page 331: Mark Edwards/Peter Arnold, Inc.; page 332: (top) NG Maps; page 332: (center) Graeme Robertson/Getty Images, Inc.; page 334: (left) Michael S. Yamashita/NG Image Collection; page 334: (right) Photo from Marco Schmidt 2006. Pflanzenvielfalt in Burkina Faso-Analyse, Modellierung und Dokumentation. http:/publikationen.ub.uni-frankfurt.de/volltexte/2006/3198/pdf/SchmidtMarco.pdf, with permission of the author; page 335: (left) ©2008 by William Robichaud; page 335: (right) ©Dr. Jack D. Ives; page 336: (left) Steve Winter/NG Image Collection; page 336: (right) Randy Olson/NG Image Collection; page 337: (top left) Susan Byrd/NG Image Collection; page 337: (top right) Justin Guariglia/NG Image Collection; page 337: (bottom left) Lynn Johnson/NG Image Collection; page 337: (bottom right) Lynn Johnson/NG Image Collection; page 338: (left) James P. Blair/NG Image Collection; page 338: (right) Stuart Forster/Alamy; page 339: Michael Melford/NG Image Collection; page 340: Ric Francis/©AP/Wide World Photos; page 341: (center) Courtesy USDA; page 341: (bottom) Mike Stewart/©AP/Wide World Photos; page 342: (left) Peter Carroll/Alamy; page 343: (right) Todd Gipstein/NG Image Collection; page 344: (bottom left inset) NG Maps; page 344: Greg Ludwig/NG Image Collection; page 345: Courtesy NASA; page 347: (bottom left) NG Maps; page 347: (bottom right) Jim Richardson/NG Image Collection; page 347: Dale R. Lightfoot; page 348: Paul Kaldjian; page 349: Todd Gipstein/NG Image Collection; page 350: Paul Kaldjian; page 351: (top) Courtesy USDA; page 351: (bottom) Patricio Robles Gil/Minden Pictures/NG Image Collection; page 352: (bottom left) Steve Winter/NG Image Collection; page 352: (bottom right) Jim Richardson/NG Image Collection; page 353: Mike Stewart/©AP/Wide World Photos.

Chapter 12

Pages 354-355: Pete Ryan/NG Image Collection; page 356: Rich Reid/NG Image Collection; page 357: (left) Lorena Ros/Getty Images, Inc.; page 357: (right) Oliver Berg/dpa/Landov LLC; page 358: (left) U.S. Fish and Wildlife Service - Nebraska Partners for Fish and Wildlife Program photo; page 358: (right) U.S. Fish and Wildlife Service - Nebraska Partners for Fish and Wildlife Program photo; page 359: (top right inset) NG Maps; page 359: (left) Danielle Lema Ngono/CIFOR; page 359: (right) Keren Su/Getty Images, Inc.; page 365: (top left) Melissa Farlow/NG Image Collection; page 365: (top right) Troy Fleece/©AP/Wide World Photos; page 365: (center right) Melissa Farlow/NG Image Collection; page 365: (center left) Melissa Farlow/NG Image Collection; page 365: (bottom left) Melissa Farlow/NG Image Collection; page 365: (bottom right) Melissa Farlow/NG Image Collection; page 367: Koji Sasahara/©AP/Wide World Photos; page 369: Peter Bennett/Ambient Images/Alamy; page 370: REUTERS/David Mdzinarishvili/Landov LLC; page 371: WEN ZHENXIAO/Xinhua /Landov LLC; page 372: NG Maps; page 373: (top) Bob Krist/NG Image Collection; page 373: (bottom) National Geographic; page 374: NG Maps; page 377: (top) U.S. Geological Survey, National Wetlands Research Center, Layette, LA; page 377: (center) US. Department of the Inteior/U.S. Geological Survey; page 377: (bottom left) Permission granted by Dr. Lionel D. Lyles, Professor and Dr. Fulbert Namwamba, Professor, Southern University Baton Rouge, LA; page 377: (bottom right) Gerald Herbert/©AP/Wide World Photos; page 378: NG Maps; page 379: NASA/UNEP/Still Pictures; page 379: (inset photo) UNEP/Levy/Topham/The Image Works; page 380: (top left) Courtesy World Resources Institute; page 380: (center) National Geographic; page 381: (left) Rich Reid/NG Image Collection; page 381: (right) Troy Fleece/©AP/Wide World Photos; page 382: (left) WEN ZHENXIAO/Xinhua /Landov LLC; page 382: (right) NG Maps; page 382: (center) Melissa Farlow/NG Image Collection; page 383: (bottom) Photo courtesy of the Sanitation Districts of Los Angeles County; page 384: (top) U.S. Fish and Wildlife Service - Nebraska Partners for Fish and Wildlife Program photo; page 384: (center) U.S. Fish and Wildlife Service - Nebraska Partners for Fish and Wildlife Program photo; page 385: (left) Melissa Farlow/NG Image Collection; page 385: (right) NASA/UNEP/Still Pictures; page 385: (right inset photo) UNEP/Levy/Topham/The Image Works.

Appendix A

Page 389: NG Maps; page 390: NG Maps; page 391: NG Maps; page 392: NG Maps.

Index

D

E

F

G

J

K

L

M

N

O

P